Lecture Notes in Computer Science 6692

Joan Cabestany Ignacio Rojas
Gonzalo Joya (Eds.)

Advances in Computational Intelligence

11th International Work-Conference
on Artificial Neural Networks, IWANN 2011
Torremolinos-Málaga, Spain, June 8-10, 2011
Proceedings, Part II

Springer

Volume Editors

Joan Cabestany
Universitat Politècnica de Catalunya (UPC)
Departament d'Enginyeria Electrònica
Campus Nord, Edificio C4, c/ Gran Capità s/n, 08034 Barcelona, Spain
E-mail: cabestan@eel.upc.es

Ignacio Rojas
University of Granada
Department of Computer Architecture and Computer Technology
C/ Periodista Daniel Saucedo Aranda, 18071 Granada, Spain
E-mail: irojas@ugr.es

Gonzalo Joya
Universidad de Málaga, Departamento Tecnologia Electrónica
Campus de Teatinos, 29071 Málaga, Spain
E-mail: gjoya@uma.es

ISSN 0302-9743 e-ISSN 1611-3349
ISBN 978-3-642-21497-4 ISBN 978-3-642-21498-1 (eBook)
DOI 10.1007/978-3-642-21498-1
Springer Heidelberg Dordrecht London New York

Library of Congress Control Number: 2011928243

CR Subject Classification (1998): J.3, I.2, I.5, C.2.4, H.3.4, D.1, D.2

LNCS Sublibrary: SL 1 – Theoretical Computer Science and General Issues

Typesetting: Camera-ready by author, data conversion by Scientific Publishing Services, Chennai, India

Printed on acid-free paper

Springer is part of Springer Science+Business Media (www.springer.com)

Preface

We are proud to present the set of final accepted papers for the eleventh edition of the IWANN conference "International Work-Conference on Artificial Neural Networks" held in Torremolinos (Spain) during June 8–10, 2011.

IWANN is a biennial conference that seeks to provide a discussion forum for scientists, engineers, educators and students about the latest ideas and realizations in the foundations, theory, models and applications of hybrid systems inspired by nature (neural networks, fuzzy logic and evolutionary systems) as well as in emerging areas related to the above items. As in previous editions of IWANN, this year's event also aimed to create a friendly environment that could lead to the establishment of scientific collaborations and exchanges among attendees. Since the first edition in Granada (LNCS 540, 1991), the conference has evolved and matured. The list of topics in the successive Call for Papers has also evolved, resulting in the following list for the present edition:

1. **Mathematical and theoretical methods in computational intelligence**: Mathematics for neural networks; RBF structures; Self-organizing networks and methods; Support vector machines and kernel methods; Fuzzy logic; Evolutionary and genetic algorithms
2. **Neurocomputational formulations**: Single-neuron modelling; Perceptual modelling; System-level neural modelling; Spiking neurons; Models of biological learning
3. **Learning and adaptation**: Adaptive systems; Imitation learning; Reconfigurable systems; Supervised, non-supervised, reinforcement and statistical algorithms
4. **Emulation of cognitive functions**: Decision making; Multi-agent systems; Sensor mesh; Natural language; Pattern recognition; Perceptual and motor functions (visual, auditory, tactile, virtual reality, etc.); Robotics; Planning motor control
5. **Bio-inspired systems and neuro-engineering**: Embedded intelligent systems; Evolvable computing; Evolving hardware; Microelectronics for neural, fuzzy and bioinspired systems; Neural prostheses; Retinomorphic systems; Brain–computer interfaces (BCI) nanosystems; Nanocognitive systems
6. **Hybrid intelligent systems**: Soft computing; Neuro-fuzzy systems; Neuro-evolutionary systems; Neuro-swarm; Hybridization with novel computing paradigms: Qantum computing, DNA computing, membrane computing; Neural dynamic logic and other methods; etc.
7. **Applications**: Image and signal processing; Ambient intelligence; Biomimetic applications; System identification, process control, and manufacturing; Computational biology and bioinformatics; Internet modeling, communication and networking; Intelligent systems in education; Human–robot interaction. Multi-agent systems; Time series analysis and prediction; Data mining and knowledge discovery

At the end of the submission process, we had 202 papers on the above topics. After a careful peer-review and evaluation process (each submission was reviewed by at least 2, and on average 2.4, Program Committee members or additional reviewer), 154 papers were accepted for oral or poster presentation, according to the recommendations of reviewers and the authors' preferences.

It is important to note that for the sake of consistency and readability of the book, the presented papers are not organized as they were presented in the IWANN 2011 sessions, but classified under 21 chapters and with one chapter on the associated satellite workshop. The organization of the papers is in two volumes and arranged following the topics list included in the call for papers. The first volume (LNCS 6691), entitled *Advances in Computational Intelligence. Part I* is divided into ten main parts and includes the contributions on:

1. Mathematical and theoretical methods in computational intelligence
2. Learning and adaptation
3. Bio-inspired systems and neuro-engineering
4. Hybrid intelligent systems
5. Applications of computational intelligence
6. New applications of brain–computer interfaces
7. Optimization algorithms in graphic processing units
8. Computing languages with bio-inspired devices and multi-agent systems
9. Computational intelligence in multimedia processing
10. Biologically plausible spiking neural processing

In the second volume (LNCS 6692), with the same title as the previous volume, we have included the contributions dealing with topics of IWANN and also the contributions to the associated satellite workshop (ISCIF 2011). These contributions are grouped into 11 chapters with one chapter on the satellite workshop:

1. Video and image processing
2. Hybrid artificial neural networks: models, algorithms and data
3. Advances in machine learning for bioinformatics and computational biomedicine
4. Biometric systems for human–machine interaction
5. Data mining in biomedicine
6. Bio-inspired combinatorial optimization
7. Applying evolutionary computation and nature-inspired algorithms to formal methods
8. Recent advances on fuzzy logic and soft computing applications
9. New advances in theory and applications of ICA-based algorithms
10. Biological and bio-inspired dynamical systems
11. Interactive and cognitive environments
12. International Workshop of Intelligent Systems for Context-Based Information Fusion (ISCIF 2011)

During the present edition, the following associated satellite workshops were organized:

1. **4th International Conference on Computational Intelligence in Security for Information Systems (CISIS 2011).** CISIS aims to offer a meeting opportunity for academic and industry-related researchers belonging to the various vast communities of computational intelligence, information security, and data mining. The corresponding selected papers are published in an independent volume (LNCS 6694).
2. **International Workshop of Intelligent Systems for Context-Based Information Fusion (ISCIF 2011).** This workshop provides an international forum to present and discuss the latest scientific developments and their effective applications, to assess the impact of the approach, and to facilitate technology transfer. The selected papers are published as a separate chapter in the second volume (LNCS 6692).
3. **Third International Workshop on Ambient-Assisted Living (IWAAL).** IWAAL promotes the collaboration among researchers in this area, concentrating efforts on the quality of life, safety and health problems of elderly people at home. IWAAL papers are published in LNCS volume 6693.

The 11th edition of IWANN was organized by the Universidad de Malaga, Universidad de Granada and Universitat Politecnica de Catalunya, together with the Spanish Chapter of the IEEE Computational Intelligence Society. We wish to thank to the Spanish Ministerio de Ciencia e Innovacion and the University of Malaga for their support and grants.

We would also like to express our gratitude to the members of the different committees for their support, collaboration and good work. We specially thank the organizers of the associated satellite workshops and special session organizers. Finally, we want to thank Springer, and especially Alfred Hofmann, Anna Kramer and Erika Siebert-Cole, for their continuous support and cooperation.

June 2011

Joan Cabestany
Ignacio Rojas
Gonzalo Joya

Organization

IWANN 2011 Organizing Committee

Honorary Chairs

Alberto Prieto	University of Granada
Francisco Sandoval	University of Malaga

Conference Chairs

Joan Cabestany	Polytechnic University of Catalonia
Ignacio Rojas	University of Granada
Gonzalo Joya	University of Malaga

Technical Program Chairs

Francisco Garcia	University of Malaga
Miguel Atencia	University of Malaga

Satellite Worshops Chairs

Juan M. Corchado	University of Salamanca
Jose Bravo	University of Castilla la Mancha

Publicity and Publication Chairs

Pedro Castillo	University of Granada
Alberto Guillen	University of Granada
Beatriz Prieto	University of Granada

IWANN 2011 Program Committee

Plamen Angelov	University of Lancaster
Cecilio Angulo	Polytechnic University of Catalonia
A. Artes Rodriguez	University of Carlos III, Madrid
Antonio Bahamonde	University of Oviedo
R. Babuska	Delft University of Technology
Sergi Bermejo	Polytechnic University of Catalonia
Piero P. Bonissone	GE Global Research
Andreu Catala	Polytechnic University of Catalonia
Gert Cauwenberghs	University of California, San Diego
Jesus Cid-Sueiro	University of Carlos III, Madrid
Rafael Corchuelo	University of Seville

Óscar Cordón	European Centre for Soft Computing
Carlos Cotta	University of Malaga
Marie Cottrell	University of Paris I
Alicia D'Anjou	University of the Basque Country
Luiza De Macedo Mourelle	State University of Rio de Janeiro (UERJ)
Dante Del Corso	Polytechnic of Turin
Angel P. del Pobil	University of Jaume I, Castellon
Richard Duro	University of A Coruña
Marcos Faundez-Zanuy	Polytechnic University of Mataro
J. Manuel Ferrández	Polytechnic University of Cartagena
Kunihiko Fukushima	Takatsuki, Osaka
Chistian Gamrat	CEA, Gif sur Yvette
Patrik Garda	University Paris Sud, Orsay
F. Javier Gonzalez Cañete	University of Malaga
Karl Goser	University of Dortmund
Manuel Graña	University of the Basque Country
Anne Guerin-Dugue	Institut National Polytechnique de Grenoble
Hani Hagras	University of Essex
Alister Hamilton	University of Edinburgh
Jeanny Hérault	GIPSA-Lab, INPG, Grenoble
Luis Javier Herrera	University of Granada
Francisco Herrera	University of Granada
Cesar Hervás	University of Cordoba
Tom Heskes	Radboud University Nijmegen
Pedro Isasi	University of Carlos III, Madrid
Simon Jones	University of Loughbourough
Christian Jutten	GIPSA-lab/DIS - CNRS - Grenoble University
Kathryn Klemic	Yale University
Amaury Lendasse	Helsinki University of Technology
Kurosh Madani	University of Paris XII
Jordi Madrenas	Polytechnic University of Catalonia
Luís Magdalena	ECSC Mieres
Dario Maravall	Polytechnic University of Madrid
Bonifacio Martín Del Brio	University of Zaragoza
Francesco Masulli	University of La Spezia, Genoa
Jose M. Molina	University of Carlos III, Madrid
Augusto Montisci	University of Cagliari
Claudio Moraga	European Centre for Soft Computing
Juan M. Moreno	Polytechnic University of Catalonia
Klaus-Robert Muller	FIRST, Berlin
Jose Muñoz	University of Malaga
Alan F. Murray	Edinburgh University
Jean-Pierre Nadal	Normal Superior School, Paris

Nadia Nedjah	State University of Rio de Janeiro
Erkki Oja	Helsinki University of Technology
Madalina Olteanu	University of Paris I
Julio Ortega	University of Granada
Kevin M. Passino	The Ohio State University USA
Witold Pedrycz	University of Alberta
Francisco Pelayo	University of Granada
Vincenzo Piuri	University of Milan
Hector Pomares	University of Granada
Carlos G. Puntonet	University of Granada
Leonardo Reyneri	Polytechnic of Turin
Eduardo Ros	University of Granada
Ulrich Rueckert	University of Paderborn
Eduardo Sanchez	LSI, EPFL
Jordi Solé-Casals	University of Vic
Peter Szolgay	Pazmany Peter Catholic University
John Taylor	Kings College London, UK
Carme Torras	Polytechnic University of Catalonia
I. Burhan Turksen	TOBB Econ Technol. University, Ankara
Mark Van Rossum	University of Edinburgh
Marley Vellasco	Pontif. Catholic University of Rio de Janeiro
Alfredo Vellido	Polytechnic University of Catalonia
Michel Verleysen	Catholic University of Louvain-la-Neuve
Thomas Villmann	University of Leipzig
Changjiu Zhou	Singapore Polytechnic
Ahmed Zobaa	University of Cairo
Pedro Zufiria	Polytechnic University of Madrid

ISCIF 2011 Program Committee

José M. Molina (Co-chair)	Universidad Carlos III (Spain)
Juan M. Corchado (Co-chair)	University of Salamanca (Spain)
Jesús García (Co-chair)	Universidad Carlos III (Spain)
Javier Bajo (Co-chair)	Pontifical University of Salamanca (Spain)
James Llinas (Co-chair)	University of Buffalo (USA)
Sara Rodríguez	University of Salamanca (Spain)
Juan F. de Paz	University of Salamanca (Spain)
Carolina Zato	University of Salamanca (Spain)
Fernando de la Prieta	University of Salamanca (Spain)
Miguel Angel Patricio	Universidad Carlos III (Spain)
Antonio Berlanga	Universidad Carlos III (Spain)
Juan Gómez	Universidad Carlos III (Spain)
José María Armingol	Universidad Carlos III (Spain)
Moises Sudit	University of Buffalo (USA)

Tarunraj Singh	University of Buffalo (USA)
Lauro Snidaro	University of Udine (Italy)
Eloi Bosse	DRDC (Canada)
Subrata Das	Xerox France (France)
Vicente Julián	Technical University of Valencia (Spain)
Eugénio Oliveira	University of Porto (Portugal)
Florentino Fdez-Riverola	University of Vigo (Spain)
Masanori Akiyoshi	Osaka University (Japan)
Juan A. Botia	University of Murcia (Spain)
Luís Lima	Polytechnic Institute of Porto (Portugal)
Pawel Pawlewski	Poznan University of Technology (Poland)
Andrew Campbell	Darthmouth College (USA)
Juan Pavón	Complutense University of Madrid (Spain)
Carlos Carrascosa	Technical University of Valencia (Spain)
Ana Cristina Bicharra Garcia	Universidade Federal Fluminense (Brazil)
Irene Díaz	University of Oviedo (Spain)
Eleni Mangina	University College Dublin (Ireland)
Luís Correia	University of Lisbon (Portugal)
Miguel Reboiro	University of Vigo (Spain)

IWANN 2011 Reviewers

Carlos Affonso	Nove de Julho University
Vanessa Aguiar	University of A Coruña
Arnulfo Alanis Garza	Instituto Tecnologico de Tijuana
Amparo Alonso-Betanzos	University of A Coruña
Juan Antonio Alvarez	University of Seville
Jhon Edgar Amaya	University of Tachira
César Andrés	Complutense University of Madrid
Anastassia Angelopoulou	University of Westminster
Plamen Angelov	Lancaster University
Davide Anguita	University of Genoa
Cecilio Angulo	Polytechnic University of Catalonia
Angelo Arleo	CNRS - University Pierre and Marie Curie Paris VI
Manuel Atencia	IIIA-CSIC
Miguel Atencia	University of Malaga
Jorge Azorin	University of Alicante
Davide Bacciu	IMT Lucca School for Advanced Studies
Antonio Bahamonde	University of Oviedo at Gijón, Asturias
Halima Bahi	University of Annaba
Javier Bajo	Pont. University of Salamanca
Juan Pedro Bandera	University of Malaga
Cristian Barrué	Polytechnic University of Catalonia
Bruno Baruque	University of Burgos
David Becerra	University of the West of Scotland

Lluis A. Belanche-Munoz	Polytechnic University of Catalonia
Sergi Bermejo	Polytechnic University of Catalonia
Nicu Bizdoaca	University of Craiova
Juan Botia	University of Murcia
Julio Bregáins	University of A Coruña
Gloria Bueno	University of Castilla-La Mancha
Joan Cabestany	Polytechnic University of Catalonia
Inma P Cabrera	University of Malaga
Tomasa Calvo	University of Alcala
Jose Luis Calvo-Rolle	University of A Coruña
Mariano Carbonero-Ruz	ETEA - Cordoba University
Carlos Carrascosa	GTI-IA DSIC Universidad Politecnica de Valencia
Luis Castedo	University of A Coruña
Pedro Castillo	University of Granada
Ana Cavalli	GET/INT
Miguel Cazorla	University of Alicante
Raymond Chiong	Swinburne University of Technology
Jesus Cid-Sueiro	University of Madrid
Máximo Cobos	Universidad Politecnica de Valencia
Valentina Colla	Scuola Superiore S. Anna
Feijoo Colomine	University of Tachira
Pablo Cordero	University of Malaga
Óscar Cordón	European Centre for Soft Computing
Francesco Corona	TKK
Ulises Cortes	Polytechnic University of Catalonia
Carlos Cotta	University of Malaga
Marie Cottrell	Universite Paris I
Mario Crespo-Ramos	University of Oviedo
Raúl Cruz-Barbosa	Universidad Tecnológica de la Mixteca
Manuel Cruz-Ramírez	Departamento de Informática y Análisis Numérico
Erzsébet Csuhaj-Varjú	Hungarian Academy of Sciences
Daniela Danciu	University of Craiova
Adriana Dapena	University of A Coruña
Alberto De La Encina	Universidad Complutense
Luiza De Macedo Mourelle	State University of Rio de Janeiro (UERJ)
Suash Deb	C.V. Raman College of Engineering
José Del Campo-Ávila	University of Malaga
Angel P. Del Pobil	Jaume-I University
Enrique Dominguez	University of Malaga
Julian Dorado	University of A Coruña
Richard Duro	University of A Coruña
Gregorio Díaz	University of Castilla-La Mancha
Marta Díaz	Polytechnic University of Catalonia

Emil Eirola	Helsinki University of Technology
Patrik Eklund	Umea University
Pablo Estevez	University of Chile
Marcos Faundez-Zanuy	Escola Universitaria Politecnica de Mataro
Carlos Fernandez	University of A Coruña
J. Fernandez De Cañete	University of Malaga
Alberto Fernandez Gil	University Rey Juan Carlos
E. Fernandez-Blanco	University of A Coruña
J.C. Fernández Caballero	University of Cordoba
M. Fernández Carmona	University of Malaga
F. Fernández De Vega	University of Extremadura
Antonio Fernández Leiva	University of Malaga
F. Fernández Navarro	University of Cordoba
J. Manuel Ferrández	Universidad Politecnica de Cartagena
Anibal R. Figueiras-Vidal	Universidad Politecnica de Madrid
Oscar Fontenla-Romero	University of A Coruña
Leonardo Franco	University of Malaga
Ana Freire	University of A Coruña
Ramón Fuentes	Universidad Publica de Navarra
Colin Fyfe	University of the west of scotland
José Gallardo	University of Malaga
Jose Garcia Rodríguez	University of Alicante
Francisco Garcia-Lagos	University of Malaga
Maite Garcia-Sebastian	University of the Basque Country
Juan Miguel García	Universidad Politecnica de Valencia
Patricio García Báez	University of La Laguna
Pablo García Sánchez	University of Granada
Maribel García-Arenas	University of Granada
Esther García-Garaluz	University of Malaga
Patrick Garda	UPMC (France)
Marcos Gestal	University of A Coruña
Peter Gloesekotter	University of Applied Sciences Münster
Juan Gomez	University of Madrid
Luis González Abril	University of Seville
Jesús González Peñalver	University of Granada
Juan Gorriz	University of Granada
Karl Goser	University of Dortmund
Bernard Gosselin	Université de Mons
Jorge Gosálbez	Universidad Politecnica de Valencia
Manuel Grana	University of the Basque Country
Bertha Guijarro-Berdiñas	University of A Coruña
Nicolás Guil	University of Malaga
Alberto Guillen	University of Granada
Pedro Antonio Gutiérrez	University of Cordoba
Vanessa Gómez-Verdejo	University of Madrid

Andrei Halanay	Polytechnic University of Bucharest
Alister Hamilton	University of Edinburgh
Francisco Herrera	University of Granada
Álvaro Herrero	University of Burgos
Cesar Hervás	University of Cordoba
Tom Heskes	Radboud University Nijmegen
M. Hidalgo-Herrero	Universidad Complutense
Rob Hierons	Brunel University
Wei-Chiang Hong	School of Management, Da Yeh University
Jeanny Hérault	GIPSA-Lab, INPG, Grenoble
José Jerez	University of Malaga
M.D. Jimenez-Lopez	University of Rovira i Virgili
J.L. Jiménez Laredo	University of Granada
Simon Jones	University of Loughbourough
Gonzalo Joya	University of Malaga
Vicente Julian	GTI-IA DSIC UPV
Christian Jutten	GIPSA-lab/DIS - CNRS - Grenoble University
Jorma Laaksonen	Helsinki University of Technology
Alberto Labarga	University of Granada
Vincent Lemaire	Orange Labs
Amaury Lendasse	HUT
Paulo Lisboa	Liverpool John Moores University
Ezequiel Lopez	University of Malaga
Rafael Luque	University of Malaga
Otoniel López	Miguel Hernandez University
Guillermo López Campos	Institute of Health "Carlos III"
M.A. López Gordo	University of Granada
Kurosh Madani	LISSI / Université PARIS XII
Jordi Madrenas	Polytechnic University of Catalonia
Luís Magdalena	ECSC Mieres
Enric Xavier Martin Rull	Polytechnic University of Catalonia
Luis Martí	University of Madrid
Mario Martín	Polytechnic University of Catalonia
Bonifacio Martín Del Brio	University of Zaragoza
José Martín Guerrero	Universiy of Valencia
José Luís Martínez	University of Castilla-La Mancha
F.J. Martínez-Estudillo	ETEA
Francesco Masulli	University of Genova
Montserrat Mateos	Pont. University of Salamanca
Jesús Medina-Moreno	University of Cadiz
Mercedes Merayo	Complutense University of Madrid
Juan J. Merelo	University of Granada
Gustavo J. Meschino	National University of Mar del Plata
Jose M. Molina	University of Madrid

David Rodríguez Rueda	University of Tachira
Ignacio Rojas	University of Granada
Fernando Rojas	University of Granada
Enrique Romero	Polytechnic University of Catalonia
Samuel Romero Garcia	University of Granada
Ricardo Ron	University of Malaga
Eduardo Ros	University of Granada
Fabrice Rossi	TELECOM ParisTech
Peter Roth	Graz University of Technology
Leonardo Rubio	University of Granada
Fernando Rubio Díez	Complutense University of Madrid
Ulrich Rueckert	University of Paderborn
Nicolás Ruiz Reyes	University of Jaen
Amparo Ruiz Sepúlveda	University of Malaga
Joseph Rynkiewicz	University of Paris I
Vladimir Râsvan	University of Craiova
Addisson Salazar	Universidad Politecnica de Valencia
Sancho Salcedo-Sanz	University of Alcalá
Albert Samà	Polytechnic University of Catalonia
Miguel A. Sanchez	Pontifical University of Salamanca
Francisco Sandoval	University of Malaga
Jose Santos	University of A Coruña
J.A. Seoane Fernández	University of A Coruña
Eduardo Serrano	Autonomous University of Madrid
Olli Simula	Helsinki University of Technology
Evgeny Skvortsov	Simon Fraser University
Sergio Solinas	Università degli studi di Pavia
Jordi Solé-Casals	Universitat de Vic
Adrian Stoica	Polytechnic University of Bucharest
José Luis Subirats	University of Malaga
Peter Szolgay	Pazmany Peter Catholic University
Javier Sánchez-Monedero	University of Cordoba
Ana Maria Tomé	Universidade de Aveiro
Carme Torras	Polytechnic University of Catalonia
Claude Touzet	Université de Provence
Gracián Triviño	University of Malaga
Ricardo Téllez	Pal Robotics
Raquel Ureña	University of Granada
Olga Valenzuela	University of Granada
Germano Vallesi	Università Politecnica delle Marche - Ancona
Agustín Valverde	University of Malaga
Pablo Varona	Autonomous University of Madrid
M.A. Veganzones	University of the Basque Country
Sergio Velastín	Kingston University

Marley Vellasco	PUC-Rio
Alfredo Vellido	Polytechnic University of Catalonia
Francisco Veredas	University of Malaga
Michel Verleysen	Université catholique de Louvain
Bart Wyns	Ghent University
Vicente Zarzoso	University of Nice Sophia Antipolis
Carolina Zato	University of Salamanca
Ahmed Zobaa	University of Exeter

IWANN 2011 Invited Speakers

Hani Hagras	The Computational Intelligence Centre School of Computer Science and Electronic Engineering, University of Essex, UK
Francisco Herrera	Head of Research Group SCI2S (Soft Computing and Intelligent Information Systems), Department of Computer Science and Artificial Intelligence, University of Granada, Spain
Tom Heskes	Head of Machine Learning Group, Intelligent Systems Institute for Computing and Information Sciences (iCIS) Faculty of Science Radboud University Nijmegen, The Netherlands

IWANN 2011 Special Sessions Organizers

New Applications of Brain–Computer Interfaces

Francisco Pelayo	University of Granada
M.A. López Gordo	University of Granada
Ricardo Ron	University of Malaga

Optimization Algorithms in Graphic Processing Units

Antonio Mora	University of Granada
Maribel García-Arenas	University of Granada
Pedro Castillo	University of Granada

Computing Languages with Bio-inspired Devices

M.D. Jimenez-Lopez	University of Rovira i Virgili
A. Ortega De La Puente	Autonomous University of Madrid

Computational Intelligence in Multimedia

Adriana Dapena	University of A Coruña
Julio Bregáins	University of A Coruña
Nicolás Guil	University of Malaga

Biologically Plausible Spiking Neural Processing

Eduardo Ros	University of Granada
Richard R. Carrillo	University of Almeria

Video and Image Processing

Enrique Domínguez	University of Malaga
José García	University of Alicante

Hybrid Artificial Neural Networks: Models, Algorithms and Data

Cesar Hervás	University of Cordoba
Pedro Antonio Gutiérrez	University of Crdoba

Advances in Machine Learning for Bioinformatics and Computational Biomedicine

Paulo J.L. Lisboa	Liverpool John Moores University
Alfredo Vellido	Polytechnic University of Catalonia
Leonardo Franco	University of Malaga

Biometric Systems for Human–Machine Interaction

Alexandra Psarrou	University of Westminster
Anastassia Angelopoulou	University of Westminster
C.M. Travieso-Gonzlez	University of Las Palmas de Gran Canaria
Jordi Solé-Casals	University of Vic

Data Mining in Biomedicine

Julián Dorado	University of A Coruña
Juan R. Rabuñal	University of A Coruña
Alejandro Pazos	University of A Coruña

Bio-inspired Combinatorial Optimization

Carlos Cotta Porras	University of Malaga
Antonio J. Fernández Leiva	University of Malaga

Applying Evolutionary Computation and Nature-Inspired Algorithms to Formal Methods

Ismael Rodríguez	Complutense University of Madrid

Recent Advances on Fuzzy Logic and Soft Computing Applications

Inma P. Cabrera	University of Malaga
Pablo Cordero	University of Malaga
Manuel Ojeda-Aciego	University of Malaga

New Advances in Theory and Applications of ICA-Based Algorithms

Addison Salazar	Polytechnic University of Valencia
Luis Vergara	Polytechnic University of Valencia

Biological and Bio-inspired Dynamical Systems

Vladimir Rasvan	University of Craiova
Daniela Danciu	University of Craiova

Interactive and Cognitive Environments

Andreu Catalá	Polytechnic University of Catalonia
Cecilio Angulo	Polytechnic University of Catalonia

Table of Contents – Part II

Video and Image Processing

Hybrid Artificial Neural Networks: Models, Algorithms and Data

Advances in Machine Learning for Bioinformatics and Computational Biomedicine

Biometric Systems for Human-Machine Interaction

Data Mining in Biomedicine

Bio-inspired Combinatorial Optimization

Applying Evolutionary Computation and Nature-inspired Algorithms to Formal Methods

Recent Advances on Fuzzy Logic and Soft Computing Applications

New Advances in Theory and Applications of ICA-Based Algorithms

Biological and Bio-inspired Dynamical Systems

Interactive and Cognitive Environments

International Workshop of Intelligent Systems for Context-Based Information Fusion (ISCIF'11)

Table of Contents – Part I

Mathematical and Theoretical Methods in Computational Intelligence

Learning and Adaptation

Bio-inspired Systems and Neuro-engineering

Hybrid Intelligent Systems

Applications of Computational Intelligence

New Applications of Brain-Computer Interfaces

Optimization Algorithms in Graphic Processing Units

Computing Languages with Bio-inspired Devices and Multi-Agent Systems

Computational Intelligence in Multimedia Processing

Biologically Plausible Spiking Neural Processing

Lossy Image Compression Using a GHSOM

E.J. Palomo, E. Domínguez, R.M. Luque, and J. Muñoz

Department of Computer Science
E.T.S.I. Informatica, University of Malaga
Campus Teatinos s/n, 29071 – Malaga, Spain
{ejpalomo,enriqued,rmluque,munozp}@lcc.uma.es

Abstract. A new approach for image compression based on the GHSOM model has been proposed in this paper. The SOM has some problems related to its fixed topology and its lack of representation of hierarchical relations among input data. The GHSOM solves these limitations by generating a hierarchical architecture that is automatically determined according to the input data and reflects the inherent hierarchical relations among them. These advantages can be utilized to perform a compression of an image, where the size of the codebook (leaf neurons in the hierarchy) is automatically established. Moreover, this hierarchy provides a different compression at each layer, where the deeper the layer, the lower the rate compression and the higher the quality of the compressed image. Thus, different trade-offs between compression rate and quality are given by the architecture. Also, the size of the codebooks and the depth of the hierarchy can be controlled by two parameters. Experimental results confirm the performance of this approach.

Keywords: Image compression, data clustering, self-organization.

1 Introduction

Image compression approaches are classified as lossy and lossless. Color quantization is one of the most useful lossy compression methods to find an acceptable set of palette color (codebook) that can be used to represent the original colors of a digital image. Generally, a full color digital image use red, green, and blue channel (each is almost 8-bit resolution) to specify the color of each pixel. The image usually is composed of a large number of distinguishable colors; although the human eye can only distinguish less than a thousand of colors (8-bit indexed color is sufficient for human perception). Therefore the color quantization problem can be modeled as the cluster analysis problem.

Color quantization (CQ) is a typical image processing task used to cluster and compress color images by selecting a small number of code vectors from a set of available colors to represent a high color resolution image with minimum perceptual distortion. CQ is used to reduce the storage requirements and the transmission bandwidth of color images. The importance of CQ is increasing due to the increase of the transmission number and storage of images. Furthermore, this task is essential for applications as videoconference, multimedia, storage

J. Cabestany, I. Rojas, and G. Joya (Eds.): IWANN 2011, Part II, LNCS 6692, pp. 1–8, 2011.

of images and transmission throughout channels limited in band. The goal of CQ is to reduce the bit rate for transmission or data storage while maintaining acceptable image fidelity.

The process of CQ requires the design of a finite set of vectors reference (codebook) that it will be used to substitute parts of the image with the minimum of error or distortion in relation to the original image. If the codebook is properly designed, the loss of visual quality will be minimum. There are several well-known codebook design algorithms such as k-means algorithm [1], fuzzy c-means [2], competitive learning [3], self-organizing map [4], and their variants. To achieve a good overall rate-distortion performance, it is important that the color quantizer possesses strong topological clustering property to preserve the neighboring pixel relationship in the mapping.

The self-organizing map (SOM) is one of the most popular algorithms in the compression of images with a good performance regarding rate of compression. The SOM is a widely used unsupervised neural network for clustering high-dimensional input data and mapping these data into a two-dimensional representation space [4]. However, it has some drawbacks. The number and arrangement of neurons (network architecture) of the SOM is static and has to be established in advance. This task can be difficult because it needs a prior study of the problem domain, especially when we have vectors with many features. Moreover, the high complexity of search is other well-known common disadvantage of the SOM. The growing hierarchical SOM (GHSOM) tries to face these problems derived from SOM. The GHSOM has a hierarchical architecture arranged in layers, where each layer is composed of different growing SOMs expanded from neurons of the upper layer maps and the number of neurons of each map are adaptively determined [5]. This way, the architecture of the GHSOM is established during the unsupervised learning process according to the input data.

The remainder of this paper is organized as follows. In Section 2, a description of the proposed GHSOM is presented. In Section 3, the performance of the proposed GHSOM is evaluated and compared to other traditional algorithms. Finally, some remarks conclude this paper in Section 4.

2 The GHSOM Model

In order to solve the limitations of the SOM related to its fixed topology and its lack of representation of hierarchical relation among data, the GHSOM was proposed in [5] as an artificial neural network with a hierarchical architecture, where SOM-like neural networks with adaptive architecture [6] build the various layers of the hierarchy. Initially, the GHSOM consists of a single SOM of 2x2 neurons. After the training process of a map has ended, this map can grow by adding neurons until is reached a certain level of detail in the representation of the data mapped onto the SOM. After growing, each neuron of the map can be expanded in a new map in the next layer of the hierarchy in order to provide a more detailed representation of the data mapped onto that SOM. Once GHSOM training has finished, the resulting architecture reflects the inherent structure of the input patterns, improving the representation achieved with a single SOM. Therefore, each neuron represents a data cluster, where data belonging to one cluster are more similar than data belonging to different clusters. An example of a GHSOM architecture is shown in Fig.1.

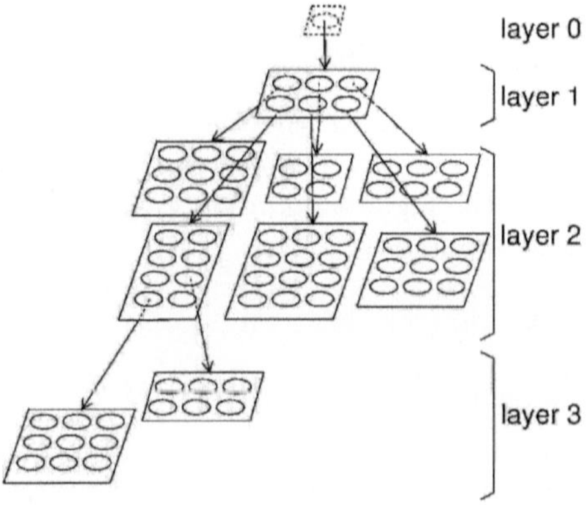

Fig. 1. An example of a GHSOM architecture

The adaptative growth process of a GHSOM is guided by two parameters, τ_1 and τ_2, which are used to control both the growth of a map and the neural expansion of the GHSOM, respectively. In the GHSOM, the growing of a map is done by inserting a row or a column of neurons between two neurons, the neuron with the highest quantization error and its most dissimilar neighbor. The quantization error (qe) is a measure of the similarity of data mapped onto each neuron, where the higher the qe is, the higher the heterogeneity of the data cluster is. The qe of a neuron i is defined as follows:

$$qe_i = \sum_{x_j \in C_i} \|w_i - x_j\|. \tag{1}$$

where w_i is the weight vector of the neuron i, C_j is the set of input patterns mapped onto the neuron i and x_j is the jth input pattern. The quantization error of a neuron can also be expressed as given in (2), which denotes the mean quantization error of a neuron i, where n_C is the number of elements of the set of input vectors C_i mapped onto the neuron i. While the qe leads to a finer representation of denser clusters, the mqe does the same with clusters with a high quantization error but regardless of their sizes. In this paper, the qe is used instead the mqe since we prefer to represent the most populated clusters in a higher level of detail, i.e. the clusters with a higher number of pixels associated.

$$mqe_i = \frac{1}{n_C} \cdot \sum_{x_j \in C_i} \|w_i - x_j\|,$$
$$n_C = \|C_i\|, C_i \neq \emptyset. \tag{2}$$

The stopping criterion for the growth of a map is defined in (3), where MQE_m is the mean of the quantization error of the map m and qe_u is quantization error of the parent neuron u in the upper layer. Thus, the growth process continues until the MQE_m of the map m reaches a certain fraction τ_1 of the quantization error of the corresponding parent neuron u in the upper layer.

$$MQE_m < \tau_1 \cdot qe_u. \tag{3}$$

If the above condition is not fulfilled, the map grows and a row or a column of neurons is inserted. These neurons are inserted between the error neuron (e), which is the neuron with the highest quantization error, and its most dissimilar neighbor (d), which is computed as

$$d = \arg\max_i \|w_e - w_i\| \qquad w_i \in \Lambda_e \tag{4}$$

where Λ_e is the set of neighbor neurons of e.

The stopping criterion for the expansion of a neuron i is defined in (5). Specifically, a neuron i is expanded into a new map at a subsequent layer unless its quantization error (qe_i) is smaller than a fraction τ_2 of the initial quantization error (qe_0).

$$qe_i < \tau_2 \cdot qe_0. \tag{5}$$

Neurons from the newly created maps are initialized in a coherent way, so the weight vectors of neurons mirror the orientation of the weight vectors of the neighbor neurons of its parent [7]. The initialization proposed computes the mean of the parent and its neighbors in their respective directions. This initialization provides a global orientation of the maps in each layer [5].

3 Experimental Results

In this section, the performance of the GHSOM for image compression is presented and discussed. Three 512x512 color images widely adopted in the image processing literature (Lena, pepper and baboon) have been selected. The images have 148,279, 230,427 and 183,525 colors, respectively, being baboon the most complex. These images are shown in Figure 2.

Each image was converted into a dataset in order to be presented to the GHSOM. This data represents pixels of the image and has three features, which correspond to the RGB components. The training was done during 5 epochs and different values of τ_1 and τ_2 parameters. The tuning of these parameters determines the minimum level of compression of the image and the maximum level of quality of the compressed image. The smaller the τ_1 and τ_2 parameters, the bigger the size of the maps and the deeper the hierarchy, respectively, which involves an increase in the number of neurons and, therefore, an increase in the number of colors used to represent the original image. The resulting architecture represents a hierarchical color structure of the image. Each layer stores a different compression of the image so the deeper the layer, the lower the compression and

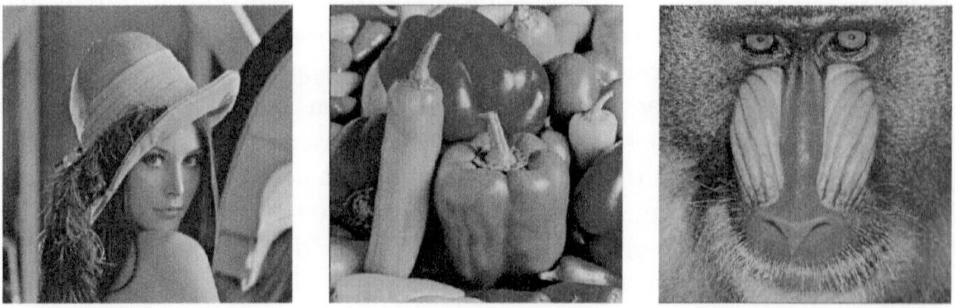

(a) (b) (c)

Fig. 2. Original 512x512 images: (a) Lena, (b) pepper and (c) baboon

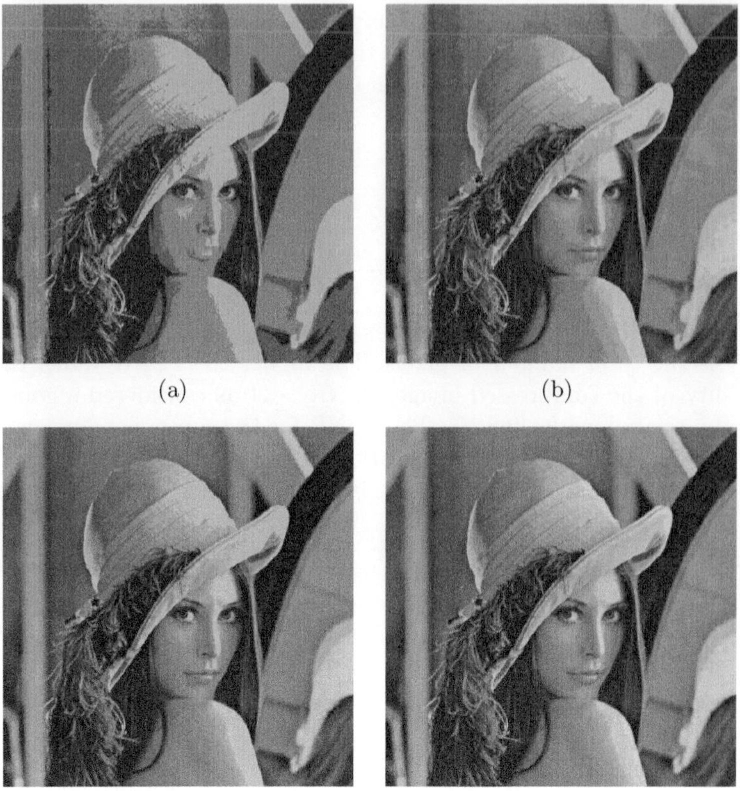

(a) (b)

(c) (d)

Fig. 3. Example of hierarchical compression for the Lena image after training with $\tau_1 = 0.5$ and $\tau_2 = 0.005$, which generated four layers. Compression achieved up to: (a) layer 1 (4 colors), (b) layer 2 (16 colors), (c) layer 3 (55 colors) and (d) layer 4 (70 colors).

the better the quality of the representation of the image. This hierarchy allows to choose among different trade-offs between compression and quality. Since it is better to have deeper hierarchies instead of bigger maps in order to have different compressions at each layer, τ_1 was set to 0.5, whereas for τ_2 the values 0.5, 0.05, 0.005 and 0.0005 were chosen. Each neuron in a layer represents the prototype of the colors of the pixels that were mapped to that neuron, i.e. a code vector of the codebook. For a layer, its codebook is composed of the leaf neurons (LNs) up to that layer. The quantization error (qe) instead of the mean quantization error (mqe) was chosen since it is better to represent the most populated neurons in a higher level of detail for image compression. Therefore, code vectors that represent a heterogeneous group of colors are represented by more code vectors in the next layer of the hierarchy, leading to a reduction of the mean square error (MSE). An example of hierarchical image compression for the Lena image is shown in Figure 3.

In order to assess the quality of compressed images, the peak signal-to-noise ratio (PSNR) [8] is the metric used in this paper. The PSNR shows how a compressed image is related to the original image. The PSNR is defined as follows:

$$PSNR = 10 \times log\left(\frac{3 \times 255^2}{MSE}\right) \tag{6}$$

$$MSE = \frac{\left(\sum_{j=0}^{N_t-1}(X_j - X_j')^2\right)}{N_t} \tag{7}$$

where X_j and X_j' are the pixel values of the original and compressed image, and N_t is the total number of pixels. The higher the PSNR value, the better the quality of the compressed image. PSNR > 30 is considered a good level of quality by related works [9,8,10]. The PSNR for the compressed image with the maximum quality from each combination of the τ_1 and τ_2 parameters for the Lena, pepper and baboon images is given in Table 1. The τ_1 parameter was

Table 1. GHSOM image compression with maximum quality for the Lena, pepper and baboon images

Image	τ_2	Layers	LNs (colors)	PSNR (dB)
Lena	0.5	1	4	28.54
	0.05	2	16	33.19
	0.005	4	70	38.03
	0.0005	7	421	43.02
pepper	0.5	1	4	25.47
	0.05	2	16	30.34
	0.005	4	61	35.32
	0.0005	6	388	40.38
baboon	0.5	1	4	24.36
	0.05	2	16	28.75
	0.005	4	85	33.62
	0.0005	6	571	39.038

Table 2. PSNR (in dB) of varios algorithms with a codebook size of 16 for the Lena, pepper and baboon images

Image	CL	LBG	SOM	GHSOM
Lena	28.80	29.65	29.61	**33.19**
pepper	26.80	26.74	26.70	**30.34**
baboon	24.78	24.91	24.85	**28.75**

fixed to 0.5 and the τ_2 parameter was progressively decreased to obtain deeper hierarchies. LN stands for leaf neurons, which are the different colors used in the representation.

The results presented in this paper have been compared with those achieved by the competetive learning (CL), the Linde, Buzo and Gray (LBG) algorithm [1] and the self-organizing map (SOM) [4] for the same color images. These results were extracted from [8]. A codebook size of 16 was selected for this comparison since it was the only one that coincides with our automatically determined codebook sizes.

4 Conclusions

A novel method for image compression using the GHSOM algorithm has been proposed in this paper. The GHSOM model solves the limitations of the SOM related to its fixed topology and its lack of representation of hierarchical relations among input data. The entire architecture (neurons, maps and layers) of the GHSOM is automatically generated after the training process according to the input data. Moreover, this architecture reflects the inherent hierarchical relations among input data.

By presenting a color image to the GHSOM, a hierarchical compression of the image is obtained encoded in the neurons of the architecture. Since each leaf neuron in the GHSOM represents a color of the compressed image, the codebook sizes are automatically determined by the GHSOM. The size of the codebook can be tuned by the τ_1 and τ_2 parameters, which control the size of the maps and the depth of hierarchy. The lower these parameters are, the larger the number of neurons and, consequently, the size of the codebooks are. Furthermore, the resulting hierarchical structure provides a different compression at each layer of the hierarchy, whereupon different trade-offs between compression rate and quality are given by the architecture. Thus, the deeper the layer is, the lower the rate compression and the higher the quality of the compressed image is.

The performance of this approach has been measured by computing the PSNR of the resulting compression of three different color images presented to the GHSOM. The obtained results have been compared with others based on unsupervised learning, overcoming the obtained results of these models.

Acknowledgements

This work is partially supported by the Ministry of Science and Innovation of Spain under grant TIN2010-15351, project name Probabilistic self organizing models for the restoration of lossy compressed images and video.

References

1. Linde, Y., Buzo, A., Gray, R.: An algorithm for vector quantizer design. IEEE Transactions on Communications 28(1), 84–95 (1980)
2. Bezdek, J.C.: Pattern Recognition with Fuzzy Objective Function Algoritms. Plenum Press, New York (1981)
3. Hertz, J., Krogh, A., Palmer, R.: Introduction to the Theory of Neural Computation. Addison-Wesley, Reading (1991)
4. Kohonen, T.: Self-organized formation of topologically correct feature maps. Biological cybernetics 43(1), 59–69 (1982)
5. Rauber, A., Merkl, D., Dittenbach, M.: The growing hierarchical self-organizing map: Exploratory analysis of high-dimensional data. IEEE Transactions on Neural Networks 13(6), 1331–1341 (2002)
6. Alahakoon, D., Halgamuge, S., Srinivasan, B.: Dynamic self-organizing maps with controlled growth for knowledge discovery. IEEE Transactions on Neural Networks 11, 601–614 (2000)
7. Dittenbach, M., Rauber, A., Merkl, D.: Recent advances with the growing hierarchical self-organizing map. In: Allinson, N., Yin, H., Allinson, L., Shek, J. (eds.) 3rd Workshop on Self-Organising Maps (WSOM 2001), pp. 140–145 (2001)
8. Chang, C.H., Pengfei, X., Xiao, R., Srikanthan, T.: New adaptive color quantization method based on self-organizing maps. IEEE Transactions on Neural Networks 16(1), 237–249 (2005)
9. Araujo, A., Costa, D.: Local adaptive receptive field self-organizing map for image color segmentation. Image and Vision Computing 27(9), 1229–1239 (2009)
10. Kanjanawanishkul, K., Uyyanonvara, B.: Novel fast color reduction algorithm for time-constrained applications. Journal of Visual Communication and Image Representation 16(3), 311–333 (2005)

Visual Features Extraction Based Egomotion Calculation from a Infrared Time-of-Flight Camera

Diego Viejo, Jose Garcia, and Miguel Cazorla

Instituto de Investigación en Informática
University of Alicante. 03080 Alicante, Spain
dviejo@dccia.ua.es, jgarcia@dtic.ua.es, miguel.cazorla@ua.es

Abstract. 3D data have been used for robotics tasks in the last years. These data provide valuable information about the robot environment. Traditionally, stereo cameras has been used to obtain 3D data, but these kind of cameras do not provide information in the lack of texture. There is a new camera, SR4000, which uses infrared light in order to get richer information. In this paper we first analyze this camera. Then, we detail an efficient ICP-like method to build complete 3D models combining Growing Neural Gas (GNG) and visual features. First, we adapt the GNG to the 3D cloud points. Then, we propose the calculation of visual features and its registration to the elements of the GNG. Finally, we use correspondences between frames, an ICP-like method to calculate egomotion. Results of mapping from the egomotion are shown.

Keywords: GNG, tof camera, visual features, 3D reconstruction.

1 Introduction

One of the central research themes in mobile robotics is the determination of the movement performed by the robot using its sensors information. The methods related with this research are called pose registration and can be used for automatic map building and SLAM [5]. Our main goal is to perform six degrees of freedom (6DoF) pose registration in semi-structured environments, i.e., man-made indoor and outdoor environments. This registration can provide a good start point for SLAM.

Using 3D information in order to get the 6DoF transformation from the robot (egomotion) is not an easy task. Although several approaches have been used (ICP [3], [1], Ransac [6], etc.) all those approaches do not work in the presence of outliers (features seen in one frame and not seen in the other). The greater the robot movement the greater the number of outliers are, and the classical methods do not provide good results. In this paper, we propose the use of visual features (like Sift [9]) from the 2D image together with a 3D representation of the scene based on a Growing Neural Gas (GNG) [7]. By means of a competitive learning, it makes an adaptation of the reference vectors of the neurons as well

J. Cabestany, I. Rojas, and G. Joya (Eds.): IWANN 2011, Part II, LNCS 6692, pp. 9–16, 2011.

as the interconnection network among them; obtaining a mapping that tries to preserve the topology of an input space. Besides, they are capable of a continuous re-adaptation process even if new patterns are entered, with no need to reset the learning. These features allows to represent fast and high quality 3D spaces, obtaining an induced Delaunay Triangulation of the input space very useful to easily obtain features like corners, edges and so on. We modify the original GNG method to be applied to sequences: the GNG is adapted sequentially, i.e. the result in a given frame is taken as input in the next frame.

The rest of the paper is organized as follows: first, a section describing SR4000 camera used for experiments; then, the GNG algorithm is explained; In section 4 the visual features used are also explained and in section 5 the method to find the egomotion is detailed; the experimental section will show our modeling results, finishing with our conclusions and future work in the last section.

2 Time-of-Flight 3D Camera

In recent years, Time-of-Flight (ToF) cameras are being developed as a new technology that delivers range (distance) and amplitude maps by the use of a modulated light source. The main advantages with respect to other 3D devices are the possibility to acquire data at video frame rates and to obtain 3D point clouds without scanning and from just one point of view.

The basic principle of ToF cameras consists of an amplitude-modulated infrared light source and a sensor field that measures the intensity of backscattered infrared light. The infrared source is constantly emitting light that varies sinusoidally. Objects that have different distances are reached by different parts of the sinusoidal wave. The reflected light is then compare to the original one, calculating the phase shift, by means of measuring the intensity of the incoming light since the phase shift is proportional to the time of flight of the light reflected by a distant object. A detailed description of the time-of-flight principle can be found in [8]. The device used in this work is the SwissRanger SR4000 ToF camera, shown in Figure 1.

In our tests all the data were acquired directly from the camera, which delivers point coordinates XYZ, amplitude data of the scene and a confidence map of the distance measurements. In particular, the confidence map is obtained using a combination of distance and amplitude measurements and their temporal variations: it represents a measure of probability that the distance measurement of each pixel is correct, so it can be useful to select regions containing high quality measurements or to reject low quality ones. In our experiments the amplitude data has low contrast so they have been equalized. Figure 1 shows an overview for the typical data obtained with the SR4000. The recorded 3D points cloud can be observed at the figure top center, corresponding amplitude by the left side and confidence by the right. Reference camera coordinates system is also shown.

ToF cameras allow to generate point clouds during real time acquisition. The accuracy of ToF cameras varies with internal components and the characteristics

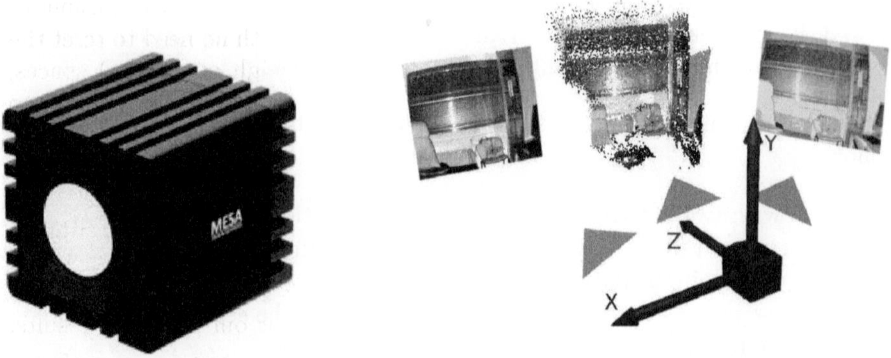

Fig. 1. Left: SR4000 camera. Right: camera data overview. The SR4000 captures both a 3D point set and two maps: amplitude (left) and confidence (right).

of the observed scene, such as objects reflectivity and ambient lighting conditions. These errors cannot be fully eliminated, but they can be reduced and optimized thanks to filtering or several techniques, such us averaging techniques or calibration procedures [4] where a distance error model was proposed which provided a reduction of distance errors in the 1.5-4m distance measurement range.

Integration Time is one of the most important camera parameters. Adjusting this value parameter can control how long each sensor pixel collects light. For lowest noise measurements the integration time should be adjusted so that all (or at least most) pixels collect as much light as possible without saturation. On the other hand if a high frame rate is more important then the integration time may be reduced to achieve the desired frame rate. The camera software allows to automatically adjust the Integration Time depending on the maximum amplitude present in the current image. This setting can be used to avoid pixel saturation and to achieve a good balance between noise and high frame rate.

3 GNG Algorithm

With Growing Neural Gas (GNG) [7] a growth process takes place from minimal network size and new units are inserted successively using a particular type of vector quantization. To determine where to insert new units, local error measures are gathered during the adaptation process and each new unit is inserted near the unit which has the highest accumulated error. At each adaptation step a connection between the winner and the second-nearest unit is created as dictated by the competitive Hebbian learning algorithm. This is continued until an ending condition is fulfilled, as for example evaluation of the optimal network topology or time deadline. The network is specified as:

- A set N of nodes (neurons). Each neuron $c \in N$ has its associated reference vector $w_c \in R^d$. The reference vectors can be regarded as positions in the input space of their corresponding neurons.

- A set of edges (connections) between pairs of neurons. These connections are not weighted and its purpose is to define the topological structure. An edge aging scheme is used to remove connections that are invalid due to the motion of the neuron during the adaptation process.

The GNG learning algorithm to map the network to the input manifold is as follows:

1. Start with two neurons a and b at random positions w_a and w_b in R^d.
2. Generate at random an input pattern ξ according to the data distribution $P(\xi)$ of each input pattern.
3. Find the nearest neuron (winner neuron) s_1 and the second nearest s_2.
4. Increase the age of all the edges emanating from s_1.
5. Add the squared distance between the input signal and the winner neuron to a counter error of s_1 such as:

$$\triangle error(s_1) = \|w_{s_1} - \xi\|^2 \tag{1}$$

6. Move the winner neuron s_1 and its topological neighbors (neurons connected to s_1) towards ξ by a learning step ϵ_w and ϵ_n, respectively, of the total distance:

$$\triangle w_{s_1} = \epsilon_w(\xi - w_{s_1}) \tag{2}$$

$$\triangle w_{s_n} = \epsilon_n(\xi - w_{s_n}) \tag{3}$$

   or all direct neighbors n of s_1.

7. If s_1 and s_2 are connected by an edge, set the age of this edge to 0. If it does not exist, create it.
8. Remove the edges larger than a_{max} . If this results in isolated neurons (without emanating edges), remove them as well.
9. Every certain number λ of input patterns generated, insert a new neuron as follows:
   - Determine the neuron q with the maximum accumulated error.
   - Insert a new neuron r between q and its further neighbor f:

$$w_r = 0.5(w_q + w_f) \tag{4}$$

   - Insert new edges connecting the neuron r with neurons q and f, removing the old edge between q and f.
10. Decrease the error variables of neurons q and f multiplying them with a consistent α. Initialize the error variable of r with the new value of the error variable of q and f.
11. Decrease all error variables by multiplying them with a constant γ.
12. If the stopping criterion is not yet achieved (in our case the stopping criterion is the number of neurons), go to step 2.

With regard to the processing of image sequences, we have introduced several improvements to the network to accelerate the representation and allow the architecture to work faster.

The main difference with the GNG algorithm is the omission of insertion/deletion actions (steps 8 to 11) after first frame. For the initial moment t_0 the representation is obtained making a complete adaptation of a GNG. However, for the following frames the previous network structure is employed. So, the new representation is obtained by performing the iteration of the internal loop of the learning algorithm of the GNG, relocating the neurons and creating or removing edges.

For the experiments, the GNG parameters used are: $N = 2000$, $\lambda = 2000$, $\epsilon_w = 0.1$, $\epsilon_n = 0.001$, $\alpha = 0.5$, $\beta = 0.95$, $\alpha_{max} = 250$. In Figure 2 a result of applying GNG to a 3D points from a SR4000 is shown.

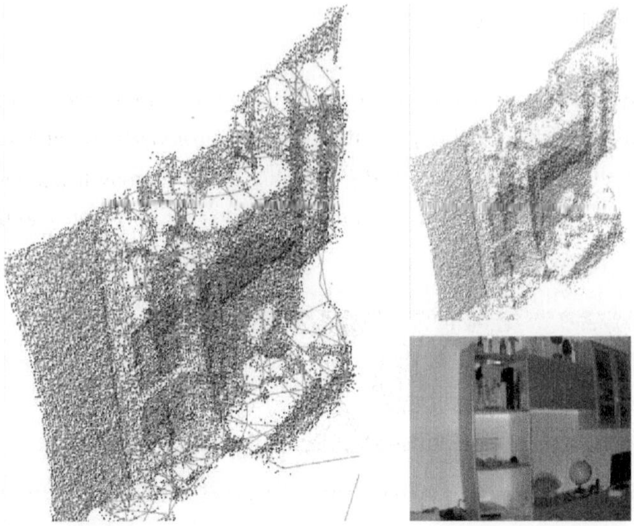

Fig. 2. Applying GNG to SR4000 data set

4 Applying a Feature Extraction Algorithm to Amplitude Images

In this section we are going to test how good are images from this camera in order to apply a feature extraction algorithm. To do that, we compute 6DoF egomotion using a ToF camera. This is an adaptation of the method proposed in [2]. In this paper a new approach for computing egomotion from stereo images is proposed. Basically, the process to obtain the transformation between two consecutive images consists of extracting features from each input image, translating the features to the 3D space using the sensor geometry, matching those 3D features using their descriptors from each input image and, finally, computing the transformation that best aligns the matched features. The main difference from a stereo vision camera and our SR4000 is that ours does not capture an image in the visible spectra but in the infrared. Therefore, we use the infrared image for extracting visual features. For the experiments presented in this paper we are using SIFT [9], described below.

One of the most used visual feature is SIFT [9], a method used in computer vision to detect and describe features in an image. It performs a local pixel appearance analysis at different scales. The SIFT features are designed to be invariant to image scale and rotation. Furthermore, it obtains a descriptor for each feature that can be used for different tasks such as object recognition. SIFT algorithm is divided into two main parts. In the first one, the location of the points of interest is extracted. The image is convolved using a Gaussian filter at different standard deviations σ. Then, the difference of Gaussians (DoG) is computed as the difference between two consecutive Gaussian-convolved images. This process is repeated in order to obtain the DoG for the input image at different scales. The localization of the points of interest starts when all DoG have been computed. A point located in a DoG is considered as a point interest if it has the maximum/minimum value compared with its 8-neighbours in the same DoG and with the 9-neighbours in the adjacent DoG at superior and inferior scale. The localization of the points of interest is then improved by interpolating nearby data, discarding low-contrast points and eliminating the edge responses. In the second part of the algorithm a descriptor vector is computed for each point of interest. Based on the image gradient around a point of interest an orientation for this point is computed. This orientation represents the starting point from where the descriptor array is computed. This is a 128-element array that holds the information about 16 histograms of 8 bins computed from the same gradient data. Finally, this descriptor vector is normalized in order to enhance invariance to changes in illumination.

We extract features for the amplitude infrared image. In the next step, we obtain GNG structure from the point cloud and hook features to the GNG. We search for the closest element of the GNG structure. We take advantage of the confidence image provided by our camera, removing those features we can not trust. This is an important difference from the stereo systems since it makes enable an important accuracy improvement as we remove erroneous points.

5 Egomotion from Visual Features

In order to get the egomotion of the vehicle, we present an ICP-like method to match 3D features. We have decided to select features close to the camera, because the longer the distance to the camera, the greater the 3D error. Thus, only features with a Z distance below a threshold are selected to match between two consecutive sets. We have to consider the movement between two consecutive frames, in order to select a set of features in both frames which intersect and have enough number of features to match. If movements are limited to, for example

1 meter, we select features from 1 to 2 meters in the first frame and from 0 to 1 in the second one. If there are not enough matches, we expand the limits from 1 to 3 meters and from 0 to 2, and so on to find a minimal number of matches or to reach a long distance (10 or 20 meters, depending on the baseline).

Once we have found matches between two consecutive frames, we apply an ICP-like algorithm to find the 3D transformation between frames. ICP is a classical algorithm used to match two 3D point sets, but it can not find a good alignment in the presence of outliers. For long movements, ICP does not give good results, because there are a lot of outliers. Using features like Sift, with an additional information, i.e. descriptors which are robust to brightness change and point of view change, are good enough for this task. So we use descriptors to find matches, instead of using Euclidean distance like the original ICP.

6 Results

We have used an indoor sequence to test the reliability of our method. The sequence has 300 frames taken at a frequency of 3Hz. Due to the low brightness of amplitude images, we first need to increase the image brightness. The mean time to calculate egomotion from two frames was 0.003s and the mean error 25mm, using 11 matches in order to get the correspondences. The mean time to get the features in a given frame was less than 0.02s. GNG process takes 1s for each frame, which provides a good method for real time problems.

Fig. 3. Mapping results using SIFT features. The red points indicate the path follow by the robot. The black points are the reconstruction once the pose registration is done.

7 Conclusions and Future Work

In this paper we have presented a method which is able to find the egomotion from a sequence of images. Data comes from a infrared camera, SR4000, which is able to obtain data in the lack of texture. The proposed method calculates a GNG over the point cloud. This fastly provides a 3D structure which has less information than the original 3D data, but keeping the 3D topology. Then, visual features, like SIFT, are calculated from the amplitude image and are attached to the GNG. The egomotion method uses an ICP-like algorithm, using correspondences among features as a matching criteria. The results presented are good enough for mapping and could be a good starting point for SLAM.

As future work, we want to extract and test other visual features.

Acknowledgments

This work has been supported by grant DPI2009-07144 from Ministerio de Ciencia e Innovacion of the Spanish Government and by the University of Alicante project GRE09-16.

References

1. Besl, P.J., McKay, N.D.: A method for registration of 3-d shapes. IEEE Trans. on Pattern Analysis and Machine Intelligence 14(2), 239–256 (1992)
2. Cazorla, M., Viejo, D., Hernandez, A., Nieto, J., Nebot, E.: Large scale egomotion and error analysis with visual features. Journal of Physical Agents 4, 19–24 (2010)
3. Chen, Y., Medioni, G.: Object modeling by registration of multiple range images. In: Medioni, G. (ed.) Proceedings of IEEE International Conference on Robotics and Automation, vol. 3, pp. 2724–2729 (1991)
4. Chiabrando, F., Chiabrando, R., Piatti, D., Rinaudo, F.: Sensors for 3d imaging: Metric evaluation and calibration of a ccd/cmos time-of-flight camera. Sensors 9(12), 10080–10096 (2009)
5. Dissanayake, M.W.M.G., Newman, P., Clark, S., Durrant-Whyte, H.F., Csorba, M.: A solution to the simultaneous localization and map building (slam) problem. IEEE Transactions on Robotics and Automation 17(3), 229–241 (2001)
6. Fischler, M.A., Bolles, R.C.: Random sample consensus: a paradigm for model fitting with applications to image analysis and automated cartography. Commun. ACM 24(6), 381–395 (1981)
7. Fritzke, B.: A Growing Neural Gas Network Learns Topologies, vol. 7, pp. 625–632. MIT Press, Cambridge (1995)
8. Burak Gokturk, S., Yalcin, H., Bamji, C.: A time-of-flight depth sensor - system description, issues and solutions. In: CVPRW 2004: Proceedings of the 2004 Conference on Computer Vision and Pattern Recognition Workshop (CVPRW 2004), vol. 3, p. 35. IEEE Computer Society, Washington, DC, USA (2004)
9. David, G.L.: Distinctive image features from scale-invariant keypoints. International Journal of Computer Vision 60, 91–110 (2004)

Feature Weighting in Competitive Learning for Multiple Object Tracking in Video Sequences

R.M. Luque, J.M. Ortiz-de-Lazcano-Lobato, Ezequiel López-Rubio, E. Domínguez, and E.J. Palomo

Department of Computer Languages and Computer Science
Bulevar Louis Pasteur, 35. 29071 Málaga. Spain
University of Málaga
{rmluque,jmortiz,ezeqlr,enriqued,ejpalomo}@lcc.uma.es

Abstract. Object tracking in video sequences remains as one of the most challenging problems in computer vision. Object occlusion, sudden trajectory changes and other difficulties still wait for comprehensive solutions. Here we propose a feature weighting method which is able to discover the most relevant features for this task, and a competitive learning neural network which takes advantage of such information in order to produce consistent estimates of the trajectories of the objects. The feature weighting is done with the help of a genetic algorithm, and each unit of the neural network remembers its past history so that sudden movements are adequately accounted for. Computational experiments with real and artificial data demonstrate the performance of the proposed system when compared to the standard Kalman filter.

1 Introduction

In computer vision system design, object tracking is a fundamental module whose performance is key to achieving a correct interpretation of the observed scene [1,2,3]. This is of paramount importance in applications such as video surveillance. One of the most commonly used approaches for visual tracking is the adaptive tracking of colored regions. It comprises proposals such as particle filtering of colored regions [4,5] and the Kalman/mean-shift hybrid scheme [6], which employs the well known mean-shift algorithm [7] to determine the search region, and the Kalman filter to predict the position of the target object in the next frame.

Many features can be added to the colour of the regions in order to improve the reliability of these algorithms. However, it is difficult for a human to manually determine which features are the most significant in order to ensure a proper tracking of the moving objects in complex environments. Here we propose a principled method to obtain a good weighting of the object features by means of a genetic algorithm. This allows us to design an appropriate scaling of the feature space, so that unsupervised learning approaches can operate efficiently. In addition to this, a growing competitive neural network (GCNN) is presented, which works in combination with the feature weighting mechanism. Each unit of the network represents a potential object in the scene, and unit creation and destruction rules are described, so as to cope with entering and exiting objects.

J. Cabestany, I. Rojas, and G. Joya (Eds.): IWANN 2011, Part II, LNCS 6692, pp. 17–24, 2011.

The structure of this paper is as follows. First the competitive neural network and the feature weighting method are presented (Sections 2 and 3). Then, some important properties of our object tracking system are discussed (Section 4). Section 5 is devoted to experimental results. Finally, the conclusions of this work are outlined in Section 6.

2 Growing Competitive Neural Network

The start point of any tracking approach requires the extracted data from object detection algorithms, which is achieved by applying the method in [8]. These extracted features represent each object in each frame and are the inputs of the tracking module, whose aim is to track the detected objects along the sequence. In this paper, the tracking stage is based on a growing competitive neural network (GCNN), which follows an online training process based on a prediction-correction scheme.

The number of neurons of the network is not fixed, and it is related to the amount of objects which must be tracked by the system in each time instant. Every object appearing in the video frame is assigned to a neuron, which is the responsible for identifying and representing that object exclusively. New neurons are created when not previously detected objects appear in the image, whereas some neurons are destroyed when the objects associated to them leave the scene.

2.1 Competition Step

In a time instant t the system is provided M D-dimensional training patterns $\mathbf{x}_i(t)$, $i \in \{1 \ldots M\}$ corresponding to M objects which were detected in the video frame sampled in time instant t. Those feature vectors are managed in sequence and for each one a competition arises.

First, every neuron j predict the new state of the object that it is assigned to it, $\hat{\mathbf{x}}_j(t)$, using the P more recent entries in its log H_j.

$$\hat{\mathbf{x}}_j(t) = \mathbf{w}_j(t-1) + \left(t - H_j^w(K)\right) \sum_{i=K-P+1}^{K-1} \frac{H_j^w(i+1) - H_j^w(i)}{H_j^f(i+1) - H_j^f(i)} \tag{1}$$

where w_j is the weight vector of the j-th neuron and $H_j^w(i)$ is the object features vector which was written down in the log of that neuron in the frame $H_j^f(i)$.

Then the neuron whose predicted vector $\hat{\mathbf{x}}_j(t)$ is nearest in the input space to the input pattern is declared the winner.

$$c(\mathbf{x}(t)) = \arg \min_{1 \le j \le N} \{\|\mathbf{m} \cdot r_j(t) \cdot (\mathbf{x}(t) - \mathbf{w}_j(t))\|^2\} \tag{2}$$

where $\cdot$ means the componentwise product, $\mathbf{m} \in [0,1]^D$ is a user defined vector which captures the importance of every object component in identifying the object regarding the user experience, and $r_j(t)$ is an automatic computed measure of reliability of those object components. (see [9] for a more detailed description)

2.2 Update Step

The neuron $c(\mathbf{x}(t))$ which wins the competition for $\mathbf{x}(t)$ is the only one which updates its weight vector including the knowledge extracted from $\mathbf{x}(t)$

$$\mathbf{w}_i(t) = \begin{cases} \mathbf{w}_i(t-1) + \alpha\,(\mathbf{x}_t - \mathbf{w}_j(t-1)) & \text{if } i = c(x(t)) \\ \mathbf{w}_i(t-1) & \text{otherwise} \end{cases} \tag{3}$$

where $\alpha \in [0 \ldots 1]$ is named the *learning rate* and determines how important is the information extracted from the current input sample with respect to the background information that the neuron already known from previous training steps. α should be fixed to a large value such as 0.9 in order to identify the object adequately in next frame.

2.3 Neurons Birth and Death

The size of the neural network layer $n(t)$ is not fixed and it can vary from frame to frame. When an unknown object appears, it cannot be asssigned to any of the existing neurons accurately and a new neuron is necessary. If the Eq. 4 holds then the birth of a neuron occurs.

$$\forall j \in \{1 \ldots n(t)\} \quad \frac{\|\mathbf{x}(t) - \mathbf{w}_j(t)\|}{\|\mathbf{x}(t)\|} > \delta \tag{4}$$

with $\delta \in [0, 1]$ a parameter fixed by the user which means the maximum relative error permitted for a neuron to represent an object.

Once the neuron is created, its memory structures are initialised. The input pattern responsible for the birth of the neuron is assigned to the weight vector of the neuron and to the first entry in the neuron log.

$$\mathbf{w}_j(t) = \mathbf{x}(t) \; ; \; H_j(1) = \mathbf{x}(t) \tag{5}$$

On the other hand, if an object leaves the scene then the neuron which represents it should be destroyed. For this purpose, each neuron has a counter C_{die} which means the *lifetime* of the neuron, measured in number of training steps, i.e., frames. Each training step, the counter value is decreased by one and, if the value reaches zero then the corresponding neuron is removed. Every time a neuron wins a competition its counter value is changed to the initial value. Therefore, only neurons associated to objects which are not longer in the scene are destroyed, since it is very unlikely for these neurons to win a competition.

3 Feature Weighting Mechanism

The use of a set of features to represent an object can help to reinforce and improve the object tracking. However, the influence which a specific feature has in the tracking depends on several factors such as the sequence quality, the behaviour of the objects in the scene and the results in the object segmentation stage. Thus, an equitable weighting mask with the same importance of all the features in the tracking process is not always suitable. In fact, this task is more and more critical when the number of features gets increase.

Algorithm 1. Main steps of the tracking algorithm.

```
Input. Time instant t and the features of the segmented objects x_i(t)
Output. Labelling of the segmented objects
foreach Segmented object x_i(t) do
    Compute winner neuron by means of Eq. (2);
    if Eq. (4) is satisfied then
        Create a new neuron. Initialize it;
    else
        Update the network using equation Eq. (3);
    end
end
Refresh the counter values belonging to the neurons which win a competition;
Decrement all neurons counter values by one;
Check out neuron counters and destroy neurons whose counter value is zero;
```

Genetic algorithms (GAs) are applied to achieve automatically a suitable weighting of the features in the tracking process. This technique has been widely used as parameter selection and tuning in combination with neural networks [10]. Although this can be a time-consuming technique, the GA is applied only at the first frames of the sequence, with the aim of getting a weighting mask for the input features of the GCNN approach. This reduces resources and time complexity, getting better tracking rates in the following frames of the sequence.

A simple encoding scheme to represent the search space as much as possible was employed, in which the chromosome is a double vector whose length is determined by the number of features extracted. Each feature i is associated with one position in the vector which represents its relevance, $R_i \in (0,1)$, with regard to the whole set of features. As initial population the relevance of the features for each chromosome is randomly generated with the requirement of $\sum_{i=1}^{N} R_i = 1$, where N is the total number of features. Scattered crossover and mutation in only one bit are used in this approach, with the only requirement of the previous equation for each chromosome. A population size of 50 individuals, an elite count value of 5 (number of chromosomes which are retained in the next generation) and a crossover rate of 0.7 are selected.

In this kind of optimisation problems, a fitness function $f(x)$ should be maximised or minimised over a given space X of arbitrary dimension. The fitness function assesses each chromosome in the population so that it may be ranked against all the other chromosomes. In this approach, the fitness function indicates how good the chosen feature weighting mask is. Because of the fact that correct trajectories of the objects in the scene are not provided, it is necessary to model a heuristic function which represents the committed error by the tracking algorithm. The objective of the GA is to minimise this function.

Let M be the incidence matrix of dimensions $p \times q$, where p is the number of tracked objects and q the number of detected objects by the segmentation phase. Each cell b_{ij} is a binary value which represents a matching between two elements. The ideal incidence matrix matches all the tracked and segmented objects as a bijective function one by one. Thus, both the number of segmented objects not associated to any tracked objects or associated to more than one (M), and the number of tracked objects with no matching or more than one to the segmented objects (N) are penalised in the fitness function. A mechanism to avoid penalising correctly predicted trajectories in the terms M and N is included.

Let H_i be the record of the centroid of an object i, defined as $H_i = \{(x_t, y_t)|t \in 1 \dots K\}$ where K is the last occurrence of the object in the scene, x_t and y_t are the coordinates of the centroid of the object i in the occurrence t. Let D_i be the derivative of the previous H_i function which represents the centroid differences frame by frame $D_i = \frac{\delta H_i}{\delta t}$. Let D_m be the median of the differences D_i, a trajectory swap (T_S) happens when it is satisfied $|D_i(K) - D_m| > T$, where T is a threshold to regulate the change variation. Finally the fitness function is defined as follows

$$fitness = N + M + \lambda * T_S \tag{6}$$

where λ reflects the penalising value of the trajectory swap term.

4 Discussion

The introduction of a genetic algorithm as a way to select the relative weights of each component of the input samples of the proposed competitive neural network can be justified as follows. The energy function of our competitive neural network can be written as

$$\mathcal{E} = \frac{1}{2T} \sum_{i=1}^{N} \sum_{t=1}^{T} u_{it} \left\| \mathbf{m} \circ \left(\mathbf{w}_i - \mathbf{x}\left(t\right) \right) \right\|^2 \tag{7}$$

where $\circ$ denotes Hadamard product, $\mathbf{m} \in [0,1]^D$ is the vector which controls the relative weight of each of the D components of the input samples $\mathbf{x}(t)$, and $u_{it} \in \{0,1\}$ are binary variables which indicate whether unit i is the winner for input vector $\mathbf{x}(t)$:

$$u_{it} = 1 \text{ iff } i = \arg \min_{k \in \{1,..,N\}} \mathbf{m} \circ \left(\mathbf{w}_k - \mathbf{x}\left(t\right) \right) \tag{8}$$

The learning rule can be obtained by considering the derivative of $\mathcal{E}$ with respect to the prototype vector $\mathbf{w}_i$ and then using gradient descent:

$$\frac{\partial \mathcal{E}}{\partial \mathbf{w}_i} = \frac{1}{T} \sum_{i=1}^{N} \sum_{t=1}^{T} u_{it} \mathbf{m} \circ \mathbf{m} \circ \left(\mathbf{w}_i - \mathbf{x}\left(t\right) \right) \tag{9}$$

$$\mathbf{w}_i\left(t+1\right) = \mathbf{w}_i\left(t\right) + \alpha \left(\mathbf{x}\left(t\right) - \mathbf{w}_i\left(t\right) \right) \tag{10}$$

where the effect of the $\mathbf{m} \circ \mathbf{m}$ factor is assumed to be integrated into the learning rate α. If we take the derivative of the energy function $\mathcal{E}$ with respect to each component of the relative weight vector $\mathbf{m}$ we get

$$\frac{\partial \mathcal{E}}{\partial m_j} = \frac{1}{T} \sum_{i=1}^{N} \sum_{t=1}^{T} u_{it} m_j \left(w_{ij} - x_j\left(t\right) \right)^2 \tag{11}$$

(a) (b)

Fig. 1. Analysis of the GA behaviour in the competitive approach. The tracking results are showed in: (a) using a mask to weight the features equitatively in the competitive approach; (b) from the mask provided by the GA. Yellow and green squares correspond to spurious and correctly tracked objects, respectively. Some objects and their trajectories are showed in red.

The above equation reveals that the variation of the energy function when we change the relative weight m_j is directly proportional to the squared differences of values in the j-th coordinate among the input samples and the prototype vectors. That is, m_j controls the effect of the j-th coordinate of the input samples on the energy function $\mathcal{E}$ which the proposed competitive neural network tries to minimize. Hence, the selection of $\mathbf{m}$ made by the genetic algorithm deeply influences the behavior of the competitive neural network.

5 Results

In this section an analysis of the tracking results obtained by our approach is done. In order to carry out the study, several sequences in which the objects are considered as rigid objects are selected. Both real and hand-generated sequences are taken into account. Thus, typical traffic sequences provided by a video surveillance online repository are used, generated by the Federal Highway Administration (FHWA) under the Next Generation Simulation (NGSIM) program[1]. Some difficulties such as occlusions or overlapping objects caused by errors in the object detection phase should be handled.

Two different approaches are applied to check the usefulness of the feature weighting mechanism. In figure 1, a qualitative comparison between a GCNN approach with all the inputs features with the same weight, and a GCNN-GA strategy is observed. Figure 1(a) highlights the amount of spurious objects which appear in the scene. Although the tracking is effective, its complexity could raise because of the increase of processed objects along the time. Unlike this approach, the GCNN-GA strategy succeeds in a better resource management, with few spurious objects and longer trajectories. Furthermore, some occlusions caused by trees in the scene are well-solved by the GCNN-GA method, as it is showed in the car trajectories with ID 1 and 19, respectively

[1] Datasets of NGSIM are available at http://ngsim-community.org/

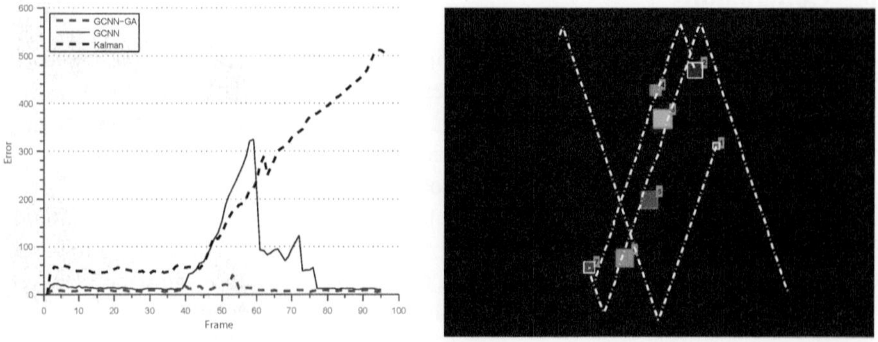

(a) (b)

Fig. 2. Comparison between different tracking approaches using a synthetic sequence. In (a), the centroid error of the tracked objects with regard to the ground truth is represented frame by frame. (b) shows a frame of the synthetic sequence and the trajectories of some objects.

Table 1. Comparative analysis of the success rate among the studied methods for the sequence observed in figure 2(b)

Method	Mean Error	Max. Error	No. spurious objs	No. mixed trajs
Kalman	26.72	59.57	2	2
GCNN	2.99	11.93	19	0
GCNN-GA	**1.26**	**2.34**	0	0

(figure 1(b)). Hand-generated sequences (figure 2(b)) are also used to perform a quantitative comparison with other tracking alternatives. The ground truth, which includes exact data about the trajectories of the objects, can be generated for these sequences in order to compare the performance of the tracking approaches, unlike the traffic sequences that do not provide this valuable information. For comparison purposes the Kalman filter [11], which is one of the main reference algorithms for tracking objects, is chosen. This method uses the centroid to predict the position of the identified object in the next frame.

In figure 2(a), the errors in the coordinates of the centroid obtained by several algorithms at each frame are shown. The closer to the x-coordinate the curve is, the better the tracking is. In table 1 the mean and maximum errors of each trajectory are calculated for each algorithm. The last two columns represent the number of the spurious objects that appear in the scene and the number of mixed trajectories. It happens when two different objects swap their trajectories. This situation is undesirable, since the analysis of each trajectory will be done incorrectly. As we can observe, the greatest errors occur in the last frames of the Kalman curve because of the confusion between two trajectories. The feature weighting mechanism (GCNN-GA) avoids the appearance of spurious objects, hereby improving considerably the results of the tracking process.

6 Conclusions

A new algorithm for tracking of rigid objects in video sequences is presented. This approach is able to take advantage of the feature set extracted from the object detection process to perform a more effective tracking. It consists in growing competitive neural networks in which the importance of the each input is computed by a feature weighting mechanism, which is based on genetic algorithms.

The combination of the two approaches is more accurate and reliable, thus diminishing the number of spurious objects and improving the resource management in terms of complexity. Both real and hand-generated sequences are used to show the viability of the system, by comparing to other alternatives such as the Kalman filter.

Acknowledgements

This work has been partially supported by the Ministry of Science and Innovation of Spain under grant TIN2010-15351, project name 'Probabilistic self organizing models for the restoration of lossy compressed images and video', and by Junta de Andalucía (Spain) under contract TIC-01615, project name 'Intelligent Remote Sensing Systems'.

References

1. Haritaoglu, I., Harwood, D., Davis, L.S.: w^4: Real-time surveillance of people and their activities. IEEE Trans. Pattern Anal. Mach. Intell 22(8), 809–830 (2000)
2. Lv, F., Kang, J., Nevatia, R., Cohen, I., Medioni, G.: Automatic tracking and labeling of human activities in a video sequence. In: Proceedings of the 6th IEEE International Workshop on Performance Evaluation of Tracking and Surveillance (2004)
3. Stauffer, C., Grimson, W.: Learning patterns of activity using real time tracking. IEEE Trans. Pattern Anal. Mach. Intell. 22(8), 747–767 (2000)
4. Grest, D., Koch, R.: Realtime multi-camera person tracking for immersive environments. In: IEEE 6th Workshop on Multimedia Signal Processing, pp. 387–390 (2004)
5. Nummiaro, K., Koller-Meier, E., Van Gool, L.: An adaptive color-based particle filter. Image Vision Comput. 21, 99–110 (2003)
6. Comaniciu, D., Ramesh, V.: Mean shift and optimal prediction for efficient object tracking. In: IEEE Int. Conf. Image Processing (ICIP 2000), pp. 70–73 (2000)
7. Comaniciu, D., Ramesh, V., Meer, P.: Real-time tracking of non-rigid objects using mean shift. In: IEEE Conference on Computer Vision and Pattern Recognition, pp. 142–149 (2000)
8. Luque, R., Dominguez, E., Palomo, E., Muñoz, J.: An art-type network approach for video object detection. In: European Symposium on Artificial Neural Networks, pp. 423–428 (2010)
9. Luque, R.M., Ortiz-de-Lazcano-Lobato, J.M., Lopez-Rubio, E., Palomo, E.J.: Object tracking in video sequences by unsupervised learning. In: Jiang, X., Petkov, N. (eds.) CAIP 2009. LNCS, vol. 5702, pp. 1070–1077. Springer, Heidelberg (2009)
10. Leung, F., Lam, H., Ling, S., Tam, P.: Tuning of the structure and parameters of a neural network using an improved genetic algorithm. IEEE Transactions on Neural Networks 14(1), 79–88 (2003)
11. Welch, G., Bishop, G.: An introduction to the kalman filter. Technical report, Chapel Hill, NC, USA (1995)

The Segmentation of Different Skin Colors Using the Combination of Graph Cuts and Probability Neural Network

Chih-Lyang Hwang [1] and Kai-Di Lu [2]

[1] Department of Electrical Engineering, National Taiwan University of Science and Technology, Taiwan, R.O.C.

[2] Department of Electrical Engineering, Tamkang University, Taiwan, RO.C.

chih-lyang_hwang@hotmail.com, try1111111@yahoo.com.tw

Abstract. It is known that fixed thresholds mostly fail in two situations as they only search for a certain skin color range: (i) any skin-like object may be classified as skin if skin-like colors belong to fixed threshold range. (ii) any true skin for different races may be mistakenly classified as non-skin if that skin colors do not belong to fixed threshold range. In this paper, a dynamic threshold of different skin colors based on the input image is determined by the combination of graph cuts (GC) and probability neural network (PNN). The compared results among GC, PNN and GC+PNN are presented not only to verify the accurate segmentation of different skin colors but also to reduce the computation time as compared with only using the neural network for the classification of different skin-colors and non-skin-color. In addition, the experimental results for different lighting conditions confirm the usefulness of the proposed methodology.

Keywords: Skin color segmentation, Graph cuts, Probability neural network, Classification.

1 Introduction

Skin segmentation means differentiating skin regions from non-skin regions in an image. A survey of different color spaces (e.g., RGB, YC_rC_b, *HSV, CIE Lab, CIE Luv* and normalized *RGB*) for skin-color representation and skin-pixel segmentation methods is given by Kakumanu et al. [1]. In [2], detected faces in videos are the basis for adaptive skin-color models, which are propagated throughout the video, providing a more precise and accurate model in its recognition performance than pure color based approaches. A method of gesture segmentation from the video image sequence based on monocular vision is presented by skin color and motion cues [3]. It is also known that fixed thresholds mostly fail as they only search for a certain skin color range [4]. Hence, in this paper instead of predefined fixed thresholds, a novel online learned dynamic threshold is employed to overcome the above drawbacks.

The dynamic thresholds for the skin color segmentation are studied by many papers. A hybrid approach based on neural networks and Bayesian classifiers is used

J. Cabestany, I. Rojas, and G. Joya (Eds.): IWANN 2011, Part II, LNCS 6692, pp. 25–33, 2011.

in the design of a computational system for automatic tissue identification in wound images [5]. In addition, a paper developed by Tsai and Yeh [6] proposes an automatic and parameter-free contrast compensation algorithm for skin detection in color face images. Inspired by stage lighting design, an active lighting system automatically adjusts the lighting so that the image looks visually appealing [7]. Based on the dynamic skin color correction method, the simple model is proposed for the convenience of hardware implementation (e.g., a 42-inch AC PDP) [8].

Min-cut-based graph partitioning has been used successfully to find clusters in networks, with applications in image segmentation as well as clustering biological and sociological networks. The central idea is to develop fast and efficient algorithms that optimally cut the edges between graph nodes, resulting in a separation of graph nodes into clusters [9]. Recently, there has been significant interest in image segmentation approaches based on graph cuts. The common theme underlying these approaches is the formation of a weighted graph, where each vertex corresponds to an image pixel or a region. The weight of each edge connecting two pixels or two regions represents the likelihood that they belong to the same segment. A graph is partitioned into components in a way that some cost function of the vertices in the components and/or the boundary between those components is minimized. A graph can also be partitioned into more than two components by recursively bi-partitioning the graph until some termination criterion is met. The termination criterion is often based on the same cost function that is used for bi-partitioning. Such a cost reduction is achieved by representing a graph using a 256×256 symmetrical weight matrix based on gray levels [9], rather than the $N\times N$ symmetrical weight matrix based on pixels, where N is the number of pixels in the image, which is typically much larger than 256. However, the skin is depicted by a 3D color space; generally, a graph cuts for the skin color segmentation needs a $256\times256\times256$ symmetrical weight matrix. Then its computational load becomes huge. Under these circumstances, we will present a subtle method to reduce the computation burden. Based on the fixed threshold for the segmentation of different skin colors, it is not necessary that each component of color space must be between 0 and 255. In addition, one of three components is reflected to illumination effect (e.g., the Y component or V component respectively for the color spaces of YC_rC_b and HSV), which slightly affects the skin-color segmentation [1]-[4]. Hence, the graph cuts of the skin color in color space (e.g., YC_rC_b) merely to compute two $Min\times Max$ weight matrices, which are generally 80×80 or 40×40. It becomes feasible to quickly obtain graph cut values for every possible threshold t from these two weight matrices.

Although the proposed graph cuts for the skin color segmentation is acceptable, a complex environment with many skin-like objects or different skin colors or different lighting conditions often results in a partial success. In this situation, a probability neural network (PNN) based on Bayesian classification is designed to classify the candidates of skin-color. The PNN is a special type of neural network that is widely used in the classification applications. It possesses a fast training process, an inherent parallel structure, and guaranteed optimal classification performance if a sufficiently large training set is provided [10]. Hence, it has the advantage of recognizing different skin colors in cluttered environments, which normally makes extracting reliable visual features difficult. Finally, the compared results among GC, PNN and GC+PNN confirm the usefulness of the proposed methodology.

2 Image Processing for Different Skin Color Segmentation

The proposed image processing for the segmentation of different skin colors is depicted in Fig. 1.

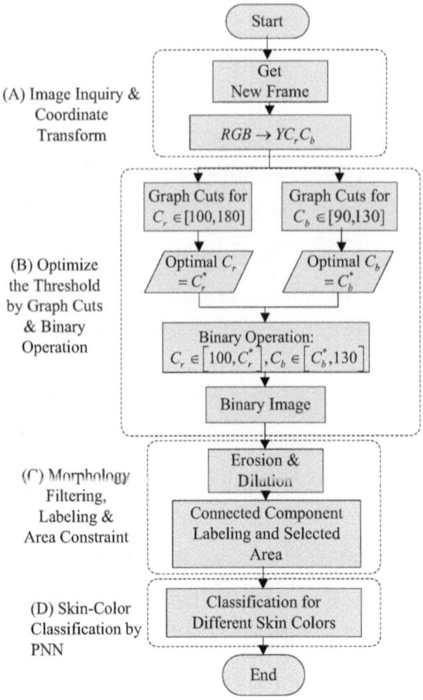

Fig. 1. Flowchart of the image processing for the skin color segmentation using the combination of GC and PNN

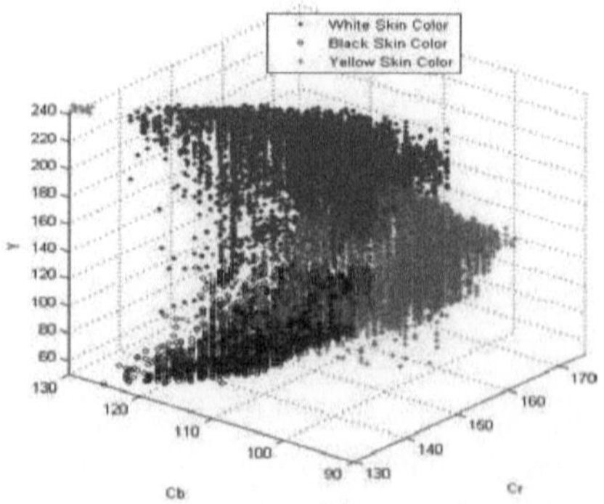

Fig. 2. The distribution of Black, White and Yellow Skin Colors in YC_rC_b

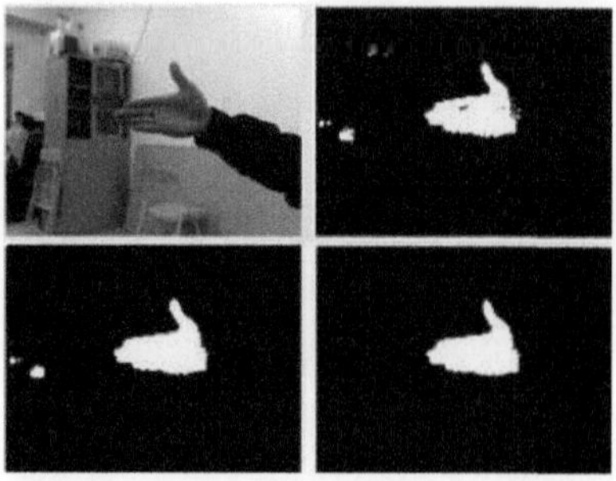

Fig. 3. The result of image processing

2.1 Image Inquiry and Coordinate Conversion

The original image from webcam is the format of *RGB*; it is easy to be affected by the illumination and also not convenient for the image processing. Based on some previous studies (e.g., [2]-[4]), the YC_rC_b format for the description of skin color is more suitable. The coordinate transform between *RGB* and YC_rC_b is given as follows:

$$\begin{bmatrix} Y \\ C_r \\ C_b \end{bmatrix} = \begin{bmatrix} 0.2990 & 0.5870 & 0.1140 \\ 0.5000 & -0.4190 & -0.0813 \\ -0.1690 & -0.3320 & 0.5000 \end{bmatrix} \begin{bmatrix} R \\ G \\ B \end{bmatrix}. \tag{1}$$

After the coordinate transform, one set of YC_rC_b, each component belongs to [0,255], is fed to the graph cuts, the other set is normalized to be the values of [0, 1] for the classification of the candidate(s) of skin color through PNN.

2.2 Graph Cuts for the Candidates of Skin Color

In the beginning, the weight of the graph edge connecting two nodes u and v as (2) is defined as follows [9]:

$$w(u,v)=\begin{cases}\exp-\left[\|F(u)-F(v)\|_2^2/d_I+\|X(u)-X(v)\|_2^2/d_X\right], & \text{if } \|X(u)-X(v)\|_2^2<r\\ 0, & \text{otherwise}\end{cases} \tag{2}$$

where $F(u)$ and $X(u)$ are the scale and spatial location of node u for the component C_r or C_b, the symbol $\|\cdot\|_2$ denotes the vector norm of the corresponding function, and d_I and d_X are positive scaling factors determining the sensitivity of $w(u,v)$ to the intensity difference and spatial location between two nodes. As a positive integer r increases, more nodes are involved in the computation of the weight and cost more time to compute.

The optimal C_r and C_b (i.e., C_r^* and C_b^*) are separately computed. At beginning, we define the set V, each component is C_r or C_b value separating from the color image of YC_rC_b .Let a threshold t to separate V into A and $B=V-A$. Threshold t is an integer and $t\in[Min,Max]$ where Min and Max are the initial threshold (or range) of C_r or C_b. Based on about 2000 images from the internet and our photographs with the skin colors of Black, White and Yellow Skin Colors, the distribution of YC_rC_b. for these images is shown in Fig. 2, which possesses $C_r\in[100,180]$ and $C_b\in[90,130]$. It implies that the C_r. component is a more important factor for the influence of the skin color segmentation than C_b. It should be noted that "Black Skin Color" more approaches the brown color but not black color because the color with very weak lighting condition is the same as black color, which is difficult to extract. Similarly, "White Skin Color" more approaches the combination of white and pink colors because the color with very strong lighting condition is the same as white color, which is also difficult to extract. If the larger range is selected for finding the optimal C_r or C_b, the larger computation time is required and however the more candidates of skin-color segmentation can be obtained.

The degree of dissimilarity between the two sets A and B can be computed as a total weight of the edges connecting the two parts as (3).

$$Cut(A,B)=\sum_{u\in A}\sum_{v\in B}w(u,v)=\sum_{i=Min}^{t}\sum_{j=t+1}^{Max}Cut(V_i,V_j) \tag{3}$$

where $Cut(V_i,V_j)=\sum_{u\in V_i,v\in V_j}w(u,v)$ is the total connection between all nodes in V_i (denoting the ith value of C_r or C_b) and all nodes in V_j (denoting the jth value of C_r or C_b). Similarly, we define $asso(A,V)$ and $asso(B,V)$ as follows:

$$asso(A,V) = \sum_{i=Min}^{t} \sum_{Min}^{Max} Cut(V_i,V_j) \text{,} \quad asso(B,V) = \sum_{i=t}^{Max} \sum_{Min}^{Max} Cut(V_i,V_j) \text{.} \tag{4}$$

The normalized cut (i.e., $Ncut$) is defined as follows [9]:

$$Ncut(A,B) = \frac{Cut(A,B)}{asso(A,V)} + \frac{Cut(A,B)}{asso(B,V)} \text{.} \tag{5}$$

Based on the definition of (3)-(5), $Ncut(A,B) \in [0,2]$. Compare all $Ncut$ corresponding to t, where $t \in [Min \quad Max]$. The optimal threshold C_r^* or C_b^* occurs when $Ncut$ is the minimum one. Before computing the $Ncut$, $M_{Min \times Max}$ in (6) is first assigned. It is much smaller than that of the previous papers for the graph cuts of a grey scale image (e.g., [9]).

$$M = \begin{bmatrix} Cut(V_{MIN},V_{MIN}) & Cut(V_{MIN},V_{MIN+1}) & \cdots & Cut(V_{MIN},V_{Max}) \\ Cut(V_{MIN+1},V_{MIN}) & \ddots & & \vdots \\ \vdots & & \ddots & Cut(V_{Max-1},V_{Max}) \\ Cut(V_{Max},V_{MIN}) & \cdots & Cut(V_{Max},V_{Max-1}) & Cut(V_{Max},V_{Max}) \end{bmatrix} \tag{6}$$

Hence, our computation time is proportional to 80×80 and 40×40, which is much smaller than that of 256×256. After the achievement of the optimal values of C_r and C_b, the corresponding binary image can be obtained as follows:

$$F = \begin{cases} 1, \text{ if } C_r \in \max\{C_r^* - 100, 180 - C_r^*\} \text{and } \max\{hist[C_r^* - 100], hist[180 - C_r^*]\}, \\ \quad C_b \in \max\{C_b^* - 90, 130 - C_b^*\} \text{and} \max\{hist[C_r^* - 100], hist[180 - C_r^*]\} \\ 0, \quad \text{otherwise} \end{cases} \tag{7}$$

where F denotes a binary image and $hist[\cdot]$ denotes the histogram of the corresponding pixel range. The result (8) is applied to ensure the larger number of possible candidate of different skin colors.

2.3 Noise Removal by Morphology Filtering, Connected Component Labeling and Area Constraint

Morphological filtering includes the erosion and dilation operations, which can be refer to the standard textbook (e.g., [11]). As to the definition of region of interest (ROI) is a rectangle including the maximum and minimum pixels of row and column in the corresponding labeling region. In addition, the unsuitable regions are removed by this constraint of area; i.e., the areas of all skin-like object must be between a low bound (e.g., $a_l = 650$ pixels) and an upper bound (e.g., $a_u = 65000$ pixels). A small region is related to high frequency noise; a large region is related to inappropriate skin-like objects. In this situation, the computation of the following classification of skin-like colors by PNN will be reduced. In addition, a typical result of image processing using GC is shown in Fig. 3: (top left) original image; (top right) after optimized segmentation via GC; (bottom left) after noise removal by morphology filtering; (bottom right) ROI after connected component labeling and the area constraint. It is satisfactory.

2.4 Classification of the Possible Skin Colors via Probability Neural Network

It is first assumed that a classification problem possesses K classes, e. g., $C_1, C_2, ..., C_K$. The corresponding rule for the classification is determined by m-dimensional feature vectors: $X=(X_1, X_2, ..., X_m)$. That is, in these m-dimensional samples space, the probability density function of each class, i.e., $f_1(X), f_2(X), ..., f_K(X)$, is function of these feature vectors. The decision formulation of Bayesian classification is then expressed as follows:

$$h_i c_i f_i(X) > h_j c_j f_j(X), \quad \forall j \neq i \tag{8}$$

where f_i denotes the *ith* probability density function, c_i is the value of cost function for misclassification of the *ith* class, and h_i denotes the prior probability of the *ith* class.

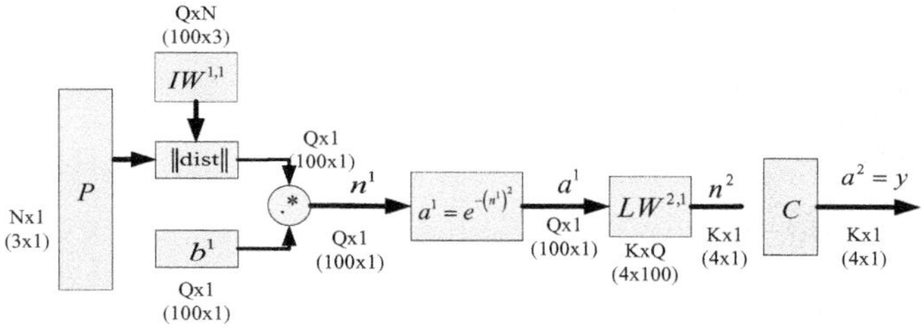

Fig. 4. Architecture of PNN

Theoretically, we can use the formula (8) to deal with the problem of classification. However, it is difficult to obtain the probability density function (PDF) of trained data in advance. Therefore, the trained data satisfying the specific condition of PDF (e.g., normal distribution) is first assumed and then are employed to train the parameters of these PDFs. In summary, the architecture of PNN is depicted in Fig. 4, where $P=[\bar{Y}, \bar{C}_r, \bar{C}_b]$ is the normalized feature vector (i.e., $\bar{Y}=Y/255$, $\bar{C}_r=C_r/\left(C_{r,\max}-C_{r,\min}\right)$, $\bar{C}_b=C_b/\left(C_{b,\max}-C_{b,\min}\right)$); the output $a^2=[C_1, C_2, C_3, C_4]$, i.e., [1,0,0,0], [0,1,0,0], [0,0,1,0], [0,0,0,1] respectively denote the Black, Yellow, White Skin Colors, and non-kin color; the number of hidden layer weights, i.e., Q=100, $IW^{1,1}$ and $LW^{1,2}$ respectively denote the input weight matrix and layer weight matrix; $\|dist\|$ denotes the Euclidean norm distance, n^i, $i=1,2$, is the activation signal of the *ith* layer, and $a^1=e^{-(n^1)^2}$ is the Gaussian distribution function.

3 Experiments

The specifications of webcam VX5000 are 130 million pixels, low light compensation, which is cheap and effective. In this paper, the resolution of image is 320×240. Our PC is the CPU of Intel i5 with 2.4GHz, memory 520MB, and the operating system of Window 7. At beginning, the skin color segmentation by GC with $d_X = 4, d_l = 625$ and $r = 2$ is shown in Fig. 5, which is also compared with fixed thresholds around the optimal threshold. The proposed GC for the skin-color segmentation is better than that using fixed threshold. However, only using the GC for the skin color segmentation in a complex environment (e.g., the 3rd case of Table 1) is not effective because at least 5 skin-like objects (i.e., one bottle of drink, a similar red paper in the left, a paper box in the right back) are also detected. This is one of important motivations for different skin-colors segmentation using the combination of the optimized threshold by GC and the classification by PNN. In this situation, some suitable data of Fig. 2 [11] will be employed to train the input weight

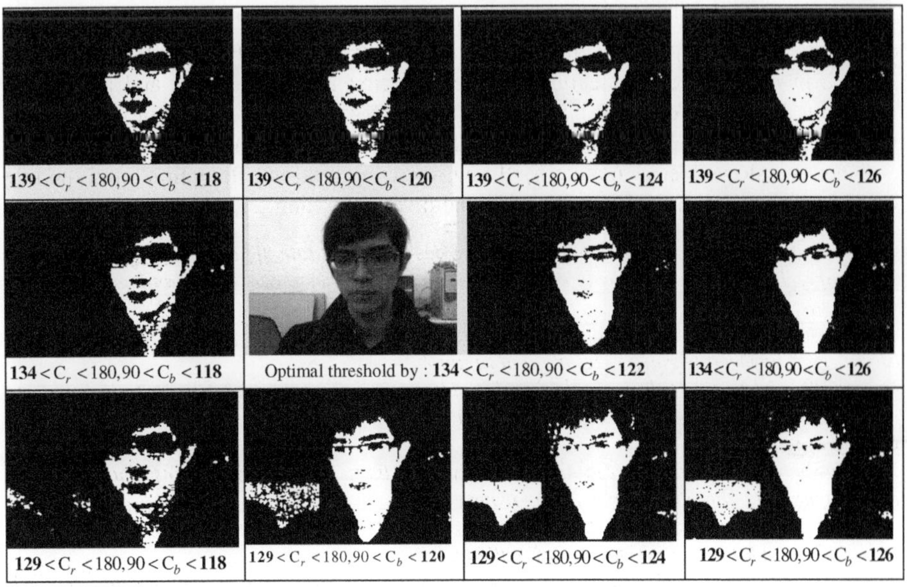

Fig. 5. The comparisons among some thresholds and the optimal threshold by the proposed GC

matrix and layer weight matrix of PNN in Fig. 4. The proposed comparisons for other cases and methods are summarized in Table 1. The 1st case is simple background and two skin colors almost in the same distance before the webcam; the 2nd case is the 1st case with the far distance from the webcam and more complex background; the 3rd case has at least six skin-like objects; and the 4th case has skin and skin-like colors are overlapped (or connected); the 5th case is a similar case of the 1st case with different lighting conditions. All these responses are acceptable and satisfactory.

Table 1. The compared results among three algorithms: GC, PNN and GC+PNN

Case	Method	GC	PNN	GC+PNN
	Time	0.048	0.407	0.121
	Skin	2		
	Result			 
	Target	2	4	2
	Time	0.045	0.390	0.058
	Skin	2		
	Result		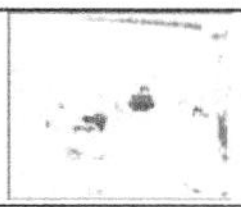	 
	Target	2	4	2
	Time	0.043	0.381	0.065
	Skin	1		
	Result	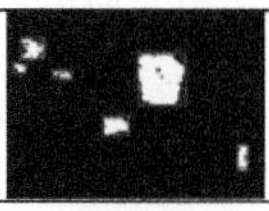		 
	Target	6	4	1
	Time	0.045	0.379	0.064
	Skin	3		
	Result			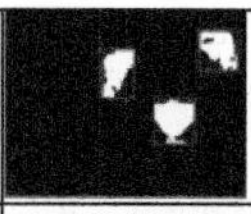 
	Target	2	3	3
	Time	0.048	0.407	0.121
	Skin	2		
	Result			 
	Target	2	4	2

4 Conclusion

In this paper, a dynamic threshold for different skin colors based on the input image is determined by the combination of the optimized threshold by graph cuts (GC) and the classification by probability neural network (PNN). The compared results among GC, PNN and GC+PNN are presented not only to verify the accurate segmentation of different skin colors but also to reduce the computation time as compared with only using the neural network for the classification of skin-colors. According to our results, the segmentation of different skin colors in complex environment and different lighting conditions is successful. The further studies include (i) the segmentation of different skin color in the other color spaces (e.g., *YUV*), (ii) an application to the visual imitation of a humanoid robot.

Acknowledgement. The authors want to thank the financial support from the project of NSC-99-2221-E-032- 066-MY3 of Taiwan.

References

1. Kakumanu, P., Makrogiannis, S., Bourbakis, N.: A Survey of Skin-Color Modeling and Detection Methods. Pattern Recognition, 1106–1122 (2007)
2. Liensberger, C., Stöttinger, J., Kampel, M.: Color-Based and Context-Aware Skin Detection for Online Video Annotation. In: IEEE International Workshop on Multimedia Signal Processing, MMSP 2009, Rio De Janeiro, pp. 1–6 (2009)
3. Cao, X.Y., Liu, H.F., Zou, Y.Y.: Gesture Segmentation Based on Monocular Vision Using Skin Color and Motion Cues. In: IEEE International Conference on Image Analysis and Signal Processing, Zhejiang China, pp. 358–362 (2010)
4. Yogarajah, P., Condell, J., Curran, K., Cheddad, A., McKevitt, P.: A Dynamic Threshold Approach for Skin Segmentation in Color Images. In: IEEE 17th International Conference on Image Processing, Hong Kong, pp. 2225–2228 (2010)
5. Veredas, F., Mesa, H., Morente, L.: Binary Tissue Classification on Wound Images Neural Networks and Bayesian Classifiers. IEEE Trans. Medical Imaging 29(2), 410–427 (2010)
6. Tsai, C.M., Yeh, Z.M.: Contrast Compensation by Fuzzy Classification and Image Illumination Analysis for Back-lit and Front-lit Color Face Images. IEEE Trans. Consumer Electronics 56(3), 1570–1578 (2010)
7. Sun, M., Liu, Z., Qiu, J., Zhang, Z., Sinclair, M.: Active Lighting for Video Conferencing. IEEE Trans. Cir. and Syst. for Video Technol. 19(12), 1819–1823 (2009)
8. Wang, Z.G., Liu, C.L.: A Method of Dynamic Skin Color Correction Applied to Display Devices. IEEE Trans. Consumer Electronics 55(3), 967–972 (2009)
9. Tao, W., Jin, H., Zhang, Y., Liu, L., Wang, D.: Image Threshold Using Graph Cuts. IEEE Trans. Syst. Man & Cybern., Pt. C 38(5), 1181–1195 (2008)
10. Psyllos, A., Anagnostopoulos, C.N., Kayafas, E.: Vehicle Model Recognition from Frontal View Image Measurements. Computer Standards & Interfaces 33, 142–151 (2011)
11. Sonka, M., Hiavac, V., Boyle, R.: Image Processing, Analysis, and Machine Vision, 3rd edn. Cengage Learning. (2008)

Reduction of JPEG Compression Artifacts by Kernel Regression and Probabilistic Self-Organizing Maps

María Nieves Florentín-Núñez, Ezequiel López-Rubio, and Francisco Javier López-Rubio

Department of Computer Languages and Computer Science
University of Málaga
Bulevar Louis Pasteur, 35. 29071 Málaga, Spain
florentin@uni.edu.py,
ezeqlr@lcc.uma.es,
xavierprof@hotmail.com

Abstract. There is a wide range of methods for lossy compression, but among those most used we find JPEG (Joint Photographic Experts Group) for still images. In this paper we present an intelligent system which is capable of restoring a compressed JPEG image by combining the knowledge extracted from the image domain and the transformed domain. It is based on probabilistic self-organizing maps and function approximation by kernel regression.

Keywords: JPEG compression artifacts, image restoration, probabilistic self-organizing maps, kernel regression.

1 Introduction

The multimedia information era has given rise to millions of image files on the Internet and stored by users. Most of them are lossy compressed, mainly in JPEG format. When the compression ratio is very high due to limitations in the transmission or the storage, the DCT codification has noticeable defects (artifacts) such as blocking effects and Gibbs' phenomenon [1]. Consequently, every algorithm which improves the quality of compressed images or videos has advantages for a number of people. This need was pointed out from the beginnings of JPEG [10] and it is attracting a growing attention in the image processing community due to the wide diffusion of wireless networks and portable devices with integrated cameras which connect to the Internet through them [1,2]. However, this task is complicated by the high dimensionality of the relevant data. A basic compression block is made up of 64 components (8×8 pixels), and in order to check the discrepancies among neighbouring blocks we must use data which have hundreds or thousands of components.

Since compressed image restoration is a complex problem, it has been considered from different perspectives. If we classify the proposals by the kind of

J. Cabestany, I. Rojas, and G. Joya (Eds.): IWANN 2011, Part II, LNCS 6692, pp. 34–41, 2011.

technique or system employed for the resolution of the problem, we obtain a wide range of non excluding possibilities:

- Transformed domain methods [4]. The process is carried out mainly on the DCT coefficients, and not on the colour values.
- Vector quantization [3]. In this case we search for a set of prototypes which best represents the image data. We must highlight the connections of this strategy with the self organizing systems that we propose, since they also learn prototypes, which implies that they can be used for vector quantization.
- Projections [10]. The observed data are projected on a subspace which captures the best solutions with respect to some quality criterion.
- Markov random fields [10,12]. This probabilistic framework is used to estimate the features of the image data. This kind of techniques reinforces the idea that the probabilistic reasoning is adequate for our goals.

In this paper we propose a restoration method which combines transformed domain and spatial domain methods. It uses adaptive kernel regression for 2D function approximation, and probabilistic self-organizing maps are used to control the restoration process.

The structure of this paper is as follows. First we present the restoration system (Subsection 2). Then we assess the quantitative and qualitative performance of our proposal (Subsection 3). Finally, Section 4 is devoted to conclusions.

2 Deblocking

Our proposed system for JPEG compressed image artifact removal consists of three subsystems which operate in sequence: DCT domain restoration (Subsection 2.1), smoothing (Subsection 2.2) and spatial domain restoration (Subsection 2.3). Next we describe each of them.

2.1 DCT Domain Restoration

Let us assume that we are given a JPEG compressed image with $8M \times 8N$ pixels which is made up of $M \times N$ compression blocks; we will assume here that the block counts in either direction (M and N) are integers for the sake of simplicity. Moreover, we will focus on one color channel only, since our method operates on each color channel independently. The i-th compression block with block coordinates $\mathbf{x}_i$, consists of 8×8 pixels with pixel coordinates $\mathbf{y}_j$, where

$$\mathbf{x}_i \in \{0, ..., M-1\} \times \{0, ..., N-1\} \tag{1}$$

$$\mathbf{y}_j \in \{8x_{i1}, ..., 8x_{i1}+7\} \times \{8x_{i2}, ..., 8x_{i2}+7\} \tag{2}$$

If we arrange the original, uncompressed values of these pixels in a 8×8 matrix $\mathbf{f}(\mathbf{x}_i)$, we can express the corresponding matrix of compressed pixel values $\tilde{\mathbf{f}}(\mathbf{x}_i)$ as

$$\tilde{\mathbf{f}}(\mathbf{x}_i) = \mathcal{T}^{-1}\left(\mathcal{Q}^{-1}\left(\mathbf{q}(\mathbf{x}_i)\right)\right) \tag{3}$$

$$\mathbf{q}(\mathbf{x}_i) = \mathcal{Q}\left(\mathbf{z}(\mathbf{x}_i)\right) \tag{4}$$

$$\mathbf{z}(\mathbf{x}_i) = \mathcal{T}\left(\mathbf{f}(\mathbf{x}_i)\right) \tag{5}$$

where $\mathcal{T}$ is the 2D-DCT transform, $\mathcal{Q}$ is the quantization operator, $\mathbf{z}(\mathbf{x}_i)$ is the 8×8 matrix of DCT coefficients, and $\mathbf{q}(\mathbf{x}_i)$ is the 8×8 matrix of quantized DCT coefficients.

The quantization operation $\mathcal{Q}$ is the only one which produces information loss. In order to alleviate its effects, we propose to consider the matrices $\mathbf{q}(\mathbf{x}_i)$ of quantized DCT coefficients as the output of a function $\mathbf{q}$ of the block coordinates $\mathbf{x}_i$:

$$\mathbf{q}: \; [0, M-1] \times [0, N-1] \rightarrow \mathbb{R}^{64} \tag{6}$$

Each of the components q_j of the function $\mathbf{q}$ is a scalar field, with $j \in \{1, ..., 64\}$. As illustrated in Figure 1, every q_j can be regarded as an image to be restored. This way, we can use a second order 2D steering kernel regression [13,14,5] to enhance each q_j separately:

$$\tilde{q}_j(\mathbf{x}_i) = \tilde{q}_j(\mathbf{x}) + \left(\nabla \tilde{q}_j(\mathbf{x})\right)^T (\mathbf{x}_i - \mathbf{x}) + \frac{1}{2} (\mathbf{x}_i - \mathbf{x})^T \left(\mathcal{H}\tilde{q}_j(\mathbf{x})\right)(\mathbf{x}_i - \mathbf{x}) + ... \tag{7}$$

where $\tilde{q}_j$ is the enhanced version of q_j, and ∇ and $\mathcal{H}$ are the gradient and Hessian operators, respectively. After that, we clip the enhanced values so that they fulfil the *narrow quantization constraint* [11,18], i.e. we ensure that the obtained image belongs to a reduced version of the set of images which could have given rise to the input JPEG file:

$$\hat{q}_j(\mathbf{x}_i) = \min\left(q_j(\mathbf{x}_i) + 0.3, \max\left(q_j(\mathbf{x}_i) - 0.3, \tilde{q}_j(\mathbf{x}_i)\right)\right) \tag{8}$$

This process leads to an enhanced version of the input image:

$$\hat{\mathbf{f}}(\mathbf{x}_i) = \mathcal{T}^{-1}\left(\mathcal{Q}^{-1}\left(\hat{\mathbf{q}}(\mathbf{x}_i)\right)\right) \tag{9}$$

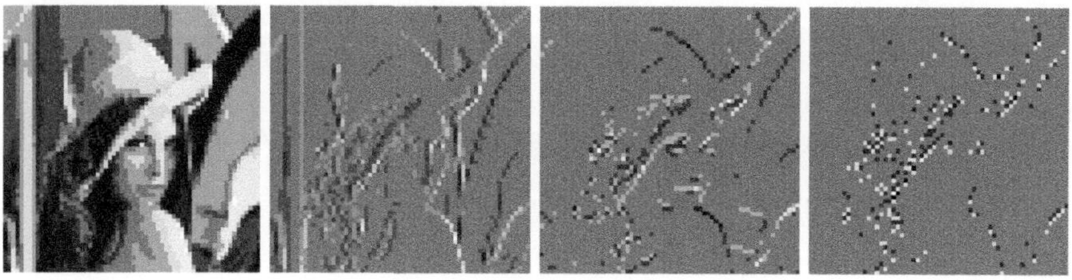

Fig. 1. Quantized DCT coefficients $q_j(\mathbf{x}_i)$ for Lena arranged as images, JPEG compression quality parameter $Q = 4$. From left to right, $j = 1, 2, 3, 4$.

In order to obtain a better reconstruction, we take advantage of the idea of shifting the DCT coefficients a half block (4 pixels) in each of the four main directions (up, down, left and right), which is explained in detail in [18]. We apply the above restoration mechanism to each of the five versions of the image (four shifted and one unshifted), and then we undo the shifts. The output of this subsystem is the average of these five images.

2.2 Smoothing

The result of the DCT domain deblocking we have just presented has considerably better quality than the input image. Nevertheless, it still suffers from blockiness. We have designed a procedure to smooth the image while ensuring that the narrow quantization constraint (8) holds.

First we generate an oversmoothed image by applying a standard Gaussian low-pass filter with a window size of 9×9 pixels and a standard deviation of 10 pixels (these values have been optimized experimentally). The oversmoothed image will not satisfy (8) in most cases. So, for each block i we consider the line segment in $\mathbb{R}^{64}$ which connects the result of the previous subsystem $\hat{\mathbf{q}}(\mathbf{x}_i)$ with the quantized DCT coefficients of the oversmoothed image $\bar{\mathbf{q}}(\mathbf{x}_i)$:

$$\dot{\mathbf{q}}(\mathbf{x}_i) = \hat{\mathbf{q}}(\mathbf{x}_i) + \lambda(\mathbf{x}_i)\left(\bar{\mathbf{q}}(\mathbf{x}_i) - \hat{\mathbf{q}}(\mathbf{x}_i)\right), \; \lambda(\mathbf{x}_i) \in [0,1] \tag{10}$$

Since we know that $\hat{\mathbf{q}}(\mathbf{x}_i)$ satisfies (8), we look for the point in the line segment closest to the oversmoothed solution $\bar{\mathbf{q}}(\mathbf{x}_i)$ which fulfils (8). This is readily computed by taking

$$\lambda(\mathbf{x}_i) = \min\left(\left\{\frac{q_j(\mathbf{x}_i) + \xi - \hat{q}_j(\mathbf{x}_i)}{\bar{q}_j(\mathbf{x}_i) - \hat{q}_j(\mathbf{x}_i)} \mid j \in \{1, ..., 64\}, \; \xi \in \{-0.3, 0.3\}\right\}\right) \tag{11}$$

where the values outside of the interval $[0,1]$ are not taken into account in the evaluation of the min function.

The obtained DCT coefficient blocks $\dot{\mathbf{q}}(\mathbf{x}_i)$ are used to build a new restored version of the input image, which both satisfies (8) and is smoother than that obtained from $\hat{\mathbf{q}}(\mathbf{x}_i)$.

2.3 Spatial Domain Restoration

Our final stage is devoted to the fine tuning of the smoothed image which comes from the previous subsystem. This is achieved by repeated application of a second order 2D steering kernel regression on the spatial domain:

$$\ddot{f}(\mathbf{y}_j) = \ddot{f}(\mathbf{y}) + \left(\nabla \ddot{f}(\mathbf{y})\right)^T (\mathbf{y}_j - \mathbf{y}) + \frac{1}{2}(\mathbf{y}_j - \mathbf{y})^T \left(\mathcal{H}\ddot{f}(\mathbf{y})\right)(\mathbf{y}_j - \mathbf{y}) + ... \tag{12}$$

After each application of the kernel regression, we obtain the corresponding DCT coefficients $\ddot{\mathbf{q}}(\mathbf{x}_i)$ and then we clip them so as to ensure that (8) holds. This way we prevent a excessive departure from the original, uncompressed image.

It is known that repeated application of kernel regression poses the fundamental problem of developing a principled criterion to stop the iteration [13]. Here we propose to solve this issue by means of a probabilistic self-organizing map model [8], namely the Probabilistic Principal Components Analysis Self-Organizing Map (PPCASOM, [7]). Other probabilistic SOM models could have also been used [6,17], but we have selected PPCASOM because of its low complexity with high dimensional data.

Our approach is based on training a PPCASOM with 64-dimensional vectors $\mathbf{q}(\mathbf{x}_i)$ of JPEG quantized coefficients. This way, the PPCASOM provides an estimation of the probability density function $p(\mathbf{q}(\mathbf{x}_i))$ of the distribution of quantized DCT blocks. The training is carried out offline, so that it does not slow down the restoration algorithm in any way.

The pre-trained PPCASOM is used to estimate the log-likelihood L of observing the restored coefficients $\ddot{\mathbf{q}}(\mathbf{x}_i)$, given the distribution of JPEG quantized DCT blocks:

$$L = \sum_{i=1}^{MN} \log p\left(\ddot{\mathbf{q}}(\mathbf{x}_i)\right) \qquad (13)$$

Successive improvements of the restored image should take the restored coefficients $\ddot{\mathbf{q}}(\mathbf{x}_i)$ away from the distribution of JPEG quantized DCT blocks. Hence, we continue the application of steering kernel regression while the likelihood L diminishes. A maximum number of iterations is also considered, which in our experiments has been 5. Consequently, the algorithm stops if L has increased or the maximum number of iterations has been reached, and then the restored image $\ddot{f}(\mathbf{y}_j)$ is given as output.

3 Experimental Results

We have tested our method over several benchmark images from USC-SIPI image repository [16]. The Tiffany image has been used to train the PPCASOM model, and then the restoration performance has been measured over six other images for JPEG compression quality parameters $Q = 2,\ 4,\ 6$. We have run comparisons with the deblocking method by re-application of JPEG [9], which we denote RDCT. The unoptimized implementation of our proposal restores a 512×512 image in about 5.4 min on a single 2GHz core of a 32-bit CPU; this could be optimized by running it in parallel for each colour channel.

For the quantitative assessment, three standard criteria have been chosen, namely the Power Signal to Noise Ratio $PSNR$ (higher is better), the Mean Absolute Error MAE (lower is better) and the Structural Similarity index $SSIM$ [15] (higher is better).

As seen in Tables 1-3, our proposal outperforms RDCT in most cases. The same situation is depicted by the qualitative results (Figures 2-4), where the zoomed sections of the benchmarks show how our method is able to reduce blockiness while preserving the original structure of the images.

Table 1. Quantitative results for JPEG compression quality parameter $Q = 2$. Best results marked in **bold**.

	JPEG			RDCT			Proposed		
	PSNR	MAE	SSIM	PSNR	MAE	SSIM	PSNR	MAE	SSIM
Baboon	18.4367	23.6358	0.3367	18.9947	22.1696	0.3372	**19.0255**	**22.0781**	**0.3514**
F-16	22.0337	14.7552	0.6797	23.1594	12.8689	**0.7478**	**23.3348**	**12.4402**	0.7462
House	20.8099	17.5578	0.6290	21.8420	15.7073	**0.6799**	**21.9247**	**15.0200**	0.6724
Lake	19.9404	19.5155	0.5087	20.9071	17.4884	0.5681	**21.1443**	**16.8388**	**0.5715**
Lena	21.5328	16.8377	0.5727	22.6748	14.7832	0.6533	**23.3979**	**13.3298**	**0.6655**
Peppers	21.0983	16.6477	0.5136	22.1759	14.7535	0.5940	**22.4607**	**14.0805**	**0.6081**

Table 2. Quantitative results for JPEG compression quality parameter $Q = 4$. Best results marked in **bold**.

	JPEG			RDCT			Proposed		
	PSNR	MAE	SSIM	PSNR	MAE	SSIM	PSNR	MAE	SSIM
Baboon	19.2794	21.1876	0.4262	19.7809	20.0671	0.4236	**19.8490**	**19.9335**	**0.4333**
F-16	23.6793	11.6518	0.7279	24.5412	10.5795	0.7762	**24.8228**	**10.1217**	**0.7854**
House	22.6299	13.9833	0.6668	23.4410	12.7988	**0.7038**	**23.4755**	**12.3913**	0.7034
Lake	21.1574	16.9985	0.5637	21.9178	15.5876	0.6091	**22.2090**	**14.8530**	**0.6212**
Lena	23.3423	13.3468	0.6445	24.2102	12.1531	0.6946	**24.8666**	**11.1658**	**0.7135**
Peppers	22.3154	14.4370	0.5539	23.1345	13.1935	0.6148	**23.4682**	**12.3659**	**0.6404**

Table 3. Quantitative results for JPEG compression quality parameter $Q = 6$. Best results marked in **bold**.

	JPEG			RDCT			Proposed		
	PSNR	MAE	SSIM	PSNR	MAE	SSIM	PSNR	MAE	SSIM
Baboon	20.3812	18.5746	0.5067	20.8458	17.5835	0.5085	**20.8816**	**17.4777**	**0.5102**
F-16	25.0458	9.6174	0.7620	25.8425	8.7440	0.8087	**26.0742**	**8.3682**	**0.8177**
House	24.4142	10.9096	0.7024	**25.2622**	**9.8721**	**0.7465**	25.1046	10.1148	0.7425
Lake	22.7284	13.7717	0.6124	23.3987	12.7154	0.6538	**23.4872**	**12.4795**	**0.6593**
Lena	25.5220	10.0455	0.6858	26.4586	9.0874	0.7369	**26.7445**	**8.6857**	**0.7484**
Peppers	23.9934	11.4644	0.6058	24.8150	10.4310	0.6621	**24.9826**	**10.0597**	**0.6753**

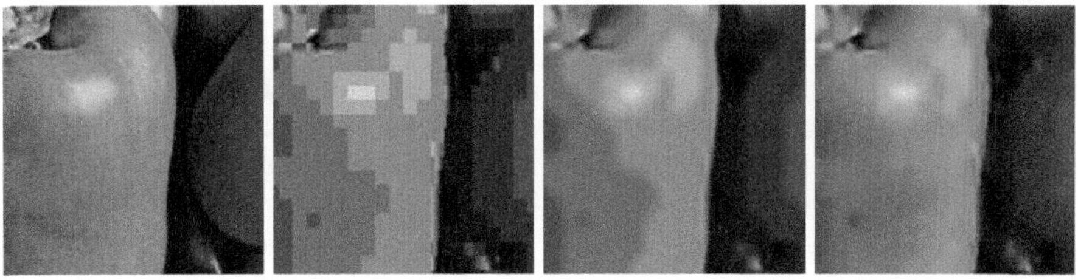

Fig. 2. Qualitative results for JPEG compression quality parameter $Q = 2$, Peppers image. From left to right: original, JPEG compressed, restored by RDCT, restored by our proposal.

Fig. 3. Qualitative results for JPEG compression quality parameter $Q = 4$, Lake image. From left to right: original, JPEG compressed, restored by RDCT, restored by our proposal.

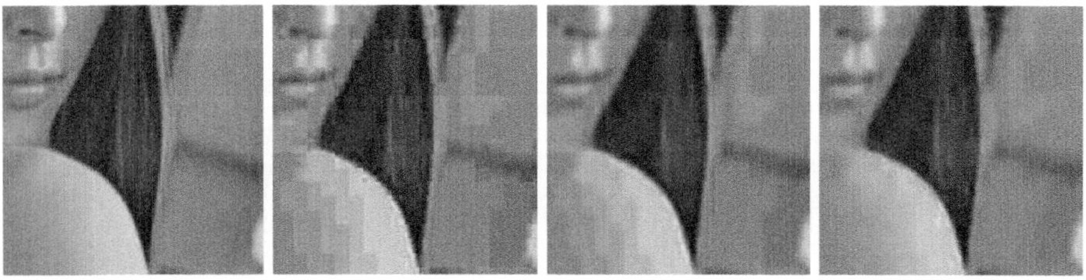

Fig. 4. Qualitative results for JPEG compression quality parameter $Q = 6$, Lena image. From left to right: original, JPEG compressed, restored by RDCT, restored by our proposal.

4 Conclusions

We have presented a new artifact reduction system for excessively compressed JPEG image files. It carries out nonparametric kernel regression approximations of 2D functions defined over the transformed and the spatial domains. The restoration is controlled by a probabilistic self-organizing map, which builds a probabilistic model of the distribution of DCT quantized coefficients. As seen in the experiments, this combination of techniques yields competitive results, both in quantitative and qualitative terms.

Acknowledgements

This work has been partially supported by the Ministry of Science and Innovation of Spain under grant TIN2010-15351, project name 'Probabilistic self organizing models for the restoration of lossy compressed images and video'.

References

1. Alter, F., Durand, S., Froment, J.: Adapted total variation for artifact free decompression of JPEG images. Journal of Mathematical Imaging and Vision 23, 199–211 (2005)

2. Chung, K.H., Chan, Y.H.: A low-complexity joint color demosaicking and zooming algorithm for digital camera. IEEE Transactions on Image Processing 16, 1705–1715 (2007)
3. Lai, J.Z., Liaw, Y.C.: Improvement of interpolated color filter array image using modified mean-removed classified vector quantization. Pattern Recognition Letters 26, 1047–1058 (2005)
4. Lim, T., Ryu, J., Kim, J., Jeong, J.: Adaptive deblocking method using a transform table of different dimension DCT. IEEE Transactions on Consumer Electronics 54, 1988–1995 (2008)
5. López-Rubio, E.: Restoration of images corrupted by Gaussian and uniform impulsive noise. Pattern Recognition 43(5), 1835–1846 (2010)
6. López-Rubio, E.: Multivariate Student-t self-organizing maps. Neural Networks 22(10), 1432–1447 (2009)
7. López-Rubio, E., Ortiz-de-Lazcano-Lobato, J.M., López-Rodríguez, D.: Probabilistic PCA self-organizing maps. IEEE Transactions on Neural Networks 20(9), 1474–1489 (2009)
8. López-Rubio, E.: Probabilistic self-organizing maps for continuous data. IEEE Transactions on Neural Networks 21(10), 1543–1554 (2010)
9. Nosratinia, A.: Enhancement of JPEG-compressed images by re-application of JPEG. The Journal of VLSI Signal Processing 27, 69–79 (2001)
10. O'Rourke, T.P., Stevenson, R.L.: Improved image decompression for reduced transform coding artifacts. IEEE Transactions on Circuits and Systems for Video Technology 5, 490–499 (1995)
11. Park, S.H., Kim, D.S.: Theory of projection onto the narrow quantization constraint set and its application. IEEE Transactions on Image Processing 8(10), 1361–1373 (1999)
12. Sun, D., Cham, W.K.: Postprocessing of low bit-rate block DCT coded images based on a fields of experts prior. IEEE Transactions on Image Processing 16, 2743–2751 (2007)
13. Takeda, H., Farsiu, S., Milanfar, P.: Kernel regression for image processing and reconstruction. IEEE Transactions on Image Processing 16(2), 349–366 (2007)
14. Takeda, H., Milanfar, P., Protter, M., Elad, M.: Super-resolution without explicit subpixel motion estimation. IEEE Transactions on Image Processing 18(9), 1958–1975 (2009)
15. Wang, Z., Bovik, A.C., Sheikh, H.R., Simoncelli, E.P.: Image quality assessment: From error visibility to structural similarity. IEEE Transactions on Image Proccessing 13(4), 600–612 (2004)
16. Weber, A.: The USC-SIPI image database (2010), http://sipi.usc.edu/database/
17. Yin, H., Allinson, N.: Self-organizing mixture networks for probability density estimation. IEEE Transactions on Neural Networks 12(2), 405–411 (2001)
18. Zhai, G., Zhang, W., Yang, X., Lin, W., Xu, Y.: Efficient deblocking with coefficient regularization, shape-adaptive filtering, and quantization constraint. IEEE Transactions on Multimedia 10(5), 735–745 (2008)

An Unsupervised Method for Active Region Extraction in Sports Videos

Markos Mentzelopoulos, Alexandra Psarrou, and Anastassia Angelopoulou

School of Electronics and Computer Science, University of Westminster,
115 New Cavendish Street, W1W 6UW, United Kingdom
{mentzem, psarroa, agelopa}@wmin.ac.uk

Abstract. In this paper, we propose a fully automatic and computationally efficient algorithm for analysis of sports videos. The goal of the proposed method is to identify regions that perform certain activities in a scene. The model uses some low-level feature video processing algorithms to extract the shot boundaries from a video scene and to identify dominant colours within these boundaries. An object classification method is used for clustering the seed distributions of the dominant colours to homogeneous regions. Using a simple tracking method a classification of these regions to active or static is performed. The efficiency of the proposed framework is demonstrated over a standard video benchmark with numerous types of sport events and the experimental results show that our algorithm can be used with high accuracy for automatic annotation of active regions for sport videos.

1 Introduction

As an important video domain, sports video has been widely studied due to its high evolution within the commercial industry. Prior work focuses on the detection of special events in sports videos such as goals, penalties or corner kicks or mutually exclusive states of the game such as *play* or *break*. Sports video segmentation could be categorized in two genre framework based on current research methodologies both aim to lead to a semantic classification. The first category, investigates the roles of semantic objects in the scenery and models their trajectories [7,10], while the second category is using structure analysis of the environment (lines in a basketball, soccer and tennis court) [1,7].

L.Duan and M. Xuan try to create an interpretation step for the analysis of such videos by using a three-level framework approach [3]. Their proposed framework used at the low layer, low level features direct extracted from the raw video. To bridge the semantic gap between the low layer and the upper layer (semantic events), they used an intermediate layer for shot classification based on audio keywords and visual pre-defined semantic shots.

Based on layer-2 functionality a numerous of approaches have been developed. Xiaofeng [9] proposed a supervised-framework for generic field-ball shots genres in which a shot is weight-characterized via three essential properties: camera shot size, subject in a scene, and video production technology. Despite the numerous

J. Cabestany, I. Rojas, and G. Joya (Eds.): IWANN 2011, Part II, LNCS 6692, pp. 42–49, 2011.

efforts in semantic sports videos analysis, it is still hard to develop a generic approach able to handle all different type sports video [9].

In this paper we present a method for the automatic segmentation of a sport video and the extraction of its dominant regions, robust to different sport scenarios, without the need of camera calibration or background extraction. The proposed model is using the Entropy Difference algorithm [5] that has a double scope: 1) To provide an automatic video segmentation down to keyframes and 2) To identify the dominant colors in the shot. The rest of the paper is utilized as follows. Section 2 describes the extraction of the Active regions in a shot. Experimental results and conclusions are presented in Sections 3 and 4 respectively.

2 Algorithm Description

In previous work [5] the entropy was used as a distance metric between consecutive frames. We considered that if we distribute the entropy among the image, then the regions that contain the important objects of the video sequence (static or moving) will contain the higher levels of distribution. After applying the video segmentation algorithm, for each extracted key-frame we keep a table of record (***DEC*** -Dominant Entropy Colour table) Figure 2(**b**), that includes the key-frame ID number in the shot sequence, the color bin values that include the highest entropy values, the pixel distributions and the entropy values.

2.1 Active Region Extraction

In terms of Content-based video analysis and indexing, a precise boundary is not essential [9] as it is has also high computational cost. This is why current research methods tend to cluster around places which are homogenous in some spatial features(color, texture, positions) [2,4]. Moreover, motion plays an important role in focusing of attention within perceptions. Therefore the main goal of the video segmentation methods should be to obtain the spatial boundaries of the regions that will be of most interest to the user. By combining the Entropy Difference Algorithm [5] with the corresponded DEC tables for each keyframe in conjunction with a simple tracking framework we can extract regions,*which at different time instances within a shot perform some action.* We define these regions as ***Active Regions***.

Our proposed method to identify the Active Regions is based on a five main step algorithm.

- Binary back-projection of the dominant colors from the DEC table extracted from the video segmentation to identify the related to the color seeds.
- Classification of the identified color seeds into regions.
- Region thresholding and labeling.
- Extraction of region properties and construction of the Minimum Bounding Box (MBB) Table (Figure 2(**b**)).
- Tracking regions in the shot sequence to distinguish static from active.

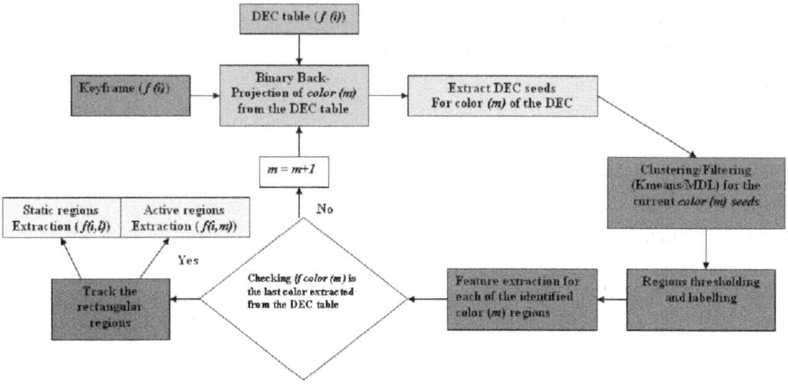

Fig. 1. Flowchart of our proposed method for Active region identification. The method is using the DEC table extracted from the Entropy Difference Algorithm in order to identify dominant colour regions in the shot.

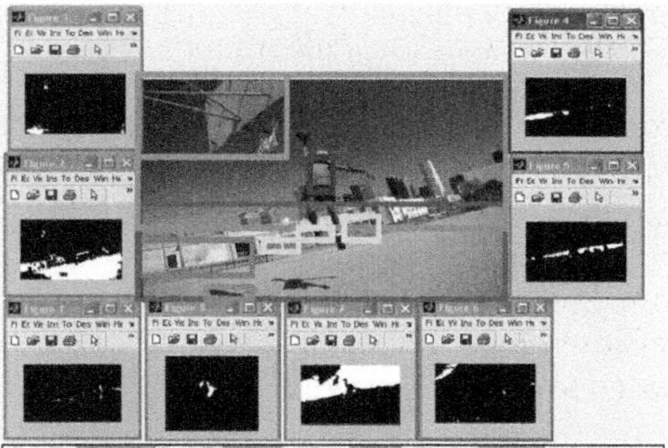

(a)

Image ID	Region ID	Binary ID	X	Y	LBP	Contrast	Area	w	h
001	001	[01...0]	12	45	111	23	122	12	10
002	002	[01...1]	23	78	12	45	34	4	7
003	003	[11...1]	45	100	45	88	212	25	7
...	...	...	...	...	...	...	...	...	...
n	j	[11...0]	55	12	65	67	78	6	10

(b)

Fig. 2. a) Example of back-projecting a keyframe using its DEC table c_i. The bi-level images $B_{i,b}$ are the ones before applying any filtering to the image (1-8 in this example). The remaining bi-level regions after applying clustering and thresholding can be seen in the centre image. (b) The MBB table shows the Region relation with attributes for the binary colour set (Binary ID), region centre (X,Y), texture descriptors (Local Binary Patterns (LBP), Contrast), number of colour seeds in the region (Area) and the width, height (w,h) of the MBB.

2.2 Back-Projection and Classification

The Binary color set back-projection [8] technique is used in order to extract color regions. Given an extracted keyframe $f_i[M,N]$ and its corresponding DEC table c_i extracted from the video sequence, then for each color bin b in the c table we generate a bi-level mode image $B_{i,b}[M,N]$:

$$B_{i,b}(x,y) = \begin{cases} 0, \text{ if } f_i(x,y) = c_{i,b}. \\ 1, \text{ otherwise.} \end{cases} \quad (1)$$

where x, y are the x,y coordinates of each pixel in the keyframe. Then we filter each $B_{i,b}[M,N]$ with a 3×3 median filter to reduce spatial noises in the areas. Figure 2(**a**) demonstrates the resulted 8 Bi-level images that have been extracted for the keyframe (colored frame in the center).

In VisualSEEk [8] there are two major drawbacks:

- A single point can simultaneously belong to two or more overlapping regions (e.g.: The wheel of a car and the car as a whole).
- When applying the binary back-projection over the key-frame for each dominant color bin, we identify seeds scattered inside the image that could possible represent more than one object with the same color information (e.g: players of the same team wearing same T-shirt(red) Figure 4,and 5, and 6).

Previous algorithms [2,4] have used the Minimum Description Length (MDL) [6] model, over a multi-dimensional feature space, to automatically estimate the optimum number of atomic components k, which is very computationally expensive. In our method we use the MDL in conjunction with the kmeans clustering, to classify each color seed, of each bi-level image $B_{i,b}[M,N]$, to a single independent homogenous region. The great advantage of our proposed algorithm compared to the other methods is that the clustering is applied on a simple feature vector making the computational cost very low. For every pixel of the bi-level image the feature vector $F_{i,b}(x,y) = \{g, x_{i,b}, y_{i,b}\}$ is consisted from the binary value g of the pixel and the corresponded x, y coordinates.

2.3 Region Thresholding and Labeling

After pixel classification, all the extracted regions n are encapsulated in a rectangular (Figure 2(**a**)) with center (X'_z, Y'_z) and width $w_{i,z}$ and height $h_{i,z}$ $(1 \leq z \leq n)$. Then in each rectangular we apply 2 criteria to verify if the identify region contains sufficient information to describe a possible active region:

$$\sum_{i=1}^{C_{i,z}} \geq 60\%(w_{i,z} \times h_{i,z}) \quad (2)$$

$$\sum_{i=1}^{Z_i} \geq 2\%(M \times N) \quad (3)$$

where: $Z_i = w_{i,z} \times h_{i,z}$ is the distribution of the pixels within the MBB and: $C_{i,z}$ is the distribution of the pixels from the corresponded MBB that belong to the specific colour.cluster (z). Then for every MBB that passes the above criteria we create a representative feature vector that will describe it in the tracking sequence:

$$D_{i,n} = g, X, Y, Lbp, C, w, h \quad (4)$$

where g is the colour value of the cluster, X, Y the centers, Lbp (Local Binary Pattern) and C (Contrast) are the texture descriptors and w, h are the width and height of the MBB. Finally we constructed a table that holds the above information (Figure 2(**b**)).

2.4 Region Tracking

The final step of the proposed method is to track each region through the video sequence in order to classify them to active regions and static. The tracking performance of our system is demonstrated in Figure 3. For each MBR we try to track to which direction (0-7 Figure 3**a**) the corresponding MBR pixel distribution is moving to. We check the pixel distribution on a 5 frames difference. Therefore after 5 frames (25-30) where the player is moving for the follow, we will see that the pixel distribution on the direction of arrow 3 (blue arrow) is higher compare to the neighborhood ones (2,4) and therefore the MBR is moving up and left. We apply the same tracking method for all the MBR's that have been identified (Figure 2(**a**)). At the end we mark only the *Active Regions* and disregard the rest.

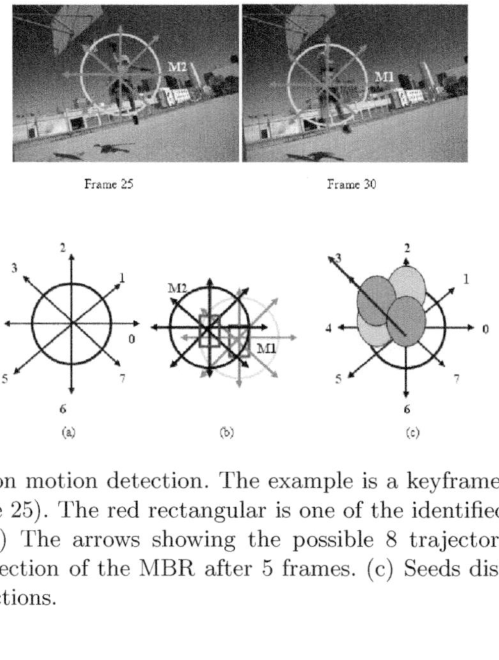

Fig. 3. Active region motion detection. The example is a keyframe taken from a basketball shot (Frame 25). The red rectangular is one of the identified regions extracted from section 3.2.(a) The arrows showing the possible 8 trajectories that the MBR will move. (b) Projection of the MBR after 5 frames. (c) Seeds distributions over the corresponding directions.

3 Experimental Results

Our proposed system has been evaluated on a data set of 11 video clips from football, squash and basketball games taken from the benchmark workshop series

PETS (2001)[1]. This specific benchmark was chosen as it includes sports activities for evaluation of our system. Table 1 shows the name, frame length and time duration of each clip of our testing data set. While the squash (Figure 4) was captured using a single camera from a single match, the football scenes where taken from the same football match between Manchester United and Tottenham but from 2 different locations in the football field (2 scenes/camera). These are: 1) a camera mounted in the centre of the stadium pointing towards the one side of the football field (Figure 5), and 2) a camera mounted again in the centre of the stadium but this time adjusted to look at the centre of the playfield (Figure 6). Table 1 shows the results from the region tracking performance. It contains

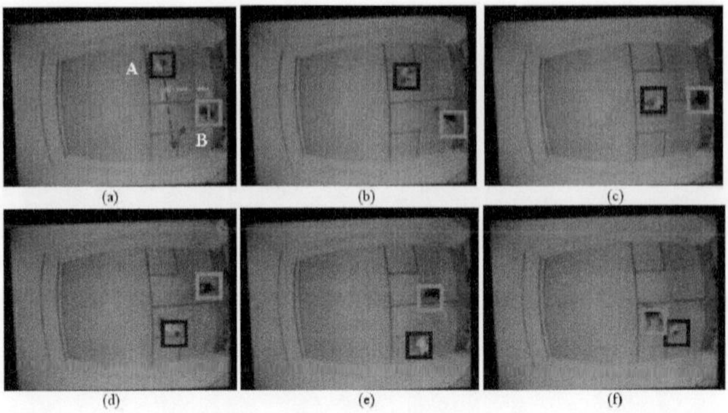

Fig. 4. Image sequence picked from the squash video with a frame distance of 20 frames/image. The red and orange dot notated arrows show the trajectory of the active regions in the duration of the 100 frames.

information regarding to how many correct active regions we have detected per video shot, how many where false and how many we missed. The % *Recall* was calculated based on how many correct regions we extracted compared to the *Total* regions detected ($Total_{Recall} = Correct + False$), while the % *Precision* was calculated how many correct regions we extracted against the *Total* number of actual active regions that our system should have detect ($Total_{Precision} = Correct + Miss$) based on *ground truth*. Figure 4,and 5, and 6 show the detected active regions bounded in the MBB's. From Table 1 results it can bee seen that the percentage tracking accuracy and recall is very high when the dominant objects are easily distinguishable from the background (eg:squash). The overall % recall for the 11 videos is at 77.32 while the precision is at 68.86%.

A problem that arises is that when the players are very close to each other, it results to be merged in a unique cluster and therefore the MBR that encapsulates the active blob will be the same for the 3 players. Finally in cases when there are regions at the background of the key-frames (Figure 5, and 6), then even if the seeds that represent the players are identified in the binary-backprojection, the encapsulated region is possible not to pass the specified threshold criteria.

[1] http://www.cvg.cs.rdg.ac.uk/VS/

Table 1. Active Region Motion Detection Results. Overall 68.86% Precision, and 77.32% Recall Rates.

Sequence Name	Camera	Length in (min:sec)	Length in Frames	Correct	False	Miss	% Recall	% Precision
Football 1	1	1:30	1999	22	5	3	81.48	88.00
Football 2	1	1:30	1999	13	3	4	81.25	76.47
Football 3	1	1:30	1999	14	3	2	82.35	87.50
Football 4	2	1:30	1999	15	6	6	71.43	71.43
Football 5	2	1:30	1999	13	4	4	76.47	76.47
Football 6	2	1:30	1999	15	5	5	75.00	75.00
Football 7	3	1:30	1999	7	4	4	63.63	63.63
Football 8	3	1:30	1999	13	5	6	72.22	68.42
Football 9	3	1:30	1999	12	3	5	80.00	70.59
Squash		0:45	888	2	0	0	100.00	0.00
Basketball		0:45	888	8	4	2	66.66	80.00

Fig. 5. Visual tracking players from a real football match using a still camera looking towards left goalpost

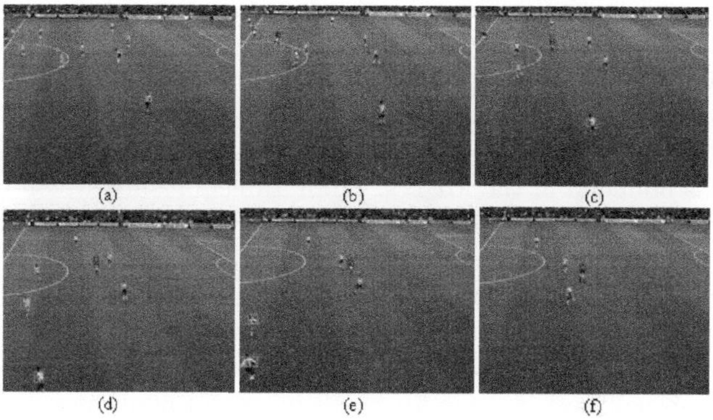

Fig. 6. Visual tracking players from a real football match using a still camera in the middle of the stadium

4 Conclusions

A new approach for *Active Region* extraction from a video sequence is presented, based on a video segmentation algorithm that uses the Entropy Difference. We applied a binary-backprojection algorithm in conjunction with K-means and MDL in order to group the colour seeds into homogenous regions (blobs). The effectiveness of the proposed algorithm was evaluated over a range of 3 different sport classes. From the results, it can be seen that the proposed model is advantageous over previous approaches because: **a)** it is capable of dealing with ambiguities among different video scenarios, **b)** there is no need for any camera calibration or background extraction and **c)** the system is very fast as it gains from the fact that information for the active regions(colours), have already been extracted using the Entropy Difference. In the future we will try to model the behavior of the active regions for semantic sport video annotation.

References

1. Assfalg, J., Bertini, M., Colombo, C., Bimbo, A.D., Nunziati, W.: Semantic annotation of soccer videos: automaitc highlights identification. Computer Vision and Image Understanding (2004)
2. Belongie, S., Carson, C., Greenspan, H., Malik, J.: Color- and Texture-Based Segmentation Using EM and Its Application to Content-Based Image Retrieval. In: IEEE International Conference Computer Vision, pp. 675–682 (1998)
3. Duan, L., Xu, M., Chua, T., Tian, Q., Xu, C.: A mid-level representation framework for semantic sports video analysis. In: Proc. of ACM MM 2003, pp. 33–44 (2002)
4. Fan, J., Gao, Y., Luo, H.: Multi-level annotation of natural scenes using dominant image components and semantic concepts. In: ACM Multimedia (ACM MM 2004), October 10-16, pp. 540–547 (2004)
5. Mentzelopoulos, M.,, P.: Key-frame Extraction Algorithm Using Entropy Difference. In: Proc. of the 6th ACM SIGMMA International Workshop on Multimedia Information Retrieval (MIR 2004), pp. 39–45 (2004)
6. Rissanen, J.: Hypothesis selection and testing by the mdl principle. The Computer Journal 42(4) (1999)
7. Jiang, S., Ye, Q., Gao, W., Huang, T.: A new method to segment playfield and its applications in match analysis in sports videos. In: ACM MM (2004)
8. Smith, J.: Image classification and querying using composite region templates. Computer Vision and Image Understanding 75 (1999)
9. Tong, X., Liu, Q., Duan, L., Lu, H., Xu, C., Tian, Q.: A unified framework for semantic shot representation of sports video. In: ACM Multimedia Information Retrieval MIR, November 10-11, pp. 127–134 (2005)
10. Yan, F., Christmas, W., Kittler, J.: A tennis ball tracking algorithm for automatic annotation of tennis match. In: BMVC 2005, vol. 2, pp. 619–628 (2005)

6DoF Egomotion Computing Using 3D GNG-Based Reconstruction

Diego Viejo, Jose Garcia, and Miguel Cazorla

Instituto de Investigación en Informática
University of Alicante. 03080 Alicante, Spain
dviejo@dccia.ua.es, jgarcia@dtic.ua.es, miguel.cazorla@ua.es

Abstract. Several recent works deal with 3D data in mobile robotic problems: mapping and SLAM related problems. Data come from any kind of sensor (time of flight cameras and 3D lasers) providing a huge amount of unorganized 3D data. In this paper we detail an efficient method to build complete 3D models from a Growing Neural Gass (GNG). The GNG obtained is then applied to a sequence. From neurons in the GNG, we propose to calculate planar patches and thus obtaining a fast method to compute the movement performed by a mobile robot by means of a 3D models registration algorithm.

Keywords: GNG, egomotion, registration, planar patches.

1 Introduction

One of the central research themes in mobile robotics is the determination of the movement performed by the robot using its sensors information. The methods related with this research are called pose registration and can be used for automatic map building and SLAM [3]. Our main goal is to perform six degrees of freedom (6DoF) pose registration in semi-structured environments, i.e., man-made indoor and outdoor environments. This registration can provide a good start point for the Simultaneous Location and Mapping (SLAM) problem. We use dense raw 3D data as input sets. Our method is developed for managing 3D point sets collected with any kind of sensor. For our experiments, we use an infrared camera SR4000, mounted on a mobile robot. The SR4000 camera is a time-of-flight camera based on infrared light. It also does not suffer from the lack of texture, as in stereo but its range is limited to 5 or 10 meters, providing gray level color. We are also interested in dealing with outliers, i.e., environments with people or none modeled objects. This task is hard to overcome because classic algorithms, like Iterative Closest Point (ICP) and its variants, are very sensitive to outliers. Furthermore, we will not use odometry information.

Furthermore, handling raw 3D data is not suitable for most of the mobile robot methodologies. In this paper we use a method for extracting and modeling planar patches from 3D raw data [8]. Using this method we achieve two main advantages: first, a complexity reduction (when comparing with raw data) is done and time and memory consumptions are improved (we obtain over 500

J. Cabestany, I. Rojas, and G. Joya (Eds.): IWANN 2011, Part II, LNCS 6692, pp. 50–57, 2011.

features from 100000 3D points); second, outliers are better overcome using those features, as points not supported by a planar patch are deleted. Planar patches are useful features as a man-made environments are easily described with them.

Nevertheless, in some situations the planar patches extraction method can not obtain a complete environment model. As we will explain later in this article, this kind of problems arise when the 3D sensor used combines both a short measurement range and a high measurement error. In those situations we propose the use of a Growing Neural Gas [4]. By means of a competitive learning, it makes an adaptation of the reference vectors of the neurons as well as the interconnection network among them; obtaining a mapping that tries to preserve the topology of an input space. Besides, they are capable of a continuous re-adaptation process even if new patterns are entered, with no need to reset the learning. These features allows to represent fast and high quality 3D spaces, obtaining an induced Delaunay Triangulation of the input space very useful to easily obtain features like corners, edges and so on. We modify the original GNG method to be applied to sequences: the GNG is adapted sequentially, i.e. the result in a given frame is taken as input in the next frame. Modeling 3D scenes using GNGs produces a more detailed result and thus further computations such as planar patches based egomotion are also improved.

The rest of the paper is organized as follows: first, the GNG algorithm is explained and results for applying GNG to our planar patches method is described; the experimental section will show our modeling results and its application on egomotion, finishing with our conclusions and future work in the last section.

2 GNG Algorithm

With Growing Neural Gas (GNG) [4] a growth process takes place from minimal network size and new units are inserted successively using a particular type of vector quantization. To determine where to insert new units, local error measures are gathered during the adaptation process and each new unit is inserted near the unit which has the highest accumulated error. At each adaptation step a connection between the winner and the second-nearest unit is created as dictated by the competitive Hebbian learning algorithm. This is continued until an ending condition is fulfilled, as for example evaluation of the optimal network topology or time deadline. The network is specified as:

- A set N of nodes (neurons). Each neuron $c \in N$ has its associated reference vector $w_c \in R^d$. The reference vectors can be regarded as positions in the input space of their corresponding neurons.
- A set of edges (connections) between pairs of neurons. These connections are not weighted and its purpose is to define the topological structure. An edge aging scheme is used to remove connections that are invalid due to the motion of the neuron during the adaptation process.

The GNG learning algorithm to map the network to the input manifold is as follows:

1. Start with two neurons a and b at random positions w_a and w_b in R^d.
2. Generate at random an input pattern ξ according to the data distribution $P(\xi)$ of each input pattern.
3. Find the nearest neuron (winner neuron) s_1 and the second nearest s_2.
4. Increase the age of all the edges emanating from s_1.
5. Add the squared distance between the input signal and the winner neuron to a counter error of s_1 such as:

$$\triangle error(s_1) = \|w_{s_1} - \xi\|^2 \tag{1}$$

6. Move the winner neuron s_1 and its topological neighbors (neurons connected to s_1) towards ξ by a learning step ϵ_w and ϵ_n, respectively, of the total distance:

$$\triangle w_{s_1} = \epsilon_w(\xi - w_{s_1}) \tag{2}$$

$$\triangle w_{s_n} = \epsilon_w(\xi - w_{s_n}) \tag{3}$$

   or all direct neighbors n of s_1.
7. If s_1 and s_2 are connected by an edge, set the age of this edge to 0. If it does not exist, create it.
8. Remove the edges larger than a_{max} . If this results in isolated neurons (without emanating edges), remove them as well.
9. Every certain number λ of input patterns generated, insert a new neuron as follows:
   - Determine the neuron q with the maximum accumulated error.
   - Insert a new neuron r between q and its further neighbor f:

$$w_r = 0.5(w_q + w_f) \tag{4}$$

   - Insert new edges connecting the neuron r with neurons q and f, removing the old edge between q and f.
10. Decrease the error variables of neurons q and f multiplying them with a consistent α. Initialize the error variable of r with the new value of the error variable of q and f.
11. Decrease all error variables by multiplying them with a constant γ.
12. If the stopping criterion is not yet achieved (in our case the stopping criterion is the number of neurons), go to step 2.

With regard to the processing of image sequences, we have introduced several improvements to the network to accelerate the representation and allow the architecture to work faster.

The main difference with the GNG algorithm is the omission of insertion/deletion actions (steps 8 to 11) after first frame. For the initial moment t_0 the representation is obtained making a complete adaptation of a GNG. However, for the following frames the previous network structure is employed. So, the new representation is obtained by performing the iteration of the internal loop of the learning algorithm of the GNG, relocating the neurons and creating or removing edges.

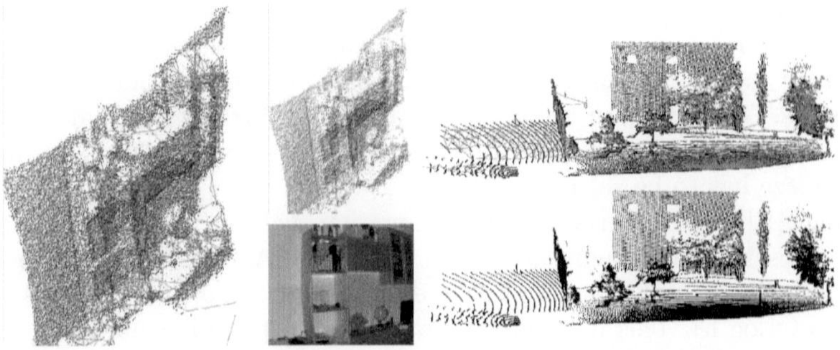

Fig. 1. Applying GNG to SR4000 data set (left) and 3D laser data (right)

For the experiments, the GNG parameters used are: $N = 2000$, $\lambda = 2000$, $\epsilon_w = 0.1$, $\epsilon_n = 0.001$, $\alpha = 0.5$, $\beta = 0.95$, $\alpha_{max} = 250$. In Figure 1 a result of applying GNG to a 3D points from a SR4000, stereo and 3D laser are shown.

3 Features Extraction Method

We can reduce more the amount of information contained in a 3D scene by modeling object surfaces included in it. Normal vectors estimated from a local area around each 3D point into the scene is a good starting point for obtaining surfaces descriptions. Some methods, such as [7] or [6] were developed for handling noisy input data sets. The basic idea consists in analyzing each point in the local neighborhood by means of a robust estimator. In [5] a singular value decomposition (SVD) based estimator is used for obtaining surface normal vectors. Using this method, when the underlying surface is a plane, the minimum singular value is quite smaller than the other two singular values, and the singular vector related to the minimum singular value is the normal vector of the surface at this point.

From this information we can label each point in a 3D scene as belonging to a planar surface, when one of the singular values is much smaller than the others or not defined objects in other case. Despite the segmentation of the scene points, we need to do some extra work to extract planar patches from the scene. We use a template matching for fitting the labeled points into a planar patch model.

This process retrieves the underlying surface normal vector of a given set of points. Furthermore, a threshold called thickness can be defined from singular values in order to determine in which situations a point, as well as its neighborhood, belong to a planar surface or not. This thickness value can be used to measure the fitting of a 3D point set to a plane. The lower thickness value we find, the better fitting between points and planar surface is. The size of the window used for obtaining neighbor points has an important impact on the results. As it is considered in [2], sample density of 3D laser range finder data presents large variations due to the divergence of consecutively sampled beams.

In general, this characteristic is present in any 3D data set, independently on the sensor used. A complete study on the impact of different window sizes was performed in [8]. Summarizing, a depth-based adaptive window provides better results. Depending on the sensor measurement error, this window has to be setup at different starting sizes. The bigger the measurement error is, the bigger window size has to be set up.

Using SVD based normal vector estimation method we can obtain a model that represents the planar surfaces in the scene. As it is also described in [8] it is possible to find planar patches descriptions for planar surfaces of a 3D scene in O(log n). This method can obtain complete and accurate models from most of the available 3D sensors. Nevertheless, some sensor characteristics may lead this method to obtain a not so complete scene model. As we stated before, the window size is a key factor. It depends on both the depth of the point and in the 3D sensor measurement error. Bigger windows will produce better results for noisy 3D sets but also will discard small objects, compared to window size, in the scene. This lack of small details may lead to problems in further computations, specially when we are using 3D sensors with a short range (up to 10 meters). Oftently, under this configurations the sensor can not obtain important information such as the end of a corridor or the rear part of a big room and then small details are really important. To overcome this problem we introduce in this paper the use of GNG in order to improve the feature extraction method. GNG produces a Delaunay Triangulation which can be used as a representation for the points neighborhood. In this way we can state the neighbor searching according to GNG and produce more detailed and accurate planar patches descriptions. Figure 2 shows planar patches extraction from a 3D image obtained with a SR4000 camera. Right image shows the results of combining GNG with the features extraction procedure. It can be compared with the left image in which no GNG has been used.

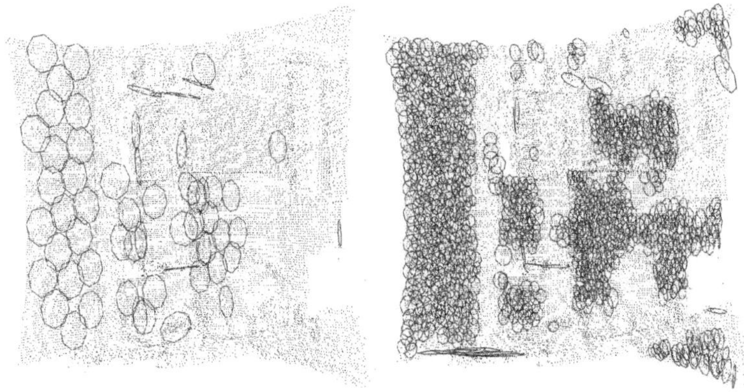

Fig. 2. Planar patches extracted from SR4000 camera. Right image uses GNG for improving planar patches extraction. As a result we obtain a more detailed planar patches descriptions than in the left image.

4 Using 3D Models: 6DoF Egomotion

In the previous section we described a method for building 3D models from scenes captured with a 3D sensor. Therefore, we want to use these models to achieve further mobile robot applications in real 3D environments. The basic idea is to take advantage of the extra knowledge that can be found in 3D models such as surfaces and its orientations. This information is introduced in a modified version of an ICP-like algorithm in order to reduce the outliers incidence in the results. ICP [1] is widely used for geometric alignment of a pair of three-dimensional points sets. From an initial approximate transformation, ICP iterates the next three steps until convergence is achieved: first, closest points between sets are stated; then, best fitting transformation is computed form paired points; finally, transformation is applied. In the mobile robotics area, the initial transformation usually comes from odometry data.

Nevertheless, our approach does not need an initial approximate transformation like ICP based methods do. We can use the global model structure to recover the correct transformation. This feature is useful for those situations where no odometry is available or it is not accurate enough, such as legged robots. In our case, we are going to exploit both the information given by the normal vector of the planar patches and its geometric position. Whereas original ICP computes both orientation and position at each iteration of the algorithm, we can take an advantage on the knowledge about planar patches orientation for decoupling the computation of rotation and translation. So we first register the orientation of planar patch sets and when the two planar patches sets are aligned we address the translation registration. In Figure 3 we can observe the steps performed for computing the alignment between two sets of planar patches. The left image shows a zenithal view of two planar patches sets computed from two consecutive 3D scenes obtained by a robot during its trajectory. The image in the center shows the result of rotation registration. Finally, right image shows the result after the translation between planar patches sets is computed.

Fig. 3. Planar patches matching example. For all the three images patches from the model are painted in dark grey whereas scene paths are represented in light grey. Left, initial situation. Middle, after rotation registration. Right, final result after translation registration is completed.

Fig. 4. Planar based 6DoF egomotion results. The sequence includes 100 3D images captured from a SR4000 camera. Left image shows map building results without using GNG while the results showed on the right are obtained after computing a GNG mesh.

In Figure 4 we show an example of 3D map building using this 6DoF egomotion approach. For this experiment, 100 3D images from a 5 meter range SR4000 camera have been used. The left image shows a 3D view of the reconstructed environment using 6DoF egomotion from planar patches. In the right image, the same scene is reconstructed but GNG have been used for improving feature extraction. While in the first experiment the registration of the sequence was almost impossible, in the second one the reconstruction was reasonably good. Computing time for obtaining planar patches descriptions after applying GNG are almost the same than without GNG and is about 100 ms per image.

5 Conclusions and Future Work

We have presented a new method for computing 3D models from unorganized raw 3D data. We do not need to know anything about the kind of sensor used for obtaining data so the method we propose can be used with the most of the 3D scanner devices. First, we have explained an algorithm for computing the planar patches that fits with the planar surfaces in the 3D scene. This is a low complexity method that can be used for obtaining online 3D models. We have also used a GNG in order to increase the level of detail for the resulting models. Results have been shown for both: time-of-flight cameras and 3D range laser. The usefulness of our models is demonstrated by applying an 6DoF egomotion algorithm that uses those models as input for computations. It have been proved that the use of GNG improves 6Dof mapping results making it possible to use this approach with any kind of 3D sensor.

Modification of GNG to represent 3D data sequences accelerates the learning algorithm and allows the architecture to work faster. This is possible because the system does not restart the map for each frame in the sequence, only readjust the network structure starting from previous map without inserting or deleting neurons.

As future work we plan to improve the accuracy and performance of our method in order to use it in 6Dof SLAM.

Acknowledgments

This work has been supported by grant DPI2009-07144 from Ministerio de Ciencia e Innovacion of the Spanish Government and by the University of Alicante project GRE09-16.

References

1. Besl, P.J., McKay, N.D.: A method for registration of 3-d shapes. IEEE Trans. on Pattern Analysis and Machine Intelligence 14(2), 239–256 (1992)
2. Harrison Cole, A.R., David, M., Newman, P.M.: Using naturally salient regions for slam with 3d laser data. In: Proc. of the IEEE International Conference on Robotics and Automation (2005)
3. Dissanayake, M.W.M.G., Newman, P., Clark, S., Durrant-Whyte, H.F., Csorba, M.: A solution to the simultaneous localization and map building (slam) problem. IEEE Transactions on, Robotics and Automation 17(3), 229–241 (2001)
4. Fritzke, B.: A Growing Neural Gas Network Learns Topologies, vol. 7, pp. 625–632. MIT Press, Cambridge (1995)
5. Martín, M., Gómez, J., Zalama, E.: Obtaining 3d models of indoor environments with a mobile robot by estimating local surface directions. Robotics and Autonomous Systems 48(2-3), 131–143 (2004)
6. Niloy, J.: Mitra and An Nguyen. Estimating surface normals in noisy point cloud data. In: SCG 2003: Proceedings of the nineteenth annual symposium on Computational geometry, pp. 322–328. ACM Press, New York (2003)
7. Page, D.L., Sun, Y., Koschan, A.F., Paik, J., Abidi, M.A.: Normal vector voting: crease detection and curvature estimation on large, noisy meshes. Graph. Models 64(3/4), 199–229 (2002)
8. Viejo, D., Cazorla, M.: 3d model based map building. In: International Symposium on Robotics. ISR 2008 (2008)

Fast Image Representation with GPU-Based Growing Neural Gas

José García-Rodríguez[1], Anastassia Angelopoulou[2], Vicente Morell[1], Sergio Orts[1], Alexandra Psarrou[2], and Juan Manuel García-Chamizo[1]

[1] Dept. of Computing Technology, University of Alicante, Ap. 99. E03080. Alicante, Spain
{jgarcia,sorts,vmorell,chamizo}@dtic.ua.es
[2] Dept. of Computer Science & Software Engineering (CSSE), University of Westminster, Cavendish W1W 6UW, United Kingdom
{agelopa,psarroa}@wmin.ac.uk

Abstract. This paper aims to address the ability of self-organizing neural network models to manage real-time applications. Specifically, we introduce a Graphics Processing Unit (GPU) implementation with Compute Unified Device Architecture (CUDA) of the Growing Neural Gas (GNG) network. The Growing Neural Gas network with its attributes of growth, flexibility, rapid adaptation, and excellent quality representation of the input space makes it a suitable model for real time applications. In contrast to existing algorithms the proposed GPU implementation allow the acceleration keeping good quality of representation. Comparative experiments using iterative, parallel and hybrid implementation are carried out to demonstrate the effectiveness of CUDA implementation in representing linear and non-linear input spaces under time restrictions.

Keywords: Growing Neural Gas, topology preservation, objects representation, Graphics Processing Units, Compute Unified Device Architecture, parallelism.

1 Introduction

Self-organising neural networks, by means of a competitive learning, make an adaptation of the reference vectors of the neurons as well as the interconnection network among them; obtaining a mapping that tries to preserve the topology of an input space. Besides, they are capable of a continuous re-adaptation process even if new patterns are entered, with no need to reset the learning.

These capacities have been applied to different tasks that deals with image segmentation [1,2] or representation of objects [3,4] among others, by means of the Growing Neural Gas (GNG) [5] that has a learning process more flexible than other self-organising models like Kohonen maps [6] and more flexible and faster than Topology Representing Networks [7].

In this work we describe an image representation system based on a fast version of the GNG implemented in parallel and accelerated with CUDA [8]. Using as input for the system an image or sequence of images we fast represent the objects that appear in the images with the GNG structure [9] with good topology preservation.

J. Cabestany, I. Rojas, and G. Joya (Eds.): IWANN 2011, Part II, LNCS 6692, pp. 58–65, 2011.

The GPU is organized in a series of multiprocessors, each of which contains a set of Stream processors and a shared cache memory which facilitates cooperation between different threads. CUDA architecture obeys the SIMD architecture (single instruction multiple data). The threads are released simultaneously in groups called "warps". Each thread within a warp can run concurrently throughout the code, thus obtaining better performance when running the same piece of code with different data sets (figure 1).

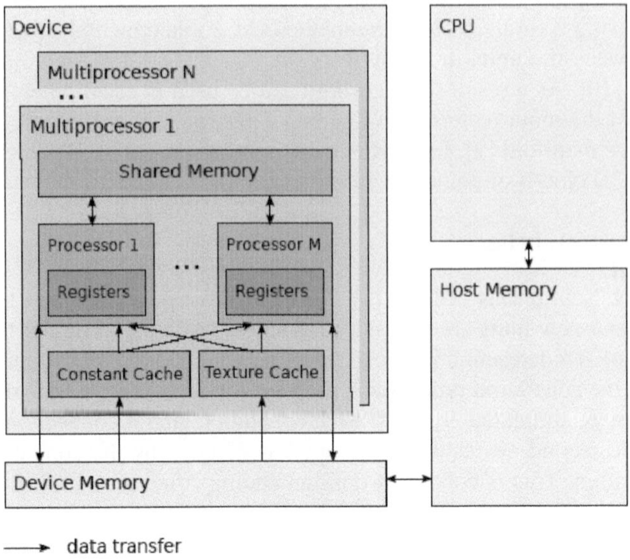

Fig. 1. CUDA architecture

Algorithms related with image processing performed small tasks on large amounts of data, hence the bandwidth tends to be a critical factor in application performance. The GPU offers another advantage in this area, allowing us to perform context switches almost instantaneously, regaining the status of the thread. Also CUDA devices offer different types of memory, the three major types of memory for our purpose are: shared memory, global memory and persistent memory. Shared memory is a shared cache to a block of threads which can be accessed with low latency for a thread group. It is very useful for implementing caches. Another important type of memory is the memory of constant and allows us the same data to be accessed by a large set of wires with low latency. These types of memory have limited space. Finally we have the global memory, which has higher latency but has up to 4 GB per device to store information.

The remainder of the paper is organized as follows: section 2 provides a detailed description of the topology learning algorithm of the GNG. Section 3 presents the GNG-based implementation with CUDA. Finally in section 4 we describe some experiments of the parallel implementation running in a GPU compared with the CPU results, followed by our major conclusions.

2 Topology Learning

One way to obtain a reduced and compact representation of 2D shapes or 3D surfaces is to use a topographic mapping where a low dimensional map is fitted to the high dimensional manifold of the shape, whilst preserving the topographic structure of the data. A common way to achieve this is by using self-organising neural networks where input patterns are projected onto a network of neural units such that similar patterns are projected onto units adjacent in the network and vice versa.

The approach presented in this paper is based on self-organising networks trained using the Growing Neural Gas learning method [5], an incremental training algorithm. The links between the units in the network are established through competitive hebbian learning [10]. As a result the algorithm can be used in cases where the topological structure of the input pattern is not known a priori and yields topology preserving maps of feature manifold [7]. In our case representing shapes of 2D objects in images or describing 3D clouds of point structure.

2.1 Growing Neural Gas

With Growing Neural Gas (GNG) [5] a growth process takes place from minimal network size and new units are inserted successively using a particular type of vector quantisation [6]. To determine where to insert new units, local error measures are gathered during the adaptation process and each new unit is inserted near the unit which has the highest accumulated error. At each adaptation step a connection between the winner and the second-nearest unit is created as dictated by the competitive hebbian learning algorithm. This is continued until an ending condition is fulfilled, as for example evaluation of the optimal network topology based on some measure. Also the ending condition could it be the insertion of a predefined number of neurons or a temporal constrain. In addition, in GNG networks learning parameters are constant in time, in contrast to other methods whose learning is based on decaying parameters.

The GNG learning algorithm to approach the network to the input manifold is a version with k=1and a predefined λ of the accelerated GNG presented in figure 2.

2.2 Accelerated Growing Neural Gas

Computer vision and image processing tasks have in many cases temporal constraints determined by the sampling rate.

To obtain a complete network, with all its neurons, in a predetermined time, the GNG learning algorithm should be modify to accelerate its conclusion. The main factor that affects the learning time is the number λ of input signals generated by iteration (step 2), since new neurons are inserted at smaller intervals, taking less time in completing the network. Another alternative is the insertion of more than a neuron by iteration, repeating k times step 9 of the learning algorithm. In this accelerated version of the GNG, each iteration step 9 is repeated, inserting several neurons in those zones where bigger accumulated error exists, creating the corresponding connections (figure 2).

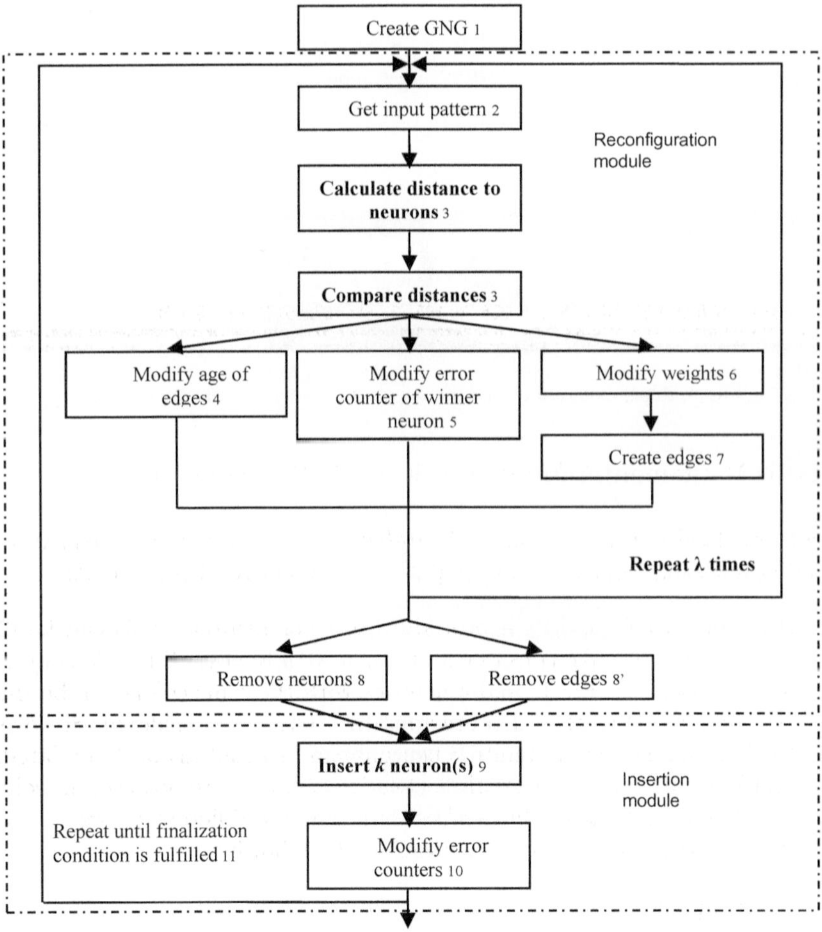

Fig. 2. Accelerated GNG learning algorithm

2.3 Representation of Objects with GNG

In the case of 2D images, given an image $I(x,y) \in \mathcal{R}$ we perform the transformation $\psi_{\mathcal{T}}(x,y) = \mathcal{T}(I(x,y))$ that associates to each one of the pixels its probability of belonging to the object, according to a property $\mathcal{T}$, for instance a threshold function.

If we consider $\xi = (x,y)$ and $\mathcal{P}(\xi) = \psi_{\mathcal{T}}(\xi)$, we can apply the learning algorithm of the GNG to the image I, so that the network adapts its topology to the object shape. This adaptive process is iterative, so the GNG represents the object during all the learning.

As a result of the GNG learning process we obtain a graph that we call Topology Preserving Graph $\mathcal{TPG} = \langle \mathcal{N}, C \rangle$, with a vertex (neurons) set $\mathcal{N}$ and an edge set C that connect them (figure 3). This $\mathcal{TPG}$ establishes a Delaunay triangulation induced by the object [9]. This consideration can be extended to 3D images adding z coordinates value.

Fig. 3. Representation of 2D/3D objects with a self-organising network

3 GPU Implementation

To address the acceleration of the GNG algorithm using CUDA, it has been used an array of threads that calculates in parallel the Euclidean distance from each neuron to the input pattern (step 3), subsequently, also in parallel we obtain the winning neuron by reducing the linear computational cost n of the sequential version to the logarithmic cost *log (n)* in the parallelized one using the algorithm Parallel Reduction [11]. Since these operations consume most of the algorithm execution time are very important to improve the global algorithm performance.

Due to the transfer of information between the memory of the GPU and CPU, initially there was no major improvement in the algorithm performance and time obtained were higher than the sequential CPU version, even increasing the number of neurons to a high number. After analyzing the flow of execution and find where the highest consumption of time occurs, we realized that acceleration of algorithms with CUDA should avoid this transfer of information.

In the initial version of the algorithm, we started copying the vector of neurons in the memory of the GPU, then calculated the distance to the input pattern and obtained the winning neuron and turned to copying the memory array of neurons to CPU to perform sequential operations as adaptation in weights of the winning neuron. This transfer of memory between CPU and GPU after each input pattern processing, increased significantly the latency each iteration making the algorithm slower than CPU version. That is why we take the decision to perform these sequenced operations in the GPU using a single CUDA thread, thus avoiding the transfer of information between CPU and GPU memory, the transfer back to CPU memory occurs only after iterate over the λ input patterns, so as the number of input patterns increase the algorithm parallelized version will increase performance.

The use of CUDA provides better performance for a large number of neurons due to the time spent in preparation of the environment to launch a CUDA kernel, for example, initially for GNG algorithm is not profitable to launch a CUDA kernel to calculate distances and get the winning neuron as the number of neurons is too low, perform these operations on CPU for vectors of 50-500 neurons is almost immediate.

However, the calculation of these operations to a number of neurons higher than 1000, obtain acceleration with respect to the sequential version of the algorithm begins to be noticeable.

The need to apply hybrid techniques comes due to the restriction that GNG algorithm initially work with a minimum number of neurons, so that the algorithm uses a sequential CPU calculation while the sequential execution time of operations is smaller than that used by the parallel version, so when it detects that the runtime version time is higher than the parallelized time, CUDA version is enabled.

4 Experiments

In this section we present some results of experiments with the GNG used to learn features in 2D and 3D images (see figure 3). In the case of 3D, the necessary number of neurons to adapt the topology is bigger than in 2D version and the parallelization of the algorithm to use the GPU obtain considerable better results than the sequential version.

After several tests we decided to make a hybrid version of this algorithm that combines the performance CPU with the GPU getting better results. For this test, we chose the parameters of 1000-5000 neurons and a maximum λ (entries per iteration) of 1000-2000. Some other parameters have been fixed ($\varepsilon_1 = 0.1$, $\varepsilon_2 = 0.01$, $\alpha = 0.5$, $\beta = 0.0005$, $a_{max} = 250$) based on our previous experience. Figure 4 shows how the algorithm evolves and what are the strengths and weaknesses of each solution.

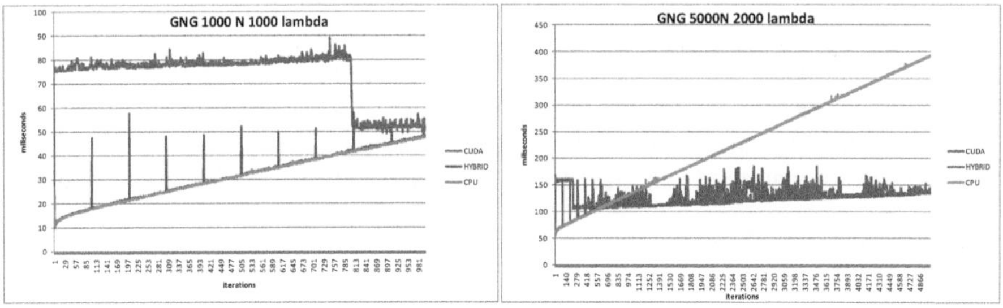

Fig. 4. 3D Image representation execution times for different GNG implementations

As we see in the figure 4, the CPU solution takes more and more time as it grows the number of neurons in the network. It can be seen how the CPU version time complexity grows much faster than the other versions and is therefore an inadequate solution for large numbers of neurons. However the parallel CUDA version increases the size of the array of neurons without degrading performance significantly. We also appreciated that the hybrid version in red is very useful because it begins with the time of the CPU version and when it is convenient switch to CUDA version decreasing the total time of calculation. Peaks in the graphs are also related to communication between CPU and GPU. Results are better for CUDA version when number of neurons and iterations raise while hybrid version is more interesting in early iterations.

Experiments were developed onto NVIDIA GeForce GTX 260 and Tesla Quadro 2000.

5 Conclusions

In this paper, we have demonstrated the capacity of Growing Neural Gas Networks neural networks to solve some computer vision an image processing tasks, demonstrating their capacity to represent 2D/3D objects in images. Establishing a suitable transformation function, the model is able to adapt its topology to represent images and 3D data.

We also propose the modification of the Growing Neural Gas learning algorithm with the objective of satisfying temporal constraints in the adaptation of the neural network to an input space, so that its process of adaptation will be determined by the number λ of input patterns and by the number N of neurons.

As it has been shown in the results, the calculation time of neural networks raised for large numbers of neurons. It is also been demonstrated that the more operations are performed in parallel with CUDA, the best results are obtained. The solutions could still be improved by using more computing on GPU implementing the operations of the external loops (almost always in relation to the neighborhood).

In conclusion, these algorithms can be accelerated using the GPU and provides a better performance than CPU at a relatively lower cost.

Acknowledgement

This work was partially supported by the University of Alicante project GRE09-16 and Valencian Government project GV/2011/034. Experiments were made possible with generous donation of hardware from NVDIA.

References

1. Ji, S., Park, W.: Image Segmentation of Color Image Based on Region Coherency. In: Proc. International Conference on Image Processing, pp. 80–83 (1998)
2. Lo, Y.S., Pei, S.C.: Color Image Segmentation Using Local Histogram and Self-organization of Kohonen Feature Map. In: Proc. International Conference on Image Processing, pp. 232–239 (1999)
3. Flórez, F., García, J.M., García, J., Hernández, A.: Representation of 2D Objects with a Topology Preserving Network. In: Proceedings of the 2nd International Workshop on Pattern Recognition in Information Systems (PRIS 2002), Alicante, pp. 267–276. ICEIS Press (2001)
4. Holdstein, Y., Fischer, A.: Three-dimensional Surface Reconstruction Using Meshing Growing Neural Gas (MGNG). Visual Computation 24, 295–302 (2008)

5. Fritzke, B.: A Growing Neural Gas Network Learns Topologies. In: Tesauro, G., Touretzky, D.S., Leen, T.K. (eds.) Advances in Neural Information Processing Systems 7, pp. 625–632. MIT Press, Cambridge (1995)
6. Kohonen, T.: Self-Organising Maps. Springer, Heidelberg (1995)
7. Martinetz, T., Schulten, K.: Topology Representing Networks. Neural Networks 7(3), 507–522 (1994)
8. NVIDIA Corporation, CUDA Programming Guide, version 3.2 (2010)
9. O'Rourke, J.: Computational Geometry. C. Cambridge University Press, Cambridge (2001)
10. Martinez, T.: Competitive hebbian learning rule forms perfectly topology preserving maps. In: ICANN (1993)
11. Harris, M.: Optimizing parallel reduction in CUDA. NVIDIA Corporation (2007)

Texture and Color Analysis for the Automatic Classification of the Eye Lipid Layer

L. Ramos[1], M. Penas[1], B. Remeseiro[1], A. Mosquera[2], N. Barreira, and E. Yebra-Pimentel[3]

[1] Departamento de Computación, Universidade da Coruña
Campus de Elviña S/N. 15071 A Coruña, Spain
[2] Departamento de Electrónica y Computación, Universidade de Santiago de Compostela, Campus Universitario Sur. 15782 Santiago de Compostela, Spain
[3] Escuela de Óptica y Optometría, Universidade de Santiago de Compostela, Campus Universitario Sur. 15782 Santiago de Compostela

Abstract. This paper describes a methodology for the automatic classification of the eye lipid layer based on the categories enumerated by Guillon [1]. From a photography of the eye, the system detects the region of interest where the analysis will take place, extracts its low-level features, generates a feature vector that describes it and classifies the feature vector in one of the target categories. We have tested our methodology on a dataset composed of 105 images, with a classification rate of over 90%.

Keywords: Eye lipid layer, Guillon categories, Butterworth filters, Lab color space, machine learning.

1 Introduction

The preocular tear film consists of an outer lipid layer, a middle aqueous layer and an inner mucous layer. The quality and thickness of each layer, as well as their adequate interaction, are important in order to have a stable tear film. Abnormalities in any of the layers can cause tear dysfunction problems. Concretely, the lipid layer plays a major role in retarding the evaporation of the tear film during the inter-blink period and, consequently, a deficit of this layer can cause the *evaporative dry eye syndrome* [2]. This condition affects a wide sector of the population, specially among contact lens users, and worsens with age.

The lipid layer thickness can be evaluated through the observation of the interference phenomena, since the color and shape of the observed patterns reflect the layer thickness. Thicker lipid layers ($\geq$ 90 nm) show color and wave patterns while thinner lipid layers ($\leq$ 60nm) are more homogeneous. The Tearscope Plus, designed by Guillon [1], is the instrument of choice for the lipid layer thickness evaluation in clinical settings. Guillon also proposed five main categories of lipid interference patterns in increasing thickness: marmoreal (open and closed

J. Cabestany, I. Rojas, and G. Joya (Eds.): IWANN 2011, Part II, LNCS 6692, pp. 66–73, 2011.

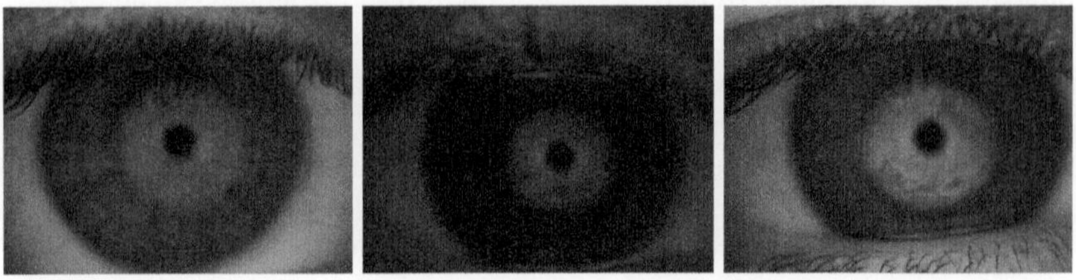

(a) (b) (c)

Fig. 1. Lipid layer thickness categories: (a) Open meshwork. (b) Flow. (c) Color fringe.

meshwork), flow, amorphous and color fringes. Figure 1 shows some representative images of each category[1].

However, the classification of the lipid layer thickness is a difficult clinical technique, especially with thinner lipid layers that lack distinct features, and is affected by the subjective interpretation of the observer. Some techniques have been designed to objectively calculate the lipid layer thickness where a sophisticated optic system was necessary [3] or an interference camera evaluated the lipid layer thickness by only analyzing the interference color [4].

In this paper we present a novel methodology of the automatic classification of the eye lipid layer. From a photography of the eye, we detect the region of interest (ROI) where the interference phenomena will be studied, we analyze its color and texture, generate a feature vector that represents it and, finally, we classify the feature vector in one of the Guillon categories. We have tested our methodology on a dataset with 105 images from healthy subjects. This paper is organized as follows: section 2 describes the methodology developed for the classification; section 3 compares the classification results depending on the frequency ranges, classifiers and color spaces used for the analysis of the images; finally section 4 enumerates our conclusions and future work.

2 Methodology

Our methodology has been tested on eye photographs acquired using a Tearscope Plus®(Keeler, Windsor, United Kingdom) attached to a Topcon SL-D4 slit lamp and a Topcon DV-3 digital camera. The magnification of the biomicroscope was set to 200X and the images stored with a resolution of 1024×768 pixels. Since the tear lipid film is not static between blinks, a video was recorded and analyzed to select the images to be processed: an image was selected only when the tear lipid film was completely expanded after the eye blink.

As we can see in figure 1 the input images include regions of the eye, like the eyelids or eyelashes, that do not contain relevant information for the classification and could mislead the results of our methodology. For this reason, in [5] we

[1] The amorphous category is not depicted in figure 1 or considered hereafter in this paper due to the lack of representative images from this category.

developed a methodology to detect the region of interest where the classification will take place. Our acquisition procedure guarantees that the region of interest corresponds to the most illuminated area of the image. In order to detect this region, first we generated a set of masks like the one depicted in figure 2(a) and selected the region of the image with the maximum normalized cross-correlation with respect to one of these masks; then, we draw a large inscribed rectangle in the lower half of the mask selected, as figure 2(b) shows.

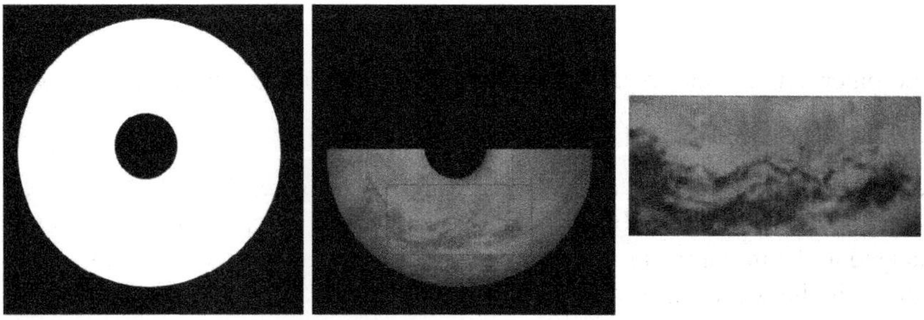

(a) (b) (c)

Fig. 2. (a) Mask used to determine the region of interest. (b) Sub-mask and region of interest. (c) Region of interest.

After extracting the regions of interest, we begin to analyze their low-level features. Color and texture seem to be two discriminant features of the Guillon categories. As we mentioned earlier, thick lipid layers show clear patterns while thinner layers are more homogeneous; therefore, the discriminant features of each category should yield stronger responses in different frequency ranges. Also, since some target categories show distinctive color features, we have analyzed the low-level features not only in grayscale but also in Lab [6] and RGB making use of the opponent color theory [7].

We have analyzed the texture of each ROI through a bank of Butterworth bandpass filters [8]. Butterworth filters have a flat response in the bandpass frequency and gradually decay in the stopband. The slope of the decay is defined by the order of the filter; the higher the order, the faster the decay, as figure 3(a) shows. We have used nine second order Butterworth filters with bands covering the whole frequency spectrum, see figure 3(b) for a 1D representation of the filters used.

In grayscale, the filter bank maps each ROI to 9 result images, one per frequency band. In Lab, each ROI is mapped to 18 result images, three per frequency band corresponding to the L, a and b components, respectively. Finally, the opponent color theory proposed by E. Hering [7] states that the human visual system interprets information about color by processing three opponent channels: red vs. green, green vs. red and blue vs. yellow; this is, either red or green can be perceived, but never both, and the same applies to blue and yellow. We have used these pairs of opponent colors to analyze our images. As in Lab,

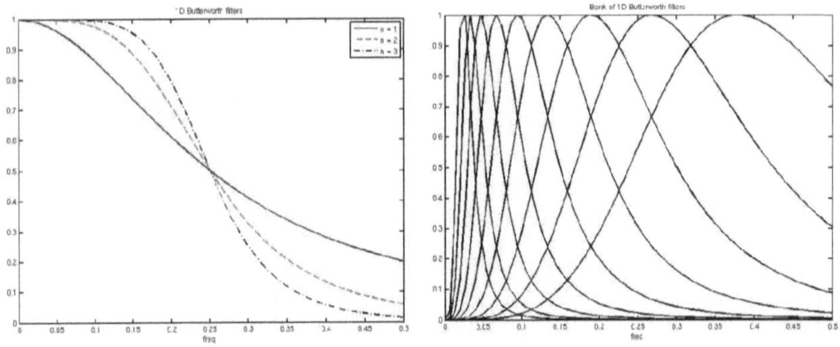

(a) (b)

Fig. 3. (a) Butterworth filters of varying orders. (b) Bank of bandpass filters.

each ROI was mapped to 18 result images, three per frequency band, computed as: $R\_G = R_F - p * G_F$, $G\_R = G_F - p * R_F$ and $B\_Y = B_F - p * (R_F + G_F)$, where p is a low pass filter and R_F, G_F and B_F are the filtered R, G, and B components, respectively.

We must now assign a feature vector to each output image in order to perform the final classification. Our first attempt was to normalize each frequency band separately and compute the histograms of its output images; such histograms concentrated most of the information in the lower bins, which made their comparison difficult. In order to boost the importance of the differences between lower values, we defined uniform histograms with non-equidistant bins; given all the output images from our dataset in a frequency band, we ordered their pixels making sure each bin contained no more than $\frac{N}{N_{bins}}$ pixels, where N is the

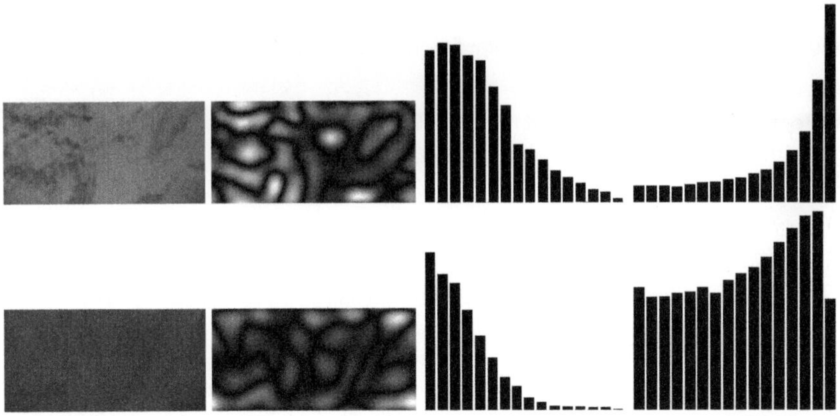

(a) (b) (c) (d)

Fig. 4. (a) Color fringe and flow regions of interest. (b) Lowest frequency band results. (c) Histograms. (d) Uniform histograms.

number of pixels in the given frequency and N_{bins} the number of histogram bins. Figure 4 shows the regular and uniform histograms of two regions of interest in the lowest frequency band.

In grayscale, the feature vector of a ROI in a frequency band is the uniform histogram of the filtering result; in Lab, the concatenation of the uniform histograms of the L, a and b component results; similarly, in RGB, the feature vector is the concatenation of the uniform histograms of the R_G, G_R and B_Y results.

The final stage of our methodology is the classification of each ROI into one of the target categories. We have analyzed and compared several machine learning algorithms like naive Bayes [9], logistic model trees [10], random decision trees [11], random decision forests [12], multilayer perceptron [13] and support vector machines (SVM) [14]. Section 3 summarizes the results obtained by these algorithms on our dataset.

3 Results

We have tested our methodology on an image dataset composed of 105 images from healthy patients with ages ranging from 19 to 33 years old. These images have been annotated by optometrists from the School of Optics and Optometry of the Universidade de Santiago de Compostela. The dataset contains 22 color fringe, 25 flow, 29 open meshwork and 29 closed meshwork images. Due to the limited size of our dataset, we have used 10-fold cross-validation on our experiments [15] to assess the generalization capability.

We have first analyzed the results on each frequency band separately in order to determine which frequency bands are more discriminative and which classification algorithms are more effective for the task at hand. Our first experiment is performed on grayscale images. Table 1 shows its results in terms of percentage accuracy, with the best results per frequency band highlighted[2].

Table 1. Classification rates (%) on separate frequency bands. The cutoff frequencies are: 1 (0.02,0.028), 2 (0.028,0.039), 3 (0.039,0.056), 4 (0.056,0.079), 5 (0.079,0.1125), 6 (0.1125,0.159), 7 (0.159,0.225), 8 (0.225,0.318) and 9 (0.318,0.45).

	Frequency bands									
Classifier	**1**	**2**	**3**	**4**	**5**	**6**	**7**	**8**	**9**	**Avg**
Naive Bayes	50.48	59.05	65.70	60	60.95	58.10	54.29	49.52	48.57	56.30
LMT	59.04	55.23	60.95	60.95	60	61.90	63.8	58.09	57.14	59.68
Random tree	43.80	51.42	55.23	50.47	60	64.76	60	55.23	40	53.43
Random forest	45.71	52.38	69.52	61.90	62.86	67.61	58.10	54.29	46.67	57.67
M. perceptron	58.09	56.19	64.77	59.05	65.71	66.67	67.62	**63.81**	**59.04**	62.33
SVM	**60.0**	**59.05**	**71.43**	**71.43**	**69.52**	**67.67**	**70.47**	62.86	52.38	**64.97**

[2] We have used the implementation of the SVM in libSVM [16] and the implementations in WEKA [17] for the remaining classifiers.

Several conclusions can be drawn from table 1. Intermediate frequencies are more discriminative than lowest or highest ones, regardless of the classifier; this information can be useful when constructing the descriptors for the regions of interest, since some of the frequency bands could be discarded. With respect to the classifiers, the SVM with radial basis kernel outperforms the other classifiers not only on average but also in most frequency bands.

We have tested the significance of the differences among classifier accuracies; first, we applied the Lilliefors test for normality [18] and then an ANOVA test. The one-way ANOVA compares the means of several distributions by estimating the variance between distributions and within a distribution. The null hypothesis, that all population means are equal, is tested using the F distribution and a p-value is computed. The results of the ANOVA test are depicted on table 2 showing how the null hypothesis would be rejected with a significance level 0.05.

Table 2. ANOVA results. SS: sum of squared deviations about the mean. df: degrees of freedom. MS: variance.

Source	**SS**	**df**	**MS**	**F**	**p-value**
Between	785.80	5	157.16	3.898	<0.05
Within	1935.36	48	40.32		
Total	2721.16	53			

Our second experiment aims at determining which frequency bands should be considered when constructing the final descriptor for a region of interest. In order to combine the adjacent frequency bands we simply concatenate their individual descriptors. The analysis will take place in grayscale using the SVM, since this is the classification algorithm that produced the best results in our previous experiment. See table 3 for a summary of the results in terms of percentage accuracy.

Table 3 shows how the combination of frequency bands clearly outperforms the individual results. We have highlighted the 7 best frequency band combinations

Table 3. SVM categorization accuracy (%) for the combination of frequency bands; cell ij depicts the results obtained combining the frequency bands from i to j

	1	**2**	**3**	**4**	**5**	**6**	**7**	**8**	**9**
1	60.00	62.86	70.48	74.28	77.14	79.05	79.05	80.0	**81.90**
2		59.05	70.48	73.33	77.14	79.04	80.0	80.0	**81.90**
3			71.43	77.14	79.05	**80.95**	**80.95**	**81.90**	80.0
4				71.43	**82.86**	**80.95**	80.00	79.05	80.0
5					69.52	75.24	78.09	75.24	75.24
6						67.67	74.29	73.33	71.43
7							70.47	69.52	71.43
8								62.86	70.47
9									52.38

with classification rates higher than 80%; we will use these combinations in our next experiment.

Our final experiment will help us decide which color space works best and, thus, should be used in the final classification. Table 4 shows the results of this experiment, where we can see how color improves the accuracy of our methodology; opponent colors outperforms grayscale in most cases and Lab produces the best results, that reach 91.43% classification rates in some cases. Table 4 also shows how the results are quite stable; there is a wide range of frequency band combinations where the results are over a 90% rate, which indicates that our low-level analysis is adequate for the classification.

Table 4. Classification rates (%) in the three color spaces considered, grayscale, Lab and opponent colors

Frequency bands	Grayscale	Lab	Opponent colors
1-9	81.90	**91.43**	84.76
2-9	81.90	**91.43**	85.71
3-6	80.95	90.48	83.81
3-7	80.95	90.48	83.81
3-8	81.90	**91.43**	84.76
4-5	82.86	90.48	80.0
4-6	80.95	90.48	80.95

4 Conclusions and Future Work

This paper presents a methodology for the automatic classification of the eye lipid layer based on the detection of a region of interest and the analysis of its low-level features through a bank of bandpass filters. We have analyzed the performance of the individual frequency bands, their combination, different color spaces and classification algorithms. These analysis show how the combination of frequency channels outperforms the individual frequency channels and how the SVM and the Lab color space produce the best classification results. The methodology has been tested on a dataset composed of 105 eye photographies with maximum classification rates of 91.43%.

We would like to test the classification rates obtained using other algorithms for texture analysis, like co-occurrence matrices, Gabor wavelets or Markov random fields, as well as their combination with the current bank of bandpass filters.

Sometimes, the eye lipid layer is very heterogeneous and cannot be classified into a single Gillon's category, which is a sign of meibomian gland abnormality. Therefore, it is also part of our future work to perform local analysis and classifications, allowing the detection of several categories in a single photograph.

Acknowledgements. We would like to thank the Escuela de Óptica y Optometría of the Universidade de Santiago de Compostela for providing us with the annotated image dataset. This paper has been partially funded by the Ministerio de Ciencia e Innovación, through the research project PI10/00578.

References

1. Guillon, J.P.: Non-invasive Tearscope Plus routine for contact lens fitting. Contact Lens & Anterior Eye 21(Suppl. 1), S31–S40 (1998)
2. Craig, J.P., Tomlinson, A.: Importance of the lipid layer in human tear film stability and evaporation. Optometry and Vision Science 74(1), 8–13 (1997)
3. King-Smith, P.E., Fink, B.A., Fogt, N.: Three interferometric methods for measuring the thickness of layers of the tear film. Optometry and Vision Science 76(1), 19–32 (1999)
4. Goto, E., Dogru, M., Kojima, T., Tsubota, K.: Computer-synthesis of an interference color chart of human tear lipid layer, by a colorimetric approach. Investigative Ophthalmology & Visual Science 44(11), 4693–4697 (2003)
5. Calvo, D., Mosquera, A., Penas, M., Garcia Resua, C., Remeseiro, B.: Color texture analysis for tear film classification: a preliminary study. In: ICIAR 2010. LNCS, vol. 6112, pp. 388–397 (2010)
6. McLaren, K.: The development of the CIE 1976 (L*a*b*) uniform colour-space and colour-difference formula. Journal of the Society of Dyers and Colourists 92(9), 338–341 (1976)
7. Hering, E.: Outlines of a Theory of the Light Sense. Harvard University Press, Cambridge (1964)
8. Gonzalez, R., Woods, R.: Digital image processing. Pearson/Prentice Hall, Englewood Cliffs (2008)
9. Zhang, H.: The Optimality of Naive Bayes. In: FLAIRS Conference (2004)
10. Landwehr, N., Hall, M., Frank, E.: Logistic Model Trees. Machine Learning 59(1-2) (2005)
11. Drmota, M.: Random Trees. In: An interplay between combinatorics and probability. Springer, New York (2009)
12. Breiman, L.: Random Forests. Machine Learning 45(1), 5–32 (2001)
13. Chauvin, Y., Rumelhart, D.: Backpropagation: Theory, architecture and applications. Lawrence Erlbaum Associates, Inc., Publishers, Mahwah (1995)
14. Burges, C.: A tutorial on support vector machines for pattern recognition. Data Mining and Knowledge Discovery 2, 121–167 (1998)
15. Rodriguez, J., Perez, A., Lozano, J.: Sensitivity analysis of k-fold cross validation in prediction error estimation. IEEE Trans. on Pattern Analysis and Machine Intelligence 32(3), 569–575 (2010)
16. Chang, C., Lin, C.: LIBSVM: a library for support vector machines. Software (2001), `http://www.csie.ntu.edu.tw/~cjlin/libsvm`
17. Hall, M., Frank, E., Holmes, G., Pfahringer, B., Reutemann, P., Witten, I.H.: The WEKA Data Mining Software: An Update. ACM SIGKDD Explorations Newsletter 11(1) (2009)
18. Lilliefors, H.W.: On the Kolmogorov-Smirnov Test for Normality with Mean and Variance Unknown. Journal of the American Statistical Association 62(318), 399–402 (1967)

Quantitative Study and Monitoring of the Growth of Lung Cancer Nodule Using an X-Ray Computed Tomography Image Processing Tool

José Luis García Arroyo, Begoña García Zapirain, and Amaia Méndez Zorrilla

Deustotech-Life, Deusto Institute of Technology, University of Deusto
Avda. de las Universidades 24, Bilbao, Spain, 48007
{jlgarcia,mbgarciazapi,amaia.mendez}@deusto.es

Abstract. Nowadays there are millions of lung cancer patients around the world, and the number is increasing each year. In this context it is essential for medical radiologists and oncologists to properly control cancer evolution and to calculate quantitative values for its characterization. This paper presents a complete system integrating a software tool to improve the monitoring of patients' lung nodules and a complete stack of algorithms for a mathematical model to accurately calculate their growth/reduction. At the current moment we have a work in progress using a database with four patients who have been successfully tested, and whose nodules have all experienced positive growth, with an average of 31.72% in area growth and 0.28% per day in area growth speed. In the future, the database is expected to be enlarged with more patients so that numerical data can be obtained for use in statistical studies and mathematical modeling.

Keywords: Image Segmentation, Otsu´s method, Lung cancer evolution, Nodule growth, Patient monitoring.

1 Introduction

Lung cancer has the highest mortality rate of any cancer in the world. According to the American Cancer Society, in the United States more than 220,500 new cases of lung cancer are diagnosed each year and the disease claims the lives of over 157,000 people annually. Against it there are several treatments, all of which are high-cost, that millions of people in the world are currently undergoing. In this context it is essential to adequately control the state of lung cancer patients and the effectiveness of the treatments.

In order to study lung cancer nodules, "X-ray computed tomography" (CT) is widely used. In [1] and [2] it is presented a description of the state of art of CT methods, some of which will be used here.

In usual medical practice, the medical radiologist uses the proprietary software tool of the scanner to analyze the lung nodules at different moments. It can be seen a summary and a comparison of the actually leading ones in [3].

J. Cabestany, I. Rojas, and G. Joya (Eds.): IWANN 2011, Part II, LNCS 6692, pp. 74–82, 2011.

The authors have analyzed some of them, with the help of radiologists specialized in their operation, and as result of that study have concluded that these tools provide a big number of useful options, but lack the functionality of a serious study of the evolution of the nodules combined with its easy monitoring. This is the contribution of the present work.

To help medical radiologists in these tasks, the authors propose a complete stack of algorithms to obtain an accurate calculation of lung nodule development. As well as the previously mentioned, in order to help the medical radiologists to control lung cancer patients, a software tool focused on three aims is presented: (1) Optimizing the monitoring of the lung cancer nodules' evolution, (2) Obtaining numerical data to be used in statistical studies, (3) and, from these results, Generating a mathematical model to look into the behavior of the lung cancer nodules.

2 Methodology

2.1 X-Ray Computed Tomography

X-ray computed tomography (CT) is a medical imaging method employing tomography created by computer processing. It consists of examining body organs by scanning them with X-rays and using a computer to construct a series of cross-sectional scans along a single axis [4]. As regards the obtained images, the standard in medical imaging is the DICOM format.

2.2 DICOM Format

DICOM (Digital Imaging and Communication in Medicine) is a standard for handling, storing, printing and transmitting information in medical imaging [5].

2.3 Image Database

The image database was created with the collaboration of the V. San Sebastián Clinic and IMQ institution in Bilbao, Spain. The DICOM images were provided by the medical radiologist Gonzalo Solís and were captured using the CT method with a General Electric GE VCT TC64. The software tool used in the parameterization and capturing of the images is CT AW Volumeshare 4.4. The collimation was 64X0.625 mm. 4 patients' images were selected from two different dates. For each one, the database has the following type of images: (1) "Pulmonary-alveolus": image of the pulmonary alveolus. (2) "Pulmonary-nodules": image of the pulmonary nodules. (3) "Nodule": image with details of the lung nodule.

2.4 Mathematical Basis Concepts and Algorithms Used

Measure of circularity

Having N points on a region boundary and its centroid, the circularity C of the region is:

$$C = \mu_R / \sigma_R \tag{1}$$

where μ_R and σ_R are the mean and standard deviation of the distances between the centroid of a shape and its boundary pixels, respectively:

$$\mu_R = (1/N)\sum_{k=1}^{N} \| (r_k, c_k) - (\bar{r}, \bar{c}) \| \tag{2}$$

$$\sigma_R = [(1/N)\sum_{k=1}^{N} (\| (r_k, c_k) - (\bar{r}, \bar{c}) \| - \mu_R)^2]^{1/2} \tag{3}$$

where (r_k, c_k) is the spatial coordinate of the kth boundary point, $(\bar{r}, \bar{c})$ is the centroid of the region and the distances are in the Euclidean metric [6].

Otsu´s method

Otsu's method is a method used to calculate the optimum threshold separating those two classes so that the intra-class variance is minimal [7]. That intra-class variance is:

$$\sigma_w^2(t) = w_1(t)\sigma_1^2(t) + w_2(t)\sigma_2^2(t) \tag{4}$$

where weights w_i are the probabilities of the two classes separated by a threshold t and σ_i^2 the variances of these classes.

Then minimizing the intra-class variance is the same as maximizing inter-class variance:

$$\sigma_b^2(t) = \sigma^2 - \sigma_w^2(t) = w_1(t)w_2(t)[\mu_1(t) - \mu_2(t)] \tag{5}$$

which is expressed in terms of class probabilities w_i and class means μ_i .

3 System design

3.1 High-Level View

In Fig. 1 a high-level view of the work is presented. As can be seen, there are two principal aims: (1) Calculating the evolution of the lung nodules, and (2) Monitoring the state of lung cancer patients. These aims result in the contribution of this work: (1) A complete stack of algorithms and (2) An innovative software tool.

Exploding both of them, the principal designed algorithms and the application modules that are part of the software tool are shown. From those modules there are invocations to the algorithms of the stack, so everything is integrated within the same platform.

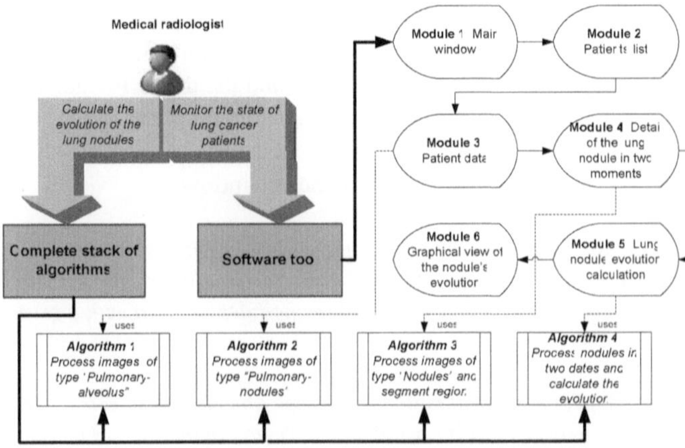

Fig. 1. High level view of the work

3.2 Description of the System

Here the authors are going to present the algorithms and the software tool in detail. To improve the clarity of the exposition the software tool's functionality is going to be used as the main thread of the whole exposition.

Modules 1 and 2

The starting point of the application is the module 1, "Main window", with all the options. The patients can be seen in the module 2, "Patients list". When the user wants to monitor a patient, selects it and can see the patient data.

Modules 3 and 4

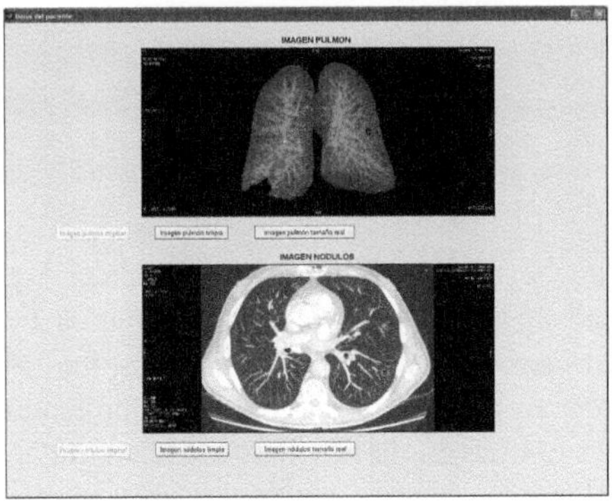

Fig. 2. Patient data

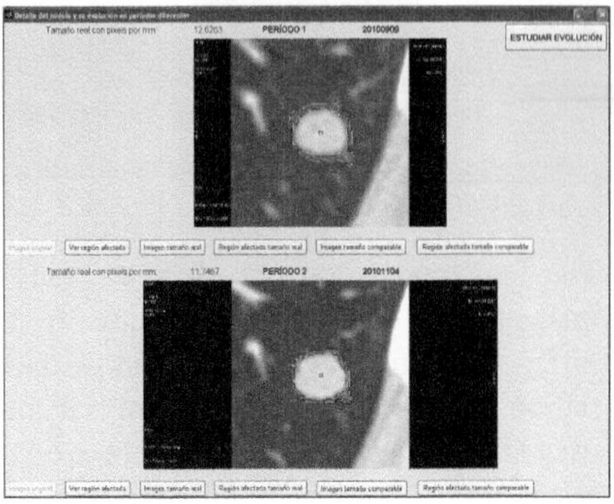

Fig. 3. Detail of lung nodule and its state in two moments

Fig.2 shows the "Pulmonary-alveolus" and "Pulmonary-nodules" images with the target regions selected with a green line. It is in module 3, "Patient data". In this window the user can segment the target regions and clean the image. The application uses algorithms 1 and 2 to process the images. After their execution, the application creates a listener that runs waiting for mouse events in the target regions. When any of them are clicked on, the user can go to the next window.

Algorithm 1. ("Process images of type "Pulmonary-alveolus"):
Summary: *The lung nodules are segmented and any irrelevant information is cleaned.*
Input: *A "Pulmonary-alveolus" image with the lung nodules selected with a green line.*
Output: *The processed image.*
Description:

To segment the image, the algorithm finds the green objects and selects those that are valid. For each one, in order to check if it is a valid nodule region in this type of image:

Step 1 fills the holes using the algorithm based on the morphological reconstruction described in [8].

Step 2 calculates whether it has a relevant area. Information on the resolution is obtained in millimeters by examining the "PixelSpacing" field that belongs to the DICOM header.

Step 3 checks if it is circular, as follows:

Step 3.1 calculates the centroid, called C.

Step 3.2 calculates the perimeter using 8 connected neighborhood-based connectivity and the Moore-Neighbor tracing algorithm modified by Jacob's stopping criteria (boundaries function defined in [9]).

Step 3.3 selects 40 points in the perimeter. To obtain them, it creates 40 straight lines from C by interpolation, each one having (for i=0 to 39) a slope value equal to the tangent of i/360 (in degrees). Each point is the result of the intersection between each line and the perimeter.

Step 3.4 calculates circularity using the formula described in section 2.

It uses erode and dilate morphological operations to clean the images. Then the algorithm marks the selected regions in green.

Algorithm 2. ("Process images of type "Pulmonary-nodules"):
Summary: *The lung nodules are segmented and the other irrelevant information is cleaned.*
Input: *A "Pulmonary-nodules" image with the lung nodules selected with a green line.*
Output: *The processed image.*
Description: *Obtaining the segmentation and cleaning the image are very similar to the previous algorithm, so it is not going to be described here.*

The lung nodule on two different dates can be seen in Fig. 3 with "Nodule"-type images. This window is in module 4, "Detail of the lung nodule and its state at two different moments". In this window the user can segment the target regions, the first region in blue and the second in red. In order to process the images, the application uses algorithm 3. Moreover, there are further options in these windows: changing the resolutions, rotating, changing the reference point, unifying the desired parameterizations and other different views. When "Study Evolution" is pressed, we can see the evolution of the nodule between those dates.

Algorithm 3. ("Process "Nodules"-type images and segment the nodule region"):
Summary: *The lung nodule is accurately segmented.*
Input: *A "Nodule" image (grayscale) called noduleImage.*
Output: *The region mask of the nodule.*
Description:

The algorithm described in [10] proposing a refinement of the Otsu method was taken into account in the design of this algorithm.

In order to segment the image, the algorithm uses a two-stage algorithm. In the first stage:

Step 1 uses a filter to add blur because this makes it is easier to obtain a good first segmentation with Otsu's method.

Step 2 uses Otsu's method (described in section 2) with some regions as the result.

Step 3 selects the region that contains the geometric center using 8-connected neighborhood based connectivit. Then it makes the first approximation to the segmentation region: regionAfterStage1. In the second stage:

Step 4 calculates the centroid of regionAfterStage1, which will be called C.

Step 5 calculates meanRegionAfterStage1, the mean value of the noduleImage in regionAfterStage1.

Step 6: calculates the threshold level called levelThreshold using the value |meanRegionAfterStage1-40|.

Step 7 calculates the binary mask using the mentioned value and selects the region that contains the geometric center, using 8-connected neighborhood based connectivity.

Step 8 uses erode and dilate morphological operations to clean the images.

Step 9 fills the holes using the algorithm based on the morphological reconstruction described in [8].

Modules 5 and 6

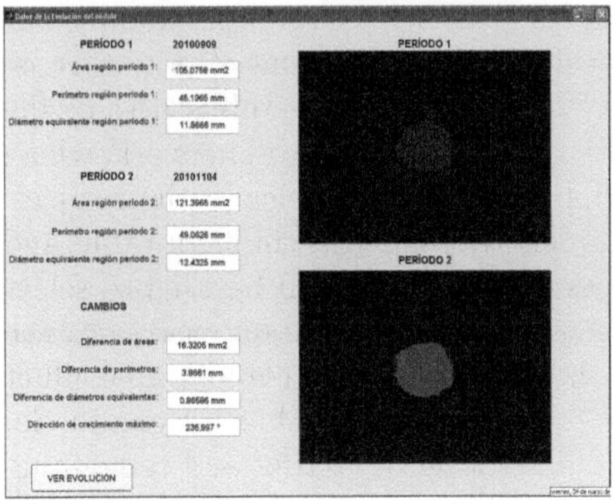

Fig. 4. Lung evolution calculations

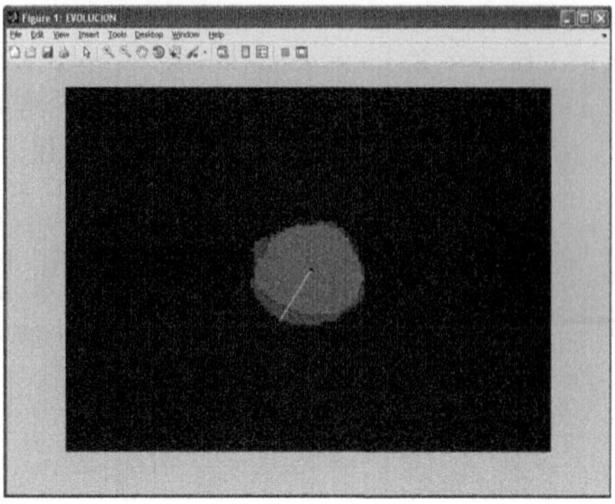

Fig. 5. Graphical view of the nodule's evolution behaviour

In Fig. 4, by invoking algorithm 4, the following can be obtained: the area and the equivalent diameter of the two shapes, the increment in these features and the direction of maximum growth, in degrees. The positive numbers are bad data, and the negative good (possibly because of the effectiveness of the treatment).

Moreover, the user can obtain a graphical view of the behavior of nodule's evolution in Fig. 5: in blue, the region on the first date but not contained in the second, in purple, the intersection between the first and the second, and in red, the region on the second date but not the first. In that view the medical radiologist can see the graphical evolution and the direction of the growth, a straight yellow line.

These windows are in module 5, "Lung nodule evolution calculation" and module 6, "Graphical view of the behavior of the nodule's evolution".

Algorithm 4 ("Process nodule images on two dates and calculate the evolution"):
Summary: *It calculates the evolution of the nodule between two dates and generates its graphical view.*
Input: *The region binary masks of the nodule on the two dates segmented by the previous algorithm. They will be called region1 and region2.*
Output: *The nodule's evolution calculated in numerical values (area, equivalent diameter and direction of maximum growth) and its graphical view.*
Description:

Initially, it calculates the area and the equivalent diameter. To generate the graphical view and calculate the direction of maximum growth:

Step 1 scales the two regions to the same resolution (as in algorithm 1).

Step 2 calculates region1's centroid, called C.

Step 3: As explained in section 2, there are two reference points in region1 and region2, having made them coincide with the geometric centers, called G1 and G2. Here it makes point C the geometric center of the generated view and translates region1 and region2 points to that center, so the translation vectors are $\overline{G1C}$ *and* $\overline{G2C}$

Step 4: To calculate the direction of maximum growth and obtain the associated graphical line:

Step 4.1 calculates the regionDiff region, the result of the difference between region2 and region1.

Step 4.2 calculates the perimeter of the abovementioned region using 8-connected neighborhood based connectivity, using the Moore-Neighbor tracing algorithm modified by Jacob's stopping criteria (boundaries function defined in [4]). It will be called regionDiffPerim.

Step 4.3 iterates for each point of regionDiffPerim's points, obtaining points P1 and P2 and the distMax distance between them, satisfying: (1) P1 is in the perimeter of region1 (2) P2 is in the perimeter of region2 (3) P1 and P2 are in the same quadrant from center C's point of view (4) C, P1 and P2 are in the same straight line (obtained by interpolation with minimal deviation) (5) distMax is the distance between P1 and P2, using Euclidean metric, and is the maximum of all pairs of points satisfying the previous four conditions.

Step 4.4 obtains a straight line between P1 and P2 (by interpolation)

Step 4.5 obtains the direction in degrees from the slope of the mentioned line. The direction is the angle from center C's point of view, taking the commonly used coordinate system with C as the origin.

Step 5 puts the region1 points not contained in region2 in blue, the intersection between region1 and region2 in purple and the regionDiff in red. The straight line that joins C, P1 and P2 is in yellow.

4 Results

Having carried out testing on the four patients contained in the database, the results are as follows:

Table 1. Dates of measures

DATES

Patient	Date 1	Date 2	N. of days
Patient 1	9 Sep 10	4 Nov 10	56
Patient 2	20 Jul 10	4 Nov 10	107
Patient 3	20 Jul 10	17 Nov 10	120
Patient 4	1 Jan 10	17 Nov 10	320

Table 2. Directions of maximum growth

DIRECTION OF MAXIMUM GROWTH

Patient	Direction of maximum growth (degrees)
Patient 1	236.897
Patient 2	142.0271
Patient 3	58.2968
Patient 4	158.3599

Table 3. Areas of nodules in the two dates and its growth

AREAS

Patient	Area1 (mm2)	Area2 (mm2)	Growth of areas (mm2)	Growth of areas p. day (mm2/day)	Growth of areas (%)	Growth of areas p. day (%/day)
Patient 1	105.0759	121.3965	16.3205	0.29	15.53	0.28
Patient 2	182.9327	276.1833	93.2506	0.87	50.98	0.48
Patient 3	25.0532	33.0828	8.0297	0.07	32.05	0.27
Patient 4	44.3236	56.8771	12.5535	0.04	28.32	0.09

Table 4. Equivalent diameters of nodules in the two dates and its growth

EQUIVALENT DIAMETERS

Patient	Diam1 (mm)	Diam2 (mm)	Growth of diams (mm)	Growth of diams p. day (mm/day)	Growth of diams (%)	Growth of diams p. day (%/day)
Patient 1	11.5666	12.4325	0.86586	0.015	7.49	0.134
Patient 2	15.2616	18.7523	3.4906	0.033	22.87	0.214
Patient 3	5.6479	6.4902	0.84228	0.007	14.91	0.124
Patient 4	7.5123	8.5099	0.99759	0.003	13.28	0.042

The growth was positive for all the patients that have experienced different values, in terms of both growth and velocity. The average of the areas' growth in absolute terms was 32.54 mm2, and in relative terms 31.72%. The average of the equivalent diameters' growth in absolute terms was 1.55 mm, and in relative terms 14.64%. The average of the of the areas' growth speed in absolute terms was 0.32 mm2/day, and in relative terms 0.28 %/day. The average of the equivalent diameters' growth speed in absolute terms was 0.0145 mm/day and in relative terms 0.1285 %/day. As regards the directions of maximum growth, it can be shown different values.

Although the number of patients is still small, the algorithms used were proved in a formal way and it is expected to take on a larger number to be able to produce better statistical treatments and to obtain relevant inferences about the mathematical behavior, in order to be modeled.

The medical radiologists that have tested the tool have found it very useful and consider it improves the monitoring process, optimizing the time employed in lung patient controlling issues.

5 Conclusion and Future Work

Most medical radiologists currently capture lung nodules' features in 2D, so this research has been performed using 2D images, using algorithms referring to two-dimensional space and obtaining measurements in terms of areas, equivalent diameters and direction of growth. From now on, the number of patients is going to be increased. Each new patient joining the research will allow regular monitoring, normally every 3 months, throughout their process control and medical treatment. Each of the cases will be incorporated into a database and processed by the implementation of signal processing system presented in this paper for characterization based on a mathematical model of growth. The current research team, consisting of radiologists, oncologists and engineers, has programmed the addition of more personnel in order to address the increased volume of patients expected.

In the future, this work is expected to be extended to three-dimensional space from 3D images (generated from large series of two-dimensional images), obtaining the measurements in terms of volume, surface, volumetric equivalent diameter and spatial direction of growth.

Acknowledgments

The authors wish to acknowledge the help of Cristina Murillo from GAIA, the Vicente San Sebastian Clinic medical team and Bilbomatica company. The support provided by the Basque Country Government Departments of Industry, Commerce and Tourism and Education also deserve mention.

References

1. Wang, G., Yu, H., Man, B.D.: An outlook on x-ray CT research and development. American Association of Physicists in Medicine (2008), doi:10.1118/1.2836950
2. Boone, J.B.: Radiological interpretation 2020: Toward quantitative image assessment. American Association of Physicists in Medicine (2007), doi:10.1118/1.2789501
3. Field, E.: Comparison of the function and performance of CT AEC systems. Cardiff and Vale University Health Board (2010)
4. Herman, G.T.: Fundamentals of computerized tomography: Image reconstruction from projection, 2nd edn. Springer, Heidelberg (2009)
5. Unknown, DICOM specification. National Electrical Manufacturers Association (NEMA), `http://medical.nema.org` (retrieved on Janaury 5, 2011)
6. Haralick, R.M.: A measure for circularity of digital figures. IEEE Transactions on Systems, Man, and Cybernetic, SMC 4(4), 394–396 (1974)
7. Otsu, N.: A Threshold Selection Method from Gray Level Histograms. IEEE T. Syst. Man Cyb. 9, 62–66 (1979)
8. Soille, P.: Morphological Image Analysis: Principles and Applications, pp. 173–174. Springer, Heidelberg (1999)
9. Gonzalez, R.C., Woods, R.E., Eddins, S.L.: Digital Image Processing Using Matlab, 2nd edn. Gatesmark Publishing (2009)
10. Hima Bindu, C.: An Improved Medical Image Segmentation Algorithm using Otsu's Method. International Journal of Recent Trends in Engineering 2(3) (2009)

A Geometrical Method of Diffuse and Specular Image Components Separation

Ramón Moreno, Manuel Graña, and Alicia d'Anjou

Computational Intelligence Group, University of the Basque Country
http://www.ehu.es/ccwintco

Abstract. Diffuse and specular image component separation is a powerful image preprocessing for image segmentation. The approach presented here is based on observed properties of the distribution of pixel colors in the RGB cube according to the Dichromatic Reflectance Model (DRM). We estimate the lines in the RGB cube corresponding to the diffuse and specular chromaticities. Then the specular component is easily removed by projection on the diffuse chromaticity line. The specular component is computed by a straightforward difference. The proposed algorithm does not need any additional information besides the image under study.

Keywords: Dichromatic reflection model, specular, diffuse, reflectance, image correction.

1 Introduction

There have been several approaches to image reflectance analysis for the separation of the diffuse and specular image components, ranging from physical methods [1,2] to computational approaches based on the Dichromatic Reflectance Model (DRM) [3,4,5], like the one presented in this letter. The diffuse component estimation is useful for color based artificial vision processes, while the specular component contains surface topological information and is required for the estimation of reflectance maps. Recent solutions [6,7] require uniform illumination and the identification of constant color regions, working on synthetic "clean" images. Color constancy analysis is a requirement for such algorithms. Our approach does not need such analysis. The process is as follows: first we estimate the chromatic lines, then we perform a dychromatization process, we estimate the diffuse image component, then we compute the specular image component. We show some computational results on well known benchmark images.

2 Dichromatic Reflectance Model (DRM)

The Dichromatic Reflectance Model (DRM) [3] explains the formation of the image of the observed surface as the addition of a diffuse component D and an specular component S, as shown in figure 1. Algebraically, the DRM is $I(x) = m_d(x)D+m_s(x)S$ where m_d and m_s the diffuse and specular component weights,

J. Cabestany, I. Rojas, and G. Joya (Eds.): IWANN 2011, Part II, LNCS 6692, pp. 83–89, 2011.

respectively, with values in the range $[0,1]$. That is, a surface with homogeneous chromatic characteristics can be expressed as the addition of the own surface color and the illumination color. In figure 1 the shaded region represents the convex region of the plane Π_c inside the RGB cube containing all the image colors resulting from the DRM equation. When there are several colors in the imaged scene, the DRM becomes $I(x) = m_d(x)D(x) + m_s(x)S$. Notice that D depends on the spatial coordinates x. The most general DRM is expressed as $I(x) = m_d(x)D(x) + m_s(x)S(x)$ where both chromaticities, the surface's and the illumination's, vary spatially. We restrict ourselves to the first model in this letter.

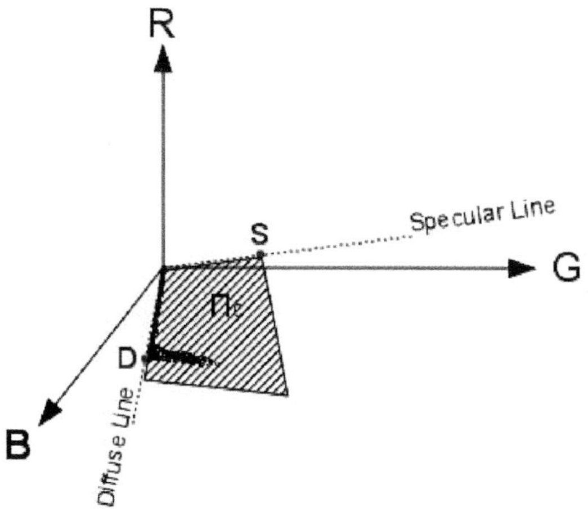

Fig. 1. Dichromatic reflection model

We have developed this work using the RGB color space, let us a brief justification of this decision. First, the DRM is defined as a vectorial sum in an euclidean space. This linearity exist in RGB however do not exist in other spaces like in the HSx family. In Fig. 2(a) we can see the distribution of the pixels in the HSV color space. On the one hand pixels with low 'Value' are very separated each others, when 'Value' is increasing they are shaping in a line (the chromatic line). On the other hand specular pixels are in a curved shape like a horn. Comparing with the Fig.3 the linearity is lost. Hence is difficult to express DRM in HSx parameters.

When working with color images is very interesting to separate color in its components; intensity, chromaticity, hue and saturation specially when we are looking for photometric invariants. We can get it expressing the RGB color space by spherical coordinates, where a pixel $p_{euclidean} = \{r, g, b\}$ can be expressed equivalently in spherical coordinates by $p_{spheric} = \{\theta, \phi, l\}$ where θ, ϕ are the angular parameters and l is the vector magnitude. Fig.2(b) shows the distribution of the image pixels in a $\theta - \phi - l$ space. As we can see the pixels distribution in this space is very close to the HSV space. But by difference with HSV, the spherical interpretation of the RGB color space let us express the DRM in spherical coordinates as $\mathbf{I}(\mathbf{x}) = (\theta_{\mathbf{D}}(\mathbf{x}), \phi_{\mathbf{D}}(\mathbf{x}), l_{\mathbf{D}}(\mathbf{x})) + (\theta_{\mathbf{S}}, \phi_{\mathbf{S}}, l_{\mathbf{S}}(\mathbf{x}))$ where the first one is the diffuse component and the second one the specular component. Then we can formulate this experiment working with the spherical interpretation, but always working in the RGB color space. In fact we trust in this approach for further works.

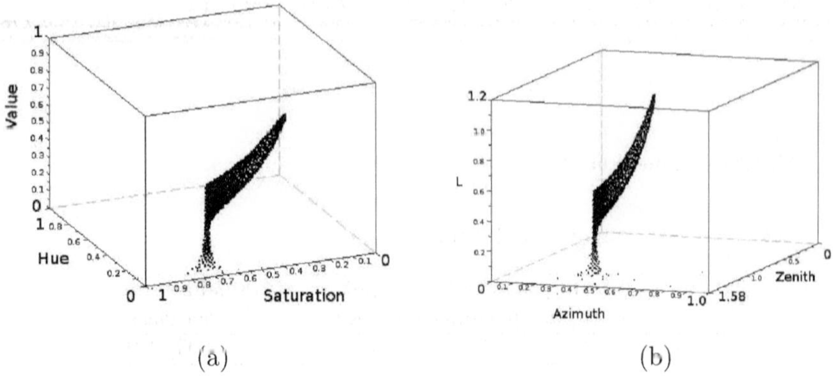

Fig. 2. Distribution of the ball image in the HSV color space (a) and in the spherical interpretation of the RGB color space (b)

3 General Description of the Method

We assume that the observed surface is decomposable into patches of homogeneous chromatic characteristics. The proposed method has the following phases:

1. Chromatic line estimation: estimate the diffuse line L_d and the specular line L_s.
2. Dichromatization: We compute the parameters of the chromatic plane Π_{dc} in the RGB cube, and we project all the pixel colors into this plane. This step involves some additive noise removal.
3. Component separation: We compute the pure diffuse image component and the specular image component.

3.1 Chromatic Line Estimation

In figure 3 we have a plot of the pixels in the image of figure 4 in the three-dimensional RGB cube. Let us denote them $\{I_i; i = 1, ..., M\}$. We can easily

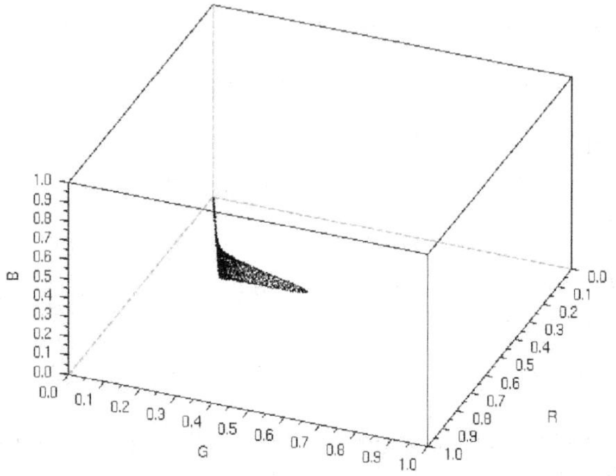

Fig. 3. Synthetic image plotted in the three-dimensional RGB space

appreciate the two main directions in the data. The most clear is the one corresponding to the diffuse line L_d which rises from the coordinate system origin. The second, less defined, appearing at the end of the diffuse elongation, is the specular direction identified by the specular line L_s.

To estimate the diffuse line, we start selecting the less bright pixels in the image region corresponding to the surface, which will have the greatest diffuse component. We plot them in the RGB cube and we estimate the best linear regression on the RGB data. In fact, we perform a Principal Component Analysis [8] (PCA) which give us the direction of the chromatic line $\overrightarrow{u}$. Therefore the diffuse chromatic line is defined as $L_d : (r, g, b) = P + s\overrightarrow{u}; \forall s \in \mathbb{R}$. Analogously, we select the brightest pixels, obtaining a mean point Q in the RGB cube and the largest eigenvector $\overrightarrow{v}$ for the specular color, therefore the specular chromaticity line is expressed as follows $L_s : (r, g, b) = Q + t\overrightarrow{v}; \forall t \in \mathbb{R}$.

3.2 Image Dichromatic Regularization

Once we know the chromatic lines, we build the dichromatic plane Π_{dc} in $\mathbb{R}^3$ which is the best planar approximation to the color distribution in RGB. It can be expressed as follows: $\Pi_{dc} : (r, g, b) = P + s\overrightarrow{u} + t\overrightarrow{v}; \forall s, t \in \mathbb{R}$, and the normal vector is $\overrightarrow{N} : \overrightarrow{u} \times \overrightarrow{v}$, where $\times$ denotes the conventional vector product. To remove noise and regularize the image colors we project the pixel's colors into this dichromatic plane Π_{dc} . For each image point color in the RGB cube I_i we compute the line $L_i : (r, g, b) = I_i + k\boldsymbol{N}; \forall k \in \mathbb{R}$, which is orthogonal to the dichromatic plane Π_{dc}, and to regularize I_i we compute its projection I_i^c as the intersection of L_i with Π_{dc}.

3.3 Component Separation

Recalling the DRM definition $I(x) = m_d(x)D + m_s(x)S$ our goal is to bring the pixels to the chromatic line, that is $\forall x : m_s(x) = 0$. We proceed as follows: for each regularized image point I_i^c lying in the plane Π_{dc} we draw the line $L_i : (r, g, b) = I_i^c + t\overrightarrow{v}; \forall t \in \mathbb{R}$ where $\overrightarrow{v}$ is the specular line vector director. The pixel diffuse component corresponds to the intersection point I_i^d of this line with the diffuse line $L_d : (r, g, b) = P + s\overrightarrow{u}; \forall s \in \mathbb{R}$ and it exists because they lie in the same plane Π_{dc} and they are not parallel lines. We have obtained $I^d(x) = m_d(x)D$ so that $\forall x, : m(x) = 0$, and the resulting image $I^d(x)$ is purely diffuse, without specular components. Obtaining the specular image component is then trivial if we recall the DRM definition: $I^s(x) = I(x) - I^d(x) = I(x) - m_d(x)D = m_s(x)S$.

4 Experimental Results

The experimental demonstration of our approach is shown in figures 4 and5. The first is a synthetic image (using Blender), and the second is a natural image. Both are monochromatic. Original image is the leftmost image in both figures. Following our approach we obtain the diffuse and specular images, shown at the center and rightmost images, respectively, in both figures. Both original images can be downloaded from http://www.ehu.es/ccwintco/index.php/Images. The natural image has been used as benchmark by several researchers [4,9]. The visual results are comparable or better than the state of the art results in [4,9]. As we know the original surface color ($r = 0.790$, $g = 0.347$ and $b = 0.221$) in the synthetic image, we can compute an estimation of the error committed by our estimation of the diffuse image. If we denote Q the original color, the error is the distance of this point to the diffuse line, computed as $d(Q, L_d) = ||\boldsymbol{PQ} - \perp(\boldsymbol{PQ}, \boldsymbol{u})||$, where $\perp(\boldsymbol{a}, \boldsymbol{b})$ denotes the projection operator. In the images shown in figure 4 the error committed is 0.0116. Variations in the error are due to the diffuse region pixel selection.

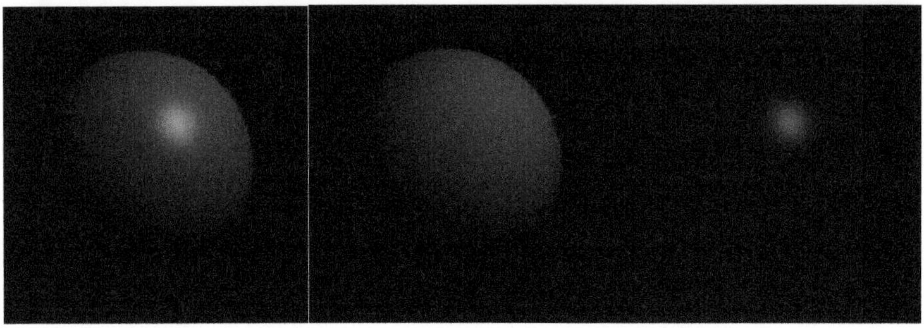

Fig. 4. Synthetic image, diffuse image and specular image

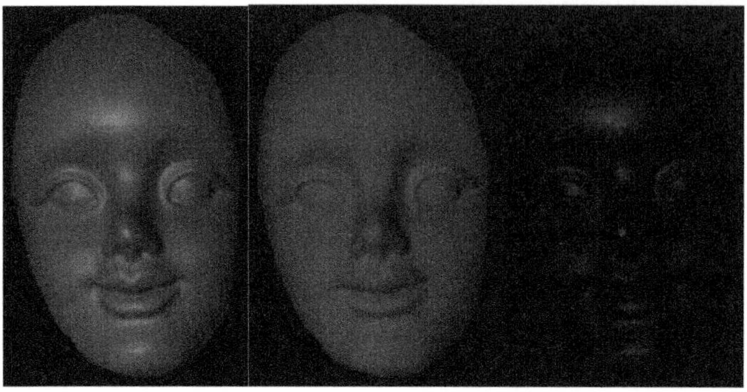

Fig. 5. Natural image, diffuse image and specular image

5 Conclusions

As conclusion of this work, we have described an image component separation for monocolor images which is very effective, fast and robust. It has been developed from the DRM and is well theoretically grounded despite its simplicity. It consists in the estimation of the diffuse and specular lines as the principal components of diffuse and specular point clouds, respectively, selected from the image by hand. Contrary to other approaches [1,2] our approach does not need specific hardware devices, and only needs one image. Our approach's complexity time is linear in the image size $O(M)$, while others [5,4] are quadratic $O(M^2)$. Our approach does not need a Specular Free image, it provides almost simultaneously both image components. On going work is addressing the extension of this approach to images containing several surface colors, i.e. $I(x) = m_d(x)D + m_s(x)S$, and to images with illumination sources of different colors, i.e. $I(x) = m_d(x)D + m_s(x)S(x)$.

References

1. Feris, R., Raskar, R., Tan, K.-H., Turk, M.: Specular reflection reduction with multi-flash imaging. In: Proceedings. 17th Brazilian Symposium on Computer Graphics and Image Processing, October 17-20, pp. 316–321 (2004)
2. Umeyama, S., Godin, G.: Separation of diffuse and specular components of surface reflection by use of polarization and statistical analysis of images. IEEE Trans. Pattern Anal. Mach. Intell. 26(5), 639–647 (2004)
3. Shafer, S.A.: Using color to separate reflection components. Color Research and Aplications 10, 43–51 (1984)
4. Tan, R.T., Nishino, K., Ikeuchi, K.: Separating reflection components based on chromaticity and noise analysis. IEEE Trans. Pattern Anal. Mach. Intell. 26, 1373–1379 (2004)
5. Yoon, K.-J., Choi, Y., Kweon, I.S.: Fast separation of reflection components using a specularity-invariant image representation. In: IEEE International Conference on Image Processing, October 8-11, pp. 973–976 (2006)

6. Li, S., Manabe, Y., Chihara, K.: Accurately estimating reflectance parameters for color and gloss reproduction. Computer Vision and Image Understanding 113, 308–316 (2009)
7. Lellmann, J., Balzer, J., Rieder, A., Beyerer, J.: Shape from specular reflection and optical flow. International Journal of Computer Vision 80, 226–241 (2008)
8. Oja, E.: Principal components, minor components, and linear neural networks. Neural Networks 5(6), 927–935 (1992)
9. Tan, R., Ikeuchi, K.: Reflection components decomposition of textured surfaces using linear basis functions. In: IEEE Computer Society Conference on Computer Vision and Pattern Recognition, CVPR 2005, June 20-25, vol. 1, pp. 125–131 (2005)

Optical Flow Reliability Model Approximated with RBF

Agis Rodrigo[1], Díaz Javier[2], Ortigosa Pilar[1], Guzmán Pablo[2], and Ros Eduardo[2]

[1] Computer Architecture and Electronics group, U. Almería,
Ctra. Sacramento s/n, La Cañada de San Urbano 04120 Almería, Spain
{rodrigo,ortigosa}@ual.es
[2] Department of Computer Architecture and Technology, CITIC, U. Granada,
C/Periodista Daniel Saucedo Aranda s/n E-18071 Granada, Spain
{jdiaz,pguzman,eros}@atc.ugr.es

Abstract. This Paper presents a new approach based on RBF NN (Radial Based Function Neural Network) in order to produce high quality optical-flow confidence estimation. The new approach is compared with a widely used confidence estimator obtaining a significant improvement. In order to evaluate the presented approach performance we have used a multi-scale version of the well known Lukas and Kanade optical flow model and widely used benchmarking optical flow sequences. The new approach aims refining optical flow representation maps but is easily applicable to other vision primitives (stereo vision, object segmentation, object recognition, object tracking, etc). Therefore, this approach represents an automatic reliability estimation model based on artificial neural networks of interest for multiple vision primitives.

Keywords: Optical flow, Confidence Estimator, Active Vision, Machine Vision, Neural Network Applications.

1 Introduction

Currently, advanced low level image processing engines allow the extraction of multi-scale low level primitives such as *local image descriptors* [1, 2], *optical flow* [3], *binocular disparity* [4, 5], etc. Optical flow can be calculated with very diverse models [5, 6] and some of them have already been implemented in real-time in specific purpose hardware [1, 2, 4]. Usually dense optical flow representation maps are obtained using different models and afterwards filtered using reliability estimators that allow discarding unreliable optical flow estimations. Traditionally, simple functions are used to extract this confidence measure, for instance, the least-squares fitting method proposed by Lucas and Kanade [7], typically uses *Eigen-values* based estimators for detection of non reliable regions [2, 8]. They rely in the spatio-temporal structure of the scene, closely related with the notion of intrinsic dimensionality of the image [9]. Nevertheless, we are going to explore if other low level primitives can be used by themselves as reliability estimators of optical flow error. This is of specific interest in the framework of multimodal vision approaches, in which scene local descriptors may be used to cross-validate different vision modalities, instead of using

J. Cabestany, I. Rojas, and G. Joya (Eds.): IWANN 2011, Part II, LNCS 6692, pp. 90–97, 2011.

Table 1. Local structure descriptors

Vector component index	Parameter
1-3	[id_0 $flat_{fs}$], [id_0 $flatp_{ms}$], [id_0 $flatp_{cs}$]
4-6	[id_1 $line_{fs}$], [id_1 $line_{ms}$], [id_1 $line_{cs}$]
7-9	[id_2 $corner_{fs}$], [id_2 $corner_{ms}$], [id_2 $corner_{cs}$]
10-12	[$energy_{fs}$], [$energy_{ms}$], [$energy_{cs}$]
13-15	[$Orientation_{fs}$], [$Orientation_{ms}$], [$Orientation_{cs}$]
16-18	[$Phase_{fs}$], [$Phase_{ms}$], [$Phase_{cs}$]
19-21	[$Imean_{fs}$], [$Imean_{ms}$], [$Imean_{cs}$]
22-24	[$Idif_{fs}$], [$Idif_{ms}$], [$Idif_{cs}$]
25	[Optical-flow velocity module]
26	[Optical-flow velocity angle]
27	[Image reconstruction error]
28	[Optical-flow velocity component X]
29	[Optical-flow velocity component Y]
30	[Pixelwise Fleet´s angular error]
31	[Pixel_index]
32	[Sequence_index]

simple local image structure estimators (such as *Eigen-values* based approaches) for filtering the optical flow representation maps.

Each pixel descriptor vector is composed by 27 components (see Table 1). The neural network aims to learn optical flow error estimator (Pixelwise angular error [14]). *id* stands for intrinsic dimensionality of local region. id_0= (flat regions), id_1= (lines), id_2= (corners). *fs*, *ms*, and *cs* stand for fine, medium and coarse scales respectively.

This study aims to obtain an efficient reliability estimator (in terms of accuracy enhancement) for optical flow maps based on low level local image descriptors. We evaluate quantitatively the quality of the confidence measures obtained from the proposed approach and we compare it with ideal confidence estimation (actual error). Furthermore, we also evaluate the impact on the accuracy improvement using Lucas and Kanade (L&K) optical flow model [7], modified including a multiscale extension [12]. Nevertheless note that the presented approach is general and could be used for other vision cues. We illustrate how this methodology efficiently allows significant accuracy enhancement on the obtained optical flow representation maps.

In order to obtain automatically an alternative optical flow confidence estimator we use a *Radial Basis Function Neural Network* (RBF) [13] to approximate the optical flow error using as training and test data sets, a set of widely used benchmarking sequences [11] (in which we known the ground truth optical-flow). We compute the optical flow with the L&K model mentioned above [12]. The error (comparing the obtained optical flow with the ground truth) is estimated based on the angular error measure [14]. Other metrics can be used [12] but similar behaviors are expected.

Optical flow errors (θ_{lk}) and vision estimator components are calculated from the image database previously for RBF training and evaluation process (see Fig. 1). θ_{gt} (ground truth optical flow), θ_{lk} (Lukas-Kanade optical-flow estimation), θ_{re} (actual optical flow error). A_E (Angular Error) is the optical-flow error, according to Fleets's

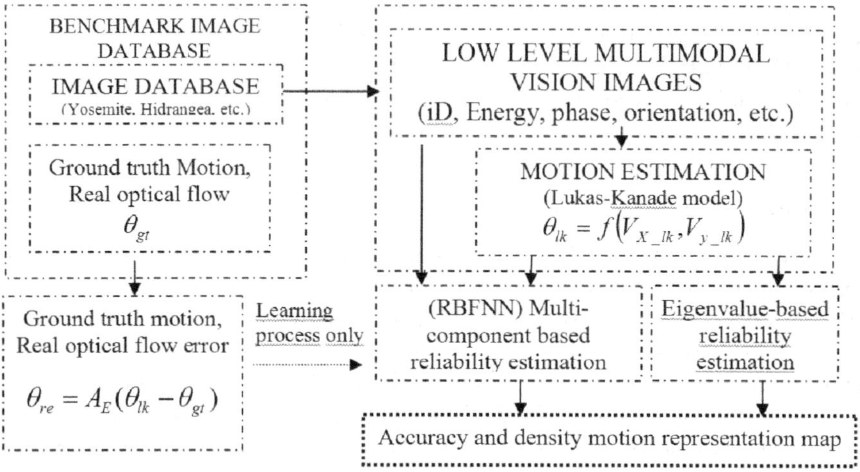

Fig. 1. Pre-processing information scheme

angular error methodology [14]. The *eigenvalue* reliability estimation methods rely on the value of the minimum *eigenvalue*, according to [5].

The RBF is trained to approximate optical flow error, we extract different local image descriptors (calculated at different spatial scales using pyramidal analysis techniques). We have "intuitively" selected a set of 27 components that allow representing relevant information of spatio-temporal properties of the scene (see Table 1). The chosen features are the intrinsic dimensionality [9], local image energy, orientation and phase as described in [1], the average intensity of the image (computed in a neighborhood of 9x9 pixels), the pixel-wise image differences (computed between the current frame and the immediately previous one), and the optical flow values themselves (velocity component, module and angle). All these values are computed at three different image pyramid levels (with decreasing resolution of one octave between pyramids values). Descriptors list are included in table 1, including also the error value and pixel/sequence index information. All these data values are pre-computed and used as input to the RBF network during the training and test processes.

2 Material and Methods

The proposed model uses a synthetic benchmarking optical flow database (in which Ground Truth optical flow is available) (see Fig. 1). This database includes a set of 11 representative sequences, specifically: Yosemite (with and without clouds), Diverging and Translating tree, Marble blocks, Grove2, Grove3, Hydrangea, Rubber-Whale, Urban2, Urban3 (see [5, 11]). They are well known database benchmark sequences widely used in the literature that cover many different image scenarios (small and large movements, textured and textureless image regions, continuous and discontinuous motions, translations, rotations, etc). The dataset represents a vector dataset of about 2 million vectors but only 1.2 million are vectors with well defined

27 features (indicated in Table 1) . The discarded vectors correspond to areas with almost zero image structure even though we use very relaxed threshold for the different local image descriptors. In this situation some features are defined as NaN (Not a Number) values because there is no meaning for some descriptors (for instance we could not assign a local orientation to a blank patch). These pixels are discarded from the training and test database. This is not a problem for the generality of the used approach, low confidence would be directly assigned to these pixels.

2.1 Optimizing the Dimensionality of the Input Database for the Function Approximation Task

The image database is composed by approximately 1.2 million vectors; i.e. it is huge and represents a computation problem when we try to apply a classical RBF approach, mainly because it needs to calculate the pseudo-inverse of a matrix with size equal to number of training vectors (see Eq. 1). The pseudo-inverse matrix is computationally expensive; therefore we simplify the process by reducing the dimensions of the input matrix. Many vector dimensions can be simplified because may not be relevant for the calculation of optical flow error estimator or contain redundant information. In order to identify the most relevant dimensions in order to reduce the complexity and computing time, an analysis by optimal local searching algorithms has been implemented. Particularly we have reduced the dimensionality by SGA (*Searching Greedy Algorithm)* [15]. Using SGA with different input vectors subsets, we obtain a set of local optimal solutions. We launch the SGA 100 experiments with different input vector subsets (500 vectors each time) and we arrive to different solutions depending on the input vectors. This is important to find the most relevant dimensions (for the whole dataset). Our SGA targets MAE *(Mean Absolute Error)* (of the test dataset) as goal function. We selected the set of dimensions (of the input vectors) with the minimum MAE.

In order to identify the most representative vector components (vector dimensions) we have calculated the histogram of used components in the different solutions found by SGA in each experiment. In this case the most frequently used dimensions were [id_0 $flat_{fs}$] and [Velocity-module]. (Note: we have selected the most relevant dimensions upon a threshold of 40% of frequency. We can select different thresholds below 40% of frequency in order to select a higher number of relevant dimensions).

This is very representative and easy to explain based on computer vision grounds. The id_0 indicates the lack of image structure at this scale. It is well known that on untextured areas, local methods (for instance the L&K gradient model) become ill-posed and the estimated flow lacks of any confidence. The most relevant id_0 values appear at the finest scale. This indicates that, although the multiscale approach uses motion information from coarser scales, the final motion vectors include the contribution of the fine scale which is likely to have a larger error. Thus id_0 at the finest scale works better for evaluating the reliability of the optical flow estimation. Regarding the other "relevant feature" (according to SGA experiments), i.e. the velocity module, it is easy to understand that large velocities modules that represent fast movements have a higher error probability. The coarse to fine process (required for estimating large movements) is prone to larger errors when motion values are higher (error propagates across the scales due to the image warping process). In addition, the hypothesis of luminance constancy and translation motion inherent to most of the optical flow models become much more difficult to satisfy. Large motions introduce modifications of the point of view and changes in illumination, which could easily fail the model hypothesis and, as consequence, are likely to produce larger errors in the estimations. The previous study concludes that the RBF network will be able to learn, from these two parameters as key base elements, to predict the error values of the optical flow.

2.2 RBF Neural Network: Inner Architecture

The RBF neural network used is based on the classic model [16] and takes a matrix representation following Eq. (1) to Eq. (3) for storing the different values of the neural weights. This allows simple mathematical processing using matrix operations, but requires large and expensive storage resources of memory. As we can see, it limits the number of training cases on the network.

$$\begin{bmatrix} Ee_1 \\ Ee_2 \\ \cdots \\ Ee_{(n)} \end{bmatrix} = \begin{bmatrix} G(d_{1,k}) & \cdots & G(d_{1,k}) \\ G(d_{2,k}) & \cdots & G(d_{2,k}) \\ \cdots & \cdots & \cdots \\ G(d_{n,k}) & \cdots & G(d_{n,k}) \end{bmatrix} \cdot \begin{bmatrix} W_1 \\ W_2 \\ \cdots \\ W_{(n\cdot k)} \end{bmatrix} ; \begin{bmatrix} W_1 \\ W_2 \\ \cdots \\ W_{(n\cdot k)} \end{bmatrix} = Pseu\left(\begin{bmatrix} G(d_{1,1}) & \cdots & G(d_{1,k}) \\ G(d_{2,1}) & \cdots & G(d_{2,k}) \\ \cdots & \cdots & \cdots \\ G(d_{n,1}) & \cdots & G(d_{n,k}) \end{bmatrix} \right) \cdot \begin{bmatrix} Ee_1 \\ Ee_2 \\ \cdots \\ Ee_{(n)} \end{bmatrix} \tag{1}$$

$$d_{i,j} = \left\| PA_{i,j} - C_{i,j} \right\| = \sqrt{\sum_{i=1}^{n} \sum_{j=1}^{k} \left(PA_{i,j} - C_{i,j} \right)^2} \tag{2}$$

$$\text{(a)} \quad G(d_{i,j}) = e^{-d_{i,j}} \quad ; \quad \text{(b)} \quad G(d_{i,j}) = \begin{cases} d_{i,j}^{\ r} & r = 1,3,5,\ldots, \\ d_{i,j}^{\ r} \cdot \ln(d_{i,j}) & r = 2,4,6,\ldots, \end{cases} \tag{3}$$

The extracted feature vectors of each pixel ($PA_{i,j}$) are used as input to the neural network together with the actual error output during the learning process. Radial function G(•) was a Gaussian function normalized between 0-1 following Eq. (3a). The distances $d_{i,j}$ between the input vector ($PA_{i,j}$) and the centers ($C_{i,j}$) are calculated following Eq. (2). Initially the training input vectors PAij are set as centers (Cij). Other (RBF) Radial Base Functions such as *poliharmonic splines* have been tested, for instance the one defined by Eq. (3b) obtaining comparable results.

Throughout the training process the weights (W_n) are calculated, the RBF receives an *input* matrix with n vectors of k components ($PA_{n,k}$ input component) and with k dimensions 1 to 27 of Table 1 (depending on the optimal dimensions estimated by SGA before). Each component corresponds to a vision primitive or feature (for example $PA_{28,25}$ corresponds to velocity module of pixel 28). The component 30 of Table 1 (actual error) is used as goal function (output) for the RBF and is approximated modifying the weight matrix according to (1) in a single stage. Nevertheless, as explained above, we have done 100 SGA experiments for optimizing the input vector dimensionality. Once the neural network is trained with a training set of the database (50000 vectors), we test the result with the test dataset (50000 vectors). During test process a new *input* matrix, without real optical flow error, feeds the RBF module, now the RBF provides the new optical flow error estimation (Ee_n) that we will use as confidence estimator.

3 Rank Order Coding Technique

Comparing two confidence or reliability functions is not straightforward. Therefore we will focus on their final aim, which is allowing the selection of most reliable estimations among the obtained optical flow dense representation map. For instance, after computing the optical flow at each pixel we will also extract the "RBF-based reliability estimator" that based on the local spatio-temporal supporting structure provides a confidence measure about the accuracy of the obtained optical flow value. A Rank Order Coding technique [17] was used to represent different error estimator's results. The RBF-based reliability estimator is different compared to the reliability estimator provided by the technique of the minimum *Eigen-value*. Therefore, a method for comparing these confidence values is needed.

We compare both approaches with respect to the ground-truth based error which would be an ideal confidence measure, i.e. if we knew the actual error at each pixel we could filter the most reliable ones (low errors) to arrive to a sparser reliable optical flow maps. We order the test vectors according to the actual optical flow error (using the ground-truth optical flow) obtaining a reference list. We also order the test vectors according to the RBF-based reliability estimator (second list), and finally, we also order the test vectors according to the Eigen-values based reliability estimator (third list). We measure the matching ratio between the ordered lists obtained with different rank-orders. For instance, we focus on the first 10% samples. We evaluate how many of the first 10% samples of the list obtained using the ground-truth values match the first 10% samples of the Eigen-values based list (third list) and how many match the first 10% samples of the RBF-based (second list). Once the test vectors are ordered according to different reliability estimators we can select the best optical flow vectors according to these different reliability estimators. And finally we can calculate the actual MAE of the selected estimations for each case.

4 Results

On this section we have compared our RBF-based confidence estimator with the Eigen-value confidence estimator using the previous methodology. The results were obtained from a random set of 100000 vectors of the test-dataset (equivalent to an image of approximately 300x300 pixels). In Fig. 2 (left graph) we point out (with arrows) the MAE of the 30% most confident vectors according to the RBF-based estimator and eigenvalue-based estimator. In 30% of density the MAE obtained using the RBF-based confidence estimator is around 0.5 whereas that the MAE obtained with *Eigen-value* based reliability estimator is 0.78. In the right plot (Fig. 2) we point out (with an indicative arrow) the MAE improvement. In 30% of density we obtain a 33% of improvement with respect to the *Eigen-value* based confidence estimator.

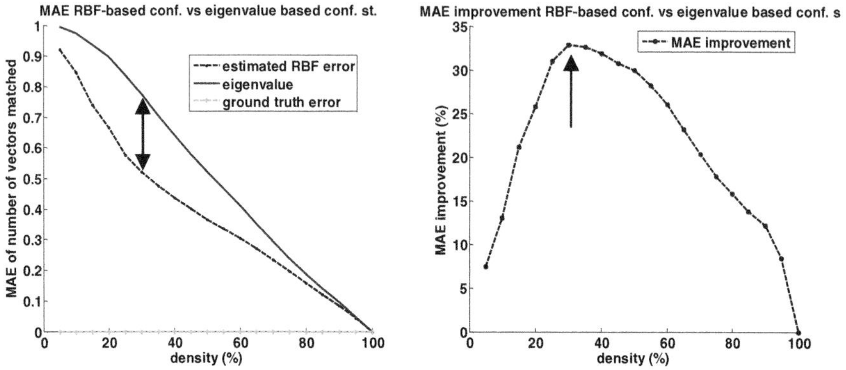

Fig. 2. Normalized improvement of the RBF optical flow confidence estimator

5 Conclusions

We have presented a new approach for optical flow error estimation based on multimodal image analysis and a RBF network for automatically learning the goal function (reliability estimator) using standard benchmarking sequences. This scheme is capable of producing a confidence optical flow estimator that leads to improvements of 33% compared with the widely used *Eigen-value* based approach (at 30% of density in optical flow maps). This is a very significant improvement compared to the widely used Eigen-value based reliability estimator.

Other optical flow error estimation may be analyzed. In [18] different optical flow confidence estimators were compared with a proposed method based on the computation of motion statistics from sample data. The method was only evaluated with three test sequences (Yosemite, Rubber-Whale, Marbel). In that work, 90% of flow vectors density was needed for reaching up goods results. Our method takes has into account 11 test images and is capable of obtaining significantly good results from only 10% of the density of vectors onwards (see Fig.2).

In addition to developing a high reliability confidence estimator capable to adapt to multiple algorithms and even specific scenarios, we have used SGA for identifying the most relevant components for the evaluation of optical flow confidence (inverse of error probability). Interestingly enough we find that the most relevant components are [id_0 $flat_{fs}$] and [Velocity module]. As discussed in section 2.1, this is reasonable and easy to explain because these two parameters encode information that indicates where the spatio-temporal structure of the image sequences violates the optical flow model hypothesis. In addition, the SGA analysis demonstrates that only 2-5 vision cues are required to provide the best results of the RBF-based reliability measure. This reduces the computing requirements significantly.

Acknowledgments

The authors thank Caja Murcia Spanish Foundation (post-doc research program) for funding support for this work also supported by the grants TIN2008-01117, P08-TIC-3518 and MULTIVISION TIC-3873 from J. de Andalucía (funded by ERDFs).

References

1. Díaz, J., Ros, E., Mota, S., Carrillo, R.: Local image phase, energy and orientation extraction using FPGAs. Int. Journal of Electronics 95(7), 743–760 (2008)
2. Bonato, V., Marques, E., Constantinides, G.: A parallel hardware architecture for scale and rotation invariant feature detection. IEEE Transactions on Circuits and Systems for Video Technology 18(12), 1703–1712 (2008)
3. Anguita, M., Diaz, J., Ros, E., Fernandez-Baldomero, F.J.: Optimization strategies for high-performance computing of optical-flow in general-purpose processors. IEEE Trans. on Circuits and Systems for Video Technology 19(10), 1475–1488 (2009)
4. Diaz, J., Ros, E., Carrillo, R.: Real-time system for high-image resolution disparity estimation. IEEE Trans. on Image Processing 16(1), 280–285 (2007)
5. Barron, J.L., Fleet, D.J., Beauchemin, S.S.: Performance of optical flow techniques. International Journal of Computer Vision 12(1), 43–77 (1994)
6. Liu, H.C., Hong, T.S., Herman, M., Camus, T., Chellappa, R.: Accuracy vs Efficiency Trade-offs in Optical Flow Algorithms. Computer Vision and Image Understanding 3, 271–286 (1998)
7. Lucas, B.D., Kanade, T.: An iterative image registration technique with an application to stereo vision. In: Proc. IJCAI., pp. 674–679 (1981)
8. Bainbridge-Smith, A., Lane, R.G.: Measuring Confidence in Optical Flow Estimation. IEE Electronic Letters 10, 882–884 (1996)
9. Felsberg, M., Kalkan, S., Krueger, N.: Continuous dimensionality characterization of image structures. Image and Vision Computing 27, 628–630 (2009)
10. Baker, S., Scharstein, D., Lewis, J.P., Roth, S., Black, M.J., Szeliski, R.: A Database and Evaluation Methodology for Optical Flow. Computer Vision, 1–8 (2011)
11. Middlebury Optical flow evaluation dataset, `http://vision.middlebury.edu/flow/eval/`
12. Bouguet, J.-Y.: Pyramidal Implementation of the Lucas Kanade Feature Tracker Description of the Aalgorithm. Intel. Corp., Microprocessor Research Labs (1999)
13. Park, J., Sandberg, I.W.: Universal Approximation Using Radial-Basis-Function Networks. Neural computation 3(2), 246–257 (1991)
14. Fleet, D.J.: Measurement of Image Velocity. In: Engineering and Computer Science. Kluwer Academic Publishers, Norwell (1992)
15. Schaback, R., Wendland, H.: Adaptive greedy techniques for approximate solution of large RBF systems. Numerical Algorithms 24, 239–254 (2000)
16. Broomhead, D.S., Lowe, D.: Multi-variable functional interpolation and adaptive networks. In: Complex Systems, pp. 269-303 (1988)
17. Thorpe, S., Gaustrais, J.: Rank Order Coding. In: Bower, J. (ed.) Computational Neuroscience: Trends in Research. Plenum Press, New York (1998)
18. Kondermann, C., Mester, R., Garbe, C.: A Statistical Confidence Measure for Optical Flows. In: ECCV 2008, Part I, pp. 290–301 (2008)

Video and Image Processing with Self-Organizing Neural Networks

José García-Rodríguez[1], Enrique Domínguez[2], Anastassia Angelopoulou[3], Alexandra Psarrou[3], Francisco José Mora-Gimeno[1], Sergio Orts[1], and Juan Manuel García-Chamizo[1]

[1] Dept. of Computing Technology, University of Alicante, Spain
{jgarcia,fjmora,sorts,juanma}@dtic.ua.es
[2] Dept. of Computer Science, University of Malaga, Spain
enriqued@lcc.uma.es
[3] Dept. of Computer Science & Software Engineering (CSSE), University of Westminster, Cavendish W1W 6UW, United Kingdom
{agelopa,psarroa}@wmin.ac.uk

Abstract. This paper aims to address the ability of self-organizing neural network models to manage video and image processing in real-time. The Growing Neural Gas networks (GNG) with its attributes of growth, flexibility, rapid adaptation, and excellent quality representation of the input space makes it a suitable model for real time applications. A number of applications are presented that includes: image compression, hand and medical image contours representation, surveillance systems, hand gesture recognition systems or 3D data reconstruction.

Keywords: Growing Neural Gas, topology preservation, real time, objects representation, parallelism.

1 Introduction

The growing computational power of current computer systems and the reduced costs of image acquisition devices allow a very large audience to have an easier access to image analysis issues.

Self-organizing models (SOM), with their massive parallelism, have shown considerable promise in a wide variety of application areas, not related to vision problems only, and have been particularly useful in solving problems for with traditional techniques have failed or proved inefficient. Accordingly, even hard video and image processing applications like some robotic operation, visual inspection, remote sensing, autonomous vehicle driving, automated surveillance, and many others, have been approached using neural networks.

The investigation in SOM in this context has received a great attention, on one hand, as imaging tasks are computationally intensive and high performances potentially can be reached in real time, and on the other hand, thanks to the versatility of

J. Cabestany, I. Rojas, and G. Joya (Eds.): IWANN 2011, Part II, LNCS 6692, pp. 98–104, 2011.

neural approaches. Versatility means that a set of interesting properties are shared by neural systems; apart from the inherent parallelism, they allow a distributed representation of the information, easy learning by examples, high generalization ability, and a certain fault tolerance. There is a vast literature on NN, it deals with both theoretical investigation of their inherent mechanisms, and solutions to real problems. SOM have been and have been used extensively for pattern recognition problems, namely classification, clustering, and feature selection.

Several works have used self-organizing models for the representation and tracking of objects. Fritzke [1] proposed a variation of the GNG [2] to map non–stationary distributions that [3] applies to the representation and tracking of people. In [4] it is suggested the use of self-organized networks for human-machine interaction. In [5], amendments to self-organizing models for the characterization of the movement are proposed.

From the works cited, only [3] represents both the local movement and the global movement, however there is no consideration of time constraints, and do not exploits the knowledge of previous frames for segmentation and prediction in subsequent frames. Neither uses the structure of the neural network to solve the problem of correspondence in the analysis of the movement.

The time constraints of the problem suggest the need for high availability systems capable to obtain representations with an acceptable quality in a limited time. Besides, the large amount of data suggests definition of parallel solutions.

In this paper we present applications of a neural architecture based on GNG that is able to adapt the topology of the network of neurons to the shape of the entities that appear in the images and can represent and characterize objects of interest in the scenes with the ability to track the objects through a sequence of images in a robust and simple way.

It has been demonstrated that using architectures based on (GNG), can be assumed the temporary restrictions on problems such as tracking objects or the recognition of gestures, with the ability to processing sequences of images offering an acceptable quality of representation and refined very quickly depending on the time available.

With regard to the processing of image sequences, it is possible to accelerate the tracking and allow the architecture to work at video frequency. In this proposal, the use of the GNG to the representation of objects in sequences solves the costly problem of matching of features over time, using the positions of neurons in the network. Furthermore, can be ensured that from the map obtained with the first image will only be required to represent a fast re-adaptation to locate and track objects in subsequent images, allowing the system to work at video frequency.

The data stored throughout the sequence in the structure of the neural network about characteristics of the entities represented as position, colour, and others, provide information on deformation, merge, paths followed by these entities and other events that may be analyzed and interpreted, giving the semantic description of the behaviors of these entities.

The remainder of the paper is organized as follows: section 2 introduces the capacities of topology learning and preservation of GNG. Section 3 presents a number of video and image processing applications of GNG followed by our major conclusions.

2 Topology Learning

One way to obtain a reduced and compact representation of 2D shapes or 3D surfaces is to use a topographic mapping where a low dimensional map is fitted to the high dimensional manifold of the shape, whilst preserving the topographic structure of the data. A common way to achieve this is by using self-organising neural networks where input patterns are projected onto a network of neural units such that similar patterns are projected onto units adjacent in the network and vice versa.

The approach presented in this paper is based on self-organising networks trained using the Growing Neural Gas learning method [2], an incremental training algorithm. The links between the units in the network are established through competitive hebbian learning. As a result the algorithm can be used in cases where the topological structure of the input pattern is not known a priori and yields topology preserving maps of feature manifold.

2.1 Growing Neural Gas

From the Neural Gas model [6] and Growing Cell Structures [7], Fritzke developed the Growing Neural Gas model, with no predefined topology of a union between neurons. A growth process takes place from minimal network size and new units are inserted successively using a particular type of vector quantisation [8]. To determine where to insert new units, local error measures are gathered during the adaptation process and each new unit is inserted near the unit which has the highest accumulated error. At each adaptation step a connection between the winner and the second-nearest unit is created as dictated by the competitive hebbian learning algorithm. This is continued until an ending condition is fulfilled, as for example evaluation of the optimal network topology based on some measure. Also the ending condition could it be the insertion of a predefined number of neurons or a temporal constrain. In addition, in GNG networks learning parameters are constant in time, in contrast to other methods whose learning is based on decaying parameters.

3 Examples

A number of applications to demonstrate the ability of GNG to manage Video and Image processing problems are presented that includes: image compression [9], hand and medical image contours representation [10], surveillance systems [11,12] and hand gesture recognition systems or 3D reconstruction [13,14].

3.1 Landmarks Extraction and Contours Representation

GNG allows to extract, in an autonomous way, the contour of any object as a set of edges that belong to a single polygon and form a topology preserving graph. The method is based on the assumption that correspondences are the nodes (the cluster centres in a high-dimensional vector space) of a network. The automatic extraction and correspondence is performed with the GNG (Figure 1).

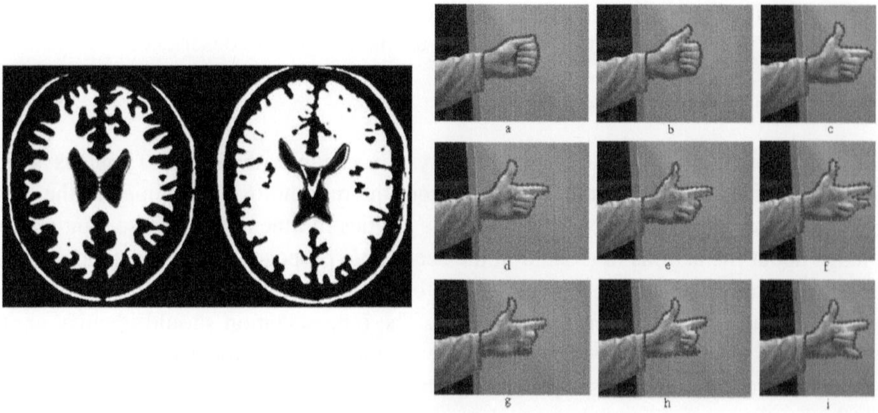

Fig. 1. 3D data representation with GNG

3.2 Image Compression and Hand Gesture Recognition

In this application we study the capacities of characterization and synthesis of objects by using a self-organizing neural model, the Growing Neural Gas. These networks, by means of their competitive learning try to preserve the topology of an input space. This feature is being used for the representation of objects and their movement with topology preserving networks. We characterize the object to be represented by means of the obtained maps and kept information solely on the coordinates and the pixel color of the neurons. With this information it is made the synthesis of the original images, applying mathematical morphology and simple filters using the available information.

The reduced structure obtained for any image in the sequence can be used as a fixed markers (figure 2) to follow along the sequence if we do not add or delete

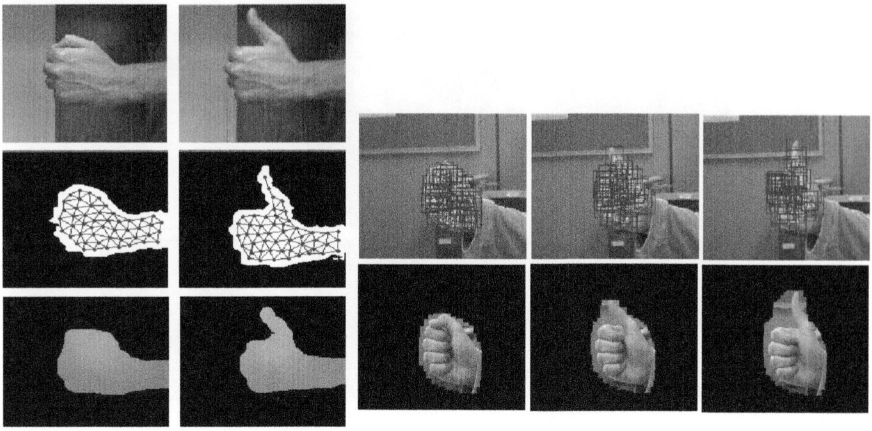

Fig. 2. Image compression and hand gesture recognition applications

neurons after first frame, only redistribute neurons. That is, we take previous map obtained from previous image as a start point for next image representation. Trajectory of neurons reference vectors can be use to define the gesture described.

3.3 Surveillance Systems

It is proposed the design of a modular system capable of capturing images from a camera, target areas of interest and represent the morphology of entities in the scene. As well as analyzing the evolution of these entities in time and obtain semantic knowledge about the actions that occur at the scene. We propose the representation of these entities through a flexible model able to characterize their morphological and positional changes along the image sequence. The representation should identify entities over time, establishing a correspondence of these during the various observations. This feature allows the description of the behavior of the entities through the interpretation of the dynamics of the model representation. The time constraints of the problem suggest the need for highly available systems capable to obtain representations with an acceptable quality in a limited time. Besides, the large amount of data suggests definition of parallel solutions.

To solve the problem, the representation of the objects and their motion is done with a modified self-growing model. We propose a neural architecture able to adapt the topology of the network of neurons to the shape of the entities that appear in the images, representing and characterizing objects of interest in the scenes. The model has also the ability to track the objects through a sequence of images in a robust and simple way.

With regard to the processing of image sequences, we have introduced several improvements to the network to accelerate the tracking and allow the architecture to

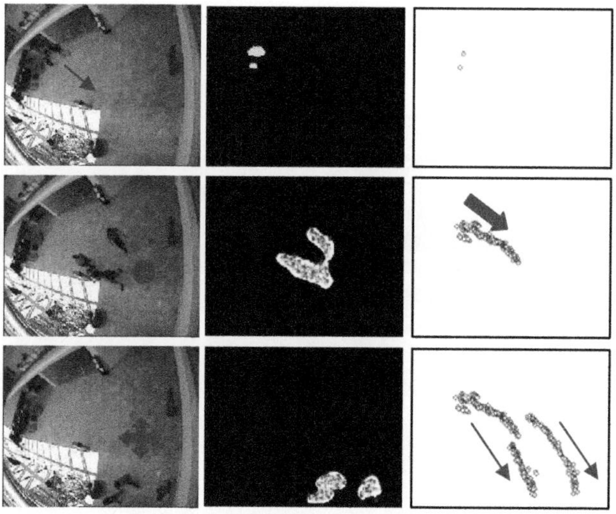

Fig. 3. GNG-based surveillance system

work at video frequency. In this paper, the use of the GNG to represent objects in image sequences solves the costly problem of matching features over time by using the positions of neurons in the network. Likewise, the use of simple prediction facilitates the monitoring of neurons and reduces the time to readapt the network between frames without damaging the quality and speed of system response.

The data stored throughout the sequence in the structure of the neural network about characteristics of the entities represented as: position, color, texture, labels or any interesting feature, provide information on deformation, merge, paths followed by these entities and other events. This information may be analyzed and interpreted, giving the semantic description of the behaviors of these entities (figure 3).

3.4 3D Data Reconstruction

The ability of GNG to represent the topology of the input space (3D in this case) permits to obtain, based on the structure of the net (neurons and edges), an induced Delaunay Triangulation that can be used to obtain several features and reduced graphs. Although, the compact and reduced representation obtain permits to deal with computationally expensive algorithms. Figure represents 3D data obtain from scanner for shoe lasts and foots in the upper images and outdoor images obtained with a 3D laser in lower one (figure 4).

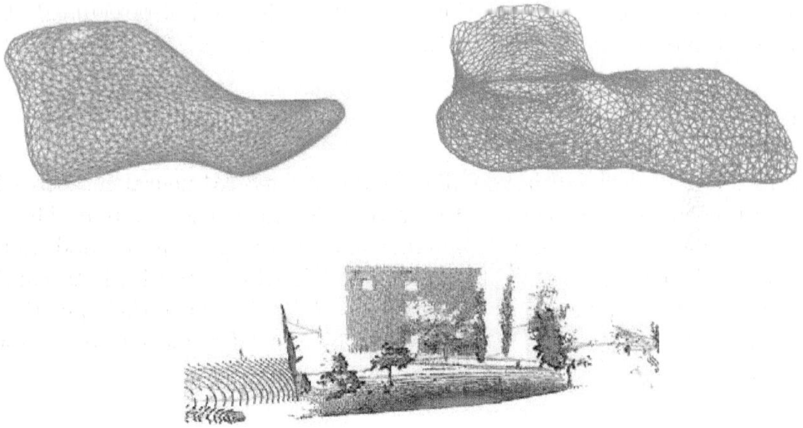

Fig. 4. 3D data representation with GNG

4 Conclusions

In this paper, we have demonstrated the capacity of Growing Neural Gas to solve some computer vision an image processing tasks. Demonstrating their capacity to segment, extract and represent 2D/3D objects in images. Establishing a suitable transformation function, the model is able to adapt its topology to images of the environment, to compress data, represent the movement and provide reduced and compact information about the structure of the entities in the images that permit real time processing video.

Acknowledgement

This work was partially supported by the University of Alicante project GRE09-16 and Valencian Government project GV/2011/034.

References

1. Fritzke, B.: A self-organizing network that can follow non-stationary distributions. In: Proc. of the International Conference on Artificial Neural Networks, pp. 613–618. Springer, Heidelberg (1997)
2. Fritzke, B.: A Growing Neural Gas Network Learns Topologies. In: Tesauro, G., Touretzky, D.S., Leen, T.K. (eds.) Advances in Neural Information Processing Systems 7. MIT Press, Cambridge (1995)
3. Frezza-Buet, H.: Following non-stationary distributions by controlling the vector quatization accuracy of a growing neural gas network. Neurocomputing 71, 1191–1202 (2008)
4. Flórez, F., García, J.M., García, J., Hernández, A.: Hand Gesture Recognition Following the Dynamics of a Topology-Preserving Network. In: Proc. of the 5th IEEE Intern. Conference on Automatic Face and Gesture Recognition, pp. 318–323. IEEE, Inc., Washington (2002)
5. Cao, X., Suganthan, P.N.: Video shot motion characterization based on hierachical overlapped growing neural gas networks. Multimedia Systems 9, 378–385 (2003)
6. Martinetz, T., Berkovich, S.G., Schulten, K.J.: Neural-Gas. Network for Vector Quantization and its Application to Time-Series Prediction. IEEE Transactions on Neural Networks 4(4), 558–569 (1993)
7. Fritzke, B.: Growing Cell Structures – A Self-organising Network for Unsupervised and Supervised Learning. In: Technical Report TR-93-026, International Computer Science Institute, Berkeley, California,
8. Martinetz, T., Schulten, K.: Topology Representing Networks. Neural Networks 7(3), 507–522 (1994)
9. García-Rodríguez, J., Flórez-Revuelta, F., García-Chamizo, J.M.: Image Compression Using Growing Neural Gas. In: Procceding of International Joint Conference on Artificial Neural Networks, pp. 366–370 (2007)
10. Angelopoulou, A., Psarrou, A., García-Rodríguez, J., Revett, K.: Automatic landmarking of 2D medical shapes using the growing neural gas network. ICCV 2005 workshop CVBIA 210-219 (2005)
11. García-Rodríguez, J., Angelopoulou, A., Garcia-Chamizo, J.M., Psarrou, A.: GNG Based Surveillance System. In: IJCNN (2010)
12. Luque, R.M., Domínguez, E., Palomo, E.J., Muñoz, J.: A Neural Network Approach for Video Object Segmentation in Traffic Surveillance. Lecture Notes in Computer Science 5112, 151–158 (2008)
13. Holdstein, Y., Fischer, A.: Three-dimensional Surface Reconstruction Using Meshing Growing Neural Gas (MGNG). Visual Computation 24, 295–302 (2008)
14. Domínguez, E., Spinola, C., Luque, R.M., Palomo, E.J., Muñoz, J.: Object recognition and inspection in difficult industrial environments. In: IEEE International Conference on Industrial Technology (ICIT), pp. 989–993 (2006)

Parallelism in Binary Hopfield Networks

José Muñoz-Pérez, Amparo Ruiz-Sepúlveda, and Rafaela Benítez-Rochel

Dept of Computer Science, Universidad de Málaga,
E.T.S. Ingenieria Informática
{munozp,amparo,benitez}@lcc.uma.es
http://www.lcc.uma.es

Abstract. Some neural networks have been proposed as a model of computation for solving combinatorial optimization problems. The ability to solve interesting classic problems has motivated the use of neural networks as models for parallel computing. In this paper the degree of parallelism of a binary Hopfield network is studied using the *chromatic number* of the graph G associated to the network. We propose a rule to coloring the vertices of the neural network associated to the Traveling Salesman Problem such that the neurons with the same color can be simultaneously updated.

Keywords: Hopfield networks, Neural networks, Combinatorial optimization

1 Introduction

Hopfield neural networks provide a powerful approaches for a wide variety of combinatorial optimization problems since Hopfield and Tank [3] applied their neural network to the travelling salesman problem (TSP). Hopfield neural networks have been used mostly in search and optimization applications [5,11,12,19]. In particular, Takefuji et al. [4,14], have obtained important solutions using this type of network and they found discrete neurons to be computationally more efficient than continuous neurons. Furthermore, discrete time dynamics are usually much easier to implement than continuous time dynamics which are described by differential equations. Some interesting modifications in the Hopfield network have been proposed [1,7,8,13,16]. In order to avoid the local minima problem, many researchers have adopted different heuristic search techniques which allow the network to escape from local minima and converge to the global optimal or near optimal state more quickly [6,9,10,18]. Moreover, different methods for selecting the penalty parameters applied to travelling salesman problem have been proposed [15].

The basic idea behind these applications is the following: An optimization problem is first mapped to a neural network in such way that the neural states correspond to possible solutions to the problem. A function of neural states, called computational energy function, is constructed on the basis of the cost function of the problem. The main property of a Hopfield neural network is that

J. Cabestany, I. Rojas, and G. Joya (Eds.): IWANN 2011, Part II, LNCS 6692, pp. 105–112, 2011.

its energy function always decreases (or remains constant) as the network evolves according to its dynamical rule that updates the state of each neuron.

If the updating is performed on one neuron at a time, then we say that the network is operating in *serial* mode; if all neurons can be updated at the same time then we say that the network is operating in *fully parallel* mode. All other cases are referred to as parallel modes of operation. In this work we explore one of the main advantages of neural networks, i.e. their capacity to perform parallel computation. Nevertheless, it is well known that fully parallel Hopfield networks guarantee a decreasing energy function only when the synaptic weight matrix is positive semidefinite. However, in many optimization problems, the weight matrix does not satisfy this property. Consequently, the question of the degree of parallelism is reduced to find those process units that can be simultaneously updated so that a decrease in the energy function is guaranteed. In this paper, the degree of parallelism is determined in terms of the *chromatic number* of the graph associated to the Hopfield network. It is demonstrated that the average number of neurons to be updated simultaneously is the total number of neurons in the network divided by the chromatic index of the graph. Although the determination of the *chromatic number* of a graph is a *NP-Complete* problem [2], we have found the *chromatic number* and the coloring graph by means of coloring rules for the traveling salesman problem. Furthermore, neurons that can be simultaneously updated will be identified as those associated to the vertices of the graph with the same color.

This paper is organized as follows: in Section 2, we prove that the energy function is reduced when the neurons with the same color are simultaneously updated. In section 3 is provided the degree of parallelism for the traveling salesman problem by a coloring rule of a graph that provide partitions of the process units so that units belonging to a same subset have the same color and can be simultaneously updated. Conclusions are presented in Section 4.

2 Theoretical Results

We consider a binary Hopfield network with N process units. Let W be the weight matrix, here $\mathbf{x}(k) = (x_1(k), x_2(k), \ldots, x_n(k))$ represents the states of the process units in the time k, and θ_i be the threshold for ith unit. This network evolves according to the following update rule (computational dynamics):

$$x_i(k+1) = \begin{cases} 1 \text{ if } & u_i \geqslant \theta_i \\ 0 \text{ if } & u_i < \theta_i \end{cases} \tag{1}$$

where $u_i = \sum_{j=1}^{N} w_{ij} x_j(k)$, is the synaptic potential of the ith process unit and the network has associated a computational energy function given by the following expression:

$$E(\mathbf{x}(k)) = -\frac{1}{2} \sum_{i=1}^{N} \sum_{j=1}^{N} w_{ij} x_i(k) x_j(k) + \sum_{i=1}^{N} \theta_i x_i(k) \tag{2}$$

It is well known that when the update rule (1) is applied in a sequential (asynchronous) mode the computational energy is reduced or, at least, is not increased. Since there is only a finite set of possible states, the network must eventually reach a state for which the energy cannot be reduced further. It is a stable state of the network (local minimum of the energy function).

Let $C_1, C_2, \ldots, C_p$ be a partition of the set of process units (nodes) of a Hopfield network into p sets, so the units belonging to C_k can be updated simultaneously, and let n_i be the number of vertices in the group C_i. From the point of view of parallel computation, a measure of the degree of parallelism in the network is given in terms of the values of $p, n_1, n_2, \ldots, n_p$.

At instant k, the degree of parallelism in the network is defined as the number of process units that are simultaneously updated and it is denoted by n_k. The maximum number of simultaneous updates obtained in the network, represented by $P_{\max}$, is the maximum degree of parallelism. Hence,

$$P_{\max} = \max\{n_1, n_2, \ldots, n_p\} \tag{3}$$

The average degree of parallelism, $\overline{P}$, is defined by the expression

$$\overline{P} = \frac{1}{p}\sum_{k=1}^{p} n_k \tag{4}$$

and the usage rate of a system is defined by

$$\mu = \frac{\overline{P}}{P_{\max}} \tag{5}$$

It is well known that simultaneous updating of all process units guarantees a decrease of the energy function when the synaptic weight matrix is positive semidefinite. However, usually the weight matrix is not positive semidefinite. Now the question is: What process units sets can we update simultaneously with a decreasing energy function? To answer this question we will determine a partition of the set of vertices of the graph in such a way that the vertices belonging to the same subset can be updated simultaneously. So, all the vertices belonging to the same subset cannot be adjacent (connected). That is, the subset is an independent set of vertices. Two adjacent vertices should be assigned to different subsets. If all the vertices of the same subset are colored in the same color, and different subsets have a different color, the degree of parallelism in the network depends on the number of colors and the size of subsets. Thus, if the vertices of graph are colored with four different colors and all the subsets have the same number of vertices, then a simultaneous updating of $N/4$ neurons is possible. Therefore, if we determine the minimum number of colors required for coloring the graph, we can find the partition which will lead to a higher level of parallelism. In this way, the degree of parallelism in the network can be defined as the number of process units that can be updated simultaneously while the subsets remain the same size. In this case, the degree of parallelism is determined in terms of the *chromatic number* of the graph G, denoted by $\gamma(G)$, i.e., the minimum number of different colors required for coloring all the vertices of G.

Let B_k be the independent set of process units that have the same color k. Let G be the graph associated to a Hopfield neural network and let $\gamma(G)$ the *chromatic number* of G. We have the following result:

Proposition 1. *If at instant $k+1$ the units of B_r are updated following the computational dynamics* (1), *then the energy function decreases every time that the state of a unit is altered. That is,*

$$E(k+1) \leq E(k), \quad k = 1, 2, \ldots, \gamma(G) \tag{6}$$

Demonstration:
The state x_i of neuron i, only depends on the synaptic potential

$$u_i = \sum_{j=1}^{N} w_{ij} x_j - \theta_i.$$

If $w_{ij} = 0$, the new state x_i of neuron i does not depend on x_j, the state of neuron j. So, a change in the state of neuron j does not affect to the state of neuron i.

This result provides a simultaneous update rule of process units, so that the units with the same color can be simultaneously updated. Now, the question is, how the colors are assigned to the process units. In many classical combinatorial optimization problems a rule for vertex coloring can be determined according to the graph properties.

3 Degree of Parallelism in the Traveling Salesman Problem

The travelling salesman problem (TSP) is a well known *NP-Complete* problem for which many algorithms have been developed, some of them are based on neural networks [6,8,15,17]. It consists of finding the shortest tour for N cities in such way that each city is visited once and only once, being the distance matrix $D = (d_{ij})$ between each pair of cities. One solution to the problem which allows it to be solved using the Hopfield network consists of minimizing the expression:

$$E = A \sum_{i=1}^{N} \left[\sum_{k=1}^{N} x_{ik} - 1 \right]^2 + \lambda \sum_{k=1}^{N} \left[\sum_{i=1}^{N} x_{ik} - 1 \right]^2 + \sum_{i=1}^{N} \sum_{\substack{j=1 \\ j \neq i}}^{N} \left(\sum_{k=2}^{N} d_{ij} x_{i(k-1)} x_{jk} + d_{ij} x_{iN} x_{j1} \right) \tag{7}$$

where A is a positive number large enough, x_{ij} is the state of the process unit (i, j) which is equal to one when the ith city is visited on jth day. Note that when the two first terms reach their minimum value, i.e.,they are equal to zero, the restrictions of the problem are verified. The first constraint says that each city appears only once on the tour, while the second says that each stop on the tour is at just one city. The third term gives us the length of the tour. This problem can be solved using a binary Hopfield network with $N \times N$ process units. The parameters of the network are:

$$w_{ik,jr} = \begin{cases} -2d_{ij} - 2\lambda \text{ for } & i \neq j; r = k+1 \\ -2\lambda \text{ for } & i = j, r \neq k \\ -2\lambda \text{ for } & k = r, j \neq i \\ 0 \text{ otherwise} & \end{cases} \tag{8}$$

$$\theta_{ij} = -2\lambda, \quad i,j = 1,2,\ldots,N$$

Now the question is: how many process units can be simultaneously updated? The following results provides the chromatic number.

Proposition 2. *The chromatic number of the graph associated to the travelling salesman problem is given by the following expression:*

$$\gamma(G) = \begin{cases} 2N \text{ if} N \text{ is even} \\ 3N \text{ if} N \text{ is odd} \end{cases} \tag{9}$$

Demonstration:
In this network all the neurons in the same row and the same column are connected. Moreover, all neurons in two consecutive columns are connected. As the $2N$ neurons of each pair of consecutive columns are mutually connected, they constitute a clique. Therefore, the *chromatic number* of the graph is greater than or equal to $2N$. If N is even then we need to find a coloring rule for the graph which has only $2N$ colors. The coloring rule assigns the consecutive values (colors) from 1 to N in the first column, $N+1$ to $2N$ in the second, $2, 3, \ldots, N, 1$, in the third, $N+2, \ldots, 2N, N+1$, in the fourth and so on, in such a way that the values from 1 to N are assigned to the odd columns and from $N+1$ to $2N$ to the even columns. This ensures that no two consecutive columns are the same color, and that a neuron will not be the same color as the others in its same row or column. Hence, $2N$ groups of $N/2$ neurons have been formed. On the other hand, when N is an odd number, the *chromatic number* must be at least $3N$, because for $N = 3$, $3N$ colors are required since each one of the nine neurons are interconnected and form a clique. Thus, in this case the *chromatic number* is $3N$. If N is odd then we need to find a coloring rule with $3N$ colors. The following rule ensures coloring the vertices in the network with only $3N$ colors. Neurons in the N rows and first $N-1$ columns are colored only with $2N$ colors, as in the case when N is even, using the values from 1 to N for the odd columns and the values from $N+1$ to $2N$, in the even columns. So, we have $2N$ groups with $(N-1)/2$ neurons that are the same color. N new colors are assigned to the N neurons in the last column, since they must be different from the N colors of the first column and of the $(N-1)$ column. So the graph can be colored with $3N$ colors.

Corollary 1. *The average degree of parallelism in the network associated to the travelling salesman problem is $N/2$ if N is even and $N/3$ if N is odd.*

Note that this proof provides a procedure to color neurons. This can be illustrated by an example. We consider a problem with 8 cities. Since N is even 16 colors can be used (see Fig.1) and 4 neurons with the same colors can be simultaneously updated. In Fig.2, a problem with 7 cities is considered. Since N is odd 21 colors can be used and until 3 neurons can be simultaneously updated. In Fig.1 and Fig.2 you can see that the colors in column k are not used in column $k-1$ and $k+1$, and in each file and column are used different colors.

So, the algorithm for updating neurons is the following:

ALGORITHM:

INPUT :

Graph $G=(V,E)$ where V is the set of vertices or nodes (process unit or neuron) and E is the set of connections (arcs or edges) between vertices. Each connection has associated a number $w_{ij} \in R$ called the synaptic weight and each vertex has associated a number i called threshold. La adjacency matrix is given by

$$a_{ij} = \begin{cases} 1 \text{ if } & w_{ij} \neq 0 \\ 0 \text{ if } & w_{ij} = 0 \end{cases}$$

Initial solution: $x_1, x_2, \ldots, x_N$ (local minimum provided by an binary Hopfield network asynchronously updated).

STEP 1:

Determine the smallest number of colors required to color the vertices of G. Let $c_1, c_2, \ldots, c_p$ be a color set.

STEP2:

Neuron-Coloring so that two adjacent neurons have different colors.

STEP3:

while $\Delta E < 0$ do
  for $i = 1 : p$
    for $j = 1 : N$
      if $C(j) = c_j$
      $x_j = g(\sum_{k=1}^{N} w_{jk} x_k - \theta_j,) \quad i = 1, 2, , N.$
      where g is the *step function*
      end
    end
  end
end

OUTPUT:

$(x_1, x_2, \ldots, x_N)$

Note that the neurons with the same color can also be updated simultaneously since they are not connected to each other.

1	9	2	10	3	11	4	12
2	10	3	11	4	12	5	13
3	11	4	12	5	13	6	14
4	12	5	13	6	14	7	15
5	13	6	14	7	15	8	16
6	14	7	15	8	16	1	9
7	15	8	16	1	9	2	10
8	16	1	9	2	10	3	11

Fig. 1. TSP graph coloring with 8 cities

1	8	15	2	9	16	10
2	9	16	3	10	17	14
3	10	17	4	11	18	12
4	11	18	5	12	19	13
5	12	19	6	13	20	14
6	13	20	7	14	21	8
7	14	21	1	8	15	9

Fig. 2. TSP graph coloring with 7 cities

4 Conclusions

Binary Hopfield networks can simultaneously update their process units and reduce the energy function when the synaptic weights matrix is positive semidefinite. This condition is very restrictive. However, in this paper we have shown the capacity of binary Hopfield networks to carry out parallel computation with a high degree of parallelism. The degree of parallelism of a binary Hopfield network has been established in terms of *chromatic number* of the graph associated to the network. Furthermore, in order to color the vertices of the graph, a simple rule has been developed for traveling salesman problem.

We provide two rules for neuron-coloring according to the number of cities is even or odd. The average degree of parallelism in the network associated to the this problem is $N/2$ if N is even and $N/3$ if N is odd and the degree of parallelism is obtained without requiring that the synaptic weight matrix be positive semidefinite. Furthermore, since the demonstration performed is constructive, a coloring rule is provided for traveling salesman problem and similar rules may be used to color neurons in many other combinatorial optimization problems.

References

1. Galán-Marín, G., Muñoz-Pérez, J.: Design and Analysis of Maximum Hopfield Networks. IEEE Trans. on Neural Networks 12(2), 329–339 (2001)
2. Garey, M.R., Johnson, D.S.: Computers and Intractability: A Guide to the Theory of NP-Completeness. Freeman, New York (1979)
3. Hopfield, J., Tank, D.W.: Neural computation of decision in optimization problems. Biol. Cyber. 52, 141–152 (1985)
4. Lee, K.C., Takefuji, Y.: Finding Knight's tours on an MxN cheesboard whith O(MN) hysteresis McCulloch-Pitts neurons. IEEE Trans. Syst. Man Cybern. 24, 300–306 (1994)
5. Looi, C.K.: Neural networks methods in combinatorial optimization. Computers Operational Research 19(3/4), 191–208 (1991)
6. Martin-Valdivia, M., Ruiz-Sepúlveda, A., Triguero-Ruiz, F.: Improving local minima of Hopfield networks with augmented Lagrange multipliers for large scale TSPs. Neural Networks 13, 283–285 (2000)

7. Matsuda, S.: "Optimal" Hopfield Network for Combinatorial Optimization with Linear Cost Function. IEEE Trans. on Neural Networks 9(6), 1319–1330 (1998)
8. Mérida-Casermeiro, E., Galán-Marín, G., Muñoz-Pérez, J.: An Efficient Multivalued Hopfield Network for the Traveling Salesman Problem. Neural Processing Letters 14, 203–216 (2001)
9. Papageorgiou, G., Likas, A., Stafilopatis, A.: Improved exploration in Hopfield network state-space through parameter perturbation driven by simulated annealing. European Journal of Operational Research 108, 283–292 (1998)
10. Peng, M., Gupta, N.K., Armitage, A.F.: An Investigation into the Improvement of Local Minima of the Hopfield Network. Neural Networks 9(7), 1241–1253 (1996)
11. Ramanujan, J., Sadayappan, P.: Mapping Combinatorial Optimization Problems onto Neural Networks. Information Sciences 82, 239–255 (1995)
12. Smith, K., Palaniswami, M., Krishnamoorthy, M.: Neural techniques for combinatorial optimization with applications. IEEE Trans on Neural Networks 9(6), 1301–1318 (1998)
13. Sun, Y.: A generalized updating rule for modified Hopfield neural network for quadratic optimization. Neurocomputing 19(1-3), 133–143 (1998)
14. Takefuji, Y., Lee, K.C.: Artificial neural networks for four-coloring map problems and K-colorability problems. IEEE Trans. on Circuits and Systems 38, 326–333 (1991)
15. Tan, K.C., Tang, H., Ge, S.S.: On Parameter Settings of Hopfield Networks Applied to Traveling Salesman Problems. IEEE Trans. on Circuits and Systems 52(5), 994–1002 (2005)
16. Jiahai, W., Tang, Z.: An improved optimal competitive Hopfield network for bipartite subgraph problems. Neurocomputing 61, 413–419 (2004)
17. Wilson, G.V., Pawley, D.: On the stability of the TSP algorithm of Hopfield and Tank. Biol. Cybern. 58, 63–70 (1988)
18. Xia, Y., Wang, J.: A General Methodology for Designing Globally Convergent Optimization. Neural Networks 9(6), 1331–1343 (1998)
19. Shih-Yi, Y., Sy-Yen, K.: A New Technique for Optimization Problems in Graph Theory. IEEE Transactions on Computers 47(2), 190–196 (1998)

Multi-parametric Gaussian Kernel Function Optimization for ϵ-SVMr Using a Genetic Algorithm

J. Gascón-Moreno, E.G. Ortiz-García, S. Salcedo-Sanz, A. Paniagua-Tineo, B. Saavedra-Moreno, and J.A. Portilla-Figueras

Department of Signal Theory and Communications, Universidad de Alcalá, 28871 Alcalá de Henares, Madrid, Spain
sancho.salcedo@uah.es

Abstract. In this paper we propose a novel multi-parametric kernel Support Vector Regression algorithm optimized with a genetic algorithm. The multi-parametric model and the genetic algorithm proposed are both described with detail in the paper. We also present experimental evidences of the good performance of the genetic algorithm, when compared to a standard Grid Search approach. Specifically, results in different real regression problems from public repositories have shown the good performance of the multi-parametric kernel approach both in accuracy and computation time.

1 Introduction

The Support Vector Regression algorithm (SVMr) [1] is a robust methodology in statistical machine learning, that has been successfully applied to real regression problems [2,3,4]. The SVMr uses *kernel theory* [1] to increase the quality of regression models and, in most cases can be solved as a convex optimization problem. Several fast algorithms can be used to carry out the SVMr training, such as the sequential minimal optimization algorithm [1]. In spite of this, the training time of a SVM model can be very high due to the SVMr performance heavily depends on the choice of several hyper-parameters, necessary to define the optimization problem and the final SVMr model. Unfortunately, there is not an exact method to obtain the optimal set of SVMr hyper-parameters, and thus search algorithms must be applied to obtain this best possible set, what usually requires a heavy computational effort.

Usually, the search algorithms used to obtain SVMr hyper-parameters are based on *grid search* (GS) [5], where the search space of parameters is divided into groups of possible parameters to be tested (usually, an uniform partition of the search space is considered). This algorithm can be easily implemented, but it has an important drawback, since if the number of combinations is large, the training time becomes high, even considering only the three standard SVMr hyper-parameters in the search, i.e, C, ϵ and γ. In the case of multi-parametric kernel optimization, the GS approach is computationally not affordable, and meta-heuristic approaches such as genetic algorithms can be a good option.

J. Cabestany, I. Rojas, and G. Joya (Eds.): IWANN 2011, Part II, LNCS 6692, pp. 113–120, 2011.

In this paper we consider the case of a multi-parametric kernel SVMr, in which $N+2$ parameters are considered in the SVMr (C,ϵ, γ_m, $m = 1, \ldots, N$). Basically the idea is to maintain the C and ϵ parameters which define the problem, and N different γ parameters, one for each dimension of the data. This way the final SVMr model should be more effective, since it has better ability to discriminate important data components. As previously mentioned, the main drawback of this model is the computation time to search the best set of hyper-parameters for the model, since the GS algorithm cannot be applied when N grows. In this paper we propose a genetic algorithm to carry out this search. We present the encoding and different operators, and the validation methodology followed to avoid overtraining. The results obtained with this multi-parametric kernel SVMr are promising in several real databases tested.

The structure of the rest of the paper is the following: next section presents the mathematical foundations of the SVMr and the multi-parametric model considered in this paper. Section 3 presents the genetic algorithm proposed as a search heuristic for the hyper-parameters of the SVMr. Section 4 shows the performance of this algorithm in several regression problems from UCI and StatLib machine learning repositories. Finally, Section 5 closes the paper giving some remarks.

2 ϵ-SVM Formulation for the SVMr

The ϵ-SVM formulation for the SVMr [1] consists of, given a set of training vectors $S = \{(\mathbf{x_i}, y_i), i = 1, \ldots, l\}$, obtaining a model of the form $y(\mathbf{x}) = f(\mathbf{x}) + b = \mathbf{w}^T\phi(\mathbf{x}) + b$, to minimize a general risk function of the form

$$R[f] = \frac{1}{2}\left\|\mathbf{w}\right\|^2 + \frac{1}{2}C\sum_{i=1}^{l} L\left(y_i, f(\mathbf{x})\right) \tag{1}$$

where $\mathbf{w}$ controls the smoothness of the model, $\phi(\mathbf{x})$ is a function of projection of the input space to the feature space, b is a parameter of bias, $\mathbf{x_i}$ is a feature vector of the input space with dimension N, y_i is the output value to be estimated and $L\left(y_i, f(\mathbf{x})\right)$ is the selected loss function. In this paper, we use the L1-SVR (L1 support vector regression), characterized by a ϵ-insensitive loss function [1]

$$L\left(y_i, f(\mathbf{x})\right) = |y_i - f(\mathbf{x_i})|_\epsilon \tag{2}$$

In order to train this model, it is necessary to solve the following optimization problem [1]:

$$\min\left(\frac{1}{2}\left\|\mathbf{w}\right\|^2 + C\sum_{i=1}^{l}(\xi_i + \xi_i^*)\right) \tag{3}$$

subject to

$$y_i - \mathbf{w}^T\phi(\mathbf{x_i}) - b \leq \epsilon + \xi_i, \quad i = 1, \ldots, l \tag{4}$$

$$-y_i + \mathbf{w}^T\phi(\mathbf{x_i}) + b \leq \epsilon + \xi_i^*, \quad i = 1, \ldots, l \tag{5}$$

$$\xi_i, \xi_i^* \geq 0, \quad i = 1, \ldots, l \tag{6}$$

The dual form of this optimization problem is usually obtained through the minimization of the Lagrange function, constructed from the objective function and the problem constraints. In this case, the dual form of the optimization problem is the following:

$$\max \left(-\frac{1}{2} \sum_{i,j=1}^{l} (\alpha_i - \alpha_i^*)(\alpha_j - \alpha_j^*) K(\mathbf{x_i}, \mathbf{x_j}) - \epsilon \sum_{i=1}^{l} (\alpha_i + \alpha_i^*) + \sum_{i=1}^{l} y_i (\alpha_i - \alpha_i^*) \right) \tag{7}$$

subject to

$$\sum_{i=1}^{l} (\alpha_i - \alpha_i^*) = 0 \tag{8}$$

$$\alpha_i, \alpha_i^* \in [0, C] \tag{9}$$

In addition to these constraints, the Karush-Kuhn-Tucker conditions must be fulfilled, and also the bias variable, b, must be obtained. We do not detail this process for simplicity, the interested reader can consult [1] for reference. In the dual formulation of the problem the function $K(\mathbf{x_i}, \mathbf{x_j})$ is the kernel matrix, which is formed by the evaluation of a kernel function, equivalent to the dot product $\langle \phi(\mathbf{x_i}), \phi(\mathbf{x_j}) \rangle$. An usual election for this kernel function is a Gaussian function, as follows:

$$K(\mathbf{x_i}, \mathbf{x_j}) = exp(-\gamma \cdot \|\mathbf{x_i} - \mathbf{x_j}\|^2) \tag{10}$$

The final form of function $f(\mathbf{x})$ depends on the Lagrange multipliers α_i, α_i^*, as follows:

$$f(\mathbf{x}) = \sum_{i=1}^{l} (\alpha_i - \alpha_i^*) K(\mathbf{x_i}, \mathbf{x}) \tag{11}$$

2.1 Extension to Multi-parametric Gaussian Kernels

In this paper we consider a SVMr with a multi-parametric Gaussian kernel function, given by the following expression:

$$K(x_i, x_j) = exp(-(x_i - x_j)^T Q (x_i - x_j)) = exp(-\sum_{m=1}^{N} \sum_{n=1}^{N} \gamma_{mn} (x_{im} - x_{jm})(x_{in} - x_{in})) \tag{12}$$

where N is the number of dimensions in the feature space, Q is the matrix with size N that represents the width of the Gaussian function in each direction in the feature space and γ_{mn} is the element of the matrix Q which associates feature m with feature n. Without loss of generality, in this work we consider that matrix Q is a diagonal matrix with different values in each position of the trace, so the simplified Gaussian kernel considered is then:

$$K(x_i, x_j) = exp(-\sum_{m=1}^{N} \gamma_m (x_{im} - x_{jm})^2) \quad (13)$$

Thus, the hyper-parameters of the proposed multi-parametric kernel SVMr are C, ϵ and γ_m, $m = 1, \ldots, N$.

3 Optimization of the Multi-parametric Gaussian Kernel Function Proposed

Grid Search (GS) is maybe the most used algorithm in SVMrs to obtain a good set of hyper-parameters. GS establishes a grid of possible points to be explored, keeping parameters which show the best performance in a validation dataset. There are recent works in the literature which have proposed the application of meta-heuristic techniques, such as evolutionary approaches, to carry out this search [7], [8], obtaining good results even in the case of the traditional SVMr algorithm, with 3 parameters (C, ϵ and γ). However, note that it would be really difficult to extend the GS to multi-parametric kernels, since the number of parameters grows to $N + 2$, so the computation time explodes using this approach. An evolutionary approach [6] should be a solution to face this problem, since it is able to look for a set of $N + 2$ parameters within a reasonable time.

We then propose an evolutionary algorithm to obtain the best possible set of parameters for the multi-parametric Gaussian kernel in SVMrs, in the following way: each individual is the evolutionary population is defined to be a vector representing the hyper-parameters of the SVMr (C,ϵ, γ_m, $m = 1, \ldots, N$). The encoding used consist of a real encoding, with a standard two-points crossover operator and a random mutation using a Gaussian function. The roulette-wheel procedure has been applied as selection mechanism [6]. As objective function, we have used the error obtained by means of a n-fold cross-validation procedure (note that this is the same scheme used as validation method in the GS). In order to avoid the random elimination of the optimal individual, an operator called *Elitism* is introduced. This operator is based on keeping in a privilege position in the population the best individuals of every generation, so that they cannot be removed or modified by other operators. Thus it is ensured that the best individuals so far in the evolution survive. In the proposed evolutionary algorithm we use three elitist individuals in the population.

We have three different stopping criteria in our algorithm. The first one consists on evaluating the differences in fitness between the best individual in the population and the third best one. Then, along the evolution, this fitness difference between the best and third best individuals are obtained. If this difference is less than 10 times the first difference found during at least 5 generations, we stop the algorithm. Note that this criterion is related to the diversity in the population of the evolutionary algorithm. This diversity-based criterion has a drawback: if the initial population is good enough, it is possible that we cannot reach to the specific reduction in fitness difference between the best and third best individuals in the population. In order to solve this point, we introduce a second stopping criterion based on the best individual in the population, in such a way that if the fitness of this individual remains constant during $\mathcal{K}$ generations, we also stop the algorithm. The third stopping criterion is based on a maximum number of generations $\aleph$, if the algorithm reaches this maximum number of generations, then it is stopped.

If the γ_m values are quite different, it means that the chromosome may be generating a model with over-fitting in some of the coordinates. In order to keep the difference between the γ_m parameters of an individual in a reasonable level, a repair function is used to correct high variations among them. To do this, the repair function adjusts the value of the different γ_m in order to keep the maximum deviation among them to be less than a given threshold, in this case 0.2.

4 Experimental Part

This section presents the experimental part of the paper, structured in several subsections: First, the following subsection describes the different data sets used in the experiments carried out, discussing their origin and main characteristics. Subsection 4.2 briefly describes the methodology followed in all the experiments carried out in the paper. Finally, Subsection 4.3 contains the results obtained with the proposed evolutionary approach.

4.1 Data Sets

The data sets considered in this paper have been obtained from the UCI machine learning [9] and the data archive of Statlib [10]. Table 1 shows the main properties of the selected data sets.

MortPollution measures the age-adjusted mortality rate, taking into account some properties of pollution. Bodyfat estimates the percentage of body fat determined by underwater weighing and various body circumference measurements. Betaplasma and Retplasma study the relationship between personal characteristics and dietary factors, and plasma concentrations of beta-carotene and retinol which might be associated with increased risk of developing cancer. Autompg concerns city-cycle fuel consumption in function of parameters of the car. Housing concerns housing values in suburbs of Boston. Concrete estimates the concrete compressive strength through the components of the mixture. Finally, Abalone set can be handled as a multi-classification or regression problem, and predicts the age of abalones from their physical measurement.

Table 1. Data sets used in experiments carried out

Data set	Samples	Attributes (N)	Repository
MortPollution	60	15	StatLib
Bodyfat	252	13	StatLib
Betaplasma	315	12	StatLib
Retplasma	315	12	StatLib
Autompg	392	7	UCI
Housing	506	13	UCI
Concrete	1030	16	UCI
Abalone	4177	8	UCI

4.2 Simulation Methodology

In an initial step, previous to the application of the SVMr, all the data sets are treated in order to homogenize them. First, the samples with missing values of each data set are eliminated, to simplify their treatment, and the non-numerical attributes are substituted by integer numbers. Note that the number of samples in Table 1 represents the number after this process. Then, each data set is divided into two sets, the training and test sets, by selecting 80% of instances for the training set and the rest of the instances for the test set, in the same order as they appear in the original data set (this ensures that the obtained results are repeatable in future works of other authors). In a final common step, the data are scaled to zero mean and unit variance. In all the experiments carried out in the paper, the solver of LIBSVM library [11] has been used to solve the different SVM optimization problems. In order to evaluate the performance of a set of parameters, we use a n-fold cross-validation with n=10, i.e, we divide into n folds the train set and evaluate each fold with the model training with the rest of the folds. The n folds are selected in the same order as they appear in the train set, in a similar way to the division of train and test sets. For speeding up the training of the models for each fold, we have modified the functions related to the matrix kernel in the LIBSVM library, in such a way that we keep in memory the complete kernel matrix K, and it is only modified when the parameters γ_m are changed. This allows calculating only one kernel matrix for all n-folds training and all the range of possible γ_m parameters.

4.3 Results

In order to show the performance of the proposed evolutionary algorithm (EA), we have applied it to obtain the optimal set of hyper-parameters in the different regression problems considered. We have run the EA with a population of 25 individuals (plus the three elitist individuals), a crossover probability $P_c = 0.7$, a mutation probability $P_m = 0.1$, we have fixed a maximum generations allowed $\aleph = 35$, a number of maximum generations with the same value of the best fitness $\mathcal{K} = 6$ and the n-fold cross-validation procedure fixed to $n = 10$. Twenty different runs of the evolutionary algorithm have been launched, and the average value have been kept.

The results obtained with the multi-parametric kernel SVMr, optimized with the proposed EA are shown in Table 2. As comparison algorithm, we use the standard SVMr (with 3 hyper-paramenters), optimized with a GS approach. In the table, the average RMSE obtained in the 20 runs of the genetic algorithm (in the test sets) are displayed, and also the computation time in the average computation time of this process. For the multi-parametric kernel SVMr, the best RMSE in the test set is also displayed for reference.

First, note that the multi-parametric kernel SVMr obtains better average accuracy values than the standard SVMr in 5 out of the 8 data sets, obtaining RMSE quite close to the standard SVMr in the other 3 data sets. The best RMSE values obtained with the genetic algorithm are always better than the RMSE obtained by the standard SVMr. The computation time is another important factor to be evaluated in this work. As the data sets are sorted by the number of samples, the increment of training time is evident as the number of samples grows. Note that the computation time with the genetic algorithm is much lower than the one provided by the GS in the standard SVMr, in all the problems tackled. Of course, this is due to the genetic algorithm evaluates less points in the search space than the GS. This point, together with the performance of the multi-parametric kernel SVMr in RMSE show that this approach is able to outperform the performance of the standard SVMr in regression problems, and it is a very good option to obtain accurate results with a reduced computation time in these kind of problems.

Table 2. SVMr performance and training time for the standard SVMr (with a Grid Search) and the multi-parametric kernel SVMr (with a genetic algorithm)

	Grid Search (standard SVMr)		Genetic algorithm (multi-parametric kernel SVMr)		
Data set	Error	Time	Error	Time	Min Error
Mortpollution	51.0935	1.50 s	44.8689	0.81 s	42.7355
BodyFat	0.01138	21.01 s	0.01129	11.05 s	0.01112
Betaplasma	172.333	32.54 s	170.032	20.72 s	167.22
Retplasma	254.8547	32.13 s	257.655	16.21 s	251.886
Autompg	4.5426	60.85 s	4.5859	24.11 s	4.5124
Housing	4.9886	113.72 s	4.9672	58.53 s	3.8493
Concrete	28.4964	669.04 s	28.5326	149.20 s	28.4826
Abalone	1.7774	1.16 h	2.1994	150.00 s	1.7589

5 Conclusions

In this paper we have proposed a novel genetic-optimized multi-parametric kernel Support Vector Regression algorithm (multi-parametric kernel SVMr). We have described the main characteristics of the multi-parametric model proposed, and the genetic algorithm considered to optimize this kernel, since standard Grid Search is computationally too expensive to be applied in this task. Results in different real regression problems have shown the good performance of the multi-parametric kernel approach against the standard SVMr with Grid Search, both in accuracy and computation time.

Acknowledgement

This work has been partially supported by Spanish Ministry of Science and Innovation, under a project number ECO2010-22065-C03-02.

References

1. Smola, A.J., Schölkopf, B.: A tutorial on support vector regression. Statistics and Computing (1998)
2. He, W., Wang, Z., Jiang, H.: Model optimizing and feature selecting for support vector regression in time series forecasting. Neurocomputing 72(1-3), 600–611 (2008)
3. Wu, C.L., Chau, K.W., Li, Y.S.: River stage prediction based on a distributed support vector regression. J. of Hydrology 358(1-2), 96–111 (2008)
4. Ortiz-Garcia, E.G., Salcedo-Sanz, S., Pérez-Bellido, A.M., Portilla-Figueras, J.A.: Improving the training time of support vector regression algorithms through novel hyper-parameters search space reductions. Neurocomputing 72(1-3), 3683–3691 (2009)
5. Akay, M.F.: Support vector machines combined with feature selection for breast cancer diagnosis. Expert Systems with Applications 36(2), 3240–3247 (2009)
6. Eiben, A.E., Smith, J.E.: Introduction to evolutionary computing. Springer, Heidelberg (2003)
7. Wu, G.H., Tzeng, G.H., Lin, R.H.: A Novel hybrid genetic algorithm for kernel function and parameter optimization in support vector regression. Expert Systems with Applications 36(3), 4725–4735 (2009)
8. Hou, S., Li, Y.: Short-term fault prediction based on support vector machines with parameter optimization by evolution strategy. Expert Systems with Applications 36(10), 12383–12391 (2009)
9. Asuncion, A., Newman, D.J.: UCI Machine Learning Repository. University of California, School of Information and Computer Science, Irvine (2007), `http://www.ics.uci.edu/~mlearn/MLRepository.html`
10. StatLib DataSets Archive, `http://lib.stat.cmu.edu/datasets`
11. Chang, C.C., Lin, C.-J.: LIBSVM: a library for support vector machines. Software, `http://www.csie.ntu.edu.tw/~cjlin/libsvm`

Face Recognition System in a Dynamical Environment

Aldo Franco Dragoni, Germano Vallesi, and Paola Baldassarri

DIIGA, Universitá Politecnica delle Marche,
via Brecce Bianche 1, 60131 Ancona, Italia
{a.f.dragoni,g.vallesi,p.baldassarri}@univpm.it

Abstract. We propose a Hybrid System for dynamic environments, where a "Multiple Neural Networks" system works with Bayes Rule to solve the face recognition problem. One or more neural nets may no longer be able to properly operate, due to partial changes in some of the characteristics of the individuals. For this purpose, we assume that each expert network has a reliability factor that can be dynamically re-evaluated on the ground of the global recognition operated by the overall group. Since the net's degree of reliability is defined as the probability that the net is giving the desired output, in case of conflicts between the outputs of the various nets the re-evaluation of their degrees of reliability can be simply performed on the basis of the Bayes Rule. The new vector of reliability will be used to establish who is the conflict winner, making the final choice. Moreover the network disagreed with the group and specialized to recognize the changed characteristic of the subject will be retrained and then forced to correctly recognize the subject. Then the system is subjected to continuous learning.

Keywords: Belief revision, face recognition, neural networks, unsupervised learning, hybrid system.

1 Introduction

Several researches in the field of Artificial Neural Networks indicated that there are problems which cannot be effectively solved by a single neural network [1]. This led to the concept of "Multiple Neural Networks" systems for tackling complex tasks improving performances w.r.t. single network systems [2]. The idea is to decompose a large problem into a number of subproblems and then to combine the individual solutions to the subproblems into a solution to the original one [2]. This modular approach can lead to systems in which the integration of expert modules can result in solving problems which otherwise would not have been possible using a single neural network [3]. The responses of the individual modules are simple and have to be combined by some integrating mechanism in order to generate the complex overall system response [4]. The combination of individual responses is particularly critical when there are incompatibilities between them. Such situations may arise for example when the system operates

J. Cabestany, I. Rojas, and G. Joya (Eds.): IWANN 2011, Part II, LNCS 6692, pp. 121–128, 2011.

in dynamic environments, where it can happen that one or more modules of the system are no longer able to properly operate [5]. In this work we propose a "Multiple Neural Networks" system to solve a face recognition problem. Where each neural network is trained to recognize a significant region of the face and to each one is assigned an arbitrary a-priori reliability. All the networks have a reliability factor that can be dynamically re-evaluated on the ground of the global recognition operated by the overall group. In case of conflicts between the outputs of the various nets the re-evaluation of their "degrees of reliability" can be simply performed on the basis of the Bayes Rule. The conflicts depend on the fact that there may be no global agreement about the recognized subject, may be for she/he changed some features of her/his face. The new vector of reliability obtained through the Bayes Rule will be used for making the final choice, by applying the "Inclusion based" algorithm [3] or another "Weighted" algorithm over all the maximally consistent subsets of the global output of the neural networks. The nets recognized as responsible for the conflicts will be automatically forced to learn about the changes in the individuals characteristics through a continuous learning process.

2 Theoretical Background

In this section we introduce some theoretical background taken from the Belief Revision (BR) field. Belief Revision occurs when a new piece of information inconsistent with the present belief set (or database) is added in order to produce a new consistent belief system [6].

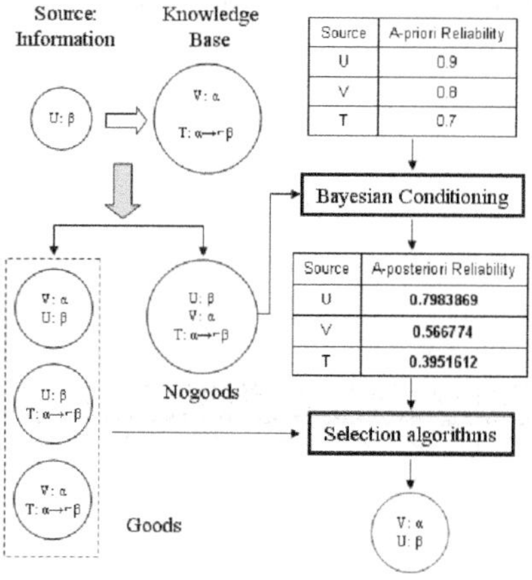

Fig. 1. "Belief Revision" mechanism

In figure 1, we see a Knowledge Base (KB) which contains two pieces of information: the α information , which come from V source, and the rule "If α, then not β" that comes from T source. Unfortunately, another piece of β

information produced by the U source , is coming, causing a conflicts in the KB. To solve the conflicts we have to found all the "maximally consistent subsets", called Goods, inside the inconsistent KB, and we choose one of them as the most believable one. In our case (figure 1) there are three Goods: $\{\alpha, \beta\}$; $\{\beta, \alpha \rightarrow \neg\beta\}$; $\{\alpha, \alpha \rightarrow \neg\beta\}$. Maximally consistent subsets (Goods) and minimally inconsistent subsets (Nogoods) are dual notions. Given an inconsistent KB finding all the Goods and finding all the Nogoods are dual processes. Each source of information is associated with an a-priori "degree of reliability", which is intended as the a-priori probability that the source provides correct information. In case of conflicts the "degree of reliability" of the involved sources should decrease after "Bayesian Conditioning" which is obtained as follows. Let $S = \{s_1, ..., s_n\}$ be the set of the sources, each source s_i is associated with an a-priori reliability $R(s_i)$. Let ϕ be an element of 2^S. If the sources are independent, the probability that only the sources belonging to the subset $\phi \subseteq S$ are reliable is:

$$R(\phi) = \prod_{s_i \in \phi} R(s_i) * \prod_{s_i \notin \phi} (1 - R(s_i)) \tag{1}$$

This combined reliability can be calculated for any ϕ providing that:

$$\sum_{\phi \in 2^S} R(\phi) = 1 \tag{2}$$

Of course, if the sources belonging to a certain ϕ give inconsistent information, then $R(\phi)$ must be zero. Having already found all the Nogoods, what we have to do is:

- Summing up into $R_{Contradictory}$ the a-priori reliability of Nogoods;
- Putting at zero the reliabilities of all the contradictory sets, which are the Nogoods and their supersets;
- Dividing the reliability of all the other (no-contradictory) set of sources by $1 - R_{Contradictory}$ we obtain the new reliability (NR).

The last step assures that the equation 2 is still satisfied and it is well known as "Bayesian Conditioning". The revised reliability $NR(s_i)$ of a source s_i is the sum of the reliabilities of the elements of 2^S that contain s_i. If a source has been involved in some contradictions, then $NR(s_i) \leq R(s_i)$, otherwise $NR(s_i) = R(s_i)$. These new or revised "degrees of reliability" will be used for choosing the most credible Goods as the one suggested by "the most reliable sources". There are three algorithms to perform this task:

1. Inclusion based (IB) Algorithm select all the Goods which contains information provided by the most reliable source.
2. Inclusion based weighted (IBW) is a variation of IB: each Good is associated with a weight derived from the sum of Euclidean distances between the neurons of the networks. If IB select more than one Good, then IBW selects as winner the Good with a lower weight.

3. Weighted algorithm (WA) combines the a-posteriori reliability of each network with the order of the answers provided. Each answer has a weight $1/n$ where $n \in [1; N]$ represents its position among the N responses.

3 Face Recognition System: An Example

In the present work ,to solve the face recognition problem [7], we use a "Multiple Neural Networks" system consisted of a number of independent modules, such as neural networks, specialized to recognize individual template of the face. We use 4 neural networks specialized to perform a specific task: eyes (E), nose (N), mouth (M) and, finally, hair (H) recognition. Their outputs are the recognized subjects, and conflicts are simple disagreements regarding the subject recognized. As an example, lets suppose that during the testing phase, the system has to recognize the face of four persons: Andrea (A), Franco (F), Lucia (L) and Paolo (P), and that, after the testing phase, the outputs of the networks are as follows: E gives as output "A or F", N gives "A or P", M gives "L or P" and H gives "L or A"', so the 4 networks do not globally agree. Starting from an undifferentiated a-priori reliability factor of 0.9, and applying the Belief revision method , for each expert network we get the following new degree of reliability: NR(E) = 0.7684, NR(N) = 0.8375, NR(M) = 0.1459 and NR(H)=0.8375. The networks N and H have the same reliability, and by applying a selection algorithm it turns out that the most credible Good is E,N,H, which corresponds to Andrea. So Andrea is the response of the system.

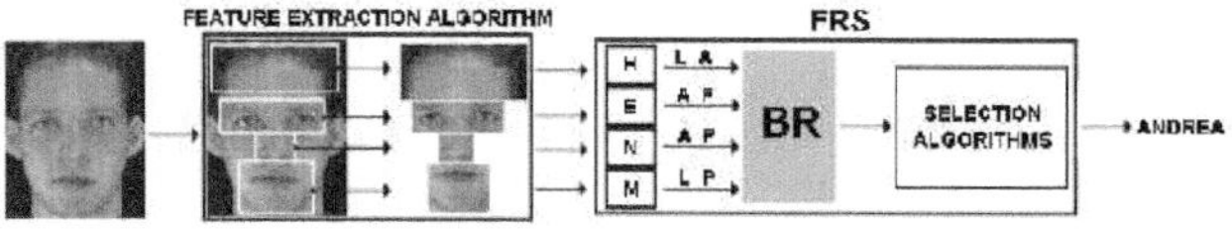

Fig. 2. Face Recognition System (FRS) representation

Figure 2 shows a schematic representation of this Face Recognition System (FRS), which is able to recognize the most probable individual even if there are serious conflicts among their outputs.

4 Face Recognition System in a Dynamical Environment

As seen in the previous Section, one or more networks may fail to recognize the subject, there can be two reasons for the fault of the net: either the task of recognizing is objectively harder, or the subject could have recently changed something in the appearance of his face (perhaps because of the grown of a goatee or moustaches). The second case is very interesting because it shows how our FRS could be useful for implementing Multiple Neural Networks able to follow dynamic changes in the features of the subjects. In a such dynamic environment, where the input pattern partially changes, some neural networks could no longer be able to recognize the input. However, if the changes are minimal, we guess that most of the networks will still correctly recognize the face. So, we force the faulting network to re-train itself on the basis of the recognition made by the overall group. Considering the a-posteriori reliability and the Goods, our idea is to automatically re-train the networks that did not agree with the others. The network that do not support the most credible Good is forced to re-train themself in order to "correctly" (according to the opinion of the group) recognize the face. Each iteration of the cycle applies Bayesian conditioning to the a-priori "degrees of reliability" producing an a-posteriori vector of reliability. To take into account the history of the responses that came from each network, we maintain an "average vectors of reliability" produced at each recognition, always starting from the a-priori degrees of reliability. This average vector will be given as input to the two algorithms, IBW and WA, instead of the a-posteriori vector of reliability produced in the current recognition. In other words, the difference with respect to the BR mechanism is that we do not give an a-posteriori vector of reliability to the two algorithms (IBW and WA), but the average vector of reliability calculated since the FRS started to work with that set of subjects to recognize. Now the subject has moustaches and goatee, while, when the system is trained, the subject did not have them. So O_M network (specialized to recognize the mouth) is no longer able to correctly indicate the tested subject. Since all the others still recognize Andrea, O_M will be retrained with the mouth of Andrea as new input pattern.

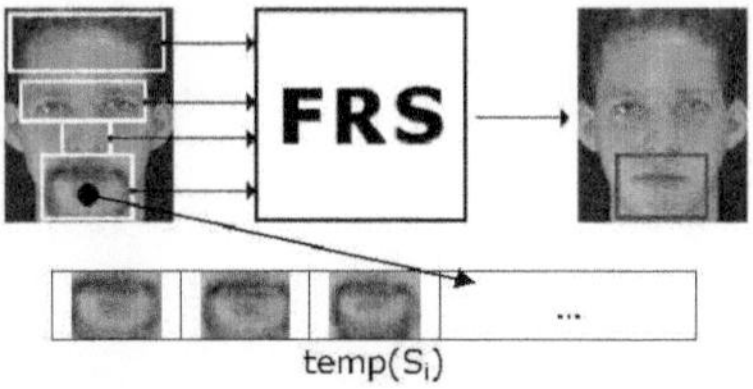

Fig. 3. Functioning of the temporal window

The re-learning procedure occurs when the changing is longer than the previously fixed temporal window (windowlength equals to 10) associated to each neural network. So we avoid the re-learning for a subject with a very variable feature. We define imm_i the portion of the image containing the feature analyzed by the network r_i; S the subject identified by the synthesis function of the FRS; s_{ik} is the subject i in the k-th position of the list ordered on the base of the distance of the LVQ output. So the re-learning procedure consists of the following steps:

1. For each network r_i the system compares S and s_{ik} used to find the Good. If $S \neq s_{ik} \forall k$ then in the temporary directory $temp(S_i)$ (that is the temporal window) of the subject S related to the network i is saved the imm_i portion,

as showed in Figure 3. On the contrary if $S = s_{ik}$ for one k the temporary directory temp(S_i) is empitied;

2. If in temp(S_i) there are windowlength samples, the *temp*(S_i) images are transferred in riadd(S_i) removing its old images, then the retraining of the r_i network begins using the riadd(S_i) images for S and the most recent images for all other subjects.

We have to highlight that the windowlength chosen strongly depends on the variability of subjects, and so on the database used for the testing. It is important also to note that there will always be a limit to the size of the windowlength beyond which for any dataset the system will be able to filter all the changes, to a value beyond this limit the system behaves as a system without re-learning. If not recognized by the networks, the introduction of re-learning in the facial recognition system allows, that the networks maintain higher reliability values than in the case without re-learning, as shown in Figures 4a and 4b. This because, the network now can recognizes a feature that could not recognize with the original knowledge acquired during the first training of all networks. If a network is no longer able to recognize one feature can not contribute to the final choice. Moreover if other networks do not recognize the subject, but indicate the same wrong subject, the whole system fails. In this case there would be a wrong Good that could be the most likely for the system but associated to the incorrect subject.

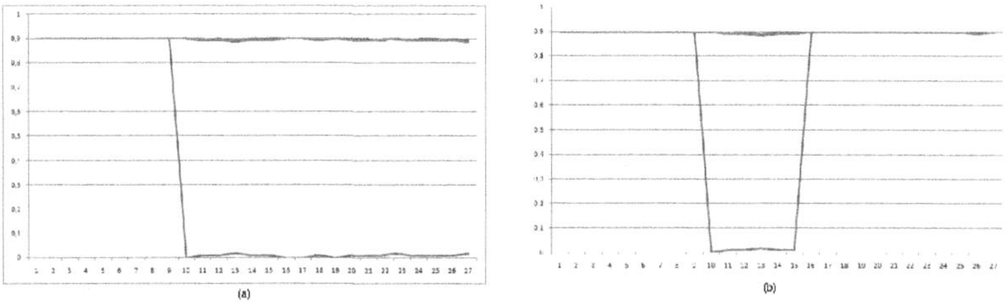

Fig. 4. Performance of the a-priori reliability (a) without and (b) with re-learning

Figure 4 shows the a-posteriori reliability trend related to the five expert neural networks concerning a particular subject. Observing the two graphs, we can see that until the networks are agree, the reliability maintains high values. While, if one of the networks (eg mouth) comes into conflict with the others giving in output another subject, since it changed its appearance, then the reliability goes down. In Figure 4a, we can see how this conflict will bring the net loser to have a low reliability, which can not be recovered even when the network agrees with the others. Conversely, in Figure 4b, we can see that if the network does not recognize the subject for a consecutive number of times corresponding to the windowlength samples the re-learning begins, after which the a-posteriori reliability back to its highest level.

5 Experimental Results

This section shows only partial results: those obtained without the feedback, discussed in the previous section. In this work we compared two groups of neural networks: the first consisting of four networks and the second with five (the additional network is obtained by separating the eyes in two distinctive networks). All the networks are LVQ 2.1, a variation of Kohonens LVQ [8], each one specialized to respond to individual template of the face. The Training Set is composed of 20 subjects (taken from FERET database [9]), for each one 4 pictures were taken for a total of 80. Networks were trained, during the learning phase, with three different epochs: 3000, 4000 and 5000. To find Goods and Nogoods, from the networks responses we use two methods:

1. Static method: the cardinality of the response provided by each net is fixed a priori. We choose values from 1 to 5, 1 meaning the most probable individual, while 5 meaning the most five probable subjects.
2. Dynamic method: the cardinality of the response provided by each net changes dynamically according to the minimum number of "desired" Goods to be searched among. In other words, we set the number of desired Goods and reduce the cardinality of the response (from 5 down to 1) till we eventually reach that number (of course, if all the nets agree in their first name there will be only one Goods).

In the next step we applied the Bayesian conditioning depending from Goods obtained with these two techniques. In this way we obtain the new reliability for each network. These new "degrees of reliability" will be used for choosing the most credible Good (then the name of subject). We use two selection algorithms to perform this task: Inclusion based weighted (IBW), Weighted algorithm (WA). To test our work, we have taken 488 different images of the 20 subjects and with these images we have created the Test Set. Using the system without re-learning the results show how the union of the Dynamic method with the selection algorithm WA and five neural networks gives the best solution to reach a 79.39% correct recognition rate of the subjects. Moreover using only one LVQ network for the entire face, we obtain the worst result. In other words, if we consider a single neural network to recognize the face, rather one for the nose, one for the mouth and so on, we have the lowest rate of recognition equals to 66%. This is because a single change in one part of the face makes the whole image not recognizable to a single network, unlike the hybrid system. Moreover, we obtain for the Hybrid system (consisted of Static method and WA selection algorithm) with re-learning a rate of recognition equals to 89.25%, while for the same Hybrid system but without re-learning a rate of recognition equals to 79.39%.

6 Conclusion

Our hybrid method integrates multiple neural networks with a symbolic approach to Belief Revision to deal with pattern recognition problems that: require the cooperation of multiple neural networks specialized on different topics and the subjects to recognize change dynamically some of their features so that some nets occasionally fail. We tested this hybrid method referring to a face recognition problem, training each network to recognize a specific region of the face: eyes, nose, mouth, and hair. Every output unit is associated with one of the persons to be recognized. Each net gives the same number of outputs. We consider a constrained environment in which the image of the face is always frontal, lighting conditions, scaling and rotation of the face being the same. We accommodated the test so that changes of the faces are partial, for example the mouth and hair do not change simultaneously, but one at a time. Under this assumption of limited changes, our hybrid system ensures great robustness to the recognition. The system assigns a reliability factor to each neural network, which is recalculated on the basis of conflicts that occur in the choice of the subject. The new "degrees of reliability" are obtained through the conflicts table and Bayesian Conditioning. These new "degrees of reliability" can be used to select the most likely subject. When the subject partially changes its appearance, the network responsible for the recognition of the modified region comes into conflict with other networks and its degree of reliability will suffer a sharp decrease. So, the overall system is engaged in a never ending loop of testing and re-training that makes it able to cope with dynamic partial changes in the features of the subjects. To maintain high values of the reliability for all the networks is very important since the choice of the right subject strongly depends on the credibility of all the experts.

References

1. Azam, F.: Biologically inspired modular neural networks. PhD Dissertation, Virginia Tech. (2000)
2. Shields, M.W., Casey, M.C.: A theoretical framework for multiple neural network systems. Neurocomputing 71(7-9), 1462–1476 (2008)
3. Sharkey, A.J.: Modularity combining and artificial neural nets. Connection Science 9(1), 3–10 (1997)
4. Li, Y., Zhang, D.: Modular Neural Networks and Their Applications in Biometrics. Trends in Neural Computation 35, 337–365 (2007)
5. Guo, H., Shi, W., Deng, Y.: Evaluating sensor reliability in classification problems based on evidence theory. IEEE Transactions on Systems, Man and Cybernetics-Part B: Cybernetics 36(5), 970–981 (2006)
6. Gárdenfors, P.: Belief Revision. In: Cambridge Tracts in Theoretical Computer Science, vol. 29 (2003)
7. Tolba, A.S., El-Baz, A.H., El-Harby, A.A.: Face Recognition a Litterature Review. International Journal of Signal Processing 2, 88–103 (2006)
8. Kohonen, T.: Learning vector quantization. In: Self Organizing Maps. Springer Series in Information Sciences, Berlin, vol. 30 (2001)
9. Philips, P.J., Wechsler, H., Huang, J., Rauss, P.: The FERET Database and Evaluation Procedure for Face-Recognition Algorithms. Images and Vision Computing Journal 16(5), 295–306 (1998)

Memetic Pareto Differential Evolutionary Neural Network for Donor-Recipient Matching in Liver Transplantation

M. Cruz-Ramírez[1,⋆], C. Hervás-Martínez[1], P.A. Gutiérrez[1], J. Briceño[2], and M. de la Mata[2]

[1] Dept. of Computer Science and Numerical Analysis, University of Córdoba, Spain
[2] Liver Transplantation Unit, Hospital Reina Sofía, Córdoba, Spain, CIBERHED

Abstract. Donor-Recipient matching constitutes a complex scenario not easily modelable. The risk of subjectivity and the likelihood of falling into error must not be underestimated. Computational tools for decision-making process in liver transplantation can be useful, despite its inherent complexity. Therefore, a Multi-Objective Evolutionary Algorithm and various techniques of selection of individuals are used in this paper to obtain Artificial Neural Network models to assist in making decisions. Thus, the experts will have a mathematical value that enables them to make a right decision without deleting the principles of justice, efficiency and equity.

1 Introduction

Liver transplantation is an accepted treatment for patients with end-stage chronic liver disease. Numerous donor and recipient risk factors interact to influence the probability of survival at 3 months after liver transplantation. It is critical to balance waitlist mortality against posttransplant mortality. Our objective was to devise a scoring system that predicts recipient survival at 3 months following liver transplantation to complement the model for end-stage liver disease score (MELD) to predicted waitlist mortality.

Most current organ allocation systems are based on the principle that the sickest patients should be treated first. Models have been developed to estimate the risk of death, considering the underlying disease and urgency of the receiving patient assuming that all donor livers carry the same risk of failure. This, however, is not the case: it has been shown in recent years that the risk of graft failure, and even patient death, after transplantation differs among recipients. While some patients may "tolerate" and overcome the initial poor function of a compromised donor organ, others may not have the same reserve. Increasing

⋆ Corresponding author at: Tel.: +34 957 218 349; Fax: +34 957 218 630; E-mail: mcruz@uco.es M. Cruz-Ramírez's research has been subsidized by the FPU Predoctoral Program (spanish Ministry of Education and Science), grant reference AP2009-0487.

J. Cabestany, I. Rojas, and G. Joya (Eds.): IWANN 2011, Part II, LNCS 6692, pp. 129–136, 2011.

awareness of the diversity in donor organ quality has stimulated the debate on matching between specific recipient and donor factors to avoid futility, but also to avoid personal and institutional differences in organ acceptance. The insufficient supply of deceased donor livers for transplantation has motivated the expansion of acceptance criteria; such organs are captured by the terms marginal and expanded criteria livers. This context of aggressive liver utilization motivated the derivation of the donor risk index, a quantitative, objective, and continuous metric of liver quality based on factors known or knowable at the time of an organ offer.

Thus, predicting the survival of liver transplant patients has the potential to play a critical role in understanding and improving the matching procedure between the recipient and graft. Although voluminous data related to the transplantation procedures is being collected and stored, only a small subset of the predictive factors has been used in modeling liver transplantation outcomes. The previous studies have mainly focused on applying statistical techniques to a small set of factors selected by the domain-experts in order to reveal the simple linear relationships between the factors and survival. Machine learning and soft computing methods offer significant advantages over conventional statistical techniques in dealing with the latter's limitations such as normality assumption of observations, independence of observations from each other, and linearity of the relationship between the observations and the output measure(s). Among these techniques, we will use Artificial Neural Network models (ANN). The use of ANNs in biomedicine as an alternative to other classification methods is baised on different approaches: a Fisher transformation [2,9], due to their flexibility and high degree of accuracy to fit to biomedical data, sigmoid functions and other types of basis functions (Multilayer Perceptron type networks) [11]. In the field of transplantation, ANNs have been designed to diagnose cytomegalovirus (CMV) disease [17] and acute rejection using data obtained from posttransplantation renal biopsies [10] after kidney transplantation. In addition, the use of ANNs was investigated in the prediction of graft failure [15] on the prediction of liver transplantation outcome [6].

ANNs can be trained with Evolutionary Computation algorithms (EC). This methodology widely used in the last few years to evolve neural-network architectures and weights. This is known as Evolutionary Artificial Neural Networks (EANNs), and it has been used in many applications [13,16]. EANNs provide a more successful platform for optimizing network performance and architecture simultaneously.

In this work, we discuss learning and generalization improvement of classifiers designed using a Multi-Objective Evolutionary learning Algorithm (MOEA) [4] for the determination of survival at 3 months after liver transplantation. The data come from eleven hospitals and we investigate the generation of neural network classifiers that achieve high classification level for each class. The methodology is based on two measures: the correct classification rate or Accuracy (C) and the Minimum Sensitivity (MS) as the minimum of the sensitivities of all classes. The aim of this study is to determine which models obtained with the

MOEA, presented the best results. In order to do this, different methods for selection of individuals and ensembles techniques [14] are used, once the execution of the MOEA is finished.

The paper is organized as follows: Section 2 describes the dataset used; Section 3 shows a description of the methodology used; Section 4 explains the experimental design; Section 5 shows the results obtained, while the conclusions and the future work are outlined in Section 6.

2 Dataset Description

A multi-centric retrospective analysis from 11 Spanish units of liver transplantation was conducted, including all the consecutive liver transplants performed between January 1, 2007, and December 31, 2008. The dataset included all transplant recipients aged 18 years or older. Recipient and donor characteristics were reported at the time of transplant. Patients undergoing partial, split or living donor liver transplantation and patients undergoing combined or multi-visceral transplants were excluded from the study. All patients were followed from the date of transplant until either death, graft loss or the first year after liver transplant. Units of liver transplantation were homogeneously distributed throughout Spain.

16 recipient characteristics, 20 donor characteristics and 3 operative factors were reported for each donor-recipient pair. The end-point variable for artificial neural network modeling was 3-month graft mortality. A total of 1031 liver transplants were initially included. The follow-up period was fulfilled in 1003 liver transplants. 28 cases were excluded because the absence of graft survival data. All losses were well distributed among the participating institutions. Donor or recipients with missing entries were not eliminated but missing values were filled using data imputation techniques. Imputation techniques employed are those commonly used. When the number of non-responses in a variable is less than 1%, we have substituted the value by the average if the variable is continuous and by the mode if the variable is discrete (categorical). In another case, we used polynomial regression models to estimate these values.

3 Methods

3.1 Accuracy and Minimum Sensitivity in Classification Problems

To evaluate a classifier, the machine learning community has traditionally used Correct Classification Rate or Accuracy (C) to measure its default performance. Actually, it suffices to realize that C cannot capture all the different behavioral aspects found in two different classifiers in multi-class problems. For these problems, two performance measures are considered: traditionally-used C, as the number of patterns correctly classified and the Minimum of the Sensitivities of all classes (MS), that is, the lowest percentage of examples correctly predicted as belonging to each class, S_i, with respect to the total number of examples in the corresponding class, $MS = \min\{S_i\}$ (for a more detailed description of these measures, see [8]). This is, we assume the premise that a good classifier should combine a high classification rate level in the testing set with an acceptable level for each class.

In [8], C and MS are presented as objectives that could be positively correlated, but, while this may be true for small values of MS and C, it is not so for values close to 1 on both MS and C, where the objectives are competitive and conflicting. This fact justifies the use of a MOEA for training ANNs optimizing both objectives.

3.2 Pareto Differential Evolution Algorithm

This paper uses the MOEA described in [5] for training ANN with sigmoid basis functions. The next section briefly explains the this algorithm. For more details about the Base Classifier Framework or Fitness Functions, see [5].

In this paper, we use one of the most prominent Multi-Objective Evolutionary Algorithms in the bibliography. This algorithm is the MPDENN (Memetic Pareto Differential Evolution Neural Network) algorithm developed by R. Storn and K. Price in [18], modified by H. Abbass to train neural networks [1] and adapted for C and MS [7]. The fundamental bases of this algorithm are Differential Evolution (DE) and the concept of Pareto dominance.

The main feature of the MPDENN algorithm is the inclusion of a crossover operator together with the mutation operator. The crossover operator is based on a random choice of three parents, where one of them (main parent) is modified using the weighted difference of the two other parents (secondary parents). The child generated by the crossover and mutation operator is included in the population if it dominates the main parent, if it has no relationship with him or if it is the best child of the rejected children. At the beginning of each generation, individuals dominated are eliminated from the population. A generation of the evolutionary process ends when the population has been completed. In three generations of the evolution (the first initially, the second in the middle and the third at the end), a local search algorithm is applied to the most representative individuals of the population. The local search algorithm used by MPDENN algorithm is *iRprop*$^+$ [12] (more details in [5]).

3.3 Automatic Selection Method Used in the Experimentation

Once the execution of MPDENN algorithm ends, various automatic selection methods of individuals are used for each run:

- **MPDENN-E:** It consists of choosing the Pareto upper extreme value in training, that is, the best individual in Entropy (E), because one of the fitness function of the MOEA is E. This method is described in [8].
- **MPDENN-MS:** This technique is similar to the previous one, but selecting the best individual in MS, i.e., the individual at the Pareto lower extreme. This method is described in [8].

- **MPDENN-CC:** This method selects all individuals from the first and second Pareto front obtained with the MPDENN algorithm. This group of individuals is divided into two subgroups by a 2-means algorithm (because there are two objective functions, C and MS). The individual that is closest to the centroid of the upper cluster (cluster that takes the C measure into account) is selected.
- **MPDENN-CMS:** This automatic method works in a similar way to the MPDENN-CC automatic method, but in this case, the individual that is closest to the centroid is selected taking the MS measure into account (lower cluster).

  Individuals selected by MPDENN-CC and MPDENN-CMS are considered the most representative individuals in the population (the fact that these individuals do not have the greatest value in any objective does not mean that they do not generalize well).

  We decided to include the second Pareto front in the clustering process, in order to expand the number of individuals and to increase diversity. In addition, individuals belonging to this front may have a high percentage of classification in generalization because it is a way to avoid over-training. In the extreme case that is only one individual in each of the fronts (there would be only two individuals), each of these individuals will be assigned to a cluster.
- **MPDENN-MV [19]:** *Majority Voting* (MV) is an ensemble technique that uses all individuals in the first Pareto front. With this technique, a pattern belong to the class that has the higher number of votes, according to the independent classification of each of the elements that make up the ensemble. To estimate the a posteriori probability of a pattern to belong to a class, the average of the output probabilities of the models who voted for this class are used. This is performed for each pattern in the training of generalization dataset so that a probability matrix is formed to obtain the $RMSE$ measure (Root Mean Square Error).
- **MPDENN-SA [19]:** The *Simple Averaging* (SA) ensemble technique uses the first Pareto front to calculate for each pattern the arithmetic mean of the probability for each Q class for each of the models in the ensemble. The assignment will take the class that has the highest average probability. For the case of the $RMSE$ measure, the arithmetic mean of the probabilities is obtained for each output of each model in the ensemble for a particular pattern. Then we use the probabilities of the output with the maximum mean probability for each model of the ensemble for that particular pattern. This is done for each pattern in the training and generalization dataset, and a probability matrix is formed to obtain the $RMSE$ measure.
- **MPDENN-WT [19]:** With the *Winner Take All* (WT) ensemble method, for each pattern the probabilities of the model with the highest probability in one of the outputs are used as the output of the ensemble. This ensemble method uses the individuals in the first Pareto front.

Table 1. Features of the dataset

#Patterns	#Training patterns	#Test patterns	#Input variables	#Classes	#Patterns per class
1003	751	252	64	2	(890,113)

4 Experimental Study

The Experimental design was conducted using a stratified holdout procedure with 30 runs, where approximately 75% of the patterns were randomly selected for the training set and the remaining 25% for the test set. During the creation of these two sets, the proportion of 75-25% for the training-testing patterns for each of the participating hospotals was also kept. Table 1 shows the features of the dataset.

In all the experiments, the population size for MPDENN is established as $M = 25$. The crossover probability is 0.8 and the mutation probability is 0.1. For *iRprop*$^+$ as local search algorithm, the adopted parameters are $\eta^+ = 1.2$, $\eta^- = 0.5$, $\Delta_0 = 0.0125$ (the initial value of the Δ_{ij}), $\Delta_{\min} = 0$, $\Delta_{\max} = 50$ and $Epochs = 10$, see [12] for the *iRprop*$^+$ parameter description.

To start processing data, each one of the input variables was scaled in the ranks $[-1.0, 1.0]$ to avoid the saturation of the signal. Addition, categorical variables have been transformed into many binary variables as possible category.

5 Results

The C and the $RMSE$ represent two most often used metrics in classification [3]. In our paper, we use these two metrics together with the MS. Table 2 presents the values of mean and Standard Deviation (SD) for C, MS and $RMSE$ in generalization in 30 runs of all the experiments performed. The analysis of the results leads us to conclude that the MPDENN-E obtained the best performance in the dataset considering C_G and the second best value in $RMSE_G$. The best result in $RMSE_G$ is obtained by the MPDENN-CC. For MS_G, the MPDENN-MS obtained the best results in the analyzed dataset. From this analysis, we can consider that the best method for C_G and $RMSE_G$ is the MPDENN-E while it is MPDENN-MS for MS_G.

The best models obtained by MPDENN-E and MPDENN-MS methods are shown in Table 2. The best MPDENN-E model has a high value on $C = 89.29$ and $RMSE = 0.3212$, while the best MPDENN-MS model produces a very acceptable value on $MS = 62.07$. The confusion matrix for the best MPDENN-E model in generalization is: True Positive (TP)=221, False Negative (FN)=2, False Positive (FP)=25, True Negative (TN)=4; and for the best MPDENN-MS model is: TP=143, FN=80, FP=11, TN=18.

These results suggest that a combination of the two models extremes of the Pareto front would provide a useful tool for the problem of donor-recipient assignment. This combination could be a rule-based system or a weighted aggregation

Table 2. Statistical results for different methods in generalization

Method	$C_G(\%)$ Mean ± SD	$MS_G(\%)$ Mean ± SD	$RMSE_G$ Mean ± SD
MPDENN-E	**88.34 ± 0.68**	1.15 ± 3.98	*0.3282 ± 0.0071*
MPDENN-MS	60.66 ± 3.04	**50.39 ± 6.56**	0.4492 ± 0.0763
MPDENN-CC	88.24 ± 0.66	1.03 ± 4.54	**0.3261 ± 0.0068**
MPDENN-CMS	60.71 ± 3.14	*50.23 ± 8.48*	0.4160 ± 0.0561
MPDENN-MV	68.33 ± 8.49	27.35 ± 11.11	0.3583 ± 0.0184
MPDENN-SA	84.26 ± 3.87	8.97 ± 7.72	0.3454 ± 0.0122
MPDENN-WT	*88.25 ± 0.67*	0.69 ± 2.29	0.3581 ± 0.0733
Method	$C_G(\%)$	$MS_G(\%)$	$RMSE_G$
Best MPDENN-E model	89.29	13.79	0.3212
Best MPDENN-MS model	63.89	62.07	0.3863

The best result is in **bold** face and the second best result in *italics*.

of the outputs of both models, although in our opinion, the rules-based system would provide a more understandable and comprehensible tool for experts. The system would receive as input a set of potential recipients and form a donor-recipient pair between each of them and donor/organ data. These pairs would be the input for these neural network models. With the results provided by these models and using a simple set of rules, the system would determine which of the recipients receiving the organ.

6 Conclusions

With the study presented in this paper, we obtain some artificial neural networks models that can help medical experts in the donor-recipient allocation. These models are obtained by a multi-objective evolutionary algorithm where Accuracy is the measure considered to evaluate model performance along with the Minimum Sensitivity measure. Minimum Sensitivity is used to avoid the design of models with high global performance but bad performance when considering the classification rate for each class (survival or not-survival).

With the two best models (obtained by MPDENN-E and MPDENN-MS methods), a rule-based system can be used to perform the matching between donor and recipient. This rule-based system must be generated by the medical experts and machine learning experts, for maintaining the principles of justice, efficiency and equity. The current allocations systems are based on thee risk of death on the waiting list and do not recognize distinctions in "donor organ quality". With the rule-based system, the "donor organ quality" would be taken into account to improve the allocation and ensure the survival of recipients.

References

1. Abbass, H.A., Sarker, R., Newton, C.: PDE: a Pareto-frontier differential evolution approach for multi-objective optimization problems. In: Proceedings of the 2001 Congress on Evolutionary Computation, Seoul, South Korea, vol. 2 (2001)

2. Bishop, C.M.: Neural networks for pattern recognition. Oxford University Press, Oxford (1996)
3. Caruana, R., Niculescu-Mizil, A.: Data mining in metric space: An empirical analysis of supervised learning performance criteria, pp. 69–78 (2004)
4. Coello Coello, C., Lamont, G., Veldhuizen, D.: Evolutionary Algorithms for Solving Multi-Objective Problems, 2nd edn. Springer, Heidelberg (September 2007)
5. Cruz-Ramírez, M., Sánchez-Monedero, J., Fernández-Navarro, F., Fernández, J., Hervás-Martínez, C.: Memetic pareto differential evolutionary artificial neural networks to determine growth multi-classes in predictive microbiology. Evolutionary Intelligence 3(3-4), 187–199 (2010)
6. Dvorchik, I., Subotin, M., Marsh, W., McMichael, J., Fung, J.: Performance of multi-layer feedforward neural networks to predict liver transplantation outcome. Methods Inf. Med. 35, 12–18 (1996)
7. Fernández, J.C., Hervás, C., Martínez, F.J., Gutiérrez, P.A., Cruz, M.: Memetic pareto differential evolution for designing artificial neural networks in multiclassification problems using cross-entropy versus sensitivity. In: Corchado, E., Wu, X., Oja, E., Herrero, Á., Baruque, B. (eds.) HAIS 2009. LNCS, vol. 5572, pp. 433–441. Springer, Heidelberg (2009)
8. Fernández, J.C., Martínez-Estudillo, F.J., Hervás-Martínez, C., Gutiérrez, P.A.: Sensitivity versus accuracy in multiclass problems using memetic Pareto evolutionary neural networks. IEEE Trans. on Neural Networks 21(5), 750–770 (2010)
9. Fisher, R.A.: The use of multiple measurements in taxonomic problems. Annals of Eugenics 7(7), 179–188 (1936)
10. Furness, P.N., Levesley, J., Luo, Z., Taub, N., Kazi, J., Bates, W., Nicholson, M.: A neural network approach to the biopsy diagnosis of early acute renal transplant rejection. Histopathology 35(5), 461–467 (1999)
11. Haykin, S.: Neural Networks: A comprehensive Foundation, 2nd edn. Prentice Hall, Upper Saddle River (1998)
12. Igel, C., Hüsken, M.: Empirical evaluation of the improved rprop learning algorithms. Neurocomputing 50(6), 105–123 (2003)
13. Kondo, T.: Evolutionary design and behavior analysis of neuromodulatory neural networks for mobile robots control. Appl. Soft Comput. 7, 189–202 (2007)
14. Löfström, T., Johansson, U., Boström, H.: Ensemble member selection using multi-objective optimization. In: IEEE Symposium on Computational Intelligence and Data Mining, pp. 245–251 (2009)
15. Matis, S., Doyle, H., Marino, I., Mural, R., Uberbacher, E.: Use of neural networks for prediction of graft failure following liver transplantation. In: IEEE Symposium on Computer-Based Medical Systems, pp. 133–140 (1995)
16. Saxena, A., Saad, A.: Evolving an artificial neural network classifier for condition monitoring of rotating mechanical systems. Appl. Soft Comput. 7, 441–454 (2007)
17. Sheppard, D., McPhee, D., Darke, C., Shrethra, B., Moore, R., Jurewitz, A., Gray, A.: Predicting cytomegalovirus disease after renal transplantation: an artificial neural network approach. Int. J. Med. Inf. 54(1), 55–76 (1999)
18. Storn, R., Price, K.: Differential evolution. A fast and efficient heuristic for global optimization over continuous spaces. J. of Global Optimization 11, 341–359 (1997)
19. Theodoridis, S., Koutroumbas, K.: Pattern Recognition, 3rd edn. Elsevier, Academic Press (2006)

Studying the Hybridization of Artificial Neural Networks in HECIC*

José del Campo-Ávila, Gonzalo Ramos-Jiménez, Jesús Pérez-García, and Rafael Morales-Bueno

Departamento de Lenguajes y Ciencias de la Computación
E.T.S. Ingeniería Informática, Universidad de Málaga
Málaga, 29071, Spain
{jcampo,ramos,morales}@lcc.uma.es

Abstract. One of the most relevant tasks concerning Machine Learning is the induction of classifiers, which can be used to classify or to predict. Those classifiers can be used in an isolated way, or can be combined to build a multiple classifier system. Building many-layered systems or knowing relation between different base classifiers are of special interest. Thus, in this paper we will use the HECIC system which consists of two layers: the first layer is a multiple classifier system that processes all the examples and tries to classify them; the second layer is an individual classifier that learns using the examples that are not unanimously classified by the first layer (incorporating new information). While using this system in a previous work we detected that some combinations that hybridize artificial neural networks (ANN) in one of the two layers seemed to get high-accuracy results. Thus, in this paper we have focused on the study of the improvement achieved by using different kinds of ANN in this two-layered system.

Keywords: Many-layered learning, Machine Learning, Multiple Classifiers Systems.

1 Introduction

Classification is one of the most relevant activities in Machine Learning and the induced models: apart from showing interesting information that is useful for the expert, it can be used to predict future examples that were not previously seen. Many different systems based on a wide variety of approaches have been proposed for a long time [1]. One of those approaches is based on the multiple classifier systems, also known as ensembles [2], which combine multiple individual classifiers (base classifiers) to get a joint classification.

The process to induce the base classifiers that constitute the ensemble is very important in order to get high-quality multiple classifier systems because getting accurate and diverse base classifiers has been revealed as a significant point [3]. Many researchers have been working on this idea but, recently some other approaches using ensembles

* This work has been partially supported by the SESAAME project number TIN2008-06582-C03-03 of the MEC, Spain.

J. Cabestany, I. Rojas, and G. Joya (Eds.): IWANN 2011, Part II, LNCS 6692, pp. 137–144, 2011.

that satisfy those requirements (accuracy and diversity) have gone one step forward. They have used those ensembles to build many-layered systems [4,5], to get a better performance when the relation between classifiers is known [6].

In our previous work [7] we simplified an existing method [8] that uses multiple layers and reduced their complexity maintaining only two layers: one multiple classifier system and an individual classifier. In that work we observed that some combinations that hybridize artificial neural networks (ANN) in one of the layers seemed to get high-accuracy results. Thus, in this paper we have focused on the study of the improvement achieved by using different kinds of ANN in this two-layer system.

In the next Section we will briefly present such two-layered system where examples not unanimously classified in the ensemble (first layer) pass to the individual classifier (second layer) that tries to solve the discrepancies. In Section 3, we will present some experimental results to show how the new method can improve the performance of isolated ensemble of classifiers, focusing our study on the hybridization with ANN and finally, in Section 4, we will offer some conclusions and future lines of research.

2 HECIC: Hybridizing Ensemble Classifiers with Individual Classifiers

In this section we summarize the basic concepts (Subsections 2.1 and 2.2) necessary to present the method that we are studying. A more detailed version can be found in [7].

2.1 Multiple Classifier Systems

When an ensemble is used to perform a classification (or prediction) task, the main objective is to improve the generalization. This improvement usually relies on different voting methods and many algorithms have been developed [9,10] and different studies have been done about them [11,12]. They can be divided into two main categories: those that change the dataset distribution depending on the previous steps of the algorithm (usually called *boosting algorithms* [10]) and those that do not change the cited distribution (usually known as *bagging algorithms* [10]).

We cite these two categories because they are among the most used ones, and they are the multiple classifier systems used in the experimental section.

2.2 Two Layered System

The basic idea of a multiple layered system [5] is the incorporation of consecutive layers in order to achieve a competitive global system. In the previous work [7] we simplified the architecture of one of those multiple layered systems [8] limiting the number of layers. We combined a multiple classifier system positioned in the first layer with an individual classifier placed in the second layer. One of the main advantages of those methods is the incorporation of new information that passes from the first layer to the second one, when an example is not unanimously classified in the ensemble. In Fig. 1 there is a schematic idea of the process that incorporates new information into a subset of the training set that passes to the second layer.

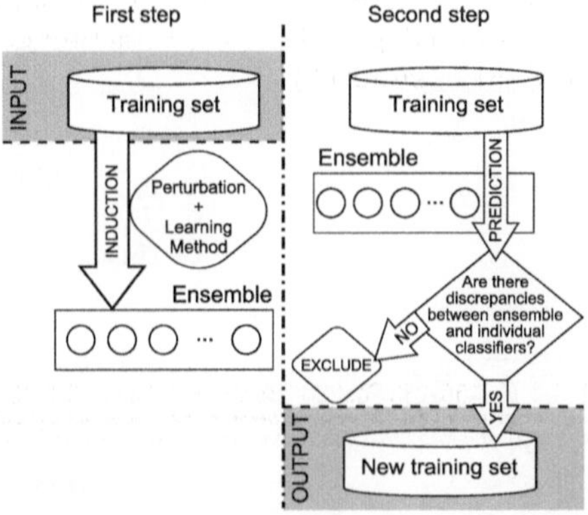

Fig. 1. Creation of the dataset for the second layer incorporating new additional information

The first layer is constituted by an ensemble of classifiers. This ensemble processes every example that is present in the dataset, and delegates a subset of them to an individual classifier that constitutes the second layer. The examples that pass from the ensemble classifier (first layer) to the individual classifier (second and last layer) are those examples that cause discrepancies in the ensemble.

In other words, after inducing the ensemble using the whole training set this set is used again and is checked by the ensemble in order to discover which examples do not get an unanimous classification. These examples pass to the next layer incorporating new information generated by the ensemble: the class estimated by each base classifier in the ensemble and the class estimated by the ensemble itself. The purpose of the individual classifier in the second layer is to learn using only the examples that do not reach a consensus. Those examples could be considered difficult to be evaluated by the ensemble, and we would like to have a second level that induces new concepts and tries to overcome the difficulties arising in the ensemble.

A similar process is used when predicting the class of an observation. It is evaluated by the ensemble and, if the classification made by every base classifier is unanimous, the observation is assigned such class. Otherwise, the observation passes to the individual classifier (including the new information induced by the ensemble) which gives a classification to such observation.

Two schematic diagrams representing the transformation of the examples that pass to the next layer and the prediction process are shown in Fig. 2. In order to make this figure more comprehensible, we have introduced some notation. Examples in the training set for the first layer ($\_{L1}$) have N attributes ($At\_1, At\_2, \ldots, At\_N$) which would have different values depending on the example. Thus, for the example e_{L1}, the value of the first attribute would be $At\_1(e\_{L1})$ and the value of the class would be $Class(e\_{L1})$. In the ensemble there are k base classifiers ($C_1, \ldots, C_k$) that combine their classifications ($C_1 - Class(e\_{L1}), \ldots, C_k - Class(e\_{L1})$) to build the ensemble classification ($Ens\_Class(e\_{L1})$). The examples used in the second layer ($\_{L2}$) are a subset of the

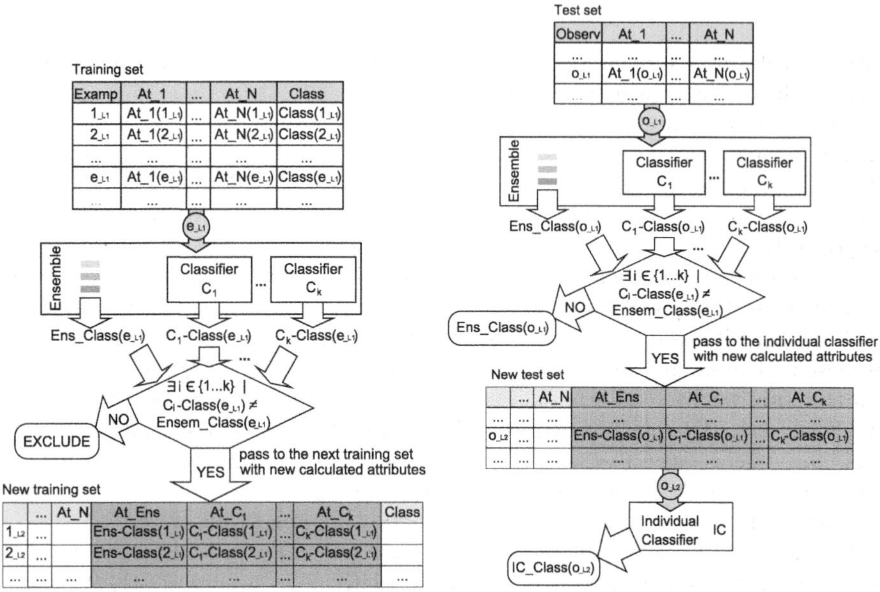

Fig. 2. Left. Details of the process to include new information for the second layer. It corresponds to the second step described in Fig. 1
Right. Details of the process to predict the class of an observation. Prediction is calculated in the first layer if there are no discrepancies, or it is calculated in the second layer when there are discrepancies.

examples used in the previous layer including new information: the classification given by the ensemble and the individual classifiers.

3 Experimental Results

One preliminary version of the system HECIC (Hybridizing Ensemble Classifiers with Individual Classifiers) was tested in a limited way using one ensemble and individual classifier designed by ourselves [7], getting promising results that revealed some kind of relation between Multilayer perceptron and other classifiers. Now we have extended that experimental section with well-known ensembles and classifiers in order to study the repercussion of using artificial neural networks in the different layers.

We have combined different multiple classifier systems like Bagging [10] and Boosting [13] with some individual classifiers: IB1 [14], C4.5 (J48) [15], Naïve bayes (NB) [16] and some artificial neural networks like Multilayer perceptron (MLP) [17], LVQ1 or LVQ3 [18]. Thus, we can show how different combinations interact and we can propose the most promising configurations for both layers. To perform these experiments we have incorporated HECIC to the Weka framework [19] as a metalearner and we have used the implementation of the classifiers implemented in Weka (including the plugin available at `http://wekaclassalgos.sourceforge.net`).

Table 1. Summary table for datasets

Abbr.	UCI name	Examples	Attributes	Type	Classes
BAL	Balance-scale	625	4	nominal	3
CAR	Car	1728	6	nominal	4
KR	KR-vs-KP	3196	36	nominal	2
TIC	Tic-Tac-Toe	958	9	nominal	3
VOT	Votes	435	16	nominal	2
ECO	Ecoli	336	7	numerical	8
ION	Ionosphere	351	33	numerical	2
PIM	Pima-indians	768	8	numerical	2
VEH	Vehicle	846	18	numerical	4
WEB	WDBC	569	30	numerical	2
YEA	Yeast	1484	10	numerical	8

The experiments we have done and the results we have obtained are now exposed. Before we go on to deal with particular experiments, let us explain some issues:

- The datasets that we have studied (11) are summarised in Table 1. It shows the number of examples, the type of attributes, and the number of values for the class. All these datasets have been taken from the UCI Machine Learning Repository and are available online. We have focused on datasets with nominal and numerical attributes in order to see if there is some difference. Nominal datasets have been previously binarised because the implementation of LVQ1 (and LVQ3) can not deal with nominal values.
- In this study we only focus our attention on the accuracy, but other parameters could be considered to make a more extensive analysis. The implementation used to do the experiments is offered by the Weka framework [19] and we have used the default configuration for all the algorithms.
- We have calculated the average values for accuracy, obtained by performing 10 x 10 fold cross-validations. The reference result, the one that we desire to improve, is the accuracy of a single ensemble (without using HECIC). A Wilcoxon test has been conducted (using the statistical package R) with the results of the cited 10 x 10 fold cross-validation. A difference is considered as significant if the significance level of the Wilcoxon test is lower than 0.05. Thus, $\oplus$ indicates that the accuracy is significantly better than the accuracy achieved when HECIC is not used, $\ominus$ signifies that the accuracy is significantly worse and $\odot$ signifies that there is no significant differences. In addition to these comparisons, the best result for each experiment (distinguishing between different ensembles used in the first layer) has been emphasized using numbers in boldface.

Because of space limitation, we can not present the details of every calculated results (we would need 22 tables), so we show in Table 2 one example of the results achieved. It is clear that the results vary between different datasets, but we have selected this dataset because it shows some of the conclusions that we have extracted. In general,

Table 2. Results for the WDBC dataset when the induction of the ensemble is done with bagging

	Bagging in the first layer						
first	**without**	**Second Layer**					
layer	**HECIC**	**IB1**	**J48**	**NB**	**MLP**	**LVQ1**	**LVQ3**
IB1	**95.62**	93.80 $\ominus$	95.48 $\odot$	95.48 $\odot$	95.45 $\ominus$	95.48 $\odot$	96.12 $\oplus$
	±0.25	±0.48	±0.36	±0.32	±0.36	±0.38	±0.29
J48	93.23	**95.80** $\oplus$	94.04 $\oplus$	94.22 $\oplus$	94.17 $\oplus$	94.06 $\oplus$	93.88 $\oplus$
	±0.27	±0.33	±0.37	±0.25	±0.35	±0.32	±0.27
NB	95.33	**96.13** $\oplus$	95.13 $\odot$	95.91 $\oplus$	95.08 $\ominus$	95.25 $\odot$	93.83 $\ominus$
	±0.43	±0.54	±0.72	±0.65	±0.43	±0.66	±0.81
MLP	97.05	96.73 $\ominus$	96.64 $\ominus$	**97.10** $\odot$	96.56 $\ominus$	96.57 $\ominus$	96.31 $\ominus$
	±0.25	±0.33	±0.37	±0.34	±0.40	±0.39	±0.60
LVQ1	92.27	92.00 $\odot$	94.67 $\oplus$	**94.76** $\oplus$	94.13 $\oplus$	95.13 $\oplus$	92.78 $\oplus$
	±0.36	±0.62	±0.41	±0.34	±0.41	±0.49	±0.46
LVQ3	92.46	91.37 $\ominus$	**94.78** $\oplus$	94.67 $\oplus$	94.76 $\oplus$	95.55 $\oplus$	92.85 $\odot$
	±0.34	±0.56	±0.66	±0.63	±0.85	±0.50	±0.76

considering all the datasets, we can summarise the following aspects attending to the impact of artificial neural networks:

- The performance is very different between nominal and numerical datasets. In half of the experiments with nominal attributes, the best results were achieved by the isolated ensembles (without using HECIC). On the other hand, when learning from numerical datasets, the accuracy is improved if HECIC creates a second layer and uses the new created information. Therefore, it seems that HECIC could be more appropriate to be used with numerical datasets.
- the best layer to use MLP is the first one, showing poorer results if it is used as isolated classifier in the second layer. On the other hand, the best place to use LVQ1 is the second layer. But it does not mean that using MLP in the second layer (or LVQ1 in the first one) worsen the results compared with those achieved without using HECIC. For example, with nominal datasets, when using isolated boosting combinations of J48 models (without HECIC), there is a poor performance, but completing it with MLP in the second layer clearly improves the accuracy.
- we have observed that LVQ methods seems to produce few discrepancies when using boosting method, because the accuracy is the same independently of the classifier selected for the second layer.
- when the ensemble is induced with the bagging method, the usage of the same ANN in both layers presents very different results depending on the kind of ANN. If we use MLP in both layers, the performance usually gets worse; but using LVQ1 in both layers generally causes improvements (sometimes relevant).
- when dealing with numerical datasets, the most accurate configuration uses a MLP classifier in the first layer (with bagging or boosting). In the case of bagging ensembles, the best results are obtained with NB in the second layer. If we are inducing ensembles with boosting method, the best classifier to be used in the second layer is J48. To show more clearly this point we have included the Table 3 where we have

Table 3. Ranking achieved in numerical datasets. Results for bagging and boosting are grouped separately to observe differences.

Layer		Bagging							Boosting						
First	Second	ECO	ION	PIM	VEH	WDB	YEA	Rank	ECO	ION	PIM	VEH	WDB	YEA	Rank
MLP	IB1	84,0	90,5	74,4	83,2	96,7	58,3	2	80,9	90,3	70,8	83,1	96,4	56,5	6
LVQ1	IB1	81,8	87,0	72,0	54,2	92,0	50,1	18	81,8	87,0	72,0	54,2	92,0	50,1	13
LVQ3	IB1	81,7	87,7	71,3	54,5	91,4	51,7	17	81,7	87,7	71,3	54,5	91,4	51,7	16
MLP	J48	82,6	90,8	73,2	84,6	96,6	56,3	4	85,0	91,4	75,1	82,3	95,3	59,6	1
LVQ1	J48	84,0	88,1	71,3	65,0	94,7	53,4	12	81,8	87,0	72,0	54,2	92,0	50,1	13
LVQ3	J48	84,3	88,3	71,1	64,6	94,8	53,8	10	81,7	87,7	71,3	54,5	91,4	51,7	16
MLP	NB	85,9	92,5	76,4	83,8	97,1	60,2	1	83,8	90,7	72,7	83,2	96,7	57,3	2
LVQ1	NB	86,5	90,7	74,6	61,3	94,8	58,8	5	81,8	87,0	72,0	54,2	92,0	50,1	13
LVQ3	NB	86,3	90,0	74,6	60,9	94,7	58,9	6	81,7	87,7	71,3	54,5	91,4	51,7	16
IB1	MLP	81,0	87,2	70,4	69,7	95,4	52,7	13	79,0	87,9	67,5	68,2	93,8	51,1	11
J48	MLP	85,0	87,7	75,4	56,7	94,2	57,4	9	84,0	92,3	73,3	45,0	95,4	58,3	4
NB	MLP	83,1	92,5	73,8	74,3	95,1	58,8	4	83,3	93,0	72,4	76,2	96,1	56,6	3
IB1	LVQ1	80,9	87,0	70,3	69,6	95,5	52,6	15	74,9	87,9	72,8	60,4	95,5	40,1	8
J48	LVQ1	85,4	87,9	75,4	56,8	94,1	56,8	8	83,1	91,7	72,7	45,0	92,7	58,0	5
NB	LVQ1	82,0	92,7	72,4	74,4	95,3	56,1	7	71,8	92,8	71,8	56,3	93,0	38,2	12
IB1	LVQ3	79,3	87,5	70,8	66,9	96,1	46,9	14	74,6	88,3	72,2	59,6	95,4	45,5	9
J48	LVQ3	83,8	88,1	75,9	48,6	93,9	54,7	11	82,5	91,7	71,9	45,0	92,7	58,2	7
NB	LVQ3	78,3	92,9	69,1	65,1	93,8	43,0	16	72,2	92,8	71,1	56,4	93,1	46,5	11

summarised the ranking for the different algorithms and datasets. In this table we have not showed all the combinations, we have focused in the usage of ANN in the first or in the second layer (and not simultaneously).

4 Conclusion

In this paper we study how some artificial neural networks works when using the HECIC system: a method that improves the performance of multiple classifier systems by including a second step that tries to solve the discrepancies that can arise in that multiple classifier system. It uses new information created by the ensemble itself to create an individual classifier in a second layer.

We have carried out an extensive experimental section where we have observed some characteristics and we have identified different configurations of HECIC with artificial neural networks that shows a good performance. The most appropriate layer to use Multilayer perceptrons is the first one, while the second layer is preferable when using LVQ1. Furthermore, combining a Multilayer perceptron in the first layer with a Naïve bayes in the second layer seems to be the best option (specially if bagging is being used).

Although we have tested different configurations on diverse datasets, we would like to find some possible characterisation of the kind of problems that best fit with the different types of classifiers. Additionally we also want to study how this approach can improve the performance in the presence of missing values.

References

1. Kononenko, I., Kukar, M.: Machine Learning and Data Mining: Introduction to Principles and Algorithms. Horwood Publishing (2007)
2. Dietterich, T.G.: Ensemble methods in machine learning. In: Kittler, J., Roli, F. (eds.) MCS 2000. LNCS, vol. 1857, pp. 1–15. Springer, Heidelberg (2000)
3. Hansen, L.K., Salamon, P.: Neural network ensembles. IEEE Transactions on Pattern Analysis and Machine Intelligence 12, 993–1001 (1990)
4. Gama, J., Brazdil, P.: Cascade generalization. Machine Learning 41, 315–343 (2000)
5. Utgoff, P.E., Stracuzzi, D.J.: Many-layered learning. Neural Computation 14(10), 2497–2529 (2002)
6. Chindaro, S., Sirlantzis, K., Fairhurst, M.C.: Modelling multiple-classifier relationships using bayesian belief networks. In: Haindl, M., Kittler, J., Roli, F. (eds.) MCS 2007. LNCS, vol. 4472, pp. 312–321. Springer, Heidelberg (2007)
7. Ramos-Jiménez, G., del Campo-Ávila, J., Morales-Bueno, R.: Hybridizing ensemble classifiers with individual classifiers. In: International Conference on Intelligent Systems Design and Applications. Workshop on Hybrid Learning for Artificial Neural Networks: Architectures and Applications, pp. 199–202. IEEE Computer Society, Los Alamitos (2009)
8. Ramos-Jiménez, G., del Campo-Ávila, J., Morales-Bueno, R.: ML-CIDIM: Multiple layers of multiple classifier systems based on CIDIM. In: Ślęzak, D., Yao, J., Peters, J.F., Ziarko, W.P., Hu, X. (eds.) RSFDGrC 2005. LNCS (LNAI), vol. 3642, pp. 138–146. Springer, Heidelberg (2005)
9. Freund, Y.: Boosting a weak learning algorithm by majority. Information and Computation 121, 256–285 (1995)
10. Breiman, L.: Bagging predictors. Machine Learning 24(2), 123–140 (1996)
11. Aslam, J.A., Decatur, S.E.: General bounds on statistical query learning and pac learning with noise via hypothesis boosting. Information and Computation 141, 85–118 (1998)
12. Bauer, E., Kohavi, R.: An empirical comparison of voting classification algorithms: bagging, boosting, and variants. Machine Learning 36, 105–142 (1999)
13. Freund, Y., Schapire, R.E.: Experiments with a new boosting algorithm. In: Proceedings of the 13th International Conference on Machine Learning (ICML-1996), pp. 146–148 (1996)
14. Aha, D.W., Kibler, D., Albert, M.K.: Instance-based learning algorithms. Machine Learning 6, 37–66 (1991)
15. Quinlan, J.R.: C4.5: Programs for machine learning. Morgan Kaufmann, San Francisco (1993)
16. John, G.H., Langley, P.: Estimating continuous distributions in bayesian classifiers. In: Proceedings of the Eleventh Annual Conference on Uncertainty in Artificial Intelligence (UAI-1995), pp. 338–345. Morgan Kaufmann, San Francisco (1995)
17. Gill, P.E., Murray, W., Wright, M.H.: Practical optimization. Academic Press, London (1981)
18. Kohonen, T.: Self-organizing maps. Springer-Verlag, New York, Inc. (1997)
19. Hall, M., Frank, E., Holmes, G., Pfahringer, B., Reutemann, P., Witten, I.H.: The WEKA data mining software: An update. In: SIGKDD Explorations, vol. 11, pp. 10–18 (2009)

Processing Acyclic Data Structures Using Modified Self-Organizing Maps*

Gabriela Andrejková and Jozef Oravec

Faculty of Science
P. J. Šafárik University in Košice, Slovakia
{gabriela.andrejkova, jozef.oravec}@upjs.sk

Abstract. The paper deals with Acyclic Graph Data Structures (AGDS) and with model of a self-organizing map (SOM) that has been modified for processing of AGDS. The motivation was found in the real world of the Academic Information System (AIS) at P. J. Šafárik University in Košice. To the modified SOM Neural Network (SOM NN), there are added contexts and counters which are built in a training phase of the neural network. The trained SOM NN in active phase can compute more information which is used to built an answer to some questions. The working application was tested on the study programs in informatics, the test results are very closed to the real values.

1 Introduction

Graphs are very important models of data structures, they give possibilities to represent facts in vertices and relations among facts in edges. The graphs as data structures are used in many applications, for example image analysis, scene description, natural language processing. In some applications, the structured data are often connected to a learning process from examples. The graph $G = (V, E)$, where V is the finite set of vertices (facts) $|V| = n$ and E is the set of edges $|E| \leq n^2$ should be represented by matrix of type $n \times n$. The problems, which we can see here, is the following: for each example - graph we can have a different type of matrix; if the biggest type of the matrix is used for all graphs then there are too many free positions, which do not represent anything. It means, it is important to find some other representations of graphs and it is necessary to do it in a connection to a model of computation.

SOM NN have been proposed by many authors [1], [2], [3], [9], [11], as models which are very good for a learning of graph data structures. It means, the graph structures in connection to neural networks should be represented in some special representation (a trained neural network) and the neural networks can be trained to graph structures. In the training, it is possible to prepare input data of the graph in some parts, it is not necessary to put full graph as one input data to the network. In the paper [10], we have prepared theoretical points of view to the presented application of using SOM neural networks to acyclic data structures.

* Supported by the Slovak Scientific Grant Agency VEGA, Grant No. 1/0035/09.

J. Cabestany, I. Rojas, and G. Joya (Eds.): IWANN 2011, Part II, LNCS 6692, pp. 145–152, 2011.

The paper is prepared in the following sections. In the first section we describe the motivation to the study of the problem, the second section is oriented to a description of acyclic data structures, in the third section we describe the model of prepared neural network. The next section contains a description of a data encoding and a training algorithm of modified SOM NN. In the last section, there is evaluated the prepared solution in some examples. In the conclusion, we summarize the work which was done and inform about a next work in the area.

2 Motivation - Prerequisites in Study Programs

Many study programs at universities are built in the style that they have compulsory subjects, compulsory elective and elective subjects. From the theoretical

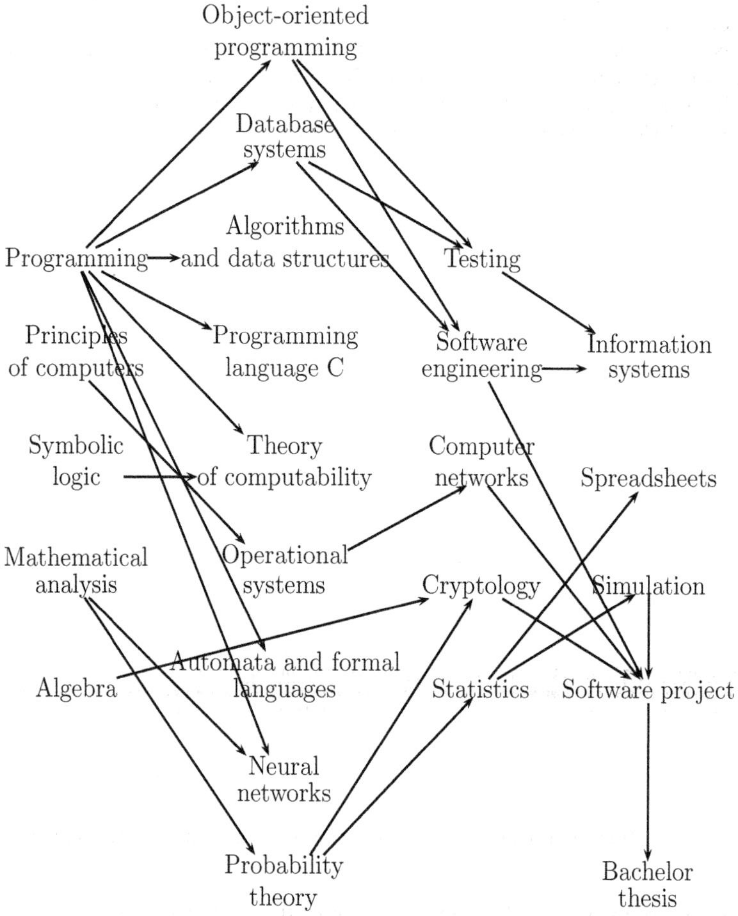

Fig. 1. Mapping of prerequisites for study program in informatics

point of view, all subjects built a partially ordered set according to the order of subject passings.

The partial ordering is defined by prerequisites of subjects. The part of study program in informatics at Faculty of Science of P. J. Šafárik university in Košice is mapped in the picture Pic. 1. At the beginning of academic year, the students choose new subjects which they will study in the current academic year. They register the chosen subjects in an Academic Information System (AIS). For example, a registration of the subject *Software engineering* can be taken after passing subjects *Object-oriented programming* and *Database systems.*

The main problem: The system has information about the structure of all subjects in the study program (some acyclic graph structure). The information about passed subjects of the student can be put to the system in two ways:

- The information is in the list of passed subjects in AIS. It is possible to obtain the acyclic structure of all student's passed subjects in automatic way.
- The way is not automatic. The student will give only some passed subjects. We will follow this way.

When the student gives some passed subjects and he chooses some subject, which he try to register for the next academic year, the system should answer him automatically *NO (you can not to register the subject, you have not fulfilled prerequisites)* in real case or it can answer *YES (it is O.K.)*.

3 Acyclic Data Structures

A directed graph or digraph is a pair $G = (V, E)$, where V is a set, whose elements are called vertices or nodes, E is a set of ordered pairs of vertices, called directed edges.

A Directed Acyclic Graph (DAG), is a directed graph with no directed cycles. That is, it is formed by a collection of vertices and directed edges, each edge connecting one vertex to another, such that there is no way to start at some vertex v and follow a sequence of edges that eventually loops back to v again.

We used the DAG to define the study program. The subjects are in the vertices and the oriented edges represents prerequisites. If the subject A is the prerequisite of the subject B then in DAG is the oriented edge from A to B. The subject A is called **right prerequisite** or the **first level prerequisite** to B. If C is the right prerequisite to A, the C is the **second level prerequisite** to B.

If we choose some vertices from DAG together with the oriented edges among them then we get a set of graphs (one or more subgraphs) and the previous information in the DAG is reduced. The reduced information is enough sometimes to compute the result, but in more cases it is not enough.

4 Modification of Self-Organizing Neural Networks

We describe a model of a neural network which has been used for the computation of the answers of AIS to given questions on passing of subjects. Our model was motivated by the model of the neural network working with tree structured data [5], [7].

The SOM for Structured Data (SOMSD) has been described for processing of labeled trees with fixed fan-out k [4], [8]. It is assumed that the neurons are arranged in rectangular d-dimensional lattice structure. Each neuron can be enumerated by tuples $\bar{i} = (i_1, \ldots, i_d)$ in $\{1, \ldots, N_1\} \times \ldots \times \{1, \ldots, N_d\}$ where $N_i \in \mathcal{N}$. The vector $\bar{i}$ describes the position of the neuron in the lattice. Let $N = N_1 * \ldots * N_d$, N be the number neurons in the lattice, n is a dimension of an input. Each neuron $n_{\bar{i}}$ is equipped with a weight $w_{\bar{i}}$ and k context vectors $c_1^{\bar{i}}, \ldots, c_k^{\bar{i}}$ in $\mathcal{R}^d$. They represent the contexts of proceed tree given by the k subtrees and the indices of their respective winners. The winning neuron with index $I(t)$ to given a tree t with root label a and subtrees $t_1, \ldots, t_k$ is recursively defined by

$$I(\epsilon) = (-1, \ldots, -1),$$

$$I(a(t_1, \ldots, t_k)) = argmin_i\{\bar{i}|\alpha\cdot\|a - w_{\bar{i}}\|^2 + \beta\cdot\|I(t_1) - c_1^{\bar{i}}\|^2 + \ldots + \beta\cdot\|I(t_k) - c_k^{\bar{i}}\|^2\}.$$

$\alpha, \beta > 0$ are constants to control mediation between the amount of pattern match versus context match. The empty tree ϵ is represented by an index not contained in the lattice, $(-1, \ldots, -1)$. The weights of the neurons consist of compact representations of trees with prototypical labels and prototypical winner indices which represent the subtrees. Starting at the leaves, the winner is recursively computed for the entire tree. The index $\bar{i}$ of the winner for the subtrees is computed. The attached weights $w_{\bar{i}}, c_1^{\bar{i}}, \ldots, c_k^{\bar{i}})$ are moved into the direction $(a, I(t_1), \ldots, I(t_k))$ after each recursive processing step, where $I(t_i)$ denotes the winning index of the subtree t_i of the currently processed part of the tree. The neighbor neurons are updated into the same direction with smaller learning rate.

The objective of a neural network learning algorithm is to find a set of weights that minimizes the error function on the given data set of input-output samples $(\mathbf{x}(t), \mathbf{d}(t))$, $t = 0, \ldots, T$,

$$\mathcal{E} = \sum_{t=1}^{T} E(t) \quad , \quad E(t) = \frac{1}{2}\sum_{i=1}^{l} e_i(t)^2 \quad , \tag{1}$$

with $e_i(t) = d_i(t) - y_i(t)$ being the difference between the desired and calculated value of the i-th neuron. The recurrent neural networks attempt to acquire the dynamics of the system, and therefore input-output pairs should be presented in causative order.

The modified model is drawn in the picture Pic. 2. The described model can be divided into three part:

A. **The lattice of neurons, contexts and counters**. The network is trained for the acyclic structure of the study program subjects. In the process are

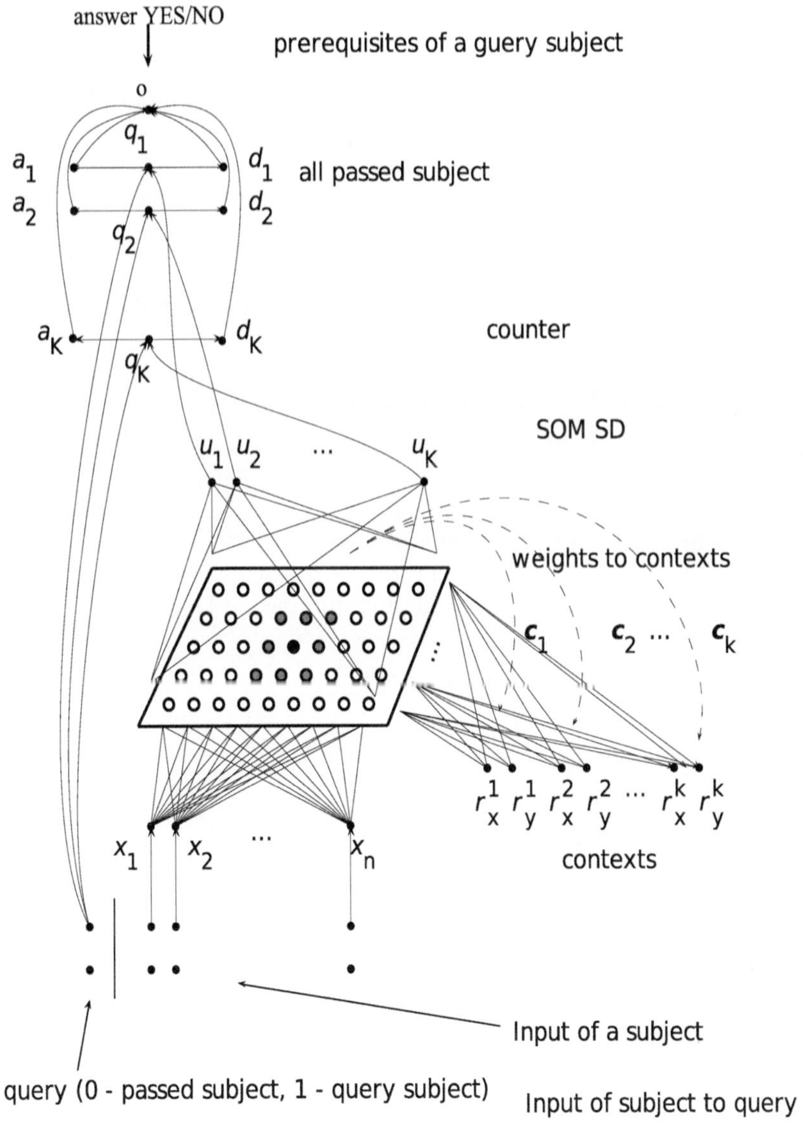

Fig. 2. The structure of the modified neural network

built contexts for prerequisites of subjects and counters for all prerequisites of a given subject

The contexts are $r_x^1, r_y^1, r_x^2, r_y^2, \ldots, r_x^k, r_y^k$. The weights $\mathbf{c_1}, \mathbf{c_2}, \ldots, \mathbf{c_k}$ are the weights to contexts. The counters are $u_1, u_2, \ldots u_K$ and each counter contains all prerequisites of one subject.

B. **Input part** - vector $\overline{x} = (x_1, \ldots x_n)$ is important in the training of the network for the acyclic data structure of the study program and in the active phase of computation, it is necessary to add one bit of information about passing the subject.

C. **Output part** is built from the vectors $\overline{a} = (a_1, a_2, \ldots, a_K), \overline{q} = (q_1, q_2, \ldots, q_K)$ and $\overline{d} = (d_1, d_2, \ldots, d_K)$. The $(0, 1)$ - vector $\overline{a}$ represents the passed subjects given on input (given by student), the $(0, 1)$ - vector $\overline{d}$ explain if the subject has prerequisites or not. From the vectors $\overline{a}, \overline{q}$ and $\overline{d}$ is computed result if the student can register the queried subject. The vector $\overline{q}$ is computed using counters and one bit information in the input.

5 Encoding of Data Structures and the Training of the Neural Network

5.1 Encoding of Training Structures

Let K be the number of subjects in the acyclic data structure of the study program. The subjects are encoded in the binary alphabet $\{0, 1\}$, it means each subject $\mathbf{p^i} = [p_1^i, p_2^i, \ldots, p_b^i] \in \{0, 1\}^b, 1 \leq i \leq K$, where $b = log_2 K + 1$ is the numbers of bits used to the encoding of each subject.

The preparation of training structures — to each subject was prepared the structure with all prerequisites on the first and second level.

5.2 Training of the Neural Network

The distance of input $\mathbf{x}$ from the neurons in the lattice and in connection to the contexts was computed according the formula (2).

$$d_i = \alpha \left\|\mathbf{x} - \mathbf{w}_i\right\|^2 + \beta \left(\left\|\mathbf{r}^1 - \mathbf{c}_i^{(1)}\right\|^2 + \ldots + \left\|\mathbf{r}^k - \mathbf{c}_i^{(k)}\right\|^2 \right), \qquad (2)$$

where the constants $\alpha, \beta > 0$. Let the winner is i^*. Then the adaptation of weights was done according the formulas (3)

$$\nabla \mathbf{w}_j = \eta(t) h(i^*, j)(t) \left\|\mathbf{x} - \mathbf{w}_j\right\|^2, \text{ where } h(t) = e^{-\frac{\|dist(i^*, j)\|}{2\sigma^2(t)}}, \qquad (3)$$

and

$$dist(i^*, j) = \|\mathbf{pos}(i^*) - \mathbf{pos}(j)\| = \sqrt{(i_x^* - j_x)^2 + (i_y^* - j_y)^2}.$$

The output layer contains K Grossberg's units with weights $\mathbf{u}_i, i = 1, \ldots, K$. The counter propagation was applied for training $\mathbf{u}_i$.

The used real values in the training: the dimension of lattices from 5×5 to 15×15: parameters α and β in interval $0.05 - 1.0$; the number of iterations 3000; the starting initial weights were chosen randomly from the interval $\langle -10; 10 \rangle$; the learning rate 0.1. The basic information on training neural networks can be found in [6].

The space complexity of the network is $n^2 \log_2 K + 2n^2 K + 6K$, where K is the number of vertices in an acyclic data structure and n^2 is the lattice dimension. The time complexity of the training is $l_s * n^2 * \log_2 K$, where l_s is the number of training cycles.

5.3 Encoding of Queries

The query (input to the neural network) is prepared in the following way:

- a is the number of all subjects in the query;
- the subjects in the query are prepared as a pairs $(\mathbf{p}_l, t_l)$, where ($\mathbf{p}_l$ is the code of subject and t_l) is one bit (for passed subject it contains the value 1, for not passed the value 0).

6 Results in the Application

In the test of prepared application we did many tests for the study of a network behavior in the comparison to the real values. The student can put in the input a set of his passed subjects (SPS) and then he put the subject Q which he would like to register for study. We analyzed the behavior of the network for the different structures of SPS. The set SPS is the set from the student point of view but in the acyclic structure of study program the subjects have their positions. The student do not put full information, it means, it is interesting to compare real information following from the study program to the information computed by network. We describe some situations:

- $|SPS| = 0, SPS = \emptyset$.
  If the student put on input any passed subjects and he put only the subject Q which he would like to register, the answer of the network is depending only on prerequisites of the subject Q. The network computed the correct values YES only for subjects without prerequisites. In the opposite case the network gave answer NO.
- $|SPS| = 1, SPS = \{A\}$.
  If the student put on input only one passed subject A and then he put the subject Q which he would like to register, the answer of the network is depending only on prerequisites of the subject A and if A is the prerequisite of Q. The network computed the correct values YES only if A is without prerequisites and A is prerequisite of Q. In the opossite case the network computed answer NO.
- $|SPS| = 2, SPS = \{A, B\}$.
  In the case that A and B are not right prerequisites of Q the case is reduced to the case $|SPS| = 0$. If one of A and B is not right prerequisite of Q then we get the case $|SPS| = 1$. If both A and B are right prerequisites of Q then the network computed answer YES if both A and B are without prerequisites, in opposite case the value NO.

We were interested mainly in the answers to the following queries:

- **Query 1:** I have passed the set of subjects $SPS = \{A, B, C\}$. Can I pass the subject Q?
- **Query 2:** I have passed the subject A. Which subjects (with the prerequisite A) could I register?

In the answer to the first query, it was necessary to analyze the acyclic substructures which can be constructed from four vertices. The best results were computed for the connected structure of four vertices, the worst results for isolated vertices.

The answer to the second question depends on the level of prerequisite to A. The trained network can answer to the query 2 very successful.

7 Conclusion

In the paper, we described the modified model of the SOM neural network which could be used in the application of an academic information system. The tests for starting data structures we evaluated and results give quite good starting point to the following work. The plan to the following activities is to continue in the work with an analysis of different sets SPS they have their positions in the acyclic data structure and to compare real values for query Q to computed values by SOM NN.

References

1. Frasconi, P.M., Gori, M., Sperduti, A.: A general framework of adaptive processing of data structures. IEEE-NN 9(5), 768–786 (1998)
2. Hagenbuchner, M., Sperduti, A., Tsoi, A.C.: A self-organizing map for adaptive processing of structured data. IEEE Transactions on Neural Networks 14(3), 491–505 (2003)
3. Hagenbuchner, M., Tsoi, A.C.: A supervised self-organizing map for structures. In: Proceedings IEEE International Joint Conference on Neural Networks, vol. 3, pp. 1923–1928. IEEE, Los Alamitos (2004)
4. Hagenbuchner, M., Tsoi, A.C., Sperduti, A.: A supervised self-organising map for structured data. In: Proc. WSOM 2001: Advances in Self-Organizing Maps, pp. 21–28. Springer, Heidelberg (2001)
5. Hammer, B., Micheli, A., Sperduti, A., Strickert, M.: A general framework for unsupervised processing of structured data. Neurocomputing (57), 3–35 (2004)
6. Haykin, S.: Neural networks: a comprehensive foundation, 2nd edn. Prentice-Hall, New Jersey (1999)
7. Sperduti, A.: Tutorial on neurocomputing of structures. In: Cloete, I., Zurada, J.M. (eds.) Knowledge-Based Neurocomputing, pp. 117–152. MIT Press, Cambridge (2000)
8. Sperduti, A.: Neural networks for adaptive processing of structured data. In: Dorffner, G., Bischof, H., Hornik, K. (eds.) ICANN 2001. LNCS, vol. 2130, pp. 5–12. Springer, Heidelberg (2001)
9. Scarselli, F., Gori, M., Tsoi, A.C., Hagenbuchner, M., Monfardini, G.: The Graph Neural Network Model. IEEE Transactions on Neural Networks 20(1), 61–80 (2009)
10. Andrejková, G., Oravec, J.: Rekurzívne neurónové siete a dátové štruktúry (Recursive neural networks and data stuctures). In: Proceedings of International Congress IMEM 2009, Catholic University, Ružomberok (2009)
11. Vančo, P., Farkáš, I.: Experimental comparison of recursive self-organizing maps for processing tree-structured data. Neurocomputing 73(7-9), 1362–1375 (2010)

On the Performance of the μ-GA Extreme Learning Machines in Regression Problems

A. Paniagua-Tineo, S. Salcedo-Sanz, E.G. Ortiz-García, J. Gascón-Moreno, B. Saavedra-Moreno, and J.A. Portilla-Figueras

Department of Signal Theory and Communications, Universidad de Alcalá, 28871 Alcalá de Henares, Madrid, Spain
sancho.salcedo@uah.es

Abstract. In this paper we carry out an statistical study of the performance of the μ-GA ELM algorithm in regression problems. Up until now, the performance of the the μ-GA ELM have not been characterized, and only a traditional evolutionary ELM have been proposed in the literature, and tested in synthetic problems. In this paper we analyze the performance of the μ-GA ELM in small 1-dimensional problems, where our results agree with the ones in previous works in the literature, and also in large real problems, where we will show that the behavior of the algorithm is worse in many cases than that of the ELM, what is a completely novel result.

1 Introduction

Extreme Learning Machines (ELM) are a class of learning scheme for neural networks, recently proposed in the literature [1], and characterized by produce an extreme fast training, with a reasonable performance in terms of accuracy. ELMs have been defined for classification [2] and regression problems [3], though in this paper we consider only the regression approach.

Briefly, an ELM is basically a learning scheme which is applied to feed forward neural networks, usually formed by an input layer, a hidden layer and an output neuron. Each neuron in the hidden and output layers is defined as a traditional neuron node, with a given activation function defined by the user. In the standard ELM algorithm the weights between the input and the hidden layer are randomly set, whereas the corresponding weights between the hidden layer and the output neuron are then calculated by means of the Penn Rose-Moore inverse matrix of the *hidden layer output matrix* $\mathbf{H}$ (output of the different neurons in the hidden layer for all the input vectors). This easy idea provides a extremely fast training algorithm (since weights in the input layer are random numbers, and the calculation of the Penn Rose-Moore inverse matrix is not a time-consuming process). Note that the number of neurons in the hidden layer is a parameter of the network that must be also set by the user.

Several attempts to improve the performance of ELMs while keeping low its training time have been done in the literature. Specifically, in regression problems there are several works in the literature dealing with methods to improve the

J. Cabestany, I. Rojas, and G. Joya (Eds.): IWANN 2011, Part II, LNCS 6692, pp. 153–160, 2011.

ELM performance, such as with two-steps methods [3], constructive hidden layer nodes [4], novel approaches to calculate the inverse of matrix **H** [5], encoding of a-priori information of the input data [6] and also evolutionary approaches [7], which is the method we further study in this paper.

The idea of including an evolutionary-type algorithm to improve the performance of the ELM was first proposed in [7] for classification and regression problems, though the regression experimental part of the paper where limited to one-dimensional benchmark functions. In [7] it was shown that evolutionary algorithms working on the search space formed by the input weights of the ELM can improve its performance. This argument has been further explored for classification problems in [8], but for regression problems there are still some points that need clarification: the performance of a micro-evolutive algorithm (just a very few generations to modify the weights) is one of these points. Also, the performance of the evolutionary ELM in real applications is of major interest, and have not been completely described in the literature.

In this paper we carry out a comprehensive statistical study of the micro-evolutionary ELM performance in regression problems. Specifically, we start from the model given in [7] for evolutionary ELM, using an Evolutionary Programming algorithm as global search technique. Then we set a very few maximum number of generations in the evolutionary algorithm (only 5), this way we are dealing with a micro-evolutionary ELM (μ-evolutive ELM). The reason for studying μ-evolutive ELM algorithms is that the computation time of these approaches is quite close to the original ELM, and in addition, with very few generations in the evolutionary algorithm, there are not overtraining problems. We study the performance of the μ-evolutive ELM in different regression problems, of increasing difficulty. We have found that the μ-evolutive ELM is better than the ELM in easy, one-dimensional regression problems, but in difficult real problems the performance of the μ-evolutive ELM reduces, and in those cases the ELM performs, in many cases, statistically better than the μ-evolutive ELM.

The rest of the paper is structured as follows: next section gives a brief summary of the main concepts about ELMs. Section 3 describes the evolutionary ELM and the specific μ-evolutive ELM studied in this paper. Section 4 contains the experimental study of the paper, and discusses about the applicability of the μ-evolutive ELM in real problems. Section 5 closes the paper by giving some final conclusions.

2 Extreme Learning Machines

Artificial neural networks (ANNs) are massively parallel and distributed information processing systems, successfully applied to model a large amount of nonlinear problems. ANNs learn from a given set of examples, by constructing an input-output mapping to perform predictions of future samples. ANN are known to be universal approximators of a large class of functions, with a high degree of accuracy. Single hidden layer neural networks are widely used in forecasting and regression problems. One of the most used ANN is the so-called Multi-Layer

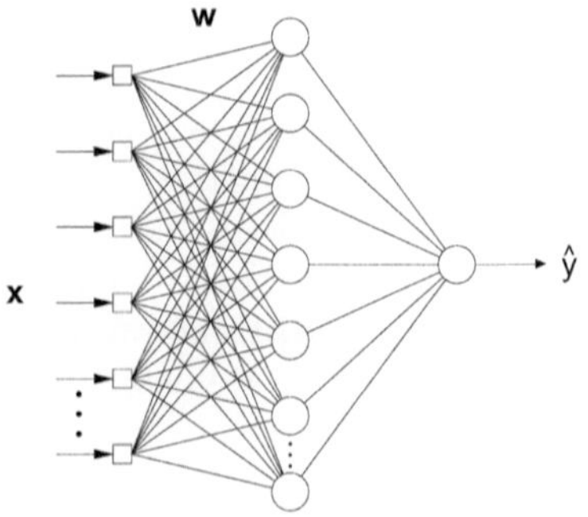

Fig. 1. MLP example diagram

Perceptron (MLP), which consists of an input layer, a hidden layer and an output layer, as shown in Figure 1.

Each neuron in the hidden layer is an autonomous processing node, which processes the arriving input information in the following way:

$$y = g\left(\sum_{j=1}^{n} w_j x_j - b\right), \tag{1}$$

where y is the output signal, x_j, $j = 1, 2, \ldots, n$ are the input signals, w_j is the weight associated with the j-th input,b is a bias variable and g stands for a transfer function, which is usually considered as a Sigmoid function:

$$g(x) = \frac{1}{1 + e^{-x}}. \tag{2}$$

The training process of a MLP consists of, given a training inputs set $\mathbf{X} = [\mathbf{x_1}, \ldots, \mathbf{t}_N]$, and a training out vector $\mathbf{T} = [\mathbf{t}_1, \ldots, \mathbf{t}_N]$, determining the optimal set of weights, and in some cases the network's structure, which minimizes a measure of error. In the network's training process the first step consists of obtaining the so-called *hidden layer output matrix* $\mathbf{H}$ (output of the different neurons for all the input vectors):

$$\mathbf{H} = \begin{bmatrix} g(\mathbf{w}_1 \cdot \mathbf{x_1} + b_1) & \cdots & g(\mathbf{w}_{\tilde{N}} \cdot \mathbf{x_1} + b_{\tilde{N}}) \\ \vdots & \cdots & \vdots \\ g(\mathbf{w}_1 \cdot \mathbf{x_N} + b_1) & \cdots & g(\mathbf{w}_{\tilde{N}} \cdot \mathbf{x_N} + b_{\tilde{N}}) \end{bmatrix}_{N \times \tilde{N}} \tag{3}$$

where $\tilde{N}$ stands for the number of neurons in the hidden layer, and N stands for the number of input vectors. Of course, the training process implies then to tune the network's weights and biases.

Several methods such as gradient-based learning approaches, Levenberg-Marquardt algorithm, evolutionary computing techniques etc., have been previously used to train MLPs. The problem of these methods is that in the majority of cases they require lots of iterations in order to obtain good training performance.

The Extreme Learning Machine (ELM) is a novel and fast learning method recently proposed in [1]. It has been applied to a large number of classification and regression problems. The beauty of this technique is its simplicity together with surprising results comparable to the best known actual techniques. The structure of an ELM is similar to a MLP, however the ideas behind ELM are completely novel. In the next subsection we describe the main concepts needed to describe the ELM approach.

2.1 The ELM Algorithm

In [1] it was proven that if the activation function is infinitely differentiable in any interval, then the hidden layer output matrix $\mathbf{H}$ is invertible and $||\mathbf{H}\beta - \mathbf{T}|| = 0$. They also proved that given any small positive value $\epsilon > 0$, there exists an integer $\tilde{N} \leq N$ that satisfies that $||\mathbf{H}_{N\times\tilde{N}}\beta_{\tilde{N}\times m} - T_{N\times m}|| < \epsilon$. Using these results, the ELM algorithm has been proposed as a very fast approach to train ANNs. It can be described as follows: given a training set $\aleph = \{(\mathbf{x}_i, \mathbf{t}_i | \mathbf{x}_i \in \mathbb{R}^n, \mathbf{t}_i \in \mathbb{R}^m, i = 1, \cdots, N\}$, an activation function $g(x)$ and number of hidden nodes ($\tilde{N}$), (note that we have followed here the notation given in [1], which is different than the one used to introduce matrix $\mathbf{H}$ in Equation (3)).

1. Randomly assign inputs weights $\mathbf{w}_i$ and bias b_i, $i = 1, \cdots, \tilde{N}$.

2. Calculate the hidden layer output matrix $\mathbf{H}$, defined as

$$\mathbf{H} = \begin{bmatrix} g(\mathbf{w_1}\cdot\mathbf{x_1}+b_1) & \cdots & g(\mathbf{w}_{\tilde{N}}\cdot\mathbf{x_1}+b_{\tilde{N}}) \\ \vdots & \cdots & \vdots \\ g(\mathbf{w_1}\cdot\mathbf{x_N}+b_1) & \cdots & g(\mathbf{w}_{\tilde{N}}\cdot\mathbf{x_N}+b_{\tilde{N}}) \end{bmatrix}_{N\times\tilde{N}} \tag{4}$$

   Usually the Sigmoid function (Equation (2)) is considered as the activation function $g(x)$.

3. Calculate the output weight vector β as

$$\hat{\beta} = \mathbf{H}^{\dagger}\mathbf{T}, \tag{5}$$

   where $\mathbf{H}^{\dagger}$ stands for the Moore-Penrose inverse of matrix $\mathbf{H}$ [1], and $\mathbf{T}$ is the training output vector, $\mathbf{T} = [\mathbf{t}_1, \ldots, \mathbf{t}_N]^T$. By calculating the output weight with the Moore-Penrose inverse matrix, we ensure that the norm of these weights is the smallest from the group of possible solutions.

Note that the number of hidden nodes ($\tilde{N}$) is a free parameter of the ELM training, and must be estimated for obtaining good results. Usually, scanning a range of $\tilde{N}$ values is the solution for this problem.

3 The Evolutionary ELM

The evolutionary ELM (E-ELM hereafter) was proposed by Zhu et al. in [7]. The idea of the E-ELM is intuitive: to use an evolutionary-based algorithm (a Differential Evolution was used in [7]) in order to improve the ELM by evolving the randomly-generated weights and bias of the network. Basically, the evolutionary algorithm is initialized with a population representing ELMs with different set of weights and bias, where each individual (representing an ELM) is encoded using a vector of real numbers, in the following way:

$$\theta = [w_{11}, \dots, w_{1\tilde{N}}, w_{21}, \dots, w_{2\tilde{N}}, \dots, w_{n1}, \dots, w_{n\tilde{N}}, b_1, \dots, b_{\tilde{N}}] \qquad (6)$$

where all w_{ij} and b_j are randomly initialized within $[-1, 1]$.

The population is then evolved by means of the application of the DE operators (selection, crossover and mutation). In order to calculate the fitness function of its individual, each associated ELM is trained in a training set, and a value of root mean square error (RMSE) in a validation set is calculated. Note that, following [7], the use of the validation set also helps to avoid overtraining of the networks. After the evolutionary process, the best individual of the population is considered the final solution to the problem, and its performance is tested in a final test set.

3.1 The μ-GA ELM

The approach proposed in this paper starts from the idea in [7], but analyzes the behavior of micro-genetic algorithms (μ-GA) as the evolutionary search approach applied to the ELM. A μ-GA is a genetic algorithm with very limited resources (small population, and very few generations are allowed). μ-GAs have several advantages over traditional evolutionary algorithms in this problem: first, they are much faster, in the case of the ELM, this is an important characteristic. In fact, in this case we have set the genetic population and number of generations to a minimum, in such a way that the computation time of the μ-GA ELM does not overpass a given threshold with respect to the computation time of the original ELM. The second important advantage of the μ-GAs in the training process of neural networks is that a validation set is usually not needed, since the fitness function can be calculated using the training set without overtraining. This allows to use a larger training set than with traditional evolutionary approaches.

Regarding specific points of the μ-GA implemented in this paper, we have used the same encoding for individuals that in [7] (see Equation (6)). We have used then a μ-GA with 50 individuals in the population, evolved during very few generations (5 generations evolution in this case). An standard two-points crossover operator has been implemented, whereas we have followed the indications in [9] to implement the selection and mutation operators. In the final generation, the best individual is considered as the solution of the problem, and its accuracy is measured in terms of the RMSE in a test set.

4 Experimental Part

This section presents the experimental part of the paper. We have structured it in two different subsections: First, a number of experiments in synthetic, one-dimensional functions, similar to the experiments carried out in [7]. Then a second subsection where we present results in more difficult problems, from public repositories ([10] and [11]). A discussion section is also included to close this experimental part of the paper.

Regarding the methodology applied, note that we are interested in obtaining statistical analysis on the comparison of the evolutionary Extreme Learning Machine with classical ELM. Thus, a significant number of experiments must be carried out. Note also that the ELM training algorithm has an important random component, so we have launched 30 times each training for the μ-GA ELM and ELM in each problem. In order to obtain more generality in the problems from public repositories, we consider different permutations of the data, in this case, 20. In each of these sets, we set the 80% of the data for training and the rest 20% for test purposes. Since we launch 30 time each permutation, we have 600 results for these problems. Note that, with this methodology, we compare the RMSE of the synthetic problems for the 30 repetitions and we can also carry out a t-test or a Fisher-test in a pool of 600 outputs in the case of public repository problems.

4.1 Experiments in Synthetic Functions

First, we compare the μ-GA ELM and ELM performances in several synthetic one-dimensional functions. Specifically, we have tested the algorithms in a Sigmoid, Hyperbolic tangent and a Sinc function. In order to obtain more generality, we have added Gaussian noise with zero-mean and unity variance to our synthetic 1-D function problems. We have used the same configuration for all the experiments: we define 5 generations in the genetic algorithm, with a 50 individuals in the population, and 10 neurons within the hidden layer of the ELM.

Table 1 shows the performance (in terms of Root Mean Square Error (RMSE)) of the compared algorithms in each 1-D problem. It can be seen that the evolutionary Extreme Learning Machine outperforms the classical ELM in most of the cases. These results agree with the ones obtained in [7], where a sigmoid function was used.

4.2 Experiments in Public Repositories

In a second round of experiments, we have carried out experiments in several data sets, obtained from popular public repositories: the UCI machine learning [10] and the data archive of Statlib [11]. The first columns of Table 2 show the main properties of the selected data sets. These regression problems are real problems more difficult than the synthetic functions tested above.

Table 1. Mean and standard deviation of the accuracy (RMSE) and computation time of evolutionary ELM for the synthetic 1-dimension problems considered

	Evolutionary ELM			ELM		
Function	Mean	Std	Time	Mean	Std	Time
Sigmoid	1.5061e-5	1.239e-5	0.6253	1.188e-4	1.5348e-4	0.001
Sigmoid with noise	0.0306	0.0017	0.6134	0.0302	0.002	0.001
Hyperbolic tangent	0.0021	0.0019	0.6288	0.0076	0.0036	0.0005
Hyperbolic tangent with noise	0.0303	0.0016	0.6995	0.0330	0.0041	0.0002
Sinc function	0.0414	0.0168	0.7966	0.1033	0.0542	0.0007
Sinc function with noise	0.0534	0.0162	0.5977	0.0840	0.0341	0.0001

Table 2. Main result in public repositories data. The table shows the data sets used in experiments carried out, Mean and Standard Deviation of the accuracy (RMSE) for both the μ-GA ELM and ELM algorithms, and finally statistical tests for the experiments carried out, where the μ-GA ELM is compared to the ELM. A symbol * stands for t-test $\alpha = 0.05$ and a symbol ** stands for Fisher-test $\alpha = 0.05$. The symbols L, W, T stand for lost, wind and tied instances (out of the total 600 runs of the algorithms.).

Data set	N	n	Rep	μ-GA ELM		ELM		μ-GA vs ELM	
				Mean	Std	Mean	Std	p-value	L-W-T
MortPollution	60	15	StatLib	136.6570	89.1044	140.4365	90.8576	0.0199*	278-322-0
Bodyfat	252	13	StatLib	0.0565	0.0403	0.0482	0.0440	0.0000**	446-154-0
Betaplasma	315	12	StatLib	177.7690	47.3315	184.0188	43.6241	0.0000**	197-403-0
Retplasma	315	12	StatLib	219.7362	29.3762	222.5146	30.4868	0.0000*	263-337-0
Autompg	392	7	UCI	3.4634	0.8035	3.4568	1.0819	0.0000**	369-231-0
Housing	506	13	UCI	5.6152	0.8786	5.2002	1.0761	0.0000**	460-140-0
Concrete	1030	16	UCI	28.7485	0.7948	28.8552	0.8756	0.0000*	261-339-0
Abalone	4177	8	UCI	2.1825	0.1068	2.1134	0.0966	0.0000*	567-33-0

Table 2 shows the main results in terms of RMSE obtained by the μ-GA ELM and ELM approaches in the different regression problems considered. Note that in this case, the ELM outperforms to the μ-GA ELM in the majority of problems. As has been mentioned before, we have carried out a t-test or a Fisher-test (depending on a previous Kolmogorov Smirnov Gaussianity test) to compare the results in a statistical way. The rightmost columns of Table 2 show the results obtained in these statistical tests. It is easy to see that in this case the ELM and the μ-GA ELM are not statistically equal. In fact, in mortpollution, betaplasma and retplasma the μ-GA ELM statistically outperforms the ELM, but in bodyfat, autompg, housing, concrete and abalone sets the ELM performs statistically better than the μ-GA ELM. This means that the μ-GA ELM does not outperform the basic ELM algorithm in large, real regression problems, on contrary it performs worse in many cases.

5 Conclusions

In this paper we have analyzed the performance of the μ-GA ELM algorithm in regression problems. The μ-GA ELM is characterized by a short evolution of the network's weights and bias, using a small-population genetic algorithm. We have shown that the μ-GA ELM performs statistically better than the basic ELM in small 1-dimensional problems, but in large real problems we have shown that the behavior of the algorithm is worse in many cases than that of the ELM.

Acknowledgement

This work has been partially supported by Spanish Ministry of Science and Innovation, under a project number ECO2010-22065-C03-02.

References

1. Huang, G.B., Zhu, Q.Y., Siew, C.K.: Extreme learning machine: Theory and applications. Neurocomputing 70(1-3), 489–501 (2006)
2. Rong, H.J., Ong, Y.S., Tan, A.H., Zhu, Z.: A fast pruned-extreme learning machine for classification problem. Neurocomputing 72(1-3), 359–366 (2008)
3. Lan, Y., Soh, Y.C., Huang, G.B.: Two-stage extreme learning machine for regression. Neurocomputing 73(16-18), 3028–3038 (2010)
4. Lan, Y., Sohand, Y.C., Huang, G.B.: Constructive hidden nodes selection of extreme learning machine for regression. Neurocomputing 73(16-18), 3191–3199 (2010)
5. Tang, X., Han, M.: Partial Lanczos extreme learning machine for single-output regression problems. Neurocomputing 72(13-15), 3066–3076 (2009)
6. Han, F., Huang, D.: Improved extreme learning machine for function approximation by encoding a priori information. Neurocomputing 69(16-18), 2369–2373 (2006)
7. Zhu, Q.Y., Qin, A.K., Suganthan, P.N., Huang, G.B.: Evolutionary extreme learning machine. Pattern Recognition 38, 1759–1763 (2005)
8. Sánchez-Monedero, J., Hervás-Martínez, C., Martínez-Estudillo, F.J., Ruz, M.C., Moreno, M.C.R., Cruz-Ramírez, M.: Evolutionary learning using a sensitivity-accuracy approach for classification. In: Corchado, E., Graña Romay, M., Manhaes Savio, A. (eds.) HAIS 2010. LNCS, vol. 6077, pp. 288–295. Springer, Heidelberg (2010)
9. Yao, X., Liu, Y., Lin, G.: Evolutionary programming made faster. IEEE Transactions on Evolutionary Computation 3, 82–102 (1999)
10. Asuncion, A., Newman, D.J.: UCI Machine Learning Repository. University of California, School of Information and Computer Science, Irvine, CA (2007), `http://www.ics.uci.edu/~mlearn/MLRepository.html`
11. StatLib DataSets Archive, `http://lib.stat.cmu.edu/datasets`

A Hybrid Evolutionary Approach to Obtain Better Quality Classifiers

David Becerra-Alonso, Mariano Carbonero-Ruz,
Francisco José Martínez-Estudillo, and Alfonso Carlos Martínez-Estudillo

Department of Management and Quantitative Methods
ETEA - University of Córdoba
{dbecerra, mariano, fjmestud, acme}@etea.com

Abstract. We present an extra measurement for classifiers, responding to the need to evaluate them with more than accuracy alone. This measure should be able to express, at least to some degree, the extent to which all classes are taken into account in a classification problem. In this communication we propose sensitivity dispersion (being as it is, the associated statistical dispersion measurement of accuracy), as the appropriate measure to have a more complete evaluation of the quality of classifiers. We use the Evolutionary Extreme Learning Machine algorithm, with a specific fitness function to optimize both measures simultaneously, and we compare it with other classifiers.[1]

1 Introduction

Accuracy has generally been used to study the performance of evolutionary classifiers. Many works extend on the idea that accuracy, although a critical piece of information, does not capture all the different aspects of how classification takes place for a given method and a given set. This becomes even more relevant when some of the classes of the dataset are shown to be predominant over others, and the system is unbalanced. The search for combined measurements as a way to evaluate a classifier is already found in a number of machine learning publications [2, 7, 9]. Here, we propose a new measure that we call sensitivity dispersion given by the dispersion between the accuracy results in each class.

In 2005, Huang et al. [5, 6] proposed the original algorithm called Extreme Learning Machine (ELM) which randomly chooses hidden nodes and analytically determines the output weights of the network. A hybrid algorithm called Evolutionary ELM (E-ELM) [11] was proposed by using the Differential Evolution algorithm [8]. In this paper, the simultaneous optimization of accuracy and sensitivity dispersion is carried out by means of the E-ELM algorithm combination. The key point of the algorithm is the fitness function considered, which tries to achieve a good balance between the classification rate level in the global

[1] This work was supported in part by the Spanish Inter-Ministerial Commission of Science and Technology under Project TIN 2008-06681-C06-03, the European Regional Development fund, and the Junta de Andaluca, Spain, under Project P08-TIC-3745.

J. Cabestany, I. Rojas, and G. Joya (Eds.): IWANN 2011, Part II, LNCS 6692, pp. 161–168, 2011.

dataset, and an acceptable level for each class. The base classifier considered is the Multilayer Perceptron (MLP) neural network.

The paper is structured as follows: Section 2 presents the two measurements we propose and the relationship between them. In Section 3 we propose the hybrid algorithm based on the Evolutionary Extreme Learning Machine with the specific fitness function built for the optimization of accuracy and sensitivity dispersion. Section 4 presents the experimental results, where 9 methods are tested for 6 datasets. Section 5 ends the paper with conclusions.

2 C vs. S^2

We consider a classification problem with Q classes and N training or testing patterns with g as a classifier obtaining a $Q \times Q$ contingency or confusion matrix $M(g) = \{n_{ij}; \sum_{i,j=1}^{Q} n_{ij} = N\}$ where n_{ij} represents the number of times the patterns are predicted by classifier g to be in class j when they really belong to class i. The accuracy C can be obtained as a weighted average of the classification rate of each class:

$$C = \sum_{Q} f_i p_i \tag{1}$$

where p_i represents the relative frequency of patterns that belong to a given class, while f_i is the rate of those correctly classified in that i class. The sensitivity dispersion (i.e. the associated statistical dispersion measurement of accuracy) is given by:

$$S^2 = \sum f_i^2 p_i - C^2 \tag{2}$$

There is also a way to relate C with S^2 in terms of how large must the latter be for every value of the former. Given our definition of S^2 as in (2), and since $\sum_i f_i^2 p_i \leq \sum_i f_i p_i = C$, we have the boundary:

$$S^2(C) \leq C - C^2 \tag{3}$$

This of course is just a tentative upper boundary for S^2. Each classifier will be represented as a point in the (C,S^2) region (see Figure 1). The worst classifier returns a confusion matrix that has $C = 0$ and $S^2 = 0$, while the best classifier returns $C = 1$, $S^2 = 0$.

Although S^2 takes on its full meaning on multi-class scenarios, it is illustrative to present its properties and boundaries in the $Q = 2$ case. For a two classes dataset we have:

$$C = f_1 p_1 + f_2 p_2 \tag{4}$$

$$S^2 = f_1^2 p_1 + f_2^2 p_2 - C^2 \tag{5}$$

Since $p_1 + p_2 = 1$, both measures can be expressed in terms of p_1, that we will call p from this point:

$$C = p f_1 + (1 - p) f_2 \tag{6}$$

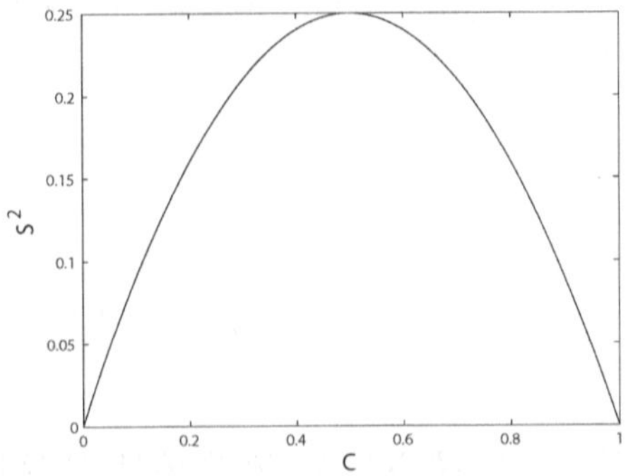

Fig. 1. Top boundary for the C vs. S^2 representation

$$S^2 = p(1-p)(f_1 - f_2)^2 \tag{7}$$

All classifiers with optimal S^2 meet on $f_1 = f_2$. Optimal S^2 is found when the relative correctly classified patterns are equal in both classes.

3 Differential Evolution and Extreme Learning Machine Algorithm

3.1 Extreme Learning Machine (ELM)

Given a dataset with Q classes, we have N patterns $D = \{(\mathbf{x}_j, \mathbf{y}_j) : \mathbf{x}_j \in R^K, \mathbf{y}_j \in R^Q, j = 1, 2, ..., N\}$ where $\mathbf{x}_j$ is a $k \times 1$ input vector and $\mathbf{y}_j$ is $Q \times 1$ target vector. Our Multilayer Perceptron system has M nodes in the hidden layer, and is given by $f = (f_1, f_2, ..., f_Q)$, where every f transforms the input $\mathbf{x}$ according to:

$$f_j(\mathbf{x}, \boldsymbol{\theta}_l) = \beta_0^l + \sum_{j=1}^{M} \beta_j^l \sigma_j(\mathbf{x}, \mathbf{w}_j), l = 1, 2, ..., Q \tag{8}$$

where $\boldsymbol{\theta} = (\boldsymbol{\theta}_1, ..., \boldsymbol{\theta}_J)^T$ is the transposed matrix that includes all the neural network weights. Thus, $\boldsymbol{\theta}_l = (\beta_0^l, \beta_1^l, ..., \beta_M^l, \mathbf{w}_1, ..., \mathbf{w}_M)$ includes the M weights from the hidden layer to the output node, and the vectors $\mathbf{w}_j = (w_{1j}, ..., w_{Kj})$ for the weights and biases between the input and the hidden layer. Each element of $\mathbf{x}$ represents a pattern of the dataset, and the activation function (sigmoidal for our case) is given by $\sigma_j(\mathbf{x}, \mathbf{w}_j)$.

Suppose we want to train a Multilayer Perceptron to learn from N patterns of a set D as proposed above. The relation $f(\mathbf{x}_j) = \mathbf{y}_j$ where $j = 1, ..., N$ can also be expressed as $\mathbf{H}\boldsymbol{\beta} = \mathbf{Y}$, where $\mathbf{H}$ represents the hidden layer output matrix of the network.

The ELM algorithm, as it is proposed in [5, 6] assigns random values to $\mathbf{w}_j = (w_{ij}, ..., w_{Kj})$. By doing so, they convert the nonlinear system into a linear one. Thus they can use these weights to analytically obtain $\{\beta_1^l, ..., \beta_M^l\}$

by finding the least square solution to the given linear system. This can be done by solving $\hat{\boldsymbol{\theta}} = \mathbf{H}^{\dagger}\mathbf{Y}$, where $\mathbf{H}^{\dagger}$ is the Moore-Penrose generalized inverse of $\mathbf{H}$.

3.2 Differential Evolution Algorithm

The Evolutionary Extreme Learning Machine (E-ELM) takes on what has been proposed in Section 3.1 and tries to improve the classifier by applying Differential Evolution (DE) to $\mathbf{w}_j$ (as proposed by Storn and Price [8, 10], and later by Zhu [11]) . This provides an opportunity to use a fitness function that involves C and S^2. DE requires a population of classifiers to compete and combine their weight chromosomes. The following is the basic strategy used for DE:

Given our $\mathbf{w}_j$ in generation G (we will define the whole as $\mathbf{w}_{j,G}$), we first apply mutation: for each target vector $\mathbf{w}_{j,G+1}$ we have a mutant vector according to:

$$\mathbf{m}_{j,G+1} = \mathbf{w}_{r_1,G} + F(\mathbf{w}_{r_2,G} - \mathbf{w}_{r_3,G}) \tag{9}$$

where r_1, r_2, r_3 are random mutually different indices, and $F \in [0, 2]$ defines the amplification of the differential variation $(\mathbf{w}_{r_2,G} - \mathbf{w}_{r_3,G})$. We then apply crossover: our population of chromosomes $\mathbf{w}_{j,G}$ has P members. For generation $G+1$, we can again define a trial vector as $\nu_{kj,G+1} = (\nu_{1j,G+1}, \nu_{2j,G+1}, ..., \nu_{Pj,G+1})$. Thus,

$$\nu_{kj,G+1} = \begin{cases} \mathbf{m}_{kj,G+1} \text{ if } rand_k \leq CR \text{ or } k = rnbr_j \\ \mathbf{w}_{kj,G+1} \text{ if } rand_k > CR \text{ and } k \neq rnbr_j \end{cases}$$

where $rand_k$ is the kth evaluation of a uniform random number generator in $[0, 1]$, $rnbr_j$ is a random chosen integer from 1 to P, and CR is the crossover constant. Finally, we apply selection: $\nu_{kj,G+1}$ will be used in generation $G+1$, as long as our fitness function returns a better outcome than that obtained from $\mathbf{w}_{kj,G}$. Otherwise, we will have $\mathbf{w}_{kj,G} = \mathbf{w}_{kj,G+1}$. Thus, a fitness function for our particular problem needs to be defined. For each individual (a set of weights and biases), ELM returns the confusion matrix, and from it we can obtain its C and S^2. Let us define a classifier that returns both C_0 and S_0^2 in generation G, whereas in generation $G+1$ (once the classifier has undergone crossover and

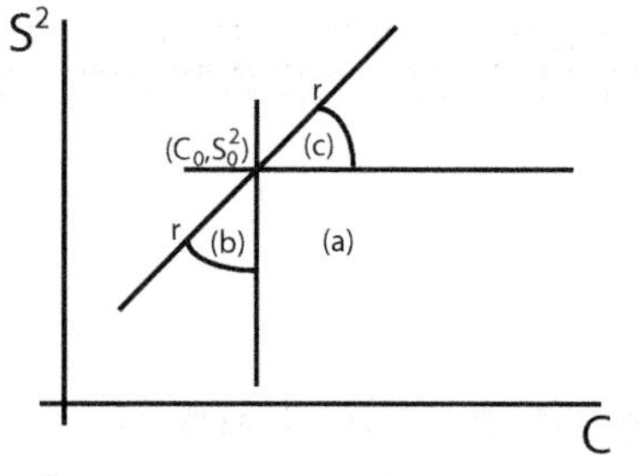

Fig. 2. C vs. S^2 scheme for the E-ELMSC2 fitness conditions

mutation) the condition for the offspring (that returns C_1 and S_1^2) to prevail over the parent is one of the following:

$$(a) \quad C_1 \geq C_0 \quad \textbf{and} \quad S_1^2 \leq S_0^2 \tag{10}$$

or

$$(b) \quad C_1 \leq C_0 \quad \textbf{and} \quad S_1^2 \leq S_0^2 \quad \textbf{and} \quad r \leq C_0/10 \tag{11}$$

when $(C_1 - C_0)^2 + (S_1^2 - S_0^2)^2 \leq r$ and $S_1^2 \leq C_1 - C_0 + S_0^2$, or

$$(c) \quad C_1 \geq C_0 \quad \textbf{and} \quad S_1^2 \geq S_0^2 \quad \textbf{and} \quad r \leq S_0^2/10 \tag{12}$$

when $(C_1 - C_0)^2 + (S_1^2 - S_0^2)^2 \leq r$ and $S_1^2 \leq C_1 - C_0 + S_0^2$.
These fitness conditions allow for small setbacks in either C or S^2 as long as these provide an opportunity for the $\mathbf{w}_j$ chromosome to thrive from a new position in the (C, S^2) space. The allowed new positions for a successful (C_1, S_1^2) can be seen in regions (a), (b) and (c) in Figure 2. We will refer to entire method with this particular fitness as E-ELMCS2.

4 Experiments

We consider six datasets with different features taken from the UCI repository [1] (see Table 1). The data was distributed in 10-fold subsets where all classes had an approximately equal representability. The method we present in this paper is compared with 7 well-known classification algorithms, as coded in Weka 3.6.3 (see [3, 4]). The methods listed include Support Vector Machines (SVM), C4.5, CART, Naive-Bayes/Decision-Tree (NBTree), Logistic, Simple Logistic and Multilayer Perceptron (MLP). Both Table 2 and the corresponding plots in Figure 3 show that we have found the best (C, S^2) results for Haberman, Pima, German and Vehicle. Glass returns an acceptable value of S^2, and our model does not compete well with others when we use Segmentation as dataset. The latter happens mostly when the dataset is very balanced, or when high accuracy C is generally easy to achieve by most methods.

Table 1. Datasets used for the experiments

Dataset	Size	#Input	#Classes	Distribution
Haberman	306	3	2	225-81
Pima	768	8	2	500-268
German	1000	24	2	700-300
Vehicle	846	18	4	212-199-218-217
Glass	214	9	6	70-76-17-13-9-29
Segmentation	2310	19	7	330-330-330-330-330-330-330

Table 2. Comparative C vs. S^2

Dataset	Algorithm	C	S^2	Dataset	Algorithm	C	S^2
Haberman	SVM	0.7353	0.1818	Vehicle	SVM	0.7435	0.0467
	C4.5	0.7288	0.0969		C4.5	0.7246	0.0410
	CART	0.7451	0.1504		CART	0.6891	0.0438
	NBTree	0.7255	0.0957		NBTree	0.7293	0.0308
	Logistic	0.7418	0.1115		Logistic	0.7979	0.0224
	Simplelog.	0.7386	0.1203		Simplelog.	0.7719	0.0302
	MLP	0.6928	0.0498		MLP	0.8168	0.0191
	ELM	0.7763	0.0380		ELM	0.8246	0.0169
	E-ELM	0.7913	0.0311		E-ELM	0.8279	0.0128
Pima	SVM	0.7734	0.0289	Glass	SVM	0.5607	0.0756
	C4.5	0.7383	0.0107		C4.5	0.6682	0.0127
	CART	0.7513	0.0254		CART	0.7056	0.0324
	NBTree	0.7435	0.0115		NBTree	0.7089	0.0257
	Logistic	0.7721	0.0217		Logistic	0.6402	0.0275
	Simplelog.	0.7747	0.0248		Simplelog.	0.6402	0.0454
	MLP	0.7539	0.0114		MLP	0.6776	0.0353
	ELM	0.8021	0.0138		ELM	0.7736	0.0524
	E-ELM	0.8035	0.0105		E-ELM	0.7759	0.0419
German	SVM	0.7640	0.0362	Segmentation	SVM	0.9307	0.0079
	C4.5	0.7390	0.0273		C4.5	0.9693	0.0009
	CART	0.7500	0.0394		CART	0.9615	0.0013
	NBTree	0.7310	0.0423		NBTree	0.9494	0.0020
	Logistic	0.7690	0.0318		Logistic	0.9580	0.0018
	Simplelog.	0.7630	0.0429		Simplelog.	0.9511	0.0028
	MLP	0.7040	0.0190		MLP	0.9606	0.0018
	ELM	0.7917	0.0270		ELM	0.9411	0.0040
	E-ELM	0.8030	0.0153		E-ELM	0.9502	0.0024

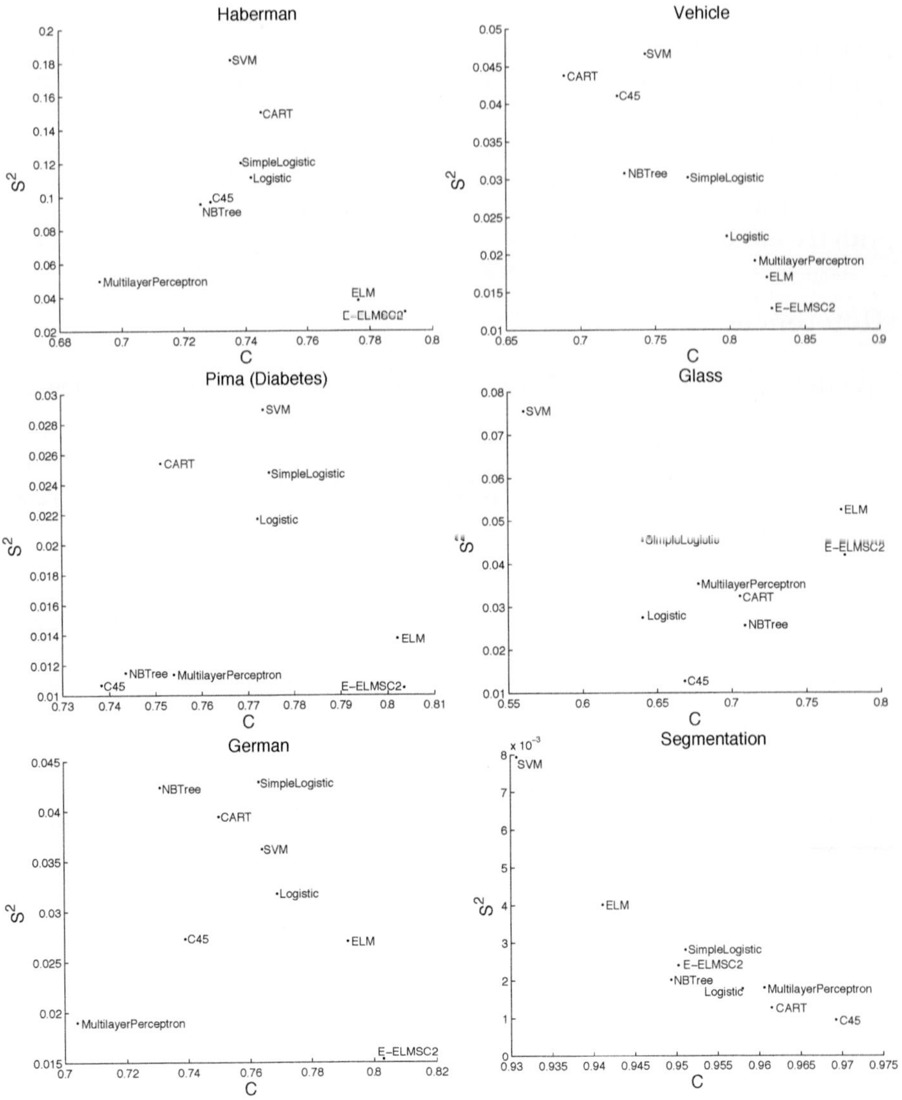

Fig. 3. C vs. S^2 for the databsets shown in Table 2

5 Conclusions

In this work we have presented a hybrid algorithm that returns results with a good balance between C and S^2. The results depicted in the (C, S^2) space show that the E-ELMCS2 algorithm is able to achieve high values of C while having a low S^2, i.e. the dispersion between the accuracy results in each class. Moreover, and looking at the methods used for comparison, the differences in S^2 between them (even when they share a similar C), indicate that this measure introduces the quality information that C did not have on its own.

For future work we will try to refine the fitness function in order to improve the robustness of the evolutionary side of the algorithm. A further analysis of a larger number of datasets will allow us to associate common features of these, with the outcomes of (C, S^2).

References

[1] Blake, C.L., Merz, C.J.: UCI repository of machine learning databases (1998)

[2] Caballero, F., Martínez, F.J., Hervás, C., Gutiérrez, P.A.: Sensitivity versus accuracy in multiclass problems using memetic pareto evolutionary neural networks. IEEE Transactions on Neural Networks 21(5), 750–770 (2010)

[3] Hall, M., Frank, E., Holmes, G., Pfahringer, B., Reutemann, P., Witten, I.H.: The WEKA data mining software: An update. ACM SIGKDD Explorations Newsletter 11(1), 10–18 (2009)

[4] Holmes, G., Donkin, A., Witten, I.H.: Weka: A machine learning workbench. In: Proceedings of the 1994 Second Australian and New Zealand Conference on Intelligent Information Systems, pp. 357–361 (2002)

[5] Huang, G.B., Zhu, Q.Y., Siew, C.K.: Extreme learning machine: A new learning scheme of feedforward neural networks. In: Proceedings 2004 IEEE International Joint Conference on Neural Networks, pp. 985–990 (2004)

[6] Huang, G.B., Zhu, Q.Y., Siew, C.K.: Extreme learning machine: theory and applications. Neurocomputing 70(1-3), 489–501 (2006)

[7] Martínez-Estudillo, F.J., Gutiérrez, P.A., Hervás, C., Fernández, J.C.: Evolutionary learning by a sensitivity-accuracy approach for multi-class problems. In: IEEE Congress on Evolutionary Computation (IEEE World Congress on Computational Intelligence) CEC 2008, pp. 1581–1588 (2008)

[8] Price, K.V., Storn, R.M., Lampinen, J.A.: Differential evolution: a practical approach to global optimization. Springer, Heidelberg (2005)

[9] Sánchez-Monedero, J., Hervás-Martínez, C., Martínez-Estudillo, F.J., Ruz, M.C., Moreno, M.C.R., Cruz-Ramírez, M.: Evolutionary learning using a sensitivity-accuracy approach for classification. In: Corchado, E., Graña Romay, M., Manhaes Savio, A. (eds.) HAIS 2010. LNCS, vol. 6077, pp. 288–295. Springer, Heidelberg (2010)

[10] Storn, R., Price, K.: Differential evolution: a simple and efficient heuristic for global optimization over continuous spaces. Journal of global optimization 11(4), 341–359 (1997)

[11] Zhu, Q.Y., Qin, A.K., Suganthan, P.N., Huang, G.B.: Evolutionary extreme learning machine. Pattern recognition 38(10), 1759–1763 (2005)

Neural Network Ensembles with Missing Data Processing and Data Fusion Capacities: Applications in Medicine and in the Environment

Patricio García Báez[1], Carmen Paz Suárez Araujo[2], and Pablo Fernández López[2]

[1] Departamento de Estadística, Investigación Operativa y Computación, Universidad de La Laguna. 38271 La Laguna, Canary Islands, Spain
[2] Instituto Universitario de Ciencias y Tecnologías Cibernéticas, Universidad de Las Palmas de Gran Canaria. 35017 Las Palmas de Gran Canaria, Canary Islands, Spain
pgarcia@ull.es, cpsuarez@dis.ulpgc.es

Abstract. An important way to reach a qualitative improvement of Artificial Neural Networks (ANNs) is to incorporate biological features in the networks. Our proposal introduces modularity at two different levels, first, at the network level and second, at the intrinsic level of the networks, generating neural network ensembles (NNEs). We designed three NNEs which incorporated new capacities with regard to the processing of missing data, introduced hybrid modularity, and also used modular ANNs for building the NNEs. We have investigated a suitable NNE design where selection and fusion are recurrently applied to a population of best combinations of classifiers. In this paper we explore the ability of the proposed NNE in different automated decision making applications, especially for those with inherent complexity in their information environment. We present some results on dementia diagnosis and on automatic pollutants detection.

Keywords: Ensemble Systems; Missing Data; Data Fusion; Artificial Neural Network; HUMANN; Dementia Diagnosis; Pollutant Detection.

1 Introduction

ANNs can be considered as information processing computational models which are biologically inspired. They are massively parallel dynamic systems based upon models of brain function. Their behaviour emerges from structural changes driven by local learning rules, resulting in the ability to generalise, and their main abilities are given by biological similarities. Thus, their qualitative improvement must be linked to approximate brain function and structure. The nature of the adaptive and open-system ANN provides not only important possibilities of working with complex problems but also with the generation of multiple neural network configurations which are close to the optimal one. The importance of this variability is evident when the biological property of the modularity at the network level is incorporated into the network design. This characteristic allows groups of ANNs to be used to improve their

J. Cabestany, I. Rojas, and G. Joya (Eds.): IWANN 2011, Part II, LNCS 6692, pp. 169–176, 2011.

accuracy and generalisation skills because the collective decision produced by the set of networks, with an appropriate collective decision strategy, is less likely to be in error than the decision made by any one of the individual networks. Furthermore they are able to tackle large and complex problems in a more efficient way than monolithic networks. This set of networks is referred to as a NNE [1].

This study introduces three NNEs: Decision Tree Based Switching Multi-net System Ensemble (DTBSE), Simple and Weighted Majority Voting Ensemble Systems (SMVE/WMVE), and Gating Neural Ensemble System (GaNEn). They incorporate modularity at the network level and at the intrinsic level of the networks. In these NNEs diversity is reached by considering whether the training data is inherently re-sampled by classifier conditions or by the generation of two levels of ensemble members. Their combination strategies use a non-trainable or a hybrid scheme. The NNEs proposed present new qualities and a new structural organization, referring to other previous ensemble-learning methods [1][2][3][4][5][6]. They contain new capabilities concerned with the processing of missing data, introduce hybrid modularity, and also use modular neural networks for building the NNE. We have investigated suitable NNE designs where selection and fusion are recurrently applied to a population of best combinations of classifiers [7].

In this paper we explore the ability of the proposed NNEs in different automated decision making applications, especially for those with inherent complexity in their information environment, without any comparative study among them. We present some results on dementia diagnosis and on automatic pollutants detection in an environmental control scenario.

2 Ensemble Systems

A NNE combines a set of neural networks which learn to subdivide the task and thereby solve it more efficiently and elegantly. The NNEs divide the data space into smaller and easier-to learn partitions, where each ANN learns only one of the simpler partitions. The underlying complex decision boundary can then be approximated by an appropriate combination of different ANNs. The NNE can perform more complex tasks than any of its components. It is more robust than a monolithic neural network. It can produce a reduction of variance and increase the confidence level of the decision, and show graceful performance degradation in situations where only a subset of neural networks in the ensemble is performing correctly [3].

Two strategies are needed to build an NNE. Firstly, a strategy for generating the ensemble members. This must seek to improve the ensemble´s diversity. Second, a combination strategy. It is necessary to combine the outputs of individual ANNs that make up the ensemble in such a way that the correct decisions are amplified, and incorrect ones are cancelled out.

We focus on the type of data classification because the set of input data will continue to reflect data fusion schemes with either independent input or different input combinations. This second case allows us to processing missing data, even though the modules do not possess this capacity.

A selection of the most diverse classifier modules (CMs) is carried out by focusing on the correlation between the validation errors of the modules, two at a time. There are essentially two combination strategies that are used, Competitive and Co-operative. The co-operative combination reveals how all of the elements that will be combined will make some contribution to the decision, even though this contribution may be weighted in some way. In the competitive combination strategy, it is assumed that for each input the most appropriate element will be selected [4]. The combination of the outputs of several classifiers does not guarantee a superior performance to the best ensemble module, but it clearly reduces the risk of making a particularly poor selection [2].

3 New Neural Network Ensembles

This section presents the three proposed NNE models, each which have the capacity to process missing data and data fusion. They are characterised by a high capability to face hard decision making problems in a more efficient way than monolithic networks. We show that each NNE has a different building strategy, different CMs and different qualities. The first NNE focuses on processing datasets with missing values using CMs with not missing data processing capability. The second NNE focuses on increasing the effectiveness of CMs using a data fusion scheme. The third NNE is the result of the fusion of the two previous NNEs, taking advantage of each one strengths. Finally, the last subsection introduces the different CMs contained in each NNE used in the Application section.

3.1 Decision Tree Based Switching Ensemble (DTBSE)

DTBSE [8] follows a competitive combination scheme. We use a Decision Tree mechanism to encode these circumstances. It faces the missing data training that all the modules with different input subgroups that can be given as variant of the input dimension. The design of the DTBSE was based on the accomplishment of training and evaluation of the generalization error of the different CMs. From the analysis of the obtained results the dominations between classifiers were obtained, that is to say, we considered that a classifier dominates another one if a first input subgroup obtains a validation error less than or equal or the other subgroup. Soon, according to the obtained dominations some modules were discarded and the suitable rules were designed to select the module with smaller generalization error according to the presence of tests in the input consultation at every moment. Then, depending on the not missing values, the module with better awaited expectations is selected, Fig. 1.

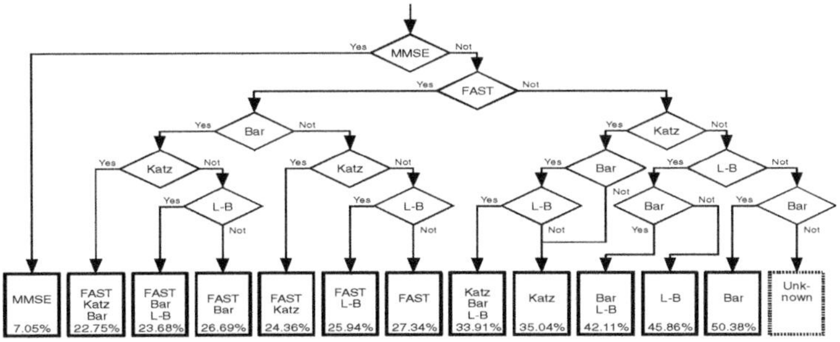

Fig. 1. DTBSE scheme and modules used in it. Each rhomboid module asks for the availability of an input. The number of each classification module (squares) is the validation error for diagnosis problem (P1) calculated using only the data without missing values.

3.2 Simple and Weighted Majority Voting Ensembles (SMVE / WMVE)

This architecture uses a co-operative method as a combination scheme. The combination strategies used were the Simple and the Weighted Majority Voting [2]. In SMVE, Fig. 2, each module emits a vote, indicating whether it considers that said input belongs to a class or is unknown. A later module is responsible for the overall count, of considering whether the input belongs to one class or another, depending on whether most of the modules consider it. Then the class with the maximum number of votes is selected as the ensemble output.

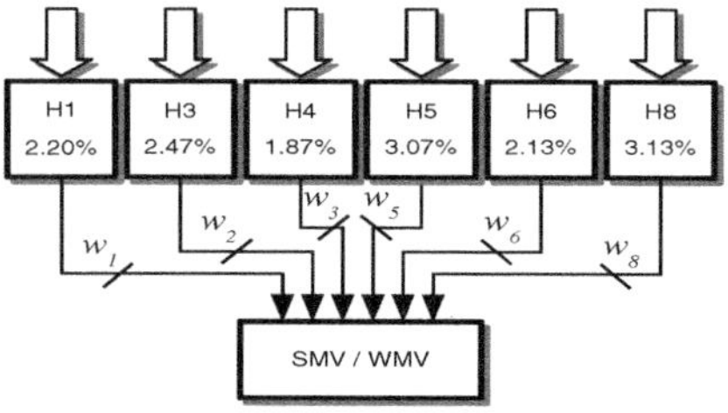

Fig. 2. SMVE/WMVE scheme and modules used in it. H*i* (classifier module *i*) and w_i (weight of module *i*). The number of each H*i* is the validation error for detection problem (P2).

The WMVE system assigns different weights to the modules, Fig. 2, according to their performance, in order to avoid the possibility of certain classifiers to be better than the other ones [2]. It is assigned a weight to each classifier in proportion to its estimated performance. Then the weights for each module are given by:

$$w_i = \log \frac{1 - e_i}{e_i} \qquad (1)$$

where e_i is the predicted error without missing classifier *i* of the ensemble.

3.3 Gating Neural Ensemble (GaNEn)

A Gating Module (GM) is another method to select the most appropriate module in a competitive combination scheme. Here expert modules are combined by means of a GM and also the GM can be used to output a set of scalar coefficients that serve to weigh the contributions of the modules in a co-operative method [4].

We design a modular intelligent system based on a gating system and neural ensemble approaches [5] called the GaNEn [9], Fig. 3. This system is a new formulation of the NNEs, where there are several GMs that take part in the combination strategy of the ensemble system. Each classifier has an associated a GM and the inputs to such a GM are their input samples [6]. The aims of the GMs are to predict the performance of the classifiers, in dependency on the position of the input sample in the input space, and to determine the missing data rate in the input sample. The most straight forward way to describe the performance of the classifiers is to use a measure which is proportional to the inverted values of the prediction error. Furthermore, to describe the participation of each classifier in the final output of the GaNEn system, according to available data in the input, we define an available data function, as follows:

$$GM_i(\mathbf{x}) = \begin{cases} w_i & if\,|P_i(\mathbf{x})| = n_i \\ 0 & otherwise \end{cases} \tag{2}$$

where w_i is the weight of the WMV of equation (1), $P_i(\mathbf{x})$ is the set of input units, j, belonging to classifier i, whose values x_j are available at time t, and n_i is the total number of input values of classifier i.

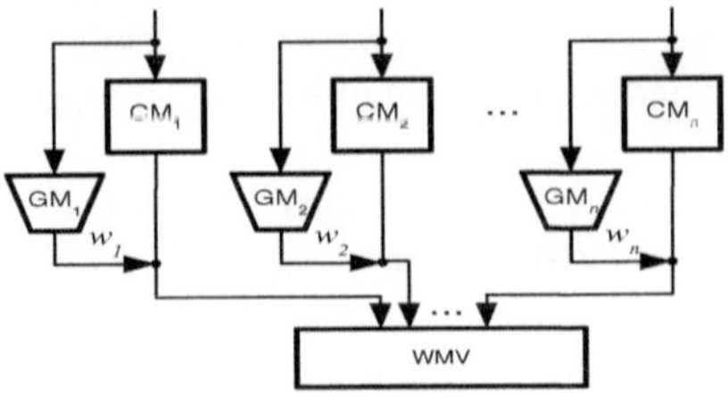

Fig. 3. Generic GaNEn diagram. GM_i (Gating Module i), CM_i (classifier module i), w_i (weight of module i) and WMV (Weighted Majority Voting).

The designed NNEs have modular ANNs as CMs. One of them is the classical Counterpropagation Network (CPN) [10] and the others are two new neural architectures which allow for an important biological plausibility, HUMANN (Hierarchical Unsupervised Modular Adaptive Neural Network) and its supervised version, HUMANN-S.

HUMANN [11] is an unsupervised modular neural structure with three modules and with different neurodynamics, connectivity topologies and learning rules. As is the case with CPN, the first layer is a variant of Kohonen´s SOM. The second is the Tolerance layer, responsible for the robustness of HUMANN against noise. Its main objective is to compare the fitting between the input patterns and the Kohonen detectors [11]. The third layer is the Labelling layer and implements the discrimination task. It maps the outputs of a neural assembly belongs to the Tolerance layer, which have been activated by a category, into different clusters represented by labelling units. This layer exhibits a dynamic dimension, which is fitted to the number of clusters detected in the data set. HUMANN-S is a supervised variant of HUMANN, where the third module is a Perceptron net [12].

In these ANNs it is possible to use a variant of the SOM architecture that processes missing data [13], although this is only required in SMVE and WMVE. This SOM variant prevents missing values from contributing when coming out or modifying weights. Even so, this approach for treating missing values is not enough on its own, especially when the proportion of missing values is excessive.

4 Applications

We present two different problems to solve using our ensemble approaches. The first one (P1) is the diagnosis of Severity Level of Dementia [8][9]. Using scores from five neuropsychological tests the system must try to classify inputs in three different classes: Mild, Moderate and Severe. The data set used for it is incomplete, that is, there are missing data features in 5.62% of the inputs. A 30-fold cross-validation method was used, grouping consultations of each of 30 patients. The second problem (P2) is the detection of Benzimidazole Fungicides pollutants [14]. Using features extracted from eight different type of spectra the system must try to detect each of the three classes of possible fungicides present in mixtures of up to four fungicides. The four different classes are Benomyl or Carbendazim (BM/MBC), Fuberidazole (FB), Thiobendazole (TBZ) and the Clean Sample (CS). Each input can be classified as belonging to none (CS), one, two or three of the classes. Then each CM must emit a vote for each category, indicating whether it considers that the said category is present in the mixture or not.

In order to tackle P1 we show the results from two different ensembles. The first ensemble is based on a DTBSE that uses 12 CPNs without missing data capacity as CMs (DTBSE-CPN). The resultant decision tree scheme is shown in Fig. 1. The chosen CMs correspond to those whose results dominate the excluded modules, as mentioned previously in section 3.1. The second ensemble includes a GaNEn that uses 20 HUMANN-S type CMs, also without the capacity to work with missing data (GaNEn-HS). The 20 CMs included in the selection of this group follow the criteria of effectiveness and diversity that are mentioned in section 2.

Table 1. P1 results for the validation set

		Best HUMANN-S	*Best DTBSE-CPN module*	*DTBSE-CPN*	*Best GaNEn-HS module*	*GaNEn-HS*
Sensit	*Mild*	88%	71%	71%	68%	68%
	Moderate	85%	93%	93%	99%	99%
	Severe	97%	72%	99%	72%	100%
Specif	*Mild*	82%	85%	85%	100%	100%
	Moderate	90%	89%	86%	89%	87%
	Severe	96%	99%	99%	99%	99%
	Error	8.61%	20.97%*	7.49%	19.48%*	5.24%

* both classifiers have 14.98% of errors originated by inputs with missing data.

In order to compare our network with an architecture with missing data processing ability, a single HUMANN-S was trained. We used (Best HUMANN-S). Table 1 presents the results from the validation set in P1. Five ensembles were tested: two cited ensembles (DTBSE-CPN and GaNEn-HS), together with the best modules of each one of these ensembles and the results from the best HUMANN-S. It is clear that both ensembles obtain improvements in accuracy that exceed the results from their modules. The best CMs have a large proportion of their errors due to missing input values, because they lack missing processing capabilities which can generate them. In addition, both improve upon Best HUMANN-S by 1.12% and 3.37%, respectively. These results reveal the effectiveness of our ensembles even when using CMs without the ability to process missing data.

P2 is carried out using two ensembles, SMVE and WMVE. Both use 6 of the 8 possible HUMANNs as CMs, Fig. 2. This selection also followed the criteria of effectivity and diversity. Table 2 presents the results of these two ensembles together with the results from the best HUMANN (Best HUMANN), which also corresponds to the best module used in the ensembles. Note that in this problem there are no significant differences between the results of the SMVE and the WMVE. On the other hand the ensembles obtain an interesting improvement, 0.47%, on the best of the CMs. These results reveal the capacity of these ensembles for automatic decision making problems with data fusion.

Table 2. P2 results for the validation set

		Best HUMANN	*SMVE*	*WMVE*
Sensit	*BM/MBC*	100%	100%	100%
	FB	100%	100%	100%
	TBZ	100%	100%	100%
	CS	100%	100%	100%
Specif	*BM/MBC*	100%	98%	98%
	FB	98%	100%	100%
	TBZ	88%	92%	92%
	CS	100%	100%	100%
	Error	1.87%	1.40%	1.40%

5 Conclusions

Our developments represent advances on NNEs. We proposed new computational intelligent tools, with the capacity for processing missing data and data fusion, which are very appropriate for automated decision making applications. We combined two NNEs to work together, specifically DTBSE and WMVE, obtaining a gating neural ensemble. Overall, the GaNEn shows excellent discrimination accuracy and it improves other previous computational solutions, ensembles or not. An important characteristic of these tools is their ability to estimate the reliability of the output, depending on the battery of available inputs. As future work we plan to enhance the capabilities of the CMs on a GaNEn system. The results could be improved by implementing other functions, perhaps based on Fuzzy Logic, or introducing learning in them. In that case the results may be less sensitive to the set of selected CMs.

The results presented in this work confirm the good performance of the proposed systems and its validity in dementia diagnosis and detection of pollutants in the environment.

One of our ensembles drawbacks occurs when there are a high number of inputs. In this case the computation of all performance combinations or types of inputs may be nearly impossible. The use of feature selection methods could help to overcome this obstacle. The search of an optimum set of CMs can also be a difficult task.

Combining our methods with other optimization techniques, such as Evolutionary ANN or Swarm optimization, could improve the module selection.

Acknowledgments. Authors would like to thank the Canary Islands Government and Science and Innovation Ministry of the Spanish Government for their support under Research Project "SolSubC200801000347" and "TIN2009-13891" respectively.

References

[1] Hansen, L.K., Salamon, P.: Neural Network Ensembles. IEEE Transactions on Pattern Analysis and Machine Intelligence 12(10), 993–1001 (1990)
[2] Polikar, R.: Ensemble Based Systems in Decision Making. IEEE Circuits and Systems Magazine 6(3), 21–45 (2006)
[3] Liu, Y., Yao, X., Higuchi, T.: Designing Neural Network Ensembles by Minimising Mutual Information. In: Mohammadian, M., Sarker, R.A., Yao, X. (eds.) Computational Intelligence in Control, pp. 1–21. Idea Group Inc., USA & London (2003)
[4] Sharkey, A.J.C.: Multi-Net Systems. In: Sharkey, A.J.C. (ed.) Combining Artificial Neural Nets, pp. 1–30. Springer, Heidelberg (1999)
[5] Jacobs, R.A., Jordan, M.I., Barto, A.G.: Task decomposition through competition in a modular connectionist architecture: The what and where vision tasks. Cognitive Science 15(2), 219–250 (1991)
[6] Kadlec, P., Gabrys, B.: Gating Artificial Neural Network Based Soft Sensors. In: Matsuo, T., Ito, T., Kobayashi, M., Minami, F. (eds.) New Challenges in Applied Intelligence Technologies, pp. 193–202. Springer, Heidelberg (2008)
[7] Ruta, D., Gabrys, D.: Classifier selection for majority voting. Information Fusion 6(1), 63–81 (2005)
[8] García Báez, P., Viadero, C.F., García, J.R., Araujo, C.P.S.: An Ensemble Approach for the Diagnosis of Cognitive Decline with Missing Data. In: Corchado, E., Abraham, A., Pedrycz, W. (eds.) HAIS 2008. LNCS (LNAI), vol. 5271, pp. 353–360. Springer, Heidelberg (2008)
[9] Suárez Araujo, C.P., García Báez, P., Fernández Viadero, C.: GaNEn: a new gating neural ensemble for automatic assessment of the severity level of dementia using neuropsychological tests. In: Broadband and Biomedical Communications (IB2Com), pp. 1–6. IEEE Xplore, Málaga (2010)
[10] Hecht Nielsen, R.: Counterpropagation networks. Applied Optics 26(23), 4979–4983 (1987)
[11] García Báez, P., Fernández López, P., Suárez Araujo, C.P.: A Parametric Study of HUMANN in Relation to the Noise. Application to the Identification of Compounds of Environmental Interest. Systems Analysis Modelling and Simulation 43(9), 1213–1228 (2003)
[12] Báez, P.G., del Pino, M.A.P., Viadero, C.F., Araujo, C.P.S.: Artificial Intelligent Systems Based on Supervised HUMANN for Differential Diagnosis of Cognitive Impairment: Towards a 4P-HCDS. In: Cabestany, J., Sandoval, F., Prieto, A., Corchado, J.M. (eds.) IWANN 2009. LNCS, vol. 5517, pp. 981–988. Springer, Heidelberg (2009)
[13] Samad, T., Harp, S.A.: Self-organization with partial data. Network 3(2), 205–212 (1992)
[14] Suárez Araujo, C.P., García Báez, P., Sánchez Rodríguez, A., Santana Rodríguez, J.J.: HUMANN-based system to determine pesticides using multifluorescence spectra: An ensemble approach. Analytical and Bioanalytical Chemistry 394(4), 1059–1072 (2009)

Hybrid Artificial Neural Networks: Models, Algorithms and Data

P.A. Gutiérrez* and C. Hervás-Martínez

Department of Computer Science and Numerical Analysis,
University of Córdoba, Spain
Tel.: +34 957 218 349; Fax: +34 957 218 630
pagutierrez@uco.es

Abstract. Artificial neural networks (ANNs) constitute a class of flexible nonlinear models designed to mimic biological neural systems. ANNs are one of the three main components of computational intelligence and, as such, they have been often hybridized from different perspectives. In this paper, a review of some of the main contributions for hybrid ANNs is given, considering three points of views: models, algorithms and data.

1 Introduction

Artificial Neural Networks (ANNs) are a very flexible modelling technique designed to mimic biological neural systems, whose computing power is developed using an adaptive learning process. Properties and characteristics of ANNs have made them a common tool when successfully solving high complexity problems from different areas, e.g. medical diagnosis, financial data modelling, predictive microbiology, remote sensing, analytical chemistry...

From a statistical point of view, one hidden-layer feed-forward ANNs are generalized linear regression models considering a linear combination of non-linear projections of the input variables (*basis functions*), $B_j(\mathbf{x}, \mathbf{w}_j)$, in the following way:

$$y(\mathbf{x}, \boldsymbol{\theta}) = \beta_0 + \sum_{j=1}^{M} \beta_j B_j(\mathbf{x}, \mathbf{w}_j) \quad (1)$$

where M is the number of non-linear combinations, $\boldsymbol{\theta} = \{\boldsymbol{\beta}, \mathbf{w}_1, ..., \mathbf{w}_M\}$ is the set of parameters associated with the model, $\boldsymbol{\beta} = \{\beta_0, ..., \beta_M\}$ are those parameters associated to the linear part of the model, $B_j(\mathbf{x}, \mathbf{w}_j)$ are each of the basis functions, $\mathbf{w}_j$ are the set of parameters associated to each basis function and $\mathbf{x} = \{x_1, ..., x_k\}$ the input variables associated to the problem. This kind of models, that include ANNs, are called *linear models of basis functions* [5]. One-dimensional polynomial regression is a concrete example of this kind of models, where there is only one input variable and each basis function is a power of this variable, $B_j(\mathbf{x}) = x^j$. One extension of this model, dividing the input space in

* Corresponding author.

J. Cabestany, I. Rojas, and G. Joya (Eds.): IWANN 2011, Part II, LNCS 6692, pp. 177–184, 2011.

different parts and approximating each part using a different polynomial, are spline regression models [28]. There are several possibilities for selecting the typology of the basis functions. One possibility is to consider functions located in subspaces of the input space, such as Radial Basis Functions, RBFs, which constitute RBF neural networks [4,42]. Other one is based on projection functions, such as sigmoidal unit basis functions, which are part of the MultiLayer Perceptron (MLP) [12], or product units which generate product unit neural networks [16]. A mixture of different kinds of basis functions [25] is an interesting alternative, which could be able to take advantage from the benefits of each one.

ANN learning consists of estimating the values for the set of parameters $\boldsymbol{\theta}$ and an architecture (i.e. number of non-linear transformations M and number of connections between the different nodes of the network). Once an architecture has been selected, supervised, unsupervised or reinforcement learning in ANNs has been usually achieved by adjusting the connection weights iteratively using a gradient descent-based optimization algorithm such as Back Propagation. The main problems associated with this kind of algorithms are the necessity of a previously defined architecture for the neural net, their sensitivity to the initial conditions of training, their local character and their restriction to only differentiable error surfaces. Much recent research has been done for obtaining neural network algorithms by combining different soft-computing paradigms, resulting in hybrid approaches with the advantages of the different paradigms considered.

One final possibility for hybridization is the use of heterogeneous datasets. Many ensembles techniques are based on reweighting datasets (bagging or boosting approaches), generating specific classifiers or regressors which complement each other depending on the specific weights of the training patterns [6,49,50].

This paper is aimed to offer a general view of the different ways in which the standard approach of ANNs has been hybridized, extending the model of Eq. 1, combining gradient based algorithms with other learning paradigms or using heterogeneous data structures.

2 Models

Hybrid models of ANNs have been proposed, where different activation/transfer functions are used for the nodes in the hidden layer. Several authors have proposed the hybridization of different basis functions, using either one single hybrid hidden layer or several connected pure layers.

According to Duch and Jankowski [15], mixed transfer functions within one network may be introduced in two ways. In the first way, a constructive method selects the most promising function from a pool of candidates in RBF-like architecture and add it to the network [14]. In the second approach, starting from a network that already contains several types of functions (such as Gaussian and sigmoidal functions), pruning or regularization techniques are used to reduce the number of functions [15].

Optimal transfer function networks were presented as a method for selecting appropriate functions for a given problem [38], creating architectures that are well matched for some given data and resulting in a very small number of hidden nodes.

In the functional link networks of Pao [47], a combination of various functions, such as polynomial, periodic, sigmoidal and Gaussian functions is used. The basic idea behind a functional link network is the use of functional links, adding nonlinear transformation of the input variables to the original set of variables, and suppressing the hidden layer, only performing a linear transformation of this derived input space. In this way, the first nonlinear mapping is fixed and the second linear mapping is adaptive.

A more complex approach considers several layers or models, each one containing a basis function structure, resulting in a modular system. For example, Iulian proposes a methodology including three distinct modules implementing a hybrid feed-forward neural network, namely a Gaussian type RBF network, a principal component analysis process, and a MLP [36]. Another proposal of Lehtokangas and Saarinen considers two hidden layers in the model [43], the first one composed of Gaussian functions and the second one made up of Sigmoidal Unit (SU) basis functions.

Neural networks using different transfer functions should use fewer nodes, enabling the function performed by the network to be more transparent. For example, one hyperplane may be used to divide the input space into two classes and one additional Gaussian function to account for local anomaly. Analysis of the mapping performed by an MLP network trained on the same data will not be so simple.

In this context, it is worth emphasizing the paper by Cohen and Intrator [11], which is based on the duality and complementary properties of projection-based functions (SU and Product Unit, PU) and kernel typology (RBF). This hybridization of models has been justified theoretically by Donoho [13], who demonstrated that any continuous function can be decomposed into two mutually exclusive functions, such as radial (kernel functions) and crest ones (based on projection). Although theoretically this decomposition is justified, in practise it is difficult to apply gradient methods to separate the different locations of a function (in order to adjust them by means of a combination of RBFs) and then to estimate the residual function by means of a functional approach based on projections, all without getting trapped in local optima in the procedure of error minimization [22].

Recently, Wedge and collaborators [53] have presented a hybrid RBF and sigmoidal neural network using a three step training algorithm for function approximation, aiming to achieve an identification of the aspects of a relationship that are universally expressed separately from those that vary only within particular regions of the input space.

An Evolutionary Algorithm (EA) has been recently proposed for the evolution of ANNs combining kernel (RBF) functions and projection ones (PUs and SUs), concluding that the combination of RBFs and PUs offers a very competitive performance [25].

Finally, ANN models can be hybridized with other kinds of simpler models. For example, a recently proposed combination of neural networks and logistic regression [26,32,33] allows the generation of hybrid linear/nonlinear classification surfaces and the identification of possible strong interactions that may exist between the attributes (also known as covariates in the Logistic Regression literature) which define the classification problem.

3 Algorithms

Feed-forward ANNs with the Back Propagation (BP) training algorithm is the standard approach when applying ANNs. BP has the advantage of the gradient-directed search, always minimizing the error.

However, there are several drawbacks when applying BP: strict *dependency* to a learning rate coefficient, which can either lead to oscillation or an indefinitely long training time; *network paralysis* might also occur, i.e. as the ANN trains, the weights tend to be quite large values; BP usually *slows down* by an order of magnitude for every extra (hidden) layer added to ANN; the error space, which can be complex and contain many deceiving local minima (multi-modal).

Therefore, the BP most likely gets trapped into a local minimum, making it entirely dependent on initial (weight) settings. There are many BP variants and extensions trying to address some or all of these problems [17,29,51]. However, all of them have one major difficulty in common: the number of hidden neurons has to be fixed. Although some solutions have been proposed for this problem [37,30], they do not generally perform well for all problems.

Adjusting the architecture of the net can be considered a search process within the architecture space containing all potential and feasible architectures. Some research in this direction include: constructive and pruning algorithms [8,21,41,48]. However, such structural hill climbing methods are susceptible to becoming trapped at structural local minima [1] because the architecture space is non-differentiable, complex, deceptive and multi-modal [45].

Evolutionary Algorithms (EAs) [2], from genetic algorithms [24], to genetic programming [40], evolutionary strategies [3] or evolutionary programming [20], are more promising candidates for both training and evolving the ANNs. This is known as Evolutionary Artificial Neural Networks (EANNs) [7,55], and it has been used in many applications [10,27,31]. In general, EANNs provide a very successful platform for optimizing network performance and architecture simultaneously.

The common point of all is that EAs are population based stochastic processes and they can avoid being trapped in a local optimum. Furthermore, they have the advantage of being applicable to any type of ANN, feed-forward or not, with any activation function, differentiable or not.

Many researchers have shown that EAs perform well for global searching because they are capable of rapidly finding and exploring promising regions in the search space, but they take a relatively long time to converge to a local optimum [44]. One possible way to improve convergence is the hybridization of these algorithms with *local optimization* algorithms, which are usually BP methods when evolving ANNs. These methods are commonly known as hybrid algorithms. A common classification of hybrid algorithms (when applied to evolve ANNs) differentiate two major types [46]: "noninvasive" which refers to approaches where EA selection is used but fitness evaluation requires BP or other gradient training [19,35]; "invasive" which refers to those approaches where the system uses EA for ANN's weight and structure evolution [1,18,23,54,56].

Other kind of algorithms have been also hybridized when learning neural networks, such as extreme learning machines (a very fast method for training one-hidden-layer MLPs [34]) have been combined with a differential evolution approach, which selects values for the weights of the hidden neurons [57].

4 Data

Another possibility for generating more complex ANNs is to train them using heterogeneous data structures. Very popular ensemble approaches are based on this idea, including bagging and boosting, which explicitly create different training sets for different ANNs by probabilistically changing the distribution of the original training data [6,50].

Heterogeneous datasets can be naturally obtained from some real-world problems and can also substantially improve quality of classifiers or regressors. Transfer learning [39] is a new learning paradigm, in which, besides the training data for the targeted learning task, data that are related to the task (often under a different distribution) are also employed to help train a better learner [9].

A very interesting framework has also been recently proposed, the learning using privileged information paradigm [52]. In this paradigm, in addition to the standard training data, $x \in X$ and $y \in \{-1, 1\}$, a teacher supplies student with the privileged information, $x^* \in X^*$. The privileged information is only available for the training examples and is never available for the test examples.

5 Conclusions

A short review of some of the different methods used for hybridizing neural networks has been presented. The methods presented in this paper have been grouped using three criteria: models (when the mathematical models include different typology functions or even different statistical learners), algorithms (when the learning procedure take advantage from traditional methods and heuristics) or data (when ANNs are obtained from heterogeneous data structures). All these methods perform reasonable well for a given set of databases, but no single methodology can be assessed as the best one in a general way.

References

1. Angeline, P.J., Sauders, G.M., Pollack, J.B.: An evolutionary algorithm that constructs recurren neural networks. IEEE Transactions on Neural Networks 5, 54–65 (1994)

2. Back, T.: Evolutionary Algorithms in Theory and Practice, Oxford (1996)
3. Back, T., Fogel, D.B., Michalewicz, Z.: Handbook of Evolutionary Computation. IOP Publishing Ltd., Bristol (1997)
4. Bishop, C.M.: Improving the generalization properties of radial basis function neural networks. Neural Computation 3(4), 579–581 (1991)
5. Bishop, C.M.: Pattern Recognition and Machine Learning. Information Science and Statistics, 1st edn. Springer, Heidelberg (2006)
6. Breiman, L.: Bagging predictors. Machine Learning 24(2), 123–140 (1996)
7. Buchtala, O., Klimek, M., Sick, B.: Evolutionary optimization of radial basis function classifiers for data mining applications. IEEE Transactions on Systems, Man, and Cybernetics, Part B 35(5), 928–947 (2005)
8. Burgess, N.: A constructive algorithm that converges for real-valued input patterns. Int. J. Neural. Syst. 5(1), 59–66 (1994)
9. Caruana, R.: Multitask learning. Machine Learning 28(1), 41–75 (1997)
10. Chaiyaratana, N., Piroonratana, T., Sangkawelert, N.: Effects of diversity control in single-objective and multi-objective genetic algorithms. Journal of Heuristics 13(1), 1–34 (2007)
11. Cohen, S., Intrator, N.: A hybrid projection-based and radial basis function architecture: initial values and global optimisation. Pattern Analysis & Applications 5(2), 113–120 (2002)
12. Cybenko, G.: Approximation by superpositions of a sigmoidal function. Mathematics of Control, Signals, and Systems 2(4), 303–314 (1989)
13. Donoho, D.L., Johnstone, I.M.: Projection-based approximation and a duality with kernel methods. The Annals of Statistics 17(1), 58–106 (1989)
14. Duch, W., Adamczak, R., Diercksen, G.: Constructive density estimation network based on several different separable transfer functions. In: Proceedings of the 9th European Symposium on Artificial Neural Networks, Bruges, Belgium, pp. 107–112 (2001)
15. Duch, W., Jankowski, N.: Transfer functions hidden possibilities for better neural networks. In: Proceedings of the 9th European Symposium on Artificial Neural Networks, Bruges, Belgium, pp. 81–94 (2001)
16. Durbin, R., Rumelhart, D.: Products units: A computationally powerful and biologically plausible extension to backpropagation networks. Neural Computation 1(1), 133–142 (1989)
17. Fahlman, S.E.: An empirical study of learning speed in back-propagation networks. Technical report, cmu-cs-88-162, Carnegie-Mellon University (1988)
18. Fogel, D.: Using evolutionary programming to greater neural networks that are capable of playing tic-tac-toe. In: International Conference on Neural Networks, pp. 875–880. IEEE Press, San Francisco (1993)
19. Fogel, D.: Evolutionary Computation: Toward a New Philosophy of Machine Intelligence. IEEE Press, New York (1995)
20. Fogel, L.J.: Artificial Intelligence through Simulated Evolution, 1st edn. John Wiley & Sons, New York (1966)
21. Frean, M.: The upstart algorithm: A method for constructing and training feedforward neural networks. Neural Computation 2, 198–209 (1990)
22. Friedman, J.: Multivariate adaptive regression splines (with discussion). Annals of Statistics 19, 1–141 (1991)
23. García-Padrajas, N., Hervás-Martínez, C., Muñoz-Pérez, J.: Covnet: A cooperative coevolutionary model for evolving artificial neural networks. IEEE Transaction on Neural Networks 14(3), 575–596 (2003)

24. Goldberg, D.E.: Genetic Algorithms in Search, Optimization, and Machine Learning. Addison-Wesley Professional, Reading (1989)
25. Gutiérrez, P.A., Hervás-Martínez, C., Carbonero, M., Fernández, J.C.: Combined projection and kernel basis functions for classification in evolutionary neural networks. Neurocomputing 27(13-15), 2731–2742 (2009)
26. Gutiérrez, P.A., Hervás-Martínez, C.,Martínez-Estudillo, F.J.: Logistic regression by means of evolutionary radial basis function neural networks. IEEE Transactions on Neural Networks 22(2), 246–263 (2011)
27. Gutiérrez, P.A., López-Granados, Peña-Barragán, J.M., Gómez-Casero, M.T., Hervás, C.: Mapping sunflower yield as affected by Ridolfia segetum patches and elevation by applying evolutionary product unit neural networks to remote sensed data. Computers and Electronics in Agriculture 60(2), 122–132 (2008)
28. Hastie, T., Tibshirani, R., Friedman, J.H.: The Elements of Statistical Learning. Springer, Heidelberg (2001)
29. Haykin, S.: Neural Networks: A comprehensive Foundation, 3rd edn. Prentice-Hall, Englewood Cliffs (2008)
30. Hecht-Nielsen, R.: Neurocomputing. Addison-Wesley, Reading (1990)
31. Hervas-Martinez, C., Garcia-Gimeno, R.M., Martinez-Estudillo, A.C., Martinez-Estudillo, F.J., Zurera-Cosano, G.: Improving microbial growth prediction by product unit neural networks. Journal of Food Science 71(2), M31–M38 (2006)
32. Hervas-Martínez, C., Martínez-Estudillo, F.J.: Logistic regression using covariates obtained by product-unit neural network models. Pattern Recognition 40(1), 52–64 (2007)
33. Hervas-Martínez, C., Martínez-Estudillo, F.J., Carbonero-Ruz, M.: Multilogistic regression by means of evolutionary product-unit neural networks. Neural Networks 21(7), 951–961 (2008)
34. Huang, G.B., Zhua, Q.Y., Siewa, C.K.: Extreme learning machine: Theory and applications. Neurocomputing 70(1–3), 489–501 (2006)
35. Islam, M.M., Yao, X., Murase, K.: A constructive algorithm for training cooperative neural network ensembles. IEEE Transactions on Neural Networks 14(4), 820–834 (2003)
36. Iulian, B.C.: Hybrid feedforward neural networks for solving classification problems. Neural Processing Letters 16(1), 81–91 (2002)
37. Jadid, M.N., Fairbairn, D.R.: Neural-network applications in predicting moment-curvature parameters from experimental data. Engineering Applications of Artificial Intelligence 9(3), 309–319 (1996)
38. Jankowski, N., Duch, W.: Optimal transfer function neural networks. In: Procedings of the 9th European Symposium on Artificial Neural Networks, Bruges, Belgium, pp. 101–106 (2001)
39. Jiang, L., Zhang, J., Allen, G.: Transferred correlation learning: An incremental scheme for neural network ensembles. In: Proceedings of the 2010 International Joint Conference on Neural Networks (IJCNN), pp. 1–8. IEEE Press, Barcelona (2010)
40. Koza, J.R., Rice, J.P.: Genetic generation of both the weights and architecture for a neural network. In: Proceedings of International Joint Conference on Neural Networks, vol. 2, pp. 397–404. IEEE Press, Seattle (1991)
41. Le Cun, Y., Denker, J.S., Solla, S.A.: Optimal brain damage. In: Advances in neural information processing systems, pp. 598–605. Morgan Kaufmann Publishers Inc., San Francisco (1990)
42. Lee, S.H., Hou, C.L.: An art-based construction of RBF networks. IEEE Transactions on Neural Networks 13(6), 1308–1321 (2002)

43. Lehtokangas, M., Saarinen, J.: Centroid based multilayer perceptron networks. Neural Processing Letters 7, 101–106 (1998)
44. Michalewicz, Z.: Genetic algorithms + data STRUCTURES = evolution programs. Springer, New York (1996)
45. Miller, G.F., Todd, P.M., Hedge, S.U.: Designing neural networks using genetic algorithms. In: Schaffer, J.D. (ed.) Proceedings of the 3rd International Conference on Genetic Algorithms and Their Applications, pp. 379–384. Morgan Kaufmann, San Mateo (1989)
46. Palmes, P.P., Hayasaka, T., Usui, S.: Mutation-based genetic neural network. IEEE Transactions on Neural Networks 16(3), 587–600 (2005)
47. Pao, Y.H., Takefuji, Y.: Functional-link net computing: theory, system architecture, and functionalities. IEEE Computer 25(5), 76–79 (1992)
48. Reed, R.: Pruning algorithms. a survey. IEEE Transactions on Neural Networks 4, 740–747 (1993)
49. Rokach, L.: Ensemble-based classifiers. Artificial Intelligence Review 33(1–2), 1–39 (2010)
50. Schapire, R.: Theoretical views of boosting. In: Proceedings 4th European Conference on Computational Learning Theory, pp. 1–10 (1999)
51. Sutton, R.S.: Two problems with backpropagation and other steepest-descent learning procedures for networks. In: Proceedings of the 8th Annual Conference of the Cognitive Science Society (1986)
52. Vapnik, V., Vashist, A.: A new learning paradigm: Learning using privileged information. Neural Networks 22(5-6), 544–557 (2009)
53. Wedge, D., Ingram, D., McLean, D., Mingham, C., Bandar, Z.: On global-local artificial neural networks for function approximation. IEEE Transactions on Neural Networks 17(4), 942–952 (2006)
54. Yao, X.: Global optimization by evolutionary algorithms. In: Proceedings of the 2nd Aizu International Symposium on Parallel Algorithms/Architecutre Synthesis (pAs 1997), Aizu-Wakamatsu, Japan, pp. 282–291 (1997)
55. Yao, X.: Evolving artificial neural networks. Proceedings of the IEEE 87(9), 1423–1447 (1999)
56. Yao, X., Liu, Y.: Making use of population information in evolutionary artificial neural networks. IEEE Transactions on Systems, Man and Cybernetics, Part B: Cybernetics 28(3), 417–425 (1998)
57. Zhu, Q.Y., Qin, A.K., Suganthan, P.N., Huang, G.B.: Evolutionary extreme learning machine. Pattern Recognition 38(10), 1759–1763 (2005)

Automatic Recognition of Daily Living Activities Based on a Hierarchical Classifier

Oresti Banos, Miguel Damas, Hector Pomares, and Ignacio Rojas

Department of Computer Architecture and Computer Technology, University of Granada, C/Periodista Daniel Saucedo Aranda s/n E-18071, Granada, Spain
{oresti,mdamas,hpomares,irojas}@atc.ugr.es

Abstract. Physical activity recognition has become an increasing research area specially on health-related fields. The amount of different postures, movements and exercises in addition to the difficulty of the individuals particular execution style determine that extremely robust efficient knowledge inference systems are extremely necessary, being classification process one of the most crucial steps. Considering the power of binary classification in contrast to direct multiclass approaches, and the capabilities offered by multi-sense environments, we define a novel classification schema based on hierarchical structures composed by weighted decision makers. Remarkable accuracy results are obtained for a particular activity recognition problem in contrast to a traditional multiclass majority voting algorithm.

Keywords: hierarchical classification, weighted classification, binary classifiers, activity recognition.

1 Introduction

Over the past decade, there has been considerable research effort directed toward monitoring and classification of physical activity patterns from body-fixed sensor information. The application on several fields such as manufacturer industry [6], sports [2] or videogames interactive entertainment [11] is clearly recognized, however a special interest is recently being focused on healthcare. Chronic disease management [15], rehabilitation systems [10] or disease prevention [14] are several topics where activity recognition potential is being revealed.

One of the most important stages on activity recognition systems is machine learning. Several paradigms such as artificial neural networks [8], support vector machines [10], Bayesian classifiers [4] or hidden Markov models [12] have been widely used, but they are less accurate as the number of classes (activities) grows [9].

Some of these schemas are originally defined through binary classification, recognized as the most interesting approach [1], but in some cases traditional multiclass generalization is not efficiently practical. Besides, the use of several monitoring systems usually improves the system accuracy rates, but to the best of our knowledge, no general models are presented for the combined use of them. We

J. Cabestany, I. Rojas, and G. Joya (Eds.): IWANN 2011, Part II, LNCS 6692, pp. 185–193, 2011.

propose a wide-ranging multiclass schema by reducing the study to multiple binary or class specialized problems, employing a weighted structure to define the decision maker. This scheme is extended to each source to define a hierarchical knowledge inference system with a two-level weighting decision framework.

The rest of the paper is organized as follows. In section 2 a brief summary of the activity recognition process is presented. Section 3 describes the hierarchical weighted classification methodology proposed, showing the fundamentals of this method and the algorithm's main steps. Finally the performance of the method is evaluated for a specific example in section 4.

2 Activity Recognition Method

The experimental setup starts from a signal set [4] corresponding to acceleration values measured by a group of sensors located in several strategic body locations (hip, wrist, arm, ankle, thigh), for four daily activities (see Fig. 1). The methodology presented from this point forward can be easily generalized to other studies related to activity recognition from a set of features.

Monitored data have some artifacts and noise associated to the acquisition data process. Considering that a 20 Hz sampling is sufficient to assess habitual daily physical activity [7], a low pass elliptic filter with 20 Hz cutoff frequency, followed by a 0.5 Hz cutoff frequency high pass elliptic filter (both 0.5dB passband ripple and 20dB stopband attenuation) are applied to respectively remove the high frequency noise and the original signal offset.

Afterward a parameter set made up of 861 features is obtained. This corresponds to a combination of statistical functions such as mode, median, variance, etc., and magnitudes obtained from a domain transformation of the original data such as energy

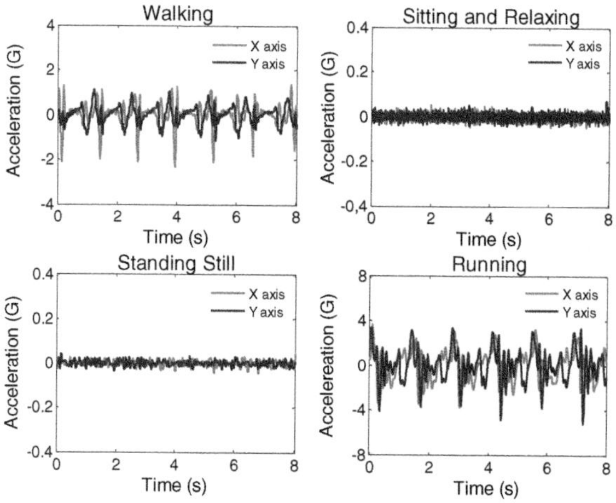

Fig. 1. Signals corresponding to four usual daily physical activities (ankle accelerometer). These activities have been considered for the daily living commonness, as well as because of the manner execution similarity between couples walking/running and sitting/standing.

Table 1. Feature set generation functions. Every statistical function is computed for each magnitude generating the set of 861 features.

Magnitudes	Statistical functions
Amplitude	4th and 5th central statistical moments
Autocorrelation function	Energy
Cepstrum	Arithmetic/Harmonic/Geometric/Trimmed mean
Correlation lags	Entropy
Cross correlation function	Fisher asymmetry coefficient
Energy spectral density	Maximum / Position of
Spectral coherence	Median
Spectrum amplitude/phase	Minimum / Position of
Histogram	Mode
Historical data lags	Kurtosis
Minimum phase reconstruction	Data range
Daubechies wavelet decomposition	Total harmonic deviation
	Variance
	Zero crossing counts

spectral density, spectral coherence or wavelet ("a1 to a5" and "d1 to d5" Daubechies levels of decomposition) among others for both signal axes. "Fisher asymmetry coefficient of the X axis signal histogram", "Y axis signal wavelet coefficients a2 zero crossing counts" or "X axis-Y axis cross correlation harmonic mean" are possible examples of features obtained from the set defined (Table 1).

Feature selection processes have the responsibility of deciding which features or magnitudes are the most important ones to infer the kind of activity the person is carrying out. Taking into account the binary class classifier approach (described in the next section), several class specialized feature selection schemas based on an 'one-against-all' strategy have been applied to the data. The methodologies implemented are based on Bhattacharyya distance, Entropy [13], and a technique (*Quality Group Ranking* or *QGR*) recently defined in a previous work [3].

3 Hierarchical Weighted Classifier (HWC)

Considering binary classification in general more accurate than direct multiclass approach, is extremely important to establish an appropriate multiclass extension schema which permits to preserve and optimize binary entities capabilities, even more when fusion of several sensors or sources is considered. A general methodology based on the combination of binary or class classifier decisions in a hierarchical structure with an special application for multisource problems is presented in this section.

The framework of the HWC is composed by three classification levels or stages related to the decision structure defined (see Fig. 2). In general, for $m=1,...,M$ sources and $n=1,...,N$ classes, a set of M x N "class classifiers" (c_{mn}) are defined. They are binary classifiers specialized in the classification of the class n by using the data acquired from the m source. Each one apply an 'one-versus-rest' strategy, so any classification paradigm can be easily applied. These define the first level or *class level classifier*. The second stage, *source level classifier*, is defined by M "source classifiers" (S_m). Source classifiers are not machine learning as class classifiers, but hierarchical decision models which define a classification entity. Source classifiers structures are composed by several class classifiers as is shown in Fig. 2, defining a decision system based on class classifiers weighted decisions. This approach is repeated for the next level, *method level classifier*, which ultimately defines a decision structure constituted by source classifiers weighted decisions.

In accordance to the structure described above, a process consisting of a few main steps is carried out to define the complete HWC. The process starts by evaluating the individual accuracy of each class classifier, defined through its corresponding feature vector (several vector lengths should be considered to find out the best results for every classifier). A 10-fold cross validation is suggested for accomplish this task and this is repeated 100 times to ensure the statistical robustness. The entirely process is repeated for each source. Considering average accuracy rates ($\overline{R_{mn}}$ for source m and class classifier n) as a measure of the pattern recognition capabilities of each classifier, an associated weight is obtained for each one:

$$\beta_{mn} = \frac{\overline{R_{mn}}}{\sum_{k=1}^{N} \overline{R_{mk}}} \tag{1}$$

These weights are a measure of the importance that every class classifier will have on the source classifier decision schema. An specific voting algorithm is considered in this point to define source classifiers decision. For a source m, given a sample x_{mk} to

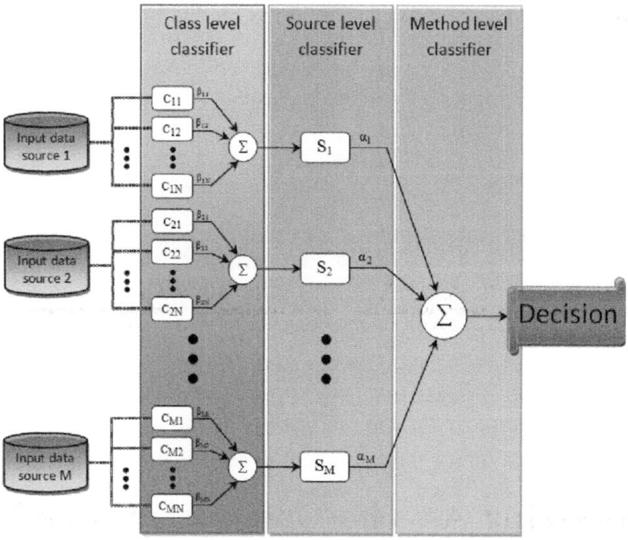

Fig. 2. HWC general structure for a problem with N classes and M sources

be classified and being q the class predicted by the classifier c_{mn}, if the sample is classified as belonging to the classifier class of specialization (q=n), the classifier will set its decision as '1' for the class n and '0' for the rest of classes. Opposite is made for ($q \neq n$). In summary, the decision from the classifier n for the class q is:

$$D_{nq}(x_{mk}) = \begin{cases} 1, & x_{mk} \ \textit{classified as } q \\ 0, & x_{mk} \ \textit{not classified as } q \end{cases} \Bigg\} \ \forall q = n \qquad \begin{cases} 1, & x_{mk} \ \textit{not classified as } q \\ 0, & x_{mk} \ \textit{classified as } q \end{cases} \Bigg\} \ \forall q \neq n \qquad \forall q = 1,\ldots,N \tag{2}$$

Once decisions have been offered by each class classifier for every class q by applying (2), it is time to compute the weighted output for the m source classifier:

$$O_{mq}(x_{mk}) = \sum_{n=1}^{N} \beta_{mn} D_{nq}(x_{mk}) \qquad \forall q = 1,\ldots, N \tag{3}$$

Finally, the class predicted for the m source classifier (q_m) is the class q for which source classifier output is maximized:

$$q_m = \arg\max_q \left(O_{mq}(x_{mk}) \right) \tag{4}$$

In this stage, the source level classifier is completely defined. Every source class classifier can be used separately, looking for the most interesting for the particular problem analyzed. If extremely accurate classifiers are found, maybe this is enough to be used as the final pattern recognition system solution. However, *fusion* or combination of sources information is in general a more robust and efficient solution. Consequently, the complete process described before is extended to a new hierarchy level, the *method level classifier*. First, source classifiers weights (α_m) are obtained by calculating the average accuracy rates for each source classifier ($\overline{R_m}$), so a cross validation process is again repeated but now focusing on the source classifiers predictions. The weight for the source m is:

$$\alpha_m = \frac{\overline{R_m}}{\sum_{k=1}^{M} \overline{R_k}} \tag{5}$$

The output is calculated taking into account the individual outputs obtained for each source classifier. For a sample x_k defined through the corresponding information obtained from each source (x_{1k},...,x_{Mk}):

$$O_q(x_k) = O_q(\{x_{1k},\ldots, x_{Mk}\}) = \sum_{p=1}^{M} \alpha_p O_{pq}(x_{pk}) \qquad \forall q = 1,\ldots, N \tag{6}$$

Similar to (4) the final class predicted q is:

$$q = \arg\max_{q} \left(O_q(x_k)\right) \qquad q \in [1,..., N] \tag{7}$$

In summary, the HWC is absolutely defined through the class classifiers (c_{mn}), class level weights (β_{mn}) and the source level weights (α_m) at this point.

4 Results

The aim of the methodology presented is to define robust and efficient pattern recognition systems based on binary class classifiers. For the activity recognition problem presented (N = 4, M = 5), two classification schemas based on majority voting (MV) and our approach (HWC) are respectively used. Naive Bayes machine learning paradigm is applied for the structure of class classifiers, using the two best features selected by each feature selector (for every source and class, see table 2), assuming the stochastical independence is satisfied. Fig. 3 shows the results.

Clearly, a comparison of MV and HWC allows us to realize the potential of HWC. All MV average accuracy rates for source classifiers are improved significantly by using HWC, up to nearly 35% in some cases as *ankle* source for QGR features selected. Furthermore, this is extensible to *fusion* of source classifiers or *method classification*, which are considerable better in all cases. In fact, no improvement appears for *fusion* approach when is used following a MV schema. Conversely, important benefits are obtained for HWC, with particular remarkable results for Bhattacharyya and Entropy, with accuracy rates above 90%. Because of QGR source classifiers are extremely accurate (>96%), results for *fusion* is in line to them (~99%).

As was mentioned in section 3, if source classifiers offer outstanding accuracy results, fusion approach may be omitted. This can be seen in Fig.3b, for QGR case. Source classifiers as based on the *thigh* or the *wrist* accelerometer define almost absolute recognition systems (~100%), so for this problem it would not be necessary to use other sensors, something especially important in wearable monitoring contexts.

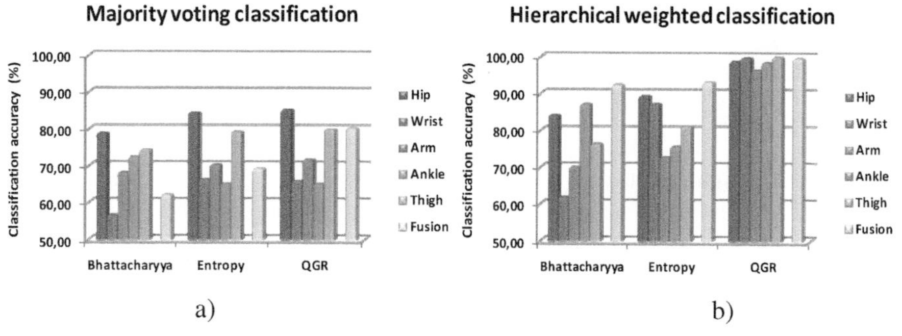

Fig. 3. Accuracy rates using a) MV and b) HWC. Results for each source classifier are identified with the corresponding sensor label. Fusion is referred to the combined use of the different source classifiers (identified as *method level classifier* in section 3).

Table 2. Two best features selected by each feature selection schema used. The features are used for the corresponding class classifier defined through the accelerometer and activity of specialization.

		Walking	*Sitting*	*Standing*	*Running*
BHATTACHARYYA	*Hip*	*Y-axis autocorr. function 5th central moment & Y-axis energy spectral density 5th central moment*	*X-axis minimum phase reconstruction position of the maximum & X-axis energy spectral density 5th central moment*	*X-axis minimum phase reconstruction position of the maximum & Y-axis minimum phase reconstruction position of the minimum*	*Y-axis autocorr. function 5th central moment & Y-axis energy spectral density 5th central moment*
	Wrist	*Y-axis autocorr. function 5th central moment & Y-axis autocorr. function 4th central moment*	*X-axis minimum phase reconstruction position of the maximum & X-axis energy spectral density 5th central moment*	*Y-axis autocorr. function entropy & Y-axis autocorr. function position of the minimum*	*Y-axis autocorr. function 5th central moment & Y-axis energy spectral density 5th central moment*
	Arm	*X-axis autocorr. function 5th central moment & Y-axis autocorr. function 5th central moment*	*Y-axis energy spectral density 5th central moment & Y-axis autocorr. function 5th central moment*	*X-axis autocorr. function entropy & X-axis autocorr. function position of the minimum*	*X-axis minimum phase reconstruction position of the maximum & X-axis autocorr. function 5th central moment*
	Ankle	*X-Y-axes cross correlation function 3th central moment & Y-axis autocorr. function 5th central moment*	*X-axis energy spectral density 5th central moment & X-Y-axes cross correlation function 5th central moment*	*X-axis autocorr. function entropy & X-axis autocorr. function position of the maximum*	*X-axis minimum phase reconstruction position of the maximum & X-Y-axes cross correlation function 5th central moment*
	Thigh	*X-axis minimum phase reconstruction position of the maximum & X-axis autocorr. function 5th central moment*	*Y-axis energy spectral density 5th central moment & Y-axis autocorr. function 5th central moment*	*X-axis autocorr. function entropy & X-axis autocorr. function position of the minimum*	*X-axis autocorr. function 5th central moment & X-axis autocorr. function 4th central moment*
ENTROPY	*Hip*	*Y-axis amplitude trimmed mean & Y-axis amplitude median*	*X-axis amplitude zero crossing counts & X-axis amplitude standard deviation*	*X-axis histogram standard deviation & X-axis histogram maximum*	*X-axis amplitude zero crossing counts & X-axis amplitude standard deviation*
	Wrist	*X-axis amplitude trimmed mean & X-axis amplitude median*	*X-axis amplitude zero crossing counts & X-axis amplitude standard deviation*	*Y-axis spectrum amplitude Fisher asymmetry coef. & Y-axis spectrum amplitude kurtosis*	*X-axis amplitude standard deviation & X-axis amplitude maximum*
	Arm	*X-axis wav. a3 coef. zero crossing counts & Y-axis wav. d2 coef. zero crossing counts*	*X-axis amplitude standard deviation & X-axis amplitude energy*	*X-axis spectrum amplitude Fisher asymmetry coefficient & X-axis spectrum amplitude kurtosis*	*X-axis amplitude zero crossing counts & X-axis amplitude standard deviation*
	Ankle	*Y-axis wav. d1 coef. zero crossing counts & Y-axis wav. d3 coef. zero crossing counts*	*X-axis amplitude standard deviation & X-axis amplitude energy*	*X-axis amplitude kurtosis & X-axis histogram Fisher asymmetry coef.*	*X-axis amplitude zero crossing counts & X-axis amplitude standard deviation*
	Thigh	*X-axis wav. d1 coef. zero crossing counts & Y-axis autocorr. function zero crossing counts*	*X-axis amplitude zero crossing counts & X-axis amplitude standard deviation*	*X-axis minimum phase reconstruction position of the minimum & Y-axis amplitude minimum*	*X-axis amplitude standard deviation & X-axis amplitude energy*
QGR	*Hip*	*X-axis wav. a3 coef. geometric mean & Y-axis wav. a3 coef. geometric mean*	*X-axis wav. a3 coef. geometric mean & Y-axis wav. a3 coef. geometric mean*	*X-axis wav. a3 coef. geometric mean & Y-axis wav. a3 coef. geometric mean*	*X-axis wav. a3 coef. geometric mean & Y-axis wav. a3 coef. geometric mean*
	Wrist	*X-axis wav. a3 coef. geometric mean & Y-axis wav. a3 coef. geometric mean*	*X-axis wav. a3 coef. geometric mean & Y-axis wav. a3 coef. geometric mean*	*X-axis wav. a3 coef. geometric mean & Y-axis wav. a3 coef. geometric mean*	*X-axis wav. a3 coef. geometric mean & Y-axis wav. a3 coef. geometric mean*
	Arm	*X-axis wav. a3 coef. geometric mean & Y-axis wav. a3 coef. geometric mean*	*X-axis wav. a3 coef. geometric mean & Y-axis wav. a3 coef. geometric mean*	*X-axis wav. a3 coef. geometric mean & Y-axis wav. a3 coef. geometric mean*	*X-axis wav. a3 coef. geometric mean & Y-axis wav. a3 coef. geometric mean*
	Ankle	*X-axis wav. a3 coef. geometric mean & Y-axis wav. a3 coef. geometric mean*	*X-axis wav. a3 coef. geometric mean & Y-axis wav. a3 coef. geometric mean*	*X-axis wav. a3 coef. geometric mean & Y-axis wav. a3 coef. geometric mean*	*X-axis wav. a3 coef. geometric mean & Y-axis wav. a3 coef. geometric mean*
	Thigh	*X-axis wav. a3 coef. geometric mean & Y-axis wav. a3 coef. geometric mean*	*X-axis wav. a3 coef. geometric mean & Y-axis wav. a3 coef. geometric mean*	*X-axis wav. a3 coef. geometric mean & Y-axis wav. a3 coef. geometric mean*	*X-axis wav. a3 coef. geometric mean & Y-axis wav. a3 coef. geometric mean*

5 Conclusions

Several advantages are obtained to traditional multiclass schemas as majority voting. Primarily only features with high binary discriminant capacity are required because of class specialized classifiers define completely the knowledge base of the model. This reduces the complexity of feature selection processes. Besides, once source and class level weights are calculated for the corresponding problem analyzed, the classification system is simply defined through a few decision rules that are easily extended from source classifiers to the complete hierarchy.

A simple activity recognition system has been defined by using solely two features for each class classifier with accuracy rates close to 100% for some cases. Results are particularly interesting for the first level of the hierarchy defined (source classifier), having remarkable accuracy rates for some specific sensors as the wrist (specially for QGR system defined), so interesting for the unobtrusively and applicability on wearable monitoring activity recognition systems.

The good results obtained for the example above are promising for applying this technique to a problem with more classes. For future work we want to test our methodology in different related problems or others (*UCI repository* [5]) with a spread range of classes.

Acknowledgments. We want to express our gratitude to Prof. Stephen S. Intille, Technology Director of the House_n Consortium with the MIT Department of Architecture for the experimental data provided. This work was supported in part by the Spanish CICYT Project TIN2007-60587, Junta de Andalucia Projects P07-TIC-02768 and P07-TIC-02906, the CENIT project AmIVital, of the "Centro para el Desarrollo Tecnológico Industrial" (CDTI- Spain) and the FPU Spanish grant AP2009-2244.

References

1. Allwein, E.L., Schapire, R.E., Singer, Y.: Reducing multiclass to binary: a unifying approach for margin classifiers. J. Mach. Learn. Res. 1, 113–141 (2001)
2. Baca, A., Dabnichki, P., Heller, M., Kornfeind, P.: Ubiquitous computing in sports: A review and analysis. Journal of Sports Sciences 27, 1335–1346 (2009)
3. Banos, O., Pomares, H., Rojas, I.: Ambient living activity recognition based on feature-set ranking using intelligent systems. In: The 2010 International Joint Conference on Neural Networks (IJCNN), pp. 1–4 (2010)
4. Bao, L., Intille, S.: Activity Recognition from User-Annotated Acceleration Data. Pervasive Computing, 1–17 (2004)
5. Frank, A., Asuncion, A.: UCI Machine Learning Repository. University of California, School of Information and Computer Science, Irvine, CA (2010), `http://archive.ics.uci.edu/ml`
6. Koskimaki, H., Huikari, V., Siirtola, P., Laurinen, P., Roning, J.: Activity recognition using a wrist-worn inertial measurement unit: A case study for industrial assembly lines. In: 17th Mediterranean Conference on Control and Automation MED 2009, pp. 401–405 (2009)

7. Mathie, M.J., Coster, A.C.F., Lovell, N.H., Celler, B.G.: Accelerometry: providing an integrated, practical method for long-term, ambulatory monitoring of human movement. Physiol. Meas. 25, 1–20 (2004)
8. Parkka, J., Ermes, M., Korpipaa, P., Mantyjarvi, J., Peltola, J., Korhonen, I.: Activity classification using realistic data from wearable sensors. IEEE Transactions on Information Technology in Biomedicine 10, 119–128 (2006)
9. Preece, S.J., Goulermas, J.Y., Kenney, L.P.J., Howard, D., Meijer, K., Crompton, R.: Activity identification using body-mounted sensors—a review of classification techniques. Physiol. Meas. 30, 1–33 (2009)
10. Sazonov, E., Fulk, G., Sazonova, N., Schuckers, S.: Automatic Recognition of postures and activities in stroke patients. In: Annual International Conference of the IEEE Engineering in Medicine and Biology Society EMBC 2009, pp. 2200–2203 (2009)
11. Schlömer, T., Poppinga, B., Henze, N., Boll, S.: Gesture recognition with a Wii controller. In: Proceedings of the 2nd international conference on Tangible and embedded interaction - TEI 2008 (2008)
12. Singla, G., Cook, D., Schmitter-Edgecombe, M.: Incorporating temporal reasoning into activity recognition for smart home residents. In: AAAI Workshop - Technical Report WS-08-11, pp. 53–61 (2008)
13. Theodoridis, S., Koutroumbas, K.: Pattern Recognition, 4th edn. Elsevier, Amsterdam (2009)
14. Warren, J.M., et al.: Assessment of physical activity – a review of methodologies with reference to epidemiological research: a report of the exercise physiology section of the European Association of Cardiovascular Prevention and Rehabilitation. European Journal of Cardiovascular Prevention & Rehabilitation 17, 127–139 (2010)
15. Zwartjes, D., Heida, T., van Vugt, J., Geelen, J., Veltink, P.: Ambulatory Monitoring of Activities and Motor Symptoms in Parkinson's Disease. IEEE Transactions on Biomedical Engineering 57, 2778–2786 (2010)

Prediction of Functional Associations between Proteins by Means of a Cost-Sensitive Artificial Neural Network

J.P. Florido, H. Pomares, I. Rojas, J.M. Urquiza, and F. Ortuño

Dept. of Computer Architecture and Computer Technology, CITIC-UGR, University of Granada, Spain
jpflorido@ugr.es

Abstract. A challenge in systems biology is to discover new functional associations between proteins and, from these new relationships, assign functions to unannotated proteins. High-throughput data can be the key to achieve this task, but they often lack the degree of specificity needed for predicting accurate protein functional associations. This improvement in specificity can be achieved through the integration of heterogeneous data sets in a suitable manner. In this paper, we assess the quality of prediction of functional associations between proteins in two cases: (1) using each data source in isolation and (2) by means of Artificial Neural Networks through the integration of different biological data sources.

Keywords: Data integration, systems biology, artificial neural networks.

1 Introduction

Nowadays, uncovering new relationships between genes or proteins is one of the major goals of biological studies [1]. Several quantities of high-throughput biological data have become available in recent years to provide diverse insights of protein functions, such as phenotypic profiles, gene expression microarrays, protein sequences, protein phylogenetic profiles or protein-protein interaction data [2]. For most of them, various analytical techniques can be applied to extract protein functional relations. However, there are a number of drawbacks to using each data type in isolation. For instance, microarray and interaction data are often noisy, of varying quality and illuminate only limited aspects of the underlying biological mechanisms [2]. Since they may lack the degree of specificity required for discovering new relationships, it is desirable to maximize the utility of several large-scale datasets to infer new associations between proteins through the integration of them by means of supervised machine learning algorithms [3].

Several methodologies have been proposed in the literature to integrate heterogeneous biological data for prediction of functional associations between proteins. One of the most used is Naïve Bayes [1],[2],[4] although its main drawback is its prediction bias when there is correlation among the data sets to be integrated [5]. Fully connected Bayes models can capture the interdependence among data

J. Cabestany, I. Rojas, and G. Joya (Eds.): IWANN 2011, Part II, LNCS 6692, pp. 194–201, 2011.

sources by directly calculating joint probabilities; however, it results in higher computational costs [5]. Linear Support Vector Machines and Artificial Neural Networks have also been used for the integration of heterogeneous biological data [2], but assuming that the cost of erroneously predicting no functional association between proteins is the same as mistakenly predicting functional association. This is an important issue and it must be taken into account that class distribution of the data set used for training is imbalanced. For example, the expected number of proteins related in any particular biological process is a small percentage of the proteome. This imbalance is particularly problematic in methods based on pairwise associations between proteins, where the expected number of protein pairs sharing functional relationships is an even smaller fraction of all possible protein combinations [6]. For instance, of the approximately 18 million possible protein pairs in yeast, it is expected that less than 1 million are functionally related. Thus, the cost of incorrectly predicting a positive case (functionally related proteins) as a negative (non-functionally related proteins) or, in other words, a false negative, must be greater than the cost of incorrectly classifying a negative as a positive (a false positive). To avoid the use of cost-sensitive learning algorithms, the relative size of positive and negative sets can be the same [2],[5], but, this way, some valuable data are then discarded from learning.

Thus, in this work, we propose a Cost-Sensitive Artificial Neural Network model to demonstrate that, combining several heterogeneous data, functional coupling of linked proteins are more accurately estimated than using individual data sources alone. Moreover, since it is a cost-sensitive model, it takes into account the class imbalance problem and the whole data set can be used during learning. We apply our framework to the prediction of functional associations (pathway sharing) between proteins in *Saccharomyces Cerevisiae.*

2 Methods and Datasets

2.1 Multi-Layer Perceptron Neural Network for Predicting Pathway Sharing

We propose a Multi-layer Perceptron (MLP) that learns to recognize whether two given proteins are in the same pathway (functionally related) based on five features: sequence similarity, phylogenetic profile correlation, correlated gene expression, protein-protein interactions and shared molecular function annotation.

Multilayer perceptrons consist of a number of interconnected processing elements (neurons) arranged in layers [7]. We start from a set of n examples $\{(x_i, y_i); i = 1, ..., n\}$, where $x_i \in \mathbb{R}^m$ (m is the number of features) represents a vector of protein pair scores for the ith example and $y_i \in \{0, 1\}$ is a label vector indicating the classes to which the ith example belongs (1 and 0 denote pathway and non-pathway sharing respectively). Assuming an MLP with one input layer, one hidden layer and a single output layer, the outputs in the hidden and output layer can be written as:

$$p_k = f\left(\sum_{j=1}^{m} w_{kj}^{(p)} s_{ji}\right), k = 1, ..., L; o_k = g\left(\sum_{j=1}^{L} w_{kj}^{(o)} p_j\right), k = 1, ..., C \quad (1)$$

where p_k and o_k are the outputs of the kth neuron in the hidden and output layers respectively. $f(\cdot)$ and $g(\cdot)$ are activation functions which transform the activation level of a neuron into an output signal. L and C are the number of neurons in the hidden and output layers respectively and s_{ji} is the score of the j-th feature for the i-th protein pair. For a given value of L and C, the network weights (i.e. $w_{kj}^{(p)}$s and $w_{kj}^{(o)}$) are adjusted or trained to achieve a desired overall behavior of the network in terms of predicting the class of protein pairs in the training set. The back-propagation algorithm is commonly used to train MLP networks [7]. Let $o_k (k \in \{1..C\})$ denote the real-value output of different output units of the neural network, $\sum_{k=1}^{C} o_k = 1$ and $0 \leq o_k \leq 1$. In standard neural classifiers, the class returned is $\arg\max_k o_k$.

As previously stated in the introduction, pathway sharing protein pairs occur with less frequency than non-pathway sharing ones and classification rules that predict pathway sharing tend to be rare in standard machine learning methods. Consequently, test samples belonging to the small class (pathway sharing) are misclassified more often than those belonging to the prevalent class (non-pathway sharing). In our case, the correct prediction of protein pairs as being in the same pathway has a greater value than the contrary case. Thus, the artificial neural network must be cost-sensitive. Although there are several ways of making a neural network cost-sensitive such as *over-* or *under-sampling* [8], we chose the method known as *threshold-moving*, which has been applied successfully to two-class problems [8]. This technique is very simple and moves the output threshold toward inexpensive classes such that examples with higher costs become harder to be misclassified. This method uses the original learning set to train a neural network and the cost-sensitivity is introduced in the test phase [8].

Concretely, in threshold-moving the class returned is $\arg\max_k o_k^*$. o_k^* is computed according to (2), where μ is a normalization term such that $\sum_{k=1}^{C} o_k^* = 1$, $0 \leq o_k^* \leq 1$ and $Cost[k, c]$ denotes the cost of misclassifying an example of the k-th class to the c-th class ($Cost[k, k] = 0$).

$$o_k^* = \mu \sum_{c=1}^{C} o_k \cdot Cost[k, c] \quad (2)$$

2.2 Gold Standard (GS)

The input set includes only protein pairs of *Saccharomyces Cerevisiae* for which both members have protein sequences in the Reference Sequence (RefSeq) collection [9] and measurements of phylogenetic profile correlation, gene expression correlation and physical interaction. This ensures the integration by providing at least four input features. Thus, the final set consists of 5079 proteins or nearly 13 million pairs. KEGG [10] is chosen as our functional ontology because its endpoint, pathway presence, is relatively well defined. We define protein pairs sharing one or more KEGG pathway as the *Gold Standard Positive (GSP)* set. To minimize bias, the two most populous informative KEGG pathways ("metabolic pathways" and "biosynthesis of secondary metabolites") are excluded from the GSP set (they are associated with the $\sim$ 82% and $\sim$ 10% respectively of the GSP). Thus, our positive set is comprised of 46,636 proteins. We define our true negative gold standard (GSN) set as the collection of pairs that: (1) are annotated in KEGG and (2) never occur in the same KEGG pathway based on current knowledge [4]. There are 993,767 such pairs.

2.3 Features and Scoring

Each of the five sources, sequence similarity, phylogenetic profile, correlated gene expression, protein-protein interactions and shared molecular function terms in Gene Ontology (GO), contributes one component to the feature vector characterizing a protein pair.

Sequence similarity. Protein sequences from RefSeq are downloaded and blastp in blast2.2.23 is used to perform an all against all blast within the proteome. As in [2], pairs are filtered by requiring that their best alignment has an E-value lower than 0.1 and the smaller protein aligns to the larger in at least 50% of its length. The E-value serves as the input feature and pairs not passing this filter take the default E-value of 1.0. A total of 3384 protein pairs pass the filter.

Phylogenetic profile. The presence and absence of a protein across a set of genomes can be represented by a binary string, its phylogenetic profile. Proteins with sufficiently similar profiles tend to be functionally related [1]. Protein pairs with correlated phylogenetic profiles and their associated correlation scores are downloaded from String database [11]. In total, 2353 pairs have non zero phylogenetic profile score among 1082 proteins. Pairs not having phylogenetic profile score take the default score of 0.

Correlated gene expression. New relationships between proteins can also be derived from functional genomics data: co-regulation of genes across diverse experimental conditions, as measured by using microarray analysis, can be a predictor of functional associations [1]. We downloaded protein pairs with correlated gene expression and their associated correlation scores from String database [11]. A total of 58,444 pairs have an score greater than zero in 2066 proteins. Pairs not having correlation score take the default value of 0.

Protein-protein interactions (PPI). Protein-protein interactions are downloaded from Biogrid database [12]. A binary value serves as the input feature denoting existence or absence of an interaction for a pair of proteins. In total, 49,536 pairs among 5073 proteins are known to interact, with self interaction and redundant interactions removed.

Shared molecular-function terms in Gene Ontology (GO-MF). Proteins with the same molecular function GO term [13] are more likely to belong to the same pathway (functionally linked) [1]. In addition, proteins sharing a more specific annotation are more likely to interact than those sharing a common less specific annotation. We downloaded yeast molecular function annotations from the *Saccharomyces Genome Database* [14]. To avoid potential circularity and achieve the strongest evidence, only IDA (Inferred from Direct Assays), IMP (Inferred from Mutant Phenotype) and TAS (Traceable Author Statement) GO terms are taken into account [3]. As a measure of functional association for two proteins with one or more shared GO terms, every shared GO term is examined for the count of the proteins annotated with that term and the smallest count is used as the score [1]. A lower score corresponds to higher degree of functional association. A total of 1,643,935 protein pairs corresponding to 3080 proteins have an score less than 5079, which is the number of proteins that are present in the root GO term. Pairs of proteins in the input set that do not share any GO term take the default value of 5079.

3 Experimental Results and Analysis

3.1 Neural Network Topology and Cross Validation of the Integration Results

We created multilayer perceptrons with three layers and different number of nodes in the hidden layer. The input layer consists of 5 nodes (one for each input feature), the hidden layer has 2, 5, 10, 15, 20 or 25 neurons and the output layer is composed of two neurons, one for each class. To evaluate the overall performance of the prediction of pathway sharing for protein pairs, we did a stratified ten-fold cross-validation. First, we randomly divided both the GSP and GSN datasets into ten separate equal sets. Then we used nine of the ten sets as the learning set to train the network and the test set to identify the positive (pathway sharing) and negatives (non-pathway sharing). It must be taken into account that the ratio of positive to negative examples is kept in both learning and test sets. We ran this process ten times so that each of the ten sets was a test set and the remaining nine constituted the learning set. The number of true positives (TP), false positives (FP), true negatives (TN) and false negatives (FN) are used to get the precision and recall rates:

$$precision = TP/(TP + FP); \; recall = TP/(TP + FN) \tag{3}$$

Since we made our artificial neural network cost-sensitive, in the testing phase of each trained network, a range of different cost values on either the positive or negative class are used to get different values of precision and recall rates. Concretely, the cost of misclassifying an example of the negative class to the positive class, $Cost[GSN, GSP]$, is varied in the range $[150, 1]$ while maintaining the cost of misclassifying an example of the positive class to the negative class, $Cost[GSP, GSN]$, equal to one. Let us call this run as *Cost(-)*. On the

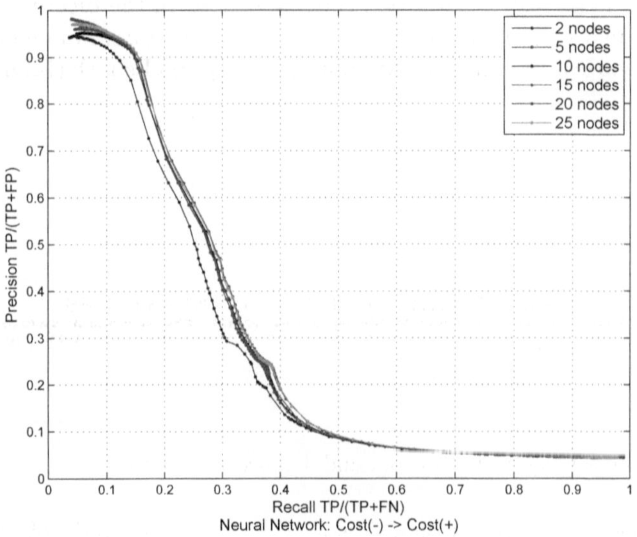

Fig. 1. Precision-recall curves for the overall performance of Neural Network with different nodes in the hidden layer. The x-axis represents the recall varying the cost from high values for negative samples, Cost(-), to high values for positive samples, Cost(+).

other hand, $Cost[GSP, GSN]$ is varied in the range [1, 150] while maintaining $Cost[GSN, GSP] = 1$. This run is named *Cost(+)*. This way, we try to cover the range from almost all predictions being true and false negatives to most of predictions being true and false positives. By the use of different cost values, a Precision-Recall (PRC) curve can be used for evaluating the performance of KEGG pathway sharing prediction. A PRC curve is chosen owing to the fact that it can provide more informative representations of performance assessment under highly imbalanced data than ROC curves [15]. Any point in the PRC curve corresponds to the average performance of the classifier for a given cost value.

3.2 Results

For a given network architecture, that is, for a given number of nodes in the hidden layer, the cross-validation procedure described above is repeated 10 times to get average results, due to the pseudo-random nature of multilayer perceptrons when they are initialized. From Fig.1, it can be observed that the performance of multilayer perceptrons with different number of neurons in the hidden layer in terms of *precision* and *recall* is very similar when the number of neurons is 5 or greater, obtaining worse results when the number of nodes is very low (2 nodes). Within the group of 5 or more neurons, the neural network with less computational cost is chosen: 5 nodes in the hidden layer.

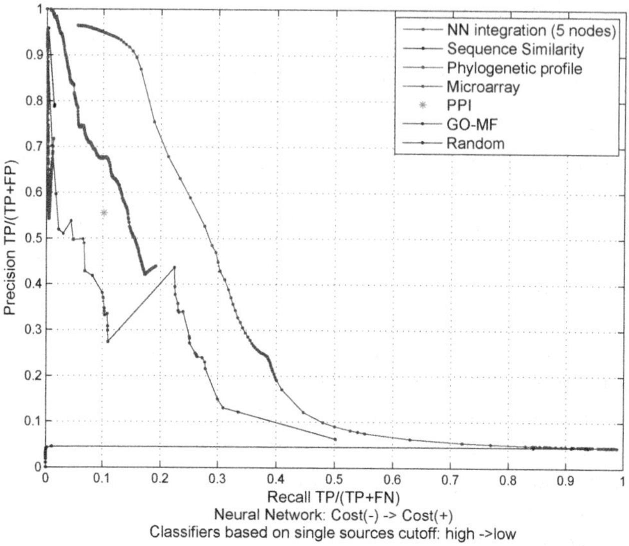

Fig. 2. Precision-recall curves for the overall performance of Neural Network and classifiers based on single sources for pathway sharing prediction. The x-axis represents the recall varying both the misclassification cost in Neural Network and the score cutoff value from high to low in classifiers based on single sources. To generate the random control curve, the class labels in the gold standard data sets are randomized and then a Neural Network with 5 nodes is applied for the integration of the features.

In order to demonstrate the power of integration, we compared the relations extracted by the Artificial Neural Network with 5 nodes in the hidden layer with those supported by evidence from each individual source on isolation: protein pairs are ranked by the score of the feature (for example, correlated gene expression score) and the ranking is then thresholded with several cutoffs to form predictions. Protein-protein interactions represents a binary relation, so, a single point in the precision-recall curve is shown. Thus, from Fig.2, it can be observed that Neural Network classifier outperforms classifiers based on single sources, indicating that a proper integration of heterogeneous datasets in a suitable manner improves the prediction of functional (pathway) relationships. Also, it can be observed that, for pathway sharing prediction, sequence similarity and phylogenetic profile are the less informative, compared to the rest of sources.

4 Conclusion and Future Work

In this paper, we have demonstrated the merit of data integration for the prediction of functional associations between proteins in *Saccharomyces Cerevisiae*. Weak evidence from multiple sources can be combined to provide strong evidence for a relation. To demonstrate this fact, a cost-sensitive Artificial Neural Network has been used since it is an efficient methodology when dealing with imbalanced data.

This is a promising research topic and we are working on a statistical study to demonstrate the advantages of using Artificial Neural Networks for the prediction of functional associations between proteins. An extension to other paradigms such as cost-sensitive support vector machines (SVMs) and Fuzzy-Rule Based Systems (FRBSs) is also under study. In the last case, a FRBS can be a valuable tool to extract knowledge (IF-THEN rules), so that researchers can gain a better understanding between the inputs (features) and outputs (functional association). Moreover, we are planning to use more features in the integration process such as genetic interactions, gene fusion, protein domain sharing, co-occurrence of protein pairs in PubMed abstracts or shared cellular component terms in Gene Ontology.

Acknowledgments. This paper has been supported by the Spanish Ministry of Education (SAF2010-20558 project) and the FPU research grant AP2007-03009.

References

1. Linghu, B., et al.: Genome-wide prioritization of disease genes and identification of disease-disease associations from an integrated human functional linkage network. Genome Biology 10 (2009)
2. Linghu, B., et al.: High-precision high-coverage functional inference from integrated data sources. BMC Bioinformatics 9 (2008)
3. Bradford, J.R., et al.: GO-At: in silico prediction of gene function in Arabidopsis thaliana by combining heterogeneous data. The Plant Journal 61, 713–721 (2010)
4. Lee, I., et al.: A probabilistic functional network of yeast genes. Science 306, 1555–1558 (2004)
5. Wu, C.C., et al.: Predicting of human functional genetic networks from heterogeneous data using RVM-based ensemble learning. Bioinformatics 26, 807–813 (2010)
6. Myers, C.L., et al.: Finding function: evaluation methods for functional genomic data. BMC Genomics 7 (2006)
7. Feng, L.H., et al.: Classification error of multilayer perceptron neural networks. Neural Computing & Applications 18, 377–380 (2009)
8. Zhou, Z., et al.: Training Cost-Sensitive NNs with Methods Addressing the Class Imbalance Problem. IEEE Trans. on Knowledge and Data Eng. 18, 63–77 (2006)
9. The Reference Sequence Collection, `http://www.ncbi.nlm.nih.gov/projects/RefSeq/`
10. Kanehisa, M., Goto, S., et al.: The KEGG resource for deciphering the genome. Nucleic Acids Res. 32, D277–D280 (2004)
11. von Mering, C., et al.: STRING 8–a global view on proteins and their functional interactions in 630 organisms. Nucleic Acids Res. 37 (2009)
12. Breitkreutz, B.J., et al.: The BioGRID Interaction Database: 2008 update. Nucleic Acids Res. 36, 637–640 (2008)
13. Ashburner, M., et al.: Gene ontology: tool for the unification of biology. The Gene Ontology Consortium. Nat.Genetc. 25, 25–29 (2000)
14. Saccharomyces Genome Database, `http://www.yeastgenome.org/`
15. He, H., et al.: Learning from Imbalanced Data. IEEE Trans on Knowledge and Data Eng. 21, 1263–1284 (2009)

Hybrid (Generalization-Correlation) Method for Feature Selection in High Dimensional DNA Microarray Prediction Problems

Yasel Couce[1], Leonardo Franco[2], Daniel Urda[2], José L. Subirats[2], and José M. Jerez[2]

[1] Universidad de Ciencias Informáticas, La Habana, Cuba
yaselc@uci.cu
[2] Universidad de Málaga, Department of Computer Science, ETSI Informática, Spain
{lfranco,jlsubirats,durda,jja}@lcc.uma.es

Abstract. Microarray data analysis is attracting increasing attention in computer science because of the many applications of machine learning methods in prediction problems. The process typically involves a feature selection step, important in order to increase the accuracy and speed of the classifiers. This work analyzes the characteristics of the features selected by two wrapper methods, the first one based on artificial neural networks (ANN) and the second in a novel constructive neural network (CNN) algorithm, to later propose a hybrid model that combines the advantages of wrapper and filter methods. The results obtained in terms of the computational costs involved and the prediction accuracy reached show the feasibility of the hybrid model proposed here and indicate an interesting research line for the near future.

Keywords: DNA Microarray, Feature Selection, Data Mining, Constructive Neural Networks.

1 Introduction

DNA microarray technology has opened up new research directions and significant opportunities in biomedical sciences. DNA microarray technology makes possible to measure simultaneously the expressions levels of thousands of genes in a single experiment, providing unique and useful data for a wide range of experimental research, e.g., predicting disease outcome in patients. However, due to the large number of features (in the order of thousands) and the small number of samples (mostly less than a hundred) in this data, microarray data analysis face the "large-p-small-n" paradigm [9] also known as the curse of dimensionality. Generally, most of these features are irrelevant to a specific study and represent noise for most of the prediction systems, therefore the application of machine learning techniques [5] are becoming increasingly needed in order to enhance the speed and accuracy in prediction systems.

The microarray data analysis usually involves a preprocessing step, which consists in the selection of features (genes) relevant for the classification step. These

J. Cabestany, I. Rojas, and G. Joya (Eds.): IWANN 2011, Part II, LNCS 6692, pp. 202–209, 2011.

feature selection algorithms are grouped into two categories, filter methods and wrapper methods [3]. Filter methods selects relevant features based on general characteristics of the training data. Wrapper methods requires an specific classifier algorithm to evaluate the suitability of each subset of features found. As a result of using the classifier in the feature subset evaluation, wrapper methods tend to outperform the prediction accuracy of the filter methods, but the constantly training of the classifier leads to a high computational cost for the wrapper methods, which makes them not much used in microarray data analysis [2]. On the other hand, filter methods are fast and computationally simple, making them more suitable for tasks on high-dimensional datasets.

A recent proposal made by Urda et al. [8] shows how constructive neural networks (CNN) algorithm can reduce time and increase accuracy in microarray data analysis, in particular in comparison to ANN. Using a novel constructive algorithm (C-Mantec) [7] to predict estrogen receptor status, Urda et al. [8] improves the results achieved by Lancashire et al. [4] using a stepwise forward selection artificial neural network approach. Despite the results obtained by [8], the solution still suffers the drawbacks of the wrapper methods making the search in feature space very time-consuming.

The combination of filter and wrapper approaches has led to alternative proposals [6][1] looking to overcome the disadvantages of the two methods separately. Hybrid models incorporate the relationship of the wrappers with the classifiers, in order to increase the accuracy prediction of the selected subset and use the analysis of the properties of the data set, performed by the filters, to achieve speed and scalability. In this work we present an hybrid wrapper-filter model using a constructive neural network algorithm to build a classifier using the information from the training patterns in order to facilitate its adaptation to a given problem. The proposal is based on the use of a recently introduced constructive neural network algorithm (C-Mantec) [7] and on the use of a simple correlation measure to rank the features in the dataset, in order to avoid redundancy in the selected features.

2 Materials and Methods

2.1 Materials

The dataset used in this work comes from the study published by West et al. [10] (`http://data.cgt.duke.edu/west.php`). This study used microarray technology to analyze primary breast tumors in relation to estrogen receptor (ER) status. The dataset contains the expression levels of a total of 7129 genes measured in 49 breast tumor samples (25 ER+ and 24 ER- cases).

2.2 Methods

We first analyze in this work two previous studies [4] [8] where feature selection is implemented and characterize the process measuring the prediction ability of using the individual genes and the correlation between the selected features, to use later these two characteristics of the dataset to test three different strategies. For measuring the generalization ability of the individual genes for the selection process and for estimating the predictive accuracy of using all the selected variables we used C-Mantec, a constructive neural network algorithm that generates very compact neural architectures with state-of-the-art generalization capabilities [7].

The well known linear correlation coefficient (r), computed for a pair of variables ($X; Y$) and shown below in equation 1, was used for estimating the redundancy among a set of variables.

$$r = \frac{\sum_i (x_i - \bar{x})(y_i - \bar{y})}{\sqrt{\sum_i (x_i - \bar{x})^2}\sqrt{\sum_i (y_i - \bar{y})^2}}, \tag{1}$$

where $\bar{x}$ is the mean of X, and $\bar{y}$ is the mean of Y. The value of r belongs to interval $[-1, 1]$. If r is zero then X and Y are totally independent. The closer is the value of r to the extremes of the interval $[-1, 1]$, the closer X and Y to a perfect linear relationship. As we need to measure the redundancy of one feature to a set of features, we computed the correlation of variable X and a set of n variables $(Y_1, Y_2, \ldots, Y_n)$ as the mean of the correlation value of each pair (X, Y_i).

3 Experiments and Results

We have first analyzed the set of features selected in the studies by Lancashire et al. [4] and Urda et al. [8], where the datasets have been selected using forward selection methods based only in measuring prediction accuracy, in order to measure the relevance of individual generalization and correlation. Table 1 shows the features selected by Lancashire et al. with their respective measures of correlation and generalization, while table 2 shows the same analysis for the variables selected by Urda et al. Both tables show first the values for the correlation measured as an average between pairs of variables, indicating the obtained correlation coefficient, then a rank position between 1 and 7129 (number of features of the dataset) and a normalized rank between 0 and 1. The three rightmost columns of the table show the generalization ability obtained when the C-Mantec algorithm is trained only with a single variable, and the columns show the absolute value, rank position and normalized rank.

The data shown in the tables indicate that the variables selected in the Lancashire analysis present relatively high individual generalization values (average normalized rank equals to 0.1579) while correlation among variables is high (average normalized rank equals to 0.7614), indicating a large redundancy between the selected variables. For the features selected in the Urda et al. study the correlation between variables is lower (average normalized rank equals to 0.2247) but the generalization measure of the individual variables is higher (0.2990 for the normalized rank). Surprisingly, even if both sets of variables were chosen using a forward selection method the characteristics of the selected variables are

Table 1. Features selected in the work of Lancashire et al. [4] ranked according to the correlation measure used in the present work (see the text for more details)

		Correlation measure			Generalization measure		
	Probe Set ID	correlation coefficient	rank position	rank [0-1]	mean %	rank position	rank [0-1]
1	X58072_at	-	-	-	86.0	4	0.0004
2	Z29083_at	0.3302	6402	0.898	83.4	9	0.0011
3	M81758_at	0.1763	5033	0.706	52.6	4550	0.6382
4	M60748_at	0.1520	3515	0.493	60.2	1419	0.1989
5	M74093_at	0.1835	4590	0.644	81.8	15	0.0020
6	U22029_f_at	0.2175	5484	0.769	58.0	2242	0.3144
7	U96131_at	0.2657	6625	0.929	73.2	131	0.0182
8	M96982_at	0.2391	6353	0.891	64.4	642	0.0899
	mean	0.2235	5429	0.761	69.9	1126	0.1579

Table 2. Features selected in the work of Urda et al. [8] ranked according to the correlation coeffficient averaged among variables and the generalization ability (see the text for more details)

		Correlation measure			Generalization measure		
	Probe Set ID	correlation coefficient	rank position	rank [0-1]	mean %	rank position	rank [0-1]
1	X76180_at	-	-	-	85.6	6	0.0007
2	HG4749-HT5197_at	0.0566	1785	0.250	59.2	1742	0.2442
3	M31520_rna1_at	0.1043	1431	0.201	55.2	3383	0.4745
4	U20325_at	0.1265	1592	0.223	55.2	3399	0.4767
	mean	0.0958	1603	0.225	63.8	2132	0.2990

different. It is worth noting that the Lancashire et al. work used ANN as classification algorithm while in the Urda et. al. analysis C-Mantec algorithm was used. Given this discrepancy between the observed characteristics of the datasets, we have decided to propose and test **three** different strategies in order to develop a selection method:

Relevance only (ROnly). It is the simplest of the three strategies, aimed to select those features with the highest generalization over the test data subset. The pseudocode of this strategy is presented below as Algorithm 1.

Algorithm 1. Pseudocode of the ROnly algorithm.

for each feature f_i in dataset D **do**
  {Create model for f_i and compute its generalization performance on D}
  $g(i) \leftarrow$ **CMantec**$(f_i, T_0, I_{max}, g_{fac})$;
end for
$n \leftarrow$ **size**(D)
for $j = 1$ **to** 10 **do**
  {Select the feature with the highest generalization performance on D}
  $Set(j) \leftarrow \{f_i \in D : \forall x \in \{1, \ldots, n\}(g(i) \geqslant g(x))\}$;
  $g(i) \leftarrow 0$
end for
return Set

Relevance first, then redundancy (RelevanceF). This strategy consists in selecting the ten percent of the most relevant features from the data set (those with the highest generalization value over the test data subset) and then within that ten percent select the feature that is less redundant with the subset of variables already selected for classification. Algorithm 2 specifies the pseudocode for this strategy.

Algorithm 2. Pseudocode of the RelevanceF algorithm.

```
for each feature f_i in dataset D do
    {Create model for f_i and compute its generalization performance on D}
    g(i) ← CMantec(f_i, T_0, I_max, g_fac);
end for
n ← size(D)
{Select the ten percent of features with the highest generalization performance on D}
for j = 1 to n ÷ 10 do
    T(j) ← {f_i ∈ D : ∀x ∈ {1, ..., n}, (g(i) ⩾ g(x))};
    g(i) ← 0
end for
Set(1) ← T(1)
for j = 2 to 10 do
    {Select the feature less redundant to features in Set}
    Set(j) ← {f_i ∈ T : ∀x ∈ {1, ..., n}, (r̄(f_i, Set) ⩽ r̄(f_x, Set))};
end for
return Set
```

Redundancy first, then relevance (RedundancyF). Contrary to the previous strategy, this algorithm select first a ten percent of the features that are less redundant to the features already selected for classification and within this ten percent the most relevant attribute (high generalization) is chosen. Algorithm 3 shows pseudocode for this strategy.

Algorithm 3. Pseudocode of the RedundancyF algorithm.

```
for each feature f_i in dataset D do
    {Create model for f_i and compute its generalization performance on D}
    g(i) ← CMantec(f_i, T_0, I_max, g_fac);
end for
n ← size(D)
{Select the feature with the highest generalization performance on D}
Set(1) ← {f_i ∈ D : ∀x ∈ {1, ..., n}, (g(i) ⩾ g(x))};
g(i) ← 0
for j = 2 to 10 do
    {Select the ten percent of features less redundant to features in Set}
    for t = 1 to n ÷ 10 do
        R(t) ← {f_i ∈ D : ∀x ∈ {1, ..., n}, (r̄(f_i, Set) ⩽ r̄(f_x, Set))};
    end for
    {Select the feature with the highest generalization performance on D}
    Set(j) ← {f_i ∈ R : ∀x ∈ {1, ..., n}, (g(i) ⩾ g(x))};
end for
return Set
```

In all cases the C-Mantec algorithm was the classification method used, with the following values for the parameters: $g_{fac} = 0.2$ (network growing factor), $I_{max} = 100,000$ (maximum number of iterations), while T_0 (initial temperature) was set equal to the number of input variables. A ten-fold cross validation approach was used and the results obtained are indicated in table 3 and in figure 1. Table 3 shows the mean and the standard deviation across the set of ten observed values for the three algorithms, with best value highlighted. We can observe that the third algorithm (Redundancy first, then relevance) outperformed the other two achieving the highest generalization and also showing the lowest standard deviation. (Figure 1), and thus we choose this algorithm as our proposal for the feature selection process.

Table 3. Mean and standard deviation observed for the three algorithms

Algorithm	Generalization C-Mantec
ROnly	89.18 ± 1.03
RelevanceF	84.72 ± 1.92
RedundancyF	92.82 ± 0.51

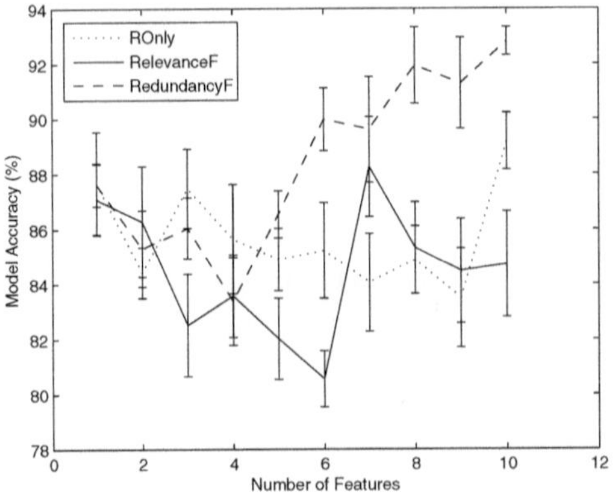

Fig. 1. Results of ten-fold cross validation of the C-Mantec algorithm applied to the test dataset using the features selected by the three algorithms. Error bars represent standard error of the mean expressed in percentage.

Table 4 shows the characteristic of the selected variables chosen by the RedundancyF algorithm. The average normalized rank is much lower for both correlation between features (0.069) and individual prediction accuracy (0.025) than in the previous analyzed studies [4] and [8].

As a final comparison, table 5 shows the generalization ability obtained using C-Mantec on a 10-fold cross validation scheme including all the selected features chosen by the three studies (Lancashire et al., Urda et al., and the present proposal) . Given that the proposed approach (RedundancyF) only uses C-Mantec in the first phase of the algorithm (estimation phase) and that the second phase (selection phase) consumes an average of 0.4 seconds of CPU time to find each new feature, the execution time remains almost invariable as the number of selected features increases, while this is not the case for the previous proposals [8] and [4]. This make our algorithm a more feasible choice to scale on high-dimensional datasets. A simple mathematical analysis shows below that for the forward selection methods used in [4] [8] the computational costs approximately scale with both the total number of features available (N_I) and with the number of features to be selected (N_V), while for the hybrid method proposed the computationally costs scale linearly only with the total number of available features. The following equations shows the calculation involved:

$$CPU_{time}(FSel) = \sum_{i=1}^{V}(N_I - i + 1)N_V T_{gen}(i) \sim N_I N_V \overline{T_{gen}} \tag{2}$$

$$CPU_{time}(Hybrid) = N_I(T_{gen}(V = 1) + N_V T_{cor}) \sim N_I T_{gen}(V = 1), \tag{3}$$

Table 4. Features selected by RedundacyF algorithm, ranked according to the correlation measure used in the present work (see the text for more details)

		Correlation measure			Generalization measure		
	Probe Set ID	correlation coefficient	rank position	rank [0-1]	mean %	rank position	rank [0-1]
1	X03635_at	–	–	–	87.6	1	0
2	M85289_at	0.0052	234	0.033	72.0	157	0.0219
3	U60269_cds2_at	0.0250	371	0.052	67.4	379	0.0530
4	L08044_s_at	0.0857	540	0.076	79.4	32	0.0043
5	U41371_at	0.0761	354	0.050	69.4	254	0.0355
6	L13278_at	0.0884	446	0.062	68.2	317	0.0443
7	X06614_at	0.1137	563	0.079	73.2	132	0.0184
8	L37199_at	0.1147	651	0.091	72.0	156	0.0217
9	HG2279-HT2375_at	0.1186	631	0.088	73.8	116	0.0161
10	S77410_at	0.1184	649	0.091	69.4	251	0.0351
	mean	0.0829	493	0.069	73.2	179	0.0250

Table 5. Percentage accuracy using C-Mantec on a 10-fold cross validation scheme using all the selected features chosen in each work

Method	Percentage accuracy
Urda et al. [8]	0.950 ± 0.103
Lancashire et al. [4]	0.946 ± 0.091
RedundancyF	0.922 ± 0.099

where $T_{gen}(i)$ is the time needed to compute the generalization ability of a given model using i input variables ($\overline{T_{gen}}$ indicates averaging). For the hybrid model T_{cor} indicates the CPU time needed to compute a correlation measure between pair of variables. In our case T_{cor} was very small and thus the total CPU time depends mainly on the computational cost of computing the generalization of the model. The validity of the previous analysis was checked for some particular values but lack of space in the present work leaves a more detailed analysis to be published elsewhere.

4 Conclusions and Further Work

In this work we have first carried out an analysis of the characteristics of the features selected by two recent publications, to further propose and test three strategies for selection of informative genes in DNA microarray experiments through applying a hybrid model of constructive neural network algorithm and a simple correlation-based algorithm. Even though the new introduced models do not reach the level of effectiveness of the reviewed approaches (Table 5), we must emphasize that a major goal of this research was firstly to reduce the computational cost of the feature selection task and on the other hand to analyze whether both correlation between features and level of generalization of individual features are important characteristics for the feature selection task, fact that the obtained results seems to confirm. Further work will be centered on extensions of the RedundancyF algorithm in order to increase its speed and prediction accuracy on DNA microarray prediction problems. Better measures for estimating feature redundancy will be tested, as they may permit to capture nonlinear correlation effects, and additionally, tests on different databases using different classifiers will be conducted to fully validate the approach.

Acknowledgements

The authors acknowledge support from CICYT (Spain) through grants TIN2008-04985 and TIN2010-16556 (including FEDER funds) and from Junta de Andalucía through grant P08-TIC-04026.

References

1. Huda, S., Yearwood, J., Strainieri, A.: Hybrid Wrapper-Filter Approaches for Input Feature Selection Using Maximum Relevance and Artificial Neural Network Input Gain Measurement Approximation (ANNIGMA). In: 4th International Conference on Network and System Security, pp. 442–449 (2010)
2. Inza, I., Larrañaga, P., Blanco, R., Cerrolaza, A.J.: Filter versus wrapper gene selection approaches in DNA microarray domains. Artificial Intelligence in Medicine 31, 91–103 (2004)
3. Kohavi, R., John, G.: The Wrapper approach. In: Feature Extraction, Construction and Selection: a data mining perspective, pp. 33–51 (1998)
4. Lancashire, L.J., Rees, R.C., Ball, G.R.: Identification of gene transcript signatures predictive for er and lymph node status using a stepwise forward selection ann modelling approach. Artif. Intell. Med. 43, 99–111 (2008)
5. Pirooznia, M., Yang, J., Yang, M.Q., Deng, Y.: A comparative study of different machine learning methods on microarray gene expression data. BMC Genomics 9, S13 (2008)
6. Sebban, M., Nock, R.: A hybrid filter/wrapper approach of feature selection using information theory. Pattern Recognition 35, 835–846 (2002)
7. Subirats, J.L., Jerez, J.M., Gómez, I., Franco, L.: Multiclass Pattern Recognition Extension for the New C-Mantec Constructive Neural Network Algorithm. Cognitive Computation 2, 285–290 (2010)
8. Urda, D., Subirats, J.L., Franco, L., Jerez, J.M.: Constructive neural networks to predict breast cancer outcome by using gene expression profiles. In: García-Pedrajas, N., Herrera, F., Fyfe, C., Benítez, J.M., Ali, M. (eds.) IEA/AIE 2010. LNCS, vol. 6096, pp. 317–326. Springer, Heidelberg (2010)
9. West, M.: Bayesian factor regression models in the "large p, small n" paradigm. Bayesian statistics 7, 723–732 (2003)
10. West, M., Blanchette, C., Dressman, H., Huang, E., Ishida, S., Spang, R., Zuzan, H., Olson, J.A., Marks, J.R., Nevins, J.R.: Predicting the clinical status of human breast cancer by using gene expression profiles. PNAS 98, 11462–11467 (2001)

Model Selection with PLANN-CR-ARD

Corneliu T.C. Arsene[1,2], Paulo J. Lisboa[2], and Elia Biganzoli[3]

[1] Project Director of National University Research Council, Bucharest, Romania
[2] School of Computing and Mathematical Sciences, Liverpool John Moores University, Liverpool, United Kingdom
[3] Istituto di Statistica Medica e Biometria, Universita degli Studi di Milano, Milano, Italy
CorneliuArsene@gmail.com, P.J.Lisboa@ljmu.ac.uk, Elia.Biganzoli@unimi.it

Abstract. This paper presents a new compensation mechanism to be used with a Partial Logistic Artificial Neural Network for Competing Risks with Automatic Relevance Determination (PLANN-CR-ARD) and tested comprehensibly on a real breast cancer dataset with excellent convergence properties and numerical stability for the non-linear model. The Model Selection is implemented for the PLANN-CR-ARD model, benefiting from a scaling of the prior error term which together with the data error term forms the total error function that is optimized. The PLANN-CR-ARD proves to be an excellent prognostic tool that can be used in regression analysis tasks such as the survival analysis of cancer datasets.

Keywords: Model Selection, PLANN-CR-ARD, Artificial Neural Networks, Convergence properties, Competing Risks.

1 Introduction

The purpose of this paper is to investigate the model selection and the convergence properties of a Bayesian neural network applied to competing risks. The theory of competing risks regards the problems in which a system (or an individual) is exposed to two or more causes of failure, but the actual first failure is attributed to only one of the causes. Typically, the competing risk data (cause-specific data) include the time of failure or censoring for each individual as well as an indicator of the type of failure or censoring. There is an extensive literature on the analysis of such data: many results and references are given by [1].

In [2][3] is presented a Partial Logistic Artificial Neural Network (PLANN) which includes regularization of the network parameters within the Bayesian framework with Automatic Relevance Determination (ARD), initially developed for single risk cases [2][3], which was then extended for the analysis of Competing Risks (PLANN-CR-ARD) in [4][5]. In this paper is presented the model selection of the PLANN-CR-ARD model in the context of a novel compensation mechanism and the convergence properties of the same model.

J. Cabestany, I. Rojas, and G. Joya (Eds.): IWANN 2011, Part II, LNCS 6692, pp. 210–219, 2011.

2 Compensation Mechanism

Because the training data is skewed, in order to preserve the numerical stability of the optimization algorithms, a compensation mechanism is necessary by PLANN-CR-ARD [2][10]. The goal of the compensation mechanism is to balance the data.

The compensation mechanism presented here is novel and it is used to scale the log-likelihood function and the Jacobian and the Hessian matrices of this function. The log-likelihood function from [2], which represents the data error term, it is scaled with the new compensation formula C_{rk} :

$$E = -\sum_{r=1}^{R+1} \sum_{i=1}^{N} \sum_{l=1}^{l_i} \left\{ d_{ril} \log(y_{ril}(x_i, l)) / C_{rk} \right\}. \tag{1}$$

where E is log-likelihood function, r is the index for the risks and the survival, i is the index for the patient, l is the index for time interval, l_i is the number of time intervals for which the patient i is observed alive, R is the number of competing risks, N is the number of patients, y is the output of the model, d is the output target for which the PLANN-CR-ARD model is trained, x is the input training covariate data of the model, k is an index which can be 1 indicating the producing of a network output or 0 otherwise.

The novel compensation mechanism is applied together with the PLANN-CR-ARD model on a realistic breast cancer dataset of 2010 patients in order to investigate the effects of the age of patients, tumour dimension, number of axillar nodes, histologic grade, tumor localization, for the cause specific hazard and the cumulative cause specific hazard for the competing events of Intra-Breast Tumour Recurrence (IBTR) and Distant Metastasis (DM). All patients suffered surgery of excision of the nodules at the breast (quadrantectomy, axillary lymph node dissection), followed by radiotherapy (QUART). Age (in years), tumour size/dimension (cm) and number of nodes involved were recorded as continuous variables scaled between 0 and 1. For the discrete variables the following classification was used: place of the tumour/ tumour site (external part, E; internal and central, IC); histologic type (Extensive Intraductal Component – EIC, Infiltrated Intraductal Component – IDC, Infiltrated Lobular Component - ILC). With a median time following the surgery of 8.5 years, 151 IBTR and 414 DM events were recorded as the first neoplazic event. The interval of time until the appearance of a neoplazic event other than the IBTR or DM was censored for the respective patient. A description of the dataset and of the previous results can be found in [6] where the effects of the covariates over the hazard functions were investigated by two different Cox regression models [2].

The patient records from the database are replicated for each time interval for which the patient is observed until the patient is censored or one of the two competing risks is producing (i.e. IBTR or DM). Time interval means that if a patient survives 10 years then the patient record is replicated from 1 to 10 with a time interval equals to 1. This results in a training dataset of 15420 training patterns.

For a model with three network outputs, the balancing of the dataset means to balance each network output to a 1/3rd of the length of the training data for the case when the event is producing or to 2/3rd of the length of training data for the case when the event is not producing. Therefore the compensation mechanism is applied to each of the 15420 training patterns with respect to the network outputs (i.e. r = 1:3) and with respect to whether a network output is producing (i.e. k =1) or the network output is not producing (i.e. k =0), where index k belongs to C_{r_k}. This results in two formulas C_{r_0} for k equals 0 and C_{r_1} for k equals 1, which implement the compensation mechanism:

$$C_{r_1} = \frac{\sum_{i=1}^{N}\sum_{l=1}^{l_i} d_{ril}}{\sum_{i=1}^{N} l_i / 3} . \tag{2}$$

$$C_{r_0} = \frac{\sum_{i=1}^{N} l_i - \sum_{i=1}^{N}\sum_{l=1}^{l_i} d_{ril}}{\sum_{i=1}^{N} l_i (2/3)} . \tag{3}$$

An example of applying the compensation mechanism to the 3rd network output (DM) is the following: the sum $\sum_{i=1}^{N} l_i$ is the number of training data patterns (i.e. 15420), the sum $\sum_{i=1}^{N}\sum_{l=1}^{l_i} d_{ril}$ is the number of DM events (i.e. 414), and the difference between $\sum_{i=1}^{N} l_i$ - $\sum_{i=1}^{N}\sum_{l=1}^{l_i} d_{ril}$ is the difference between 15420 and 414 (i.e. 15420-414 = 15006). Therefore C_{r_1} for r equals 3 is 414*3/15420 that is 0.0805, while C_{r_0} is (3/2)*(15006/15420) that is 1.46. These values are used to scale the log-likelihood function, the Jacobian and the Hessian matrices of this function.

In Fig. 1b is shown a sequence from the training data of 15420 patterns for patients 4 and 5, in Fig. 1c is shown an example of balanced network output and in Fig 1d is shown the compensation values which are used to scale the log-likelihood function, the Jacobian and the Hessian matrices.

Patient	Sequence of training data from the 15420 patterns	Balanced network output	Compensation values applied to the log-likelihood function, Jacobian and Hessian matrices of the function
Patient 4	1 0 0	1 0 0	2.90 1.48 1.46
Patient 4	1 0 0	1 0 0	2.90 1.48 1.46
Patient 4	0 1 0	0 1 0	0.0495 0.0247 1.46
Patient 5	1 0 0	0 1 0	2.90 1.48 1.46
Patient 5	1 0 0	0 0 1	2.90 1.48 1.46
Patient 5	0 0 1	0 0 1	0.0495 1.48 0.0805
a)	b)	c)	d)

Fig. 1. Compensation mechanism calculated per time interval for each patient and for each network output: a) patient; b) unbalanced training data; c) example of balanced network output usually not met in real datasets; d) compensation factors calculated from equations (2) and (3)

3 Model Convergence

The determination of the parameters of the PLANN-CR-ARD consists of determination of two sets of variables: the weights of the network and the hyper-parameters. The weights (*w*) are calculated based on the scaled conjugate gradient method [7][9].

The total error function that is optimized by the scaled conjugate gradient method:

$$fnew = E + eprior \ . \qquad (4)$$

where *eprior* is the prior error term, *E* is the log-likelihood function.

The prior error term regularizes the weights in order to avoid overfitting [8][9].

The convergence of the scaled gradient algorithm (i.e. scg.m file in NETLAB [9]) is obtained with the following two convergence criteria:

$$\max(abs(fnew - fold)) \ < \ \varepsilon_1 \ . \qquad (5)$$

$$\max(abs(alpha * srch)) \ < \ \varepsilon_2 \ . \qquad (6)$$

where ε_1 and ε_2 are error thresholds (e.g. 10^(-16)), *alpha* a parameter of the optimization procedure, *srch* is the Jacobian of the total error function with respect to the weights, *fnew* and *fold* are the new and the old values of the total error function calculated at two successive iterations of the scaled conjugate gradient algorithm.

In Fig. 2 is shown two different numerical results of the marginalized and the non-marginalized outputs [2][3][4] of the PLANN-CR-ARD compared to the Nelson-Aalen non-parametric estimates [11] with 95% confidence intervals [12]. Each numerical result is obtained for different initialization points of the network parameters, which are used by the optimization algorithms of the PLANN-CR-ARD.

The model converges in both simulations to values of the IBTR and DM cause specific cumulative hazards situated within the 95% confidence intervals of the non-parametric estimates for the entire follow-up. Similar results, as the ones shown below, within the 95% confidence intervals were obtained in all numerical simulations realized on this dataset and it demonstrates that the PLANN-CR-ARD model has a very good power of prediction.

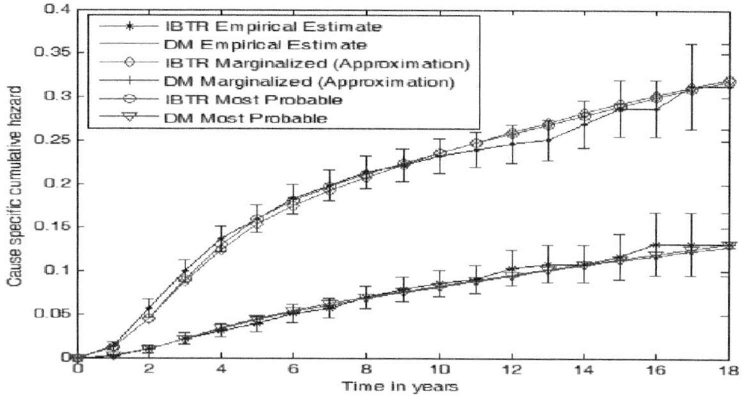

a)

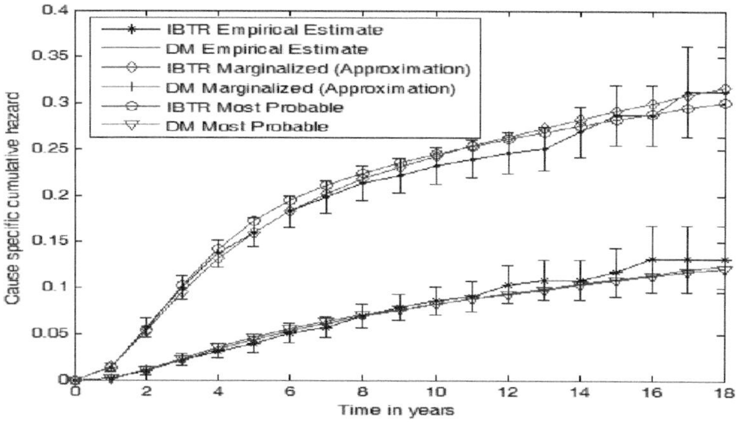

b)

Fig. 2. Comparison between the marginalized and the non-marginalized outputs of the PLANN-CR-ARD compared to the non-parametric estimation Nelson-Aalen with confidence intervals: a) first simulation; b) second simulation

In Fig. 3 is shown the convergence criteria $\max(abs(fnew - fold)) < \varepsilon_1$ (continuous line) and $\max(abs(alpha * srch)) < \varepsilon_2$ (dash line) in two different runs of the PLANN-CR-ARD model with the two convergence criteria reaching values of 10^(-16).

As described in [8][10], the effect of the *eprior* error term is to move the minimum of the total error function *fnew* from a set of network weights w^* to a different set of network weights $\tilde{w}$. The non-monotonic behavior shown in Fig. 3 for the two criteria is due to the shift of the network weights from w^* to $\tilde{w}$.

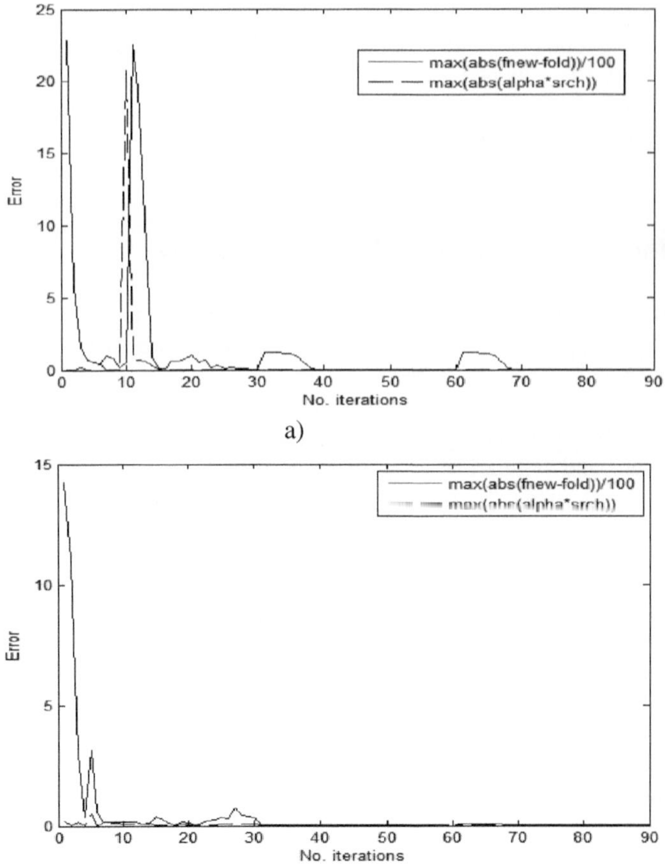

a)

b)

Fig. 3. Convergence criteria $\max(abs(fnew - fold)) < \varepsilon_1$ (continuous line) and $\max(abs(alpha * srch)) < \varepsilon_2$ (dash line) in two different running of the PLANN-CR-ARD model on the same dataset: a) first simulation; b) second simulation

The determination of the weights w^* is realized with the scaled conjugate gradient method. The calculation of the weights is followed by the estimation of the hyper-parameters based on the re-calculation of the *eprior* error term and a new set of network weights will result $\tilde{w}$. The new set of weights $\tilde{w}$ may not be a point of minimum for the total error function and for the two convergence criteria until later during the convergence of the model that is the iterations 70 to 90 in Fig. 3.

For illustrative purposes, there were used 3 main iterations in order to achieve the convergence of the PLANN-CR-ARD model. Each of the 3 main iterations included 30 iterations for the re-estimation of the network weights, followed by 30 iterations for the re-estimation of the hyper-parameters. This resulted in 3*60 iterations in order to obtain a very good convergence. The 3 main iterations took 2 minutes on a Pentium(R) Dual-Core E5400 with 3 GB of RAM which computational time is very good. In Fig. 3 is shown the 90 iterations (3*30) used to re-estimate the weights. The results were obtained with a model with 18 hidden nodes. With a less number of hidden nodes is obtained a decreasing of the computational time. For example, the model converged with 2 main iterations and 20 iterations for the estimation of weights and 20 iterations for the estimation of the hyper-parameters which coupled with 8 hidden nodes resulted in a computational time well under 1 minute.

4 Model Selection with PLANN-CR-ARD

The Bayesian framework can be utilized to carry out model selection once the PLANN-CR-ARD parameters have been estimated with ARD regularization, in order to soft-prune irrelevant variables in the model.

The evidence in support of a particular model hypothesis *H* requires a distinct level in the ARD methodology, beyond estimation of the evidence for the weight parameters and regularization of another set of network parameters called hyper-parameters. A model *H* consists of the input variables to be used with the PLANN-CR-ARD model. By using Bayes theorem the following equation stands:

$$P(H \mid D) = \frac{P(D \mid H)P(H)}{P(D)} \cdot \tag{7}$$

where *P(H)* is the prior probability in the support of model *H*, *P(D)* is the probability distribution of the input data, *P(D/H)* is the probability of observing data D given the model *H*.

Assuming a flat prior for the space of possible models, considered here as the available set of explanatory input variables, the evidence *P(D/H)* for a particular model selection is obtained.

$$P(H \mid D) \propto P(D \mid H) = \int P(D \mid \alpha, H)P(\alpha \mid H)\mathrm{d}\alpha \cdot \tag{8}$$

Approximating the integral in (8) by its mode multiplied by the width of the prior, it is obtained an analytical approximation to the evidence in support of a candidate PLANN-CR-ARD model [2][8]. The accurate evaluation of the evidence can be difficult on real examples. The evaluation of the evidence requires the calculation of the determinant of the Hessian matrix of the total error function with respect to the weights. The determinant is given by the product of the eigenvalues and there could be encountered difficulties since the respective determinant, which measures the volume of the posterior distribution of weights, will be dominated by the small eigenvalues. These small values correspond to directions in which the weight distributions are relatively broad. Some approaches to avoid such problems are presented in [9]. It consists in the use of a set of network parameters called hyper-parameters which are also employed to avoid the model overfitting through the *eprior* error term. The hyper-parameters depend on the sum of the eigenvalues of the Hessian matrix and so are less sensitive to errors in the small eigenvalues.

The model selection can be realized based on the values of the hyper-parameters which are calculated by maximizing the posterior probability of the hyper-parameters:

$$P(\alpha \mid D, H) \propto \frac{\exp(-fnew(w^{MP}, \alpha))}{Z_W(\alpha)} (2\pi)^{N_W/2} \det(A)^{-1/2} . \tag{9}$$

where α is the hyper-parameter of interest, A is the Hessian matrix of the total error function *fnew* with respect to the weights, N_w is the total number of weights, w^{MP} is the most probable weight vector and $Z_w(\alpha)$ is a normalization constant [8].

By maximizing (9) with respect to the weights results the hyper-parameters:

$$\begin{aligned} \gamma_m &= N_m - \alpha_m Tr_m(A^{-1}) = \frac{N_m \sum_{n=1}^{N_m} (W_{mn}^{MP})^2}{\sum_{n=1}^{N_m} (W_{mn}^{MP})^2 + Tr_m(A^{-1})} \\ \frac{1}{\alpha_m} &= \frac{2\,eprior}{\gamma_m} = \frac{\sum_{n=1}^{N_m} (W_{mn}^{MP})^2}{\gamma_m} = \frac{\sum_{n=1}^{N_m} (W_{mn}^{MP})^2 + Tr_m(A^{-1})}{N_m} . \end{aligned} \tag{10}$$

where $Tr_m(A^{-1})$ is the trace of the inverse of the Hessian matrix of the total error function with respect to the weights, α_m and γ_m are hyper-parameters, N_m is the number of weights associated with α_m and m is the input variable.

There are two separate (α_m, γ_m) hyper-parameters for each input covariate including the Time input covariate, the bias terms in the hidden units, the weights to the multiple output units, and the output node biases. For the model selection, the hyper-parameter α_m alphas are of interest because the smallest values correspond to the input variables with prognostic significance. The hyper-parameters γ_m which are calculated based on α_m, represent a measure of the uncertainty in the estimation of the weights and values close to 1 correspond to weights controlled by the data while values close to 0 correspond to weights controlled by the prior.

In the training stage of the neural network, the record of a patient is repeated with respect to the Time variable, which results in the case of the breast cancer dataset in training the PLANN-CR-ARD model with 15420 patterns. In order to express the relationship between the Time variable which stretches over the 15420 patterns and the other 5 input variables (i.e. Number of nodes involved, Tumour size, Age, Histologic grade, Tumour Site) a scaling is needed for the *eprior* error term and the Jacobian matrix of this error term:

$$eprior = \alpha_m \frac{\| w \|^2}{2} \; 0.000000005 . \tag{11}$$

The scaling value used for the breast cancer dataset is 0.000000005, which is the inverse of the squared number of training pattern of data (i.e. 1/(15420*15420)). The scaling value was applied to the *eprior* vector except for the *eprior* variable which corresponds to the Time input variable. This is a general procedure which can be applied in the survival analysis of any type of cancer dataset for the purpose of the model selection with the scaling value being calculated from the number of training pattern of data. The PLANN-CR-ARD model was simulated numerous times with the results for the model selection shown at Fig. 4, which are in agreement with the findings from the literature [6]. The hyper-parameters are shown normalized by the order of the smallest hyper-parameter.

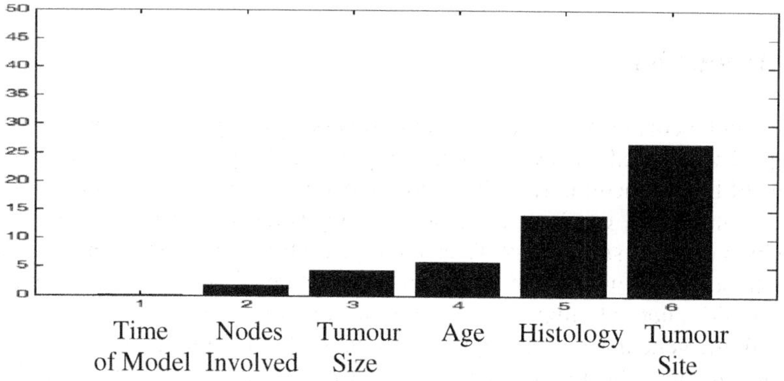

Fig. 4. Hyper-parameters with the most important variables having the smallest values from left to right on the x axis: Time of Model, Number of Nodes Involved, Tumour Size, Age, Histology and Tumour Site

5 Conclusions

A novel compensation mechanism was implemented for the PLANN-CR-ARD model and tested successfully on a real breast cancer dataset. The numerical results showed that the predictions of the PLANN-CR-ARD model are within the 95% confidence intervals for the entire follow-up described in the dataset. The convergence properties of the model were in agreement with what was described in the literature [8][10] for such type of models.

The results for the model selection were in conformity with previous studies realized on the same dataset by using other methods of investigation [6], which showed that the order of the important prognostic variables is Time, Nodes Involved, Tumour Size, Age, Histologic grade and Tumour Site. The model selection was applied to the full model with the three network outputs. A further refinement of this method will be to obtain three lists of prognostic input variables for each of the three network outputs (IBTR, DM and survivorship) and to identify whether there are any changes in the order of the variables in the three lists. Finally, although it can be done simultaneously, a robust procedure is to run the PLANN-CR-ARD model and to calculate the network outputs followed by a subsequent run of the model with the scaling of the eprior terms in order to obtain the model selection.

Acknowledgment

This work was supported by CNCSIS-UEFISCU, project number PN II-RU 246/2010. This work was also supported by the Biopattern Network of Excellence under FP6/2002/IST/1 and IST-2002-508803.

References

1. Collett, D.: Modelling survival data in medical research. Chapman & Hall, London (1994)
2. Lisboa, P.J.G., Wong, H., Harris, P., Swindell, R.: A Bayesian neural network approach for modelling censored data with an application to prognosis after surgery for breast cancer. Artificial Intelligence in Medicine 28(1), 1–25 (2003)
3. Wong, H.A.: A Bayesian neural network for censored survival data, PhD thesis, Liverpool John Moores University (2001)
4. Lisboa, P.J.L., Etchells, T., Jarman, I., Arsene, C.T.C., Aung, M.S.H., Eleuteri, A., Taktak, A.F.G., Ambrogi, F., Boracchi, P., Biganzoli, E.: Partial Logistic Artificial Neural Network for Competing Risks Regularized With Automatic Relevance Determination. IEEE Transactions on Neural Networks 20(9), 1403–1416 (2009)
5. Arsene, C.T.C., Lisboa, P.J.C., Borrachi, P., Biganzoli, E., Aung, M.S.H.: Bayesian Neural Networks for Competing Risks with Covariates. In: The Third International Conference in Advances in Medical, Signal and Information Processing, MEDSIP 2006. IET (July 2006)
6. Veronesi, U., Marubini, E., Del Vecchio, M., Manzari, A., Andreola, S., Greco, M., Luini, A., Merson, M., Saccozzi, R., Rilke, F., Salvadori, B.: Local recurrences and distant metastases after conservative breast cancer treatments: partly independent events. Journal of the National Cancer Institute 87, 19–27 (1995)
7. Moller, M.: A scaled conjugate gradient algorithm for fast supervised learning. Neural Networks 6(4), 525–533 (1993)
8. Bishop, C.: Neural Networks for Pattern Recognition (1995)
9. Nabney, I.: NETLAB: algorithms for pattern recognition. Springer, Heidelberg (2002)
10. Bishop, C.: Pattern recognition and Machine Learning. Springer, Heidelberg (2006)
11. Aalen, O.O.: Non parametric inference in connection with multiple decrement models. Scandinavian J. Statist. 3, 15–27 (1976)
12. Hosmer, D., Stanley, L.: Applied survival analysis: regression modeling of time to event data. John Wiley & Sons, New York (1999)

Gender Recognition Using PCA and DCT of Face Images

Ondrej Smirg[1], Jan Mikulka[1], Marcos Faundez-Zanuy[2], Marco Grassi[3], and Jiri Mekyska[1]

[1] Brno University of Technology, UTKO, Purkynova 118, 612 00, Czech Republic
{smirg,mikulka,mekyska}@feec.vutbr.cz
[2] Escola Universitària Politècnica de Mataró (UPC), Tecnocampus 08302 Mataró, Spain
faundez@eupmt.es
[3] Department of Biomedical, Electronic and Telecommunication Engineering, Politecnica delle Marche, Ancona, Italy
margra75@hotmail.com

Abstract. In this paper we propose a gender recognition algorithm of face images. We have used PCA and DCT for dimensionality reduction. The algorithm is based on a genetic algorithm to improve the selection of training set of images for the PCA algorithm. Genetic algorithm helps to select the images, which best represent each gender, from the image database. We have evaluated a nearest neighbor classifier as well as a neural network. Experimental results show a correct identification rate of 85,9%.

1 Introduction

In the last decade significant advances have been achieved on biometrics, especially on face recognition [10]. This has been possible due to the increase of computational power of the computers. Specific area of face recognition is gender recognition. Algorithms for gender recognition are mainly focused on the human body, because there is more specific information. However, we can have a situation, where the human silhouette is not shown completely or when people is so close to the camera that it can only acquire the face. These cases could be covered by an algorithm that can recognize gender from a single face image. These algorithms are less efficient because the input data (faces) contain less information than the whole body. The advantage of these algorithms lies in the classification of gender in a real time. For the classifier is important to choose the most important characteristics of each group. In the case of the gender classification this is particularly important.

Gender recognition may find application in areas such as advertising. An example might be an advertising screen, which uses a camera for gender recognition of a human who looks at advertising and displays the content accordingly.

2 General Gender Recognition

Gender recognition belongs to the category of pattern recognition. This implies that a large amount of methods exist. Usually, a pattern recognition system consists of three main blocks: information source, feature extraction and classifier. Fig. 1 summarizes

J. Cabestany, I. Rojas, and G. Joya (Eds.): IWANN 2011, Part II, LNCS 6692, pp. 220–227, 2011.

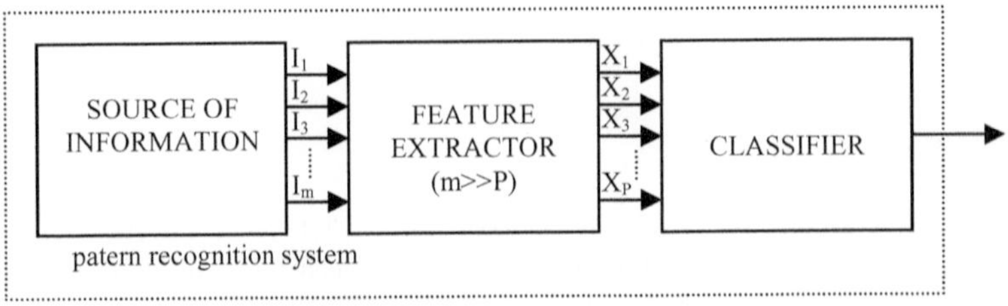

Fig. 1. General pattern recognition system

this scheme. In our case an RGB camera is the source of information, used for acquiring the face image. The image from camera is transformed to grayscale and continues to next block. Because the image contains a large amount of redundant information, system must include a block for the selection of relevant information that best describe the pattern. Features extraction is a very important block of the pattern recognition system. The *m* is a much lower value of *P*, because Features extractor selects from *I* only significant feature.

There are two main approaches for features extraction:

a) Statistical approaches are based on statistical analysis to choose the most important characteristics of the image. For the statistical approach is necessary to have a sufficient number of input data that adequately characterizing the pattern. The main problem is that the images contain a large amount of redundant information; hence the image is usually transferred from a high dimensional space into a small set of measurements. Dimensionality reduction algorithms using statistical methods search for the most common features in the input data. Typically the Karhunen-Loeve transform (KLT) is applied with a simplified algorithm known as eigenfaces [9]. However, eigenfaces algorithm is a suboptimal approximation to KL transform. Nowadays, with the improvements on computational speed and memory capacities, it is possible to compute the KLT directly. Generalized eigenfaces algorithm is called Principal Component Analysis (PCA) and is used in speech processing for example. Another dimensionality reduction method is based on DCT. DCT transforms data from the spatial to the frequency space. DCT is used, for instance, for JPEG compression algorithm. JPEG algorithm uses the fact that the largest amount of relevant information is concentrated in the low frequencies. Both algorithms are used for face recognition very often [2].
b) Geometry-features-based methods try to identify the position and relationship between face parts, such as eyes, nose, mouth etc., and the extracted parameters are measures of textures, shapes, sizes, etc. of these regions. Geometrical methods are used less frequently because it requires studies of the geometric variations of each recognized groups [8].

In this paper, we mainly focus on the study of the feature extraction for the gender recognition using statistical approaches. Classifier is the last block of the pattern recognition system. Its function is to establish rules for separating the groups. Examples of classifiers can be Nearest Neighbor, or a more advanced classifier with neural network. Because we have only two groups, which we want distinguished (male, female), we can use a Nearest Neighbor classifier.

3 Feature Extractors

DCT and PCA are most widely used algorithms for face recognition. These algorithms have been published with different types of classifiers, and achieved very good results in face recognition. PCA is considered the norm among the features extractors and is implemented in many libraries, including open source (for example: OpenCV).

3.1 PCA and Eigenfaces

Eigenfaces are a set of eigenvectors PCA methods used in the computer vision problem of human face recognition. The first image processing system using the eigenfaces was developed by Sirovich and Kirby in 1987 and first face recognition was used by Matthew Turk and Alex Pentland in 1991. A more detailed description can be found in [2] and [9] respectively.

3.2 Discrete Cosine Transform

The Discrete Cosine Transform (DCT) is an invertible linear transform and is similar to the Discrete Fourier Transform (DFT). The original signal is converted to the frequency domain by applying the cosine function for different frequencies. After the original signal has been transformed, its DCT coefficients reflect the importance of the frequencies that are present in it. The very first coefficient refers to the signal's lowest frequency, and usually carries the majority of the relevant (the most representative) information from the original signal [5]. The last coefficients represent the component of the signal with the higher frequencies. These coefficients generally represent greater image detail or fine image information, and are usually noisier. DCT has an advantage when compared with DFT: the coefficients are real values while DFT produces complex values [6]. We used 2D-DCT (two dimensional DCT) in this articke. The 2D-DCT definition is shown [6] in Equations (1) and (2).

$$X[k,l] = \frac{2}{N} c_k c_l \sum_{m=0}^{a-1} \sum_{n=0}^{b-1} x[m,n] \cos\left[\frac{(2m+1)k\pi}{2N}\right] \cos\left[\frac{(2n+1)l\pi}{2N}\right] \tag{1}$$

where, in Equation (1) :

$$c_k, c_l = \begin{cases} \sqrt{\left(\frac{1}{2}\right)} \quad to \quad k=0, l=0 \\ 1 \quad to \quad k=1,2,.....a-1 \quad and \quad l=1,2,...b-1 \end{cases} \tag{2}$$

Features extraction using DCT consists of selecting the coefficients around the X[0,0] coefficient (DC coefficient), where the highest energy of the image is. This is a zonal coding, because the coefficients around a predefined zone are retained. Another approach for feature extraction using DCT coefficients is to find the low frequencies that best describe each group [5]. This is a threshold coding, because only the coefficients over a threshold are selected. For this approach is necessary to have a training set of images for each group (male and female). The principle can be described by the following steps:

1. The images are transferred to the frequency domain by DCT (equation 1). The coefficients are chosen from area around the DC. This is done for each group separately ($X_{male}(x,y)$, $X_{female}(x,y)$.
2. This step calculates the average of coefficients E_{male}, E_{female} for each group by

$$E_{male}(x, y) = \frac{1}{n}\sum_{i=1}^{n} X_{male}(x, y, i), \tag{3}$$

where x, y determine the position of the frequency coefficient and i is the order of image input set of faces male or female group.
3. All coefficients of the groups are merged into one big group and the average value E_{all} for all coefficients are calculated:

$$E_{all}(x, y) = \frac{1}{n}\sum_{i=1}^{n} X_{all}(x, y, i), \tag{4}$$

where X_{all} contains all elements of the X_{male} and X_{female} groups.
4. Mean values of E_{male}, E_{female} are compared with the mean values E_{all} by equation (12). The highest values of the coefficients of the equations (5) and (6) represent the most important factors in the group.

$$X_M(x, y) = \frac{E_{male}(x, y)}{E_{all}(x, y)} \tag{5}$$

$$X_F(x, y) = \frac{E_{female}(x, y)}{E_{all}(x, y)} \tag{6}$$

4 Database

The training stage performs various attribute selection steps and decides which one produces the best result. With this in mind, the training algorithm for PCA tests various images from database. Choosing the best input data for each group was performed using a genetic algorithm.

In this paper, we have used our own database of faces to train and test the proposed method. This database contains photos from faces of 185 female and 215 male, each person have 4 sets of pictures and each set contains 4 photo with different poses, totalizing 6400 photos. The faces were photographed at different moments (different sets), with varying lighting, facial expressions (eyes closed/opened, smiling/not smiling), facial poses and facial details (with/without glasses, with/without beard), among other types of variations. The images are in grayscale, with 256 different levels and a dimension of 92x112 pixels. This resolution was chosen for correspondence with the AT&T and ORL database of faces. The database was obtained from BiosecurID database [11]. Examples of used images are shown in fig. 2.

Fig. 2. Images from database

5 Results of PCA and DCT

This article discusses the possibility of using face recognition techniques for the gender recognition. For testing, we chose the most common methods used for features extraction described in Chapter 3. For classification, we chose the nearest neighbor classifier, which is the simplest classifier. Nearest neighbor classifier using the mean square error (MSE) or the mean absolute difference (MAD) [2] defined as

$$MSE(\vec{x},\vec{y}) = \sum_{i=1}^{P}(x_i - y_i)^2 \tag{7}$$

$$MAD(\vec{x},\vec{y}) = \sum_{i=1}^{P}|x_i - y_i| \tag{8}$$

For the DCT method we also tested the classifier using the neural network described in [12]. The results of the effectiveness of each method are shown in Table 1. Number of coefficients for each method was chosen according to a study published in [2].

Table 1. Recognition rates

Transform	Number of images for learning (male/female)	Number of images for testing (male/female)	The effectiveness of recognition % (male/female)	The effectiveness of All recognition %
PCA	1600/1600	100/100	83/63	73
DCT with neural network classifier	1600/1600	100/297	76/57	57
DCT with the selection of the most significant frequency components	1600/1600	100/100	60/60	60

This experimental results show that the PCA algorithm provides the best result. However, PCA is dependent on the choice of images for the train set. For this reason we have also tested a genetic algorithm for selecting the best input train set.

5.1 PCA with Genetic Algorithm

The genetic algorithm operates on the principle of making each generation of individuals with genotype. These individuals are crossed among themselves and create a new generation. In addition to cross, the genetic algorithm uses a mutation, which randomly changes the genetic code of certain individuals. Every individual has the genotype, which consists of the number of images of input data for training, in this article. Each individual is evaluated by the so-called fitness function that determines the quality of the individual. According to this function is determined by the probability with which an individual participates in the creation of new generation. [8] Fitness function was calculated by using the following equation:

$$fitness = \frac{(T_{tm} + T_{tf})}{(T_{tm} + T_{tf} + T_{fm} + T_{ff})} \tag{9}$$

where T_{tm} is the number of male correctly identified, T_{tf} is the number of female correctly identified, T_{fm} is the number of male incorrectly identified and T_{ff} is the number of female incorrectly identified. The test was carried out of 100 photos which were not used for training. The genetic algorithm process of the best fitness of individuals in 500 generations is shown in fig. 3.

From the all generations the best result is chosen. This result is an outcome of genetic algorithm. Termination of the genetic algorithm is given either by reaching the desired fitness function, or if it reaches the maximum number of generations.

Results of using a combination of genetic algorithm with the PCA algorithm are shown in table 2. Testing takes place in 20000 randomly selected images, from that 10000 male and 10000 female. Images may be repeated. For testing, we excluded images used for training.

Comparing table 1 and table 2 we can see that experimental results are improved from 73% to 83,8% when using the genetic algorithm.

In addition, we also observe that the male recognition is slightly higher than the female one.

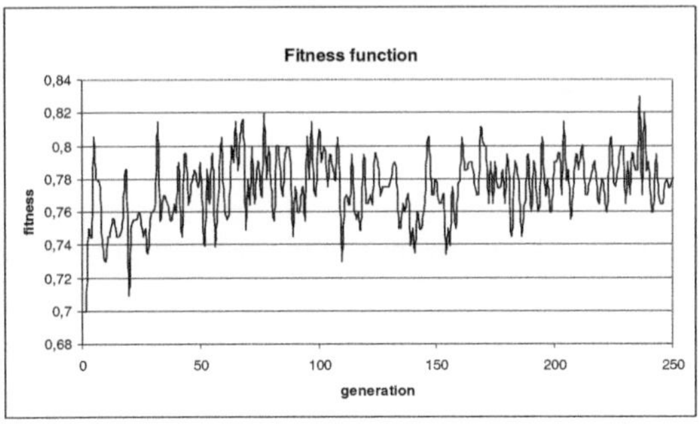

Fig. 3. Process of fitness function for 250 generations

Table 2. Recognition rates for PCA with genetic algorithm for various numbers of training set

Number of training images (male/female)	The effectiveness of recognition % (male/female)	The effectiveness of all recognition %
80/80	80,3/70,0	75,2
160/160	89,1/73,7	81,4
800/800	88,7/79,1	83,8
1600/1600	91,0/80,7	85,9

6 Conclusion

The article focused on the gender recognition by the methods used for face recognition. Comparison of different methods was evaluated. Experimental results show that the most efficient method is PCA. We found that the effectiveness of PCA depends both on the number of trained set of images and also on the specific images used for training. For training, we developed genetic algorithm for finding the images that contain the best information about each gender. As shown in Table 2, We achieved more than 85,9% efficiency when using 1600 images. The time required to find the optimal set is dependent on the number of training images. We needed more than 72 hours to find an optimal set of PCA algorithm when using 1600 images as training set.

The performance of the system is comparable to others. Brunelli and Poggio [13] trained a hyper basis function network on automatically extracted geometrical features. They achieved a correct gender classification rate of 87.5%. Golomb et al. [14] used a template-based approach. They trained a backpropagation network on a compressed representation (40 units) of low resolution face images of 30 x 30 pixels and achieved a performance of 91.9%. They used limited hair information and aligned the faces under manual control. Laurenc Wiskott et al. used classifier, which used to typical facial features of a gender-specific, such as beard for example. They achieved a correct gender classification rate of 90.2% [15]. The advantage of the system is in the requirements for performance, which is low for used algorithm.

Acknowledgment

This work has been supported by FEDER and MEC, TEC2009-14123-C04-04, COST-2102, and Research project SIX (CZ.1.05/2.1.00/03.0072).

References

1. Faundez-Zanuy, M.: On the vulnerability of biometric security systems. IEEE Aerospace Electron. System Mag. 19(6), 3–8 (2004)
2. Faundez-Zanuy, M., Roure, J., Espinosa-Duró, V., Ortega, J.A.: An efficient face verification method in a transformed domain. Pattern Recognition Letters 28, 854–858 (2007)
3. Faundez-Zanuy, M.: Data fusion in biometrics. IEEE Aerospace Electron. System Mag. 20(1), 34–38 (2005)
4. Faundez-Zanuy, M., Espinosa, V., Ortega, J.A.: A low-cost webcam and personal computer opens door. IEEE Aerospace Electron. Systems Mag. 20(11), 23–26 (2005)
5. Do, T.T., Le, T.H.: Facial Feature Extraction Using Geometric Feature and Independent Component Analysis, pp. 231–241. Springer, Heidelberg (2009), ISBN 978-3-642-01714-8
6. Omaia, A., Poel, J.K., Batista, L.V.: 2D-DCT Distance Based Face Recognition Using a Reduced Number of Coefficients. In: Computer Graphics and Image Processing (SIBGRAPI 2009), pp. 291–298 (2009), ISSN 1550-1834
7. Jain, A.K.: Fundamentals of Digital Image Processing. Prentice-Hall, Englewood Cliffs (1989)
8. Roth, G., Levine, M.D.: Geometric primitive extraction using a genetic algorithm. IEEE Transactions on Pattern Analysis and Machine Intelligence, 901–905 (1994), ISSN 0162-8828
9. Turk, M., Pentland, A.: Eigenfaces for recognition. J. Cognitive Neurosci. 3(1), 71–86 (1991)

10. Amornraksa, T., Tachaphetpiboon, S.: Fingerprint recognition using DCT features. Electronics Letters 42, 522–523 (2006), ISSN 0013-5194
11. Fierrez, J., Galbally, J., Ortega-Garcia, J., Freire, M.R., Alonso-Fernandez, F., et al.: BiosecurID: a multimodal biometric database. In: Pattern Analysis & Applications, vol. 13, pp. 235–246. Springer-Verlag London Limited, Heidelberg (2009)
12. Grassi, M., Faundez-Zanuy, M.: Protecting DCT templates for a face verification system by means of pseudo-random permutations. In: Cabestany, J., Sandoval, F., Prieto, A., Corchado, J.M. (eds.) IWANN 2009. LNCS, vol. 5517, pp. 1216–1223. Springer, Heidelberg (2009)
13. Brunelli, R., Poggio, T.: Caricatural effects in automated face perception. Biological Cybernetics 69, 235–241 (1993)
14. Golomb, B.A., Lewrence, D.T., Sejnowski, T.J.: A neural network identifies sex for human faces. In: Touretsky, D.S., Lippman, R. (eds.) Advances in Neural Information Processing System, vol. 3, Morgan Kaufmann, SantMateo (1991)
15. Wiskott, L., Fellous, J.-M., Krüger, N., von der Malsburg, C.: Face Recognition and Gender Determination (1995)

Efficient Face Recognition Fusing Dynamic Morphological Quotient Image with Local Binary Pattern

Hong Pan, Siyu Xia, Lizuo Jin, and Liangzheng Xia

School of Automation, Southeast University, Nanjing, China, 210096
enhpan@seu.edu.cn, xia081@gmail.com, jinlizuo@gmail.com,
lzxia@seu.edu.cn

Abstract. In this paper, we propose a novel illumination normalized Local Binary Pattern (LBP)-based algorithm for face recognition under varying illumination conditions. The proposed DMQI-LBP algorithm fuses illumination normalization, using the Dynamic Morphological Quotient Image (DMQI), into the current LBP-based face recognition system. So it makes full use of advantages of illumination compensation offered by the quotient image, estimated with a dynamic morphological close operation, as well as the powerful discrimination ability provided by the LBP descriptor. Evaluation results on the Yale face database B indicate that the proposed DMQI-LBP algorithm significantly improve the recognition performance (by 5% for the first rank) of the original raw LBP-based system for face recognition with severe lighting variations. Furthermore, our algorithm is efficient and simple to implement, which makes it very suitable for real-time face recognition.

Keywords: face recognition, local binary pattern, quotient image.

1 Introduction

Face Recognition is a major biometric research topic that uses automated methods to recognize the identity of a person based on one's facial characteristics. In recent several decades, many methods have been proposed and successfully applied in some practical applications. Nevertheless, there are still many challenging problems, such as facial expressions, pose variations and illumination changes, exist in current face recognition systems. Those variations inevitably degrade the performance of face recognition systems in real-world scenarios.

Generally speaking, current face recognition methods can be divided into two broad categories, i.e. holistic-based methods and local-based methods. Holistic-based methods use the whole face region as the input to a recognition system. The principle of holistic methods is to construct a subspace using principle component analysis (PCA)[1], linear discriminant analysis (LDA)[2] or independent component analysis (ICA)[3]. Face images are then projected and compared in a low-dimensional subspace to avoid the curse of dimensionality. Whereas, local-based methods first locate several facial components, and then classify faces by comparing and combining the corresponding local statistics. Since local representation provides robustness to partial occlusion, local-based methods have dominated face recognition from the mid 1990s.

J. Cabestany, I. Rojas, and G. Joya (Eds.): IWANN 2011, Part II, LNCS 6692, pp. 228–235, 2011.

For example, Pentland et al.[4] extended the eigenface technique to a layered representation by combining eigenfaces and other eigenmodules, such as eigeneyes, eigennoses, and eigenmouths. A similar approach, named subpattern PCA (SpPCA), was studied by Chen and Zhu [5]. Considering the fact that different parts of the human face may contribute differently to recognition, Tan and Chen [6] extended SpPCA to adaptively weighted subpattern PCA. Wiskott et al. [7] achieved good performance with the elastic bunch graph matching (EBGM) method on the FERET dataset. The elastic bunch graph is a graph-based face model with a set of Gabor wavelet jets attached to each node of the graph. The algorithm recognizes new faces by first locating a set of facial features (graph nodes) to build a graph, which is then used to compute the similarity of both jets and topography. Local binary pattern (LBP), proposed by Ojala et al.[8-11], was introduced to face recognition in 2004 [10]. In LBP-based approaches, the face area was divided into a couple of small windows. LBP operators on each window were extracted and the weighted chi square (χ^2) statistic was adopted to compare LBP histograms.

Compared with other local features such as local PCA feature, Gabor feature, SIFT feature etc., LBP operator achieves superior discriminative power on face images with expression, aging, pose variations as well as partial occlusion [12]. By definition, LBP is invariant under any monotonic transformation of the pixel intensity. However, it is significantly affected by non-monotonic intensity transformations. Unfortunately, in contrast to the flat surfaces where texture images are usually captured, faces are not flat and non-monotonic intensity transformations occur in face images. Therefore, LBP may have problems to deal with illumination variations in face recognition.

In order to reduce the performance degradation of LBP-based face recognition methods, caused by illumination variations, some approaches [13-14] suggest integrating Gabor transforms with the LBP representation, which applies Gabor filtering before calculating LBP features. However, Gabor filtering at multi-scale and multi-orientation are too much computation-demanding for practical applications. As a remedy against the illumination-variant characteristic of LBP operators and these time-consuming Gabor-LBP approaches, in this paper, we propose an effective DMQI-LBP algorithm for real-time face recognition. The DMQI-LBP algorithm fuses illumination normalization, using the Dynamic Morphological Quotient Image (DMQI), with the current LBP-based face recognition system, so it makes full use of advantages of the illumination compensation offered by the quotient image, estimated with a dynamic morphological close operation, as well as the powerful discrimination ability provided by the LBP representation. Fig.1 shows the framework of the proposed DMQI-LBP algorithm.

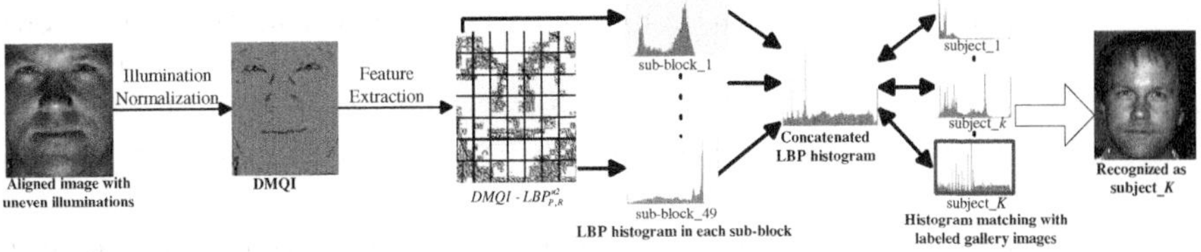

Fig. 1. Framework of the proposed DMQI-LBP algorithm for face recognition

2 Illumination Normalization

It is known that illumination variations is one of the most important factors that change the appearance of a face image because of the intensity fluctuation of the face image, caused by shadow cast from different light source directions. Usually, variations between images of the same face due to illumination and viewing directions are almost always larger than image variations due to the change in face identity. Therefore, the robustness of face recognition system must be improved, under varying illumination conditions.

Numerous algorithms [15-19] have been proposed to deal with face recognition under varying illumination conditions. Shashua et al. [15] proposed a quotient image model in which all face images were defined as the same shape but differ in their surface albedo. The quotient image is a ratio between a probe image and any linear combination of three different illumination images. Hence, the quotient image is invariant to illumination variations. Wang et al. [16] proposed a self quotient image based only on a single image. The Self Quotient Image (SQI) was obtained by using the Gaussian function as a smoothing kernel function. Srisuk and Petpon [17] extended it to Gabor based SQI where the 2D Gabor filter is applied instead of weighted Gaussian filter to increase the efficiency for face recognition. Classified Appearance-based Quotient Image (CAQI) [18] is proposed by Nishiyama et al. They classified facial appearances caused by illumination into four main components and learnt a statistical model to extract the illumination invariant ratio of albedo. More recently, Wang et al. [19] proposed a morphological quotient image in which the illumination variations can be estimated easily by utilizing the morphological close operation.

2.1 Reflection Model

In Retinex theory[20], an image $I(x, y)$ can be modeled as the product of the reflectance function $R(x, y)$ and luminance function $L(x, y)$, as shown in Eq.(1)

$$I(x, y) = R(x, y)\, L(x, y) \tag{1}$$

The reflectance $R(x, y)$, relating to the characteristics of objects in the scene of the image, is dependent on the albedo of the scene's surfaces, whereas the luminance $L(x, y)$, relating to the amount of illumination falling on the observed scene, is determined by the illumination source. Since the reflectance $R(x, y)$ only relates to the objects in the image, it is obvious that, when successfully estimated, it can be an illumination invariant representation of the input image. Thus, in order to obtain an illumination invariant image representation, the luminance $L(x, y)$ of an image is commonly estimated first. Then, the reflectance is formulated as the quotient of the image $I(x, y)$ and its luminance $L(x, y)$, as shown in Eq.(2).

$$R(x, y) = I(x, y) \,/\, L(x, y) \tag{2}$$

As already emphasized [21], the luminance $L(x, y)$ is considered to vary slowly with the spatial position, and therefore can be estimated as a smooth version of the original image $I(x, y)$.

2.2 Dynamic Morphological Quotient Image (DMQI)

Various smoothing filters and smoothing methods have been proposed to estimate the luminance $L(x, y)$, which results in different illumination normalization algorithms. Inspired by the good performance and low-complexity offered by the MQI [19] for luminance estimation, we adopt the MQI with an adaptive template size, i.e. DMQI, to compensate the illumination fluctuation, and improve the LBP-based face recognition performance in severe illumination variations.

In the MQI-based illumination normalization method, the luminance $L(x, y)$ is estimated by applying a morphological close operation, with a fixed size template, to an image $I(x, y)$. In [19], it has been shown that, with a suitable template structure, close operation can preserve some particular patterns while attenuating other patterns in an image, without blurring the original image edges. Specially, when processing an image under uneven illuminations, such close operation can be used to obtain a smooth version of the original image, and the region with pixels having the same intensity is expected to share the same light intensity.

The template size is a key parameter that impacts the performance of the MQI method. With a large size template, the close operation focuses on large scale features, and results in poor performance on local illumination compensation, especially in shadow regions. On the other hand, for a small size template, it results in good local illumination normalization, but simultaneously misses large scale features. To further improve the estimation accuracy of the luminance $L(x, y)$, the DMQI is applied in our algorithm to make the template size adaptive to the given image. The scheme of DMQI can be formulated as

$$DClose(x,y)=\begin{cases} Close^{l}(x,y), & \alpha\times Close^{s}(x,y)<Close^{l}(x,y) \\ Close^{m}(x,y), & \beta\times Close^{s}(x,y)<Close^{l}(x,y)<\alpha\times Close^{s}(x,y) \\ Close^{s}(x,y), & Close^{l}(x,y)<\beta\times Close^{s}(x,y) \end{cases} \quad (3)$$

$$DMQI(x,y)=\frac{I(x,y)}{DClose(x,y)}$$

where α, β denote feature scales, with $\alpha>\beta>1.0$; l, m, and s denote the optional sizes for large, medium and small templates, respectively, with $l > m > s > 1$. If $\alpha\times Close^{s}(x, y)<Close^{l}(x, y)$, which means that the large size template is very different from the small size one, the area around point (x, y) would be brow, eye, nose, mouth, or the boundary of illumination variation, where pixel intensity changes significantly. Therefore, large size template is chosen to preserve features. If $Close^{l}(x, y) < \beta\times Close^{s}(x, y)$, which indicates that large size template and small size template have almost the same effect, the point (x, y) is in a smooth region (such as cheek and forehead) with even illumination conditions. Pixel intensities in such areas have little changes. In this case, small size template is a good choice for the close operation.

3 Feature Extraction Using Local Binary Pattern (LBP)

LBP [8-11] is an efficient and powerful texture descriptor which labels pixels of an image by thresholding the neighborhood of each pixel with the value of its center pixel and considers the result as a binary string. Due to its discriminative power and low complexity, LBP descriptor has become a popular solution for many applications.

The idea behind the LBP descriptor is that 2D surface textures can be described by two complementary measures: local spatial patterns and grayscale contrast. The basic LBP descriptor [8] is calculated in a 3x3 neighborhood around each pixel. Formally, given a pixel at (x_c, y_c), the resulting LBP can be expressed in the decimal form as

$$LBP = \sum_{p=0}^{7} 2^p \times S(g_p - g_c) \tag{4}$$

where p runs over the 8 neighbors of the central pixel (x_c, y_c); g_c and g_p are the gray-scale values of the central pixel (x_c, y_c) and the surrounding pixel (x_p, y_p), respectively; and $S(x)$ is 1 if $x \geq 0$ and 0 otherwise. The histogram of these 2^8 =256 different labels can then be used as a texture descriptor. Ojala later made several extensions [9] to the basic descriptor. Firstly, the descriptor was extended to use neighborhoods with different sizes and capture dominant features at different scales. Using a circular sampling structure and bilinearly interpolating the neighboring pixel values, the basic LBP descriptor is extended to $LBP_{P,R}$ with local neighborhood of P equally-spaced sampling points on a circle of radius of R. Fig 2. gives an example for the calculation of $LBP_{8,2}$ descriptor. Then, in order to remove the effect of rotation, each LBP descriptor is circularly shifted to a unique and rotation invariant code $LBP_{P,R}^{ri}$, which gives a minimum decimal value. Finally, they proposed to use a small subset of the whole 2^P patterns to describe image textures. These patterns, called uniform patterns – $LBP_{P,R}^{u2}$, contain at most two bitwise transitions from 0 to 1 or vice versa, when considered as a circular binary string. The uniform patterns represent local primitives such as edges and corners. It was observed that most of the texture information was contained in the uniform patterns.

Ahonen et al. [10-11] extended the LBP representation to face recognition. In their scheme, a face image is divided into local regions and LBP descriptors are extracted from each region independently. The occurrences of the LBP codes in each local region are collected into the corresponding histograms which are then concatenated to form a global description of the face image.

4 Fusing DMQI with LBP

In order to utilize the excellent discriminative power and computational simplicity of the LBP descriptor, while abating the performance degradation due to varying illumination conditions for face recognition. We propose an enhanced LBP-based face recognition algorithm that fuses illumination invariant DMQI with discriminative LBP descriptors. In our DMQI-LBP algorithm, illumination variations in face images are first normalized with the DMQI, then the DMQI is segmented into 7×7 sub-blocks, and uniform patterns $LBP_{8,2}^{u2}$ are extracted in these sub-blocks to form the LBP feature histograms. Finally, LBP histograms from all sub-blocks are concatenated into face feature vectors, and a weighted chi-square distance metric (χ_w^2) which considers the different roles of each sub-block for face recognition, is evaluated to measure the similarity between a probe face feature and the stored subject face features.

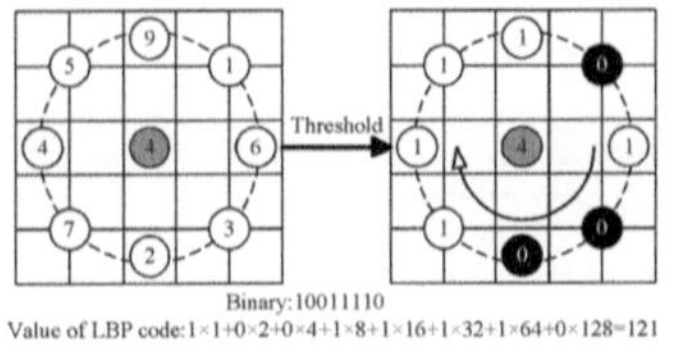

Fig. 2. Calculation of the $LBP_{8,2}$ operator

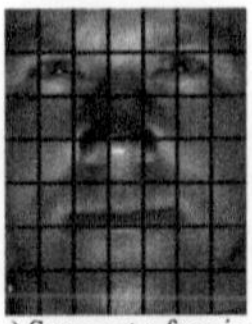

(a) Segment a face image into 7×7 sub-blocks

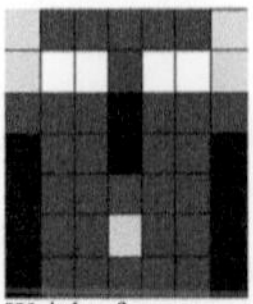

(b) Weights for squares with black, red, yellow and white are 0, 1.0, 2.0 and 4.0, respectively

Fig. 3. Weight coefficients for a segmented face image

(a) Original face images with uneven illumination conditions

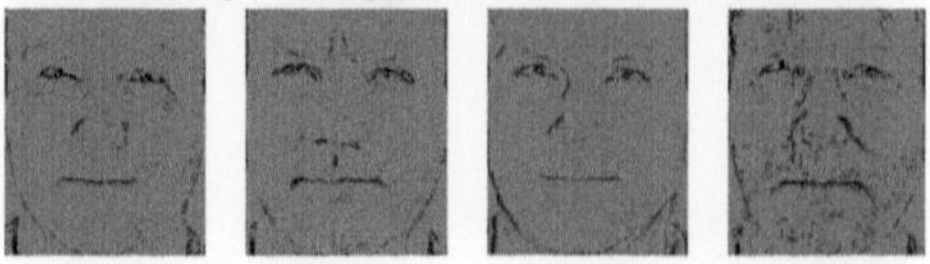

(b) Normalized images using the DMQI technique

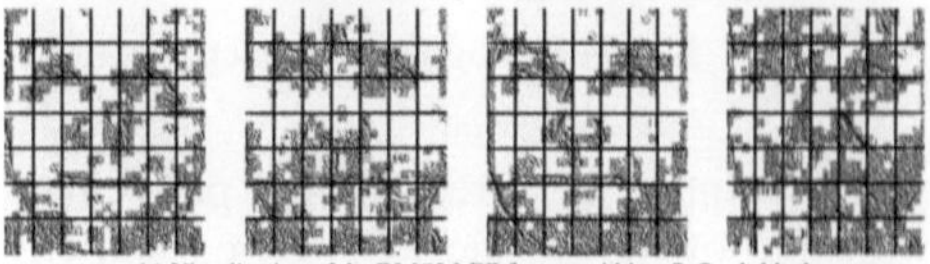

(c) Visualization of the DMQI-LBP feature within a 7×7 sub-block

Fig. 4. Comparison of original images with the DMQI and DMQI-LBP images

$$\chi^2_\omega(P,G)=\sum_{i,j} w_j \frac{(P_{i,j}-G_{i,j})^2}{P_{i,j}+G_{i,j}} \quad (5)$$

where P and G denote the probe face image and the labeled subject's gallery face image, respectively. w_j is the weight coefficient for the jth sub-block in a face image. $P_{i,j}$ denotes the probability of the probe LBP feature, from the jth sub-block, which drops in the ith histogram bin, and $G_{i,j}$ denotes the probability of the gallery LBP feature, from the jth sub-block, which drops in the ith histogram bin. The probe image is categorized to the subject that has the largest similarity value with the probe image. Fig.3 illustrates the segmentation of a face image into 7×7 sub-blocks, as well as the weight coefficients for each sub-block. Fig.4 compares original images under uneven illumination conditions with results of the normalized DMQI and DMQI-LBP images.

5 Experimental Results

We evaluated the proposed DMQI-LBP algorithm on Yale face database B [22].This database contains 5,760 single light source images of 10 subjects each captured under 576 viewing conditions (9 poses × 64 illumination conditions). The whole database was divided into 5 subsets according to the angle of the light source directions. In this experiment, we used a total of 640 images, taken from 10 individuals in a frontal pose. Face images were aligned according to the coordinates of two eyes and cropped to 210×252 pixels, so that only face regions are contained in the cropped images. We selected a single image of each individual as the gallery image, in which the lighting conditions are the same for all subjects. Other images taken under the remaining 63 lighting conditions are used as the probe images. For illumination normalization using the DMQI, the parameters are set as: α=1.8, β=1.35, s=5, m=7, l=9.

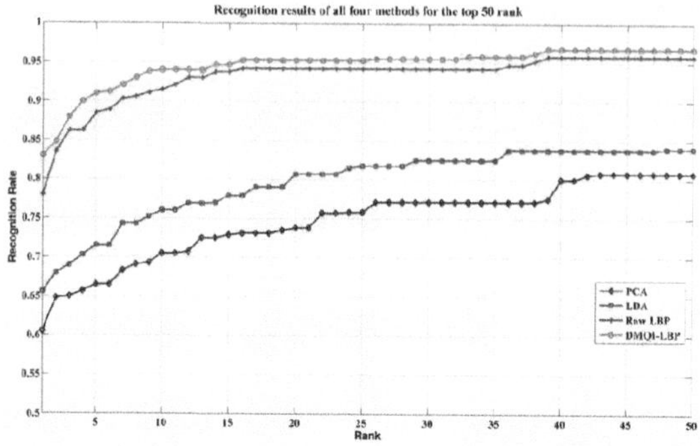

Fig. 5. Recognition rate curves of all four methods for the top 50 ranks

We compared the recognition performance of the proposed DMQI-LBP algorithm with that of the classic PCA-based and LDA-based method as well as the approach using raw LBP descriptors without illumination normalization. In particular, the dimension of face features is reduced to 200 for PCA-based and LDA-based method. In the raw LBP-based method, we follow the similar configuration in [11]. Fig.5 plots the recognition rate curves of all four methods for the top 50 ranks. It can be seen that the raw LBP-based method achieves a recognition rate of 78% for the first rank, which outperforms that of the PCA-based and LDA-based method by 17.5% and 12.4%, respectively. The proposed DMQI-LBP algorithm further improves the recognition rate to 83% for the first rank, and the recognition rate for the top 5 rank reaches 91%. It is interesting to note that the fusion of DMQI really helps the raw LBP-based method alleviate the challenges caused by uneven illuminations. Compared with other illumination compensation techniques, our method achieves the similar normalization quality, but with a high efficiency in terms of the computational complexity. In particular, it takes only 0.20 seconds for illumination normalization, 0.16 seconds for DMQI-LBP feature extraction and 0.12 for feature matching on average, on a platform with a Pentium IV 3.0G CPU.

6 Conclusion

We propose an enhanced LBP feature-based algorithm for face recognition under varying illumination conditions. The proposed DMQI-LBP algorithm integrates the dynamic morphological quotient image technique to compensate the lighting variations that greatly degrades the discriminability of LBP descriptors. By fusing the illumination invariant DMQI and the discriminative LBP descriptors into a common framework, we significantly improve the recognition performance of the current raw LBP feature-based face recognition system. Another attracting feature of the DMQI-LBP algorithm is the computational simplicity, which make our method very suitable for real-time face recognition.

Acknowledgments. This work was supported in part by the National Natural Science Foundation of China under Grant 60805002 and 90820009, Foundation for Young Scholars of Southeast University under Grant 4008001015.

References

1. Turk, M., Pentland, A.: Eigenfaces for Recognition. J. Cogn. Neurosci. 13, 71–86 (1991)
2. Etemad, K., Chellappa, R.: Discriminant Analysis for Recognition of Human Face Images. Journal of the Optical Society of America A 14, 1724–1733 (1997)
3. Bartlett, M., Movellan, J., Sejnowski, T.: Face Recognition by Independent Component Analysis. IEEE Trans. Neural Networks 13(6), 1450–1464 (2002)
4. Pentland, A., Moghaddam, B., Starner, T.: View-Based and Modular Eigenspaces for Face Recognition. In: Proc. IEEE Conf. CVPR, pp. 84–91 (1994)
5. Chen, S.C., Zhu, Y.L.: Subpattern-Based Principle Component Analysis. Pattern Recognition 37(5), 1081–1083 (2000)
6. Tan, K., Chen, S.: Adaptively Weighted Sub-pattern PCA for Face Recognition. Neurocomputing 64, 505–511 (2005)
7. Wiskott, L., Fellous, J., Kruger, N., Malsburg, C.: Face Recognition by Elastic Bunch Graph Matching. IEEE Trans. Pattern Anal. Mach. Intell. 19(7), 775–779 (1997)
8. Ojala, T., Pietikäinen, M., Harwood, D.: A Comparative Study of Texture Measures with Classification Based on Feature Distributions. Pattern Recognition 29(1), 51–59 (1996)
9. Ojala, T., Pietikäinen, M., Mäenpää, T.: Multiresolution Gray-scale and Rotation Invariant Texture Classification with Local Binary Patterns. IEEE Trans. Pattern Anal. Mach. Intell. 24, 971–987 (2002)
10. Ahonen, T., Hadid, A., Pietikäinen, M.: Face recognition with local binary patterns. In: Pajdla, T., Matas, J(G.) (eds.) ECCV 2004. LNCS, vol. 3021, pp. 469–481. Springer, Heidelberg (2004)
11. Ahonen, T., Hadid, A., Pietikäinen, M.: Face Description with Local Binary Patterns: Application to Face Recognition. IEEE Trans. Pattern Anal. Mach. Intell. 28, 2037–2041 (2006)
12. Zou, J., Ji, Q., Nagy, G.: A Comparative Study of Local Matching Approach for Face Recognition. IEEE Trans. Image Processing 16(10), 2617–2628 (2007)
13. Tan, X., Triggs, B.: Fusing Gabor and LBP Feature Sets for Kernel-based Face Recognition. In: Proc. of Analysis and modeling of faces and gestures, pp. 235–249 (2007)
14. Wang, J.G., Yau, W.Y., Wang, H.L.: Age Categorization via ECOC with Fused Gabor and LBP Features. In: Proc. Workshop on WACV, pp. 1–6 (2009)
15. Shashua, A., Riklin-Raviv, T.: The Quotient Image: Class-Based Re-Rendering and Recognition with Varying Illuminations. IEEE Trans. Pattern Anal. Mach. Intell. 23(2), 129–139 (2001)
16. Wang, H., Li, S.Z., Wang, Y.: Face Recognition under Varying Lighting Conditions Using Self Quotient Image. In: Proc. of Face and Gesture Recognition, pp. 819–824 (2004)
17. Srisuk, S., Petpon, A.: A Gabor Quotient Image for Face Recognition under Varying Illumination. In: Bebis, G., Boyle, R., Parvin, B., Koracin, D., Remagnino, P., Porikli, F., Peters, J., Klosowski, J., Arns, L., Chun, Y.K., Rhyne, T.-M., Monroe, L. (eds.) ISVC 2008, Part II. LNCS, vol. 5359, pp. 511–520. Springer, Heidelberg (2008)
18. Nishiyama, M., Kozakaya, T., Yamaguchi, O.: Illumination Normalization using Quotient Image-based Techniques. In: Recent Advances in Face Recognition, I-Tech, Vienna, Austria, pp. 97–108 (2008)
19. Wang, J., Wu, L., He, X., Tian, J.: A new method of illumination invariant face recognition. In: Proc. of International Conference on Innovative Computing, pp. 139–142 (2007)
20. Land, E.H., McCann, J.J.: Lightness and Retinex Theory. Journal of Optical Society of America 61(1), 1–11 (1971)
21. Park, Y., Park, S., Kim, J.: Retinex Method Based on Adaptive Smoothing for Illumination Invariant Face Recognition. Signal Processing 88(8), 1929–1945 (2008)
22. Georghiades, A., Belhumeur, P., Kriegman, D.: From Few to Many: Illumination Cone Models for Face Recognition under Variable Lighting and Pose. IEEE Trans. Pattern Anal. Mach. Intell. 23(6), 643–660 (2001)

A Growing Neural Gas Algorithm with Applications in Hand Modelling and Tracking

Anastassia Angelopoulou[1], Alexandra Psarrou[1], and José García Rodríguez[2]

[1] School of Electronics and Computer Science, University of Westminster, UK
{agelopa,psarroa}@wmin.ac.uk
[2] Department of Computing Technology, University of Alicante, Spain
{jgarcia}@dtic.ua.es

Abstract. Growing models have been widely used for clustering or topology learning. Traditionally these models work on stationary environments, grow incrementally and adapt their nodes to a given distribution based on global parameters. In this paper, we present an enhanced Growing Neural Gas (GNG) model for applications in hand modelling and tracking. The modified network consists of the geometric properties of the nodes, the underline local feature of the image, and an automatic criterion for maximum node growth based on the probability of the objects in the image. We present experimental results for hands and T1-weighted MRI images, and we measure topology preservation with the topographic product.

Keywords: Unsupervised Learning, Topology preservation, Self-organising networks, Nonrigid Shapes.

1 Introduction

When using shape or feature information or combination of the two to segment and track nonrigid objects in video sequences, the most effective models are the Active Contour Models ('Snakes') [11] and their extensions [12], the Active Shape Models (ASMs) [3], and the Active Appearance Models (AAMs) [5]. In recent years, there have been a number of papers that have used self-organising models in applications related to computer vision, medical imaging, and man-machine interaction like: image compression [8], segmentation and representation of objects and medical shapes [14], objects tracking [1,6], recognition of gestures [9], or 3D reconstruction [10,4]. From the cited works, only [6] represents both the local as well as the global movement, however there is no consideration of time constraints, and the condition of finalisation for the GNG algorithm is commonly defined by the insertion of a predefined number of nodes. The automatic election of this number and other parameters that constitutes our enhanced GNG algorithm, accelerates the learning process and makes it suitable for modelling and tracking an object in a sequence of k frames.

Considering the work in the area and previous studies about representation capabilities of self-growing neural models, this paper presents (1) the automatic

J. Cabestany, I. Rojas, and G. Joya (Eds.): IWANN 2011, Part II, LNCS 6692, pp. 236–243, 2011.

extraction and correspondence of landmark points using only topological relations derived from competitive hebbian learning; and (2) the adaptive learning of the number of nodes needed to represent an object.

The remainder of the paper is organised as follows. Section 2 introduces the framework for object modelling using topological relations. Section 3 discusses the extentions of GNG to various images. A set of experimental results are presented in Section 4, before we conclude in Section 5.

2 Characterising 2*D* Objects Using GNG

One way of extracting landmark points along the contour of shapes is to use a topographic mapping where a low dimensional map is fitted to the high dimensional manifold of the contour, whilst preserving the topographic structure of the data. GNG [7] allows us to extract in an autonomous way the contour of any object as a set of edges that belong to a single polygon and form a topology preserving graph. Identifying the points of the image that belong to objects allows the network to obtain an induced Delaunay triangulation of the objects.

Let an object $O = [O_G, O_A]$ be defined by its geometry and its appearance. The geometry provides a mathematical description of the object's shape, size, and parameters such as translation, rotation, and scale. The appearance defines a set of object's characteristics such as colour, texture, and other attributes.

Given $\mathbf{S} \subseteq \mathbb{R}^2$, an image intensity function $\mathbf{I}(x, y) \in \mathbb{R}$ such that $\mathbf{I} : \mathbf{S} \rightarrow [0, \mathbf{I}_{max}]$, and an object O, we perform the mapping $\Psi_T(x, y) = f_T(I(x, y))$ that associates to each point $(x, y) \in \mathbf{S}$ based on the texture T the probability of belonging to the object O. Considering:

- The input distribution as the set of points in the image:

$$\mathbf{A} = \mathbf{S} \tag{1}$$

$$x_w = (x, y) \in \mathbf{S} \tag{2}$$

- The probability density function obtained for each point of the image:

$$p(x_w) = p(x, y) = \Psi_T(x, y) \tag{3}$$

Learning takes place with the GNG algorithm. By doing this, from the appearance O_A of the object is obtained an approximation to its geometry O_G.

Figure 1 shows two different Topology Preserving Graphs $TPG = \langle N, A \rangle$ obtained for the same object in two and one dimensions respectively (Figure 1*left* and 1*right*). When obtaining the contour, the result of the learning process is a list of non-ordered nodes. The list of nodes define a graph. To normalise the graph that represents the contour we must define a starting point, for example the node on the left-bottom corner. Taking that node as the first we must follow the neighbours until all the nodes had been added to the new list. If necessary we must apply a scale and a rotation to the list with respect to the centre of gravity of the list of nodes. The non-ordered nodes and the normalised nodes can be seen in Figure 2A, and 2B. Figure 3 shows another example of normalised nodes in a T1-weighted MRI medical shape.

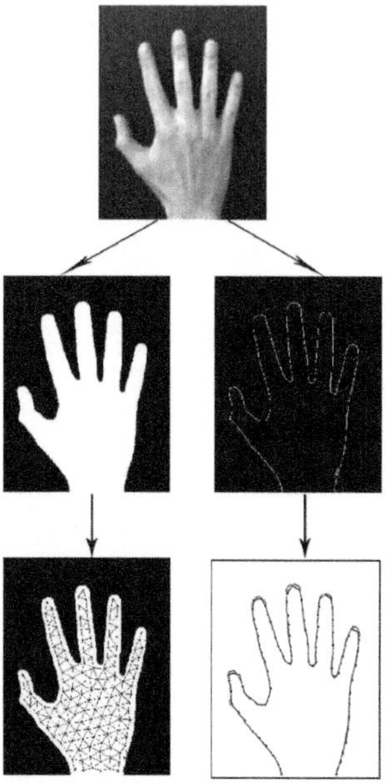

Fig. 1. Different adaptations of the growing neural gas

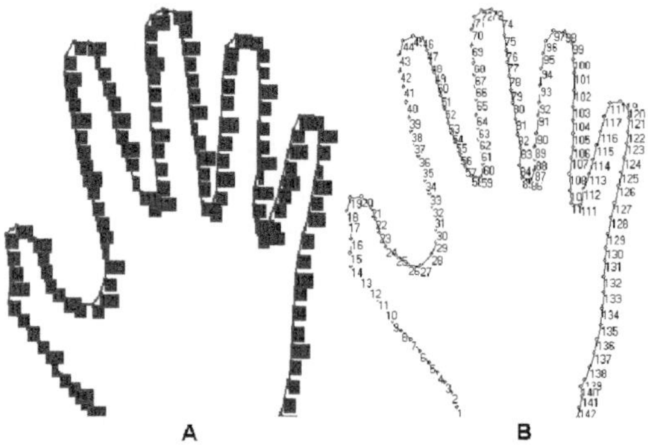

Fig. 2. Image A shows the automatic node extraction and position before any reordering is applied. B shows the nodes after normalisation is performed.

3 Adaptive Learning

In the GNG algorithm, the stopping criterion can either be a predefined number of neurons manually inserted or time constrains specified by the user. Both methods have problems because the quality of the network depends either on the arbitrary selection of the maximum number of nodes or the available time of the network to converge. For example, Figure 4 shows topology preservation for

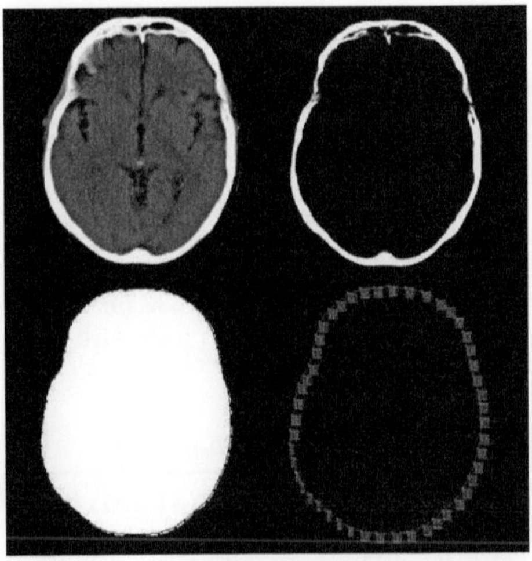

Fig. 3. Normalisation of nodes in a T1-weighted MRI medical image

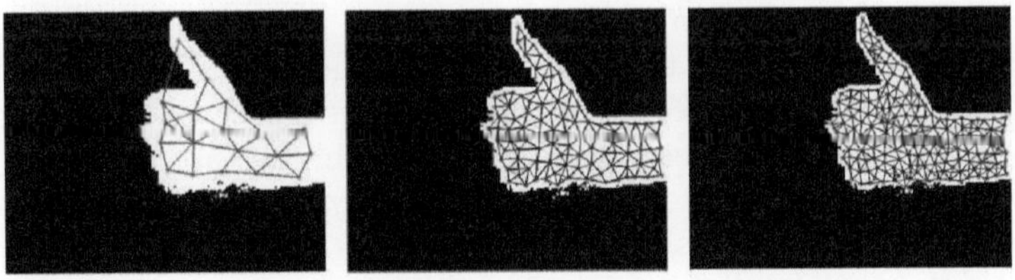

Fig. 4. Mapping of the same object with image resolution $200x160$ and nets of 20, 101 and 181 nodes

variant number of nodes of the same object. It is evident that 20 nodes are not enough to describe the topology of the object. However, the mapping with 101 nodes is good enough and has no difference, in terms of topology representation, with the map obtained inserting 181 nodes. We introduce an automatic method of the stopping criterion that defines the insertion of maximum nodes by calculating the image size and the probability of the objects in the image. In GNG and RGNG [13] the maximum number of nodes (*prenumnode*) to grow is set manually and chosen according to the scale of the clustering tasks. In our case, the maximum number of nodes is defined automatically by the system based on equation (4). In our examples we use hand configurations and we model the colour distribution p_{skin} of skin pixels by a single Gaussian in RGB space with mean and covariance estimated from hand-selected training patches. We assume that non-skin pixels have a uniform distribution p_{bkgd}.

Let $\Omega(x)$ denote the set of pixels in the objects of interest based on the configuration of x (e.g. colour, texture, etc.) and Υ the set of all image pixels. The likelihood of the required number of nodes to describe the topology of an image y is:

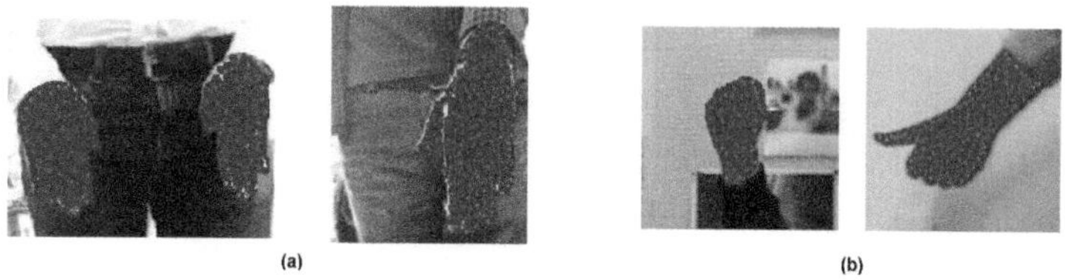

Fig. 5. Likelihood node ratios for images with same image resolution but different skin distribution. (a) Net adaptation to images of 46,332 pixels with maps of 102 and 162 nodes. (b) Net adaptation to images of 21,903 pixels with maps of 46 and 132 nodes.

$$p(y|x) = \frac{\prod_{u \in \Omega(x)} p_{skin}(u)}{\prod_{u \in \Omega(x)} p_{skin}(u) + \prod_{v \in \Upsilon \setminus \Omega(x)} p_{bkgd}(v)} * e_T \quad (4)$$

and $e_T \leq \prod_{u \in \Omega(x)} p_{skin}(u) + \prod_{v \in \Upsilon \setminus \Omega(x)} p_{bkgd}(v)$. Figure 5 plots the likelihood node ratios for different images. e_T is a similarity threshold and defines the accuracy of the map. If e_T is low the topology preservation is lost and more nodes need to be added. On the contrary, if e_T is too big then nodes have to be removed so that Voronoï cells become wider. For example, let us consider an extreme case where the total size of the image is $I = 100$ pixels and only one pixel represents the object of interest. Let us suppose that we use $e_T = 100$ then the object can be represented by one node. In the case where $e_T \geq I$ then overfit occurs since twice as many nodes are provided. In our experiments the numerical value of e_T ranges from $100 \leq e_T \leq 900$ and the accuracy depends on the size of the objects' distribution. The difference between choosing manually the maximum number of nodes and selecting e_T as the similarity threshold, is the preservation of the object independently of scaling operations.

4 Experiments

We tested our method on an image data set composed of 56 images from various participants with different hand posture. We obtained the data set from the University of Alicante, Spain and the University of Westminster, UK. Also, we tested our method with 40 images from Mikkel B. Stegmann online data set (http://www2.imm.dtu.dk/~aam/). For computational efficiency, we have resized the image data sets to $124x123$ and $120x160$ pixels. In these experiments, the insertion of maximum nodes per object distribution and similarity threshold (e_T) determines the topology preservation. The test consists of two processes: learning and evaluation. During learning, we choose different similarity thresholds taken from an image vector set, then input the results to the GNG, and report number of nodes per similarity threshold, computational time of the network, and Mean Square Error (MSE) as the results of the adaptation. In the evaluation process, we measure the topology preservation with the topographic product [2].

Table 1. Number of nodes for different similarity thresholds on two data sets

Data sets		Number of nodes	Time (sec)	MSE
$Set_{120x160}$	$H_{e_T=100}$	19	5.03	12.06
	$H_{e_T=200}$	37	12.42	4.56
	$H_{e_T=300}$	56	23.87	2.42
	$H_{e_T=400}$	75	40.23	1.58
	$H_{e_T=500}$	**94**	**60.13**	**0.98**
	$H_{e_T=600}$	112	88.25	0.87
	$H_{e_T=700}$	131	123.21	0.68
	$H_{e_T=800}$	150	160.96	0.56
	$H_{e_T=900}$	168	204.9	0.47
$Set_{124x123}$	$H_{e_T=100}$	22	5.77	13.26
	$H_{e_T=200}$	44	16.03	4.67
	$H_{e_T=300}$	66	31.37	2.55
	$H_{e_T=400}$	88	52.91	1.66
	$H_{e_T=500}$	**110**	**79.63**	**0.90**
	$H_{e_T=600}$	132	120.73	0.81
	$H_{e_T=700}$	154	164.94	0.72
	$H_{e_T=800}$	176	219.76	0.59
	$H_{e_T=900}$	198	289.25	0.49

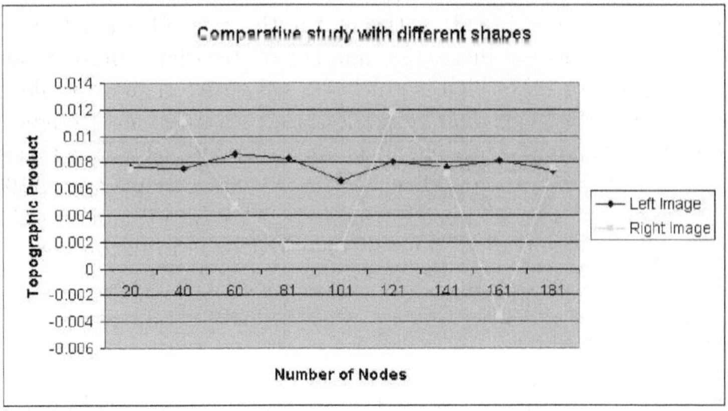

Fig. 6. Comparative study for images with different shapes

Table 1 details specifications of the test images. The maximum number of nodes depends on the similarity threshold e_T with range $100 \leq e_T \leq 900$. It is worth noting that the network stabilises and can represent sufficiently the object when $e_T = 500$. This is an optimum number obtained during testing that maximises topology learning versus adaptation time and MSE. The insertion of more nodes as e_T increases makes no difference in terms of topographic representation to the object's topology as can be seen in Figure 6. The graph shows that the topology is best preserved with maps containing 101 nodes while fewer nodes are not enough to the recognition of the object (Figure 7). Furthermore,

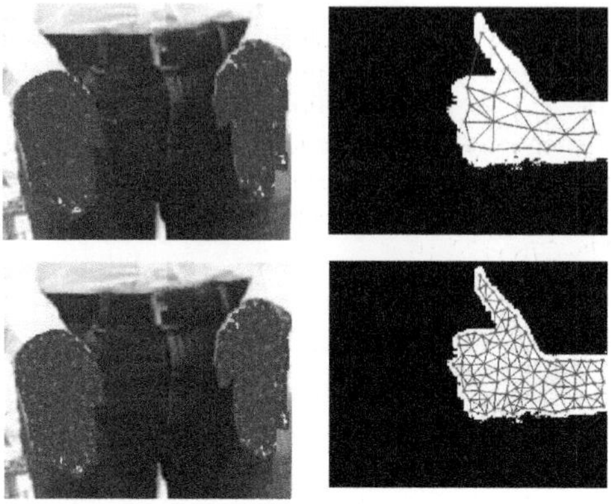

Fig. 7. Images with different shapes and GNG maps with 20 and 101 nodes

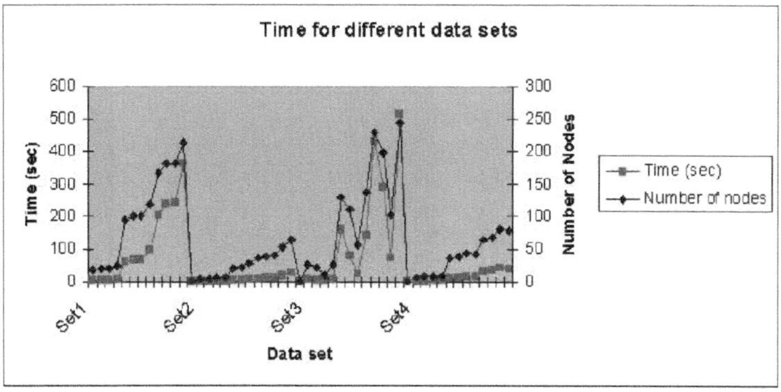

Fig. 8. Time taken to insert the maximum number of nodes per data set

the more nodes added during the learning process the more time it takes for the network to growth (Figure 8).

5 Conclusions and Future Work

We developed an approach to automatically extract, re-order and label the contour of an MRI and hand-gesture module using only topological relations derived from competitive hebbian learning. We introduced an automatic criterion for the insertion of maximum nodes based on the object's distribution and the similarity threshold (e_T) which determines the preservation of the topology. During testing we found that for different shapes there exists an optimum number that maximises topology learning versus adaptation time and MSE. In the future, we would use a more systematic approach for the selection of the optimal number by using a model selection criterion like the Minimum Description Length (MDL) approach.

References

1. Angelopoulou, A., Psarrou, A., Gupta, G., García, J.: Robust Modelling and Tracking of Nonrigid Objects Using Active-GNG. In: IEEE Workshop on Non-rigid Registration and Tracking through Learning, NRTL 2007, in conjuction with ICCV 2007, pp. 1–7 (2007)
2. Bauer, H., Pawelzik, K.: Quantifying the neighbourhood preservation of self-organizing feature maps. IEEE Transactions on Neural Networks 3(4), 570–579 (1992)
3. Cootes, T.F., Taylor, C.J., Cooper, D.H., Graham, J.: Active Shape Models - Their Training and Application. Comp. Vision Image Underst. 61(1), 38–59 (1995)
4. Cretu, A., Petriu, E., Payeur, P.: Evaluation of growing neural gas networks for selective 3D scanning. In: Proc. of IEEE International Workshop on Robotics and Sensors Environments, pp. 108–113 (2008)
5. Edwards, G., Taylor, C., Cootes, T.: Interpreting face images using active appearance models. In: Proc. of the International Conference on Face And Gesture Recognition, pp. 300–305 (1998)
6. Frezza-Buet, H.: Following non-stationary distributions by controlling the vector quantisation acccuracy of a growing neural gas network. Neurocomputing 71(7-9), 1191–1202 (2008)
7. Fritzke, B.: A growing Neural Gas Network Learns Topologies. In: Advances in Neural Information Processing Systems 7, NIPS 1994, pp. 625–632 (1995)
8. García-Rodríguez, J., Flórez-Revuelta, F., García-Chamízo, M.: Image Compression Using Growing Neural Gas. In: Proc. of the International Joint Conference on Artificial Neural Networks, pp. 366–370 (2007)
9. Hans-Joachim, B., Anja, B., Ulf-dietrich, B., Markus, K., Horst-michael, G.: Neural Networks for Gesture-based Remote Control of a Mobile Robot. In: Proceedings of the IEEE World Congress on Computational Intelligence, vol. 1, pp. 372–377 (1998)
10. Holdstein, Y., Fischer, A.: Three-dimensional surface reconstruction using meshing growing neural gas (MGNG). The Visual Computer: International Journal of Computer Graphics 24(4), 295–302 (2008)
11. Kass, M., Witkin, A., Terzopoulos, D.: Snakes: Active Contour Models. In: Proc. of the 1st Internationl Conference on Computer Vision, pp. 259–268. IEEE Computer Society Press, Los Alamitos (1987)
12. Lankton, S., Tannenbaum, A.: Localizing region-based active contours. IEEE Transactions on Image Processing 17(11), 2029–2039 (2008)
13. Qin, A.K., Suganthan, P.N.: Robust growing neural gas algorithm with application in cluster analysis. Neural Networks 17(8-9), 1135–1148 (2004)
14. Rivera-Rovelo, J., Herold, S., Bayro-Corrochano, E.: Object segmentation using growing neural gas and generalized gradient vector flow in the geometric algebra framework. In: Martínez-Trinidad, J.F., Carrasco Ochoa, J.A., Kittler, J. (eds.) CIARP 2006. LNCS, vol. 4225, pp. 306–315. Springer, Heidelberg (2006)

Object Representation with Self-Organising Networks

Anastassia Angelopoulou[1], Alexandra Psarrou[1], and José García Rodríguez[2]

[1] School of Electronics and Computer Science, University of Westminster, UK
{agelopa,psarroa}@wmin.ac.uk
[2] Department of Computing Technology, University of Alicante, Spain
{jgarcia}@dtic.ua.es

Abstract. This paper, aims to address the ability of self-organising networks to automatically extract and correspond landmark points using only topological relations derived from competitive hebbian learning. We discuss, how the Growing Neural Gas (GNG) algorithm can be used for the automatic extraction and correspondence of nodes in a set of objects, which are then used to built statistical human brain MRI and hand gesture models.

Keywords: Unsupervised Learning, Self-organising networks, Nonrigid Shapes.

1 Introduction

Accurate nonrigid shape modelling and tracking is a challenging problem in machine vision with applications in human-computer interaction, motion capture, nonlinear registration, image interpretation and scene understanding. One objective of modelling is the construction of decision boundaries based on unlabeled training data that can solve for correct correspondences between a set of shapes. Such correspondences can be classified as the problem of finding homogeneous landmark points in a multidimensional data set.

One way of extracting landmark points along the contour of shapes is to use a topographic mapping where a low dimensional map is fitted to the high dimensional manifold of the contour, whilst preserving the topographic structure of the data. A common way to achieve this is by using self-organised networks where input patterns are projected onto a network of nodes such that similar patterns are projected onto nodes adjacent in the network and vice versa. As a result of this mapping a representation of the input patterns is achieved that in postprocessing stages allows one to exploit the similarity relations of the input patterns.

Such models have been successfully used in applications such as speech processing [8], robotics [14,10,9], biology [13], medicine [1,2,5,4], and image processing [12]. These models share in common the attributes of dynamically generating and removing processing elements (vectors of a network) and dynamically generating and removing synaptic links (neighbourhood connections). Furthermore,

J. Cabestany, I. Rojas, and G. Joya (Eds.): IWANN 2011, Part II, LNCS 6692, pp. 244–251, 2011.

these models make no assumptions about the global structure of the shape to be modelled or more generally of the problem to be learned.

The remainder of the paper is organised as follows. Section 2 discusses the statistical shape models and the main parts of the GNG algorithm. Section 3 presents a number of examples in modelling and tracking, before we conclude in Section 4.

2 Point Correspondence

When analysing biological shapes it is convenient and usually effective to describe them using statistical shape models. The most well known statistical shape models are Cootes *et al.* [3] 'Point Distribution Models' (PDMs) that models the shape of an object and its variation by using a set of n_p landmark points from a training set of S_i shapes. In order to generate flexible shape models the S_i shapes are aligned and normalised to a common set of axes. The modes of variations of the ventricles are captured by applying principal component analysis (PCA). The ith shape in the training set can be back-projected to the input space by a linear model of the form:

$$\mathbf{x} = \overline{\mathbf{x}} + \Phi\beta_i \tag{1}$$

where $\overline{\mathbf{x}}$ is the mean shape, Φ describes a set of orthogonal modes of shape variations, and β_i is a vector of weights for the ith shape. PCA works well as long as good correspondences exist. To obtain the correspondences and represent the contour of the ventricles and hands we use the GNG algorithm.

2.1 Review of Growing Neural Gas (GNG) Algorithm

GNG [7] is an unsupervised incremental self-organising network independent of the topology of the input distribution or space. It uses a growth mechanism inherited from the Growth Cell Structure [6] together with the Competitive Hebbian Learning (CHL) rule [11] to construct a network of the input date set. In some cases the probability distribution of the input data set is discrete and is given by the characteristic function $x_w : \mathbb{R}^n \to \{0, 1\}$ with x_w defined by

$$x_w = \left\{ \begin{matrix} 1 \text{ if } x \in W \\ 0 \text{ if } x \in W^c \end{matrix} \right\} \tag{2}$$

In the network x_w represents the random input signal generated from the set $W \subseteq \mathbb{R}^n$ and W^c is the complement of $W \in \mathbb{R}^n$. The growing process starts with two nodes, and new nodes are incrementally inserted until a predefined conditioned is satisfied, such as the maximum number of nodes or available time. During the learning process local error measures are gathered to determine where to insert new nodes. New nodes are inserted near the node with the highest accumulated error and new connections between the winner node and its topological neighbours are created.

The GNG algorithm consists of the following:

- A set A of cluster centres known as nodes. Each node $c \in N$ has its associated reference vector $\{\boldsymbol{x}_c\}_{c=1}^{N} \in \mathbb{R}^q$. The reference vectors indicate the nodes' position or *receptive field centre* in the input distribution. The nodes move towards the input distribution by adapting their position to the input's geometry using a winner take all mapping. Generating $\boldsymbol{x}_{\boldsymbol{w}}$ input signals from the random vector $\boldsymbol{W}$, we want to find a mapping $G : \mathbb{R}^n \longrightarrow \mathbb{R}^q$ and its inverse $F : \mathbb{R}^q \longrightarrow \mathbb{R}^n$ such that $\forall c = 1, ..., \mid N \mid$,

$$f(x) = E_{\boldsymbol{W}|g(\boldsymbol{W})}\{\boldsymbol{W}|g(\boldsymbol{W}) = x\}, \forall x \in \{\boldsymbol{x}_c\}_{c=1}^{N} \subseteq \mathbb{R}^q \tag{3}$$

$$g(\boldsymbol{W}) = \arg \min_{\nu \in \{\boldsymbol{x}_c\}_{c=1}^{N}} \|\boldsymbol{W} - \boldsymbol{x}_\nu\| \tag{4}$$

  where E is the distance operator of the data points from the random vector $\boldsymbol{W}$ projecting onto $f(x)$, $g(\boldsymbol{W})$ is the projection operator, $\{\boldsymbol{x}_c\}_{c=1}^{N} \subseteq \mathbb{R}^q$ are the reference vectors of the network and $\boldsymbol{x}_\nu$ represents the reference vector closest to the input signal $\boldsymbol{x}_{\boldsymbol{w}}$, called the winner node. Equations (3) and (4) show that while the forward mapping G is approximated as a projection operator, the reverse mapping F is nonparametric and depends on the unknown latent variable $\boldsymbol{x}$. In order to compute $f(x)$ the GNG algorithm evaluates (3) and (4) in an iterative manner. n and q denote the dimensionality of the original and the reduced latent topology. In this work, current experiments include topologies of a line which is the contour of the object ($q = 1$) and triangular grid which is the topology preserving graph ($q = 2$). Figure 1 shows an example of a 2D GNG network with its associated Voronoi diagram in 2D data distribution.
- Local accumulated error measurements. Each node $c \in N$ with its associated reference vector $\{\boldsymbol{x}_c\}_{c=1}^{N} \in \mathbb{R}^q$ has an error variable $E_{\boldsymbol{x}_c}$ which is updated at every iteration according to:

$$\Delta E_{\boldsymbol{x}_\nu} = \|\boldsymbol{x}_{\boldsymbol{w}} - \boldsymbol{x}_\nu\|^2 \tag{5}$$

  where $\boldsymbol{x}_\nu$ refers to the winner node. The local accumulated error is a statistical measure and is used for the insertion and the distribution of new nodes. Nodes with larger errors will cover greater area of the input probability distribution, since their distance from the generated signal is updated by the squared distance. Knowing where the error is large a new node is inserted near the nodes with the largest local accumulated error. A global decrease according to:

$$\Delta E_{\boldsymbol{x}_c} = -\beta E_{\boldsymbol{x}_c} \tag{6}$$

  is performed to all local errors by a constant β. This is important since new errors will gain greater influence in the network resulting in a better representation of the topology.

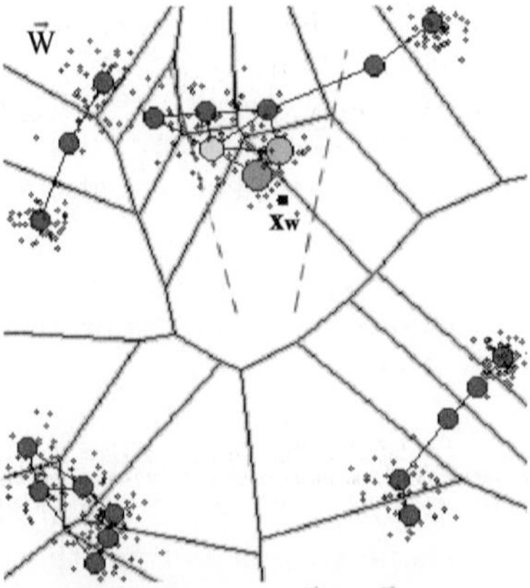

Fig. 1. A random signal $\boldsymbol{x}_w$ on the discrete input distribution and the best matching within the topological neighbourhood of $\{\boldsymbol{x}_c\}_{c=1}^{|N|} \subseteq \mathbb{R}^q$. In this example, the green node is the winner node of the network among its direct topological neighbours (orange and yellow nodes). The orange node is the second nearest node to the random signal $\boldsymbol{x}_w$.

- A set A of edges (connections) between pair of nodes. These connections are not weighted and its purpose is to define the topological structure. The edges are determined using the competitive hebbian learning method. The updating rule of the algorithm is expressed as:

$$\Delta \boldsymbol{x}_\nu = \epsilon_x(\boldsymbol{x}_w - \boldsymbol{x}_\nu) \tag{7}$$

$$\Delta \boldsymbol{x}_c = \epsilon_n(\boldsymbol{x}_w - \boldsymbol{x}_c), \forall c \in N \tag{8}$$

where ϵ_x and ϵ_n represent the constant learning rates for the winner node $\boldsymbol{x}_\nu$ and its topological neighbours $\boldsymbol{x}_c$. An *edge aging scheme* is used to remove connections that are invalid due to the activation of the node during the adaptation process. Thus, the network topology is modified by removing edges not being refreshed by a time interval α_{max} and subsequently by removing the nodes connected to these edges.

3 Examples

This section presents results for two different data sets, hands and ventricles. A number of examples to demonstrate the ability of GNG to model objects are presented with applications in biometrics and computer vision.

3.1 Ventricles

The data that we used in this study was obtained from the MNI BIC Centre for Imaging at McGill University, Canada. These images are 1 mm thick, $181x217$

pixels per slice ($1.0mm^2$ in-plane resolution), 3% noise and 20% INU. These images are used as our gold standard for segmentation, as every voxel in the entire volume has been correctly labelled to a tissue class by the McGill Institute. The entire brain volume consisted of 181 slices, from which we extracted those that contained ventricles (slices $49-91$). The images are 16 bit grey scale, which were threshold out manually to remove all but the outline of the ventricles. Since most typical clinical MRI volumes are on average 5mm thick, we selected 4 groups of 5 contiguous slices to produce our point distribution model.

In Figure 2 the modes of variation for all four groups are displayed by varying the first shape parameter $\beta_i\{\pm 3\sigma\}$ over the training set. The qualitatively results show that GNG leads to correct extraction of corners of anatomical shapes and are compact when the topology preservation of the network is achieved (Figure 4). In Figure 3 two shape variations from the automatically generated landmarks were superimposed to groups 4 and 3 from the training set. These modes effectively capture the variability of the training set and present only valid shape instances.

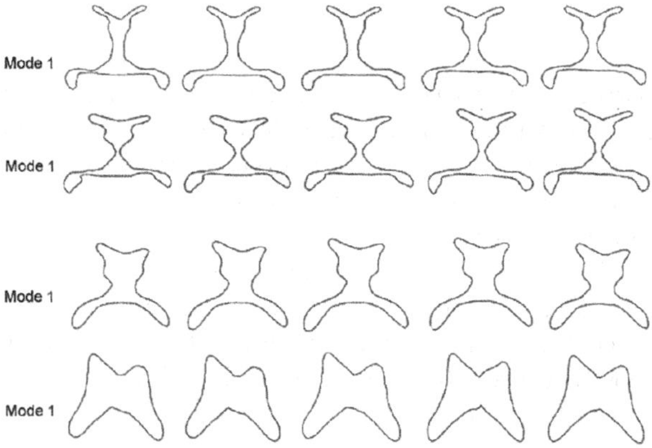

Fig. 2. The first mode ($m = 1$) of variation for the four groups of 5 contiguous slices taken from MR brain data. Range of variation $-3\sqrt{\lambda} \leq \beta_i \leq 3\sqrt{\lambda}$.

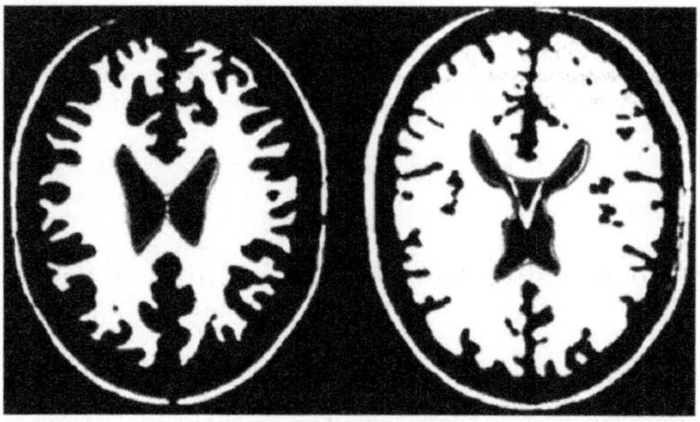

Fig. 3. Superimposed shape instances to groups 4 and 3 from the training set

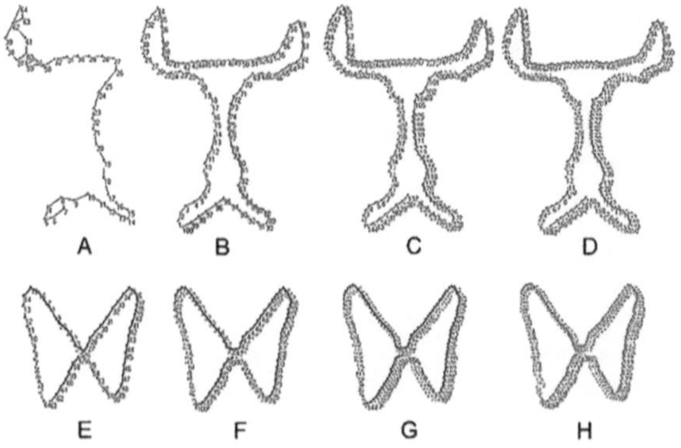

Fig. 4. Automatic annotation with network size of 64 (Image A, E), 100 (Image B, F), 144 (Image C, G) and 164 (Image D, H) nodes for two groups of the MRI volumes of the ventricles

3.2 Hands

The comparison was made by taking two reference models, a manually hand built model with 60 landmarks manually located around the boundaries, and an automatically hand built model with 144 nodes automatically generated around the boundaries (Figure 5). In Figure 6 the modes are displayed by varying the first three shape parameters $\beta_i\{\pm 3\sigma\}$ over the training set. The first mode β_1 varies the shape of the thumb and increases the distance between the middle and the index finger. The second mode β_2 varies the distance between the thumb and the index finger, and bends the middle finger. The third mode β_3 varies the shape of the middle finger and the thumb. In Figure 7 two shape variations from the

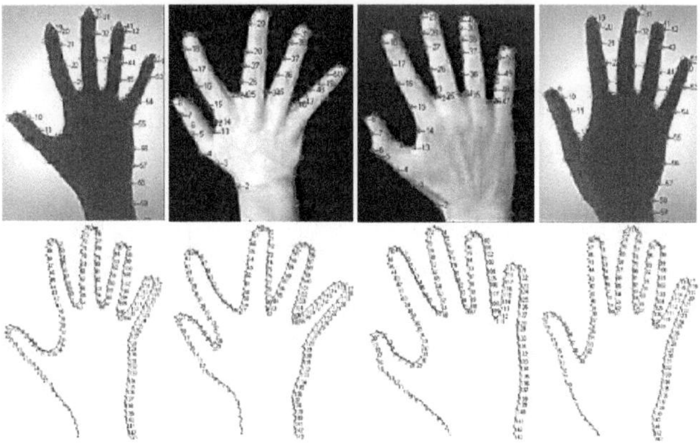

Fig. 5. First row manually annotated landmarks. All landmarks are major and have been located manually. Second row hand adaptation with 144 nodes.

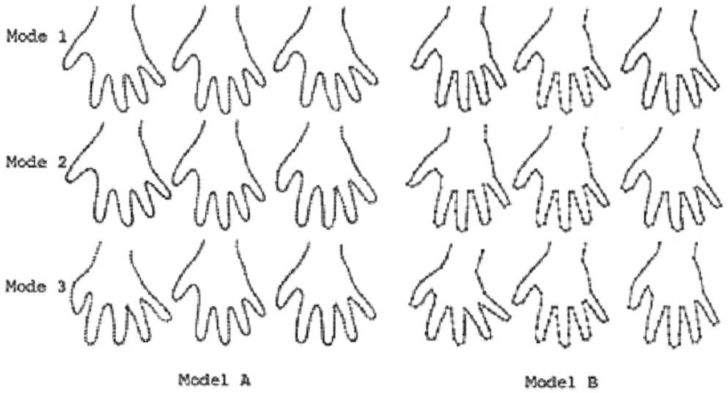

Fig. 6. Model A shows the first three modes of variation of the automatically hand built model. Model B shows the first three modes of variation of the manually hand built model. Range of variation $-3\sqrt{\lambda} \leq \beta_i \leq 3\sqrt{\lambda}$.

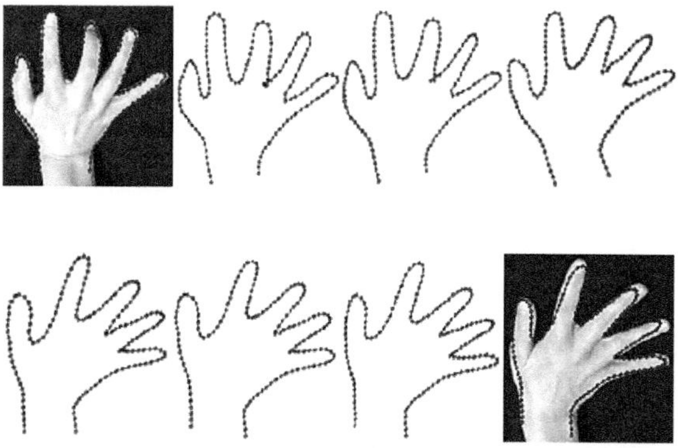

Fig. 7. Superimpose instances to the training set and taking the in-between steps

automatically generated landmarks were superimposed to the training set and the in between shape instances are drawn which shows the flexing of middle finger and hand rotation. These modes effectively capture the variability of the training set and present only valid shape instances.

4 Conclusions and Future Work

In this paper, we have demonstrated the capacity of Growing Neural Gas to perform some computer vision and biomedical tasks. Establishing a suitable transformation function the model is able to adapt its topology to the the high dimensional manifold of the ventricles and the hands, allowing good eigenshape models to be generated completely automatically from the training sets.

References

1. Angelopoulou, A., Psarrou, A., Rodríguez, J.G., Revett, K.R.: Automatic landmarking of $2D$ medical shapes using the growing neural gas network. In: Liu, Y., Jiang, T.-Z., Zhang, C. (eds.) CVBIA 2005. LNCS, vol. 3765, pp. 210–219. Springer, Heidelberg (2005)
2. Cheng, G., Zell, A.: Double growing neural gas for disease diagnosis. In: Proc. of Artificial Neural Networks in Medicine and Biology Conference (ANNIMAB-1), pp. 309–314 (2000)
3. Cootes, T.F., Taylor, C.J., Cooper, D.H., Graham, J.: Training Models of Shape from Sets of Examples. In: Proc. of the 3rd British Machine Vision Conference, pp. 9–18 (1992)
4. Cselényi, Z.: Mapping the dimensionality, density and topology of data: the growing adaptive neural gas. Computer Methods and Programs in Biomedicine 78(2), 141–156 (2005)
5. Fatemizadeh, E., Lucas, C., Soltania-Zadeh, H.: Automatic Landmark Extraction from Image Data Using Modified Growing Neural Gas Network. IEEE Transactions on Information Technology in Biomedicine 7(2), 77–85 (2003)
6. Fritzke, B.: Growing Cell Structures - a self-organising Network for Unsupervised and supervised Learning. The Journal of Neural Networks 7(9), 1441–1460 (1994)
7. Fritzke, B.: A growing Neural Gas Network Learns Topologies. In: Advances in Neural Information Processing Systems 7 (NIPS 1994), pp. 625–632 (1995)
8. Kohonen, T.: Topology Representing Networks. Springer, Heidelberg (1994)
9. Marsland, S., Nehmzow, U., Shapiro, J.: A real time novelty detector for a mobile robot. In: Proc. of EUREL European Advanced Robotics Systems Masterclass and Conference (2000)
10. Martinez, T., Ritter, H., Schulten, K.: Three dimensional neural net for learning visuomotor-condination of a robot arm. IEEE Transactions on Neural Networks 1, 131–136 (1990)
11. Martinez, T., Schulten, K.: Topology Representing Networks. The Journal of Neural Networks 7(3), 507–522 (1994)
12. Nasrabati, M., Feng, Y.: Vector Quantisation of images based upon Kohonen self-organizing feature maps. In: Proc. IEEE Int. Conf. Neural Networks., pp. 1101–1108 (1988)
13. Ogura, T., Iwasaki, K., Sato, C.: Topology representing network enables highly accurate classification of protein images taken by cryo electron-microscope without masking. Journal of Structural Biology 143(3), 185–200 (2003)
14. Ritter, H., Schulten, K.: Topology conserving mappings for learning motor tasks. In: AIP Conf. Proc. Neural Networks for Computing, pp. 376–380 (1986)

SNP-Schizo: A Web Tool for Schizophrenia SNP Sequence Classification

Vanessa Aguiar-Pulido, José A. Seoane, Cristian R. Munteanu, and Alejandro Pazos

Information and Communication Technologies Department, Faculty of Informatics, University of A Coruña,
Campus de Elviña s/n, 15071 Spain
vanesa.aguiar@udc.es, jseoane@udc.es, muntisa@gmail.com, apazos@udc.es

Abstract. This work presents a tool which is an online implementation of the best machine learning-based model obtained after an exhaustive computational study. Twelve techniques were applied to schizophrenia data to obtain the results of this study and, with these, Quantitative Genotype – Disease Relationships (QDGRs) for disease prediction. Thus, the tool offers the possibility to introduce SNP sequences (which contain the SNPs considered in the study) in order to classify a patient. In the future, QDGR models could be extended to other diseases. The model implemented online is a linear neural network.

Keywords: SNP, schizophrenia, machine learning, neural networks, data mining, bioinformatics.

1 Introduction

Quantitative Structure – Activity Relationships (QSARs) are widely used for predicting protein properties [1] and Quantitative Protein (or Proteome) – Disease Relationships (QPDRs) [2-8] for disease prediction. In a similar way, a Quantitative Genotype – Disease Relationship (QGDR) can be established in order to automatically evaluate schizophrenia DNA sequences using SNP data.

A SNP (Single Nucleotide Polymorphism) [9] is a single nucleotide variation in a genetic sequence that occurs at appreciable frequency in the population, that is, at least in 1%. Thus, SNPs can be used as inputs in disease computational studies such as pattern searching or classification models.

This work presents a tool that implements the machine learning-based method which obtains the best results in [10]. In this paper, a computational study of machine learning disease classification models using only single nucleotide polymorphisms (SNPs) at the HTR2A and DRD3 genes from Galician (Northwest Spain) schizophrenic patients is presented. Methods such as artificial neural networks [11], support vector machines [12], evolutionary computation [13, 14] and other machine learning techniques [15] have been used to find the best classification models.

J. Cabestany, I. Rojas, and G. Joya (Eds.): IWANN 2011, Part II, LNCS 6692, pp. 252–259, 2011.

2 Materials and Methods

2.1 Schizophrenia Data

For the computational study, schizophrenia data from Galician patients [16] were used. This data contained 48 SNPs at the DRD3 and HTR2A genes, genes which are associated to schizophrenia. These SNPs were encoded taking different values:

- 0 if homozygous (both copies of a given gene have the same allele) for the first allele (one of a number of alternative forms of the same gene occupying a given position on a chromosome),
- 1 if heterozygous (the patient has two different alleles of a given gene),
- 2 if homozygous for the second allele or
- 3 if unknown.

The original dataset contained 260 positive subjects (genetically predisposed to schizophrenia) and 354 negative subjects (not predisposed), a total of 614 patients.

To perform more tests, six other datasets were obtained from the original one. This was done adding negative subjects generated using the HAP-SAMPLE [17] simulation tool. This data was modified to include genotyping errors (represented as the value 3) taking into account the error frequencies of the real data, but choosing randomly which positions were modified.

Thus, these datasets included 307, 614, 1228, 1842, 2456 and 3070 simulated negative subjects. Datasets were named following the pattern 1:N, where this label represents the proportion between the real subjects (positive and negative) and the simulated negative subjects.

2.2 LNN Model

252 QDGR classification models were obtained after applying machine learning techniques to the data described previously. Twelve methods and seven datasets were used.

The method implemented in the tool presented in this work is a linear neural network (LNN) [18] with 40 neurons in the first layer, 152 in the second layer and 1 neuron as output. These neurons were selected according to the results obtained from several feature selection methods (Best First [19], Linear Forward Selection [20], FCBF Search [21], Genetic Search [22], Scatter Search [23] and Random Search [24]). It was proved that taking only 40 neurons as input the method achieved good results. This technique obtained 78.2% in test accuracy when the 1:0.5 dataset was used as input.

In addition to LNN, the following techniques were applied to the datasets: Multilayer Perceptron (MLP) [25], Radial Base Function (RBF) [26], Evolutionary Computation (EC) [27], Multifactor Dimensionality Reduction (MDR) [28, 29], Naive Bayes [30], Bayes Networks (Bayes Nets.) [31], Support Vector Machine (SVM) [32], Decision Tables (Decis. Tb.) [33], Decision Table Naive Bayes Hybrid Classifier (DTNB) [34], Best-First decision Tree classifier (BFTree) [35] and Adaptative Boosting (AdaBoost) [36].

After carrying out this computational study, the LNN model was implemented online.

2.3 Single Nucleotide Polymorphism Schizophrenia Processing (SNP-Schizo)

Bio-AIMS (http://bio-aims.tic.udc.es/, Biomedical Artificial Intelligence Model Server) [37-40] is a portal that offers theoretical models based on Artificial Intelligence, Computational Biology and Bioinformatics to study Complex Systems in OMICS (Genomics, Transcriptomics, Metabolomics, Reactomics) that are relevant for Cancer, Neurosciences, Cardiovascular diseases, Parasitology, Microbiology and other Biomedical research in general. It is the result of the collaboration between several scientific institutions. This portal includes three parts: aMDAPred (antiMicrobial Drug Action Prediction), TargetPred (Target Prediction) and DiseasePred (Disease Prediction).

The DiseasePred part includes biomedicine applications for predicting human diseases from different data sources, such as EEG recordings, blood proteome mass spectra or genotypes.

SNPSchizo (http://miaja.tic.udc.es/Bio-AIMS/SNPSchizo.php, Single Nucleotide Polymorphism Schizophrenia Processing) (figure 1) is the result of an online implementation of the previously described machine learning method which takes as input SNPs from two different genes related to schizophrenia and performs a classification [10].

The interface of this tool was implemented using PHP, XHTML and Python, and the method was implemented using Java and Weka's [41] APIs. The tool is running on Apache HTTP – Server.

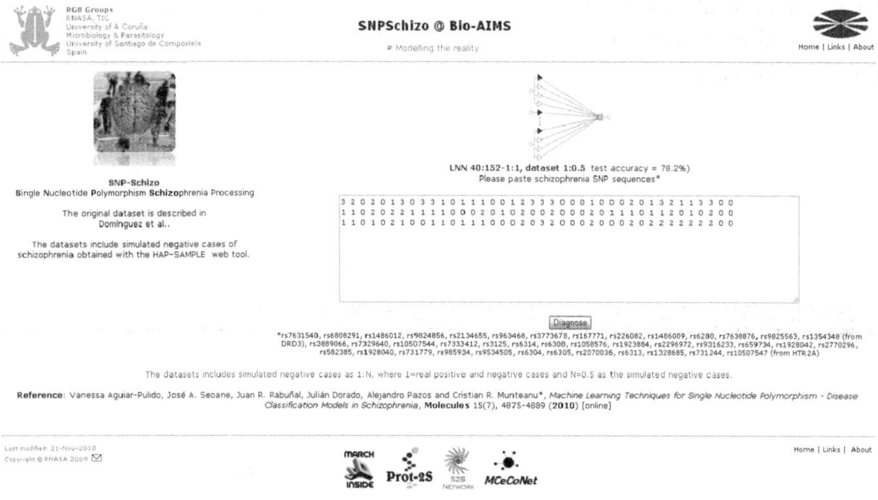

Fig. 1. SNP-Schizo web tool

This tool is simple and easy to use. To get a classification result, the user has to introduce a list of sequences of SNPs in the format used by the tool and click on the "Diagnose" button. A new window will pop up with information about the results. These results can be saved as a text file and include the following information:

- For each sequence: the classification result (genetically predisposed to schizophrenia or not) and the SNP sequence.
- Information about the original dataset.
- Information about the method implemented online and its test accuracy.
- Input SNP order.
- Reference to the article of the computational study with this data.

To test this tool three example sequences are provided following the coding described above.

3 Results and Discussion

A graphical representation of the evolution of the different methods is shown in figure 2. As said before, 1:N represents the proportion between the real subjects and the simulated negative subjects. Thus, the first dataset does not include any simulated subject and the last dataset includes 5 simulated subjects per real one.

For each method, the percentage of correctly classified subjects is shown for each dataset. It can be observed that classification percentages do not increase significantly after adding five parts of simulated subjects.

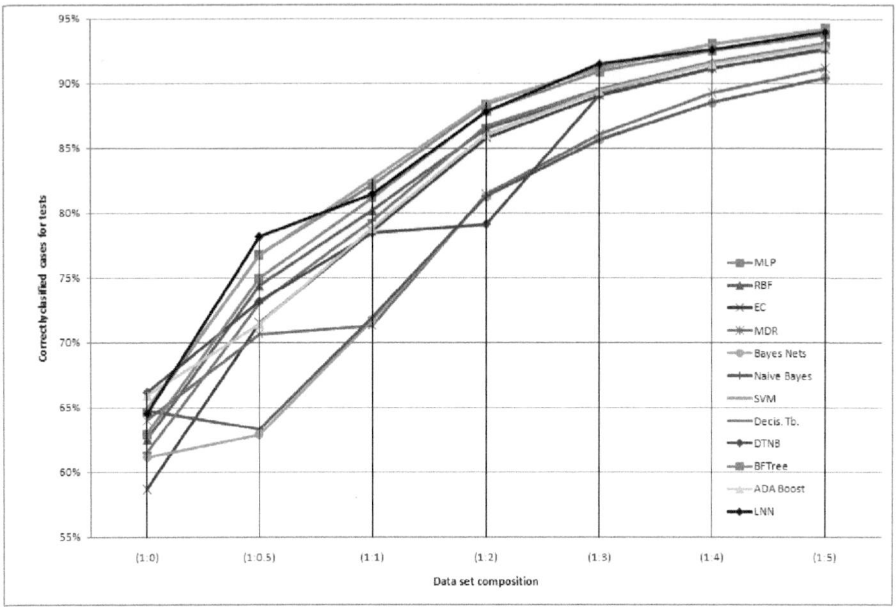

Fig. 2. Classification results of the different methods

Classification accuracy percentages range from 56.6-66.6% for 1:0, which is the original dataset. For the datasets which included simulated subjects, these percentages range from 60.5-78.2% for 1:0.5, 69.8-83.0% for 1:1, 76.2-88.8% for 1:2, 84.8-91.5% for 1:3, 87.4-93.2% for 1:4 and 88.4-94.3% for 1:5.

Among the best models, the LNN described in the previous section is proposed. This QGDR model includes only a minimum of simulated subjects (1:0.5). Thus, this dataset is composed of 921 subjects: 260 real positive subjects, 354 real negative subjects and 307 simulated negative subjects for schizophrenia. As mentioned previously, this neural network is based on 40 SNPs which are taken as an input and has a hidden layer of 152 neurons.

In figure 3, the area under the receiver operating curve (AUC-ROC) for the cross-validation group (0.8405) demonstrates that this model is not a random one. In addition, for this model, the threshold is 0.8.

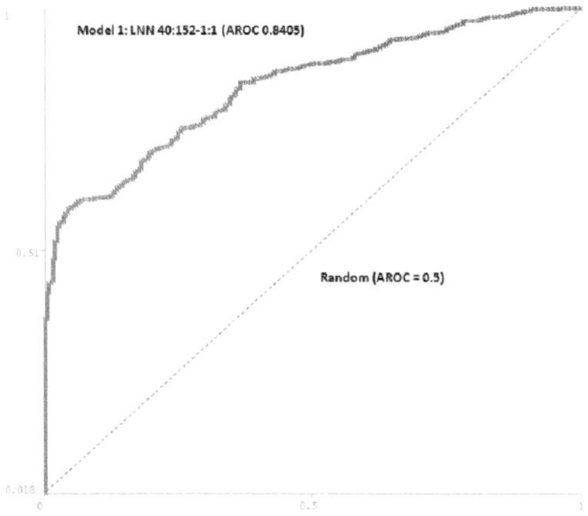

Fig. 3. Area under the receiver operating characteristic curve (AUC-ROC)

4 Conclusions

This work presents an online implementation of a model obtained as a result of a computational study of schizophrenia based on SNP sequences. This model is a linear neural network with 40 neurons in the first layer, 152 in the second layer and 1 neuron as output. This tool offers the possibility to test the system with new sequences of the SNPs considered or to try it using three examples. Thus, this tool enables the user to classify a patient according to some of his/her DNA molecules, based on QGDR (Quantitative Genotype – Disease Relationship) models. In future work, these models will be extended to other diseases in which there may be a genetic predisposition to its development.

Acknowledgements

José A. Seoane and Cristian R. Munteanu acknowledge the funding support for a research position by “Isabel Barreto” grant and an “Isidro Parga Pondal” Program from Xunta de Galicia (Spain), respectively. This work is supported by the following projects: "Galician Network for Colorectal Cancer Research" (REGICC, Ref. 2009/58) from the General Directorate of Research, Development and Innovation of Xunta de Galicia, "Ibero-American Network of the Nano-Bio-Info-Cogno Convergent Technologies", Ibero-NBIC Network (209RT-0366) funded by CYTED (Spain), grant Ref. PIO52048, RD07/0067/0005 funded by the Carlos III Health Institute and "PHR2.0: Registro Personal de Salud en Web 2.0" (Ref. TSI-020110-2009-53) funded by the Spanish Ministry of Industry, Tourism and Trade and the grant (Ref. PIO52048), "Development of new image analysis techniques in 2D Gel for biomedical research " (ref.10SIN105004PR) funded by Xunta de Galicia and RD07/0067/0005, funded by the Carlos III Health Institute.

References

1. Devillers, J., Balaban, A.T.: Topological Indices and Related Descriptors in QSAR and QSPR. Gordon and Breach, The Netherlands (1999)
2. Barabasi, A.L., Bonabeau, E.: Scale-free networks. Sci. Am. 288, 60–69 (2003)
3. Balaban, A.T., Basak, S.C., Beteringhe, A., Mills, D., Supuran, C.T.: QSAR study using topological indices for inhibition of carbonic anhydrase II by sulfanilamides and Schiff bases. Mol. Divers 8, 401–412 (2004)
4. Barabasi, A.L., Oltvai, Z.N.: Network biology: understanding the cell's functional organization. Nat. Rev. Genet. 5, 101–113 (2004)
5. Barabasi, A.L.: Sociology. Network theory-the emergence of the creative enterprise. Science 308, 639–641 (2005)
6. González-Díaz, H., Vilar, S., Santana, L., Uriarte, E.: Medicinal Chemistry and Bioinformatics – Current Trends in Drugs Discovery with Networks Topological Indices. Curr. Top Med. Chem. 7, 1025–1039 (2007)
7. Ferino, G., Gonzalez-Diaz, H., Delogu, G., Podda, G., Uriarte, E.: Using spectral moments of spiral networks based on PSA/mass spectra outcomes to derive quantitative proteome-disease relationships (QPDRs) and predicting prostate cancer. Biochem. Biophys. Res. Commun. 372, 320–325 (2008)
8. Gonzalez-Diaz, H., Gonzalez-Diaz, Y., Santana, L., Ubeira, F.M., Uriarte, E.: Proteomics, networks and connectivity indices. Proteomics 8, 750–778 (2008)
9. den Dunnen, J.T., Antonarakis, S.E.: Mutation nomenclature extensions and suggestions to describe complex mutations: a discussion. Hum. Mutat. 15, 7–12 (2000)
10. Aguiar-Pulido, V., Seoane, J.A., Rabunal, J.R., Dorado, J., Pazos, A., Munteanu, C.R.: Machine learning techniques for single nucleotide polymorphism - disease classification models in schizophrenia. Molecules 15, 4875–4889
11. Diederich, J.: Artificial neural networks: concept learning. IEEE Press, Piscataway (1990)
12. Byvatov, E., Schneider, G.: Support vector machine applications in bioinformatics. Appl. Bioinformatics 2, 67–77 (2003)
13. Eberbach, E.: Toward a theory of evolutionary computation. Biosystems 82, 1–19 (2005)
14. Rowland, J.J.: Model selection methodology in supervised learning with evolutionary computation. Biosystems 72, 187–196 (2003)
15. Tan, P.-N., Steinbach, M., Kumar, V.: Introduction to Data Mining. Pearson Addition Wesley, Boston (2006)

16. Dominguez, E., Loza, M.I., Padin, F., Gesteira, A., Paz, E., Paramo, M., Brenlla, J., Pumar, E., Iglesias, F., Cibeira, A., Castro, M., Caruncho, H., Carracedo, A., Costas, J.: Extensive linkage disequilibrium mapping at HTR2A and DRD3 for schizophrenia susceptibility genes in the Galician population. Schizophr. Res. 90, 123–129 (2007)
17. Wright, F.A., Huang, H., Guan, X., Gamiel, K., Jeffries, C., Barry, W.T., de Villena, F.P., Sullivan, P.F., Wilhelmsen, K.C., Zou, F.: Simulating association studies: a data-based resampling method for candidate regions or whole genome scans. Bioinformatics 23, 2581–2588 (2007)
18. Rosenblatt, F.: Principles of neurodynamics; perceptrons and the theory of brain mechanisms. Spartan Books, Washington (1962)
19. Russel, S., Norvig, P.: Artificial Intelligence: A Modern Approach, 2nd edn. Prentice Hall, Upper Saddle River (2003)
20. Gutlein, M., Frank, E., Hall, M., Karwath, A.: Large-scale attribute selection using wrappers. In: Proceedings of Symposium on Computational Intelligence and Data Mining, pp. 332–339. IEEE Computer Society, Nashville (2009)
21. Yu, L., Liu, H.: Feature Selection for High-Dimensional Data: A Fast Correlation-Based Filter Solution. In: Proceedings of the Twentieth International Conference on Machine Learning, pp. 856–863 (2003)
22. Goldberg, D.: Genetic Algorithms in Search, Optimization and Machine Learning. Addison-Wesley Longman Publishing Co., Boston (1989)
23. Garcia Lopez, F., Garcia Torres, M., Melian Batista, B., Moreno Perez, J.A., Moreno-Vega, J.M.: Solving feature subset selection problem by a Parallel Scatter Search. European Journal of Operational Research 169, 477–489 (2006)
24. Liu, H., Setiono, R.: A probabilistic approach to feature selection - A filter solution. In: 13th International Conference on Machine Learning, Bari, Italy, pp. 319–327 (1996)
25. Bishop, C.: Neural Networks for pattern recognition. Oxford University Press, New York (1995)
26. Buhmann, M.D.: Radial Basis Functions: Theory and Implementations. Cambridge University Press, Cambridge (2003)
27. Aguiar, V., Seoane, J.A., Freire, A., Munteanu, C.R.: Data Mining in Complex Diseases Using Evolutionary Computation. In: Cabestany, J., Sandoval, F., Prieto, A., Corchado, J.M. (eds.) IWANN 2009. LNCS, vol. 5517, pp. 917–924. Springer, Heidelberg (2009)
28. Moore, J.H., Gilbert, J.C., Tsai, C.T., Chiang, F.T., Holden, T., Barney, N., White, B.C.: A flexible computational framework for detecting, characterizing, and interpreting statistical patterns of epistasis in genetic studies of human disease susceptibility. J. Theor. Biol. 241, 252–261 (2006)
29. Cordell, H.J.: Detecting gene-gene interactions that underlie human diseases. Nat. Rev. Genet. 10, 392–404 (2009)
30. John, G.H., Langley, P.: Estimating Continuous Distributions in Bayesian Classifiers. In: 11th Conference on Uncertainty in Artificial Intelligence, pp. 338–345. Morgan Kaufman, Quebec (1995)
31. Bouckaert, R.R.: Bayesian Networks in Weka. Computer Science Department. University of Waikato, Tauranga, New Zealand (2004)
32. Vapnik, V.: Statistical Learning Theory. John Wiley and Sons, New York (1998)
33. Kohavi, R.: The Power of Decision Tables. In: 8th European Conference on Machine Learning, pp. 174–189. Springer, Heidelberg (1995)
34. Mark Hall, E.F.: Combining Naive Bayes and Decision Tables. In: 21st Florida Artificial Intelligence Society Conference (FLAIRS). AAAI Press, Florida (2008)

35. Shi, H.: Best-first Decision Tree Learning. MsC. University of Waikato, New Zealand, Hamilton (2007)
36. Freund, Y., Schapire, R.E.: Experiments with a new boosting algorithms. In: Thirteenth International Conference on Machine Learning, pp. 148–156. Morgan Kaufman, Desenzano sul Garda (1996)
37. Gonzalez-Diaz, H., Prado-Prado, F.J., Garcia-Mera, X., Alonso, N., Abeijon, P., Caamano, O., Yanez, M., Munteanu, C.R., Pazos Sierra, A., Dea-Ayuela, M.A., Gomez-Munoz, M.T., Garijo, M.M., Sansano, J., Ubeira, F.M.: MIND-BEST: web server for drugs & target discovery; design, synthesis, and assay of MAO-B inhibitors and theoretic-experimental study of G3PD protein from Trichomona gallineae. J. Proteome Res. (2010)
38. Rodriguez-Soca, Y., Munteanu, C.R., Dorado, J., Pazos, A., Prado-Prado, F.J., Gonzalez-Diaz, H.: Trypano-PPI: a web server for prediction of unique targets in trypanosome proteome by using electrostatic parameters of protein-protein interactions. J. Proteome Res. 9, 1182–1190 (2010)
39. Munteanu, C.R., Vazquez, J.M., Dorado, J., Sierra, A.P., Sanchez-Gonzalez, A., Prado-Prado, F.J., Gonzalez Diaz, H.: Complex network spectral moments for ATCUN motif DNA cleavage: first predictive study on proteins of human pathogen parasites. J. Proteome Res. 8, 5219–5228 (2009)
40. Concu, R., Dea-Ayuela, M.A., Perez-Montoto, L.G., Prado-Prado, F.J., Uriarte, E., Bolas-Fernandez, F., Podda, G., Pazos, A., Munteanu, C.R., Ubeira, F.M., Gonzalez-Diaz, H.: 3D entropy and moments prediction of enzyme classes and experimental-theoretic study of peptide fingerprints in Leishmania parasites. Biochim. Biophys. Acta. 1794, 1784–1794 (2009)
41. Hall, M., Frank, E., Holmes, G., Pfahringer, B., Reutemann, P., Witten, I.A.. The WEKA Data Mining Software: An Update. SIGKDD Explorations 11 (2009)

MicroRNA Microarray Data Analysis in Colon Cancer: Effects of Normalization

Guillermo H. López-Campos[2,*], Alejandro Romera-López[1,*], Fernando Martín-Sánchez[2], Eduardo Diaz-Rubio[1], Victoria López-Alomso[2], and Beatriz Pérez-Villamil[1]

[1] Molecular Oncology Laboratory, Medical Oncology Department1, Hospital Clinico San Carlos, C/ Martin Lagos s/n, Madrid 28040 Spain
[2] Medical Bioinformatics Dept. Institute of Health "Carlos III", Ctra. Majadahonda-Pozuelo Km.2 28220, Majadahonda, Spain
{aromera.hcsc,ediazrubio.hcsc, bperezvillamil.hcsc}@salud.madrid,org, {glopez,fms,victorialopez}@isciii.es

Abstract. Although gene expression microarray data analysis has been developed and used during the last 15 years, microRNA (miRNA) analysis is still under development and important aspects affecting the final results such as normalization still remain unclear. In this work we have studied and compared the effect of non normalization and four different normalization strategies (75th percentile, hsa-mir103 and hsa-mir-let7a control based and median of all expressed miRNA) on a data set of colon cancer miRNA microarrays. Different subsets of samples were used to study normalization effects comparing data distributions before and after normalization with the different strategies. Median of expressed miRNA behaved the best among the normalizations studied. Class comparison based analysis showed that differentially expressed miRNA are highly dependent on the normalization techniques applied.

Keywords: Microarray, Data analysis, microRNA, Cancer, Normalization.

1 Introduction

In the last decade high throughput molecular techniques have become a popular and a very useful tool in the study of complex biological processes and diseases. In these massive approaches, evolution of biomedical informatics has played a key role as an enabling technology that has supported and provided tools and methods for the analyses of these new complex datasets.

It is in this context where microarrays have become a popular analysis technique in biology and biomedical research. These techniques allow the researchers to quantitatively and simultaneously interrogate a vast amount of features in a sample. The most popular application of microarrays has been the study of gene expression and more particularly, the study of different cancers. In these genomic studies, gene expression (messenger RNA or mRNA) is analyzed, trying to understand the

* Both Authors contributed equally to this article.

J. Cabestany, I. Rojas, and G. Joya (Eds.): IWANN 2011, Part II, LNCS 6692, pp. 260–267, 2011.

molecular mechanisms underlying the disease, and trying to identify those genes differentially expressed between tumors and normal samples or among different cancer subtypes (1). It is very important to understand the genes affected and differentially expressed in cancer, but also the mechanisms related with the regulation of those changes. MicroRNAs (miRNA) are one of the different species of RNA present in the cells. miRNAs act by regulating or regulate the effect of mRNA by means of interactions that may lead either to the degradation of the mRNA or to the inhibition of its translation. miRNA expression is tissue dependent and therefore different tissues may express different miRNAs at different levels. Because of their regulatory role, analysis of miRNA is becoming of greater importance in cancer studies (2) and microarray based analysis are one of the most commonly used techniques for their detection and study.

Several works have been done in gene expression analysis with the aim of studying and standardizing the effects of different data analysis techniques and approaches (3,4). Despite these efforts, when microarrays are applied in micro RNA (miRNA) studies, there are some aspects involved with data analysis that still remain unclear, especially those related with data normalization. Normalization is a set of mathematical transformations used to allow the comparison of the experiments. An important difference in the study of whole genome mRNA and miRNA by microarrays is that while in mRNA microarrays most of the analyzed genes are expected to remain unchanged the expression of miRNA is highly variant and very dependent of the analyzed tissue. For this reason some of the assumptions accepted in some of the whole genome microarray analysis cannot be directly exported into the miRNA analysis.

2 Colon Cancer MicroRNA Microarrays Data Analysis

The main objective of this study is to evaluate whether normalization in miRNA microarray datasets is necessary or not and in case of being needed to recommend the proposal of a normalization algorithm based on the analysis of the performance (considered as the effect on the studied dataset) that different algorithms have after applying them in a colon cancer miRNA microarrays dataset. A secondary goal of the work is to analyze the differences in the expression of miRNA between normal colon and colorectal tumor tissues. The data set studied was generated in our laboratory using 116 Agilent's Human miRNA v.3 microarrays and 110 samples.. Samples included were 19 normal tissue samples and 91 tumor samples.

2.1 Study of Normalization Algorithms for miRNA Microarrays

This analysis was carried out using only a subset of the 116 reactions comprising: 5 pairs of replicated samples, 19 normal samples and 13 pairs of normal-tumor samples (normal and tumor tissues obtained from the same patient).

There are several normalization methods described in the bibliography that can be classified as either based on control miRNAs or on a global miRNA dataset. Among these methods we have selected 4 different normalization algorithms based on both strategies and compared them against non-normalized data: hsa-miR-let-7a, hsa-miR-103 as control miRNA, normalization to 75th percentile and the median of the 103 miRNAs that were identified in all the reactions as global based methods. When miR-let-7a or miR-103 were selected as control genes, normalization was carried out independently to each sample by subtracting the value of the control miRNA from all miRNAs in the sample. Same strategy was used for the 75th percentile normalization. In this case, for each array, the 75th percentile of the expression values (including all miRNAs) was calculated and then subtracted from the expression value of each single miRNA. Finally, to use the mean of expressed miRNAs as a normalization factor, the means of the 103 miRNAs that were present in all the reactions was calculated. Then, this value was subtracted from all miRNAs in the sample. Again, each sample was treated independently.

hsa-miR-let-7a has already been employed as control gene for normalization in other studies such as Monzo et al. (5) and hsa-miR-103 has been proposed as an universal housekeeping miRNA (6). No normalization has also been applied in different projects such as Wang et al. (7). In global miRNA datasets based normalizations, 75th percentile has been the default normalization method in Agilent's Genespring data analysis software for many years, instead of median normalization commonly applied in mRNA datasets, and it has recently been changed to the 90th percentile. We propose an alternative global method based on the median of those miRNAs identified in all the analyzed samples.

The comparison among the proposed normalization algorithms was carried out using different approaches:

1. A graphical approach based on the analysis of the scatter-plots generated in the comparison of different normalized samples with themselves or with other samples of the same category (normal or tumor)
2. Analysis of the histograms of the ratios of the normalized subset including only replicated samples.
3. Analysis of the accumulated variability in the normal subset using different normalization strategies.

For the two latter strategies, it was necessary to perform a re-scaling of the normalized datasets setting a minimal value of 0.

The scatter-plot analysis summarizes the effects that normalization or the lack of it may have in a dataset. For this approach, three different sample combinations were studied: one based on the comparison of the results coming from replicates using the same sample (CT3, which is a colorectal cancer sample), using two colorectal normal samples (CN100 vs CN103) and finally using two different colorectal tumor samples (CT24 vs CT105). The use of the normalization algorithms should distribute the data around the diagonal of the graphic (Fig. 1) which represents ratio = 1. Therefore, the aim of this approach is to identify those normalizations methods that generate datasets that differ from the ideal diagonal, with a narrow dispersion, and distribution.

Due to the fact that single color arrays were used, non corrected technical variation is shown in the final fluorescence leading to skewed scatter plots and therefore it could lead to artificial up-regulated or down-regulated miRNAs when comparing conditions. Under these conditions, results show that normalization improves the comparability of the results and also that among the normalization methods hsa-miR-103 and median based on the constantly detected miRNAs (103 miRNAs) performs the best. These methods are the ones that better adjust the results to the diagonal in the scatter-plots. Obviously, the dispersion increases from replicates of the same sample to the replicates of the normal tissue and from these to the replicates of the tumor tissue. The increase of variation is explained by the underlying biology of the samples, being the tumor samples more heterogeneous in their expression than the normal samples.

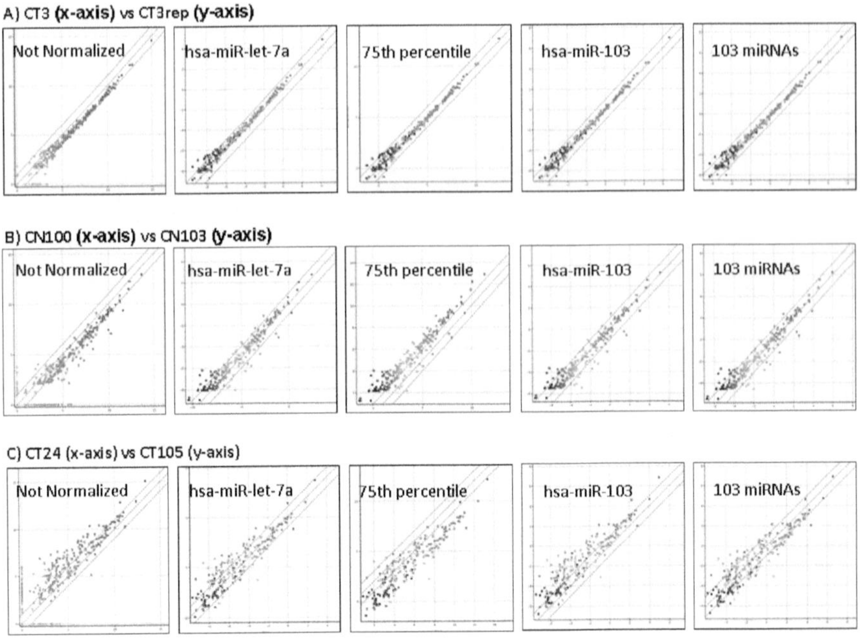

Fig. 1. Scatter-plots for log2 fluorescence after applying five different normalization strategies in: A) replicated sample (CT3-CT3rep). B) Normal sample vs. Normal sample (CN100 vs CN103). C) Tumor sample vs Tumor Sample (CT24 vs CT105). It is possible to appreciate that normalizations based on hsa-miR-103 and median (103miRNAs) perform the best.

Part of the underlying idea of the normalization of datasets is that ratio between replicated samples should be equal to 1. Therefore comparison of replicated samples, where ratios along all range of intensity should be equal to 1, is a useful tool to evaluate if technical biases are present or not.

Five pairs of replicated samples were available and a histogram was generated based on the ratio of the 203 miRNAs analyzed between replicated samples (Fig. 2). More than 50% of miRNAs provided ratios equal to 1.1 when 75th percentile, hsa-miR-103 or median of 103 miRNA normalization was applied, while in the other strategies, only around 30% of ratios seem to reach this value. Moreover, more than 38% and 22% of the miRNAs of the dataset had a ratio equal or superior to 1.4 when no normalization or hsa-miR-let-7a strategies were applied in comparison with the other three normalization procedures where no more than 10% of the data reached this limit.

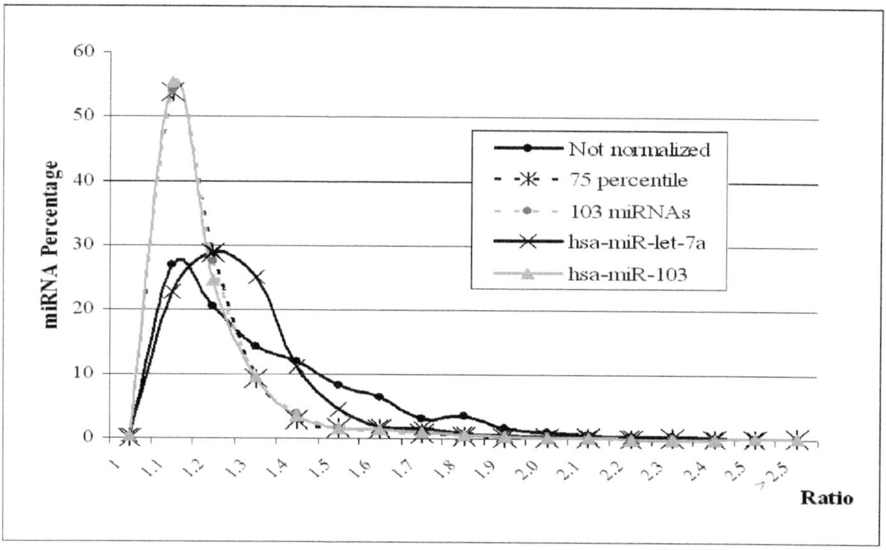

Fig. 2. Histogram of the ratios obtained when replicated samples are compared. 203 miRNAs present in all the 5 different pairs of replicates were used to generate the histograms. It is easy to appreciate that not normalized and has-miR-107 methods performs worse than the others, showing a higher population of ratio values higher than 1.2.

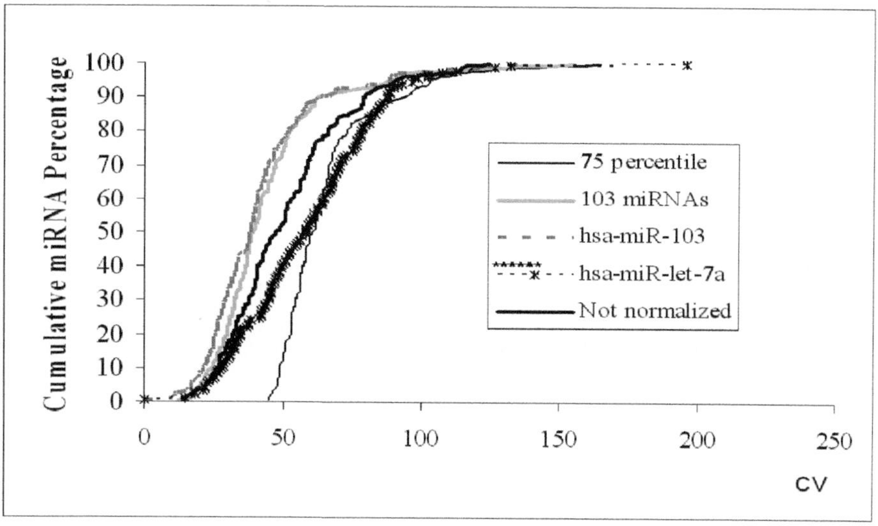

Fig. 3. Graphic representing the cumulative percentage of miRNAs according to their coefficient of variation (CV) when 19 normal samples are analyzed. In this figure it is possible to appreciate that hsa-miR107 and 75th percentile normalized dataset show a poorer performance than the non-normalized dataset.

Finally, a third strategy to evaluate the performance of the normalization methods is based on the analysis of variation of normal samples. Variation in miRNA expression between normal samples can be explained as the combination of biological and technical variation. It is expected that appropriate normalization methods decrease the coefficient of variation between normal samples while inappropriate methods do not.

When variation between 19 normal colorectal samples, using the 150 miRNAs present in all samples, was analyzed, only hsa-miR-103 and 103miRNAs normalization methods showed a decrease in the coefficient of variation compared to no normalization (Fig. 3). Contrary to our expectations, hsa-miR-let-7a and 75th percentile normalization performed incorrectly adding even more technical variation in their attempt to normalize the dataset.

2.2 Effect of Different Normalization Strategies on Comparison between Normal and Colorectal Cancer miRNA Expression

Finally, the effect of the different normalization strategies was analyzed in the context of a class comparison study between normal and colorectal cancer miRNA expression. The purpose of these kinds of studies is to identify the miRNA differentially expressed between both conditions.

A class comparison between healthy and tumor samples was carried out using 13 paired normal-tumor colorectal samples. The analysis was carried out using a paired T-Test and applying the Benjamini-Hochberg method to correct for multiple comparisons.

Eventually, only those miRNAs with a p-value ≤ 0.01 and a fold change ≥ 2 were seen as statistically significant. Data obtained shows a great imbalance in the results found between up-regulated and down-regulated identified miRNAs, table 1. When no normalization was applied, 18 miRNAs were described as differentially expressed with only 1 out of 18 being up-regulated in normal samples compared to tumor samples.

Normalization with hsa-miR-let-7a or 75th percentile showed the opposite with all miRNAs (11 and 13 respectively) being up-regulated when normal tissues were compared to colorectal cancer samples. Moreover, when the list obtained after hsa-miR-let-7a or 75th percentile normalization was compared with the one obtained after no normalization, only 1 miRNA (hsa-miR-145) was in common.

When hsa-miR-103, one of the universal miRNAs proposed by Peltier et al., was used as normalizer, a better balance was found between the miRNAs up-regulated (9 miRNAs) and down-regulated (4 miRNAs). However, it was the strategy of using the median of the 103 miRNAs expressed in all samples as a housekeeping data set, which showed the best balance. 8 miRNAs were found as up-regulated while 6 miRNAs were found as down-regulated. Moreover or furthermore, only when the combination of the 103 miRNAs were used as normalizer, miRNAs generally accepted as altered in colorectal cancer such as hsa-miR-143, hsa-miR-145, hsa-miR-20a and hsa-miR-21, were found as differentially expressed.

Table 1. Class comparison between 13 paired normal-tumor samples after using 5 different normalization strategies. (+) → up-regulated in normal tissue when compared to tumor tissue. (-) → down-regulated in normal tissue compared to tumor tissue.

Systematic name (hsa-miR-)	No Normalization	hsa-miR-let7a Normalization	75th percentile Normalization	hsa-miR-103 Normalization	103miRNAs Normalization
106b	(-) 2.1				
1246	(-) 4.6			(-) 3.3	(-) 3.5
1275	(-) 2.1				
130b	(-) 2.7				
140-3p			(+) 3.0		
143		(+) 5.7	(+) 8.2	(+) 5.1	(+) 4.8
145	(+) 4.8	(+) 7.6	(+) 10.8	(+) 6.8	(+) 6.3
146b-5p			(+) 2.7		
150		(+) 3.6	(+) 5.2	(+) 3.2	(+) 3.0
17	(-) 2.6				
186			(+) 2.9		
195		(+) 4.3	(+) 6.2	(+) 3.9	(+) 3.6
19b	(-) 2.4				
203	(-) 4.1			(-) 2.8	(-) 3.1
20a	(-) 3.0				(-) 2.2
21	(-) 3.0			(-) 2.1	(-) 2.3
210	(-) 2.9				
221	(-) 3.3			(-) 2.3	(-) 2.5
28-5p		(+) 2.1	(+) 3.1		
29a	(-) 2.0				
29b-1*	(-) 2.9				
29c		(+) 2.1			
30a		(+) 4.2	(+) 6.0	(+) 3.7	(+) 3.5
30c		(+) 2.3	(+) 3.3	(+)) 2.1	
30e*		(+) 2.4	(+) 3.5	(+) 2.2	(+) 2.0
342-3p		(+) 2.4	(+) 3.5	(+) 2.2	(+) 2.0
34a	(-) 2.4				
455-3p	(-) 2.7				(-) 2.1
497		(+) 4.7	(+) 6.8	(+) 4.3	(+) 4.0
92a	(-) 2.3				
93	(-) 2.4				

3 Conclusions

Microarray data analysis results are highly influenced by different aspects, being one of them the transformations done to allow the comparison of the different experiments that belong to the same dataset. The new miRNA microarrays are especially influenced by these artifacts since there is not a standard for normalization.

We have compared the results of 4 different normalization strategies and non normalization in a dataset generated in our laboratory. These results showed that the best normalization method is the one based on the use of the subset of microRNA that can be detected in all the analyzed microarrays. The use of this normalization approach led in further class comparison analysis to results making more sense from a biological point of view than the other methods studied in this work. Despite two different normalization algorithms (hsa-miR-103 and median of all the detected miRNA) have been identified as the better performers in our colon cancer dataset the nature of the miRNA and their variability in their level of expression suggests that further work is necessary in order to identify a general normalization approach that might be suitable for every possible condition or tissue.

Acknowledgments. Isabel Hermosilla for her collaboration with this manuscript. This work is partially funded by the RETICS COMBIOMED Network (FIS, Ministry of Science and Innovation of Spain) and by the European Commission Support Action INBIOMEDvision and the CYTED Network IBERO-NBIC.

References

1. Garcia-Hernandez, O., et al.: Microarray data analysis and management in colorectal cancer. In: Proceedings on Biological and Medical Data Analysis, vol. 3745, pp. 391–400 (2005)
2. Cummins, J.M., Velculescu, V.E.: Implications of micro-RNA profiling for cancer diagnosis. Oncogene 25, 6220–6227 (2006)
3. Chen, J.J., et al.: Reproducibility of microarray data: a further analysis of microarray quality control (MAQC) data. BMC Bioinformatics 8, 412 (2007)
4. Shi, L., et al.: The MicroArray Quality Control (MAQC)-II study of common practices for the development and validation of microarray-based predictive models. Nat. Biotechnol. 28, 827–838 (2010)
5. Monzo, M., et al.: Overlapping expression of microRNAs in human embryonic colon and colorectal cancer. Cell Res. 18, 823–833 (2008)
6. Peltier, H.J., Latham, G.J.: Normalization of microRNA expression levels in quantitative RT-PCR assays: identification of suitable reference RNA targets in normal and cancerous human solid tissues. RNA 14, 844–852 (2008)
7. Wang, H., Ach, R.A., Curry, B.: Direct and sensitive miRNA profiling from low-input total RNA. RNA 13, 151–159 (2007)

Automatic Handling of Tissue Microarray Cores in High-Dimensional Microscopy Images

G. Bueno[1], M. Fernández[1], O. Déniz[1], and M. García-Rojo[2]

[1] E.T.S.I. Industriales, Universidad de Castilla-La Mancha, Spain
[2] Hospital General de Ciudad Real, Spain
Gloria.Bueno@uclm.es

Abstract. This paper describes a specific tool to automatically perform the segmentation and archiving of tissue microarray (TMA) cores in microscopy images at different magnification, that is, 5x, 10x, 20x and 40x. TMA enables researchers to extract small cylinder of single tissues (core sections) from histological sections and arrange them in an array on a paraffin block such that hundreds can be analyzed simultaneously. A crucial step to improve the speed and quality of these analyses is the correct recognition of each tissue position in the array. However, usually the tissue cores are not aligned in the microarray, the TMA cores are broken and the digital images are noisy. We develop a robust framework to handle core sections under these conditions. The algorithms are able to detect, stitch and archive the TMA cores. Once the TMA cores are segmented they are stored in a relational database allowing their location and classification for further studies of benign-malignant classification. The method was shown to be reliable for handling the TMA cores and therefore enabling further large scale molecular pathology investigations.

1 Introduction

The tissue Microarray (TMA) represents a powerful new technology designed to assess the expression of proteins or genes across large sets of tissue specimens [1]. A TMA is an ordered array of up to several hundreds of small cylinders of single tissues (core sections) in a paraffin block from which sections can be cut and treated like any other histological section, using immunohistochemistry (IHC) for protein targets and in situ hybridization (ISH) to detect gene expressions or chromosomal alterations [2], [3]. TMA allows rapid and reproducible investigations of biomarkers. The integration of TMA and clinical pathology data is emerging as a powerful approach to molecular profiling of human cancer [4], [5].

Another use of the TMA is to provide random samples of a representative lesion, which may be evaluated by automated methods (image recognition: nuclear density, colour density, results of immunohistochemical staining) in order to achieve an objective diagnosis in pathology , which still is nowadays one of the diagnostic laboratories with more human intervention (manual work), and more subjective assessment [6],[7],[8].

However, the TMA also has their drawbacks, both in data acquisition and in its management and interpretation. The use of IHC with TMA generates

J. Cabestany, I. Rojas, and G. Joya (Eds.): IWANN 2011, Part II, LNCS 6692, pp. 268–275, 2011.

large amounts of information, which requires careful analysis. Currently, this analysis is done manually by microscope, which besides being a tedious job that hinders the workflow, is subject to errors due to subjective interpretations of the specialists. The automatic analysis of TMA data and multicenter studies is still a challenge [2], [9], [10], [11]. The use of automatic acquisition systems for various digital imaging and tissue staining, as well as the development of tools for processing these images, will help to improve these problems.

Another difficulty in TMA analysis is that usually the cores are neither aligned nor regular, or they do not have tumour tissue, besides the typical problems of digital images such as noise, distortion, etc .... This leads to lost cores in the detection process. Thus, there is a need to develop reliable tools to acquire, share and assess microarrays and correlated data.

At the moment, and as far as the authors know, there are only 4 systems described in the literature for TMA handling, Della Mea [9], Demichelis [10], Shaknovich [12] and Liu [13], [6] and a preliminary report by Morgan *et al.* [14]. The systems by Liu [6] and Shaknovich [12], are based on commercial products, particularly Microsoft Excel and Adobe Photoshop software together with some additional basic image processing tools developed by the authors. The main drawback is that they are developed for a specific study. Besides this, it seems they work just with low resolution images, i.e., magnification lower than 10x. Moreover, in the case of Liu et al. the rigid registration or stitching problem is not solved and the TMA images appear with overlapping regions. The system by Demichelis [10] is web-based and it is developed using proprietary commercial software. This system favours the automatic TMA handling compared to the manual one to facilitate subsequent archiving and processing, which is one of the biggest problems in commercial systems. The main drawback of the system is the segmentation process. The system provides many false negatives, that is, many cores are missed, even in TMA images without noise nor overlapping. The systems by Morgan and Della Mea are similar. They are web-based and open source. Della Mea presents a website dedicated to management, including archiving and retrieval. The system includes more tools for this management than the above mentioned systems, however, it does not fully cover data analysis and automatic image processing.

Thus, there is a need of tools for automatic TMA analysis including tissue core locations, segmentation and rigid registration of digital microscopic images acquired at different magnification (5x, 10x, 20x, 20x and 40x) from different devices. This is the aim of the present work, which describes the implemented tools for the detection and automatic storage of the TMA cores and their pathology information for further studies of their benign or malignant character.

Section 2 describes the methods and materials used for this work. This includes the segmentation procedure applied to the TMA images at different magnifications, the storage, as well as the experimental database used. Section 3 describes the results obtained with the proposed method and finally in Section 4 the main conclusions are drawn.

2 Methods

The first objective of this work is the automatic segmentation of the TMA cores previous to their archiving and processing. The segmentation is applied over TMA images acquired at different magnification, 5x, 10x, 20x and 40x. This process includes the TMA core detection, selection and extraction. Once the TMA cores are segmented they are archived. The archiving process preserves all the pathological information in a relational database for further classification of the selected cores. The whole segmentation process is ilustrated in Fig. 1 and the methods are described as follows.

2.1 Detection

The algorithm developed for the detection of the tissue cores depends on the availability of a thumbnail of the TMA. If a thumbnail of the microscopic image is available then the cores are detected on this image. In the case where there is not a thumbnail, the cores are obtained in the 5x images. Then, the coordinates of each core are calculated for different magnifications (5x, 10x, 20x, 40x). It must be kept in mind that the 5x images are bigger than 500MB and therefore some processing operations may lead to memory errors. To avoid this, the 5x image is divided into 6 pieces and the algorithm is applied to the 6 pieces. Thus, the algorithm proceeds as follows:

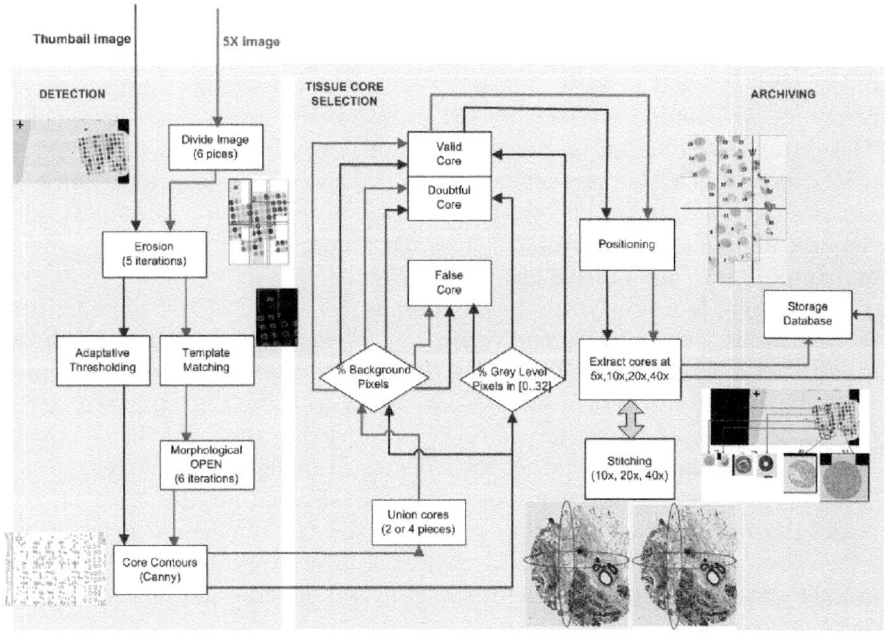

Fig. 1. TMA core segmentation process

1. The colour image is converted into a gray image.
2. (a) If there is a thumbnail image then an erosion of 5 iterations with a $3x3$ kernel is performed in the image.
2. (b) If there is not a thumbnail image then a template matching is applied on the 5x image. The template matching uses a normalized correlation coefficient method. The correlation coefficient indicates the extent to which the input template coincides with the image. The template corresponds to a core sample obtained from one of the TMA at 5x. A perfect coincidence with the input template has value 1, whereas no coincidence gives -1 and the value 0 indicates no correlation.

$$R_{ccoeff}(x,y) = \sum_{x',y'} [T'(x',y').I'(x+x',y+y')]^2 \quad (1)$$

$$T'(x',y') = T(x',y') - \frac{1}{(w.h)\sum_{x'',y''} T(x'',y'')} \quad (2)$$

$$I'(x+x',y+y') = I(x+x',y+y') - \frac{1}{(w.h)\sum_{x'',y''} I(x+x'',y+y'')} \quad (3)$$

The correlation coefficient method can be normalized for better results, helping to reduce illumination differences between the template and the image. Normalization is always performed in the same manner.

$$Z(x,y) = \sqrt{\sum_{x',y'} T(x',y')^2 . \sum_{x'',y''} I(x+x',y+y')^2} \quad (4)$$

The normalized correlation coefficient method is therefore represented as:

$$R_{ccoeff}(x,y) = \frac{R_{ccoeff}(x,y)}{Z(x,y)} \quad (5)$$

3. An adaptive thresholding is performed.
4. A morphological closing of 5 and 6 iterations on the thumbnail and on the 5x image, respectively, with a $3x3$ kernel, is applied.
5. A Canny operator is applied to find the core contours. This algorithm is used for the detection of the contour pixels that divide each segment of the image, allowing to store them through sequences and in a way such that they can be later manipulated individually. This operation identifies external and internal contours of the region of interest, that is the tissue cores.

If the thumbnail is available the erosion operation performs better than the matching (step 2). It was observed that an initial erosion in the thumbnail image (instead of a matching operation) can eliminate artifacts that may be considered as cores, thus reducing the number of false positives.

2.2 Tissue Core Selection

A selection of the previously detected cores is done before performing the extraction of the tissue cores. Thus, those cores that do not fulfill the requirements to be considered as valid are discarded. In this way false positive detections due to noise and imperfections on the glass slide (fragmented, spots and stains produced in the preparation process or missing core sections) are avoided. The condition to considered a valid core is based on the amount of segmented tissue. This condition is fulfill if 1st) the percent of background pixels measured after the adaptative threshold (step 3) is lower than 92% and 2nd) if the percent of pixels with grey values similar to the grey value of stains is lower than 32%. The latest condition is measured on a squared region of $80x80$ in the center of the core. The core is classified as doubtful if the percent of background pixels is between 82% and 92% or if the percent of pixels with grey values similar to stains is between 12% and 32%.

2.3 Tissue Core Extraction

Once the cores that compound the TMA are selected, they go through the positioning and extraction phase. For the extraction process, the minimum bounding rectangle of the core is selected. This bounding box indicates the position coordinates of the core over the whole TMA image. These coordinates allow to control the core position and enumerate the core tissues.

Then, the tissue cores are extracted at different magnification, that is 5x, 10x, 20x and 40x. The relationship that exists between the pixels of the thumbnail of an image of one core and the rest of the images at different magnifications is straightforward, because each image in ascending order is always the double of the previous one apart from the change between the thumbnail and the magnification at 5x. The image at 5x is eight times greater than the thumbnail, the image at 10x is the double of the 5x and so on.

One of the main problems when extracting the tissue core at 10x and 20x is that they are not located in the same tile, but it is divided into several image tiles. Thus, when performing the extraction it is necessary to perform previously the stitching between different tiles to form the image with the tissue core. This stitching is done by mean of a rigid registration with a cost function based on square mean error and a Levenberg-Marquardt optimizer. After the stitching the tissue core is segmented from the image using the coordinates obtained in the positioning process.

2.4 Tissue Core Archiving

Each segmented core is saved as an individual image and their information is archived in a relational database. This allows further consultations or modification of their information. The database created is composed of five interrelated tables, where the main link is the table *cores_thumbnail_TMA* which as its name

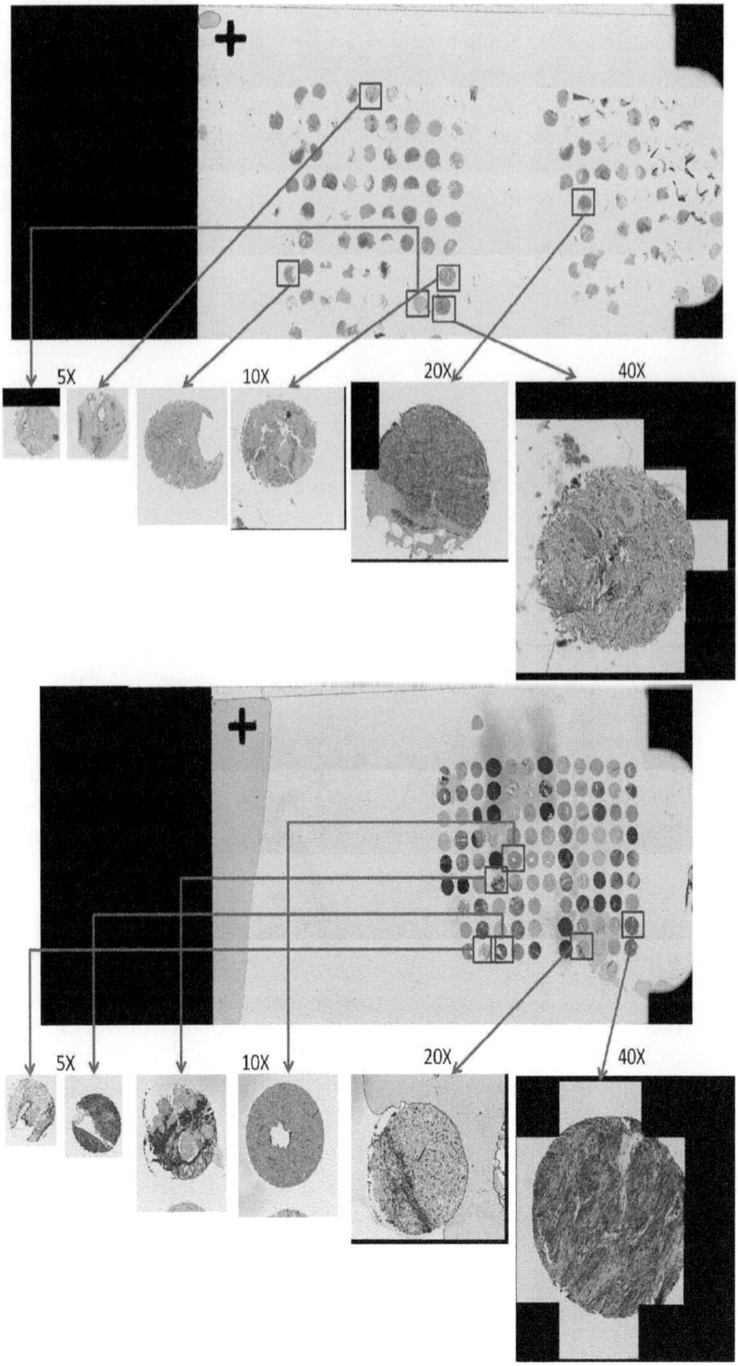

Fig. 2. TMA core segmentation at different magnifications

suggests has information about the thumbnail image. This table has five attributes that indicate: its identity (TMA core), the validity of the core (a core can be considered as valid or doubtful), the minimum bounding rectangle or bounding box coordinates of the core and its classification (malignancy or benignity). The other tables correspond to the tissue cores belonging to the images at different magnifications 5x, 10x, 20x and 40x. They have four attributes: its identity, the minimum bounding rectangle coordinates and the relationship with the thumbnail.

3 Results

Two datasets were considered. One composed of 9 TMAs acquired by the motorized microscope ALIAS II (LifeSpan Biosciences Inc.) at different magnification, 5x, 10x, 20x and 40x. And the other one composed of 15 TMAs acquired by Aperio ScanScope T2. These TMAs have an average of 36 cores per TMA, that is 3456 cores in total were processed. The scanning method to acquire the digital image is different for the scanner than for the microscope. The acquisition of microscopic fields is square-by-square, from the slide's upper left corner to the lower right one. Thus, the final image is a mosaic composed of multiple files of $2000 * 2000$ size each one. The Aperio ScanScope uses a linear camera, where the acquisition file corresponds to a strip set of $1000 * D$, where D variates between 72098 to 87891 pixels length.

Some examples of segmentation results are shown in Fig.2. We carried out a study on the overall percentage of TMA cores that are extracted from the TMA. Over the whole database an average of 90% of the cores were properly segmented. This result is quite promissing if we take into account that most of the glass slides have disorders.

4 Conclusions

This paper has described an specific tool to automatically perform the segmentation and archiving of tissue microarray (TMA) cores in microscopy images at different magnification, that is, 5x, 10x, 20x and 40x. The tool shows promissing results segmenting different microscopic images from TMA glass slides with different disorders and image quality. The tool improve previous systems described in the literature and address the problem of handling high dimensional microscopic TMA images at different magnification.

Acknowledgements

The authors acknowledge partial financial support from the Spanish Research Ministry and Junta de Comunidades de Castilla-La Mancha through projects RETIC COMBIOMED, DPI2008-06071 and PAI08-0283-9663.

References

1. Chen, W., Reiss, M., Foran, D.: A prototype for unsupervised analysis of tissue microarrays for cancer research and diagnostics. IEEE Trans. on Information Technology in Biomedicine 8(2), 89–96 (2004)
2. Dell'Anna, R., Demichelis, F., Sboner, A., Barbareschi, M.: An automated procedure to properly handle digital images in large scale tissuemicroarray experiments. Comput. Methods and Programs in Biomedicine 79(3), 197–208 (2005)
3. Rimm, D., Camp, R., Charette, L., Olsen, D., Reiss, M.: Tissue microarray: A new technology for amplification of tissue resources. Cancer 7(1), 24–31 (2001)
4. Dhanasekaran, S.M., Barrette, T., Ghosh, D., Shah, R., Varambally, S., Kurachi, K., Pienta, K., Rubin, M., Chinnalyan, A.: Delineation of prognostic biomarkers in prostate cancer. Nature 412, 822–826 (2001)
5. Kuraya, K.A., Simon, R., Sauter, G.: Tissue microarrays for high-throughput molecular pathology. Ann. Saudi. Med. 24, 169–174 (2004)
6. Liu, C., Montgomery, K., Natkunam, Y., West, R., Nielsen, T., Cheang, M., Turbin, D., Marinelli, R., de Rijn, M.V., Higgins, J.: Tma-combiner, a simple software tool to permit analysis of replicate cores on tissue microarrays. Mod. Pathol. 18, 1641–1648 (2005)
7. Nohle, D., Hackman, B., Ayers, L.: The tissue micro-array data exchange specification: a web based experience browsing imported data. BMC Med. Inform. Decis. Mak. 5(25) (2005)
8. Rabinovich, A., Krajewski, S., Krajewska, M., et al.: Framework for parsing, visualizing and scoring tissue microarray images. IEEE Tran. on Information Technology in Biomedicine (2), 209–219 (2006)
9. Mea, V.D., Bin, I., Pandolfi, M., Loreto, C.D.: A web-based system for tissue microarray data management. Diagnostic Pathology 1, 31–36 (2006)
10. Demichelis, F., Sboner, A., Barbareschi, M., Dell'Anna, R.: Tmaboost: An integrated system for comprehensive management of tissue microarray data. IEEE Trans. on Information Technology in Biomedicine 10(1), 19–27 (2006)
11. Strömberg, S., Björklund, M., Asplund, C., Sköllermo, A., et al.: A high-throughput strategy for protein profiling in cell microarrays using automated image analysis. Proteomics 7, 2142–2150 (2007)
12. Shaknovich, R., Celestine, A., Yang, L., Cattoretti, G.: Novel relational database for tissue microarray analysis. Archives of Pathology Laboratory Medicine 127 (2003)
13. Liu, C., Prapong, W., Natkunam, Y., Alizadeh, A., Montgomery, K., Gilks, C., Rijn, M.: Software tools for high-throughput analysis and archiving of ihc staining data obtained with microarrays. Am. J. Pathol. 161(5), 1557–1565 (2002)
14. Morgan, J., Iacobuzio-Donahue, C., Razzaque, B., Faith, D., Marzo, A.D.: Tmaj: Open source software to manage a tissue microarray database. Proc. of APIII Meeting (2003)

Visual Mining of Epidemic Networks

Stéphan Clémençon[1,*], Hector De Arazoza[2,3], Fabrice Rossi[1], and Viet-Chi Tran[3]

[1] Institut Télécom, Télécom ParisTech, LTCI - UMR CNRS 5141
46, rue Barrault, 75013 Paris – France
[2] Facultad de Matemática y Computación, Universidad de la Habana, La Habana, Cuba
[3] Laboratoire Paul Painlevé UMR CNRS No. 8524, Université Lille 1, 59 655 Villeneuve d'Ascq Cedex, France
{stephan.clemencon,fabrice.rossi}@telecom-paristech.fr,
arazoza@matcom.uh.cu, chi.tran@math.univ-lille1.fr

Abstract. We show how an interactive graph visualization method based on maximal modularity clustering can be used to explore a large epidemic network. The visual representation is used to display statistical tests results that expose the relations between the propagation of HIV in a sexual contact network and the sexual orientation of the patients.

1 Introduction

Large graphs and networks are natural mathematical models of interacting objects such as computers on the Internet or articles in citation networks. Numerous examples can be found in the biomedical context from metabolic pathways and gene regulatory networks to neural networks [10]. The present work is dedicated to one type of such biomedical network, namely epidemic networks [7]: such a network models the transmission of a directly transmitted infectious disease by recording individuals and their contacts, other individuals to whom they can pass infection.

Understanding the dynamic of the transmission of diseases on real world networks can lead to major improvements in public health by enabling effective disease control thanks to better information about risky behavior, targeted vaccination campaigns, etc. While transmissions can be studied on artificial networks, e.g., some specific types of random networks [7], such networks fail to exhibit all the characteristics observed in real social networks (see e.g.[10]). It is therefore important to get access to and to analyze large and complex real world epidemic networks. As pointed out in [7], the actual definition of the social network on which the propagation takes place is difficult, especially for airborne pathogens, as the probability of disease transmission depends strongly on the

* This work was supported by the French Agency for Research under grant ANR Viroscopy (ANR-08-SYSC-016-03) and by AECID project D/030223/10.

J. Cabestany, I. Rojas, and G. Joya (Eds.): IWANN 2011, Part II, LNCS 6692, pp. 276–283, 2011.

type of interaction between persons. This explains partially why sexually transmitted diseases (STD) epidemic networks have been studied more frequently than other networks [7,9].

We study in this paper a large HIV epidemic network that has some unique characteristics: it records almost 5400 HIV/AIDS cases in Cuba from 1986 to 2004; roughly 2400 persons fall into a single connected component of the infection network. STD networks studied in the literature are generally smaller and/or do not exhibit such a large connected component and/or contain a very small number of infected persons. For instance, the Manitoba study (in Canada, [15]) covers 4544 individuals with some STD, but the largest connected component covers only 82 persons. The older Colorado Springs study [13] covers around 2200 persons among which 965 falls in connected component (the full network is larger but mixes sexual contacts and social ones; additionally, the sexual networks contains only a very small number of HIV positive persons).

While the large size and coverage of the studied network is promising, it has also a main negative consequence: manual analysis and direct visual exploration, as done in e.g. [9], is not possible. We propose therefore to analyze the network with state-of-the-art graph visualization methods [3].

We first describe the epidemic network in Section 2 and give an example of the limited possibilities of macroscopic analysis on this dataset. Then Section 3 recalls briefly the visual mining technique introduced in [3] and shows how it leads to the discovery of two non obvious sub-networks with distinctive features.

2 Cuban HIV/AIDS Database

The present work studies an anonymized national dataset which lists 5389 Cuban residents with HIV/AIDS, detected between 1986 and 2004. Each patient is described by several variables including gender, sexual orientation, age at HIV/AIDS detection, etc. (see [1] for details.)

2.1 Data Collection

The Cuban HIV/AIDS program produces this global monitoring using several sources that range from systematic testing of pregnant women and all blood donations to general practitioner testing recommendations. In addition, the program conducts an extended form of infection tracing that leads to the epidemic network studied in this work.

Indeed, each new infected patient is interviewed by health workers and invited to list his/her sexual partners from the last two years. The primary use of this approach is to discover potentially infected persons and to offer them HIV testing. An indirect result is the construction of a network of infected patients. Sexual partnerships are indeed recorded in the database for all infected persons. Additionally, a probable infection date and a transmission direction are inferred from other medical information, leading to a partially oriented infection network. While this methodology is not contact tracing *stricto sensu* as non infected patients are not included in the database (contrarily to e.g. [13]), the program records the total number of sexual partners declared for the two years period as well as a few other details, leading to an extended form of infection tracing. (see [7] for differences between contact and infection tracing.)

2.2 Macroscopic Analysis

The 5389 patients are linked by 4073 declared sexual relations among which 2287 are oriented by transmission direction. A significant fraction of the patients (44 %) belong to a giant connected component with 2386 members. The rest of the patients are either isolated (1627 cases) or members of very small components (the second largest connected component contains only 17 members).

As the sexual behavior has a strong influence on HIV transmission, it seems important to study the relations between the network structure and sexual orientation of the patients. In the database, female HIV/AIDS patients are all considered to be heterosexual as almost no HIV transmission between female has been confirmed [8]. Male patients are categorized into heterosexual man and "Man having sex with Men" (MSM); the latter being men with at least one male sexual partner identified during their interview.

The distributions of genders and of sexual orientations is given in Table 1: the giant component contains proportionally more MSM than the full population; this seems logical because of the higher probability of HIV transmission between men [14]. Additionally, as shown on Figure 1, the sexual orientation of the newly detected patients changes through years (note that year 2004 is incomplete and therefore has not been included in this analysis): the percentage of heterosexual men seems to decrease. This is also the case for women, but to a lesser extent.

Table 1. Gender and sexual orientation distributions in the whole network and in the giant component

	full network		**giant component**	
	absolute	relative	absolute	relative
woman	1109	0.21	472	0.20
heterosexual man	566	0.11	110	0.05
MSM	3714	0.69	1804	0.76

However, the evolution of the percentage of MSM through time is difficult to analyze at a macroscopic level. For instance, multiple explanations can be offered to explain the influence of the sexual orientation of the patients on their average shortest path distances in the graph (see Table 2). One the one hand, temporal distance should be reflected by distance in the graph as the infection tracing has a short time span. Then one might expect short paths between heterosexual men as they should be concentrated in the early phase of the epidemic and long chains should be rare. However, on the other hand, heterosexual men could also be linked to the network mostly through women. Then their relative short distances could be explained via the relative short distances between women themselves and the time aspect would play no role in the observed distances.

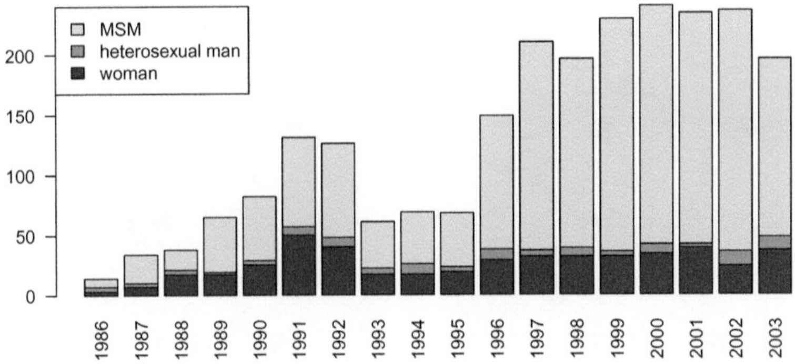

Fig. 1. Yearly sexual orientation distribution in the giant component

Table 2. Average shortest path distances in the giant component of the infection network, conditioned by the sexual orientation of the extremal points (the global average is 10.24)

	MSM	heterosexual man	woman
MSM	10.30	10.83	10.24
heterosexual man	10.83	9.87	9.30
woman	10.24	9.30	8.76

3 Visual Mining

Because of the size of the giant component, a direct visual analysis is impossible and the interplay between sexual orientation and infection is difficult to analyze. We show in this section how the difficulty can be circumvented by combining *clustered graph* visualization techniques [4] with efficient maximal modularity graph clustering [12], as proposed in [3]. The methodology introduced here can be used to analyze the relation between other quantities and the infection structure.

3.1 Methodology

This section briefly explains the graph visualization method used to analyze the infection network. Details can be found in [3].

The classical strategy used to display a large network (e.g., with more than a hundred nodes) is to coarsen the network via a clustering method, leading to the so-called *clustered graph* visualization problem [4]. To implement this strategy, we use a maximal modularity graph clustering approach [12], as maximizing the modularity leads in general to meaningful clusters [5] which are additionally well adapted to visualization [11]. Then rather than displaying the original network, we use the standard Fruchterman Reingold algorithm [6] to display the network of clusters. Figure 2 gives concrete examples of the results: in the present context, each cluster consists of a group of patients linked by sexual relationships (the size

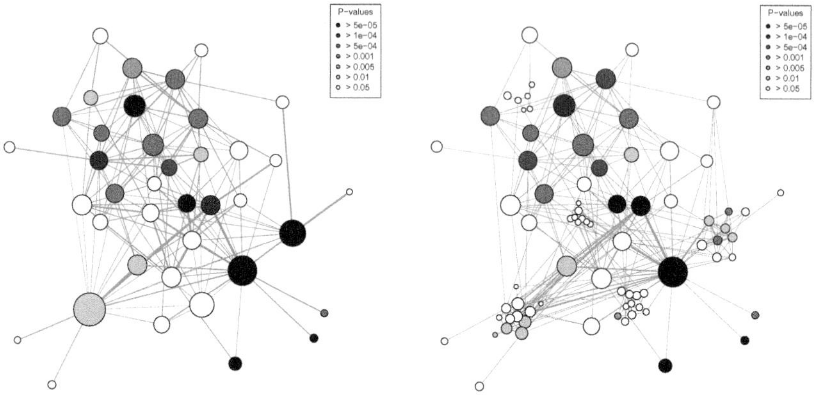

(a) Best clustering **(b)** Maximally refined clustering

Fig. 2. Clustered graph visualization

of the group is represented by the surface of the disk used in the figures). Two groups are linked in the display when there is at least one sexual partnership between patients of the two groups. The thickness of the link encodes the number of between group sexual contacts.

We implemented the hierarchical principle used in [4] by providing interactive coarsening and refining of the clustering. Indeed the best clustering of the network might be too coarse to give insights on the structure of network or too fine to lead to a legible drawing. Coarsening is implemented by a greedy merging of clusters (as is used in [12]) while refinement is obtained by applying maximal modularity clustering to each sub-cluster, taken in isolation from the rest of the network. We keep only statistically significant coarsenings and refinements: the modularity of the selected clusterings must be higher than the maximal modularity obtained on random graphs with the same degree distribution (see [3] for details). Figure 2 (b) gives an example of a refinement for the clustering used in Figure 2 (a), while Figure 3 is based on a coarsening of the clustering.

3.2 Results

Using [12], we obtain a partition of the giant component into 39 clusters, with a modularity of 0.85. This is significantly higher than the modularities of random graphs with identical sizes and degree distributions: the highest value among 50 random graphs is 0.74. The corresponding layout is given by Figure 2 (a). We use this layout as a support for visualization exploration of the sexual orientation distribution: nodes are darkened according to the p value of a chi squared test conducted on the distribution of the sexual orientation of persons in each cluster versus the distribution of the same variable in the full connected component. It appears clearly that some clusters have a specific distribution of the sexual orientation variable.

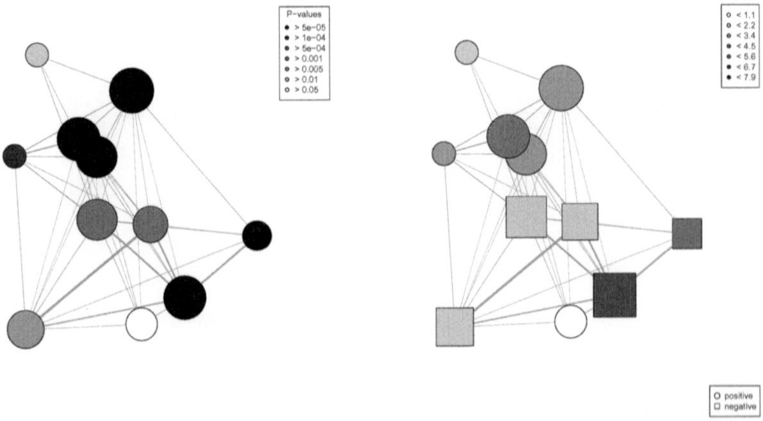

(a) Chi square P values **(b)** Pearson's residuals for MSM

Fig. 3. Coarsened clustered graph visualization

The possibilities for refining the clustering in this case are quite limited: only 5 of the 39 clusters have a significant substructure. Nevertheless, Figure 2 (b), which shows the fully refined graph (with modularity 0.81) gives interesting insights on the underlying graph. For instance, an upper left gray cluster is split into 6 white clusters: while the best clustering of those persons leads to an atypical sexual orientation distribution, this is not the case of each sub-cluster. This directs the analyst to a detailed study of the corresponding persons: it turns out that the cluster consists mainly in MSM patients. Sub-clusters are small enough ($\sim$ 7 patients) for MSM dominance to be possible by pure chance, while this is far less likely for the global cluster with 41 patients (among which 39 are MSM).

Coarsening can be done more aggressively on this graph: clusterings down to 8 clusters have modularity above the random level. With 11 clusters, the modularity reaches 0.81, a similar value as the maximally refined graph. While Figure 2 (a) is legible enough to allow direct analysis, the coarsening emphasizes the separation of the graph into two sparsely connected structures with mostly atypical sexual orientation distributions in the associated clusters, as shown in Figure 3 (a). Figure 3 (b) represents the Pearson's residuals of the chi square tests for the MSM sexual orientation: it clearly shows that a part of the largest connected component contains more than expected MSM (circle nodes) while the other part contains less than expected (square nodes).

This finding directs the analyst to a sub-population study. The original 39 clusters are merged into three groups: MSM clusters (atypical clusters in the upper part of the graph which contain more MSM than expected), Mixed clusters (atypical clusters in the lower part of the graph, which contain less MSM than expected) and typical clusters. Then the geodesic analysis summarized in Table 2 is done at this group level, leading to Table 3.

Table 3. Average geodesic distances between members of the three cluster groups. Paths are restricted to patients belonging to the groups under consideration

	MSM clusters	Mixed clusters	Typical clusters
MSM clusters	9.79	12.28	11.93
Mixed clusters	12.28	7.56	9.24
Typical clusters	11.93	9.24	12.04

This analysis shows that the two groups made of atypical clusters are far from each other compared to their internal distances. This is confirmed by the detection date analysis displayed on Figure 4. It appears that the epidemic in the giant component has two separated components. One mostly male homosexual component tends to dominate the recent cases (note that even typical clusters contain at least 57 % of MSM), while a mixed component with a large percentage of female patients was dominating the early epidemic, but tends to diminish recently. It should also be noted that this mix component is dominated by the growth of the homosexual component, but seems to decay only slightly in absolute terms. In other words, the reduction should be seen as an inability to control the growth homosexual epidemic rather than as a success in eradicating the heterosexual epidemic.

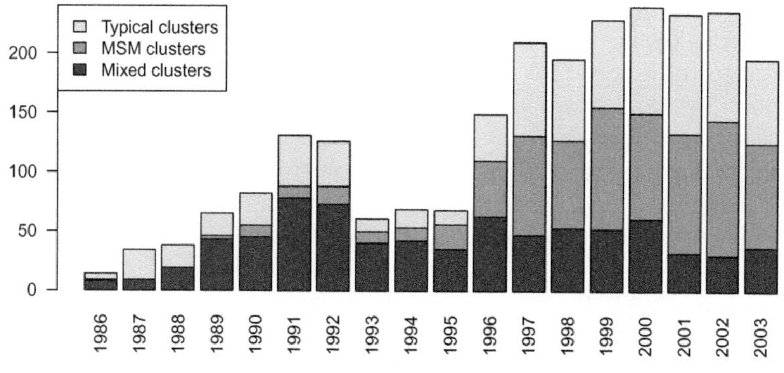

Fig. 4. Yearly distribution of the three groups of clusters

4 Conclusion

The proposed visual mining method for graphs has been shown to provide valuable insights on the epidemic network. It is based on links between modularity and visualization and leverages recent computationally efficient modularity maximizing methods. Future works include the integration of the proposed methods in graph mining tools such as [2] and its validation on other aspects of epidemic networks analysis.

References

1. de Arazoza, H., Joanes, J., Lounes, R., Legeai, C., Clémençon, S., Perez, J., Auvert, B.: The HIV/AIDS epidemic in Cuba: description and tentative explanation of its low prevalence. BMC Disease (2007)
2. Bastian, M., Heymann, S., Jacomy, M.: Gephi: An open source software for exploring and manipulating networks. In: International AAAI Conference on Weblogs and Social Media (2009)
3. Clémençon, S., De Arazoza, H., Rossi, F., Tran, V.C.: Hierarchical clustering for graph visualization. In: Proceedings of XVIIIth European Symposium on Artificial Neural Networks (ESANN 2011), Bruges, Belgique (April 2011) (to be published)
4. Eades, P., Feng, Q.W.: Multilevel visualization of clustered graphs. In: Proceedings of the Symposium on Graph Drawing, GD 1996, Berkeley, California, USA, pp. 101–112 (September 1996)
5. Fortunato, S.: Community detection in graphs. Physics Reports 486(3-5), 75–174 (2010)
6. Fruchterman, T.M., Reingold, E.M.: Graph drawing by force-directed placement. Software - Practice and Experience 21(11), 1129–1164 (1991)
7. Keeling, M.J., Eames, K.T.: Networks and epidemic models. Journal of The Royal Society Interface 2(4), 295–307 (2005)
8. Kwakwa, H.A., Ghobrial, M.W.: Female-to-female transmission of human immunodeficiency virus. Clinical infectious diseases: an official publication of the Infectious Diseases Society of America 36(3) (February 2003)
9. Liljeros, F., Edling, C.R., Nunes Amaral, L.A.: Sexual networks: implications for the transmission of sexually transmitted infections. Microbes and Infection 5(2), 189–196 (2003)
10. Newman, M.E.J.: The structure and function of complex networks. SIAM Review 45, 167–256 (2003)
11. Noack, A.: Modularity clustering is force-directed layout. Physical Review E 79(026102) (February 2009)
12. Noack, A., Rotta, R.: Multi-level algorithms for modularity clustering. In: Proceedings of the 8th International Symposium on Experimental Algorithms SEA 2009, pp. 257–268. Springer, Heidelberg (2009)
13. Rothenberg, R.B., Woodhouse, D.E., Potterat, J.J., Muth, S.Q., Darrow, W.W., Klovdahl, A.S.: Social networks in disease transmission: The colorado springs study. In: Needle, R.H., Coyle, S.L., Genser, S.G., Trotter II, R.T. (eds.) Social Networks, Drug Abuse, and HIV Transmission Research Monographs, National Institute on Drug Abuse, vol. 151, pp. 3–18 (1995); Research Monographs, National Institute on Drug Abuse
14. Varghese, B., Maher, J., Peterman, T., Branson, B., Steketee, R.: Reducing the risk of sexual hiv transmission: Quantifying the per-act risk for hiv on the basis of choice of partner, sex act, and condom use. Sexually Transmitted Diseases 29(1), 38–43 (2002)
15. Wylie, J.L., Jolly, A.: Patterns of chlamydia and gonorrhea infection in sexual networks in manitoba, canada. Sexually Transmitted Diseases 28(1), 14–24 (2001)

Towards User-Centric Memetic Algorithms: Experiences with the TSP

Ana Reyes Badillo, Carlos Cotta, and Antonio J. Fernández-Leiva

Dept. Lenguajes y Ciencias de la Computación, ETSI Informática,
Campus de Teatinos, Universidad de Málaga,
29071 Málaga – Spain
{ccottap,afdez}@lcc.uma.es

Abstract. User-centric evolutionary computation is an optimization paradigm that tries to integrate the human user and the evolutionary algorithm in a smooth way, favoring bi-directional communication and establishing synergies among these two actors. We explore the possibilities for such an approach in the context of memetic algorithms, with application to the travelling salesman problem. Some ways to canalize this cooperation via the introduction of dynamic constraints and selective local search are hinted, and implementation and interfacing issues are discussed. The reported experiments on TSPLIB instances provide encouraging results for these techniques.

1 Introduction

One of the lessons learned in the last years in the metaheuristics community, and most prominently in the area of evolutionary computation (EC), is the need of exploiting problem knowledge in order to come up with effective optimization tools [1,2,3]. This problem-knowledge can be provide in a variety of ways: ad hoc representations, specialized operators, or combination with other problem-specific techniques, just to cite a few [4]. There are however situations in which endowing the optimization algorithm with this knowledge is a much more elusive task. This may be the case when this problem-awareness is hard to encapsulate within a specific algorithmic description, e.g., they belong more to the space of human-expert's intuition than elsewhere. An extreme case of this situation can take place when the evaluation itself of solutions is not algorithmic, but needs the introduction of a human to critically assess the quality of solutions, e.g., see [5]. Other forms of interaction are possible though.

The above use of a combined human-user/evolutionary-algorithm approach is commonly termed interactive evolutionary computation (IEC) [6,7]. The term user-centric evolutionary computation [8] is however more appropriate since it hints possibilities for the system to be proactive rather than merely interactive, i.e., to anticipate some of the user behavior and/or exhibit some degree of creativity. Granted, such features constitute ambitious goals that require a good grasp of the basic underlying issues surrounding interactive optimization. In this sense, we believe that –while several EC flavors have been already tackled from

J. Cabestany, I. Rojas, and G. Joya (Eds.): IWANN 2011, Part II, LNCS 6692, pp. 284–291, 2011.

the point of view of IEC– full exploration of this topic is already required in the area of memetic algorithms (MAs) [9,10,11].

MAs constitute a metaheuristic optimization paradigm based on the systematic exploitation of knowledge about the problem being solved, and the synergistic combination of ideas taken from other population-based and trajectory-based metaheuristics. In many respects, MAs are a natural paradigm to incorporate problem-specific components, and IEC is no exception. Actually, some works have already highlighted the benefits attainable via the use of the human-user interaction with the MA, in particular in the context of multi-objective optimization [12,13]. We explore some of these capabilities in this work, focusing in particular in the dynamic management of user-defined constraints, and in user-controlled local search.

2 A User-Centric Approach to Memetic Algorithms

As already mentioned in Sect. 1, memetic algorithms (MAs) are particularly suited to integrate different sources of problem-knowledge into a single optimization tool. We refer to [14] for an up-to-date review of the state-of-the-art in MAs. In the following we shall describe how we have integrated user-centric capabilities in MAs.

2.1 Rationale

Some of the of the most common themes in IEC are using a human-expert to provide subjective evaluation information, or to perform subjective selection of solutions for breeding, among many others. We defer to [7] for an overview of the area. One of the recurring issues in this context is dealing with human fatigue, i.e., coping with the fact that the human expert cannot be forced to provide a continuous supply of information, and hence the search algorithm has to exhibit a degree of autonomy. This is particularly feasible in domains in which some objective optimization measure is already available, and therefore the human expert is a source on knowledge that can improve results, but is not necessarily required for obtaining some solutions (even if just low-quality ones). In this sense, we adhere to this vision of having an human expert overseeing the evolution of resolution process, and providing hints [15] on which directions the search should proceed but only sporadically (and asynchronously if possible).

More precisely, we have considered three particular ways to put the user in the loop, biasing the search dynamics:

– Allowing her to change dynamically some parameters of the algorithm, including the application probability and choice of operators (in order to change the way solutions are generated and thus direct the exploration process). Note in this sense that there are many works focusing in

self-parameterization of evolutionary algorithms [16]. Thus, the human expert would here act as a high-level controller that would exert direct control of these parameters, or supervise the procedure of self-adaptation, superseding the latter if necessary.
- Allowing her to provide search bias via the dynamic introduction (and removal) of additional constraints, i.e., constraints that are not a part of the problem definition, but are forced by the user in order to drive the search towards-to/away-from specific regions of solution space. Such constraints are handled as soft-constraints, i.e., their violation results in a penalty term being added to the raw fitness of solutions.
- Allowing her to selectively use local-search add-ons. This is particularly relevant in the case of MAs, in which several studies exist focusing on which solutions should undergo local improvement, and how this local improvement should be done (i.e., which local search –LS– operator to use, how intense this local improvement has to be, etc.) – e.g., see [17,18]. Allowing the user to interfere in this regard allows further possibilities such as applying local-improvement just to particular portions of solutions rather than undergoing a full-fledged local optimization.

Next section will describe how we have accommodated the above capabilities in a memetic solver for the Traveling Salesman Problem (TSP).

2.2 Implementation and Management of User Input: The TSP Case

We have built a prototype of user-centric MA on the basis of the ECJ library[1]. ECJ is an evolutionary computation framework written in java available under the Academic Free License (AFL) version 3.0, and it has been chosen due to its high flexibility and modularity among other reasons. Our implementation comprises problem-specific classes (corresponding to the representation of solutions and variation operators used) and interaction-specific classes (providing the functionality for supplying information to the user and accepting feedback from her). Among the latter we can cite:

- Output: this class has been modified in order to allow the user select specific actions, e.g., modify parameters, introduce constraints, etc.
- VectorSpecies: a derived class PermutationVectorSpecies has been defined for the TSP in order to store problem-specific parameters and dynamic constraints.
- Statistics: a derived class from the former is responsible for controlling when user interaction takes place. In this prototype we have opted for two interaction possibilities: a pre-scheduled mechanism (interacting every certain number of generations; this is dynamically reconfigurable by the user, who can effectively set up when the next interaction will take place), and a trigger mechanism (interacting when the algorithms fulfills some condition, i.e., diversity drops below a certain threshold).

[1] `http://www.cs.gmu.edu/~eclab/projects/ecj/`

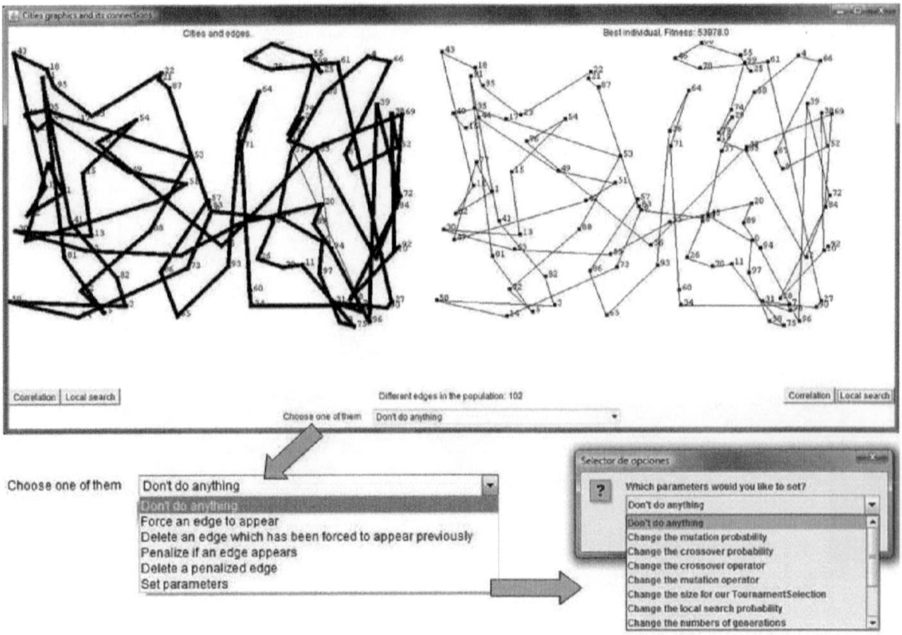

Fig. 1. General depiction of the user interface for interacting with the memetic solver in the context of the TSP

- Canvas: several problem-specific classes are derived from the latter in order to provide the means to display sensible information to the user.

The latter aspect is particularly important if the interaction with the user is to be fruitful. The user needs being provided with relevant (yet not overwhelming) information upon which to base her decisions on the course the search has to take. In this sense, the TSP has been chosen as test-suite precisely because of its amenability for graphical depiction, and intuitive visual nature. Fig. 1 shows the basic interface. The left panel provides a description of the population: a graph is built by merging all tours in the population, subsequently, it is drawn making edge-width be proportional to the frequency of that edge in the population. As to the right panel, it provides a description of the best solution found and its quality. At the bottom, a drop-down menu provides the user a list of available actions (some of which can in turn result in additional lists of options and/or text inputs). An important feature is the possibility of selectively applying local-improvement to a specific portion of a solution. This is shown in Fig. 2. As it can be seen, the user can select a subset of the solution upon which 2-opt local search will be applied (i.e., only edges adjacent to selected cities can be modified).

Our prototype is available under the same license –AFL v3.0– as ECJ. It can be downloaded from `http://nemesis.lcc.uma.es/?page_id=17`

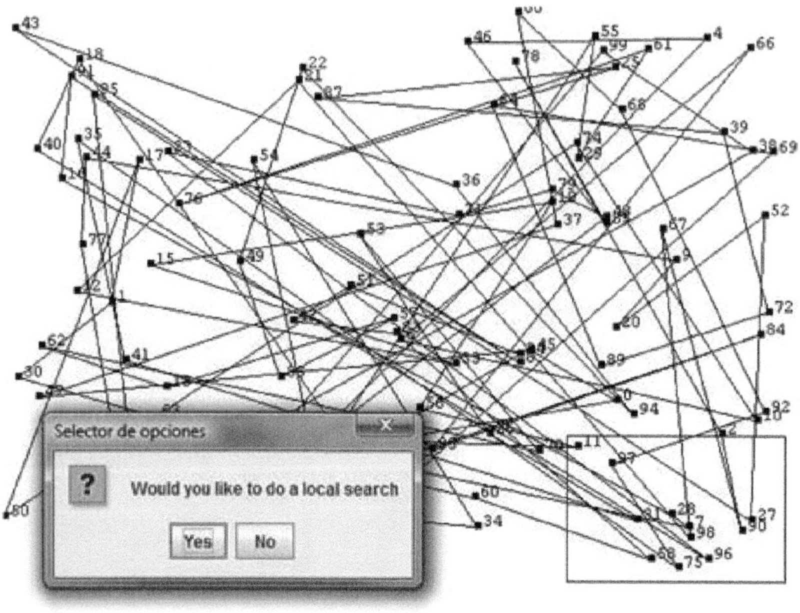

Fig. 2. The user can control the application of local search to specific portions of the current best solution

3 Experiments

The experiments have been done using an elitist evolutionary algorithm (*popsize* = 100, binary tournament selection) with edge-recombination crossover ($p_X = 1.0$), and subtour-inversion mutation ($p_M = 0.005$). Two TSP instances

Table 1. User interaction in the `kroA100` instance

# interactions	Action performed
1	forbid $\langle 15-50 \rangle$, $\langle 25-65 \rangle$, $\langle 4-72 \rangle$ and $\langle 43-68 \rangle$
2	forbid $\langle 43-79 \rangle$, $\langle 14-89 \rangle$ and $\langle 62-73 \rangle$
	2-opt LS in the bottom right corner
4	forbid $\langle 65-98 \rangle$, $\langle 50-56 \rangle$ and $\langle 50-60 \rangle$
	forbid $\langle 21-82 \rangle$, $\langle 22-68 \rangle$ and $\langle 22-48 \rangle$
	forbid $\langle 13-50 \rangle$, $\langle 64-82 \rangle$ and 2-opt LS in the bottom right corner
	forbid $\langle 57-62 \rangle$ and 2-opt LS in the top left corner
8	forbid $\langle 14-30 \rangle$, $\langle 13-46 \rangle$ and $\langle 18-61 \rangle$
	forbid $\langle 3-50 \rangle$ and $\langle 43-54 \rangle$
	forbid $\langle 23-71 \rangle$ and $\langle 55-71 \rangle$
	forbid $\langle 17-47 \rangle$
	2-opt LS in the bottom right corner
	2-opt LS in the top left corner
	2-opt LS in the top right corner
	2-opt LS in the bottom left corner

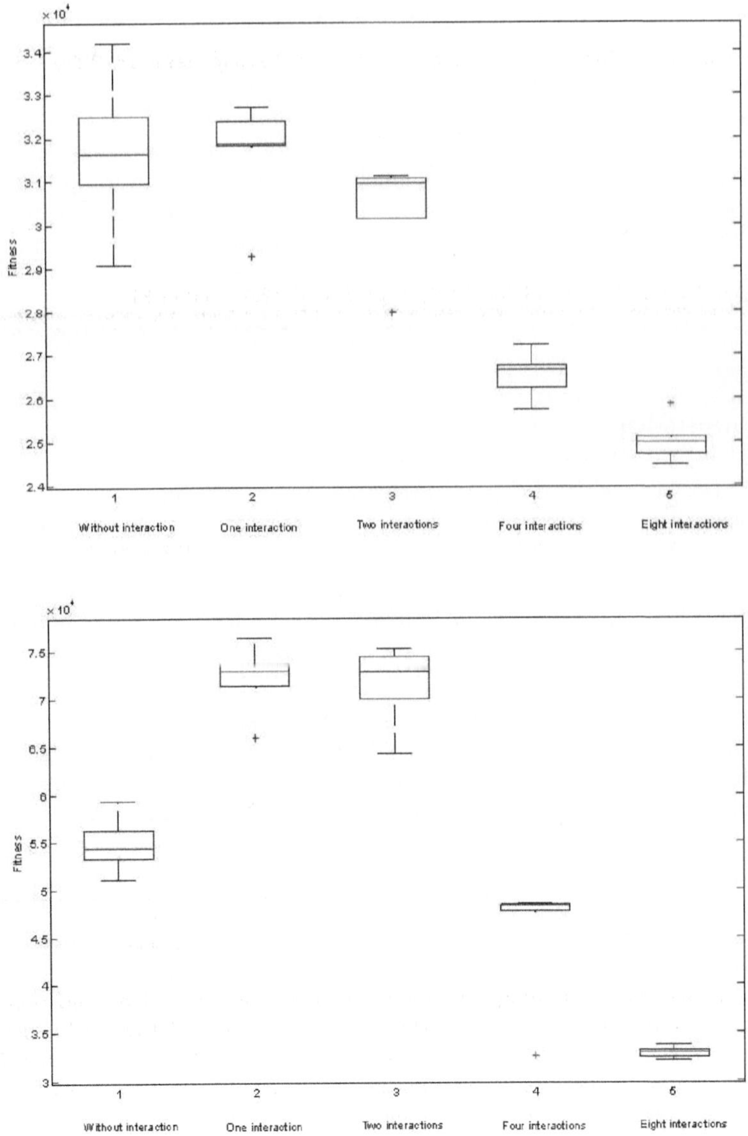

Fig. 3. Results obtained by interactive and non-interactive algorithms on the `kroA100` instance (top) and on the `kroA200` instance (bottom)

from the TSPLIB[2], namely `kroA100` and `kroA200` have been used. In order to obtain baseline results, 20 runs of the algorithm have been done without user interaction. Subsequently, we have done single runs with 1, 2, 4 and 8

[2] `http://comopt.ifi.uni-heidelberg.de/software/TSPLIB95/`

user-interactions. These interactions have been logged (specific actions and time at which they are done), and are subsequently replicated in automatic runs of the algorithm in order to determine their general goodness. Table 1 shows an example of the kind of actions performed on the `kroA100` instance.

The results are shown in Fig. 3. Notice how in the case of the `kroA100` instance the results are better for an increasing number of iterations, mostly due to the selective application of LS (which is much less expensive than a full-fledged LS, and whose cost is already accounted in the total computational budget). In the case of the `kroA200` such improvement is only attained for a larger number of interactions (which is where LS is effectively deployed). Except in `kroA100` and 1 interaction, in all cases the differences with respect to the autonomous algorithm are statistically significant at 5% level using a Wilcoxon ranksum test.

4 Conclusions

User-centric EC is an thriving research topic in the confluence of areas such as metaheuristic optimization and machine learning. Paving the way for further extensions, we have conducted in this work a study on the deployment of interactive capabilities in a memetic algorithm, with application to the TSP. The results have been encouraging, since it has been shown that even some forms of limited interaction are capable of improving the results of a baseline autonomous algorithm. While the computational scenario is not a tough one, these results indicate that these techniques are capable of profiting from adequately-crafted human feedback, not merely as a carrier of subjective information but as a source of problem-aware perturbations that can drive/focus the algorithm towards specific regions of the search space. At any rate, much remains to be done. As mentioned before, IEC is merely the tip of the iceberg; full-fledged user-centric optimization will also imply proactivity in the search heuristic, anticipating the needs of the user, or trying to follow her preferences in order to provide hints in the direction she is headed to. We are currently working on some related user-modelling areas in the context of videogames, from which some common lessons can be learned. Additionally, we are approaching other combinatorial problems, and trying to incorporate the user in a fully asynchronous way.

Acknowledgements. This work is supported by project TIN-2008-05941 of the Spanish MICINN, and project TIC-6083 of Junta de Andalucía.

References

1. Hart, W.E., Belew, R.K.: Optimizing an arbitrary function is hard for the genetic algorithm. In: Belew, R.K., Booker, L.B. (eds.) 4th International Conference on Genetic Algorithms, pp. 190–195. Morgan Kaufmann, San Francisco (1991)
2. Wolpert, D.H., Macready, W.G.: No free lunch theorems for optimization. IEEE Transactions on Evolutionary Computation 1(1), 67–82 (1997)

3. Bonissone, P., Subbu, R., Eklund, N., Kiehl, T.: Evolutionary algorithms + domain knowledge = real-world evolutionary computation. IEEE Transactions on Evolutionary Computation 10(3), 256–280 (2006)
4. Davis, L.D.: Handbook of Genetic Algorithms. Van Nostrand Reinhold Computer Library, New York (1991)
5. Herdy, M.: Evolutionary optimisation based on subjective selection – evolving blends of coffee. In: 5th European Congress on Intelligent Techniques and Soft Computing, pp. 640–644 (1997)
6. Banzhaf, W.: Interactive evolution. In: Back, T., Fogel, D., Michalewicz, Z. (eds.) Evolutionary Computation 1: basic algorithms and operators, pp. 228–234. IoP, Bristol (2000)
7. Takagi, H.: Interactive evolutionary computation: Fusion of the capabilities of EC optimization and human evaluation. Proceedings of the IEEE (9), 1275–1296 (2001)
8. Parmee, I.C., Abraham, J.A.R., Machwe, A.: User-centric evolutionary computing: Melding human and machine capability to satisfy multiple criteria. In: Knowles, J., Corne, D., Deb, K., Chair, D.R. (eds.) Multiobjective Problem Solving from Nature. Natural Computing Series, pp. 263–283. Springer, Berlin Heidelberg (2008)
9. Moscato, P.: On Evolution, Search, Optimization, Genetic Algorithms and Martial Arts: Towards Memetic Algorithms. Technical Report Caltech Concurrent Computation Program, Report. 826, California Institute of Technology, Pasadena, California, USA (1989)
10. Hart, W., Krasnogor, N., Smith, J.E.: Recent Advances in Memetic Algorithms. STUDFUZZ, vol. 166. Springer, Heidelberg (2005)
11. Moscato, P., Cotta, C.: A gentle introduction to memetic algorithms. In: Glover, F., Kochenberger, G. (eds.) Handbook of Metaheuristics, pp. 105–144. Kluwer Academic Publishers, Boston (2003)
12. Dias, J., Captivo, M., Clímaco, J.: A memetic algorithm for multi-objective dynamic location problems. Journal of Global Optimization 42, 221–253 (2008)
13. Jaszkiewicz, A.: Interactive multiple objective optimization with the pareto memetic algorithm. In: Gottlieb, J., et al. (eds.) 4th EU/ME Workshop: Design and Evaluation of Advanced Hybrid Meta-heuristics, Nottingham, UK (2004)
14. Moscato, P., Cotta, C.: A modern introduction to memetic algorithms. In: Gendreau, M., Potvin, J.-Y. (eds.) Handbook of Metaheuristics, 2nd edn. International Series in Operations Research and Management Science, vol. 146, pp. 141–183. Springer, Heidelberg (2010)
15. Abu-Mostafa, Y.S.: Hints and the VC dimension. Neural Computation 5, 278–288 (1993)
16. Smith, J.E.: Self-adaptation in evolutionary algorithms for combinatorial optimisation. In: Cotta, C., Sevaux, M., Sörensen, K. (eds.) Adaptive and Multilevel Metaheuristics. SCI, vol. 136, pp. 31–57. Springer, Heidelberg (2008)
17. Ong, Y.S., Keane, A.: Meta-lamarckian learning in memetic algorithms. IEEE Transactions on Evolutionary Computation 8(2), 99–110 (2004)
18. Ong, Y., Lim, M., Zhu, N., Wong, K.: Classification of adaptive memetic algorithms: a comparative study. IEEE Transactions on Systems, Man, and Cybernetics, Part B 36(1), 141–152 (2006)

A Multi-objective Approach for the 2D Guillotine Cutting Stock Problem

Jesica de Armas, Gara Miranda, and Coromoto León

Universidad de La Laguna, Dpto. Estadística, I. O. y Computación,
Avda. Astrofísico Fco. Sánchez s/n, 38271 La Laguna, Spain
{jdearmas,gmiranda,cleon}@ull.es

Abstract. This work presents a multi-objective approach to solve the Constrained 2D Cutting Stock Problem. The problem targets the cutting of a large rectangle of fixed dimensions in a set of smaller rectangles using orthogonal guillotine cuts. Although the problem is usually focused on a single objective, in this work we want to optimise the layout of rectangular parts on the sheet of raw material so as to maximise the total profit, as well as minimise the number of cuts to achieve the final demanded pieces. For this, we apply Multi-Objective Evolutionary Algorithms given its great effectiveness when dealing with other types real-world multi-objective problems. For the problem solution, we have implemented an encoding scheme which uses a post-fix notation. According to the two different optimisation criteria the approach provides a set of solutions offering a range of trade-offs between the two objectives, from which clients can choose according to their needs.

Keywords: Cutting Stock Problems, Multi-objective Optimisation, Evolutionary Algorithms.

1 Introduction

Cutting Stock Problems (CSPs) arise in many production industries where large stock sheets (glass, textiles, pulp and paper, steel, etc.) must be cut into smaller pieces [1]. Here we have focused on a general guillotine problem which does not introduce constraints about the number of cutting stages. The studied problem is named Constrained Two-Dimensional Cutting Stock Problem (2DCSP). It targets the cutting of a large rectangle of fixed dimensions in a set of smaller rectangles using orthogonal guillotine cuts. That means that any cut must run from one side of the rectangle to the other end and be parallel to the other two edges (Fig. 1). This is possible only generating vertical or horizontal builds of pieces [2]. The produced rectangles must belong to one of a given set of rectangle types. Associated with each type of rectangle there is a profit and a demand constraint. Usually, the main goal is to find a feasible cutting pattern maximising the total profit, and therefore minimising the total trim loss when pieces have profit proportional to their area. However, in some industrial fields, the raw material is either very cheap or can be easily recycled, so in such cases, a more

J. Cabestany, I. Rojas, and G. Joya (Eds.): IWANN 2011, Part II, LNCS 6692, pp. 292–299, 2011.

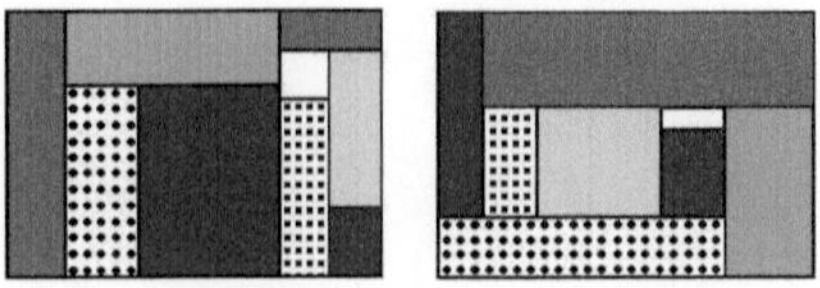

Fig. 1. Guillotine and non-guillotine cuts

important criterion for the pattern generation may be the speed at which the pieces can be obtained, thus minimising the production times and maximising the usage of the cutting equipment. This cutting process is specifically limited by the features of the machinery available but, in general, it is determined by the number of cuts involved in the packing pattern. Moreover, the number of cuts required for the cutting process is also crucial to the life of the industrial machines. Therefore, in this study, the number of cuts is taken as a second design objective. This way, the problem can be posed as a *multi-objective optimisation problem* (MOP) for optimising the layout of rectangular parts so as to maximise the total profit as well as minimise the number of cuts to achieve the final demanded pieces. MOPs [3] arise in most real-world disciplines where different and usually contrary objectives must be simultaneously optimised. In the 2DCSP, the maximisation of the total profit implies a better usage of the raw material. This usually involves compact cutting patterns containing little internal trim loss. In most cases, filling all these gaps implicitly produces a higher number of cuts. So, we can state that in general, for this problem, we will obtain a set of non-dominated solutions instead of a single and optimal solution.

A large number of exact algorithms [2,4,5,6] and heuristics [7,8] have been proposed to solve the single-objective formulation of the problem. Exact algorithms are based on post-fix notations and allow to deal with the complete solution space. However, the existing heuristics deal with a reduced part of the solution space and thus, they don't guarantee the achievement of the optimal solution. On the other hand, works dealing with such a multi-objective formulation of the problem are unknown. Some previous works [9,10] prove the effectiveness of multi-objective evolutionary algorithms (MOEAs) when applied to other kind of cutting problems. For this reason, we have developed an approach which applies MOEAs and uses a codification of solutions based on a post-fix notation, which try to maximize the total profit as well as minimize the number of cuts to achieve the final demanded pieces. So we obtain solutions which take into account both criteria independently, and then clients choose according to their needs.

The remaining content of this paper is organised as follows. In section 2, we present the approach designed to deal with the multi-objective 2DCSP. The experimental results of this approach are presented in section 3. Finally, the conclusions and some lines of future work are given in section 4.

2 Multi-objective Approach

The approach was evolved using three different MOEAs: NSGA-II, SPEA2, and an adaptive version of IBEA. Making use of the hypervolume and ϵ-indicator,

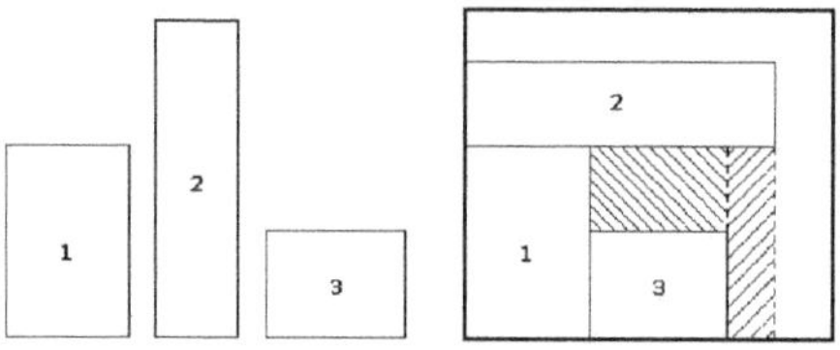

Fig. 2. Layout on the mother sheet for the chromosome '1 3 H 2 V'

NSGA-II showed a better behaviour than the other algorithm alternatives, as in previous cutting related works [9,10]. For this reason we have focused on results obtained using NSGA-II.

2.1 Representation

A post-fix notation is used to represent the candidate solutions. The operands are the piece identifiers, while the operators are 'V' and 'H' (Fig. 2). The operator 'H' concatenates its two operands in the horizontal, while the 'V' operator concatenates them in the vertical. If the problem width or length constraints are violated, the corresponding operator behaves like the opposing operator. Using such a representation based on vertical and horizontal composition of pieces, all the layouts obtained can be cut in guillotine mode [2]. In order to constitute a valid chromosome, a piece of type T_i can't appear more than b_i times. Moreover, for any operator, if the number of pieces to its left is n_p and the number of operators to its left plus itself is n_o, then the following condition must hold: $1 \leq n_o \leq n_p - 1$. Using such a representation, the chromosome size remains constant when all the available pieces are placed on the surface and no parentheses are required to uniquely represent a solution. For the generation of the initial individuals, a random order of the pieces is established and an uniform probability is applied to determine the operators. Each individual is created until the application of an operator provides a combination of pieces which does not fit on the sheet of raw material, i.e., the chromosome does not satisfy the mother sheet width and length constraints. Then, the last operator is exchanged. If even though the combination of pieces does not fit in the material, the last piece is exchanged with the following piece. If it still does not fit, the last operator is changed again. This process is repeated until a valid solution is obtained or until a maximum number of changes had been applied. When no valid solution is reached, the method is applied again with the previous piece in the chromosome. Finally, if this procedure does not work, the chromosome is cut in the right size, and a filling operator is applied trying to fill the remaining space by adding pieces vertically or horizontally at the end.

2.2 Evaluation of the Objectives

The chosen codification gives information on how pieces must be combined or placed on the raw material. Based on this information, both optimisation objectives considered - the total profit and the number of necessary cuts - can be evaluated. For this purpose, the methods applied here are based on the usage of stacks and the post-fix notation, which represents the chromosome [11].

For the evaluation of the second objective - the number of cuts required - an iterative method is applied. The chromosome is traversed from left to right, interpreting every element and creating the indicated constructions, thus calculating the partial widths, lengths, and profit. At least one cut is necessary for each implied vertical or horizontal combination of pieces. If the combined rectangles do not match in length (for vertical builds) or in width (for horizontal builds), an extra cut is required for the construction. At the end of the process, the complete final pattern is obtained. In this case, the value of the first objective - total profit - is immediately given by the profit of the resulting final pattern.

2.3 Operators

As we have used a codification implicitly representing solutions to the problem, the type of operators to be applied must deal with the problem specific features. We have proved several crossover operators, and finally, the *Partially Mapped Crossover* (PMX) [12] was selected, given its better behaviour. The technique is based on the recombination of two chromosomes chains where only the information of the pieces is considered, i.e. the operators are not taken into account for the application of this operator. Considering this type of chain, first, two crossing points inside each of the given parents are randomly chosen. Then, the segments of the parents inside the crosspoints are swapped in order to generate the offspring. The remaining chains in the offspring are obtained by mapping between the two parents. If a chromosome value outside the swapped segment is not contained in the swapped segment, it remains the same, but if it is already contained, it must be replaced by a value contained in the original segment of the chromosome but not contained in the new segment under consideration.

The mutation applied [10,11] operates as follows. First, two chromosome elements, p_1 and p_2, are picked at random. Both elements represent piece numbers or operators and p_1 is closer to the left of the chromosome. If both are piece numbers or operators, or p_1 is an operator and p_2 is a piece, they are swapped. If p_1 is a piece number and p_2 is an operator, they are swapped only when, after performing the swap, condition $1 \leq n_o \leq n_p - 1$ still holds for any operator. Finally, an operator of the chromosome is randomly chosen and flipped based upon the mutation probability.

After applying each operator, crossover and mutation, a repair operator is used to ensure that only new valid chromosomes are generated. This operator cuts the chromosome in the right size. Moreover, depending on a probability, the chromosome is traversed from left to right exchanging a piece for another unused and an operator for the other one, checking if the combination of pieces fits in the material and provides better profit or less number of cuts than the original chromosome. In this case, the original chromosome is replaced with the new improved one. Lastly, a filling operator is applied trying to fill the remaining space by adding pieces vertically or horizontally at the end.

3 Computational Results

The experimental evaluation was performed on a dedicated Debian GNU/Linux cluster of 20 dual-core nodes. Each node consists of two Intel® Xeon 2.66 Ghz and has 1Gb RAM and a Gigabit Ethernet interconnection network. The framework and the approach for the problem were implemented in C++ and compiled with *gcc 4.1.3* and MPICH *1.2.7*. For the computational study, some test instances available in the literature [13,14] have been used. These test instances indicate the number of different pieces, the dimensions of the raw material, and, for each piece, its length, its width, its profit (proportional to its area), and the number of pieces of its kind.

We have defined an individual for the solution of the 2DCSP: a direct codification based on post-fix representation of pattern layouts For this approach, we defined the corresponding representation and implemented the evaluation, generation, and operator methods involved. The approach was evolved using NSGA-II, which has showed a better behaviour than the other algorithm alternatives - as in previous cutting related works [9,10] - and the parameter were fixed to the following values: crossover probability = 0.7, mutation probability = 0.3, and population size = 50. So, for all the following experiments we have applied this evolutionary algorithm with these parameters.

In previous works [9] we have proved the effectiveness of applying MOEAs to others cutting problems, providing quality solutions in acceptable computational time. So, now we want to know if we can obtain feasible solutions when we considering both objectives - total profit and total number of needed cuts - in this 2DCSP. To start with the testing of the multi-objective approach, thirty repetitions of ten minutes each were performed for each test problem. If we want to identify the search space areas being explored by this multi-objective approach, directly plotting Pareto fronts could be rather messy since we are dealing with the results of thirty executions, so as an alternative we have used the summary attainment surfaces [15]. Fig. 3 shows the summary attainment surface 1, 15, and 30, for four different instances of the problem. As we can see, the attainment surfaces are uniformly distributed, and cover a large area of the solution space. Besides, two different optimisation criteria has a major advantage for potential customers: the multi-objective approach provide a set of solutions offering a range of trade-offs between the two objectives, from which clients can choose according to their needs, e.g. cost associated with the raw material or even times imposed for the production process. So, we can choose solutions from those which minimise the number of cuts and have an associated low profit, to those which maximise the profit and have an associated higher number of cuts, depending on the particular problem at hand.

Moreover, the single-objective optimal solutions (maximisation of total profit) are known in the literature for the selected instances and using them, we can evaluate the number of cuts required to obtain the pieces. A solution indicates the way in which pieces are placed on the material, so any solution gives the profit and the number of cuts implicitly, whether the problem has been solved using single-objective techniques. These results have been taken as a reference to measure the

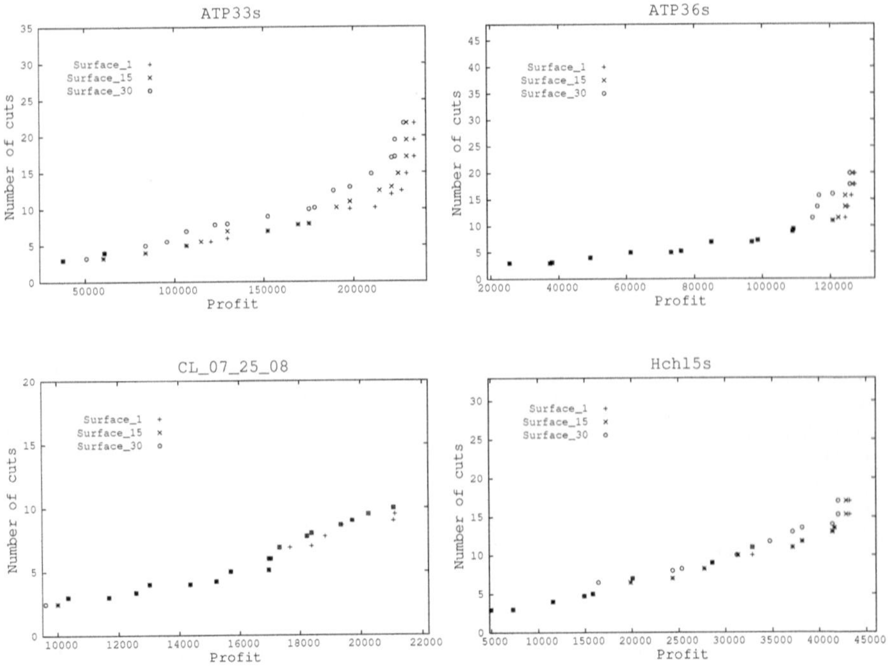

Fig. 8. Attainment surfaces for the multi-objective approach

Table 1. Comparison of single and multi-objective approaches

	Single-objective solutions		Multi-objective solutions	
Instance	**Profit**	**Cuts**	**Profit**	**Cuts**
ATP33s	236611	34	230072.23	17.20
ATP36s	130744	45	126703.50	17.46
CW6	12923	26	11780.76	17.70
CL_07_100_08	22443	30	21721.50	7.73
Hchl2	9954	21	9125.73	16.13
Hchl5s	45410	31	42603.16	15.10
CL_07_50_09	22088	14	21752.23	5.63
CL_07_25_08	21915	18	21134.96	10.00

quality of our multi-objective solution. In Table 1 we present, for each test instance, the optimal single-objective solution and two solutions obtained by the multi-objective approach: the solution with best (average) total profit and the solution with best (average) number of cuts. If we compare the average values for the best profits of the multi-objective approach with the single-objective optimal profit values, we realise that the multi-objective approach is not able to reach the optimal profit values. However, it provides profit values rather close to the optimal profit, but involving a considerable lower number of cuts. The single-objective solutions involve a quite small increase in profit but an important difference when considering the second objective (the number of cuts). For example, to achieve a 3% of growth in the total profit, the single-objective solution has increased by

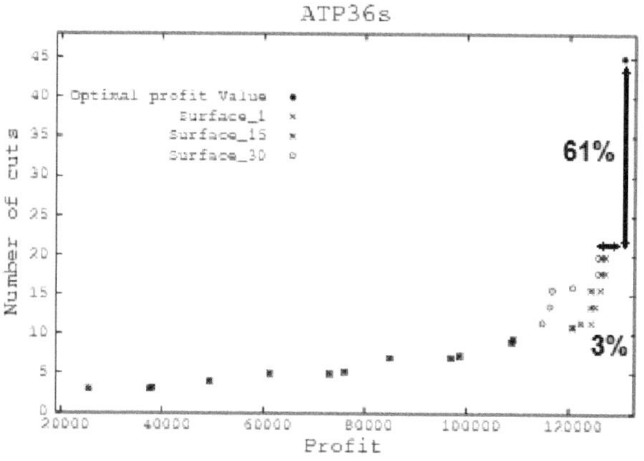

Fig. 4. Attainment surfaces and single-objective solution percentages

61% the number of cuts necessary to generate the pieces placed on the raw material (Fig. 4). It means that, although the optimum profit is not achieved, the proposed multi-objective approach provides a set of solutions with good compromise between the two objectives, thus allowing to offer to the clients a wide range of solutions from which they can choose according to their needs.

4 Conclusions

In this work we have presented a multi-objective approach to solve the Constrained 2D Cutting Stock Problem. The problem goal is to optimise the layout of rectangular parts so as to maximise the total profit as well as minimise the number of cuts to achieve the final demanded pieces. In the literature, works dealing with such a multi-objective formulation of the problem are unknown. For this purpose, we have selected the NSGA-II algorithm and an encoding scheme which is based on a post-fix notation. The obtained results demonstrate the great effectiveness of MOEAs when applied to such kind of problems. According to two different optimisation criteria, the implemented approach provides a set of solutions offering a range of trade-offs between the two objectives, from which clients can choose according to their needs. Although the multi-objective approach doesn't reach the profit values provided by the single-objective method, the obtained solutions are very close to such values and involve quite lower values for the other objective (the number of cuts). This way, we have designed an approach which provides a wide range of solutions with a fair compromise between the two objectives. Moreover, we have achieved good quality solution without having to implement an exact algorithm which involves an important associated difficulty and cost, and is just focused on one objective without considering the possible negative effects on other features of the solutions. As future work, it would be interesting to test the behaviour of other kind of encoding schemes. For example, it would be interesting to check some kind of hyperheuristic-based encoding schemes.

Acknowledgements

This work was funded by the EC (FEDER) and the Spanish Ministry of Science and Technology as part of the 'Plan Nacional de I+D+i' (TIN2008-06491-C04-02). The Canary Government has also funded this work through the PI2007/015 research project. The work of Jesica de Armas was funded by grant FPU-AP2007-02414.

References

1. Wäscher, G., Haußner, H., Schumann, H.: An improved typology of cutting and packing problems. European Journal of Operational Research 183(3), 1109–1130 (2007)
2. Wang, P.Y.: Two Algorithms for Constrained Two-Dimensional Cutting Stock Problems. Operations Research 31(3), 573–586 (1983)
3. Steuer, R.E.: Multiple Criteria Optimization: Theory, Computation and Application. John Wiley, New York (1986)
4. Viswanathan, K.V., Bagchi, A.: Best-First Search Methods for Constrained Two-Dimensional Cutting Stock Problems. Operations Research 41(4), 768–776 (1993)
5. Hifi, M.: An Improvement of Viswanathan and Bagchi's Exact Algorithm for Constrained Two-Dimensional Cutting Stock. Computer Operations Research 24(8), 727–736 (1997)
6. Cung, V.D., Hifi, M., Le-Cun, B.: Constrained Two-Dimensional Cutting Stock Problems: A Best-First Branch-and-Bound Algorithm. Technical Report 97/020, Laboratoire PRiSM, Université de Versailles (1997)
7. Burke, E.K., Kendall, G., Whitwell, G.: A New Placement Heuristic for the Orthogonal Stock-Cutting Problem. Operations Research 52(4), 655–671 (2004)
8. Ntene, N., Van Vuuren, J.: A survey and comparison of guillotine heuristics for the 2D oriented offline strip packing problem. Discrete Optimization 6(2), 174–188 (2009)
9. de Armas, J., Miranda, G., Leon, C., Segura, C.: Optimisation of a Multi-Objective Two-Dimensional Strip Packing Problem based on Evolutionary Algorithms. International Journal of Production Research 48(7), 2011–2028 (2009)
10. Tiwari, S., Chakraborti, N.: Multi-objective optimization of a two-dimensional cutting problem using genetic algorithms. Journal of Materials Processing Technology 173, 384–393 (2006)
11. Ono, T., Ikeda, T.: Optimization of two-dimensional guillotine cutting by genetic algorithms. In: Zimmermann, H.J. (ed.) European Congress on Intelligent Techniques and Soft Computing, vol. 1, pp. 7–10 (1998)
12. Goldberg, D.E., Lingle, J.R.: Allelesloci and the traveling salesman problem. In: Proceedings of the 1st International Conference on Genetic Algorithms, pp. 154–159. Lawrence Erlbaum Associates, Inc., Mahwah (1985)
13. DEIS - Operations Research Group: Library of Instances: Bin Packing Problem, `http://www.or.deis.unibo.it/research_pages/ORinstances/2CBP.html`
14. Hifi, M.: 2D Cutting Stock Problem Instances, `ftp://cermsem.univ-paris1.fr/pub/CERMSEM/hifi/2Dcutting/`
15. Knowles, J.: A summary-attainment-surface plotting method for visualizing the performance of stochastic multiobjective optimizers. In: Proceedings of the 5th International Conference on Intelligent Systems Design and Applications, pp. 552–557. IEEE Computer Society, Los Alamitos (2005)

Ant Colony Optimization for Water Distribution Network Design: A Comparative Study

C. Gil[1], R. Baños[1], J. Ortega[2], A.L. Márquez[1], A. Fernández[1], and M.G. Montoya[1]

[1] Dept. Arquitectura de Computadores y Electrónica, Universidad de Almería, La Cañada de San Urbano s/n, 04120 Almería (Spain)
{cgilm,rbanos,almarquez,afdezmolina,dgil}@ual.es
[2] Dept. Arquitectura y Tecnología de Computadores, Universidad de Granada, C/Periodista Daniel Saucedo s/n, 18071 Granada (Spain)
julio@atc.ugr.es

Abstract. The optimal design of looped water distribution networks is a major environmental and economic problem with applications in urban, industrial and irrigation water supply. Traditionally, this complex problem has been solved by applying single-objective constrained formulations, where the goal is to minimize the network investment cost subject to pressure constraints. In order to solve this highly complex optimization problem some authors have therefore proposed using heuristic techniques for their solution. Ant Colony Optimization (ACO) is a meta-heuristic that uses strategies inspired by real ants to solve optimization problems. This paper presents and evaluates the performance of a new ACO implementation specially designed to solve this problem, which results in two benchmark networks outperform those obtained by genetic algorithms and scatter search.

Keywords: ant colony optimization, heuristic optimization, combinatorial optimization, water distribution network design.

1 Introduction

The optimal design of water distribution networks is a combinatorial optimization problem that consists of finding the best way of conveying water from the sources (tanks and reservoirs) to the users (demand nodes) satisfying some requirements. It is a non-linear, constrained and multi-modal problem included in the category of NP-hard problems [1]. As a result of the extensive research effort made to solve this problem a large number of methods have been applied, including heuristic algorithms. Heuristic methods are procedures that provide approximate solutions to complex problems in a quick way. In the last decades the research interest in the design and application of heuristics and meta-heuristics (extensions of heuristics to tackle general problems) has grown remarkably, including implementations for solving the water distribution network design problem [2]. One of these meta-heuristics is Ant Colony Optimization (ACO) [3], which has been used by some authors to solve the water distribution network

J. Cabestany, I. Rojas, and G. Joya (Eds.): IWANN 2011, Part II, LNCS 6692, pp. 300–307, 2011.

design [4,5,6], although their performance has not been evaluated sufficiently in comparison to other meta-heuristics. This paper presents an ACO implementation which is compared with other meta-heuristics using benchmark water supply networks of different size and topology often used in the literature [7].

The remainder of the paper is organized as follows. Section 2 defines the looped water distribution network problem and gives a brief overview of how ACO has been applied to this problem. Section 3 offers a description of the ant colony optimization algorithm presented here. Section 4 presents the empirical study using the benchmark networks, while the conclusions of this paper are provided in Section 5.

2 Problem Description and Related Work

The problem consists of minimizing the network investment cost with pipe diameters as decision variables, while minimum pressure is the constraint, and pipe layout, minimum and maximum flow velocities are input data [7]. Equation 1 shows the cost function (C), where c_i is the cost of the pipe with diameter i per unit length, L_i is the total length of pipe with diameter i in the network, n_d is the number of available pipe diameters, ha_j is the pressure available at node j, and hr_j is the pressure required at node j.

$$\begin{aligned} &C = \textstyle\sum_{i=1}^{n_d} c_i L_i \\ &subject\ to: ha_j \geq hr_j,\ \forall j \in [1...n_n] \end{aligned} \tag{1}$$

Therefore, the size of the search space depends on the number of links and the number of available pipe diameters. Formally, if n_l is the number of links, the size of the search space (number of possible network configurations) is ${n_d}^{n_l}$.

In the literature it is possible to find an important number of papers dealing with the application of heuristics and meta-heuristics for solving the water distribution network design problem [4,7,8,9]. An interesting meta-heuristic sucessfully applied to this problem is the Ant Colony Optimization (ACO). ACO is a meta-heuristic inspired by the behaviour of real ants, which are almost blind and so rely very little on sight for communication [3,10,11]. An artificial ant in ACO is a stochastic constructive procedure that incrementally builds a solution by adding opportunely defined solution components to a partial solution under construction [11]. The first implementation of this meta-heuristic was the Ant System (AS) [3], which uses a decision policy based on the relative weighting of pheromone intensity and desirability of each option at a decision point. In each iteration the ants add pheromone to their path, which is updated with time according to an evaporation index. Since this meta-heuristic was first proposed by Dorigo, it has been extended in order to improve its performance. Some extensions include AS_{elite} [12], Ant Colony System (ACS) [13], AS_{rank} [14], etc.

Some authors have analysed the performance of ant colony optimization for water distribution network design. Maier et al. [4] proposed an ACO implementation that outperformed a genetic algorithm (GA). Zecchin et al. [5] adapted the Max-Min Ant System (MMAS) [15], which aims to avoid the premature convergence problem often encountered with elitist ACO implementations, and demonstrated that results obtained by MMAS outperformed those obtained by a basic Ant System algorithm.

3 Description of the ACO Implementation

The parameters of our implementation are: the number of ants that form the ant colony (P_{size}); the evaporation rate ($Evap_{rate}$) that determines the speed of dissipation of the pheromone; the exploration rate ($Expl_{rate}$) which is the probability assigned to the ants to explore random ways, omitting the indications obtained from the pheromone and heuristic; the importance of the heuristic in the decision process ($H_{importance}$); the maximum penalty coefficient (MAX_{penal}) for poor quality ants; the pheromone importance ($Ph_{importance}$) that indicates the probability of considering information from the pheromone of other ants; the probability of using local optimizer (LO_{rate}); and the recompense multiplicator ($R_{multiplicator}$) that determines the recompense for good quality ants.

The operation of this ACO implementation is now described. The algorithm starts by initializing the population of P_{size} ants, all of them satisfying the pressure constraints. In each iteration of the algorithm the ants are modified by a mutation operator that changes pipe diameters in links taking into account the freedom assigned to the ants to explore random ways, $Expl_{rate}$, and also the pheromone $Ph_{importance}$ and heuristic importance $H_{importance}$. These modified solutions are improved with the local search optimizer with a probability LO_{rate}. This local optimizer is based on modifying a pipe diameter to the next diameter value available until no improvement in reached. The ants are then evaluated and ranked according to their fitness function values (equation 1), and the ant with the best fitness is then stored. Later, all the ants are recompensed or penalized according to their relative position in the previous ranking. Thus, the better 50% of solutions are recompensed using the $R_{multiplicator}$ value, while the worst 50% of solutions are penalized according to MAX_{penal}. Before passing to the following iteration, the best path is reinforced, while the pheromone is evaporated in the other paths according to the parameter $Evap_{rate}$. Finally, when the stop condition is fulfilled, the ant with the fitness value is returned.

4 Empirical Analysis: Results and Discussion

The performance of the ACO implementations and of the other meta-heuristics is evaluated in the following two gravity-fed looped water distribution networks:

- Alperovits and Shamir network [16] is a simple two-loop network, with 7 nodes and 8 pipes arranged in 2 loops. A total of 14 commercial pipe diameters can be selected, i.e. there exist $14^8 = 1,4758*10^9$ possible configurations;
- Hanoi network [17] consists of 32 nodes, 34 pipes, and 3 loops. A set of 6 available commercial-diameter pipes is used, which implies a total of $6^{34} = 2,8651 * 10^{26}$ possible configurations.

A minimum pressure limitation of 30 meters above ground level for each node ($hr_j \geq 30$) is established in both networks. The interface of the program and the memetic algorithm have been programmed in the Visual-Basic programming language. Database management system has been implemented using a relational database and the ActiveX Data Objects (ADO) model. EPANET network solver (Version 2.00.07) [18] has been used considering its default values.

In order to analyse the performance of the single-objective ACO implementation in this problem, it has been compared with two other methods: Genetic Algorithms (GA) and Scatter Search (SS). Genetic algorithms [19] use stochastic search techniques that guide a population of solutions using the principles of evolution and natural genetics, including selection, mutation and recombination. Some authors have proposed GA to solve this problem [7]. The GA used in this analysis is an adaptation of that proposed by [7]. On the other hand, Scatter Search (SS) [20] is a method based on diversifying the search through the solution space. It operates on a set of solutions, named reference set (RS), formed by good and sparse solutions of the main population (P). These solutions are periodically combined with the aim of improving their fitness, while maintaining diversity. A further improvement phase using local search is also applied.

To compare the results of different executions, the stop criterion in the experiments has been that all the methods perform the same number of evaluations of the fitness function. That number of evaluations, n_e, should depend on the complexity of the network, i.e. it depends of the size of the search space. This number of evaluations has been established according to the equation 2, which is a criteria previously proposed in [21]. Considering a multiplication constant K_m=1000, the resulting fitness function evaluations are 9161 and 26457 for the Alperovits-Shamir and Hanoi networks, respectively. These common number of fitness function evaluations involve that the runtimes for all the methods range from few seconds to one minute according to the test network.

$$n_e = K_m * n_l * \log_{10}(n_d) \qquad (2)$$

When applying heuristic methods to optimization problems it is advisable to perform a sensitivity analysis, i.e. to determine to what extent the output of the model depends upon the inputs. The ACO implementation has a large number of parameters, as commented in Section 3, which is why the effect of modifying a subset of these parameters it has been analysed, while another subset has been fixed to certain values. Thus, fixed values have been established for the following parameters: $Ph_{importance}$ and $H_{importance}$ use the maximum values (100%), which means that the previous information from ants (pheromone) and the heuristic is applied in all cases. $R_{multiplicator}$ and MAX_{penal} have been set to 0.25, which means that the ants are recompensed or penalized in a factor of 25%, respectively. The probability of applying the local search optimizer has been fixed to LO_{rate}=40%.

On the other hand, a sensibility analysis has been performed to other parameters, such as the number of ants (P_{size}), the exploration probability ($Expl_{rate}$), and the evaporation rate ($Evap_{rate}$). The population size (P_{size}) in this empirical study depends on the number of pipes of the network to be optimized, i.e. P_{size} has been set according to the problem size, using P_{size}=$\{n_d/2,\ n_d,\ 2{*}n_d\}$ ants. Another important parameter to be analysed is the probability a given ant has to perform a random exploration, which has been set in an interval ranging from 5% to 45%. Finally, the evaporation rate has also been analysed with values within an interval ranging from 1% to 15%.

Table 1. Parameters used in the empirical executions

Technique	Parameters	Values
GA	P_{size}	100
	$cross_{rate}$	0.95
	mut_{rate}	0.05
SS	P_{size}	100
	RS_{size}	10
ACO	$Ph_{importance}$	1
	$H_{importance}$	1
	MAX_{penal}	0.25
	$R_{multiplicator}$	0.25
	LO_{rate}	0.4
	P_{size}	$n_d/2$,n_d,$2{*}n_d$
	$Expl_{rate}$	0.05,0.10,0.15,0.25,0.45
	$Evap_{rate}$	0.01,0.1,0.15

Firstly we should consider the results obtained after performing the sensitivity analysis (30 runs of each parametric configuration), which will determine accurate values of P_{size}, $Expl_{rate}$, and $Evap_{rate}$, using the values described in Table 1. In particular, Table 2 describes the results obtained when using an ant colony of $n_p/2$ ants in the Alperovits-Shamir network, which shows that the best configuration is that formed by $Expl_{rate}$=0.45, and $Evap_{rate}$=0.01, i.e. there is high probability of randomly exploring the search space, while the pheromone evaporates slowly. The same analysis is performed using P_{size}=$\{n_d,\ 2{*}n_d\}$ in Alperovits-Shamir and Hanoi networks.

Table 3 shows the summary of the results obtained by GA, SS and ACO in Alperovits-Shamir network. It is observed that all the methods are able to reach the best known result (419000 monetary units) in any of the runs. However, considering the average cost obtained by these methods with 30 runs, the ACO implementation obtains the best average cost, while GA and SS obtain slightly worse results. Figure 1(a) displays the evolution of the cost as the search advances, using the best parametric configuration. It can be seen that all the methods converge to 419000 monetary units.

Table 4 shows the average and minimum cost obtained by all the methods in the Hanoi network. Here, the best result is also obtained by ACO (6081127),

Table 2. Results obtained by ACO system in Alperovits-Shamir using different parameters

P_{size}	$Expl_{rate}$	$Evap_{rate}$	AVG cost	MIN cost (runs)	Deviation
		0.01	419467	419000 (16)	0.0011
$n_d/2$	0.05	0.10	419920	419000 (14)	0.0022
		0.15	419767	419000 (13)	0.0018
		0.01	419033	419000 (20)	0.0001
$n_d/2$	0.10	0.10	419780	419000 (13)	0.0019
		0.15	419680	419000 (13)	0.0016
		0.01	419033	419000 (29)	0.0001
$n_d/2$	0.15	0.10	419266	419000 (17)	0.0007
		0.15	419397	419000 (14)	0.0009
		0.01	419067	419000 (28)	0.0002
$n_d/2$	0.25	0.10	419250	419000 (17)	0.0013
		0.15	419133	419000 (18)	0.0003
		0.01	419000	419000 (30)	0.0000
$n_d/2$	0.45	0.10	419066	419000 (24)	0.0002
		0.15	419100	419000 (23)	0.0002

Table 3. Results obtained in Alperovits-Shamir

Algorithm	AVG Cost	Deviation	MIN cost	Deviation	Best config.
GA	423200	1.0095	419000	1.0000	0,95/0,05
SS	426100	1.0169	419000	1.0000	Ti=50
ACO	419228	1.0000	419000	1.0000	N/2, 0.45, 0.01

while GA and SS obtain more expensive solutions, but with a difference of less than 4.5%. Taking into account the average result, ACO system also obtains the best performance (6274123 monetary units), while the other methods obtain a higher cost, but with a difference of less than 5.1%. In this case, ACO obtains the best configuration using $Expl_{rate}$=0.15, and $Evap_{rate}$=0.1. Figure 1(b) shows the tendency of each algorithm using the best parametric configuration. In this case, the GA and SS do not converge to that obtained by ACO.

Table 4. Results obtained in Hanoi

Algorithm	AVG Cost	Deviation	MIN cost	Deviation	Best config.
GA	6575682	1.0481	6388319	1.0505	0.95/0.05
SS	6688675	1.0661	6272752	1.0315	Ti=25
ACO	6274123	1.0000	6081127	1.0000	N/2, 0.15, 0.1

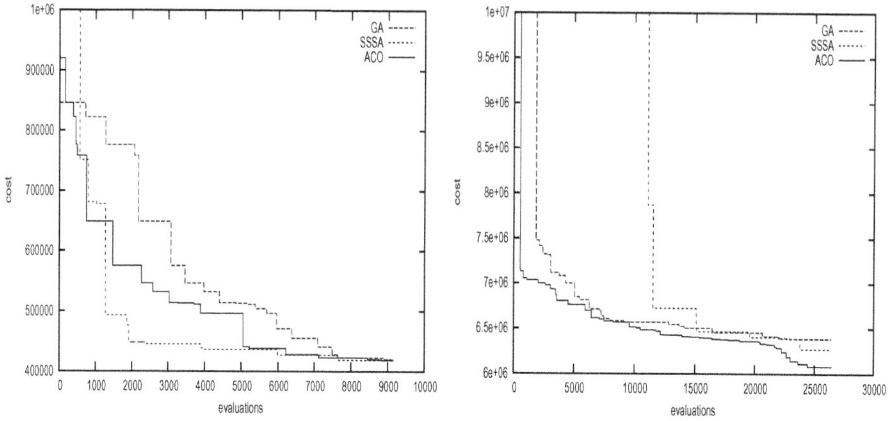

Fig. 1. Comparing GA, SS and ACO in Alperovits-Shamir and Hanoi networks

5 Conclusions

This paper presents and evaluates the performance of an ant colony optimization algorithm to solve the water distribution network design problem, which aims to reduce the total investment cost by modifying the pipe diameters, which are the decision variables subject to pressure constraints. The sensibility analysis has shown that the ACO implementation obtains better results when using a small population of ants that perform a higher number of iterations than using larger populations of ants which perform less iterations. No clear conclusions are obtained about the probability of randomly exploring the search space, although values of over 15% seem to be more suitable. The performance is often better in both networks when the pheromone evaporates slowly. The results obtained by the ACO implementation are compared with a genetic algorithm and scatter search implementations, and it is observed that all three methods obtain good results when applied to two benchmark water distribution networks, but especially the ACO algorithm. The global results obtained in both formulations reinforce the previous conclusions of other authors about the good performance of ant colony optimization to solve this problem.

Acknowledgements

Work supported by the Excellence Project of Junta de Andalucia (P07-TIC-02988), in part financed by the European Regional Development Fund (ERDF).

References

1. Gupta, I., Bassin, J.K., Gupta, A., Khanna, P.: Optimization of Water Distribution System. Environmental Software 8(4), 101–113 (1993)
2. Perelman, L., Krapivka, A., Ostfeld, A.: Single and multi-objective optimal design of water distribution systems: application to the case study of the Hanoi system. Water Science and Technology Water Supply 9(4), 395–404 (2009)

3. Dorigo, M.: Optimization, learning and natural algorithms (in Italian). PhD Thesis, Dipartamento di Electtronica, Politecnico di Milano, Milan (1992)
4. Maier, H.R., Simpson, A.R., Zecchin, A.C., Foong, W.K., Phang, K.Y., Seah, H.Y., Tan, C.L.: Ant colony optimization for design of water distribution systems. J. of Water Resources Planning And Management, ASCE 129(3), 200–209 (2003)
5. Zecchin, A.C., Simpson, A.R., Maier, H.R., Nixon, J.B.: Parametric study for an ant algorithm applied to water distribution system optimisation. IEEE Transactions on Evolutionary Computation 9(2), 175–191 (2005)
6. Zecchin, A.C., Simpson, A.R., Maier, H.R., Leonard, M., Roberts, A.J., Berrisford, J.M.: Application of two ant colony optimisation algorithms to water distribution system optimisation. Mathematical and Computer Modelling 44, 451–468 (2006)
7. Reca, J., Martínez, J.: Genetic algorithms for the design of looped irrigation water distribution networks. Water Resources Research 42, W05416 (2006)
8. Cunha, M.D., Sousa, J.: Water distribution network design optimization: simulated annealing approach. J. of Water Resources Planning And Management, ASCE 125(4), 215–221 (1999)
9. Suribabu, C.R.: Differential evolution algorithm for optimal design of water distribution networks. Journal of Hydroinformatics 12(1), 66–82 (2010)
10. Cordón, O., Herrera, F., Stutzle, T.: A review on the ant colony optimization metaheuristic: Basis, models and new trends. Mathware and Soft Computing 9(2-3), 141–175 (2002)
11. Dorigo, M., Stutzle, T.: Ant colony optimization. Bradford Book (2004)
12. Dorigo, M., Maniezzo, V., Colorni, A.: The ant system: optimization by a colony of cooperating agents. IEEE Transactions on Systems, Man, and Cybernetics - Part B 26(1), 1–13 (1996)
13. Dorigo, M., Gambardella, L.M.: Ant colony system: a cooperative learning approach to the travelling salesman problem. IEEE Transactions on Evolutionary Computation 1(1), 53–66 (1997)
14. Bullnheimer, B., Hartl, R.F., Strauss, C.: A computational study A new rank based version of the Ant System. Water Resources Research 7(1), 23–38 (1999)
15. Stutzle, T., Hoos, H.H.: MAX-MIN ant system. Future Generation Computer Systems 16, 889–914 (2000)
16. Alperovits, A., Shamir, U.: Design of optimal water distribution systems. Water Resources Research 13(6), 885–900 (1977)
17. Fujiwara, O., Khang, D.B.: A two-phase decomposition method for optimal design of looped water distribution networks. Water Resources Research 26(4), 539–549 (1990)
18. Rossman, L.A.: EPANET 2 user's manual. EPA/600/R-00/057 (September 2000)
19. Holland, J.: Adaptation in Natural and Artificial Systems. MIT Press, Cambridge (1975)
20. Marti, R., Laguna, M., Glover, F.: Principles of scatter search. European Journal of Operations Research 169(2), 359–372 (2006)
21. Baños, R., Gil, C., Reca, J., Martínez, J.: Implementation of scatter search for multi-objective optimization: a comparative study. Computational Optimization and Applications 42(3), 421–441 (2009)

A Preliminary Analysis and Simulation of Load Balancing Techniques Applied to Parallel Genetic Programming

F. Fernández de Vega[1], J.G. Abengózar Sánchez[2], and C. Cotta[3]

[1] Universidad de Extremadura
Mérida, España
fcofdez@unex.es
[2] Junta de Extremadura
Mérida, España
jg.abengozar@extremadura.es
[3] Universidad de Málaga
Málaga, España
ccottap@lcc.uma.es

Abstract. This paper addresses the problem of Load-balancing when Parallel Genetic Programming is employed. Although load-balancing techniques are regularly applied in parallel and distributed systems for reducing makespan, their impact on the performance of different structured Evolutionary Algorithms, and particularly in Genetic Programming, have been scarcely studied. This paper presents a preliminary study and simulation of some recently proposed load balancing techniques when applied to Parallel Genetic Programming, with conclusions that may be extended to any Parallel or Distributed Evolutionary Algorithm.

Keywords: Parallel Genetic Programming, Load Balancing, Distributed Computing.

1 Introduction

Evolutionary Algorithms are nowadays routinely applied for solving search and optimization problems. They are based in Darwinian principles: By means of progressive refinement of candidate solutions, evolution can provide useful solutions in a number of generations.

Nevertheless, EAs and, particularly those employing variable size chromosomes, such as GP, have a problem when facing hard optimization problems: they require large computing resources and time to reach a solution. Researchers have demonstrated for GP, that individuals tend to grow progressively as generations are computed, thus featuring the well known bloat phenomenon [4]. Therefore, a number of factors have led researchers to making use of some degree of parallelism: the large number of candidate solutions -individuals from the population- that must be evaluated every generation; the

J. Cabestany, I. Rojas, and G. Joya (Eds.): IWANN 2011, Part II, LNCS 6692, pp. 308–315, 2011.

large number of generations frequently required to reach a solution, and the high computing cost due to fitness evaluations.

Although researchers have deeply studied parallel models when applied to EAs [7][2], few have considered the need of specifically designed Load Balancing techniques. This could be particularly relevant for GP, given the differences in complexity and time required for evaluating each of the individuals of the population -featuring different sizes and structures [4].

This paper addresses these questions for GP using the well known Master Slave model. Using standard tests problems for GP, and by means of simulations, we analyze different load-balancing techniques and their usefulness when running GP on parallel or distributed infrastructures.

The rest of the paper is organized as follows: Section 2 presents Parallel Genetic Programming and Load Balancing principles. Section 3 describes our methodology and Section 4 presents the simulations and results obtained. Finally, Section 5 includes the conclusions.

2 Parallel Genetic Programming and Load Balancing

Genetic Programming was popularized by John Koza in the nineties [3], and rapidly grew with the work of researchers that not only employed it for solving problems, but also developed their mathematical foundations [4].

The main difference with GAs also leads to one of their main drawbacks: the variable size of chromosomes encoding candidate solutions. The size increase that usually happens when the evolutionary process takes place, as well as the difficulty of problems usually addressed, makes frequently necessary the use of some degree of parallelization.

Among the parallel models described in the literature, and analyzed for GAs and GP [7], we are particularly interested in the Master-Slave model. Basically, It tries to simultaneously compute the fitness function for a number of individuals of the population -tasks assigned to slaves- and then evolve the next generation in the master, so that the distribution of new fitness evaluations can proceed. The advantage of this parallel model is that it doesn't introduce any change in the main algorithm. The distribution of tasks -fitness evaluations- must follow some load-balancing policy.

Load Balancing aims at properly distributing computing tasks among processors, so that all of them employ similar time when computing their assigned tasks, therefore reducing *makespan*, i.e., time elapsed from the beginning of the first task to the completion of the last one. It is not always easy to reach that goal: differences in processor architectures and uncertainty in task sizes are some of the factors that influences the problem.

If we refer to Parallel GP, some detailed analysis of Parallel GP has been published in the last decade, particularly for the Island models [2], but no specific study on load-balancing techniques has been recently published. We must go back to 1997 to find the first papers considering the importance of Load-Balancing when using Master-slave versions of Parallel GP [1]. Usually authors have considered the application of Load Balancing techniques when addressing other problems [8], [9].

This paper tries to continue this area of research, by analyzing new load-balancing techniques that has been successfully developed recently. In this context, it is relevant the work by Yang and Casanova, that defines new load-balancing policies that are based in task sizes and different ordering principles [11], [12]. Next section considers the application of the proposed tasks ordering to GP.

3 Methodology

In our study, we will consider the use of a Master-Slave GP model. Tasks to be distributed and run simultaneously will consist of the fitness evaluation for each of the individuals. Therefore, we will have as many tasks as individuals in the population. The main goal is to analyze the application of different load-balancing policies. We must be aware that in GP two individuals with the same size may feature different complexities: this is due to the use of different functions within the program structure [3]. Measuring sizes or complexities may thus lead to different results when using load-balancing techniques.

When evaluating load-balancing techniques, a number of factors must be considered. As described by Yang and Casanova [11], [12], equation (1) describes the communication time for the master with a given slave *i*:

$$Tcomm_i = nLat_i + \frac{chunk_i}{B_i} + tLat_i \tag{1}$$

where $nLat_i$ refers to the time required for beginning the communication, $chunk_i$ is the amount of information including in task *i*, B_i is the communication rate, and $tLat_i$ is the time elapsed since the master finishes the sending of $chunk_i$ until slave *i* *receives* the last byte. In the meanwhile, the master can begun another communication with a different slave. Both $nLat_i$ and B_i are independent on the data size that is sent. On the other hand, computing time for a given slave ($Tcomp_i$) can be evaluated as described in equation 2:

$$Tcomp_i = cLat_i + \frac{chunk_i}{S_i} \tag{2}$$

where $cLat_i$ is the time required for the slave to begin the running of the task, and S_i the speed of the processor. These values do not depend on the size of data to be processed.

As described below, some simplifications will be considered for this preliminary analysis. Specifically, we will focus on computing time, given that all the simulations and analysis will be performed on a single processor. The processor speed will be used as the basis for a simulated homogeneous distributed system, with all of the processors sharing the same features.

We have employed for the simulation two well-known GP problems: the artificial ant on the Santa Fe trail, and the even parity-12. A complete description of both problems can be found in [3][5][6]. The experiments have been run using *Evolutionary Computation in Java* (ECJ), and the basic parameter configuration included in the tool. ECJ has been developed by ECLab[1] (Evolutionary Computation Laboratory), George Mason University, Washington DC.

As stated above, all the simulations have been run on a single computer: Intel Centrino Duo 1,7 Ghz. For both problems 100 individuals have been employed in the population, and 50 generations have been computed. All the remaining parameters have been employed as defined in ECJ for both problems, so that the replication of the experiments can be easily performed. Some changes in the source code have been applied so that the computing time -the only information of interest for the simulation- can be computed. Therefore, we obtain the computing time for each individual evaluation. This basic information obtained in a run, is then considered when evaluating the performance that a given load-balancing policy will obtain in a parallel or distributed infrastructure, whose processors would share exactly the same features as the one employed for the simulation. Of course, with the data obtained, conclusions that may be drawn could be easily extrapolated to other infrastructures whose features are known.

4 Simulation and Results

We have computed the evaluation time for each of the individuals, and then the evaluation time per generation. This is the total computing time required for running experiments in a single processors. Moreover, given that task completion time in a single processors heavily depends on other tasks that are run on the background -due to the operating system, cron tasks, etc- we have performed each of the experiments 10 times using the same random seed, so that we know that exactly the same individuals are generated every generation, every run. We have then computed the average time per individual, which provides a good approximation for their actual computing time.

Figure 1 shows computing time required for each of the experiments along the 50 runs. First of all, we notice that Even Parity-12 is harder than the Ant problem. Although this is not new, this information is relevant when considering the effect of load balancing policies for task distribution. The Figure also shows the maximum depth of individuals. We see that the Ant problem quickly reaches the maximum depth allowed (17 levels, as described in the literature). Again, this information is of interest if a relationship between size, depth and computing time is to be used for deciding tasks distribution and the load balancing technique to be used.

4.1 Analysis of Different Load-Balancing Policies

Let us consider now the situation on a homogeneous distributed system when the Master-Slave topology for Parallel GP is employed. We will analyze the results that would be obtained for different Load-Balancing policies when the main goal is avoiding processors idle time, consequently improving makespan.

---

[1] http://cs.gmu.edu/~eclab/projects/ecj/

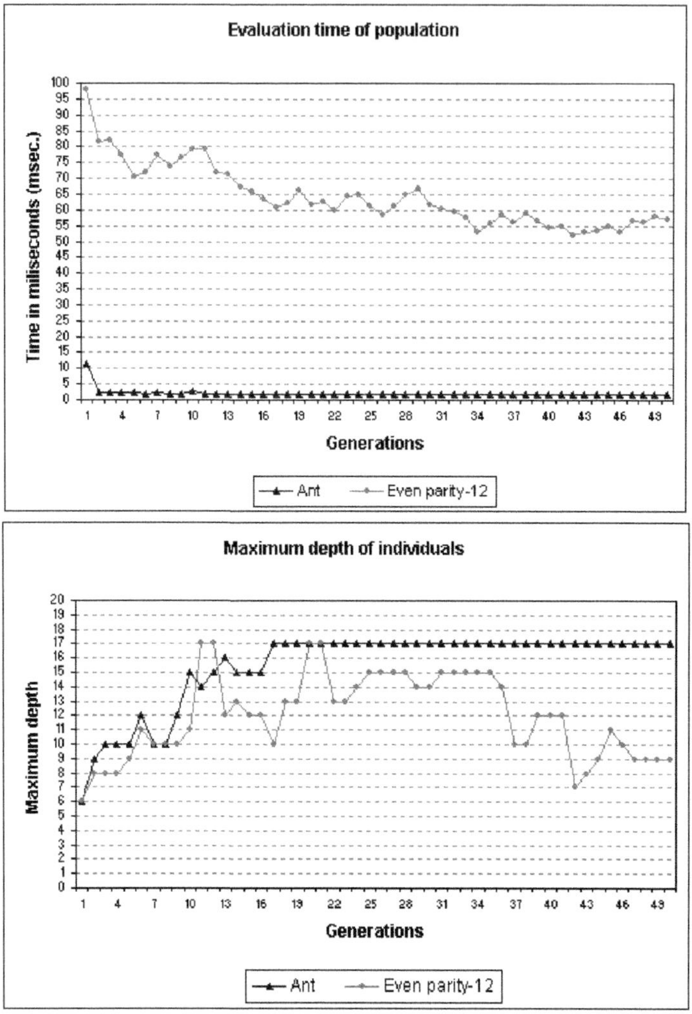

Fig. 1. Evaluation time and maximum depth per generation

Figure 2 shows a typical diagram with different steps required for sending individuals, computing fitness in the slaves, and returning back results. This is useful to see how communication and computing time can overlap, thus reducing makespan. The relationship between communication time and computing time is also relevant when deciding the policy to be used.

Some preliminary conclusions can be drawn from the figure. When the total time required for evaluating the whole generation is short -this happens when fitness is computed quickly-, when compared with the latencies and total communication times of tasks (see the case of the ant problem, with low computing time per individual) the best choice would be to send as much individuals as possible in a single task. This way communication time is reduced.

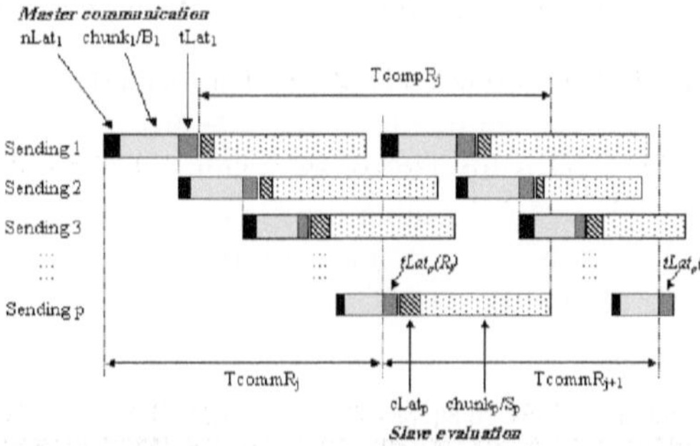

Fig. 2. Transmission and evaluation times in a parallel infrastructure

On the other hand, if fitness evaluation takes long time, it is better to send individuals to processors in a round-robin fashion, so that communication time overlaps as much as possible with computing time. In this case, the decision about the size of tasks, and therefore the number of individuals to be included in every task has to be decided. A number of alternatives are available for generating tasks: (i) Balanced Tasks: All of the task will require the same computing effort and time. (ii) Unbalanced Tasks: according to Yang and Casanova [11] and [12], unbalanced tasks may be of interest in some circumstances for reducing makespan.

We will now analyze the computing time obtained for each of the individuals in both problems considered, ant and even-parity-12, and considering that the same number of individuals are sent to every processor. We consider a distributed system with 5 slave processors. We have 100 individuals per generation, so 20 rounds are required, sending 1 individual per round per slave. Of course, other possibilities are available.

If we compute the total time required for all the fitness evaluations, we obtain 1,9130 milliseconds for the ant problem and 58,6960 milliseconds for the Even Parity 12. This is the time employed by a sequential system, and the basis for the analysis.

4.2 Analyzing Task Ordering and Submission

When all the tasks are balanced -requiring the same computing effort-, a round-robin mechanism will send tasks in the following way: the first round task 1 -first individual from the population- is sent to slave 1, task 2 -second individual- to slave 2, and so on. Second round will proceed again by sending task n+1 to slave 1, n+2 to slave 2, etc. Every chunk submitted -task- requires initiating a communication operation with a slave. Therefore, the total communication time will strongly depend on the number of rounds and the number of slaves.

Regarding GP, notice that communication time of a task will be influenced by individuals size, while computing time by both size and complexity of individuals. If this complexity is low, then total computing time will be dominated by communication time. Processors will be idle long time. In this case, an infrastructure with low communication latencies will be a must: both supercomputer platforms or commodity clusters using optimized network connections will be required. Researchers could also consider the possibility of increasing population size, so that more individuals are available, larger tasks can be conformed and processors will thus spent more time computing. Given that there are idle processors, no reason for using larger populations should keep us from using the resources available.

The second possibility is that computing time is much longer than communication time. Processors would never be idle. In this case, other kind of distributed platforms could be used, such as Grids and Desktop Grids and Cloud Computing infrastructure.

4.3 Applying Ordering

Another important factor is the order in which individual are sent. Several possibilities exist:

Random ordering: If we randomly pack individuals in tasks every generation, then, there will be random differences among completion time of tasks every round. The total time for a round is given by the task that takes longer. After considering the computed time for each of the individuals in the experiment, and computing the total time for every task in a round (5 tasks per round, given that 5 processors are considered), we have computing the total time of the experiment as the addition of all the rounds' largest task. We have thus obtained 0,4746 milliseconds for the ant problem and 17,6341 for the even-parity-12. This is better than the sequential time, but in can be even improved with better balancing techniques, as described below.

Weighted Factoring: Hummel et al. describe in [10] *Weighted Factoring* model. They consider task sizes and apply a descending ordering when submitting them to slaves. Therefore, we will consider first that the most complex tasks are is sent firstly. The advantage of this model is that for each of the rounds, all the tasks are similar, so the differences between computing time will be smaller. In the case of GP, this is only useful if we allow the algorithm to perform several rounds per generation. If a single round is to be performed per generation, the algorithm cannot work.

Table 1. Comparing Load Balancing Techniques

Model	Ant – Computing Time	EP-12 Comp. Time.
Sequential	1,9130	58,6960
Random distribution	0,4746	17,6341
Weight. Fact. Cmpx.	**0,4082**	12,1201
Weight. Fact. Size	0.4715	14.6687

If we perform a simulation using the computing time for each of the individuals in both problems tested -5 processors, 20 rounds per generation, 1 individual sent to each processor per round- ordering them and computing the time for the largest individual in the round- we obtained for the ant problem 0,4082 milliseconds and 12,1201 milliseconds for the even-parity-12 problem. Nevertheless, if we use the size of individuals for the ordering instead of computing time, we would obtain 0,4715 and 14,6687 respectively. This confirms that even when using size for balancing is positive, it is better to use complexity – a kind of computing time estimation. Table 1 summarizes results and shows the differences obtained with each of the models.

5 Conclusions

This paper has presented a preliminary analysis on the application Load-balancing techniques to Parallel Genetic Programming. By analyzing the time required for evaluating each of the individuals in a population, we have studied differences between load-balancing methods that could be applied when using the master-slave model. This preliminary analysis allows us to reach some conclusions of interest. Firstly, problems with short fitness evaluation time must be run on supercomputers or commodity clusters with optimized network connections, and should never be run on Grid infrastructures. Second, weighted factoring approach allows to reduce makespan when compared to previously employed more standard Load Balancing Techniques. Results are sensitive to the use of Complexity or Size during the ordering process.

Acknowledgments. Spanish Ministry of Science and Technology, project TIN2008-05941, and Junta de Extremadura project GR10029 and European Regional Development Fund.

References

1. Oussaidène, M., Chopard, B., Pictet, O.V., Tomassini, M.: Parallel Genetic Programming: an application to Trading Models Evolution, pp. 357–362. MIT Press, Cambridge (1996)
2. Fernández, F., Tomassini, M.,Vanneschi,L.: An empirical study of multipopulation genetic programming. In: GPEM, vol. 4(1), pp. 21–51 (2003)
3. Koza, J.R.: Genetic programming III. Morgan Kaufmann, San Francisco (1999)
4. Poli, R., Langdon, W.B., McPhee, N., Koza, J.: A field guide to genetic programming. Lulu Enterprises Uk Ltd (2008)
5. Koza, J.R.: Evolution and co-evolution of computer programs to control independently-acting agents. In: First International Conference on Simulation of Adaptive Behavior, p. 11. MIT Press, Cambridge (1991)
6. Koza, J.R.: Genetic programming: on the programming of computers by means of natural selection. MIT Press, Cambridge (1992)
7. Cantú-Paz, E.: A survey of parallel genetic algorithms. Calculateurs Paralleles, Reseaux et Systems Repartis 10(2), 141–171 (1998)
8. Folino, G., Pizzuti, C., Spezzano, G.: A scalable cellular implementation of parallel genetic programming. IEEE Transactions on Evolutionary Computation 7(1), 37–53 (2003)
9. Wang, N.: A parallel computing application of the genetic algorithm for lubrication optimization. Tribology Letters 18(1), 105–112 (2005)
10. Hummel, S.F., Schmidt, J., Uma, R.N., Wein, J.: Load-sharing in heterogeneous systems via weighted factoring. In: 8th annual ACM Symposium on Parallel Algorithms and Architectures, pp. 318–328 (1996)
11. Yang, Y., Casanova, H.: UMR: a multi-round algorithm for scheduling divisible workloads. In: 17th IEEE (IPDPS), p. 24 (2003)
12. Yang, Y., Casanova, H.: RUMR: Robust Scheduling for Divisible Workloads. In: Proceedings 12th IEEE HDPC 2003, p. 114 (2003)

A Study of Parallel Approaches in MOACOs for Solving the Bicriteria TSP*

A.M. Mora, J.J. Merelo, P.A. Castillo, M.G. Arenas,
P. García-Sánchez, J.L.J. Laredo, and G. Romero

Dpto. de Arquitectura y Tecnología de Computadores,
Universidad de Granada, Spain
{amorag,jmerelo,pedro,maribel,pgarcia,juanlu,gustavo}@geneura.ugr.es

Abstract. In this work, the parallelization of some Multi-Objective Ant Colony Optimization (MOACO) algorithms has been performed. The aim is to get a better performance, not only in running time (usually the main objective when a distributed approach is implemented), but also improving the spread of solutions over the Pareto front (the ideal set of solutions). In order to do this, colony-level (coarse- grained) implementations have been tested for solving the Bicriteria TSP problem, yielding better sets of solutions, in the sense explained above, than a sequential approach.

1 Introduction

When a classical method is redesigned for a parallel setup, the aim is usually to yield good solutions by improving the running time. Moreover, the parallelization may imply a different searching scheme in some metaheuristics, as in the case of Ant Colony Optimization (ACO) [1] algorithms. These metaheuristics are based in a set of artificial agents (ants) which explore the search space cooperating to get the solution for a problem. In addition, the main feature of a good multi-objective (MO) algorithm [2] (devoted to find solutions for more than one objective function) is to get the maximal set of non-dominated solutions, the so-called Pareto Set (PS), which includes those solutions that optimize all the functions in the problem. The ACO algorithms implemented to deal with several objectives are known as MOACOs (see [3] for a survey).

The idea addressed in this study has been the distribution of the ants (grouped in colonies) into several computing nodes, being each one of these nodes focused into a different area of the searching space. Such an structure contributes to yield a better set of results (including a bigger amount of non-dominated solutions), by promoting the explorative behaviour.

* This work has been supported in part by HPC-Europa 2 project (with the support of the European Commission - Capacities Area - Research Infrastructures), by the CEI BioTIC GENIL (CEB09-0010) Programa CEI del MICINN (PYR-2010-13) project, the Junta de Andalucía TIC-3903 and P08-TIC-03928 projects, and the Jaén University UJA-08-16-30 project.

J. Cabestany, I. Rojas, and G. Joya (Eds.): IWANN 2011, Part II, LNCS 6692, pp. 316–324, 2011.

Two different parallelization approaches have been tested, considering in the study two of the most known MOACOs in the literature: Iredi et al. *BIANT* [4], and Barán et al. *MOACS* [5].

What we have done and present in this paper is the parallelization of these two problems at the colony level. In addition, the proposed models have been applied to solve the same problem: a *Bicriteria Travelling Salesman Problem* (Bi-TSP) [6], which is the transformation into a multicriteria problem of the classical TSP. In the single-objective TSP the target is to minimize distance, while in this version there is a set of different costs between each pair of connected cities, which could correspond, for instance to distance and travel time. Those distributed implementations have been compared with the correspondent sequential approach, and the profits in running time and in the quality of the solutions yielded have been analyzed.

This work presents a novel study, since as far as we know there are in the literature some distributed or parallel ant colony algorithms [7], but no one which deals with multi-objective problems.

2 Parallel Approaches

Since an ACO algorithm works using independent agents (ants), it can be adapted to a parallel architecture in a direct way. The ants communicate with each other through the so called pheromone matrix (which simulates the real environment for the stigmergy effect), which can be updated asynchronously, so they do not require to pass continuous synchronization nor information data through the net, as many of the parallel implementations of other methods.

There are several parallel ACO approaches [7] which mainly distribute the ants into several computing nodes following a different parallelization grain. In fine-grained implementations, every ant goes to its own node, while in coarse-grained implementations every node contains a set of ants.

Typically, these implementations are centralized (following a master/slave architecture), that means there is one node, called master process, which collects the solutions or the pheromone information from all the other nodes. After this, it performs the pheromone updating and computes the new pheromone matrix, which is then sent to the other nodes (called slaves processes). In a decentralized approach every node has to compute the pheromone update by itself, using information that it has received from other nodes. The main goal of these approaches is to improve the running time without changing the optimization behaviour of the algorithm.

In contrast, the specifically-designed parallel ACO algorithms try to change the standard ACO algorithm, so that the parallel version works more efficiently. One approach is to divide the whole population of ants into several subsets which do information exchange between them every few iteration (not in every one). This can also have a positive effect on the optimization behaviour because the subset in each node may specialize in different regions of the searching space.

ACO algorithms composed by several colonies of ants, where each of them uses their own (and different) pheromone matrix, are called *multi colony ACO algorithms*. They are suitable for parallelization, since a processor can host just a colony of ants, and normally there will be less information exchange among the colonies as would have between groups of ants in standard ACO. They are typically decentralized.

The aim in this work is to get mainly one profit: improve the quality of the solutions obtained solving the Bi-TSP problem, rather than just improve the running time as usual in parallel approaches. That is, obtaining a good (large) set of non-dominated solutions with a good distribution along the Pareto Front, which is the main task of any MO algorithm.

With respect to this question, it has been demonstrated in the literature [4,8,5] that in MOACOs, the use of specialized colonies (or ants) for each objective or even each area of the searching space, yields very good results [3].

This way, the proposal in this paper implies adapting some models to a parallel environment with the advantages that can be expected, and taking a *coarse-grain* parallelization approach, that is, a parallelization at colony level, so every computation node will contain a set of ants. We propose two different distributed approaches:

- *Space Specialized Colonies* (SSC): it consists on a group of independent colonies, each of them searching in a different area of the space of solutions. At the end of the process they merge their solutions (their Pareto sub-sets) to constitute a single Pareto Set (considering dominance criteria to build it, since the non-dominated solutions of a colony may be dominated by the solutions yielded by another colony). The split of the space is made through the use of some parameters which weights the objectives in the search for each ant in every colony.
- *Objective Specialized Colonies* (OSC): it also consists on a group of independent colonies, but this time, each one tries to optimize only one of the objectives. Every colony does not consider the others in the search, but all of them are taken into account when the solutions are evaluated, so the colonies search, as in the previous model, in a multi-objective space of solutions. Again, at the end, all the PSs are merged (considering the dominance criterion) into the final (or global) PS.

3 MOACOS to Study

As previously stated, the approaches commented in the previous section have been applied to two state of the art MOACO algorithms from the literature.

In both methods we have used a key parameter in the search (inside the State Transition Rule), $\lambda \in [0,1]$, which let us to focus in a concrete area of the search space to explore.

The first algorithm is ***BIANT*** (BiCriterion Ant), which was proposed by Iredi et al. [4] as a solution for a multi-objective problem with two criteria (the Single Machine Total Tardiness Problem, SMTTP). It is an Ant System (AS) which uses just one colony, and two pheromone matrices and heuristic functions (one per objective).

The State Transition Rule, STR (the main element in an ACO algorithm), is as follows:

$$P(i,j) = \begin{cases} \dfrac{\tau_1(i,j)^{\alpha\cdot\lambda}\cdot\tau_2(i,j)^{\alpha\cdot(1-\lambda)}\cdot\eta_1(i,j)^{\beta\cdot\lambda}\cdot\eta_2(i,j)^{\beta\cdot(1-\lambda)}}{\sum\limits_{u\in N_i}\tau_1(i,u)^{\alpha\cdot\lambda}\cdot\tau_2(i,u)^{\alpha\cdot(1-\lambda)}\cdot\eta_1(i,u)^{\beta\cdot\lambda}\cdot\eta_2(i,u)^{\beta\cdot(1-\lambda)}} & if\ j\in N_i \\ 0 \quad otherwise \end{cases} \tag{1}$$

Where α and β are weighting parameters to set the relative importance of pheromone and heuristic information respectively, and N_i is the current feasible neighbourhood for the node i. These terms and parameters are the same as in classical Ant System equations, but this time there are one τ and one η per objective. In addition, the rule considers the λ parameter to weight the objectives in the search. This expression calculates the probability for the feasible nodes; then, the algorithm uses a roulette wheel to choose the next node in the path of the solution that is being built. Since BIANT is an AS, it is just performed a global pheromone updating, including evaporation in all nodes and contribution just in the edges of the best paths to the moment (those included in the Pareto Set (PS)).

The second algorithm is ***MOACS*** (Multi-Objective Ant Colony System), which was proposed by Baran et al. [5], to solve the Vehicle Routing Problem with Time Windows (VRPTW). It uses a single pheromone matrix for both objectives (instead of one per objective, as usual in other approaches).

The STR is defined this time as:

```
If (q ≤ q0)
```

$$j = \arg\max_{j\in N_i}\left\{\tau(i,j)\cdot\eta_f(i,j)^{\beta\cdot\lambda}\cdot\eta_s(i,j)^{\beta\cdot(1-\lambda)}\right\} \tag{2}$$

```
Else
```

$$P(i,j) = \begin{cases} \dfrac{\tau(i,j)\cdot\eta_1(i,j)^{\beta\cdot\lambda}\cdot\eta_2(i,j)^{\beta\cdot(1-\lambda)}}{\sum\limits_{u\in N_i}\tau(i,u)\cdot\eta_1(i,u)^{\beta\cdot\lambda}\cdot\eta_2(i,u)^{\beta\cdot(1-\lambda)}} & if\ j\in N_i \\ 0 & otherwise \end{cases} \tag{3}$$

In this expression q is a random number in [0,1] and q_0 is a parameter which set the balance between exploration and exploitation. If $q \leq q_0$, the best node is chosen as next (exploitation), on the other hand one of the feasible neighbours is selected, considering different probabilities for each one (exploration). The rest of the terms and parameters are the same as in Equation 1, but this time there are two heuristic functions: η_1 and η_2. This rule applies again λ to balance the relative importance of the objectives in the search. Since *MOACS* is an ACS, there are two levels of pheromone updating, local and global.

Both algorithms were initially defined with a policy for λ, which consists in assign a different value for the parameter to each ant h, following the expression:

$$\lambda_h = \frac{h-1}{m-1} \qquad \forall h \in [1,m] \tag{4}$$

Considering that there are m ants, the parameter takes an increasing value that goes from 0 for the first ant to 1 for the last one. This way, the algorithms search in all the possible areas of the space of solutions (each ant is devoted to a zone of the Pareto Front, PF).

In this work, this parameter has been used to determine the area of the searching space that each colony has to explore, so it is a constant for all the ants in a colony (and different to the rest of the colonies). In addition, both approaches have been improved by means of a local search (LS) application, the 2-OPT method.

4 Experiments and Results

We have performed some experiments to test the validity of the methods, so firstly, a sequential implementation of each of them have been tested (in just one processor). Then, their parallel versions (and the two approaches) have been run in a different number of processors from 2 to 16. The parallelization has been implemented using MPI [9] and tested in a 16 processors cluster, with shared memory. All the experiments have been performed for solving the *Kroa100* problem [6] (a 100 cities TSP), with two different instances (a and b) corresponding to the two objectives. In addition, the same random seed have been considered. The experimental setup can be found in Table 1, having obtained the set of configuration parameters through systematic experimentation. These parameters have been used by all the algorithms.

Table 1. Parameters of the algorithms

Number of ants	35
Number of iterations	500
Number of iterations in LS (2-OPT)	15

The experiments have been run in different numbers of processors being: 1) the sequential approach is run in one processor, considering the variable λ policy (the one proposed by the authors); 2) the two processor considers the OSC approach, since there are two objectives, so two colonies are used; 3) the approaches for 4, 8 and 16 processors apply the SSC approach, considering different number of colonies (one per processor), with a different value for λ in each one, but the same for all the ants in the colony.

The results for the BIANT approach are showed in Figure 1. As it is shown, each colony explores a different area of the search space yielding a set of solutions different from the rest. In addition, it can be seen that the distribution of the solutions in the approaching to the Pareto Front (PF) is better with a higher number of colonies (and processors), yielding smaller values in both objectives, and covering a higher area. Sometimes an approach with a smaller number of colonies gets some better solutions in a concrete zone, since they explore in a more restricted area of the space of solutions (due to the λ value), so a higher exploitation factor is applied in that area. The OSC approach (in two processors)

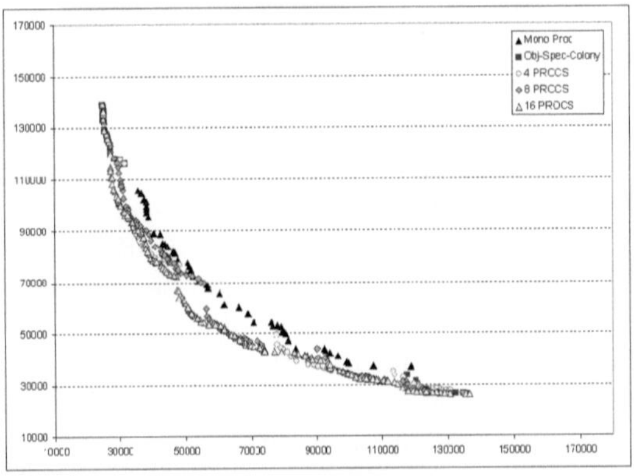

Fig. 1. Results for BIANT algorithm solving the Bi-TSP (100 cities). They are distributed in objective specialized colonies, and in space specialized colonies, from 2 to 16 processors. They are compared with those obtained by a mono-processor approach.

does not yield very good solutions, since these colonies explore just in the edges of the PF, obtaining good solutions there, but not in the central area.

The MOACS experiments, are represented in Figure 2. The figure shows that

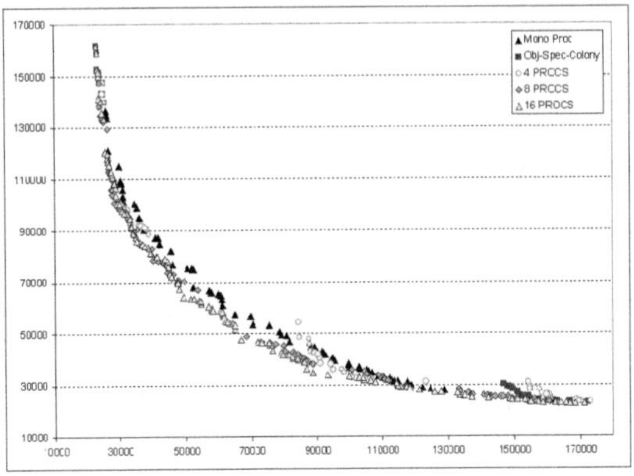

Fig. 2. Results for MOACS algorithm solving the Bi-TSP (100 cities). They are distributed in objective specialized colonies, and in space specialized colonies, from 2 to 16 processors. They are compared with those obtained by a mono-processor approach.

the results distributions yielded by each of the approaches are quite similar to the BIANT experiment. This time the PSs are closer between them than in the previous case, but again, the 16 processors approach yields the best set: better solutions and better distribution along the PF. The mono-processor run is quite good with this algorithm, but it shows a flaw in the distribution in some areas. The results yielded by MOACS are much better than those obtained by BIANT, since the PSs are closer to the minimum, are wider (they reach more values in the edges), and also shows a better distribution in every case (they cover a bigger area of the PF).

The final idea when one perform these kind of parallelization is to get a unique PS, so as an example, we have made an experiment to achieve this. The MOACS approach have been distributed into 11 processors using the SSC scheme, yielding 11 PSs as can be seen in Figure 3 (left) . Then, all these sets are merged (just considering the whole set of non-dominated solutions), getting a global PS, as it is show in Figure 3 (right). It is more diverse (overcoat in the edges), and closer to the ideal PF, than the set obtained by the sequential run (Mono-Proc), which have been run again considering a variable value for λ (one value per ant). Then, in Table 2 the number of non-dominated solutions in each of the global PSs per experiment is shown.

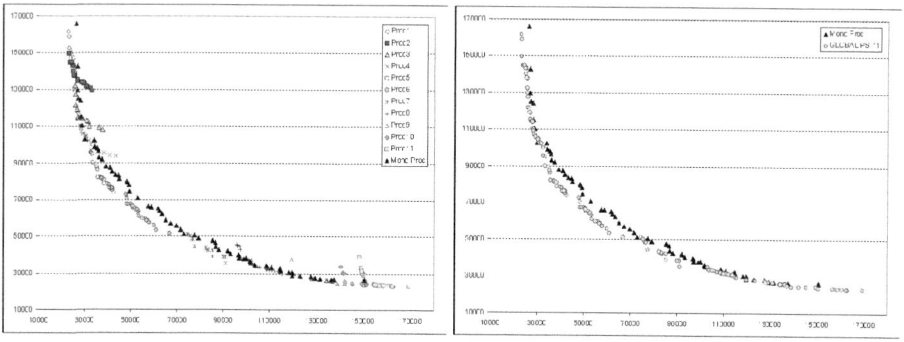

Fig. 3. Example of the results for MOACS distributed in space specialized colonies (11 processors). In the left subfigure each colony pareto set is shown in a different colour. The global pareto set is shown in the right subfigure.

Table 2. Number of solutions in the whole Pareto Set in each of the experiments

	Mono	2 Procs	4 Procs	8 Procs	16 Procs
BIANT	46	50	89	158	202
MOACS	69	28	63	97	142

Looking at those results, it can be noticed that BIANT yields more solutions in each case. The reason is that it is more explorative (is an AS) than MOACS. But if we look at the previous figures (1, 2), the latter shows a better distribution as stated, while BIANT concentrates a big amount of solutions in some concrete areas, so it can be considered that MOACS performs better.

Finally, the last analysis performed has been the running time profit due to the parallelization, usually the main objective when a distributed approach is implemented. The conclusion is the expected one: it is necessary a bit more time when we consider a higher amount of processors, but the performance in respect

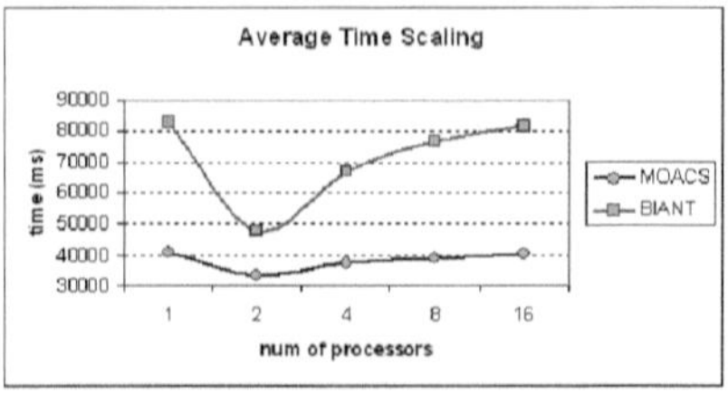

Fig. 4. Time scaling in average for MOACS and BIANT in a different number of processors

to the value of the solutions is worthwhile. The time scalability functions are shown in Figure 4.

As can be seen, both algorithms follow the same progression, taking much less time (in average) to get the results for a number of processors smaller than 16. In the last case, the average time is closer (but smaller) to the time taken by the sequential approach, but the quality of the set of solutions justifies this distribution. Again MOACS shows a better performance than BIANT.

5 Conclusions and Future Work

In this work, two Multi-Objective Ant Colony Optimization Algorithms (BIANT [4] and MOACS [5]) have been implemented in a distributed shape.

Two different parallelization approaches have been tested, one considering a different colony specialized in a concrete area of the search space (SSC), and another where there is a colony specialized in each one of the objectives of the problem (OSC). Both of them use a parameter named λ which sets the relative importance of the objectives in the search, aiming all the ants in a colony to the same zone of the space.

Some experiments have been performed distributing the colonies into a different number of processors (from 1 to 16), showing that the best set of solutions is obtained for the 16 processors approach, being a very good distributed (and crowded of solutions) set, which is the aim of the multi-objective algorithms.

In the comparison between them, MOACS yields better results (both in value and in distribution along the ideal set of solutions) than BIANT.

Finally, looking at the running time profit, it is as good as expected, being improved when the process is distributed along a bigger number of processors until 16, when it is closer (but lower) than a mono-processor run time, but it is worthwhile because of the quality of the yielded solutions.

The results yielded in this work are very promising, thus several future lines of work arise. Firstly we would like to test these approaches to solve some other multi-objective problems (such as the Vehicle Routing Problem with Time Window). Other line guides us to implement other approaches, such as heterogeneous colonies (a different algorithm running in each processor), to compensate the flaws of one with the solutions of another.

The next objective could be to implement a fine-grained parallelization approach (at ant level), in order to improve the performance in time. The aim is to deal with very large instances of multi-objective problems.

References

1. Dorigo, M., Stützle, T.: The ant colony optimization metaheuristic: Algorithms, applications, and advances. In: Glover, F. (ed.) Handbook of Metaheuristics, pp. 251–285. Kluwer, Dordrecht (2002)
2. Coello, C.A.C., Veldhuizen, D.A.V., Lamont, G.B.: Evolutionary Algorithms for Solving Multi-Objective Problems. Kluwer Academic Publishers, Dordrecht (2002)
3. García-Martínez, C., Cordón, Ó., Herrera, F.: An empirical analysis of multiple objective ant colony optimization algorithms for the bi-criteria TSP. In: Dorigo, M., Birattari, M., Blum, C., Gambardella, L.M., Mondada, F., Stützle, T. (eds.) ANTS 2004. LNCS, vol. 3172, pp. 61–72. Springer, Heidelberg (2004)
4. Iredi, S., Merkle, D., Middendorf, M.: Bi-criterion optimization with multi colony ant algorithms. In: Zitzler, E., Deb, K., Thiele, L., Coello Coello, C.A., Corne, D.W. (eds.) EMO 2001. LNCS, vol. 1993, pp. 359–372. Springer, Heidelberg (2001)
5. Barán, B., Schaerer, M.: A multiobjective ant colony system for vehicle routing problem with time windows. In: IASTED International Multi-Conference on Applied Informatics. Number 21 in IASTED IMCAI, 97–102 (2003)
6. Reinelt, G.: Tsplib software/TSPLIB95/ (2004), `http://www.iwr.uni-heidelberg.de/groups/comopt/`
7. Janson, S., Merkle, D., Middendorf, M.: 8. Parallel Metaheuristics. In: Parallel ant algorithms, Wiley, London (2005)
8. Gambardella, L., Taillard, E., Agazzi, G.: Macs-vrptw: A multiple ant colony system for vehicle routing problems with time windows. In: Corne, D., Dorigo, M. (eds.) New Ideas in Optimization, pp. 73–76. McGraw-Hill, New York (1999)
9. Gropp, W., Lusk, E., Doss, N., Skjellum, A.: A high-performance, portable implementation of the MPI message passing interface standard. Parallel Computing 22(6), 789–828 (1996)

Optimizing Strategy Parameters in a Game Bot*

A. Fernández-Ares, A.M. Mora, J.J. Merelo,
P. García-Sánchez, and C.M. Fernandes

Depto. de Arquitectura y Tecnología de Computadores, U. of Granada
{antares,amorag,jmerelo,pgarcia,cfernandes}@geneura.ugr.es

Abstract. This paper proposes an Evolutionary Algorithm for fine-tuning the behavior of a bot designed for playing Planet Wars, a game that has been selected for the the Google Artificial Intelligence Challenge 2010. The behavior engine of the proposed bot is based on a set of rules established by means of heuristic experimentation, followed by the application of an evolutionary algorithm to set the constants, weights and probabilities needed by those rules. This bot eventually defeated the baseline bot used to design it in most maps, and eventually played in the Google AI competition, obtaining a ranking in the top 20%.

1 Introduction and Problem Description

In a computer game environment, a *Bot* is usually designed as an autonomous agent which tries to play under the same conditions as a human player, cooperating or competing with the human or with other bots.Real-time strategy (RTS) games is a sub-genre of strategy video games in which the contenders control units and structures, distributed in a playing area, in order to beat the opponent (usually in a battle). In a typical RTS, it is possible to create additional units and structures during the course of a game, although usually restrained by a requirement to expend accumulated resources. These games, which include Starcraft™and Age of Empires™ , typically work in real time: the player does not wait for the results of other players' moves.

Google chose Planet Wars, a game of this kind that is a simplified version of the classic Galcon game (`http://galcon.com`), for their Artificial Intelligence Challenge 2010 (GAIC) (`http://ai-contest.com`), that pits user-submitted players against each other. The aim of this research is to design the behavioral engine of a bot that plays this game, trying to maximize its efficiency. A Planet Wars match takes place on a map which contains several planets, each of them with a number on it that represents the number of starships it hosts At a given time, each planet has a specific number of starships, and it may belong to the player, to the enemy, or it may be neutral (i.e., it belongs to nobody). Ownership

* Supported in part by Andalusian Government grant P08-TIC-03903, by the CEI BioTIC GENIL (CEB09-0010) Programa CEI del MICINN (PYR-2010-13) project, the Junta de Andalucía TIC-3903 and P08-TIC-03928 projects, and the Portuguese Fellowship SFRH /BPD / 66876 / 2009.

J. Cabestany, I. Rojas, and G. Joya (Eds.): IWANN 2011, Part II, LNCS 6692, pp. 325–332, 2011.

is represented by a colour, being blue for the player, red for the enemy, and grey for neutral (a non-playing character). In addition, each planet has a growth rate that indicates how many starships are generated during each round of action and added to the starship fleet of the player that owns the planet.

The objective of the game is to conquer all of the opponent's planets. Although Planet Wars is a RTS game, the implementation has transformed it in a turn-based one, with each player having a maximum number of turns to accomplish the objective. The player with more starships at the end of the match (set to 200 actions in the challenge) wins. Each **planet** has some properties: *X and Y Coordinates*, *Owner's PlayerID*, *Number of Starships* and *Growth Rate*. Players send fleets to conquer other planets (or to reinforce its own), and every **fleet** also has a set of properties: *Owner's PlayerID*, *Number of Starships*, *Source PlanetID*, *Destination PlanetID*, *Total Trip Length*, and *Number of turns remaining until arrival*. A simulated turn is implemented by a second in duration. The bot only has this maximum time to order the next actions list. Moreover, a peculiarity of the problem is that the bot is unable to store any kind of knowledge about their actions in previous turns, the actions of his opponent or the game map, for instance. In short, every time elapses the simulation of a turn (second) the bot gets to meet again with a unknown map, like a new game. This inability to store knowledge about the gameplay makes the creation of the bot an interesting challenge.

In fact, each autonomous bot is implemented as a function that takes as an input the list of planets and fleets (the current status of the game), each one with its properties' values, and outputs a text file with the actions to perform. In each simulated shift, a player must choose where to send fleets of starships, departing from one of the player's planets and heading to other planet on the map. This is the only type of actions that the bot can do. The fleets can take some time steps to reach their destination. When a fleet reaches a planet, it fights against the existing enemy's forces (losing one starship for each one at the planet) and, in the case of outnumbering the enemy's units, the player becomes owner of that planet. If the planet already belongs to the player, the incoming fleet is added as reinforcement. Each planet owned by a player (but not the "neutral" ones) will increase the forces there according to that planet's growth rate. Therefore, the goal is to design/evolve a function that considers the state of the map in each simulated shift and decides the actions to perform in order to get an advantage over the enemy, and, at the end, win the game.

This paper proposes an evolutionary approach for generating the decision engine of a bot that plays *Planet Wars* (or Galcon), the RTS game that has been chosen for the Google AI Challenge 2010. This decision engine has been implemented in two steps: first, a set of rules, which, depending on some parameters, models the behavior of the bot, is defined by means of exhaustive experimentation; the second step applies a Genetic Algorithm (GA) to evolve (and improve) these parameters off-line, i.e., not during a match, but previously. Next we will present the state of the art in RTS games like this one.

2 State of the Art

RTS games show an emergent component [1] as a consequence of the two level AI (making decisions on the set of units, and one devoted to the each of these small units), since the units behave in many different (and sometimes unpredictable) ways. This feature can make a RTS game more entertaining for a player, and maybe more interesting for a researcher. In addition, in many RTS games, traditional artificial intelligence techniques fail to play at a human level because of the vast search spaces that they entail. In this sense, Ontano et at. [2] proposed to extract behavioral knowledge from expert demonstrations in form of individual cases. This knowledge could be reused via a case based behavior generator that proposed advanced behaviors to achieve specific goals.

So, recently a number of soft-computing techniques and algorithms, such as co-evolutionary algorithms [3] or multi-agent based methods [4], just to cite a few, have already been applied to handle these problems in the implementation of RTS games. For instance, there are many benefits attempting to build adaptive learning AI systems which may exist at multiple levels of the game hierarchy, and which co-evolve over time. In these cases, co-evolving strategies might be not only opponents but also partners operating at different levels [5] Other authors propose using co-evolution for evolving team tactics [6], but the problem is how tactics are constrained and parametrized and how compute the overall score.

Evolutionary algorithms have also been used in this field [7,8], but they involve considerable computational cost and thus are not frequently used in on-line games. In fact, the most successful proposals correspond to EAs' off-line applications, that is, the EA works (for instance, to improve the operational rules that guide the bot's actions) while the game is not being played, and the results or improvements can be used later during the game. Through offline evolutionary learning, the quality of bots' intelligence can be improved, and this has been proved to be more effective than opponent-based scripts.

This way, in this work, an offline EA is applied to a parametrized tactic (set of behavior model rules) inside the Planet Wars game (a "simple" RTS game), in order to build the decision engine of a bot for that game, which will be considered later in the online matches. The process of designing this bot is presented next.

3 GeneBot: The Galactic Conqueror

As previously stated in Section 1, the main constraint in the environment is the limited processing time available to perform the correspondent actions (1 second). In addition, there is another key constraint: no memory is allowed, i.e., the bot cannot maintain a register of the results or efficiency of previous actions. These restrictions strongly limit the design and implementation possibilities for a bot, since many metaheuristics are based on a memory of solutions or on the assignment of payoffs to previous actions in order to improve future behavior, and most of them are quite expensive in running time; running an Evolutionary Algorithm in each time-step of 1 second, for instance, or a Monte Carlo method [9], is almost impossible. Besides, only the overall result of the strategy can be evaluated. It is not possible to optimize individual actions due to the lack of feedback from one turn to the next. These are the reasons why we have decided to define a set of rules which models the on-line (during the game) bot's AI. The rules have been formulated through exhaustive experimentation, and are strongly dependent on some key parameters, which ultimately determine the behavior of the bot. Anyway, there is only one type of action: move starships from one planet to another. The action is very simple, so the difficulty lies in choosing which planet creates a fleet to send forth, how many starships will be included in it and what will the target planet be.

The main example of this type of behavior is the Google-supplied baseline example, which we will call GoogleBot, included as a Java program in the game kit that can be downloaded from the GAIC site. GoogleBot works as follows: for a specific state of the map, the bot seeks for the planet owned by him that hosts the most ships and uses it as the base for the attack; The target will be chosen by calculating the ratio between the growth-rate and the number of ships for all enemy and neutral planets. Then it waits until the expeditionary attack fleet has reached its target; then it goes back to *attack mode*, selecting another planet as base for a new expedition.

Despite its simplicity, GoogleBot manages to win enough maps if its opponent is not good enough or is geared towards a particular situation or configuration. In fact the Google AI Contest recommends that any candidate bot should be able to win the GoogleBot every time in order to have any chance to get in the hall of fame; this is the baseline to consider the bot as a challenger, and the number of turns it needs to win is an indicator of its quality.

AresBot was designed to beat GoogleBot, and it works as follows: at the beginning of a turn, the bot tries to find its own *base planet*, decided on the basis of a score function. The rest of the planets are designed *colonies*. Then, it determines which *target planet* to attack (or to reinforce, if it already belongs to it) in the next turns (since it can take some turns to get to that planet). If the planet to attack is neutral, the action is designed *expansion*; however, if the planet is occupied by the enemy, the action is designed *conquest*. The base planet is also reinforced with starships coming from colonies; this action is called *tithe*, a kind of tax that is levied from the colonies to the imperial see. The rationale for this behavior is first to keep a stronghold that is difficult to conquer by the enemy, and at the same time to easily create a staging base for attacking the enemy. Furthermore, colonies that are closer to the target than to the base also send fleets to attack the target instead of reinforcing the base. This allows starships to travel directly to where they are required instead of accumulating at the base and then be sent. Besides, once a planet is being attacked it is marked so that it is not targeted for another attack until it is finished; this can be done straightforwardly since each attack fleet includes its target planet in its data.

The set of parameters is composed by weights, probabilities and amounts, that have been included in the rules that model the bot behavior. These parameters have been adjusted by hand, and they obviously totally determine the behavior of the bot. Its value and meaning are:

- $tithe_{perc}$ and $tithe_{prob}$: percentage of starships the bot sends (regarding the number of starships in the planet) and probability it happens.
- ω_{NS-DIS} and ω_{GR}: weight of the number of starships and planet growth rate hosted at the planet and the distance from the base planet to the target planet; it is used in the score function of target planet.
- $pool_{perc}$ and $support_{perc}$: proportion and percentage of extra starships that the bot sends from the base planet to the target planet.
- $support_{prob}$: probability of sending extra fleets from the colonies to the target planet.

Each parameter takes values in a different range, depending on its meaning, magnitude and significance in the game. These values are used in expressions used by the bot to take decisions. For instance, the function considered to select the *target planet* is defined this way:

$$Score(p) = \frac{p.NumStarships \cdot \omega_{NS-DIS} \cdot Dist(base, p)}{1 + p.GrowthRate \cdot \omega_{GR}} \tag{1}$$

where ω_{NS-DIS} and, ω_{GR} are weights related to the number of starships, the growth rate and the distance to the target planet. *base*, as explained above, is the planet with the maximum number of starships, and p is the planet to evaluate. The divisor is added 1 to protect against division by zero.

Once the target enemy planet is identified, a particular colony (chosen considering the tithe probability) can provide a part of its starships to the base planet. Moreover, if the distance between the colony and the target planet is less than the distance between the base and target planet, there is a likelihood that the colony also sent a number of troops to the target planet. When these movements are scheduled, a fleet is sent from the base planet with enough starships to beat the target.

All parameters in AresBot are estimated; however, they can be left as variable and optimized using an evolutionary algorithm [10] before sending out the bot to compete; we called the result *GeneBot*. The proposed GA uses floating point array to codify for all parameters shown in the previous versions, and follows a *generational* [10] scheme with *elitism* (the best solution always survives). The genetic operators include a *BLX-alpha* crossover [11] (with α equal to 0.5) and a *gene mutator* which mutates the value of a random gene by adding or subtracting a random quantity in the $[0, 1]$ interval. Each operator have an application rate (0.6 for crossover and 0.02 for mutator). These values were set by hand; since each run of the algorithm took a whole day.

The *selection mechanism* implements a *2-tournament*. Several other values were considered, but eventually the best results were obtained for this one, which represents the lowest selective pressure. The elitism has been implemented by replacing a random individual in the next population with the global best at the moment. The worst is not replaced in order to preserve the diversity.

The evaluation of one individual is performed by setting the correspondent values in the chromosome as the parameters for GeneBot's behavior, and placing the bot inside a scenario to fight against a GoogleBot in five maps that were chosen for its significance.The bots then fights five matches (one in each map). The result of the match is not deterministic, but instead of doing several matches over each map, we consider that the different results obtained for a single individual in each generation will make only those that consistently obtain good results be kept within the population.

The performance of the bot is reflected in two values: the first one is the number of turns that the bot has needed to win in each arena (WT), and the second is the number of games that the bot has lost (LT). Every generation bots are ranked considering the LT value; in case of coincidence, then the WT value is also considered, as shown above: the best bot is the one that has won every single game; if two bots have the same WT value, the best is the one that needs less turns to win. A multi-objective approach would in principle be possible here; however, it is clear that the most important thing is to win the most games, or all in fact, and then minimize the number of turns; this way of ranking the population can be seen as an strategy of implementing a constrained optimization problem: minimize the number of turns needed to win *provided that* the individual is able to win every single game. Finally, in case of a complete draw (same value for LT and WT), zero is returned.

4 Experiments and Results

To test the algorithm, different games have been played by pitting the standard bot (AresBot) and the optimized bot (GeneBot) versus the GoogleBot. The parameters considered in the GA are a population of 400 individuals, with a crossover probability of 0.6 (in a random point) and a mutation rate of 0.02. A 2 individuals elitism has been implemented.

The evaluation of each individual takes around 40 seconds, that is why we had time to make single run in time to enter the bot in the competition. In this run we obtained the values shown in Table 1.

Table 1. Initial behavior parameters values of the original bot (AresBot), and the optimized values (evolved by a GA) for the best bot obtained using the evolutionary algorithm (GeneBot)

	$tithe_{perc}$	$tithe_{prob}$	ω_{NS-DIS}	ω_{GR}	$pool_{perc}$	$support_{perc}$	$support_{prob}$
AresBot	0.1	0.5	1	1	0.25	0.5	0.9
GeneBot	0.294	0.0389	0.316	0.844	0.727	0.822	0.579

Results in Table 1 show that the best results are obtained by strategies where colonies have a low probability of sending tithe to the base planet (only 0.3), and those tithes send a few hosted starships, which probably implies that colonies should be left on its own to defend themselves instead of supplying the base planet. On the other hand, the probability for a planet to send starships to attack another planet is quite high (0.58), and the proportion of units sent is also elevated, showing that it is more important to attack with all the available starships than wait for reinforcements. Related to this property is the fact that, when attacking a target planet, the base one also sends a large number of extra starships (72.7 % of the hosted ships). Finally, to define the target planet to attack, the number of starships hosted in the planet is not as important as the growth range, but being the distance also an important value to consider.

After these experiments, the value of the obtained parameters has been tested considering 100 different games (matches), where the 'evolved' GeneBot, and the AresBot have fought against a standard GoogleBot. The results are shown in Table 2.

Table 2. Results after 100 games for our standard bot (AresBot) and the best optimized bot (GeneBot) versus the Google Standard Bot

	Turns			Victories
	Average and Std. Dev	Min	Max	
AresBot	210 ± 130	43	1001	99
GeneBot	159 ± 75	22	458	100

The number of turns a bot needs to win in a map is the most important factor of the two considered in the fitness, since it needs to beat GoogleBot in all maps for having any kind of chance in the challenge. In the first turns, the two bots handle the same number of starships so making a difference in a few turns implies the bot knows what to do and is able to accrue many more ships (by conquering ship-growing planets) fast. If it takes many turns, the actions of the bot have some room for improvement, and it would be even possible, if the enemy is a bit better than the one Google issues as a baseline, to be defeated.

In general, the improvement to the original AresBot offered by the algorithm could seem small from the purely numeric point of view; GeneBot is able to win in one of the maps where AresBot was beaten, which was one of the 5 selected to perform evolution, and the aggregate number of generations is around 10%. However, this small advantage confers some leverage to win more battles, which in turn will increase its ranking in the Google AI challenge. This indicates that an evolutionary algorithm holds a lot of promise in optimizing any kind of behavior, even a parametrized behavior like the one programmed in GeneBot. However, a lot of work remains to be done, either to compete in next year's challenge, or to explore all the possibilities the genetic evolution of bot behavior can offer.

5 Conclusions and Future Work

The Google AI Challenge 2010 is an international programming contest where game-playing programs (bots) fight against others in a RTS game called Planet Wars. In this paper we wanted to show how evolutionary algorithms can be applied to obtain good results in a real-world challenge, by submitting to the competition a bot whose behavioral parameters are obtained using a Genetic Algorithm, and it has been shown that using this kind of algorithms increases the efficiency in playing versus hand-coded bots, winning more runs in a lower number of turns. Results obtained in this work show that it is important to attack planets with all available ships hosted in these planets, instead of storing these ships for future attacks.

The bot described here eventually finished 1454[1], winning nine matches and losing six; this placed it among the 32% best, which means that at least this technique of fine-tuning strategy parameters shows a lot of promise; however, it can only take you as far as the strategy allows. This was an improvement of more than 1000 positions over the non-optimized version.

In the future we will try to improve the baseline strategy, and even make the evolutionary process choose between different possible strategies; we will also try to make evolution faster so that we can try different parametrizations to obtain bots as efficient as possible.

References

1. Sweetser, P.: Emergence in Games. In: Game development, Charles River Media, Boston (2008)
2. Ontañón, S., Mishra, K., Sugandh, N., Ram, A.: Case-based planning and execution for real-time strategy games. In: Weber, R.O., Richter, M.M. (eds.) ICCBR 2007. LNCS (LNAI), vol. 4626, pp. 164–178. Springer, Heidelberg (2007)
3. Keaveney, D., OŔiordan, C.: Evolving robust strategies for an abstract real-time strategy game. In: International Symposium on Computational Intelligence in Games, Milano. Italy, pp. 371–378. IEEE Press, New York (2009)
4. Hagelbäck, J., Johansson, S.J.: A multiagent potential field-based bot for real-time strategy games. Int. J. Comput. Games Technol., 2009, 4:1–4:10 (2009)
5. Livingstone, D.: Coevolution in hierarchical ai for strategy games. In: IEEE Symposium on Computational Intelligence and Games (CIG 2005), pp. 190–194. IEEE, Colchester (2005)
6. Avery, P., Louis, S.: Coevolving team tactics for a real-time strategy game. In: Proceedings of the 2010 IEEE Congress on Evolutionary Computation (2010)
7. Ponsen, M., Munoz-Avila, H., Spronck, P., Aha, D.W.: Automatically generating game tactics through evolutionary learning. AI Magazine 27(3), 75–84 (2006)
8. Jang, S.H., Yoon, J.W., Cho, S.B.: Optimal strategy selection of non-player character on real time strategy game using a speciated evolutionary algorithm. In: Proceedings of the 5th IEEE Symposium on Computational Intelligence and Games (CIG 2009), pp. 75–79. IEEE Press, Piscataway (2009)
9. Lucas, S.: Computational intelligence and games: Challenges and opportunities. International Journal of Automation and Computing 5(1), 45–57 (2008)
10. Michalewicz, Z.: Genetic Algorithms + Data Structures = Evolution Programs. Springer, Heidelberg (1996)
11. Herrera, F., Lozano, M., Sánchez, A.M.: A taxonomy for the crossover operator for real-coded genetic algorithms: An experimental study. International Journal of Intelligent Systems 18, 309–338 (2003)

[1] Final ranking at: `http://ai-contest.com/profile.php?user_id=8220`

Implementation Matters: Programming Best Practices for Evolutionary Algorithms*

J.J. Merelo, G. Romero, M.G. Arenas, P.A. Castillo, A.M. Mora, and J.L.J. Laredo

Dpto. de Arquitectura y Tecnología de Computadores. Univ. of Granada, Spain
{jmerelo,gustavo,mgarenas,pedro,amorag,juanlu}@geneura.ugr.es

Abstract. While a lot of attention is usually devoted to the study of different components of evolutionary algorithms or the creation of heuristic operators, little effort is being directed at how these algorithms are actually implemented. However, the efficient implementation of any application is essential to obtain a good performance, to the point that performance improvements obtained by changes in implementation are usually much bigger than those obtained by algorithmic changes, and they also scale much better. In this paper we will present and apply usual methodologies for performance improvement to evolutionary algorithms, and show which implementation options yield the best results for a certain problem configuration and which ones scale better when features such as population or chromosome size increase.

1 Introduction

The design of evolutionary algorithms (EAs) usually includes a methodology for making them as efficient as possible. Efficiency is measured using metrics such as the number of evaluations to solution; implicitly seeking to reduce running times. However, the same amount of attention is not given to designing an implementation as efficient as possible, even as small changes in it can have a much bigger impact in the overall running time than any algorithmic improvement. This lack of interest, or attention, in the actual implementation of algorithms proposed results in the quality of scientific programming being, on average, worse than what is usually found in companies [1] or released software.

It can be argued that the time devoted to an efficient implementation can be better employed pursuing scientific innovation or a precise description of the algorithm; however, the methodology for making improvements in program running time is well established in computer science: there are several static or dynamic analysis tools which look at memory and running time (called <u>monitors</u>), and thus, it can be established how much memory and time the program takes, and then which parts of it (variables, functions) are responsible for that, for which

* This work has been supported in part by the CEI BioTIC GENIL (CEB09-0010) Programa CEI del MICINN (PYR-2010-13) project, the Junta de Andalucía TIC-3903 and P08-TIC-03928 projects, and the Jaén University UJA-08-16-30 project.

J. Cabestany, I. Rojas, and G. Joya (Eds.): IWANN 2011, Part II, LNCS 6692, pp. 333–340, 2011.

profilers are used. Once this methodology has been included into the design process of scientific software, it does not need to take much more time than, say, running statistical tests. In the same way that these tests establish scientific accuracy, an efficient implementation makes results better and more easily reproducible and understandable.

Profiling the code that implements an algorithm also allows to detect potential bugs, see whether code fragments are executed as many times as they should, and detect the which parts of the code can be optimized in order to obtain the most impact on performance. After profiling, the deepest knowledge on the structure underlying the algorithm will allow a more efficient redesign, balancing algorithmic with computational efficiency; this deep knowledge also allows to find out computational techniques that can be leveraged in the search for new evolutionary techniques. For instance, knowing how a sorting algorithm scales with population size would allow the EA designer to choose the best option for a particular population size, or eliminate sorting completely using a methodology that avoids sorting altogether, possibly finding new operators or selection techniques for EAs.

In this paper, we will comment the enhancements applied to a program written in Perl [2–4] which implements an evolutionary algorithm, and also a methodology for its analysis, proving the impact of the identification of bottlenecks in a program, and its elimination through common programming techniques. This impact can go up to several orders of magnitude, but of course it depends on the complexity of the fitness function and the size of the problem it is applied to, as has been proved in papers such as the one by Laredo et al. [5]. In principle, the methodology and tools that have been used are language-independent, and can be found in any programing language, however the performance improvements and the options for changing a program will depend on the language implied.

From a first baseline or straightforward implementation of an EA, we will show techniques to measure the performance obtained with it, and how to derive a set of rules that improve its efficiency. Given that research papers are not commonly focused on detailing such techniques, best programming practices for EAs use to remain hidden and can not benefit the rest of the community. A typical research paper do not detail these techniques, so that this knowledge remains hidden and can not benefit the rest of the community. This work is an attempt to highlight those techniques and encourage the community to reveal how published results are obtained.

The rest of this paper is structured as follows: Section 2 presents a comprehensive review of the approaches found in the bibliography. Section 3 briefly describes the methodology followed in this study and discusses the results obtained using different techniques and versions of the program. Finally, conclusions and future work are presented in Section 4.

2 State of the Art

EA implementation has been the subject of many works by our group [6–10] and by others [11–14]. Much effort has been devoted looking for new hardware platforms to run EAs as GPUs [14] of specialized hardware [15]) than trying to maximize the potential of usual hardware.

As more powerful hardware is available every year researchers have pursuit the invention of new algorithms [16–18] forgiving how important efficiency is. There has been some attempts to calculate the complexity of EAs with the intention of improving it: by avoiding random factors [19] or by changing the random number generator [20].

However, even on the most modern systems, EA experimentation can be a extremely long process because every algorithm run can last several hours (or days), and it must be repeat several times in order to obtain accurate statistics. And that just in the case of knowing the optimal set of parameters. Sometimes the experiments must be repeated with different parameters to discover the optimal combination (systematic experimentation). So in the following sections we pay attention to implementation details, making improvements in an iterative process.

3 Methodology, Experiments and Results

The initial version of the program is taken from [2], and it is shown in Tables 1 and 2. A *canonical EA* with proportional selection, two individual elite, mutation and crossover is implemented. The problem used is MaxOnes (also called OneMax)[21], where the function to optimize is simply the number of ones in a bit-string, with chromosomes changing in length from 16 to 512.

The initial population has 32 individuals, and the algorithm runs for 100 generations. The experiments are performed with different chromosome and population sizes, since the algorithms implemented in the program have different complexity with respect to those two parameters. These runs have been repeated 30 times for statistical accuracy reasons.

Running time in user space (as opposed to wallclock time, which includes time spent in other user and system processes) is measured each time a change is made.

In these experiments, the first improvement tested is to include a *fitness cache* [16, 2], that is, a data structure called hash which remembers the values already computed for the fitness function.

This change trades off memory for fast access, as has been mentioned above, increasing speed but also the memory needed to store the precomputed values. This is always a good option if there is plenty of memory available, but if this aspect is not checked and swapping (virtual memory in other OSs) is activated, it might imply a huge decrease in performance: parts of program data will start to be swapped out to disk, resulting in a huge performance decrease. However, a quick calculation beforehand will tell us if we should worry about this and turn cache off if that is the case.

It is also convenient to look for the fastest way of computing the fitness function, using language-specific data structures, functions and expressions[1].

[1] Changes can be examined in the code repository at `http://bit.ly/b0k3z3`

Table 1. First version of the program used in the experiments (main program). An evolutionary algorithm is implemented.

```
my $chromosome_length = shift || 16;
my $population_size = shift || 32;
my $generations = shift || 100;
my @population = map(random_chromosome($chromosome_length), 1..$population_size);
map( compute_fitness( $_ ), @population );
for ( 1..$generations ) {
  my @sorted_population =sort{$b->{'fitness'}<=>$a->{'fitness'}}@population;
  my @best = @sorted_population[0,1];
  my @wheel = compute_wheel( \@sorted_population );
  my @slots = spin( \@wheel, $population_size );
  my @pool;
  my $index = 0;
  do {
    my $p = $index++ % @slots;
    my $copies = $slots[$p];
    for (1..$copies) {
      push @pool, $sorted_population[$p];
    }
  } while ( @pool <= $population_size );
  @population = ();
  map( mutate($_), @pool );
  for ( my $i = 0; $i < $population_size/2 -1 ; $i++ )  {
    my $first = $pool[rand($#pool)];
    my $second = $pool[ rand($#pool)];
    push @population, crossover( $first, $second );
  }
  map( compute_fitness( $_ ), @population );
  push @population, @best;
}
```

As can be seen in the results shown in Figure 1-left, running time of the initial version grows more than linearly. For big instances of the problem run time can became to long to be practical. For example, for chromosomes of length 16 running time of the program version used as baseline is half of the third version. For chromosomes of length 256 the run time difference is an order of magnitude greater. This lead us to think that optimizing the EA implementation is more valuable than any other algorithmic change. Also it must be remarked than code changes have been minimal.

In order to obtain further improvements, a profiler (one of the tools described in the introduction to this paper) has to be used. In Perl, `Devel::DProf` carries out an analysis of the different subroutines; however, a sentence-by-sentence analysis has to be done. Thus, the `Devel::NYTProf` Perl module (developed at the New York Times) is used. The results of applying this profiler show that operators like crossover and mutation are used quite a lot, and some improvements can be made over them; however, the function which takes the most time is the one that sorts the population. Changing the default Perl sort, which implements quicksort [22], to another version using *mergesort* [23] meant a small improvement, but using `Sort::Key`, the best one available in the Perl language yields the best results.

Table 2. First version of the program used in the experiments (subrutines)

```
sub compute_wheel {
  my $population = shift;    my $total_fitness;
  map( $total_fitness += $_->{'fitness'}, @$population );
  my @wheel = map( $_->{'fitness'}/$total_fitness, @$population);
  return @wheel;
}
sub spin {
  my ( $wheel, $slots ) = @_;    my @slots = map( $_*$slots, @$wheel );
  return @slots;
}
sub random_chromosome {
  my $length = shift;    my $string = '';
  for (1..$length) {     $string .= (rand >0.5)?1:0;    }
  { string => $string,      fitness => undef };
}
sub mutate {
  my $chromosome = shift;
  my $clone = { string => $chromosome->{'string'}, fitness => undef };
  my $mutation_point = rand( length( $clone->{'string'} ));
  substr($clone->{'string'}, $mutation_point, 1,
  ( substr($clone->{'string'}, $mutation_point, 1) eq 1 )?0:1 );
  return $clone;
}
sub crossover {
  my ($chrom_1, $chrom_2) = @_;
  my $chromosome_1 = { string => $chrom_1->{'string'} };
  my $chromosome_2 = { string => $chrom_2->{'string'} };
  my $length = length( $chromosome_1 );
  my $xover_point_1 = int rand( $length -1 );
  my $xover_point_2 = int rand( $length -1 );
  if ( $xover_point_2 < $xover_point_1 )  {
    my $swap = $xover_point_1;
    $xover_point_1 = $xover_point_2;      $xover_point_2 = $swap;
  }
  $xover_point_2 = $xover_point_1 + 1 if  ( $xover_point_2 == $xover_point_1
);
  my $swap_chrom = $chromosome_1;
  substr($chromosome_1->{'string'}, $xover_point_1, $xover_point_2 -
$xover_point_1 + 1,
  substr($chromosome_2->{'string'}, $xover_point_1, $xover_point_2 -
$xover_point_1 + 1) );
  substr($chromosome_2->{'string'}, $xover_point_1, $xover_point_2 -
$xover_point_1 + 1,
  substr($swap_chrom->{'string'}, $xover_point_1, $xover_point_2 -
$xover_point_1 + 1) );
  return ( $chromosome_1, $chromosome_2 );
}

sub compute_fitness {
  my $chromosome = shift;    my $unos = 0;
  for ( my $i = 0; $i < length($chromosome->{'string'}); $i ++ ) {
      $unos += substr($chromosome->{'string'}, $i, 1 );
  }
  $chromosome->{'fitness'} = $unos;
}
```

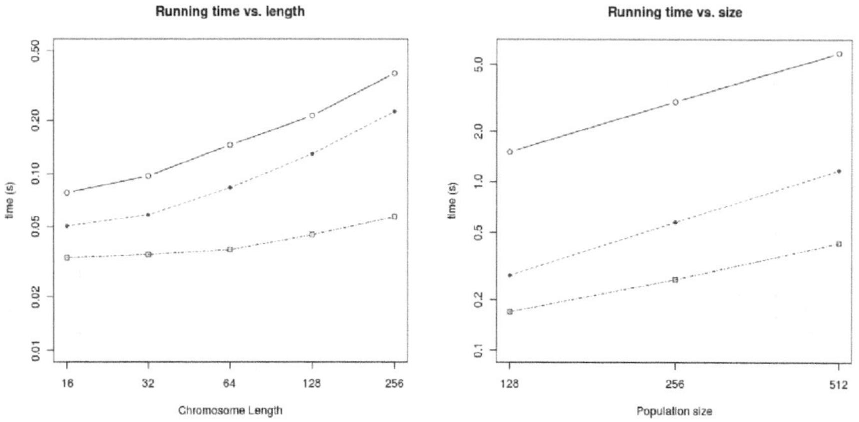

Fig. 1. Log-log plot of running time for different chromosome (left) and population sizes (right). Solid-line corresponds to the baseline version. (Left) Dashed version uses a cache, and dot-dashed one changes fitness calculation. (Right) Dashed version changes fitness calculation, while dot-dashed one uses best-of-breed sorting algorithm for the population. Values are averages for 30 runs.

Figure 1-right shows how run time grows with population size for a fixed chromosome size of 128. The algorithm is run 100 times regardless of whether the solution is found or not. The EA behavior is similarly to the previous analysis. The most efficient version, using `Sort::Key`, is an order of magnitude more efficient than the first attempt and the difference grows with the population size. Adding up both improvements, for the same problem size, almost two order of magnitude better results are obtained without changing our basic algorithm.

It should be noted that since these improvements are algorithmically neutral, they do not have a noticeable impact on results, being statistically indistinguishable from the one obtained by the baseline program.

4 Conclusions and Future Work

This work shows how good programming practices and a deep knowledge of data and control structures of a programming language can yield an improvement of up to two orders of magnitude in an evolutionary algorithm (EA). Our tests consider a well known problem whose results can be easily extrapolated to others. An elimination of bottlenecks after the profiling of the implementation of an evolutionary algorithm can give better results than a new algorithm with different, and likely more complex algorithms or a change of parameters in the existing algorithm. A cache of evaluations can be used on a wide variety of EA problems. Moreover, a profiler program can be applied on every implementation, to detect bottlenecks and concentrate efforts on solving them.

From these experiments, we conclude that applying profilers to identify the bottlenecks of evolutionary algorithm implementations, and then careful and informed programming to optimize those fragments of code, greatly improves running time of evolutionary algorithms without degrading algorithmic performance.

Several other techniques can improve EA performance; for instance mutithreading can be used to take advantage of symmetric multiprocessing and multicore machines; message passing techniques can be applied to divide the work for execution on clusters, and vectorization for execution on a GPU, are three of the more well known and usually employed, but almost every best practice in programming can be applied successfully to improve EAs. In turn, these techniques will be incorporated to the ALGORITHM::EVOLUTIONARY [16] Perl library. A thorough study of the interplay between implementation and the algorithmic performance of the implemented techniques will also be carried out.

References

1. Merali, Z.: Computational science: Error, why scientific programming does not compute. Nature 467(7317), 775–777 (2010)
2. Merelo-Guervós, J.J.: A Perl primer for EA practitioners. SIGEvolution 4(4), 12–19 (2010)
3. Wall, L., Christiansen, T., Orwant, J.: Programming Perl, 3rd edn. O'Reilly & Associates, Sebastopol (2000)
4. Schwartz, R.L., Phoenix, T., foy, B.D.: Learning Perl, 5th edn. O'Reilly & Associates (2008)
5. Laredo, J., Castillo, P., Mora, A., Merelo, J.: Exploring population structures for locally concurrent and massively parallel evolutionary algorithms. In: Computational Intelligence: Research Frontiers, pp. 2610–2617. IEEE Press, Los Alamitos (2008)
6. Merelo-Guervós, J.J.: Algoritmos evolutivos en Perl. Ponencia presentada en el V Congreso Hispalinux, disponible en (November 2002), http://congreso.hispalinux.es/ponencias/merelo/ae-hispalinux2002.html
7. Merelo-Guervós, J.J.: OPEAL, una librería de algoritmos evolutivos en Perl. In: Alba, E., Fernández, F., Gómez, J.A., Herrera, F., Hidalgo, J.I., Merelo-Guervós, J.J., Sánchez, J.M. (eds.) Actas primer congreso español algoritmos evolutivos, AEB 2002, Universidad de Extremadura, pp. 54–59 (February 2002)
8. Arenas, M., Foucart, L., Merelo-Guervós, J.J., Castillo, P.A.: JEO: a framework for Evolving Objects in Java. In: [24], pp. 185–191, http://geneura.ugr.es/pub/papers/jornadas2001.pdf
9. Castellano, J., Castillo, P., Merelo-Guervós, J.J., Romero, G.: Paralelización de evolving objects library usando MPI. In: [24], pp. 265–270
10. Keijzer, M., Merelo, J.J., Romero, G., Schoenauer, M.: Evolving objects: A general purpose evolutionary computation library. In: Collet, P., Fonlupt, C., Hao, J.-K., Lutton, E., Schoenauer, M. (eds.) EA 2001. LNCS, vol. 2310, pp. 231–244. Springer, Heidelberg (2002)
11. Fogel, D., Bäck, T., Michalewicz, Z.: Evolutionary Computation: Advanced algorithms and operators. Taylor & Francis, Abington (2000)
12. Setzkorn, C., Paton, R.: JavaSpaces–An Affordable Technology for the Simple Implementation of Reusable Parallel Evolutionary Algorithms. Knowledge Exploration in Life Science Informatics, 151–160

13. Rummler, A., Scarbata, G.: eaLib – A Java Frameword for Implementation of Evolutionary Algorithms. Theory and Applications Computational Intelligence, 92–102
14. Wong, M., Wong, T.: Implementation of parallel genetic algorithms on graphics processing units. Intelligent and Evolutionary Systems, 197–216 (2009)
15. Schubert, T., Mackensen, E., Drechsler, N., Drechsler, R., Becker, B.: Specialized hardware for implementation of evolutionary algorithms. In: Genetic and Evolutionary Computing Conference, Citeseer, p. 369 (2000)
16. Merelo-Guervós, J.J., Castillo, P.A., Alba, E.: Algorithm: Evolutionary, a flexible Perl module for evolutionary computation. Soft Computing (2009), `http://sl.ugr.es/000K` (to be published)
17. Ventura, S., Ortiz, D., Hervás, C.: JCLEC: Una biblioteca de clases java para computación evolutiva. In: Primer Congreso Español de Algoritmos Evolutivos y Bioinspirador, pp. 23–30. Mérida, Spain (2002)
18. Ventura, S., Romero, C., Zafra, A., Delgado, J., Hervás, C.: JCLEC: a Java framework for evolutionary computation. Soft Computing-A Fusion of Foundations, Methodologies and Applications 12(4), 381–392 (2008)
19. Salomon, R.: Improving the performance of genetic algorithms through derandomization. Software - Concepts and Tools 18(4), 175 (1997)
20. Digalakis, J.G., Margaritis, K.G.: On benchmarking functions for genetic algorithms. International Journal of Computer Mathematics 77(4), 481–506 (2001)
21. Muhlenbein, H.: How genetic algorithms really work: I. mutation and hillclimbing. In: Munner, R., Manderick, B. (eds.) Proceedings of the Second Conference on Parallel Problem Solving from Nature (PPSN II). pp. 15–25. North-Holland, Amsterdam (1992)
22. Hoare, C.: Quicksort. The Computer Journal 5(1), 10 (1962)
23. Cole, R.: Parallel merge sort.In: 27th Annual Symposium on Foundations of Computer Science 1985, pp. 511–516 (1986)
24. UPV. In: Actas XII Jornadas de Paralelismo, UPV, Universidad Politécnica de Valencia (2001)

Online vs. Offline ANOVA Use on Evolutionary Algorithms

G. Romero, M.G. Arenas, P.A. Castillo, J.J. Merelo, and A.M. Mora

Dep. of Computer Architecture and Technology, University of Granada, Spain
gustavo@atc.ugr.es

Abstract. One of the main drawbacks of evolutionary algorithms is their great amount of parameters. Every step to lower this quantity is a step in the right direction. Automatic control of variation operators application rates during the run of an evolutionary algorithm is a desirable feature for two reasons: we are lowering the number of parameters of the algorithm and making it able to react changes in the conditions of the problem. In this paper, a dynamic breeder able to adapt the operators application rates over time following the evolutionary process is proposed. The decision to raise or to lower every rate is based on ANOVA to be sure of statistical significant.

1 Introduction

Evolutionary algorithms (EAs) usually need a great number of components and parameters (i.e. population size, time and resources allowed, kinds of transformations and recombinations and their application rates, selective pressure). Every EA component may have one or more parameters. The values of this parameters affects the probability of finding the optimum problem solution and the efficiency of the algorithm.

Although many authors state that an optimum parameter configuration will be very difficult or impossible to be found [5], most researchers look for it. Several methods have been used to try to discover the best configuration possible, supposing that one exists:

- With some luck initial election of values will be good enough to produce solutions within given time and resources constrains.
- One of the most used methods is to choose the values based on previous experience from us or some expert in the field. Thus, for a genetic algorithm (GA), configurations proposed by De Jong [4] or Grefenstette [6] might be useful.
- Other method is to repeat an experiment with and without some variation operator to determine how the algorithm is affected by an individual operator. This method as stated by Jones [9] is not significative enough.
- If good operators are known, only its application rates should be found. Again experimentation has been widely used. The same experiment is repeated varying only the application rate of one operator at a time. Best

J. Cabestany, I. Rojas, and G. Joya (Eds.): IWANN 2011, Part II, LNCS 6692, pp. 341–347, 2011.

fitness average will be an indicator of optimal value. Recently, analysis of variance (ANOVA) [12,10,2] is employed to assert that averages are significative [3,13]. This approach has two problems: parameters values are not independent and it is very time consuming.
- The most recent trend is self-adaptation [1,15,5]. The algorithm starts its execution with known good default parameters values that are modified over time following some variations in the evolutive process [8,16].

In this paper we center our attention in two problems: what are the best variation operators and what application rates will be optimal. We also will try to avoid the high computational cost of experimental methods. This will be obtained through the use of a self-adaptative method for parameters application rates.

The remainder of this paper is structured as follows: Section 2 describes how ANOVA is used to look for an optimal configuration of operator application rates. Section 3 describes a new method to self-adapt the operator application rates over time based on ANOVA. Section 4 present a set of test using some well known test problems solved using the two methods previously seen on sections 2 and 3. Finally some conclusions and future work directions are presented in section 5.

2 Offline ANOVA

Many researchers try to determine the optimal parameter set for their EAs through experimentation. Lately, averages over a certain number of repetitions of an experiment are not considered valid without a more serious statistical analysis. One of the most common and reliable is ANOVA. This way you can discover if the mean behaviour is true or if it is not statistically significant.

Let's see an example with several well known test functions: Ackley, Griewangk, Rastrigin and Schwefel. To solve them a simple genetic algorithm as the one described in [11] is used. If optimum variation operators for them are unknown some generic ones will be tested. This way three kinds of mutation and another three types of crossover will are introduced in the algorithm. Remember that not only their rates but if they are appropriate is what we want to discover.

Experimentation will be expensive because if 6 operators with r values, from 0 to 1, should be tested for every parameter, and every experiment is repeated n times, the total amount of runs for every problem is $r^6 \times n$. After this process data should be processed with ANOVA to discover if it is statistically significant. This way optimal parameter values can be discovered but this process can easily led us to a dead end results are not significant. What are optimum rates them? If no optimum can be chosen, lower application rate values can be chosen for better performance.

For the cited problems the tested operators were: random mutation (uniform random number over the search space), additive mutation (normal with center 0 and sigma 1), multiplicative mutation (normal with center 1 and sigma 1), one point crossover, two point crossover and blend crossover. The ANOVA study

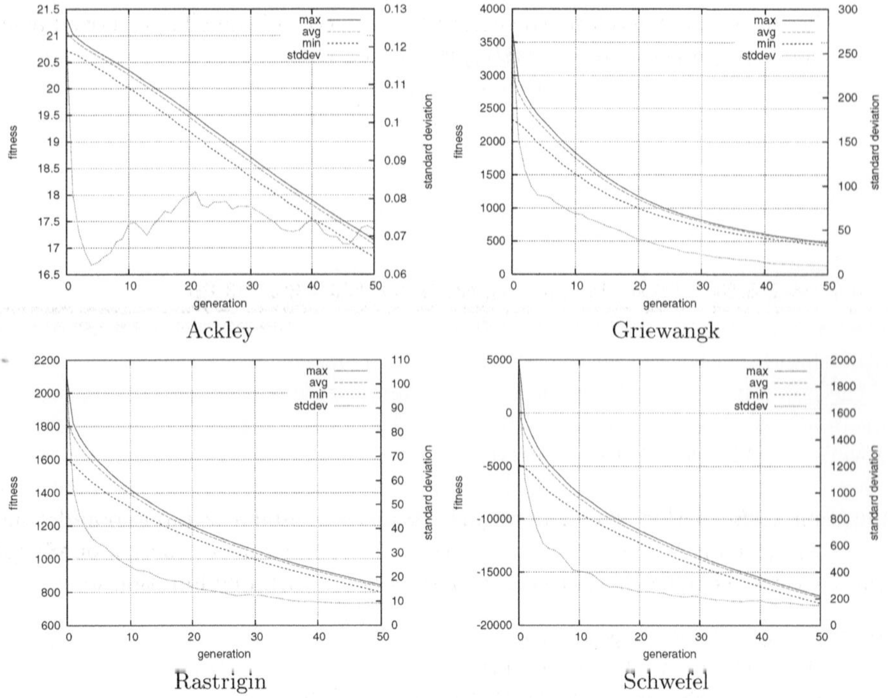

Fig. 1. Averaged results over 100 runs for every test problem with 100 dimensions with optimal parameter rates obtained with an ANOVA analysis

show that the best variation operators are multiplicative mutation and one point crossover. Every problem has different but near optimum rates, all of them are near 0.15 for mutation and 0.85 for crossover. Averaged results over 100 runs for every test problem with 100 dimensions with optimal parameter rates obtained with an ANOVA analysis can be seen on figure 1.

3 Online ANOVA

For many problems one operator can be good at the beginning and bad at the end of the process. There are other were conditions change over time for which optimum parameter rates does not exist. For these reasons not always an optimum set of parameter rates does exist. That is why an alternative method is necessary. Our algorithm is less complex than others proposed in the literature as those by C. Igel [8] or M. Hüsken [7]. The proposed method adapts the application rate of an operator by a fixed amount considering it success in one generation. The dynamic breeder, executed inside the reproductive part of the EA, is specified in the following pseudo code:

```
for every operator
    for every individual
        apply operator to individual
    if ANOVA judges population change significative
        if population is better than before
           increment operator application rate
        else
           decrement operator application rate
```

The adaptation of application rates follows an easy process. Every generation a fixed amount, α, is added or substracted to the previous value. If the population after the application of an operator is better than before α is added. On the other hand, if the population get worse than before, substract α. To determine if the population change produced by the application of one operator, instead of mean fitness, ANOVA analysis will tell us if the change is statistically significant. Several α values were tested but 0.05 usually produces best results. The operator rates are limited to a range of application. Maximum rate is 1. The minimum must be great enough to be applied to some individuals every generation. A good value is 1% although higher values can be used if population size is very small. Using our method three important question can be addressed:

- What are the best operators? If an operator is convenient its rate will grow. Otherwise it will fall to near 0. It will be almost the same as if it were not used in the algorithm. It should not reach 0 because conditions can change and it can became good in another phase of the evolution.
- What application rates are best? If an operator introduces positive changes its rate will grow, otherwise it will go down.
- If the conditions of the EA change, operator rates may change and not remain unaffected as when the evolutionary precess start.

4 Experimental Results

The same simple genetic algorithm described in section 2 is used again. Some others important parameters not mentioned before are: population size 100, 50 generations, tournament selection with size 2, inclusion replacement (type "+"). The previous experiments were repeated 100 times with identical initial conditions. This time, instead of the optimum parameters discovered offline with ANOVA, the proposed dynamic breeder method was used. The new results can be seen on figure 2.

Comparing offline ANOVA (Fig.1) with online ANOVA (Fig. 2) results shows an improvement using our proposed method. Table 1 summarize this comparison with numbers. Not only better fitness values are reached but with much less work and without the need of expert knowledge or a lengthly process of experimentation to adjust the operator application rates. Results are better for all the test cases. For new problems without a priory knowledge, this method can be an advantage as it can guess the optimum operator application rates. If no known good operators exists, many can be introduced an it will be able to raise application rates for the best and lower it for bad ones.

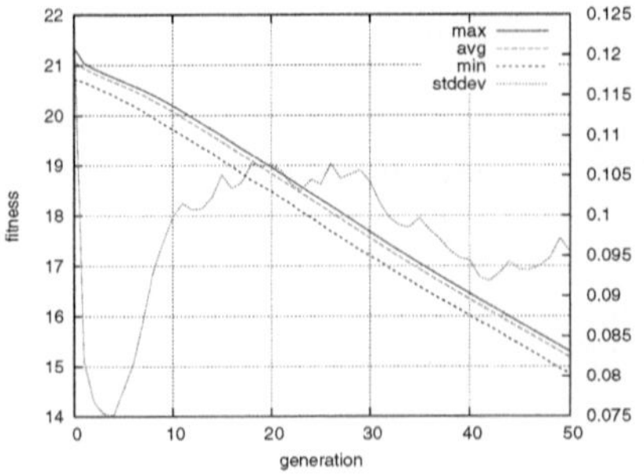

Ackley: fitness

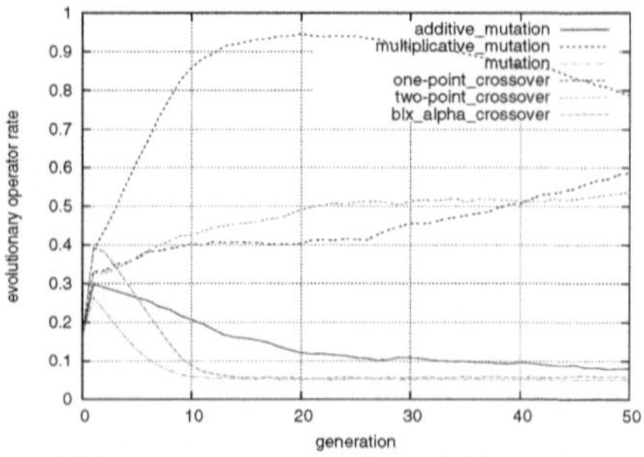

Ackley: evolutionary operator rates

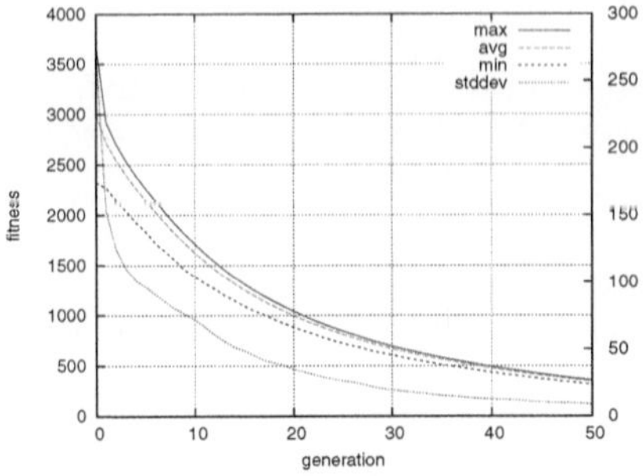

Griewangk: fitness

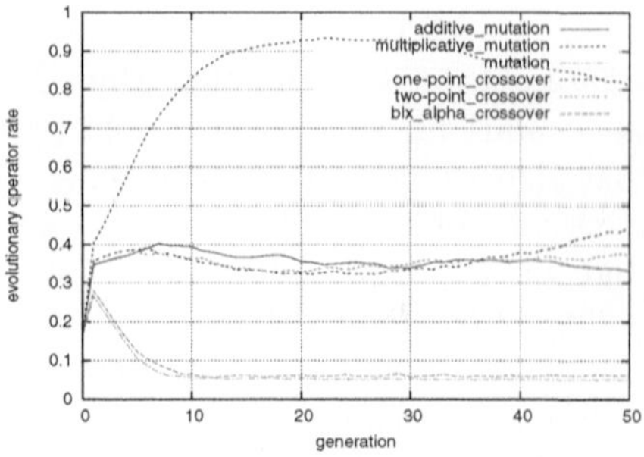

Griewangk: evolutionary operator rates

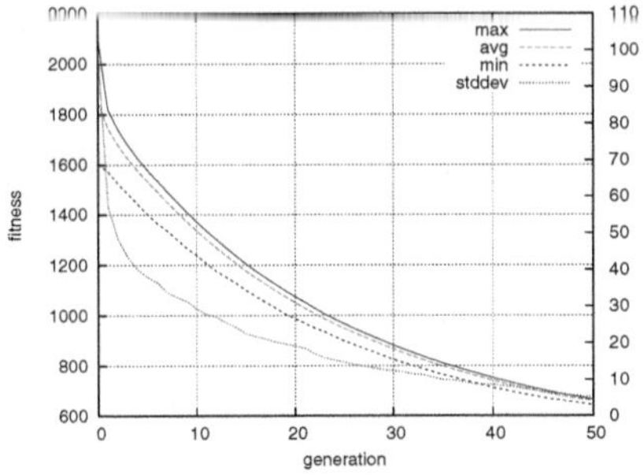

Rastrigin: fitness

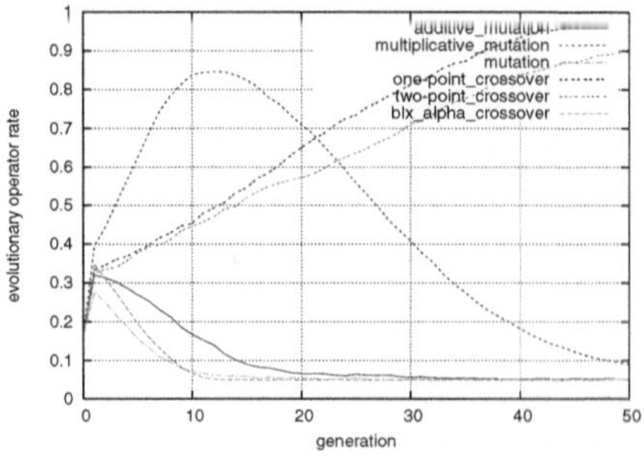

Rastrigin: evolutionary operator rates

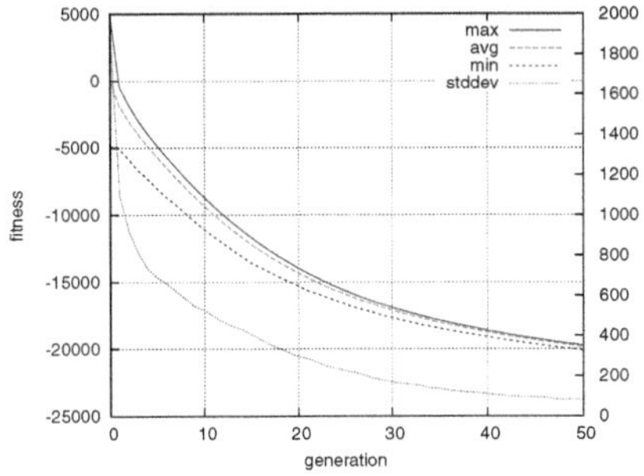

Schwefel: fitness

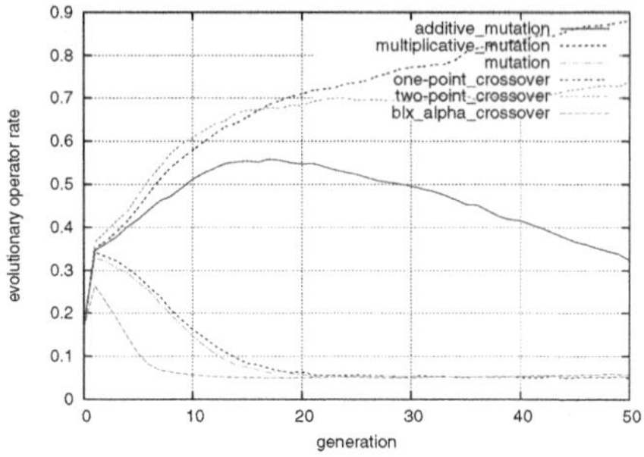

Schwefel: evolutionary operator rates

Fig. 2. Averaged results over 100 runs of every test problem for 100 dimensions using the proposed dynamic breeder based on ANOVA

Table 1. Offline vs Online ANOVA comparison. Mean fitness and standard deviation values for the test problems averaged over 100 experiments.

	offline ANOVA	online ANOVA
Ackley	16.825 ± 0.072	14.851 ± 0.096
Griewangk	429.534 ± 9.866	315.173 ± 8.871
Rastrigin	799.772 ± 9.197	642.847 ± 5.297
Schwefel	-17939.3 ± 153.559	-20114.5 ± 83.432

5 Conclusion and Future Work

Using ANOVA inside the reproductive section of an EA improves the fitness values obtained by the algorithm. The algorithm is able of react to changes in the conditions of the evolutionary process considering its state. This is one thing that algorithms with fixed parameters can't achieve.

Despite of its simplicity, the proposed method, is able to obtain better results than the application of ANOVA offline in all of the test problems.

One of the main drawbacks of evolutionary computation in general is the big amount of parameters that must be fixed. Every effort to lower this number is a step in the right direction.

As future work we are implementing a new version of the proposed dynamic breeder using n-way ANOVA instead of the one-way version used for the experiments shown in this work. This way not only the operator success will be considered but also the interactions between them.

It would be interesting to test the algorithm with real word problems, specially with those that change its conditions over time, to discover if our method is good adapting the operator application rates to changes in the environment.

For practical reasons and time constrains the statistical analysis is not as accurate as it should. Some advices and hits about a more thorough comparison of Evolutionary Algorithms can be found in [14]. This will be followed in a future review of this work.

This work just want to probe that auto-configuration of parameters is a doable task if time is not a constraint. Because once the optimal parameter set is discovered it can be used many times for free, losing time just one in the process can be a valuable inversion.

Acknowledgements

This work has been supported in part by the CEI BioTIC GENIL (CEB09-0010) Programa CEI del MICINN (PYR-2010-13) project, the Junta de Andalucía TIC-3903 and P08-TIC-03928 projects, and the Jaén University UJA-08-16-30 project.

References

1. Angeline, P.J.: Adaptive and self-adaptive evolutionary computations. In: Palaniswami, M., Attikiouzel, Y. (eds.) Computational Intelligence: A Dynamic Systems Perspective, pp. 152–163. IEEE Press, Los Alamitos (1995)

2. Casella, G., Berger, R.L.: Statistical Inference. Duxbury Press (1990)
3. Castillo, P.A., Merelo, J.J., Prieto, A., Rojas, I., Romero, G.: Statistical analysis of the parameters of a neuro-genetic algorithm. IEEE Transactions on Neural Networks 13(6), 1374–1394 (2002)
4. De Jong, K.A.: An analysis of the behavior of a class of genetic adaptive systems. PhD thesis, University of Michigan, Ann Arbor (1975)
5. Eiben, Á.E., Hinterding, R., Michalewicz, Z.: Parameter control in evolutionary algorithms. IEEE Trans. on Evolutionary Computation 3(2), 124–141 (1999)
6. Grefenstette, J.J.: Optimization of control parameters for genetic algorithms. IEEE Transactions on Systems, Man, and Cybernetics 16(1), 122–128 (1986)
7. Hüsken, M., Igel, C.: Balancing learning and evolution. In: Langdon, W.B., Cantu-Paz, E., Mathias, K., Roy, R., Davis, D., Poli, R., Balakrishnan, K., Honavar, V., Rudolph, G., Wegener, J., Bull, L., Potter, M.A., Schultz, A.C., Miller, J.F., Burke, E., Jonoska, N. (eds.) Proceedings of the Genetic and Evolutionary Computation Conference (GECCO 2002), pp. 391–398. Morgan Kaufmann, San Francisco (2002)
8. Igel, C., Kreutz, M.: Operator adaptation in structure optimization of neural networks. In: Spector, L., Goodman, E.D., Wu, A., Langdon, W.B., Voigt, H.-M., Gen, M., Sen, S., Dorigo, M., Pezeshk, S., Garzon, M.H., Burke, E. (eds.) Proceedings of the Genetic and Evolutionary Computation Conference (GECCO-2001), p. 1094. Morgan Kaufmann, San Francisco (2001)
9. Jones, T.: Crossover, macromutation, and population-based search. In: Eshelman, L. (ed.) Proceedings of the Sixth International Conference on Genetic Algorithms, pp. 73–80. Morgan Kaufmann, San Francisco (1995)
10. Mead, R.: The design of experiments. Statistical principles for practical application. Cambridge University Press, Cambridge (1988)
11. Michalewicz, Z.: Genetic Algorithms + Data Structures = Evolution Programs, third revised and extended edition. Springer, Heidelberg (1999)
12. Montgomery, D.C.: Design and Analysis of Experiments. Wiley, New York (1984)
13. Rojas, I., González, J., Pomares, H., Merelo, J.J., Castillo, P.A., Romero, G.: Statistical Analysis of the Main Parameters Involved in the Design of a Genetic ALgorithm. IEEE Transactions on Systems, Man and Cybernetics 32(1), 31–37 (2002)
14. Shilane, D., Martikainen, J., Dudoit, S., Ovaska, S.J.: A general framework for statistical performance comparison of evolutionary computation algorithms. Information Sciences 178(14), 2870–2879 (2008)
15. Smith, J., Fogarty, T.C.: Operator and parameter adaptation in genetic algorithms. Soft Computing 1(2), 81–87 (1997)
16. Toussaint, M.: Self-adaptive exploration in evolutionary search. Technical Report IRINI-2001-05, Institut für Neuroinformatik, Ruhr-Universität Bochum (2001)

Bio-inspired Combinatorial Optimization: Notes on Reactive and Proactive Interaction

Carlos Cotta and Antonio J. Fernández-Leiva

Dept. Lenguajes y Ciencias de la Computación, ETSI Informática,
Campus de Teatinos, Universidad de Málaga,
29071 Málaga – Spain
{ccottap,afdez}@lcc.uma.es

Abstract. Evolutionary combinatorial optimization (ECO) is a branch of evolutionary computing (EC) focused on finding optimal values for combinatorial problems. Algorithms ranging in this category require that the user defines, before the process of evolution, the fitness measure (i.e., the evaluation function) that will be used to guide the evolution of candidate solutions. However, there are many problems that possess aesthetical or psychological features and as a consequence fitness evaluation functions are difficult, or even impossible, to formulate mathematically. Interactive evolutionary computation (IEC) has recently been proposed as a part of EC to cope with this problem and its classical version basically consists of incorporating human user evaluation during the evolutionary procedure. This is however not the only way that the user can influence the evolution in IEC and currently one can find that IEC has been been successfully deployed on a number of hard combinatorial optimization problems. This work examines the application of IEC to these problems. We describe the basic fundament of IEC, present some guidelines to the design of interactive evolutionary algorithms (IEAs) to handle combinatorial optimization problems, and discuss the two main models over which IEC is constructed, namely reactive and proactive search-based schemas. An overview of the existing literature on the topic is also provided. We conclude with some reflections on the lessons learned, and the future directions that research might take in this area.

1 Introduction

Combinatorial optimization is ubiquitous and comprises an enormous range of practical applications. Problems arising in this area are typically hard to solve –due to both the size of the associated search spaces and the intrinsic complexity of efficiently traversing them in order to find the optimal solution– and thus the use of powerful solving methodologies is required. Among these, bio-inspired algorithms emerge as cutting-edge tools, due to their search power. Bio-inspired algorithms (including, evolutionary computation methods, swarm intelligence, and metaheuristics) have been shown to be adequate tools for combinatorial optimization in many different areas, and one of their most important characteristics, particularly inherent to evolutionary computation (EC), is their flexibility

J. Cabestany, I. Rojas, and G. Joya (Eds.): IWANN 2011, Part II, LNCS 6692, pp. 348–355, 2011.

to be adjusted to different problem domains, that is to say, in certain form these methods represent generic procedures that can be used with certain adjustments guided by the user to handle a plethora of combinatorial optimization problems. However, in spite of their proved efficacy as optimization methods, recently the need to exploit problem knowledge in order to obtain solutions of better quality as well as accelerate the optimization process has become evident [1–3]. In this sense, the programmer (i.e., the user) has usually incorporated specific information about the problem to guide the search; this has been done for instance via the hybridization with other techniques [4, 5], by designing specific genetic operators or by defining intelligent representations with inherent information in them. Other forms of adding problem-knowledge to an EC algorithm are possible though.

However, there still exists one main complication that lies precisely in the difficulty to characterize the subjective interest through a certain mathematical expression or algorithm that can be optimized. This difficulty is generally common to those problems in which the search has to be conducted (directly or indirectly, completely or partially) in a psychological space. Within the framework of the metaheuristics –and more specifically of evolutionary computing– the solution that has been proposed is the so-called interactive evolutionary computing (IEC). In a broad sense, IEC is an approach based on the optimization of a certain target system, using evolutionary computing and interacting with a human user. Traditionally, this interaction was based on the subjective assessment of the solutions generated by the algorithm; in this line see for instance the seminal work of Dawkins [6] as well as different applications in artistic fields (see, e.g., the proceedings of EvoMUSART), industrial design, processing of audiovisual information, data mining or robotics, among other fields [7]. The common nexus of classical IEC is the existence of a reactive search-based mechanism in which the user provides some feedback to the demands of the running evolutionary algorithm.

Even though they represent a powerful advance for the optimization of problems requiring some kind of subjective evaluation, classical IEC methods also have an important limitation: the fatigue of the human user that is produced by the continuous feedback that the subjacent EC technique demands to the user. Advanced IEC techniques smooth this drawback by employing proactive algorithms that are able to guess the further user interactions and thus reduce the requirement of user interventions.

The aim of this paper is to provide a general overview on user-centric evolutionary computation, and more specifically on both interactive EC and proactive search when they are applied to combinatorial optimization problems.

2 Interactive Evolutionary Computation

Generally speaking, interactive evolutionary computation (also termed indistinctly as user-centric evolutionary computation) is an optimization paradigm that incorporates the user intervention during the search process in an evolutionary algorithm; more specifically, IEC basically consists of a computational model that promotes the communication between a human user and an automated evolutionary algorithm (EA). In a classic view of IEC, the user usually acts as the fitness evaluation function in standard evolutionary algorithms (EA) and is continuously required by the EA to provide the assessment of candidate solutions; in other words, the EA is responsible for evolving the population of individuals in the evolutionary optimization process where as the user evaluates the outputs generated by the EA. As shown further on in this paper, more modern models of IEC propose different ways to attain the collaboration between the human user and the EA.

IEAs have already been implemented in all the standard types of EC as for instance in genetic programming [8, 9], genetic algorithms [10], evolution strategies [11], and evolutionary programming [12] just to name a few. Interactivity has also been added to a number of cooperatives models (e.g., [13–15]).

3 Design Principles for Effective IEAs

It is important to underline a fundamental fact in relation to the use of IEC and the context over which it is applied: the applications of IEC are often conducted on domains in which there is no exact (or reasonably good) way to assess the quality of the solutions, and this is precisely the reason why it is necessary to involve the user in the loop (the applications of aesthetic nature are appropriate examples, although they are not unique). In any case, no general approach for the design of effective interactive evolutionary algorithms exists in a well-defined sense, and hence this design phase must be addressed from an intuitive point of view as well. However, in order to help the reader to understand the mechanisms of IEAs, let us consider in what follows the main forms that exist in the literature to design IEAs. In general the user can influence the optimization process in several ways that basically can be summarized as follows:

- Allowing the user to evaluate –even if just sporadically– the candidate solutions that are generated by the evolutionary algorithm during the search process.
- Allowing the user to order (or classify) the set of candidates according to some (possibly psychological) criteria provided specifically by them. For instance the user might select solutions to be maintained in further generations (in case of elitist evolutionary algorithms).
- Allowing the user to reformulate the objective function. For instance the user interaction might consist of refining this function by adjusting the weighing of specific terms.
- Allowing the user to modify the set of constraints attached originally to the definition of the problem. More specifically, the user might provide new (or remove existing) constraints dynamically; this means that the user can impose additional constraints in the form of hard constraints (i.e., their satisfaction is mandatory) or soft constraints (i.e., they can be violated but in this case the violation generates an extra penalty value to be added to the fitness value of the solution).

- Allowing the user to change dynamically some parameters of the algorithm; for instance, a soft change might be to assign new values for genetic operators application probability; also, the user might determine the choice of the genetic operators (playing thus the role of a hyper-heuristics selector that works inside the subjacent EA mechanism).
- Allowing the user to incorporate additional mechanisms for improving the optimization process. This is for instance the case when the user decides dynamically to add a restarting mechanism during the optimization process in case of stagnation (or premature convergence) or add some local-search methods with the aim of obtaining a memetic version of running evolutionary algorithm. Moreover, the user might decide wether the phases of restarting/local-improvement should be performed only on a reduced subset of the population (e.g., the user can play the role of a selector that decides which solutions should undergo local improvement what in other words means the decision of using *partial Lamarckism* i.e., not using local search on every new solution computed, but only on some of them).

From a global perspective, the basic idea is to let the user affect the search dynamics with the objective of driving (resp. deviating) the search towards (resp. from) specific regions of the solution space. There are alternative ways to reach this objective. For instance, [16] proposes using techniques of dimensionality reduction to project the population of the EA to a bidimensional plane that is displayed to the user and over which the user selects the most promising candidates. It is also worthwhile to mention the work conducted in the area of multi-objective IEC [17, 18] in which the aim is to direct the exploration toward particular regions of the Pareto front. Again this kind of participation only represents one of the manifold forms that exist to fix search priorities.

4 IEC Limitations: The Fatigue of the Human User

The classical IEC as described previously can be catalogued as a reactive interaction procedure in which the user operates under the demand of feedback from the subjacent evolutionary computation technique. In this context, as already mentioned, one of the major concerns of classic IEC approaches is the fatigue that they cause in the human user; this fatigue is the result of demanding continuously feedback from the subjacent evolutionary algorithm. This section is devoted to discussing different mechanisms already described in the literature to mitigate this fatigue.

For instance, this can be done by reducing the accuracy of the judgements required of them [19], via the use of micropopopulations [20], or via the use of prediction methods of quality. These methods can be used for example, to make a pre-selection of promising solutions, presenting to the user (or to a set of users if a cooperative model is considered) just a reduced number of solutions to evaluate. Typically, this pre-selection can be attained from metrics that measure the distance between the tentative solutions and those for which a subjective assessment exists.

Another more sophisticated approach to reduce the human user fatigue in IEAs is by replacing the reactive collaboration by a proactive reaction in which the intervention of the user is optional and the algorithm runs autonomously [11]. In this context IEC is usually identified as *proactive user-centric evolutionary search/optimization*. This is the case when the IEC model employs computational learning techniques to predict the adequacy of the solutions still to be evaluated. If the prediction model of this adequacy is sufficiently adjusted, then alternating phases between optimization via IEC and optimization via the predictive model can be conducted [21]. In general, an approach of this type has the difficulty of finding a measure of the adequate distance that captures the subjective preferences of the human user. Another potential difficulty is the inherent noise that often exists in the human response (due for instance to the fatigue of user, to the evolution of their subjective perception, or to an adjustment of its response to the characteristics of the solutions in the current generation).

Additionally, [22] approaches the problem of the fatigue in IEC and proposes an interactive genetic algorithm with an individual fitness not assigned by a human; the basic idea is to automatically compute the fitness value of randomly selected individuals from the population by recording the time employed to mark them as valid or invalid candidates and then performing a transformation from the time space to the fitness space. The proposal was applied to a problem of fashion design.

[13] describes a mixed-initiative interaction technique for an interactive GA in which a simulated expert created by using a machine learning model (i.e., in particular via fuzzy logic modeling) can share the workload of interaction with the human expert; in addition the human user preferences are constantly been learnt. This collaborative framework also allows the system to observe the learning behaviors of both the human and simulated expert, while utilizing their knowledge for search purposes.

In [23] an interactive genetic algorithm applied over a nurse scheduling problem does not only generate an offspring that is further evaluated by the user but also includes a mechanism to learn the criteria of the user evaluation; this knowledge is then applied to construct schedules under the learn criteria. The best schedules are displayed for the user with the aim of providing them with a decision mechanism to choose which parts of the schedules to adopt, or to improve.

5 Conclusions

This paper tries to provide a general overview on interactive evolutionary computation (IEC) identifying its main components, advantages and disadvantages. One of the main conclusions that can be drawn from the literature on IEC is that it constitutes a versatile and effective optimization paradigm to tackle combinatorial optimization problems whose candidate solutions have to be evaluated in a psychological space (for instance because it is difficult to translate the fitness evaluation function into a mathematical formulation). Indeed, IEC represents one of the main paradigms to cope with this kind of problem, and provides an appropriate framework to seamlessly integrate human knowledge into evolutionary computation techniques. Traditionally IEC was based on a reactive optimization model in which the subjacent evolutionary algorithm demands the intervention of the user by requiring from them some kind of feedback; in the most classic models, the user just acts as a mere fitness evaluator. We have though discussed other models in which the user interacts with the running EA in a number of ways, as for instance adding constraints to bias the search, reformulating the objective function by changing the weights of its parameters, identifying the best solution candidates, or redesigning (parts of) the chromosome representation, among others.

We have also discussed the main drawback of IEC, that is to say, the fatigue that affects the human user as a consequence of the continuous requirement of feedback. A model that mitigates this problem consists of replacing the reactive answer of the user by a proactive approach in which the subjacent running algorithm usually infers the user's answer before the feedback demand. A number of proactive schemas, that have in common the incorporation of a specific learning mechanism, have also been discussed throughout the paper.

The flexibility of the proactive approach makes it helpful in cases in which the user wants to obtain an added value, but makes it also useful in complex optimization problems with perfectly well defined evaluation functions; in these cases the inherent skills of perception and information processing of the human user can help to both lead the search towards suboptimal regions of the search space and avoid the stagnation (or even premature convergence) of the algorithm in specific parts of this space.

User-centric optimization should be considered as a natural mechanism to cope with combinatorial optimization problems in which subjective evaluation of candidates is required. The interested reader is also referred to [7] and [24] that present surveys for interactive evolutionary computation and human-guided search respectively.

Acknowledgements. This work is supported by project NEMESIS (TIN-2008-05941) of the Spanish Ministerio de Ciencia e Innovación, and project TIC-6083 of Junta de Andalucía.

References

1. Hart, W.E., Belew, R.K.: Optimizing an arbitrary function is hard for the genetic algorithm. In: Belew, R.K., Booker, L.B. (eds.) Proceedings of the 4th International Conference on Genetic Algorithms, pp. 190–195. Morgan Kaufmann, San Mateo (1991)
2. Wolpert, D.H., Macready, W.G.: No free lunch theorems for optimization. IEEE Transactions on Evolutionary Computation 1(1), 67–82 (1997)
3. Bonissone, P., Subbu, R., Eklund, N., Kiehl, T.: Evolutionary Algorithms + Domain Knowledge = Real-World Evolutionary Computation. IEEE Transactions on Evolutionary Computation 10(3), 256–280 (2006)

4. Puchinger, J., Raidl, G.R.: Combining metaheuristics and exact algorithms in combinatorial optimization: A survey and classification. In: Mira, J., Álvarez, J.R. (eds.) IWINAC 2005. LNCS, vol. 3562, pp. 41–53. Springer, Heidelberg (2005)
5. Moscato, P., Cotta, C.: A modern introduction to memetic algorithms. In: Gendreau, M., Potvin, J.-Y. (eds.) Handbook of Metaheuristics, 2nd edn. International Series in Operations Research and Management Science, vol. 146, pp. 141–183. Springer, Heidelberg (2010)
6. Dawkins, R.: The BlindWatchmaker. Longman, Essex (1986)
7. Takagi, H.: Interactive evolutionary computation: Fusion of the capabilities of EC optimization and human evaluation. Proceedings of the IEEE (9), 1275–1296 (2001)
8. Lim, S., Kim, K.-M., Hong, J.-H., Cho, S.-B.: Interactive genetic programming for the sentence generation of dialogue-based travel planning system. In: 7th Asia-Pacific Conference on Complex Systems, Cairns, Australia, Asia-Pacific Workshops on Genetic Programming, pp. 6–10 (2004)
9. Lim, S., Cho, S.-B.: Language generation for conversational agent by evolution of plan trees with genetic programming. In: Torra, V., Narukawa, Y., Miyamoto, S. (eds.) MDAI 2005. LNCS (LNAI), vol. 3558, pp. 305–315. Springer, Heidelberg (2005)
10. Kosorukoff, A.: Human-based genetic algorithm. In: 2001 IEEE International Conference on Systems, Man, and Cybernetics, pp. 3464–3469. IEEE Press, Tucson (2001)
11. Breukelaar, R., Emmerich, M.T.M., Bäck, T.: On interactive evolution strategies. In: Rothlauf, F., Branke, J., Cagnoni, S., Costa, E., Cotta, C., Drechsler, R., Lutton, E., Machado, P., Moore, J.H., Romero, J., Smith, G.D., Squillero, G., Takagi, H. (eds.) EvoWorkshops 2006. LNCS, vol. 3907, pp. 530–541. Springer, Heidelberg (2006)
12. Kubota, N., Nojima, Y., Sulistijono, I.A., Kojima, F.: Interactive trajectory generation using evolutionary programming for a partner robot. In: 12th IEEE International Workshop on Robot and Human Interactive Communication (ROMAN 2003), Millbrae, California, USA, pp. 335–340 (2003)
13. Babbar, M., Minsker, B.: A collaborative interactive genetic algorithm framework for mixed-initiative interaction with human and simulated experts: A case study in long-term groundwater monitoring design. In: World Environmental and Water Resources Congress (2006)
14. Quiroz, J.C., Banerjee, A., Louis, S.J.: Igap: interactive genetic algorithm peer to peer. In: Proceedings of the 10th Annual Conference on Genetic and Evolutionary Computation, GECCO 2008, pp. 1719–1720. ACM, New York (2008)
15. Quiroz, J.C., Louis, S.J., Banerjee, A., Dascalu, S.M.: Towards creative design using collaborative interactive genetic algorithms. In: IEEE Congress on Evolutionary Computation (CEC 2009), pp. 1849–1856. IEEE, Singapore (2009)
16. Takagi, H.: Active user intervention in an ec search. In: 5th Joint Conf. Information Sciences (JCIS2000), Atlantic City, NJ, pp. 995–998 (2000)
17. Deb, K., Chaudhuri, S.: I-mode: An interactive multi-objective optimization and decision-making using evolutionary methods. KanGal report 2007003, Kanpur Genetic Algorithms Laboratory (2007)
18. Deb, K., Kumar, A.: Interactive evolutionary multi-objective optimization and decision-making using reference direction method. KanGal report 2007001, Kanpur Genetic Algorithms Laboratory (2007)
19. Ohsaki, M., Takagi, H., Ohya, K.: An input method using discrete fitness values for interactive ga. Journal of Intelligent and Fuzzy Systems 6(1), 131–145 (1998)

20. Sáez, Y., Viñuela, P.I., Segovia, J., Castro, J.C.H.: Reference chromosome to overcome user fatigue in IEC. New Generation Comput. 23(2) (2005)
21. Dozier, G.: Evolving robot behavior via interactive evolutionary computation: From real-world to simulation. In: 16th ACM Symp. Applied Computing (SAC2001), Las Vegas, NV, pp. 340–344 (2001)
22. Gong, D., Yao, X., Yuan, J.: Interactive genetic algorithms with individual fitness not assigned by human. Journal of Universal Computer Science 15(13), 2446–2462 (2009), http://www.jucs.org/jucs_15_13/interactive_genetic_algorithms_with
23. Inoue, T., Furuhashi, T., Fujii, M., Maeda, H., Takaba, M.: Development of nurse scheduling support system using interactive ea. In: IEEE Int. Conf. Systems, Man, and Cybernetics, vol. 5, pp. 533–537 (1999)
24. Klau, G., Lesh, N., Marks, J., Mitzenmacher, M.: Human-guided search. Journal of Heuristics 16, 289–310 (2010)

A Preliminary General Testing Method Based on Genetic Algorithms*

Luis M. Alonso, Pablo Rabanal, and Ismael Rodríguez

Dept. Sistemas Informáticos y Computación
Facultad de Informática
Universidad Complutense de Madrid, 28040 Madrid, Spain
luismaal@estumail.ucm.es, prabanal@fdi.ucm.es, isrodrig@sip.ucm.es

Abstract. We present a testing methodology to find suitable test suites in environments where the application of each test to the implementation under test (IUT) might be very expensive in terms of cost or time. The method is general in the sense that it keeps very low the dependence on the underlying model (e.g. finite state machines, timed automata, Java programs, etc). A genetic algorithm (GA) is used to find optimal test suites according to cost and distinguishability criteria.

1 Introduction

Formal testing techniques [9,15] allow testers to (semi-)automatically perform some or all of the following testing tasks: Extracting a set of tests from a specification, applying tests to the *implementation under test* (IUT), collecting the responses given by the IUT, and providing IUT (in-)correct diagnoses by assessing the observations. There exist methods to extract test suites for systems defined as *finite state machines* (FSMs) [9], *extended finite state machines* (EFSMs) [11], labeled transition systems [17], temporal systems [16,8], and probabilistic systems [10], among others. For instance, if the system specification is given by means of a temporal machine, then extracting tests from this specification consists in composing some interaction plans where inputs are produced at some specific times, and/or specific delays are allowed/forbidden after each interaction. Ideally, test suites derived from a given specification should be *complete*, i.e. such that, if the IUT passes all tests in the set, then the IUT is necessarily correct. However, in most of the cases, finite complete test suites do not exist or, if they exist, some strong hypotheses about the IUT must be assumed.

Testing methodologies can be abstracted from the selected underlying model. In [14], a general testing theory, allowing testers to reason about testing regardless of the kind of systems we wish to check, is presented. The properties presented in that work allow to classify testing problems in terms of the (in-)existence of complete test suites of different sizes (finite, countable infinite, infinite but *finitely approachable*, etc). Each behavior of the IUT that is *possible* according to our assumptions is defined by a function relating received inputs

* Work partially supported by project TIN2009-14312-C02-01.

J. Cabestany, I. Rojas, and G. Joya (Eds.): IWANN 2011, Part II, LNCS 6692, pp. 356–363, 2011.

and possible responses. Thus, the possible behaviors that the IUT could actually have (according to our assumptions about it) are defined by a set of functions. A subset of this set defines those behaviors that we consider as *correct* (i.e. the *specification*). Thus, the ideal purpose of testing consists in finding a set of tests (i.e. inputs) such that the responses of any possible IUT to these tests allow us to *precisely* determine whether the IUT belongs to the subset of correct behaviors.

Given this general framework, the problem of finding the *minimum complete test suite* (provided that there exists a *finite* complete test suite) is defined in that work. In addition, given a set of hypotheses about the IUT that the tester may assume or not, the problem of finding the minimum set of hypotheses that have to be *assumed* to make a given test suite become *complete* is defined as well in several variants. Finding out the minimum set of hypotheses such that they make a given test suite become complete allows comparing incomplete test suites: We prefer those test suites requiring weaker hypotheses to be complete, because their completeness is more *feasible*. In [14], it is shown that these problems are NP-complete in their more general forms. However, the problem of solving these problems in practice is not considered.

Evolutionary Computation methods (EC) such as *genetic algorithms* (GA) [1], *ant colony optimization* (ACO) [4], or *river formation dynamics* (RFD) [12] have been applied to solve optimization problems related to testing [2,3,13]. In this paper, we introduce a methodology to solve the general testing problems proposed in [14] by means of GA. Following the spirit of [14], the methodology can be applied to any kind of systems under test. However, the methodology is *efficient* only if an additional condition is assumed: We consider that the cost required to find out the response(s) of the IUT to a given input is much more expensive (in terms of money, time, risks, etc) than the cost required to find out the response of a *model* representing that behavior (e.g. a finite state machine, a timed automata, etc) to that input. For instance, this is the case if the IUT is a temporal system where transitions take very long times (e.g. hours or days) but these transitions can be simulated almost instantaneously by a timed automata simulator (let us note that a model simulator can trivially pretend that a timeout is reached in a *model*). Similarly, breaking a fragile component of the IUT, or shutting down a system that is running with real customers in order to test it (thus, the company business is stopped for a while), is much more expensive than *simulating* these activities in a model. Though the purpose of our testing method will be making real experiments with the IUT, the task of selecting *a priori* a test suite with high fault-detection capability on the IUT will be based on using models instead. The structure of this paper is straightforward.

2 General Testing Model

We briefly introduce some formal concepts appearing in [14]. Abundant examples showing that the framework actually allows testers to define very different testing scenarios (e.g. testing FSMs, temporal systems, Java programs, etc) can be found in [14] together with many properties and additional details.

We present a general notion to denote implementations and specifications in our framework. Since testing consists in studying systems in terms of their observable behavior, the behavior of a system can be defined by a function relating inputs with their possible outputs. Let us assume that 2^S denotes the power set of the set S. Let I be a set of *input symbols* and O be a set of *output symbols*. A *computation formalism* C for I and O is a set of functions $f : I \rightarrow 2^O$ where for all $i \in I$ we have $f(i) \neq \emptyset$. Given a function $f \in C$, $f(i)$ represents the set of outputs we can obtain after applying input $i \in I$ to the computation artifact represented by f. Since $f(i)$ is a set, f may represent a non-deterministic behavior. Besides, C, I, and O can be infinite sets. For us, an *input* is a complete plan to interact with the IUT (e.g. a *sequence* of buttons to press, a sequence of buttons together with *delays*, etc).

Computation formalisms will be used to represent the set of implementations we are considering in a given testing scenario. Implicitly, a computation formalism C represents a *fault model* (i.e. the definition of what can be wrong in the *implementation under test*, IUT) as well as the *hypotheses* about the IUT the tester is assuming. Computation formalisms will also be used to represent the subset of specification-compliant implementations. Let C represent the set of possible implementations and $E \subseteq C$ represent the set of implementations fulfilling the specification. The goal of testing is interacting with the IUT so that, according to the collected responses, we can decide whether the IUT actually belongs to E or not. For us, a *specification* of a computation formalism C is any set $E \subseteq C$. If $f \in E$ then f denotes a *correct* behavior, while $f \in C \backslash E$ denotes that f is incorrect. Thus, a specification implicitly denotes a correctness criterion. In addition, testers can define when two IUT observations can be distinguished from each other. Let O be a set of outputs. A *distinguishing relation* for O is an anti-reflexive symmetric binary relation $\mathcal{D}$ over O. $\not{\mathcal{D}}$ denotes the complementary of $\mathcal{D}$. A trivial distinguishing relation $\mathcal{D}$, where $o_1 \ \mathcal{D} \ o_2$ iff $o_1 \neq o_2$, may be considered in many cases. However, different distinguishing relations might have to be considered in specific scenarios (e.g. if systems may not terminate and non-termination is not observable). See [14] for details.

Let us identify *complete test suites*, i.e. sets of inputs such that, if they are applied to the IUT, then collected outputs allow us to precisely determine if the IUT fulfills the considered specification or not. Let C be a computation formalism for I and O, $E \subseteq C$ be a specification, $\mathcal{D}$ be a distinguishing relation, and $\mathcal{I} \subseteq I$ be a set of inputs. Let $f \in C$. We denote by $\texttt{pairs}\,(f, \mathcal{I})$ the set of all pairs $(i, f(i))$ such that $i \in \mathcal{I}$. We say that $f \in E$ and $f' \in C \backslash E$ are *distinguished* by $\mathcal{I}$, denoted by $\texttt{di}\,(f, f', \mathcal{I})$, if there exist $i \in \mathcal{I}$, $(i, outs) \in \texttt{pairs}\,(f, \mathcal{I})$, and $(i, outs') \in \texttt{pairs}\,(f', \mathcal{I})$ such that for all $o \in outs$ and $o' \in outs'$ we have $o \ \mathcal{D} \ o'$. We say that $\mathcal{I}$ is a *complete test suite* for C, E, and $\mathcal{D}$ if for all $f \in E$ and $f' \in C \backslash E$ we have $\texttt{di}\,(f, f', \mathcal{I})$.

Let us define the problem of finding minimum complete test suites when computation formalisms are *finite*. Let C be a finite computation formalism for the finite sets of inputs and outputs $I = \{i_1, \ldots, i_k\}$ and O, respectively, $E \subseteq C$ be a finite specification, and $\mathcal{D}$ be a finite distinguishing relation. Let C' and

$E' \subseteq C'$ be sets of tuples representing the behavior of functions of C and E respectively; formally, for all $f \in C$ we have $(f(i_1), \ldots, f(i_k)) \in C'$ and vice versa, and for all $g \in E$, we have $(g(i_1), \ldots, g(i_k)) \in E'$ and vice versa. Given C', E', I, O, $\mathcal{D}$, and some $K \in \mathbb{N}$, the *Minimum Complete Suite* problem (MCS) is defined as follows: Is there any complete test suite $\mathcal{I}$ for C, E, and $\mathcal{D}$ such that $|\mathcal{I}| \leq K$?

Theorem 1. [14] MCS $\in$ NP-complete. □

Let us introduce a notion to measure the coverage of an incomplete test suite. Let C be a finite computation formalism for I and O, $E \subseteq C$ be a specification, $\mathcal{D}$ be a distinguishing relation, and $\mathcal{I} \subseteq I$ be a set of inputs. We define the *distinguishing rate* of $\mathcal{I}$ for $(C, E, \mathcal{D})$, denoted by $\texttt{d-rate}(\mathcal{I}, C, E, \mathcal{D})$, as $\frac{|\{(f, f')/f \in E, f' \in C\backslash E, \texttt{di}(f, f', \mathcal{I})\}|}{|E| \cdot |C\backslash E|}$.

The problem of finding the weakest hypothesis that makes a test suite *complete* is defined next in three alternative forms. An *hypothesis* H denotes a set of functions the tester could *assume* not to be the actual IUT. Let us consider the same notation preliminaries as when we defined problem MCS before. In addition, let $\mathcal{I} \subseteq I$ be a set of inputs and $\mathcal{H} = \{H_1, \ldots, H_n\}$, where for all $1 \leq i \leq n$ we have $H_i \subseteq C'$. Let $K \in \mathbb{N}$. Given C', E', I, O, $\mathcal{D}$, $\mathcal{I}$, and K, the *Minimum Function Removal* problem (MFR) is defined as follows: Is there any set $R \subseteq C$ with $|R| \leq K$ such that $\mathcal{I}$ is a complete test suite for $(C\backslash R, E\backslash R, \mathcal{D})$? Given C', E', I, O, $\mathcal{D}$, $\mathcal{I}$, $\mathcal{H}$, K, the *Minimum Function removal via Hypotheses* problem (MFH) is defined as follows: Is there any set of hypotheses $R \subseteq \mathcal{H}$ with $|\bigcup_{H \in \mathcal{H}} H| \leq K$ such that $\mathcal{I}$ is a complete test suite for $(C\backslash(\bigcup_{H \in R} H), E\backslash(\bigcup_{H \in R} H), \mathcal{D})$? Given C', E', I, O, $\mathcal{D}$, $\mathcal{I}$, $\mathcal{H}$, and K, the *Minimum Hypotheses Assumption* problem (MHA) is defined as follows: Is there any set $R \subseteq \mathcal{H}$ with $|R| \leq K$ such that $\mathcal{I}$ is a complete test suite for $(C\backslash(\bigcup_{H \in R} H), E\backslash(\bigcup_{H \in R} H), \mathcal{D})$? In MFR, hypotheses consist in any set of functions the tester believes (i.e. assumes) not to be the IUT. In MFH and MHA, hypotheses must be taken from a given repertory, and the assumption cost is measured in terms of the number of removed functions and the number of assumed hypotheses, respectively.

Theorem 2. [14] We have the following properties:
(a) MFR $\in$ P. MFR can be solved in time $\mathcal{O}(|C'|^{5/2} + |C'|^2 \cdot |\mathcal{I}| \cdot |O|^2)$.
(b) MFH $\in$ NP-complete
(c) MHA $\in$ NP-complete □

3 General Testing Methodology

In this section we introduce our general methodology based on GA to find good test suites. The methodology consists of the following steps.

First, we compose a finite set of inputs, called *ideal set*, which is representative of a big number of possible ways to interact with the IUT (recall that, for us, an input is a *complete* plan to interact with the IUT; e.g. a sequence of FSM inputs). We assume that the ideal set is much longer than the set we will actually be able to apply to the IUT (called *real set*) due to the high cost of applying each test. From now on, we will assume that the *ideal set* contains the only inputs we consider as available for doing testing, so the ideal set coincides with the set denoted by I in the framework presented in the previous section. On the other hand, our goal will be finding a good real set containing the inputs to be applied to the IUT, that is, a good test suite $\mathcal{I} \subseteq I$.

Next, we define a finite representative set of the behaviors we believe the IUT could actually have – according to our assumptions about the IUT. That is, each behavioral model represents a *possible IUT*. According to the framework given in the previous section, the behavior of each model for all inputs is represented by a *function* relating inputs with possible outputs, and the set of behaviors of all models is a *computation formalism* C (a set of functions). One or more of these behaviors are correct behaviors. The set of these correct behaviors is represented in the previous section by a set of functions $E \subseteq C$. The rest of behaviors represent possible *incorrect* definitions of IUT.

There are several ways to construct and represent these correct and incorrect functions (behaviors). For instance, rather than working with functions, we may work with *models* (e.g. FSMs, timed automata, or even Java programs serving as models). Given a specification model, we may create modified versions of this model by introducing, in each one, one or several mistakes that could be made by a typical programmer (e.g. a call leads to the wrong part of the program, a transition produces a wrong signal, a timeout triggers sooner than expected, etc), following a standard *mutation testing* approach [6]. Another possibility consists in systematically applying a given *fault model* (i.e. a definition of what can be wrong in the IUT with respect to the specification) to extract the set of all possible IUT definitions. Let us note that, in this case, the set of behaviors of these alternative models could be constructed without considering the models themselves, as we could define functions belonging to C as modifications of other functions belonging to E. Each function could be extensionally represented by listing all pairs *(input, set of possible outputs)* (recall that the ideal set I is finite). Thus, the underlying model (FSMs, timed automata, etc) could be ignored in this step. In cases where the set of possible wrong behaviors to be detected is very small, all models of alternative IUTs could be manually defined.

Next, if the behavior of functions in C and E is still given in the form of models, then all correct and incorrect models are executed for all inputs belonging to the ideal set (recall that we are running the models, not the IUT). In this way, all functions $f \in E$ and $f' \in C \backslash E$, defining the possible responses of each possible IUT for all inputs in the ideal set I, can be extensionally defined. Alternatively, we can also run only a sample of combinations *(model, input)* if executing all of these combinations takes a very long time even working with models. On the contrary, if behaviors were already represented by functions in the last step, then there is no model to execute now.

Next, we seek for a good subset of the ideal set of inputs such that (a) it is *cheap*; and (b) it has high capability to distinguish incorrect models from correct models. Regarding (a), we may consider that a test suite is cheap if, for instance, it is small and/or it contains small tests (e.g. short sequences of inputs, short temporal delays, etc). Regarding (b), we may assess the distinguishing capability of a test suite by either considering the distinguishing rate as given in the previous section, or by identifying the *weakest required hypothesis* that would make the suite complete. In this case, let us note that measuring the kind of metric calculated by the MFR problem is feasible because, according to Theorem 2, it can be solved polynomially. Thus, this metric can be efficiently introduced as part of a fitness function. However, it is not the case for the metrics calculated by MFH and MHA problems, which are NP-complete problems (see Theorem 2). A fitness function combining factors (a) and (b) is defined according to the previous alternatives, and next a GA is executed to find a good subset of inputs $\mathcal{I} \subseteq I$ according to this criterion.

Next we describe how we can implement the genetic algorithm for finding good test suites according to either (a) their distinguishing rate; or (b) the weakest hypothesis they require for being complete. First we have to decide how to encode the individuals (i.e. candidate solutions). Let us note that a solution $\mathcal{I}$ is a subset of the ideal set I; Thus, we may represent individuals by means of bit vectors. A vector $b_1 \cdots b_n$ denotes a test suite where input $i_j \in I$ is included in the test suite iff $b_j = 1$.

We specify the fitness functions for each problem. Given a test suite $\mathcal{I}$, we consider $fitness(\mathcal{I}) = \frac{distinguish(\mathcal{I})^\alpha}{cost(\mathcal{I})}$ where $cost(\mathcal{I})$ is the cost of $\mathcal{I}$ (e.g. the addition of costs of all tests in $\mathcal{I}$) and $distinguish(\mathcal{I})$ is the *distinguishing capability* of $\mathcal{I}$. This value is either the *distinguishing rate* of $\mathcal{I}$ according to the previous section (i.e. a measure of how many pairs of correct-incorrect candidate definitions of the IUT the test suite $\mathcal{I}$ actually distinguish), or the inverse of the *minimum function removal* also given in the previous section (i.e. a measure of how small is the weakest hypothesis that would make $\mathcal{I}$ complete). Finally, the parameter α controls the relative weight of the distinguishing capability of test suites against their cost. Alternatively, if the cost/time available for testing is *fixed* then we may consider an alternative where, if the cost of $\mathcal{I}$ is under the cost threshold, then the fitness is $distinguish(\mathcal{I})$, else it is 0.

We discuss how the initial population of the GA is selected. Rather than using a totally random set of individuals, we can use some specific test suites in order to speed up the convergence of the GA to satisfactory solutions. If the distinguishing rate is considered as the basis for measuring the distinguishing capability, we can compose an individual (test suite) as follows. We take an empty set of inputs, and we extend the set step by step by iteratively adding the input that adds the largest number of newly distinguished correct-incorrect pairs. On the other hand, if the weakest hypothesis strategy is followed, we can also compose a test suite input by input, this time by adding the input that enables the existence of the smallest function removal to get completeness (compared with the rest of inputs we could add). In both cases, we stop adding inputs when the fitness decays for the first time (recall that the suite size reduces its fitness).

In order to illustrate the application of GA to our methodology, we have implemented a GA and we have applied it to find good test suites in a few simple abstract examples. Next we present one of them. The next table shows the behavior of five correct machines A, B, C, D, E and five incorrect machines V, W, X, Y, Z in response to six possible inputs a, b, c, d, e, f. For each machine and input, the table shows the possible responses of the machine to that input. E.g. we have $X(e) = \{a, e, g\}$, that is, if e is given to the (wrong) machine X, it may non-deterministically reply a, e, or g.

	a	b	c	d	e	f
A	a, b	b, c	e	d, f	a, e	h
B	b	c	e	d, f	e	h
C	a	b	g	d, f	a, g	h
D	b	b, c	e	d, f	e	h
E	a	b, c	g	d	g	h

	a	b	c	d	e	f
V	a, b	b, c	e, g	d, f	a, e	h
W	b	c	e	d, f	b, e	h
X	a, b	b, c	e	d, f	a, e, g	h
Y	a	b, c, d	g	d, f	a, g	h
Z	a, b	b, c	e, g	d, f	a, e	h

We assign a cost to the task of testing each input. The costs of testing a, b, c, d, e, f are $1.1, 1.6, 0.8, 1.6, 2, 1.4$, respectively. We run the GA for computing the best test suites (in terms of distinguishing rate) whose cost is under some given threshold. Only random solutions are initially given to the GA, due to the simplicity of this example. The selection operator is the *roulette wheel* selection operator [5], and we use single-point crossover. The mutation operator flips the (in-)existence of some input with low probability (0.1).

If the maximum cost of test suites is set to 2, then the best solution found by the GA is the test suite $\{e\}$, which has cost 2 and fitness 0.4 (i.e. it distinguishes 40% of all correct-incorrect pairs). If the maximum cost is 3 or 4, then the test suite $\{c, e\}$, with cost 2.8 and fitness 0.48, is found by the GA. Finally, for all examples where the cost is 5 or higher, the GA finds the suite $\{b, c, e\}$, which has cost 4.4 and fitness 0.52. Let us remark that, in this example, many correct-incorrect pairs are not distinguishable by *any* of the considered inputs, so this is the best test suite we can find indeed.

4 Conclusions and Future Work

In this paper we have presented a general testing methodology to construct good test suites. The method is kept as independent as possible from the kind of models used for defining the possible behaviors of the IUT. Depending on the case, it could be feasible to work directly with functions in the methodology, or we might need to work with models for some steps. In the latter case, model-dependant tools might be required to construct the set C of possible behaviors of the IUT as well as the set $E \subseteq C$ of correct behaviors, and these models might have to be executed for (some or all) inputs in I. This is the case if the set C of possible IUT behaviors is constructed by manipulating the models, rather than by manipulating the functions denoting the behavior of these models. In any case, from this point on, the task of constructing good test suites is independent from the underlying models, because it relies on directly manipulating functions in C and E to construct optimal test suites according to the criteria given in Section 2 (in particular, by running a GA). This high independence from underlying models enables a high reusability of the method and, in particular, a high reusability of fitness functions.

References

1. Davis, L. (ed.): Handbook of genetic algorithms. Van Nostrand Reinhold, New York (1991)
2. Derderian, K., Hierons, R.M., Harman, M., Guo, Q.: Automated unique input output sequence generation for conformance testing of FSMs. The Computer Journal 49(3), 331–344 (2006)
3. Derderian, K., Merayo, M.G., Hierons, R.M., Núñez, M.: Aiding test case generation in temporally constrained state based systems using genetic algorithms. In: Cabestany, J., Sandoval, F., Prieto, A., Corchado, J.M. (eds.) IWANN 2009. LNCS, vol. 5517, pp. 327–334. Springer, Heidelberg (2009)
4. Dorigo, M., Stützle, T.: Ant Colony Optimization. The MIT Press, Cambridge (2004)
5. Goldberg, D.E., Deb, K.: A comparative analysis of selection schemes used in genetic algorithms. In: Foundations of Genetic Algorithms, pp. 69–93. Morgan Kaufmann, San Francisco (1991)
6. Howden, W.E.: Weak mutation testing and completeness of test sets. IEEE Transactions on Software Engineering 8, 371–379 (1982)
7. De Jong, K.A.: Evolutionary computation: a unified approach. MIT Press, Cambridge (2006)
8. Krichen, M., Tripakis, S.: Black-box conformance testing for real-time systems. In: Graf, S., Mounier, L. (eds.) SPIN 2004. LNCS, vol. 2989, pp. 109–126. Springer, Heidelberg (2004)
9. Lee, D., Yannakakis, M.: Principles and methods of testing finite state machines: A survey. Proceedings of the IEEE 84(8), 1090–1123 (1996)
10. López, N., Núñez, M., Rodríguez, I.: Specification, testing and implementation relations for symbolic-probabilistic systems. Theoretical Computer Science 353(1-3), 228–248 (2006)
11. Petrenko, A., Boroday, S., Groz, R.: Confirming configurations in EFSM testing. IEEE Transactions on Software Engineering 30(1), 29–42 (2004)
12. Rabanal, P., Rodríguez, I., Rubio, F.: Using river formation dynamics to design heuristic algorithms. In: Akl, S.G., Calude, C.S., Dinneen, M.J., Rozenberg, G., Wareham, H.T. (eds.) UC 2007. LNCS, vol. 4618, pp. 163–177. Springer, Heidelberg (2007)
13. Rabanal, P., Rodríguez, I., Rubio, F.: A formal approach to heuristically test restorable systems. In: Leucker, M., Morgan, C. (eds.) ICTAC 2009. LNCS, vol. 5684, pp. 292–306. Springer, Heidelberg (2009)
14. Rodríguez, I.: A general testability theory. In: Bravetti, M., Zavattaro, G. (eds.) CONCUR 2009. LNCS, vol. 5710, pp. 572–586. Springer, Heidelberg (2009)
15. Rodríguez, I., Merayo, M.G., Núñez, M.: $\mathcal{HOTL}$: Hypotheses and observations testing logic. Journal of Logic and Algebraic Programming 74(2), 57–93 (2008)
16. Springintveld, J., Vaandrager, F., D'Argenio, P.R.: Testing timed automata. Theoretical Computer Science 254(1-2), 225–257 (2001); Previously appeared as Technical Report CTIT-97-17, University of Twente (1997)
17. Tretmans, J.: A Formal Approach to Conformance Testing. PhD thesis, University of Twente, Enschede, The Netherlands (1992)

Tackling the Static RWA Problem by Using a Multiobjective Artificial Bee Colony Algorithm

Álvaro Rubio-Largo, Miguel A. Vega-Rodríguez, Juan A. Gómez-Pulido, and Juan M. Sánchez-Pérez

Department of Technologies of Computers and Communications, University of Extremadura, Polytechnic School, Cáceres, 10003 Spain
{arl,mavega,jangomez,sanperez}@unex.es

Abstract. Nowadays, the most promising technology for designing optical networks is the Wavelength Division Multiplexing (WDM). This technique divides the huge bandwidth of an optical fiber link into different wavelengths, providing different available channels per link. However, when it is necessary to interconnect a set of traffic demands, a problem comes up. This problem is known as Routing and Wavelength Assignment problem, and due to its complexity (NP-hard problem), it is very suitable for being solved by using evolutionary computation. The selected heuristic is the Artificial Bee Colony (ABC) algorithm, an heuristic based on the behavior of honey bee foraging for nectar. To solve the Static RWA problem, we have applied multiobjective optimization, and consequently, we have adapted the ABC to multiobjective context (MOABC). New results have been obtained, that significantly improve those published in previous researches.

Keywords: Artificial Bee Colony, Routing and Wavelength Assignment, WDM networks, Multiobjective Optimization.

1 Introduction

Nowadays, the most promising technique to exploit the huge bandwidth of optical networks is based on Wavelength Division Multiplexing (WDM). This technology multiplies the available capacity of an optical fiber link by adding new channels, each channel on a new wavelength of light. The aim of WDN is to ensure fluent communications between several devices, avoiding bottlenecks [4].

However, a problem happens when it is necessary to establish a set of traffic demands. This problem is known in the literature as Routing and Wavelength Assignment (RWA) problem. There are two varieties of RWA problem, depending on the establishment of the demands, we could refer to a Static problem when the demands are given in advance (Static RWA problem), and a Dynamic problem when the demands are given in real time (Dynamic RWA problem).

In this paper we have developed a new Multiobjective Evolutionary Algorithm (MOEA) for solving the Static RWA problem (the most usual one). The selected algorithm is the Artificial Bee Colony (ABC) algorithm, due to the

J. Cabestany, I. Rojas, and G. Joya (Eds.): IWANN 2011, Part II, LNCS 6692, pp. 364–371, 2011.

promising results it has obtained in other studies. Since we tackle the Static RWA problem as a Multiobjective Optimization Problem (MOOP), we have to adapt the heuristic to multiobjective context (MOABC). To demonstrate the proper functioning of our proposal, we present several comparisons with other approaches published in the literature. After performing these comparisons, we conclude that the MOABC overcomes the results obtained by almost twenty different approaches published in previous researches.

The rest of this work is organized as follows. The Static RWA problem is presented in a formal way in Section 2. A description of the Multiobjective Artificial Bee Colony (MOABC) algorithm appears in Section 3. In Section 4, we present several comparisons with previous approaches developed by us. A comparison with other algorithms published in the literature by other authors appears in Section 5. Finally, the conclusions and future work are left for Section 6.

2 Static RWA Problem

In this paper, an optical network is modeled as a direct graph $G = (V, E, C)$, where V is the set of nodes, E is the set of links between nodes and C is the set of available wavelengths for each optical link in E.

- $(i, j) \in E$: Optical link from node i to node j.
- $c_{ij} \in C$: Number of channels or different wavelengths at link (i, j).
- $u = (s_u, d_u)$: Unicast request u with source node s_u and destination node d_u, where $s_u, d_u \in V$.
- U : Set of demands, where $U = \{\ u \mid u \text{ is an unicast request}\}$.
- $|U|$: Cardinality of U.
- $u^{\lambda}_{i,j}$: Wavelength (λ) assigned to the unicast request u at link (i, j).
- l_u : Lightpath or set of links between a source node s_u and a destination node d_u; with the corresponding wavelength assignment in each link (i, j).
- L_u : Solution of the RWA problem considering the set of U requests.

Notice that $L_u = \{l_u | l_u$ is the set of links with their corresponding wavelength assignment$\}$. Using the above definitions, the RWA problem may be stated as a Multiobjective Optimization Problem (MOOP) [2], searching the best solution L_u that simultaneously minimizes the following two objective functions:

1. *Number of hops* (y_1):

$$y_1 = \sum_{u \in U} \sum_{(i,j) \in l_u} \Phi_j where \begin{Bmatrix} \Phi_j = 1 & \text{if} & (i,j) \in l_u \\ \Phi_j = 0 & & otherwise \end{Bmatrix} \tag{1}$$

2. *Number of wavelength conversions* (y_2):

$$y_2 = \sum_{u \in U} \sum_{j \in V} \varphi_j where \begin{Bmatrix} \varphi_j = 1 & \text{if} & j \in V switches\ \lambda \\ \varphi_j = 0 & & otherwise \end{Bmatrix} \tag{2}$$

Furthermore, we have to fulfill *the wavelength conflict constraint*: Two different unicast transmissions must be allocated with different wavelengths when they are transmitted through the same optical link (i, j).

3 Multiobjective Artificial Bee Colony Algorithm

The Artificial Bee Colony (ABC) algorithm is an algorithm created by Dervis Karaboga [6]. It is a population-based algorithm inspired by the intelligent behavior of honey bees.

In this paper, we have used the Multiobjective Artificial Bee Colony (MOABC) algorithm. This multiobjective version is based on the ABC algorithm [6], but adapted to multiobjective problems. The definition of the individuals used in the MOABC algorithm for solving the Static RWA problem is the same as in [7], for further information, please refer to [7]. We have incorporated the fast non-dominated sort procedure from the well-known *Fast Non-Dominated Sorting Genetic Algorithm* (NSGA-II).

In Algorithm 1, we show an outline of the pseudocode for the MOABC algorithm. In first place, we fill the first half of the colony with random employed bees and after that, we calculate for each employed bee its respective value of MOFitness (lines 1-2). For further information about how calculating the value of MOFitness, please consult [9]. After that, we initialize the Pareto Front (line 3). Every generation of the algorithm can be divided up in three main steps. Firstly, we try to improve the first half of the colony (employed bees). To perform this step (lines 5-11), we apply to each employed bee a mutation (the amount of mutation is defined by the parameter F or mutation factor). In case of obtaining a new bee with better value of MOFitness, we replace the old employed bee

Algorithm 1. Pseudocode for MOABC Algorithm

```
1.  /* Generate the initial first half of the colony (employed bees) C = {X_1, X_2, ..., X_{N/2}}. */
2.  C ⇐ generateAndEvaluateRandomEmployedBees(N/2)
3.  ParetoFront ⇐ ∅
4.  while not time-limit do
5.     /* Improve the first half of the colony (employed bees) C = {X_1, X_2, ..., X_{N/2}}. */
6.     for i=1 to N/2 do
7.        X_{newEmployedBee} ⇐ generateAndEvaluateNewEmployedBee(X_i, F)
8.        if X_{newEmployedBee}.MOFitness > X_i.MOFitness then
9.           X_i ⇐ X_{newEmployedBee}
10.       end if
11.    end for
12.    /* Generate the probability vector using the employed bees */
13.    probabilityVector ⇐ calculateProbabilityVector(C,N/2)
14.    /* Generate the second half of the colony (onlooker bees) C = {X_{N/2}, ..., X_N}. */
15.    for i=N/2 to N do
16.       X_{employedBee} ⇐ selectEmployedBee(probabilityVector, C)
17.       X_{newOnlookerBee} ⇐ generateAndEvaluateNewOnlookerBee(X_{employedBee}, F)
18.       if X_{newOnlookerBee}.MOFitness ≥ X_{employedBee}.MOFitness then
19.          X_i ⇐ X_{newOnlookerBee}
20.       else
21.          X_i ⇐ X_{employedBee}
22.       end if
23.    end for
24.    /* Generate NS scout bees C = {X_{N+1}, ..., X_{N+NS}}. */
25.    for i=N to N + NS do
26.       X_i ⇐ generateAndEvaluateNewScoutBee()
27.    end for
28.    /* Sort the colony by quality */
29.    C ⇐ fastNonDominatedSort(C,N + NS)
30.    ParetoFront ⇐ updateParetoFront(ParetoFront,C)
31. end while
```

by the new one. After improving the employed bees, we generate a probability vector (line 13), which contains the probability of each employed bee to be selected in the next step. Secondly, we generate the second half of the colony (onlookers bees) (lines 14-23). To generate an onlooker bee, we have to select an employed bee, according to the probability vector. After applying a mutation to the employed bee selected, we check if this new bee obtains a higher or equal value of MOFitness than the bee selected, in that case, we store this new bee. By contrast, we store the employed bee selected. Finally, we add to the colony SN scout bees and sort the colony by quality (using the fast non-dominated sort procedure of the NSGA-II) for obtaining the new employed bees in the next generation (lines 24-29) . Finally, we update the Pareto front by using the best bees in the colony (line 30).

4 Experimental Results

In this section we describe the methodology followed for tuning the parameters of each algorithm and we show a comparison among their performance.

To conduct the parameter tuning of the algorithms, we have used a real-world network topology, the Nippon Telegraph and Telephone (NTT, Japan) network, and six sets of demands, for further information, refer to [7] (see Table 1). For each experiment we have performed 30 independent runs and a statistical analysis using ANOVA tests, in this way, we can say that the parameter tuning of each algorithm was statistically relevant.

Table 1. Runtimes, reference points (r_{min} and r_{max}) to calculate the hypervolume and Short-Names for each data set

	$\|U\|$	c_{ij}	Runtime (s)	r_{min}	r_{max}	Short Name
NTT	10	10	6	(0, 0)	(220, 20)	NTT1
		20	65	(0, 0)	(530, 70)	NTT2
		40	110	(0, 0)	(790, 190)	NTT3
NTT	8	10	6	(0, 0)	(230, 20)	NTT4
		20	65	(0, 0)	(520, 110)	NTT5
		30	70	(0, 0)	(560, 80)	NTT6

In order to make a comparison, we have selected two novel multiobjective approaches: *Differential Evolution with Pareto Tournaments* (DEPT) [7] and *Multiobjective Variable Neighborhood Search* (MO-VNS) [8]. Furthermore, we have chosen the following well-known algorithm: *Fast Non-Dominated Sorting Genetic Algorithm* (NSGA-II) [3]. To compare these approaches with the MOABC, we have used two multiobjective metrics, *Hypervolume* [11] and *Coverage Relation* [10]. To calculate the hypervolume, it is necessary to use two reference points, $r_{min}(x_{min}, y_{min})$ and $r_{max}(x_{max}, y_{max})$, where x is the number of hops (y_1) and y is the number of wavelength switchings (y_2). In Table 1 we show the different reference points for each data set. The r_{max} point for every data set was calculated from the experience. Every algorithm uses the same value of K-shortest-paths (10) and 25 individuals as population size (only for population-based algorithms). The DEPT algorithm used a 20% of crossover probability,

50% of mutation factor and Best/1/Binomial as selection Schema. The NSGA-II presents a 70% of crossover probability, a single point crossover schema, a 75% of elitism probability and a 10% of mutation probability. For the MOABC, we obtain a 7.5% of mutation probability and 3 scout bees ($SN = 3$). The only parameter of the MO-VNS is the K-shortest-paths, that is 10 for all approaches.

As we can see in Table 2, the MOABC obtains equal values of hypervolume in the easiest data sets (NTT1 and NTT4), however, it overcomes the results obtained by the other approaches in the rest of data sets. In Figure 1, we present a visual comparison among the algorithms, where we can see that the MOABC pareto front dominates all non-dominated solutions achieved by the other approaches. To create the plot, we have used the pareto front that has a value of hypervolume closer to the mean hypervolume obtained in the 30 runs by each algorithm (see Table 2).

To confirm the proper functioning of our proposal, we have compared the algorithms using the Coverage Relation. This metric measures the fraction of non-dominated solutions evolved by an algorithm B, which are covered by the non-dominated points achieved by an algorithm A in average. As we can see in Table 3, the MOABC covers the 100% of the surface of all the other approaches in all data sets. However, we can notice that the DEPT, MO-VNS and NSGA-II only covers a mean surface of 33.33%, 38.89% and 33.33% of the MOABC non-dominated solutions, respectively.

Table 2. Comparison among the algorithms DEPT, MO-VNS, NSGA-II and MOABC by using average *Hypervolume* of 30 runs

	DEPT	MO-VNS	NSGA-II	MOABC
NTT1	**69.55%**	**69.55%**	**69.55%**	**69.55%**
NTT2	69.43%	68.81%	69.81%	**70.75%**
NTT3	63.48%	62.73%	62.54%	**66.98%**
NTT4	**70.87%**	**70.87%**	**70.87%**	**70.87%**
NTT5	68.66%	67.92%	68.02%	**69.42%**
NTT6	64.31%	61.79%	64.62%	**67.36%**
Mean	67.72%	66.95%	67.57%	**69.15%**

Fig. 1. Non-Dominated Solutions obtained by the Algorithms (NTT3)

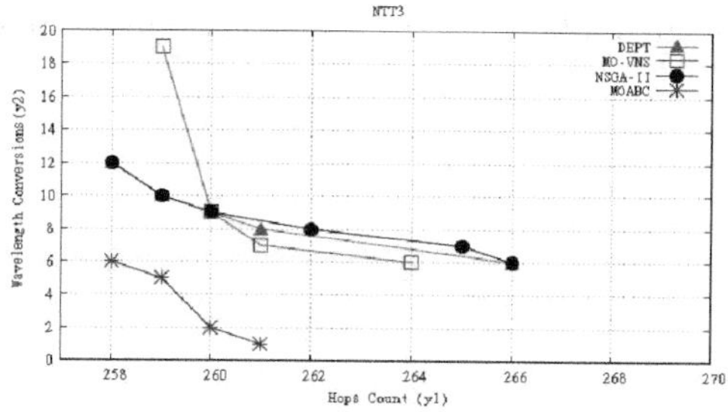

Table 3. Comparison among the algorithms DEPT, MO-VNS, NSGA-II and MOABC using the Coverage Relation ($A \succeq B$)

A	DEPT			MO-VNS			NSGA-II			MOABC		
B	MO-VNS	NSGA-II	MOABC	DEPT	NSGA-II	MOABC	DEPT	MO-VNS	MOABC	DEPT	MO-VNS	NSGA-II
NTT1	100%	100%	100%	100%	100%	100%	100%	100%	100%	100%	100%	100%
NTT2	0%	0%	0%	100%	80%	0%	100%	80%	0%	100%	100%	100%
NTT3	50%	83.33%	0%	60%	66.67%	0%	80%	50%	0%	100%	100%	100%
NTT4	100%	100%	100%	100%	100%	100%	100%	100%	100%	100%	100%	100%
NTT5	0%	100%	0%	100%	100%	33.33%	66.67%	0%	0%	100%	100%	100%
NTT6	100%	100%	0%	50%	50%	0%	100%	100%	0%	100%	100%	100%
Mean	58.33%	80.56%	33.33%	85%	82.78%	38.89%	91.11%	71.67%	33.33%	100%	100%	100%

We can conclude that the MOABC obtains equal or better performance that the DEPT, MO-VNS and NSGA-II in all data sets. In a global view, the MOABC seems to be a very promising approach for solving the Static Routing and Wavelength Assignment problem.

5 Comparison with Other Authors

Other authors have also tackled the Static RWA problem, so the aim of this section is to show several comparisons with other approaches published in the literature. In Table 4 we present the different heuristics (typical in telecommunication field) and varieties of MOACOs proposed in [1] and [5].

Table 4. Heuristics (typical in telecommunication field) and MOACOs proposed in [1] and [5]

	Wavelength Assignment			
Routing	First-Fit (FF)	Least-Used (LU)	Most-Used (MU)	Random (RR)
3-Shortest Path (3SP)	3SPFF	3SPLU	3SPMU	3SPRR
Shortest Path Dijsktra (SP)	SPFF	SPLU	SPMU	SPRR

Multiobjective Ant Colony Optimization Algorithms (MOACOs)			
BIANT	Bicriterion Ant	M3AS	Multiobjective Max-Min Ant System
COMP	COMPET Ants	MAS	Multiobjective Ant System
MOAQ	Multiple Objective Ant Q Algorithm	PACO	Pareto Ant Colony Optimization
MOACS	Multiple Objective Ant Colony System	MOA	Multiobjective Omicrom ACO

To make these comparisons, we have used the same methodology that we explained in Section 4. However, in [1], the authors only present the non-dominated solutions obtained by the best typical heuristics and MOACOs in the following data sets: NTT2, NTT3, NTT4 and NTT5. Therefore, we will compare the MOABC algorithm with the best approaches for each data set used in [1].

As we can see in Table 5, the MOABC obtains higher value of hypervolume than the best typical heuristic and MOACO for almost all data sets. We can notice that the suggested typical heuristics and MOACOs obtain the same value of hypervolume as the MOABC in NTT4 (70.87%). In Figure 2, we can see that

Table 5. Comparison among the best approaches suggested in [1] and [5] for each data set and the MOABC algorithm (*Hypervolume* metric)

	NTT2		NTT3
3SPLU	62.96%	3SPLU	63.18%
MOA	56.01%	BIANT	57.52%
MOABC	**70.75%**	MOABC	**66.98%**
	NTT4		NTT5
3SPLU, 3SPRR, SPLU, SPRR, M3AS, MOA and MOABC	**70.87%**	3SPLU	66.81%
		MAS	64.79%
		MOA	63.37%
		MOABC	**69.42%**

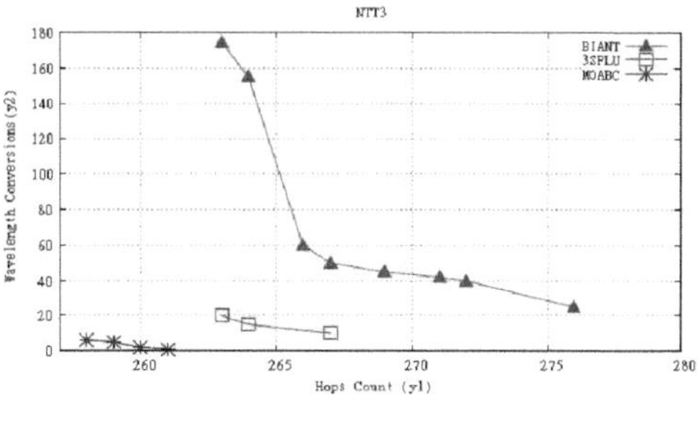

Fig. 2. Non-Dominated Solutions obtained by the approaches (NTT3)

Table 6. Comparison among the best approaches suggested in [1] and [5] for each data set and the MOABC algorithm using the *Coverage Relation* metric ($A \succeq B$)

	NTT2						NTT3					
A	MOA		3SPLU		MOABC		BIANT		3SPLU		MOABC	
B	3SPLU	MOABC	MOA	MOABC	MOA	3SPLU	3SPLU	MOABC	BIANT	MOABC	BIANT	3SPLU
	0%	0%	88.89%	0%	100%	100%	0%	0%	100%	0%	100%	100%

	NTT5											
A	MAS			MOA			3SPLU			MOABC		
B	MOA	3SPLU	MOABC	MAS	3SPLU	MOABC	MAS	MOA	MOABC	MAS	MOA	3SPLU
	16.67%	0%	0%	87.50%	0%	0%	87.50%	83.33%	0%	62.50%	50%	0%

the front achieved by MOABC clearly dominates the front obtained by the best approaches suggested in [1] and [5].

Secondly, we present a direct comparison (Coverage Relation) of the outcomes achieved by the algorithms presented above. In this case, we are going to discard the NTT4 data set, due to all approaches have obtained the same pareto front. In Table 6, we can notice that the pareto front obtained by the MOABC dominates the fronts obtained by the best MOACOs and by the best typical heuristics in data sets NTT2 and NTT3. On the one hand, in NTT5, the MOABC has a better coverage relation than the best MOACOs. On the other hand, the non-dominated solutions provided by 3SPLU are not able to dominate the non-dominated solutions obtained by the MOABC and viceversa.

To sum up, after performing an exhaustive comparison with the best typical heuristics proposed in [1] and the best MOACOs proposed in [1] and [5]; we can say that the MOABC algorithm obtains very promising results. It obtains better results than the best approaches suggested in [1] and [5], so it performs better than sixteen different heuristics.

6 Conclusions and Future Work

In this work, we have proposed the use of a new multiobjective approach of the Artificial Bee Colony (ABC) algorithm (MOABC) for solving the Static Routing and Wavelength Assignment (RWA) problem in WDM networks. To ensure the effectiveness of our proposal, we have made several comparisons with other approaches published in the literature. To make these comparisons, we have used a real-world network topology, the Nippon Telegraph and Telephone network (NTT, Japan) and six sets of demands. Furthermore, to decide which of the approaches performs better, we have used two well-known metrics in multiobjective field: *Hypervolume* and *Coverage Relation*. After performing the comparisons, we can conclude that the MOABC overcomes the results obtained by almost twenty different heuristics. As future work, we intend to apply other multiobjective versions of Swarm Intelligence algorithms for the Static RWA problem with the aim of comparing with the results achieved by the MOABC.

Acknowledgements. Álvaro Rubio-Largo is supported by the research grant PRE09010 from Junta de Extremadura. This work has been partially funded by the Spanish Ministry of Education and Science and ERDF (the European Regional Development Fund), under contract TIN2008-06491-C04-04 (the M* project).

References

1. Arteta, A., Barán, B., Pinto, D.: Routing and Wavelength Assignment over WDM Optical Networks: a comparison between MOACOs and classical approaches. In: LANC 2007: Proceedings of the 4th international IFIP/ACM Latin American conference on Networking, pp. 53–63. ACM, New York (2007)
2. Deb, K.: Multi-Objective Optimization Using Evolutionary Algorithms. John Wiley & Sons, Inc., New York (2001)
3. Deb, K., Pratap, A., Agarwal, S., Meyarivan, T.: A Fast Elitist Multi-Objective Genetic Algorithm: NSGA-II. IEEE Transactions on Evolutionary Computation 6, 182–197 (2000)
4. Hamad, A.M., Kamal, A.E.: A survey of multicasting protocols for broadcast-and-select single-hop networks. IEEE Network 16, 36–48 (2002)
5. Insfrán, C., Pinto, D., Barán, B.: Diseño de Topologías Virtuales en Redes Ópticas. Un enfoque basado en Colonia de Hormigas. In: XXXII Latin-American Conference on Informatics 2006 - CLEI 2006, vol. 8, pp. 173–195 (2006)
6. Karaboga, D., Akay, B.: A survey: algorithms simulating bee swarm intelligence. Artificial Intelligence Review 31, 61–85 (2009)
7. Rubio-Largo, A., Vega-Rodríguez, M.A., Gómez-Pulido, J.A., Sánchez-Pérez, J.M.: A Differential Evolution with Pareto Tournaments for solving the Routing and Wavelength Assignment Problem in WDM Networks. In: Proceedings of the 2010 IEEE Congress on Evolutionary Computation (CEC 2010), vol. 10, pp. 129–136 (2010)
8. Rubio-Largo, A., Vega-Rodríguez, M.A., Gómez-Pulido, J.A., Sánchez-Pérez, J.M.: Solving the Routing and Wavelength Assignment Problem in WDM Networks by Using a Multiobjective Variable Neighborhood Search Algorithm. In: 5th International Workshop on Soft Computing Models in Industrial and Environmental Applications, SOCO 2010, vol. 73, pp. 47–54 (2010)
9. Weicker, N., Szabo, G., Weicker, K., Widmayer, P.: Evolutionary Multiobjective Optimization for base station transmitter placement with Frequency Assignment. IEEE Transactions on Evolutionary Computation 7(2), 189–203 (2003)
10. Zitzler, E., Deb, K., Thiele, L.: Comparison of Multiobjective Evolutionary Algorithms: Empirical Results. Evolutionary Computation 8, 173–195 (2000)
11. Zitzler, E., Thiele, L.: Multiobjective optimization using evolutionary algorithms - A comparative case study. In: Eiben, A.E., Bäck, T., Schoenauer, M., Schwefel, H.-P. (eds.) PPSN 1998. LNCS, vol. 1498, pp. 292–301. Springer, Heidelberg (1998)

Applying a Multiobjective Gravitational Search Algorithm (MO-GSA) to Discover Motifs

David L. González-Álvarez, Miguel A. Vega-Rodríguez, Juan A. Gómez-Pulido, and Juan M. Sánchez-Pérez

University of Extremadura,
Dep. of Technologies of Computers and Communications, ARCO Research Group,
Escuela Politécnica. Campus Universitario s/n, 10003. Cáceres, Spain
{dlga,mavega,jangomez,sanperez}@unex.es

Abstract. Currently there are a large number of Bioinformatics problems that are tackled using computational techniques. The problems discussed range from small molecules to complex systems where many organisms coexist. Among all these issues, we can highlight genomics: it studies the genomes of microorganisms, plants and animals. To discover common patterns, motifs, in a set of deoxyribonucleic acid (DNA) sequences is one of the important sequence analysis problems and it is known as Motif Discovery Problem (MDP). In this work we propose the use of computational Swarm Intelligence for solving the MDP. A new heuristic based on the law of gravity and the notion of mass interactions, the Gravitational Search Algorithm (GSA), is chosen for this purpose, but adapted to a multiobjective context (MO-GSA). To test the performance of the MO-GSA, we have used twelve real data sets corresponding to alive beings. After performing several comparisons with other approaches published in the literature, we conclude that this algorithm outperforms the results obtained by others.

Keywords: Swarm Intelligence, Gravitational Search Algorithm, DNA, motif finding, multiobjective optimization.

1 Introduction

Bioinformatics arises from the need to work specifically with a large amount of deoxyribonucleic acid (DNA) and protein sequences stored in databases. This information is currently used in many research domains [1], ranging from multiple sequence alignment, DNA fragments assembly, or genomic mapping, to the prediction of DNA motifs, the search of these motifs in sequences of other species, or protein folding. In this paper we predict motifs using Swarm Intelligence, solving the Motif Discovery Problem (MDP). The MDP aims to maximize three conflicting objectives: support, motif length, and similarity. So we have to apply multiobjective optimization (MOO) to obtain motifs in the most efficient way. To solve the MDP we have designed and implemented a multiobjective version of an innovative algorithm, the Gravitational Search Algorithm (GSA), that

J. Cabestany, I. Rojas, and G. Joya (Eds.): IWANN 2011, Part II, LNCS 6692, pp. 372–379, 2011.

we have named the Multiobjective Gravitational Search Algorithm (MO-GSA). Over the last years, there has been growing interest in algorithms inspired by the observation of natural phenomena. It has been shown by many researches that these algorithms are good replacements as tools to solve complex computational problems. To demonstrate the effectiveness and efficiency of our approach we have performed experiments by using twelve real data sets. The results obtained improve other well-known methods for discovering motifs such as AlignACE, MEME, and Weeder, as well as achieve better performance than the results from other major researchers in the field.

This paper is organized as follows. In Section 2 we describe the MDP. Section 3 details the adaptation and modifications made on the Swarm Intelligence algorithm used. Section 4 shows the results obtained by our proposal, including comparisons with other multiobjective algorithms implemented previously. Furthermore, we compare the algorithm results with those achieved by other techniques and algorithms for discovering DNA motifs in Section 5. Finally, some conclusions and future lines are included in Section 6.

2 The Motif Discovery Problem

We use the following objectives to discover many long and strong motifs: motif length, support, and similarity, proposed by [2]. Given a set of sequences $S = \{S_i | i = 1, 2, ..., D\}$ of nucleotides defined on the alphabet $B = \{A, C, G, T\}$. $S_i = \{S_i^j | j = 1, 2, ..., w_i\}$ is a sequence of nucleotides, where w_i is the sequence width. The motif length is l nucleotides long and it is the first objective to maximize. In motif discovery, motifs are usually very short, so that, after conducting various studies, we have restricted the minimum and maximum motif length to 7 and 64, respectively. The set of all the subsequences contained in S is $\{s_i^{j_i} | i = 1, 2, ..., D, j_i = 1, 2, ..., w_i - l + 1\}$, where j_i is the binding site of a possible motif instance s_i^j on sequence S_i. To obtain the values of the other two objectives we have to build the Position Indicator Matrix (PIM) $A = \{A_i | i = 1, 2, ..., D\}$ of the motif, where $A_i = \{A_i^j | j = 1, 2, ..., w_i\}$ is the indicator row vector with respect to a sequence S_i. A_i^j is 1 if the position j in S_i is a binding site, and 0 otherwise. We refer to the number of motif instances as $|A| = \sum_{i=1}^{D} \sum_{j=1}^{w_i} A_i^j$. We also require to find the consensus motif, which is a string abstraction of the motif instances. In this work we consider a single motif instance per sequence. Only those sequences that achieve a motif instance of certain quality with respect to the consensus motif are taken into account when we perform the final motif. This is indicated by the second objective, the *support*.

$S(A) = \{S(A)_1, S(A)_2, ..., S(A)_{|A|}\}$ is a set of $|A|$ motif instances, where $S(A)_i = S(A)_i^1 S(A)_i^2 ... S(A)_i^l$ is the ith motif instance in $|A|$. $S(A)$ can also be expanded as $(S(A)^1, S(A)^2, ..., S(A)^l)$, where $S(A)^j = S(A)_1^j S(A)_2^j ... S(A)_{|A|}^j$ is the list of nucleotides on the jth position in the motif instances. Then, we build the Position Count Matrix (PCM) $N(A)$ with the numbers of different nucleotide bases on each position of the candidate motifs (A) who have passed the threshold marked by the support. $N(A) = \{N(A)^1, N(A)^2, ..., N(A)^l\}$, and

$N(A)^j = \{N(A)_b^j | b \in B\}$, where $N(A)_b^j = |\{S(A)_i^j | S(A)_i^j = b\}|$. The dominant nucleotides of each position are normalized in the Position Frequency Matrix (PFM) $\widehat{N} = \frac{N(A)}{|A|}$. Finally, we calculate the third objective, the similarity, averaging all the dominance values of each PFM column. As is indicated in the following expression:

$$Similarity(Motif) = \frac{\sum_{i=1}^{l} max_b\{f(b,i)\}}{l} \tag{1}$$

where $f(b,i)$ is the score of nucleotide b in column i in the PFM and $max_b\{f(b,i)\}$ is the dominance value of the dominant nucleotide in column i. For further information, refer to [3].

3 The Multiobjective Gravitational Search Algorithm

The Gravitational Search Algorithm (GSA) is an heuristic recently introduced by Rashedi et al. [4]. It uses the Newtonian physics theory and its searcher agents are the collection of masses. In the GSA, we have an isolated system of masses. Using the gravitational force, every mass in the system can see the situation of other masses. All these objects attract each other by the gravity force, and this force causes a global movement of all objects towards the objects with heavier masses (better solutions). Hence, masses cooperate using a direct form of communication, through gravitational force. To solve the MDP we have defined the Multiobjective Gravitational Search Algorithm (MO-GSA). It incorporates features of two standard algorithms in multiobjective computation such as NSGA-II [5] and SPEA2 [6]. The definition of the individuals in the algorithm is the same as in [3] and [7]. In Algorithm 1 we include the algorithm pseudocode.

First, we create and evaluate the initial population randomly (line 4 of Algorithm 1). Then, the algorithm execution starts, finishing when it reaches the time limit set for the execution. The algorithm classifies the population into different Pareto fronts, ranking the individuals by using the Pareto front and the Crowding Distance concepts from the NSGA-II algorithm. To calculate the MOFitness (line 9), we apply a linear bias b_r to the rth ranked element by using the expression: $b_r = 1/r$, obtaining values from 1 to 1/N. Then, we update the Gravitational Constant (G), the best solution, and the worst solution. To update G (line 11) and K_{best} (lines 2 and 36), we have used the equations and the parameter values (G0, α, ε, and K_{best}) proposed in [4]. At this point, we have to calculate the mass assigned to each idividual (lines 15 and 16). Then we may calculate the force acting on each individual X_i from all other K_{best} individuals (line 22) for each dimension. The total force (line 24) that acts on X_i in a dimension d is a randomly weighted sum of dth components of the forces exerted from the K_{best} individuals. Finally, we calculate the values of the accelerations (line 25) and the velocities (line 31) to update the new positions (line 32) of the chromosomes of the population. This process is repeated until the execution time expires. Note that MO-GSA archives all generated solutions (a feature that we have obtained from the SPEA2 algorithm).

Algorithm 1. Pseudocode for MO-GSA

```
NDSarchive ⇐ ∅
K_best ⇐ N
/* Generate Initial Population P = {X_1,X_2,...,X_N} */
P ⇐ generateInitialPopulation(N)
while not time limit do
   /* Evaluate the fitness for each individual */
   P ⇐ fastNondominatedSort(P)
   P ⇐ crowdingDistanceAssignment(P)
   P ⇐ calculateMOFitnessUsingBias(P)
   /* Update G, best, and worst */
   G ⇐ G_0 * e^(-αt/T)
   best ⇐ X_1.MOFitness
   worst ⇐ X_N.MOFitness
   /* Calculate masses for each individual */
   X_i.m ⇐ (X_i.MOFitness - worst ) / ( best - worst ) i = 1, ... ,N
   X_i.M ⇐ (X_i.m) / (Σ_{j=1}^{N} X_j.m) i = 1, ... ,N
   /* Calculate forces and accelerations for each individual */
   for d = 1 to chromosomeNumber do
      for i = 1 to N do
         for j = 1 to K_best do
            R_ij ⇐ ||X_i, X_j||_2
            F_ij^d ⇐ G * ((X_i.M * X_j.M) / (R_ij + ε)) * (X_j^d.chromosome - X_i^d.chromosome)
         end for
         X_i.F^d ⇐ Σ_{j∈K_best, j≠i}^{N} rand[0, 1] * F_ij^d
         X_i.acceleration^d ⇐ X_i.F^d / X_i.M
      end for
   end for
   /* Update velocities and positions of every chromosome of each individual */
   for d = 1 to chromosomeNumber do
      for i = 1 to N do
         X_i.velocity^d ⇐ rand[0,1] * X_i.velocity^d + X_i.acceleration^d
         X_i.chromosome^d ⇐ X_i.chromosome^d + X_i.velocity^d
      end for
   end for
   NDSarchive ⇐ updateNDSarchive(P)
   K_best ⇐ decrease(K_best)
   P ⇐ applyMutationFunction(P) /* If we detect stagnation */
end while
```

4 Experimental Results

We have conducted several experiments to configure our algorithm, for each of them we have performed 30 independent runs to assure its statistical relevance. The results are measured by using the hypervolume (HV) indicator to facilitate the comparison of performances, and they are displayed by using the average values of the HVs and their standard deviations. The reference volume is calculated using the maximum values of each objective in each data set, for example, a data set with seven sequences will have: support=7, motif length=64, and similarity=1 (100%). The experiments are organized taking into account the influence of each parameter. Finally, to compare the algorithms we have used, besides the HV indicator, the Coverage Relation [8], that is useful to analyze which algorithms get the best Pareto fronts. For comparison with other authors [2], we have used the same population size and the same runtimes as them. We have used twelve real sequence data sets as a benchmark for discovering motifs, which were selected from

Table 1. Data sets properties

Data set	Seq.	Len.	Time (sec.)
dm01r	4	1500	15
dm04r	4	2000	15
dm05r	5	2500	15
hm03r	10	1500	25
hm04r	13	2000	25
hm16r	7	3000	15
mus02r	9	1000	15
mus07r	4	1500	15
mus11r	12	500	25
yst03r	8	500	15
yst04r	7	1000	15
yst08r	11	1000	25

Table 2. Comparison of Algorithm Hypervolumes

	MO-GSA	MO-VNS	DEPT	NSGA-II
	Mean±Std. dev.	Mean±Std. dev.	Mean±Std. dev.	Mean±Std. dev.
dm01r	**82.39%±0.01**	81.21%±0.00	83.43%±0.01	82.66%±0.01
dm04r	**83.94%±0.01**	81.58%±0.00	85.37%±0.00	83.91%±0.01
dm05r	**82.88%±0.01**	83.30%±0.00	83.92%±0.00	83.38%±0.00
hm03r	**69.42%±0.02**	49.92%±0.02	70.18%±0.05	59.82%±0.06
hm04r	**55.84%±0.02**	33.50%±0.03	35.39%±0.04	37.78%±0.05
hm16r	**84.30%±0.03**	68.22%±0.02	78.52%±0.03	79.51%±0.03
mus02r	**69.35%±0.01**	54.64%±0.01	71.62%±0.01	66.04%±0.01
mus07r	**84.76%±0.03**	84.27%±0.01	87.11%±0.01	86.40%±0.01
mus11r	**60.77%±0.02**	40.12%±0.04	59.37%±0.09	57.09%±0.03
yst03r	**75.85%±0.00**	63.66%±0.02	75.92%±0.00	73.83%±0.01
yst04r	**78.85%±0.01**	70.36%±0.02	80.21%±0.00	77.37%±0.01
yst08r	**75.69%±0.01**	59.22%±0.03	77.04%±0.05	72.53%±0.01
mean	**75.336244%**	64.165783%	74.007811%	71.694822%

the TRANSFAC database [9]. The twelve data sets correspond to alive beings and have different number of sequences and sizes (nucleotides per sequence) to ensure that our algorithms work with several types of instances. The established runtimes and the data set properties are shown in Table 1.

This first comparison is performed by using the hypervolumes obtained by two novel algorithms, the Differential Evolution with Pareto Tournaments [3], and the Multiobjective Variable Neighborhood Search [7]. We also compare our results with those obtained by NSGA-II, a standard algorithm in multiobjective optimization. To improve its results is a first step to demonstrate the proper functioning of our proposals. The results of this comparison are shown in Table 2. We notice how MO-GSA achieves the best average results (last row in Table 2). For the case of NSGA-II and MO-VNS, MO-GSA obtains better hypervolumes in almost all the data sets. As the results of the MO-GSA and DEPT are more even, we have made a second comparison by using the Coverage Relation. The Coverage Relation is another indicator to measure the performance of the results accomplished by the algorithms [8]. More precisely, considering the dominance concept, the Coverage Relation indicator considers that x_1 covers x_2 if $x_1 \succ x_2$ or $x_1 = x_2$. It is applied to all nondominated solutions obtained by the algorithms, and it is used as a comparison criterion. In Table 3 we include the results of this comparison. In this table we can see how the nondominated solutions of MO-GSA cover the 64.04% of the nondominated solutions of DEPT, while DEPT covers the 59.09% of the nondominated solutions of MO-GSA. This means that many of the motifs discovered by MO-GSA dominate the motifs found by DEPT,

Table 3. Coverage Relation ($A \succeq B$)

A	B	dm01r	dm04r	dm05r	hm03r	hm04r	hm16r	mus02r	mus07r	mus11r	yst03r	yst04r	yst08r	**mean**
MO-GSA	DEPT	79.76%	62.62%	80.77%	65.91%	100.00%	99.09%	40.76%	83.00%	12.60%	80.75%	49.67%	13.57%	**64.04%**
MO-GSA	MOVNS	89.04%	91.04%	75.00%	90.91%	77.59%	87.80%	94.79%	100.00%	79.49%	88.98%	93.59%	65.98%	**86.18%**
MO-GSA	NSGAII	71.26%	67.92%	72.58%	78.40%	50.00%	57.45%	62.98%	86.60%	72.87%	44.50%	94.87%	92.55%	**71.00%**
DEPT	**MO-GSA**	73.26%	88.79%	39.34%	42.31%	0.00%	6.90%	73.08%	91.58%	89.14%	37.09%	79.59%	88.04%	**59.09%**
MOVNS	**MO-GSA**	29.07%	22.43%	44.26%	6.59%	11.06%	8.05%	8.17%	23.16%	11.76%	15.96%	7.48%	23.92%	**17.66%**
NSGAII	**MO-GSA**	79.07%	82.24%	57.38%	28.02%	37.17%	49.43%	47.60%	86.32%	32.13%	57.28%	14.97%	12.44%	**48.67%**

so that the Pareto fronts obtained by MO-GSA are of better quality. It should be noted that, although it is not included in the tables due to space constraints, the MO-GSA covers more solutions of the other two algorithms (MO-VNS and NSGA-II) than DEPT, resulting in a rate of 78.59% compared to the 71.15% obtained by DEPT.

5 Comparisons with Other Author Approaches

In this section we analyze the motifs obtained by the MO-GSA algorithm. To that end we compare the motifs discovered by our algorithm with the solutions predicted by the best configuration of MOGAMOD [2], another multiobjective algorithm for discovering motifs, and with other well-known methods in the bioinformatics field such as AlignACE [10], MEME [11], and Weeder [12]. Each method has its own operating principles: AlignACE uses a Gibbs sampling algorithm that returns a series of motifs as weight matrices that are overrepresented in the input set, MEME optimizes the e-value of a statistic related to the information content of the motif, and Weeder is a consensus-based method. We could not perform this comparison using the hypervolume indicator because unfortunately, we do not have this information for the other methods. In order to compare with [2], we focus our comparisons on yst04r, yst08r, and hm03r data sets (the only data sets that appear in [2]).

The comparison with the MOGAMOD algorithm is done in two ways. In the first one we compare the similarities obtained by both algorithms, maintaining fixed values of the other two objectives (motif length and support). In the second one, we compare the sizes of the motifs discovered by the two algorithms, also keeping fixed the other two objectives, in this case the support and the similarity. In Table 4(a), 4(b), and 4(c) we include this comparison. In these tables we can see how the solutions of MO-GSA get higher similarities than those obtained by MOGAMOD, and how the MO-GSA algorithm discovers larger motifs than MOGAMOD. These comparisons show the superiority of our proposal in two ways, demonstrating that the motifs discovered by our algorithm are more robust than those obtained by the other methods.

In addition to compare with MOGAMOD, we have compared our algorithm with other well-known methods in the Bioinformatics field such as AlignACE, MEME, or Weeder. Tables 4(d), 4(e), and 4(f) give the results of this comparison. In these tables we include some of the solutions obtained by these methods (besides some of MOGAMOD). A key point here is that while MO-GSA finds solutions with different support, other methods, except MOGAMOD, extract only one motif per run. Moreover, MO-GSA gets very long motifs that also have a good value of support and similarity. As we can notice in these tables, the solutions always maintain a balance between the values of the three objectives. We see how as the support and the motif length values increase, the similarity value decreases. However, with the same value of support, as the motif length decreases, the similarity value raises.

Table 4. Comparison of the motifs predicted for yst04r, yst08r, and hm03r

(a)

		Sim. Comparison		Len. Comparison		
yst04r		MO-GSA	MOGAMOD	MO-GSA	MOGAMOD	
Sup.	Len.	Sim.	Sim.	Len.	Len.	Sim.
4	24	**0.854167**	0.76	**47**	24	0.76
	20	**0.881266**	0.78	**39**	20	0.78
	15	**0.933333**	0.87	**22**	15	0.87
5	15	**0.907692**	0.82	**25**	15	0.82
	14	**0.915385**	0.84	**22**	14	0.84
6	14	**0.869048**	0.77	**28**	14	0.77
	13	**0.910256**	0.81	**22**	13	0.81
7	9	**0.920635**	0.80	**21**	9	0.80
	8	**0.946429**	0.84	**15**	8	0.84

(b)

		Sim. Comparison		Len. Comparison		
yst08r		MO-GSA	MOGAMOD	MO-GSA	MOGAMOD	
Sup.	Len.	Sim.	Sim.	Len.	Len.	Sim.
7	20	**0.828571**	0.75	**36**	20	0.75
	15	**0.857143**	0.84	**18**	15	0.84
8	15	**0.841667**	0.79	**23**	15	0.79
	14	**0.866071**	0.83	**18**	14	0.83
	13	**0.884615**	0.85	**16**	13	0.85
9	13	**0.871795**	0.82	**15**	13	0.82
	12	**0.907407**	0.84	**14**	12	0.84
10	12	**0.875000**	0.79	**20**	12	0.79
	11	**0.881818**	0.82	**15**	11	0.82
11	11	**0.876033**	0.80	**15**	11	0.80

(c)

		Sim. Comparison		Len. Comparison		
hm03r		MO-GSA	MOGAMOD	MO-GSA	MOGAMOD	
Sup.	Len.	Sim.	Sim.	Len.	Len.	Sim.
6	25	**0.786667**	0.71	**46**	25	0.71
	22	**0.809524**	0.76	**32**	22	0.76
7	22	**0.782313**	0.74	**27**	22	0.74
	18	**0.793651**	0.82	**15**	18	0.82
8	18	**0.770833**	0.76	**21**	18	0.76
	13	**0.836538**	0.81	**15**	13	0.81
9	13	**0.811966**	0.77	**16**	13	0.77
	11	**0.838384**	0.78	**15**	11	0.78
10	11	**0.809091**	0.74	**17**	11	0.74
	10	**0.830000**	0.79	**12**	10	0.79
	9	**0.844444**	0.81	**11**	9	0.81

(d)

Method	Sup.	Len.	Sim.	Predicted motif (hm03r)
AlignACE	N/A	13	N/A	TGTGGATAAAAAA
MEME	N/A	20	N/A	AGTGTAGATAAAAGAAAAAC
Weeder	N/A	10	N/A	TGATCACTGG
MOGAMOD	7	22	0.74	TATCATCCCTGCCTAGACACAA
	7	18	0.82	TGACTCTGTCCCTAGTCT
	10	11	0.74	TTTTTTCACCA
	10	10	0.79	CCCAGCTTAG
	10	9	0.81	AGTGGGTCC
MO-GSA	**7**	**24**	**0.779762**	TTAGTGCCTGACACACAGAGGTGC
	10	**11**	**0.809091**	TCTGAGACTCA

(e)

Method	Sup.	Len.	Sim.	Predicted motif (yst04r)
AlignACE	N/A	10	N/A	CGGGATTCCA
MEME	N/A	11	N/A	CGGGATTCCCC
Weeder	N/A	10	N/A	TTTTCTGGCA
MOGAMOD	5	14	0.84	CGAGCTTCCACTAA
	6	14	0.77	CGGGATTCCTCTAT
MO-GSA	**5**	**21**	**0.847619**	TGGCATCCACTAATTGAAAGA
	6	**16**	**0.854167**	GTTACACCTAGACACC

(f)

Method	Sup.	Len.	Sim.	Predicted motif (yst08r)
AlignACE	N/A	11	N/A	CACCCAGACAC
	N/A	12	N/A	TGATTGCACTGA
MEME	N/A	11	N/A	CACCCAGACAC
Weeder	N/A	10	N/A	ACACCCAGAC
MOGAMOD	7	15	0.84	GCGACTGGGTGCCTG
	8	14	0.83	GCCAGAAAAAGGCG
	8	13	0.85	ACACCCAGACATC
MO-GSA	**7**	**18**	**0.841270**	TTCTAAGACAATCTTTTT
	9	**14**	**0.849206**	TTCTTGCATAAATT

6 Conclusions

In this paper we have discovered quality motifs by using an innovative multiobjective version of the Gravitational Search Algorithm (GSA), named Multiobjective Gravitational Search Algorithm (MO-GSA). Real data sets of alive beings such as fly, human, mouse, or yeast have been used in the experimental section. After performing several comparisons with other approaches published in the literature, we can conclude that the MO-GSA overcomes the results obtained by the other approaches. For future work, we will apply this innovative multiobjective algorithm to other extensions of the Motif Discovery Problem. Furthermore, it would be interesting to develop different Swarm Intelligence algorithms with the aim of making comparisons with MO-GSA.

Acknowledgements. Thanks to the Fundación Valhondo Calaff for the economic support offered to David L. González-Álvarez to make this research. This work was partially funded by the Spanish Ministry of Science and Innovation and ERDF (the European Regional Development Fund), under the contract TIN2008-06491-C04-04 (the M* project).

References

1. Dopazo, J., Zanders, E., Dragoni, I., Amphlett, G., Falciani, F.: Methods and approaches in the analysis of gene expression data. Journal of immunological methods 250(1-2), 93–112 (2001)
2. Kaya, M.: MOGAMOD: Multi-objective genetic algorithm for motif discovery. Expert Systems with Applications: An International Journal 36(2), 1039–1047 (2009)
3. González-Álvarez, D.L., Vega-Rodríguez, M.A., Gómez-Pulido, J.A., Sánchez-Pérez, J.M.: Solving the Motif Discovery Problem by Using Differential Evolution with Pareto Tournaments. In: Proceedings of the 2010 IEEE Congress on Evolutionary Computation (CEC 2010), pp. 4140–4147. IEEE Computer Society, Los Alamitos (2010)
4. Rashedi, E., Nezamabadi-pour, H., Saryazdi, S.: GSA: A Gravitational Search Algorithm. Information Sciences 179(13), 2232–2248 (2009)
5. Deb, K., Pratap, A., Agarwal, S., Meyarivan, T.: A fast and elitist multi-objective genetic algorithm: NSGA II. IEEE Transactions on Evolutionary Computation 6, 182–197 (2002)
6. Zitzler, E., Laumanns, M., Thiele, L.: SPEA2: Improving the Strength Pareto Evolutionary Algorithm for Multiobjective Optimization. Evol. Methods for Design Optimization and Control with Applications to Industrial Problems, 95–100 (2001)
7. González-Álvarez, D.L., Vega-Rodríguez, M.A., Gómez-Pulido, J.A., Sánchez-Pérez, J.M.: A Multiobjective Variable Neighborhood Search for Solving the Motif Discovery Problem. AISC, vol. 73, pp. 39–46. Springer, Heidelberg (2010)
8. Zitzler, E., Deb, K., Thiele, L.: Comparison of multiobjective evolutionary algorithms: empirical results. IEEE Transactions on Evolutionary Computation 8(2), 173–195 (2000)
9. Wingender, E., Dietze, P., Karas, H., Knüppel, R.: TRANSFAC: a database on transcription factors and their DNA binding sites. Nucleic Acids Research 24(1), 238–241 (1996)
10. Roth, F.P., Hughes, J.D., Estep, P.W., Church, G.M.: Finding DNA regulatory motifs within unaligned noncoding sequences clustered by whole genome mRNA quantitation. Nature Biotechnology 16(10), 939–945 (1998)
11. Bailey, T.L., Elkan, C.: Fitting a mixture model by expectation maximization to discover motifs in biopolymers. In: Proceedings of the Second International Conference on Intelligent Systems for Molecular Biology, pp. 28–36. AAAI Press, Menlo Park (1994)
12. Pavesi, G., Mereghetti, P., Mauri, G., Pesolev, G.: Weeder Web: discovery of transcription factor binding sites in a set of sequences from co-regulared genes. Nucleic Acids Research 32, 199–203 (2004)

Looking for a Cheaper ROSA

Fernando L. Pelayo*, Fernando Cuartero, and Diego Cazorla

Departamento de Sistemas Informáticos
Escuela Superior de Ingenieria Informática
Universidad de Castilla-La Mancha
02071-Albacete, Spain
{FernandoL.Pelayo,Fernando.Cuartero,Diego.Cazorla}@uclm.es

Abstract. Process Algebras, PAs, are formalisms able to capture the behaviour of a computing system by, for example, giving the labelled transition system, LTS, where states are nodes and where all possible evolutions of the system are arcs; The drawing of the complete LTS is a NP-complete task, so that, the reaching of a particular 'desired' state is a problem which deserves some heuristic for improving the amount of resources to be carried out. In this line, Artificial Intelligence by means of Genetic Algorithms (GA's), provides metaheuristic techniques that have obtained good results in problems in which exhaustive techniques fail due to the size of the search space, as it is the exploration of a LTS.

In this paper, we try to avoid this problem, so only unfolding the most promising (for the task of reaching a 'goal' state) branches within the LTS.

Keywords: Process Algebra, Genetic Algorithm, Complexity.

1 Introduction

Artificial intelligence, AI, can be seen as the intelligence of machines and the branch of computer science that aims to create it. AI studies and designs intelligent agents, i.e., systems that perceive its environment and take actions that maximize its chances of success.

These agents can be categorized into several kinds according to the type of problems to solve or according to the strategies to follow. One of the typical problems to work in, is searching for a particular state among a lot of them. Genetic Algorithms, GAs [5,4], are strategies to be followed in order to solve AI problems, specially when the knowledge of the environment is not strong enough to easily guide the searching process. In fact, although they have been widely used to solve problems in the fields of combinatorial and numerical optimization, it is very rare to find them used dealing with the problem of improving the computational cost of analyzing via Process Algebras [10].

ROSA is a Markovian process algebra "functionally" close to PNAL [2]. Markovian time is added by means of the inclusion of actions whose duration is

* Research partially supported by projects TIN2009-14312 & CGL2007-66440-C04-03.

J. Cabestany, I. Rojas, and G. Joya (Eds.): IWANN 2011, Part II, LNCS 6692, pp. 380–387, 2011.

modelled by Exponentially distributed random variables of parameters $\lambda \in \mathbb{R}^+ - \{0\}$ and immediate actions, whose duration can be modelled by $Exp[\infty]$. There are some other differences between **ROSA** and PNAL as the order when solving the non-deterministic choices against the probabilistic ones or the inclusion of non-determinism when cooperating some type of actions.

ROSA [8] does not impose any syntactical restrictions on the components of a parallel operator, and thus, the specification labour becomes easier than in some other models.

The usefulness of **ROSA**, as well as of so many PAs is out of any doubt, but as exposed, the computational cost of the unfolding of the whole LTS is unbroachable from a practical perspective, so that we propose a way to only unfold the more promising states among the reachable (through a single transition) set of states from a given one. This, of course, would mean a saving on the computational cost of producing the LTS by the operational semantics of **ROSA**, in this sense we entitled this paper *Looking for a Cheaper ROSA*.

This paper is structured as follows: next 2 sections provide rough descriptions of the Markovian process Algebra **ROSA** and of a generic Genetic Algorithm, respectively. Then a topology structure over the set of **ROSA** processes is defined and that *promising function* which is claimed to make **ROSA** a "cheaper formalism" is finally presented.

2 The Markovian Process Algebra ROSA

Let $\Delta = \{a, b, c, \ldots\}$ be an ordered finite set of action types.

Let $Id = \{X, Y, Z, \ldots\}$ be a finite set of variables of process.

We will denote by the latest letters of the latin alphabet $r, s, t, \ldots$ probabilities.

We will denote by greek letters $\alpha, \beta, \gamma, \ldots$ time parameters for actions.

Terms of **ROSA** are defined by the following BNF expression:

$$P ::= \mathbf{0} \mid X \mid a.P \mid \langle a, \lambda\rangle.P \mid P \oplus P \mid P + P \mid P \oplus_r P \mid P||_A P \mid recX : P$$

where $\lambda \in \mathbb{R}^+ - \{0\}$, $A \subseteq \Delta$, $a \in \Delta$, $X \in Id$, . is concatenation, $\oplus$, $+$ and $\oplus_r$ are internal, external and probabilistic choices, $r \in [0, 1]$, $||$ is parallel, *rec* stands for recursion and P is a process of **ROSA**.

The Algebra induced by this expression makes up the set of **ROSA** processes.

A detailed description of the operational semantics and the performance evaluation algorithm of **ROSA** can be found in [9], where with the aim of making **ROSA** a more usable formalism, some steps have been done in the line of fully automatize its analyzing skills.

3 A Basic Genetic Algorithm

Although there are different types of GA's, they all share the following three processes: selection, reproduction and evaluation. The algorithm repeats these processes cyclically until a stop condition is reached. In [7], the authors have developed a first approximation to the problem we are dealing with, including:

- A generic description of a basic GA
- A formal definition of the reproduction operators
- A **ROSA** specification of the referred GA
- A complete performance study of this GA

In this paper we are concerned with the proper definition of the evaluation (of population in GAs) process. So that, we propose a metric on the states space, to be taken as basis for the selection (of the more promising individuals to conform the new population in GAs) process and therefore preventing to generate all branches of the LTS of **ROSA**.

4 Towards a Cheaper ROSA

Our main goal is to improve **ROSA** to be able to solve problems in a cheaper way, even automatically as in the line followed by [9]. In order to do this, our next step towards a Genetic Process Algebra is to define a function that, given a final state, associates to each state/process a measure of how promising is such state as being path to reach the final one. This function will be named *promising function, p-f* (we hope that the definition of the definitive fitness function could take this as basis).

We adopt the Means-End policy which tries to minimize the distance between the present state and the final one. In order to do that, following the reference [11], given P and Q a pair of **ROSA** processes, our metric takes as basis the Bayre metric and is defined as follows

$$d(P,Q) = \frac{1}{2^{l(P \sqcap Q)}} - \frac{1}{2^n}$$

- $l(P)$ is the length of the process P and is defined inductively over the syntactic structure of **ROSA** processes, as follows

$$\begin{array}{lcl}
\mathbf{l}: \{\textbf{ROSA}\ \mathit{procs.}\} & \longrightarrow & \mathbb{N} \\
\mathbf{0} & \mapsto & 0 \\
X & \mapsto & 1 \\
a.P & \mapsto & 2 + l(P) \\
\langle a, \lambda \rangle.P & \mapsto & 2 + l(P) \\
P \oplus Q & \mapsto & l(P) + 1 + l(Q) \\
P + Q & \mapsto & l(P) + 1 + l(Q) \\
P \oplus_r Q & \mapsto & l(P) + 1 + l(Q) \\
P||_A Q & \mapsto & l(P) + 1 + l(Q) \\
recX : P & \mapsto & 2 + l(P) \\
(P) & \mapsto & 2 + l(P)
\end{array}$$

- $n = max\{l(P), l(Q)\}$
- $P \sqcap Q$ is the longest common initial part of processes P and Q

Theorem 1. *The function d so defined is a metric.*

Proof. "d is a metric over $\{\textbf{ROSA}\ processes\} \Leftrightarrow d$ holds $(1) \wedge (2) \wedge (3)$" where:

1. $\forall P, Q \in \{\textbf{ROSA}\ processes\}.d(P,Q) = 0 \Leftrightarrow P = Q$
2. $\forall P, Q \in \{\textbf{ROSA}\ processes\}.d(P,Q) = d(Q,P)$
3. $\forall P, Q, T \in \{\textbf{ROSA}\ processes\}.d(P,Q) \leq d(P,T) + d(T,Q)$

1. $d(P,Q) = 0 \Leftrightarrow \frac{1}{2^{l(P \sqcap Q)}} - \frac{1}{2^n} = 0 \Leftrightarrow l(P \sqcap Q) = n \Leftrightarrow P = Q$
2. $d(P,Q) = d(Q,P) \Leftrightarrow \frac{1}{2^{l(P \sqcap Q)}} - \frac{1}{2^n} = \frac{1}{2^{l(Q \sqcap P)}} - \frac{1}{2^n} \Leftrightarrow$
   $\Leftrightarrow P \sqcap Q = Q \sqcap P \Leftrightarrow \sqcap\ symmetry$
3. $d(P,Q) \leq d(P,T) + d(T,Q) \Leftrightarrow d(P,T) + d(T,Q) - d(P,Q) \geq 0 \Leftrightarrow$
   $\Leftrightarrow \frac{1}{2^{l(P \sqcap T)}} - \frac{1}{2^m} + \frac{1}{2^{l(T \sqcap Q)}} - \frac{1}{2^o} - \frac{1}{2^{l(P \sqcap Q)}} + \frac{1}{2^n} \geq 0 \Leftrightarrow$
   $\Leftrightarrow (\frac{1}{2^n} - \frac{1}{2^m} - \frac{1}{2^o}) + (\frac{1}{2^{l(P \sqcap T)}} + \frac{1}{2^{l(T \sqcap Q)}} - \frac{1}{2^{l(P \sqcap Q)}}) \geq 0$
   where:
   - $m = max\{l(P), l(T)\}$
   - $o = max\{l(T), l(Q)\}$
   - $n = max\{l(P), l(Q)\}$

   where either one can be less than the other two (**A**), or all the same (**B**)
   A : Let's assume $n < (m = o)$

$$n < m \Leftrightarrow n \leq m - 1 \Leftrightarrow 2^n \leq 2^{m-1} \Leftrightarrow \frac{1}{2^n} > \frac{1}{2^{m-1}} \Leftrightarrow$$

$$\Leftrightarrow \frac{1}{2^n} - \frac{1}{2^{m-1}} \geq 0 \Leftrightarrow \frac{1}{2^n} - \frac{1}{2^m} - \frac{1}{2^m} \geq 0 \Leftrightarrow \frac{1}{2^n} - \frac{1}{2^m} - \frac{1}{2^o} \geq 0$$

$$\sqcap\ transitivity \Leftrightarrow l(P \sqcap Q) \geq min\{l(P \sqcap T), l(T \sqcap Q)\} \Leftrightarrow$$

$$\Leftrightarrow (l(P \sqcap Q) \geq l(P \sqcap T)) \vee (l(P \sqcap Q) \geq l(T \sqcap Q)) \Leftrightarrow$$

$$\Leftrightarrow (2^{l(P \sqcap Q)} \geq 2^{l(P \sqcap T)}) \vee (2^{l(P \sqcap Q)} \geq 2^{l(T \sqcap Q)}) \Leftrightarrow$$

$$\Leftrightarrow (\frac{1}{2^{l(P \sqcap Q)}} \leq \frac{1}{2^{l(P \sqcap T)}}) \vee (\frac{1}{2^{l(P \sqcap Q)}} \leq \frac{1}{2^{l(T \sqcap Q)}}) \Leftrightarrow$$

$$\Leftrightarrow (\frac{1}{2^{l(P \sqcap T)}} - \frac{1}{2^{l(P \sqcap Q)}} \geq 0) \vee (\frac{1}{2^{l(T \sqcap Q)}} - \frac{1}{2^{l(P \sqcap Q)}} \geq 0) \Rightarrow$$

$$\Rightarrow \frac{1}{2^{l(P \sqcap T)}} + \frac{1}{2^{l(T \sqcap Q)}} - \frac{1}{2^{l(P \sqcap Q)}} \geq 0$$

   The case where $m < (n = o)$ or equivalently $o < (m = n)$, has a very similar proof.
   B : The proof is also valid here

Once it has been checked, some considerations must be made mainly over the property (1) $\forall P, Q \in \{\textbf{ROSA}\ processes\}.d(P,Q) = 0 \Leftrightarrow P = Q$.

Since both P and Q are just **ROSA**-syntactical expressions denoting processes, some distinctions on these syntactical expressions can affect processes with the same meaning, i.e., two syntactical-different processes not always represent two different processes in terms of their behaviours, let us see some examples:

Example 1. Let P and Q be a pair of **ROSA** processes, we need that processes $P \oplus Q$ and $Q \oplus P$ have distance 0, because in whatever interpretation of the semantics of processes, they should be equivalent. The same could be said about the processes $P + Q$ and $Q + P$, so this commutative property should be preserved. Moreover, the weighted commutative property of $\oplus_r$ should be also fulfilled, thus $P \oplus_r Q$, has to be equivalent to $Q \oplus_{1-r} P$, or more precisely the distance between them must be 0 in a correct definition of distance. □

Example 2. Furthermore, the definition of distance should also respect the associativity of the processes so that, given P, Q and R three **ROSA** *processes* we want that $d((P \oplus Q) \oplus R, P \oplus (Q \oplus R)) = 0$. In this line the associativity of $+$ and the weighted associativity of $\oplus_r$ has to be preserved. □

Example 3. Also, there are some cases in which distributive property must be satisfied. For instance, let us take P, Q and R as **ROSA** *processes*, then we want that $d((P \oplus Q) + R, (P + R) \oplus (Q + R)) = 0$.

Distributive is a difficult property to be studied and guaranteed, thus, we will follow the results presented in [2], and the corresponding distributive laws. □

Example 4. Finally, we want that derivative operators could be removed, and then, the equivalent expression without them should have distance 0 with the previous one. For instance we want that $d(a.0||_\emptyset b.0, a.b.0 + b.a.0) = 0$. □

In fact, we want that in an appropriate semantics, two equivalent processes would have distance 0 between them. The main objective of this paper is not the study of a theoretical semantics, such as denotational or axiomatic semantics. Of course, with the basis of our operational semantics, we could define a notion of bisimulation ([1,3]), and take this equivalence as the basis. But this is a considerable amount of effort, and this work have been already done. In fact, in [2] a **Proof System** is defined, and it is demonstrated the equivalence of a denotational semantics and a set of axioms and inference rules, in the sense that this system is sound and complete. That is, if two processes have the same denotational semantics, then, it can be proved by using the proof system that they are equivalent, and on the contrary, if the equivalence may be proved in the proof system, then, the processes have the same denotational semantics.

In order to solve all the cases shown in the above examples we need to introduce normal forms for **ROSA** processes.

Normal Forms

In the line of [2], we can define normal forms in a very natural way. They consist in a generalized probabilistic choice at the top, followed by a generalized internal choice between a set of states, which is followed by a generalized prefixed external choice between the actions (timed and immediate) in this set, whose continuations are also in normal form.

Definition 1. *(Normal forms)*

- *Process* $\mathbf{0}$ *is in normal form.*
- *If* $\mathcal{A}_i$ *is a convex set of sets of* $\Delta\times(0,+\infty)\cup\{\infty\}$ *and for every* $a \in \mathbf{Type}(A_j)$ *(see [8]) where* $A_j \in \mathcal{A}_i$ *there is a normal form* $nf(P_{A_j,a})$, *then*

$$\bigotimes_i [q_i] \bigoplus_{A_j \in \mathcal{A}_i} \sum_{a \in \mathbf{Type}(A_j)} \langle a, \lambda_{A_j}\rangle.nf(P_{A_j,a})$$

is a normal form.

Notice that immediate action a is denoted in normal form as $\langle a, \infty\rangle$ and $\otimes_i[q_i], i \in \{1,\ldots,n\}$ represents the n-extension of $\oplus_r$ in this way:

- $P \oplus_r Q$ will be represented by $[r]P \otimes [1-r]Q$
- $P \oplus_r (Q \oplus_s T)$ will be represented by $[r]P \otimes [(1-r)*s]Q \otimes [(1-r)*(1-s)]T$

As usual, normal forms are unique modulo associativity and commutativity. Nevertheless we need to impose more restrictions in order to have one and only one normal form for every process, i.e., we want that two processes such as $a+b$ and $b+a$ have the same normal form, for instance, $a+b$.

We need then to impose some restrictions related to the order in which actions, sets and probabilities appear in the normal form. These restrictions are the following:

- At external choice level, actions must appear in alphabetical order.
- At internal choice level, sets must appear in the induced lexicographic order.
- At probabilistic level, probabilities must appear in decreasing order. If two would have the same probability then the lexicographic order of their already ordered internal choice level processes will determinate.

Let us see an example. The longest common initial part of the following two processes is 0:

$$((\langle d,1\rangle.0 + \langle a,2\rangle.0) \oplus (\langle b,1\rangle.0 + \langle a,3\rangle.0)) \oplus_{0.3} (\langle f,\infty\rangle.0 \oplus \langle e,1\rangle.0)$$
$$(\langle e,1\rangle.0 \oplus \langle f,\infty\rangle.0) \oplus_{0.7} ((\langle a,3\rangle.0 + \langle b,1\rangle.0) \oplus (\langle d,1\rangle.0 + \langle a,2\rangle.0))$$

Nevertheless both processes share the same *ordered* normal form:

$$[0.7]\langle e,1\rangle.0 \oplus \langle f,\infty\rangle.0 \otimes [0.3]\langle a,3\rangle.0 + \langle b,1\rangle.0 \oplus \langle a,2\rangle.0 + \langle d,1\rangle.0$$

so their distance must be 0.

Notation: The *ordered* normal form of process P will be denoted by $\|P\|$

We assume as equal **ROSA** processes, every pair of them which have the same corresponding ordered normal forms:

$\forall P,Q \in \{\textbf{ROSA}\ \textit{processes}\}.P = Q \Leftrightarrow \|P\| = \|Q\|$

It is a sound assumption since in [2] an equivalent proof of the soundness of the pure functional behaviour of **ROSA** can be found, and in [6] a complete Proof System for Timed Observations is presented.

In this section we will omit the treatment of recursion, because it implies an important mathematical apparatus so requiring a considerable amount of space, and the result does not justify this effort. This is due to the fact that for defining correctly a normal form for infinite processes, we need a power domain, as well as an order relation, so that, an infinite process would be the limit of a chain of ascending finite processes, each of them, an approximation of this limit. In order to guarantee the existence of this limit, we need both to introduce a fixed point theory, and to proof that every operator is continuous.

Since we think that this considerable work is not interesting in our study, we leave for a future work the completion of this operator, and we address the interested reader to the paper [2], where it is defined the semantics for infinite processes in a similar syntax to **ROSA**. Thus, from now on, operator $recX : P$ is not considered.

Once the notion of *ordered normal form* is defined it is time to provide the metric which solve all the problems previously stated.

Definition 2. *Given a pair of **ROSA** processes P and Q the distance between them is $D(P,Q)$*

$$D(P,Q) = \frac{1}{2^{l(\|P\| \sqcap \|Q\|)}} - \frac{1}{2^N}$$

where:

- *$l(\|P\|)$ is the length of the process $\|P\|$ and is defined inductively over the syntactic structure of ordered normal forms of **ROSA** processes, as follows*

$$\begin{array}{lcl}
l : \|\{\textbf{\textit{ROSA}}\ procs.\}\| & \longrightarrow & \mathbb{N} \\
\mathit{0} & \mapsto & 0 \\
a.P & \mapsto & 2 + l(P) \\
\langle a, \lambda \rangle.P & \mapsto & 2 + l(P) \\
+_{a \in A_j} P_a & \mapsto & m - 1 + \sum_{a \in A_j} l(P_a)(m = |A_j|) \\
\bigoplus_{A_j \in \mathcal{A}_i} P_j & \mapsto & k - 1 + \sum_{A_j \in \mathcal{A}_i} l(P_j)(k = |\mathcal{A}_i|) \\
\bigotimes_{i \in \{1 \ldots n\}} [q_i] P_i & \mapsto & 2n - 1 + \sum_{i \in \{1 \ldots n\}} l(P_i) \\
(P) & \mapsto & 2 + l(P)
\end{array}$$

- *$N = max\{l(\|P\|), l(\|Q\|)\}$*

Promising Function, $\mathbf{p - f}$, gives higher values to the more promising states to be followed for reaching S_F:

$$\begin{array}{lcl}
\mathbf{p - f} : \{\textbf{ROSA}\ procs.\} & \longrightarrow & (0, 1] \\
P & \mapsto & 1 - D(P, S_F)
\end{array}$$

Finally, our proposal is, given an initial state S_0 and a final one S_F, to apply all the rules of the operational semantics of **ROSA** to S_0 so generating a set of processes, and only follow on, with the state of this set that maximizes $p - f$ (associated to S_F). Therefore, the computational cost of the LTS, is moved from exponential to polynomial, so making it cheaper.

5 Conclusions and Future Work

In this paper we have provided the set of **ROSA** processes with a metric structure which allows to define a *promising function* $p - f$ for the sake of (computationally) improving the searching for 'a goal node' by means of this heuristic. This *promising function* establishes the first step towards a Genetic Process Algebra definition, since a slight variation of it, could be a $fitness$ function.

Our future work in this line is also concerned with the translation of the former operational semantics rules of **ROSA** to those rules which capture the same behaviour but over the domain of Ordered Normal Form processes.

References

1. Bloom, B.: Ready Simulation, Bisimulation, and the Semantics of CCS-like Languages. PhD thesis, Department of Electrical Engineering and Computer Science, MIT (1989)
2. Cazorla, D., Cuartero, F., Valero, V., Pelayo, F.L., Pardo, J.J.: Algebriac Theory of Probabilistic and Nondeterministic Processes. Journal of Logic and Algebraic Programming 55(1-2), 57–103 (2003)
3. Cleaveland, R., Hennessy, M.: Testing equivalence as a bisimulation equivalence. Formal Aspects of Computing 5, 1–20 (1993)
4. Goldberg, D.E.: Genetic algorithms in search, optimization, and machine learning. Addison-Wesley, Reading (1989)
5. Holland, J.H.: Adaptation in natural and artificial systems. The University of Michigan Press, Ann Arbor (1975)
6. Ortega-Mallén, Y., de Frutos-Escrig, D.: A complete proof system for timed observations. In: TAPSOFT 1991, vol. 1, pp. 412–440 (1991)
7. Pelayo, F.L., Cuartero, F., Ossa, L., Pelayo, M.L., Guirao, J.L.G.: Towards the evolutionary process algebra. In: 8th IEEE Int. Conf. on Cognitive Informatics, pp. 69–76. IEEE Computer Society Press, Los Alamitos (2009)
8. Pelayo, F.L., Cuartero, F., Valero, V., Cazorla, D.: Analysis of the MPEG-2 encoding algorithm with ROSA. Electronic Notes on Theoretical Computer Science 80(1), 185–202 (2003), `http://www.elsevier.nl/locate/entcs/volume80.html`
9. Pelayo, F.L., Pelayo, M.L., Guirao, J.G.: Generating the syntactic and semantics graphs for a markovian process algebra. Journal of Computational and Applied Mathematics 204, 38–47 (2007)
10. Godefroid, P., Khurshid, S.: Exploring very large state spaces using genetic algorithms. In: Procedings of the Conference on Tools and Algorithms for Construction and Analysis of Systems, Grenoble, France, pp. 266–280 (2002)
11. Rodriguez-Lopez, J., Romaguera, S., Valero, O.: Denotational semantics for programming languages, balanced quasi-metrics and fixed points. International Journal of Computer Mathematics 85(3), 623–630 (2008)

A Parallel Skeleton for Genetic Algorithms*

Alberto de la Encina[1], Mercedes Hidalgo-Herrero[2],
Pablo Rabanal[1], and Fernando Rubio[1]

[1] Dpto. Sistemas Informáticos y Computación
Facultad Informática, Universidad Complutense de Madrid, Spain
{albertoe,prabanal,fernando}@sip.ucm.es
[2] Dpto. Didáctica de las Matemáticas
Facultad Educación, Universidad Complutense de Madrid, Spain
mhidalgo@edu.ucm.es

Abstract. Nowadays, most users own multicore computers, but it is not simple to take advantage of them to speedup the execution of programs. In particular, it is not easy to provide a parallel implementation of a concrete genetic algorithm. In this paper we introduce a parallel skeleton that given a sequential implementation automatically provides a corresponding parallel implementation of it. In order to do it, we use a parallel functional language where skeletons can be defined as higher-order functions. Thus, the parallelizing machinery is defined only once, and it is reused for any concrete application of the skeleton to a concrete problem.

1 Introduction

Due to its higher-order nature, functional languages provide elegant strategies to implement generic solutions to a family of problems. This advantage is especially useful in the case of parallel programming, because its higher-order programming level allows to define the coordination of subcomputations in terms of the same constructions used in the rest of the program, which enables the definition and use of skeletons [2] to develop simpler parallel programs. During the last years, several parallel functional languages have been proposed (see e.g. [14,9,13,11,8]). In this paper we present how to use one of them to simplify the development of parallel versions of Evolutionary Computation methods [4,3,1]. In particular, we use the language Eden [9,10] to create a generic skeleton dealing with the parallelization of genetic algorithms [5], but the main ideas presented in the paper could also be applied to deal with other evolutionary methods.

One advantage of pure functional languages is that the absence of side-effects allow them to offer a clear semantic framework to analyze the correctness of programs. In particular, the semantics of the parallel language we will use is clearly defined in [6], and it is simple to relate it with the concrete parallel programs developed by the user [7]. The core notion of functional programming

* Research partially supported by projects TIN2009-14312-C02-01, TIN2009-14599-C03-01, S2009/TIC-1465, and UCM-BSCH GR58/08 - group number 910606.

J. Cabestany, I. Rojas, and G. Joya (Eds.): IWANN 2011, Part II, LNCS 6692, pp. 388–395, 2011.

is the mathematical function, that is, a program is a function. Starting with simple basic functions and by using functional composition, complex programs are created. Haskell [12] is the *de facto* standard of the lazy-evaluation functional programming community. It is a strongly typed language including polymorfism, higher-order programming facilities and lazy order of evaluation of expressions.

As it can be expected, the language provides large libraries of predefined functions, and also predefined data types for the most common cases, including lists. Let us remark that Haskell provides polymorfism. Thus, data types can depend on other types. New functions can be defined by analyzing cases on the structure of the data types. For instance, the following function computes the total number of elements of any list (of any concrete type) by using *pattern matching*:

```
length :: [a] -> Int
length [] = 0
length (x:xs) = 1 + length xs
```

The first line of the definition is optional, and it represents the type declaration of the function: Given a list of any concrete type a, it returns an integer. The rest of the definition represents the definition of the behavior of the function: If it receives an empty list, then it returns 0; otherwise, it adds 1 to the length of the tail of the list.

Other powerful characteristic of Haskell is higher-order. It means that functions can be arguments of functions. For instance, the following predefined function `map` receives as input a function `f` and a list, and then it applies function `f` to every element of the list:

```
map :: (a -> b) -> [a] -> [b]
map f [] = []
map f (x:xs) = f x : map f xs
```

Notice that the type declaration of function `map` indicates that its first parameter has type `a->b`, denoting that the first parameter is a function that receives values of type `a` and returns values of type `b`. The second parameter is a list of elements of type `a`, and the result is again a list, but in this case of elements of type `b`.

Notice that in higher-order languages like Haskell, it is also possible to deal with partially applied functions. For instance, `map` can take as its functional argument a partial application of function `(+)`:

```
mapPlusOne :: [Int] -> [Int]
mapPlusOne xs = map (1+) xs
```

Thus, it adds one to each element of the list.

The rest of the paper is structured as follows. In the next section we briefly describe the parallel language Eden. Then, in Section 3, we present how to develop a higher-order sequential Haskell function dealing with genetic algorithms. Afterwards, in Section 4 we introduce two different parallel versions of such higher-order function. Finally, Section 5 contains our conclusions and lines for future work.

2 Introduction to Eden

Eden [10,7] is a parallel extension of Haskell. It introduces parallelism by adding syntactic constructs to define and instantiate processes explicitly. It is possible to define a new *process abstraction* `p` by using the following notation that relates the inputs and the outputs of the process: `p = process x -> e` , where variable `x` will be the input of the process, while the behavior of the process will be given by expression `e`. Process abstractions are similar to functions – the main difference is that the former, when instantiated, are executed in parallel. From the semantics point of view, there is no difference between process abstractions and function definitions. The differences between processes and functions appear when they are invoked. Processes are invoked with a *process instantiation* (`e1 # e2`), while functions are invoked with an *application* (`e1 e2`).

Therefore, when we refer to a *process* we are not referring to a syntactical element but to a new *computational environment*, where the computations are carried out in an autonomous way. Thus, when a *process instantiation* (`e1 # e2`) is invoked, a new *computational environment* is created. The new process (the child or instantiated process) is fed by its creator by sending the value for e_2 via an input channel, and returns the value for $e_1 e_2$ (to its parent) through an output channel.

Let us remark that, in order to increase parallelism, Eden employs pushing instead of pulling of information. That is, values are sent to the receiver before it actually demands them. In addition to that, once a process is running, only fully evaluated data objects are communicated. The only exceptions are *streams*, which are transmitted element by element. Each stream element is first evaluated to full normal form and then transmitted. Concurrent threads trying to access not yet available input are temporarily suspended. This is the only way in which Eden processes synchronize. Notice that process creation is explicit, but process communication (and synchronization) is completely implicit.

Eden Skeletons. Process abstractions in Eden are not just annotations, but first class values which can be manipulated by the programmer (passed as parameters, stored in data structures, and so on). This facilitates the definition of skeletons as higher order functions. Next, we illustrate, by using a simple example, how skeletons can be written in Eden. More complex skeletons can be found in [10].

The most simple skeleton is `map`. Given a list of inputs `xs` and a function `f` to be applied to each of them, the sequential specification in Haskell is as follows:

```
map f xs  =  [f x | x <- xs]
```

that can be read as *for each element* `x` *belonging to the list* `xs`*, apply function* `f` *to that element.* This can be trivially parallelized in Eden. In order to use a different process for each task, we will use the following approach:

```
map_par f xs =  [pf # x | x <- xs] `using` spine
  where pf = process x -> f x
```

The process abstraction pf wraps the function application (f x). It determines that the input parameter x as well as the result value will be transmitted through channels. The spine strategy (see [15] for details) is used to eagerly evaluate the spine of the process instantiation list. In this way, all processes are immediately created. Otherwise, they would only be created on demand.

Let us remark that Eden's compiler has been developed by extending the GHC Haskell compiler. Hence, it reuses GHC's capabilities to interact with other programming languages. Thus, Eden can be used as a coordination language, while the sequential computation language can be, for instance, C.

3 Generic Genetic Algorithms in Haskell

In order to implement a generic scheme to deal with genetic algorithms, we start by identifying the list of input parameters that the scheme requires. In addition to the initial population given to the algorithm, we need parameters dealing with mutation, crossover, and fitness, as well as a maximum number of iterations. Moreover, in order to implement it in a pure functional language like Haskell, we need an additional parameter to introduce randomness. Note that Haskell functions cannot produce side-effects, so they need an additional input parameter to be able to obtain different results in different executions. Taking into account these considerations, the type of the higher-order Haskell function dealing with genetic algorithms is the following:

```
ga :: Int ->                                      -- Number of iterations
      [Rands] ->                                  -- Random generators
      Float ->                                    -- Mutation probability
      ([Rands] ->Float ->Genome ->Genome)->       -- Mutation function
      Float ->                                    -- Selection proportion
      ([Rands] -> [Genome] -> [Genome]) ->        -- Crossover function
      (Genome -> Float) ->                        -- Fitness function
      [Genome] ->                                 -- Initial population
      [Genome]                                    -- Final population
```

Note that we use two parameters to deal with mutation: the probability of a mutation to take place in a genome, and a function describing how a genome is modified by a mutation with a given probability. The first parameter of this function allows to introduce randomness in pure functions, as commented before.

Regarding crossover, we use a numerical parameter describing the proportion of the population that is selected for crossover in the next stage. Moreover, we also use a function describing how to perform the crossover among a given population. Note that this function also need an additional parameter to introduce randomness. The concrete value passed through this parameter will be obtained from the same list passed to the corresponding parameter of the ga function, but we need to handle both parameters independently to be able to introduce randomness in both functions.

Once the input parameters are defined, we can implement the body of the function. This is done recursively on the number of iterations. If there isn't any iteration to be performed then we just return as result the input population. Otherwise, we apply one iteration step (by using an auxiliary function `oneStep`) and then we recursively call to our generic scheme with one iteration less:

```
ga 0 _ _ _ _ _ _ _ pop = pop
ga n rl mp mf bests cf fitness pop
  = ga (n-1) newRl mp mf bests cf fitness newPop
  where (newRl,newPop) = oneStep rl mp mf bests cf fitness pop
```

Regarding how to perform each step, we start by applying the fitness function to each element of the population, and then we use the result to sort the population. Thus, we can trivially select the first elements of this list in a variable `newBests` that will be used to generate the new `offspring` by applying the crossover function on them. Finally, we use the mutation parameters to introduce the corresponding mutations to the new population. The code dealing with all these issues is as follows:

```
oneStep rl mp mf bests cf fitness pop = (newRl,newMuts)
  where withFitness = sort (zip (map fitness pop) pop)
        newBests = take bests' (map snd withFitness)
        bests' = round (bests * fromIntegral (length pop))
        nbests' = length pop - bests'

        offspring = take nbests' (cf rl newBests)
        rl' = drop (length pop) rl

        newPop = newBests ++ offspring
        (newRl,newMuts) = mmf rl' mf mp newPop
```

Note that the higher-order nature of Haskell makes it simple to define a generic function dealing with genetic algorithms. It is also easy to provide a library describing the most common strategies to perform crossover and mutation.

4 Parallel Skeleton in Eden

In order to parallelize any program, we need to identify the most time-consuming tasks appearing inside it. In our concrete case, when the computation of the fitness function is very time-consuming, we should try to parallelize its application to each population. Note that this can be done in Eden by using the skeleton `map_par` to substitute function `map` in the application of the fitness function. This is done by modifying function `oneStep` as follows:

```
oneStep rl mp mf bests cf fitness pop = (newRl,newMuts)
  where withFitness = sort (zip (map_par noPe fitness pop) pop)
        ...
```

By doing so, we are creating a new process to compute the fitness function of each element of the population. Obviously, in the most common case we will have many elements in the population, and not so many processors available. Thus, it would be more efficient to create only as many processes as processors available, and to fairly distribute the population among them. This can be done as follows:

```
where withFitness = sort (zip (map_farm noPe fitness pop) pop)
```

where `noPe` is an Eden variable equals to the number of available processors in the system, while `map_farm` implements the idea of distributing a large list of tasks among a reduced number of processes. The implementation firstly distributes the tasks among the processes, producing a list of lists where each inner list is to be executed by an independent process. Then, it applies `map_par`, and finally it collects the results joining the list of lists of results into a single list of results. Notice that, due to the laziness, these three tasks are not done sequentially, but in interleaving. As soon as any worker computes one of the outputs it is computing, it sends this subresult to the main process, and it goes on computing the next element of the output list. Notice that the communications are asynchronous, so that it is not necessary to wait for acknowledgments from the main process. When the main process has received all the needed results, it finishes the computation. The Eden source code of this skeleton is shown below, where not only the number `np` of processors but also the distribution and collection functions (`unshuffle` and `shuffle` respectively) are also parameters of the skeleton:

```
map_farmG np unshuffle shuffle f xs
  = shuffle (map_par (map f) (unshuffle np xs))
```

Different strategies to split the work into the different processes can be used provided that, for every list `xs`, `(shuffle (unshuffle np xs)) == xs`. In our case, we will use a concrete version of `map_farmG` called `map_farm` where the functions used to unshuffle/shuffle distribute the tasks in a round-robin way.

4.1 Improved Skeleton

In many situations, the computation of the fitness function is not expensive enough to obtain good speedups when using our previous approach. The solution implies to increase the granularity of the tasks to be performed by each process. In order to increase it, a typical approach to deal with genetic algorithms consists in splitting the population into groups. After that, each group evolves in parallel during a given number of iterations. Then, they are combined again in a sequential step of the evolution process to enable crossover among the groups. Afterwards, the mechanism is repeated again, that is, they are splitted into groups, they evolve in parallel, they are combined in a sequential step, and then the parallel process start again.

In order to implement in Eden a generic skeleton dealing with this idea, we will add a new parameter `nip` to the main function. This parameter will indicate how many iterations are to be performed in parallel before recombining the groups into a single population. The body of the main function will also change a little bit to deal with this parameter. In case the number of iterations to be performed is smaller than the number of parallel iterations, we perform a parallel step and we finish. Otherwise, we perform a parallel step and we go on again with the main process performing the rest of iterations:

```
gaPar nip ni rl mp mf bests cf fitness pop
  | ni <= nip  = snd (oneStepPar ni rl mp mf bests cf fitness pop)
  | otherwise = gaPar nip (ni-nip-1) newRl mp mf bests cf fitness newPop
  where (newRl,newPop) = oneStepPar nip rl mp mf bests cf fitness pop
```

Regarding the parallel step, we split into noPe groups both the population and the random generators, and then for each group we apply an independent genetic algorithm. In order to run in parallel each of them, we substitute the sequential main function ga by a parallel version of it called pga. The only difference between both functions is that the second one is encapsulated inside a process abstraction, and it is applied by using a process instantiation. The source code is the following:

```
oneStepPar nip rl mp mf bests cf fitness pop
  = oneStep newRl mp mf bests cf fitness newPop
  where pops = unshuffle noPe pop
        rls = distribute noPe rl
        newRl = rls!!noPe
        newPop = shuffle ((zipWith f pops rls) `using` spine)
        f newpop newrl = (pga nip newrl mp mf bests cf fitness) # newpop

pga ni rl mp mf bests cf fitness
  = process pop -> ga ni rl mp mf bests cf fitness
```

Let us remark that in order to convert a sequential genetic algorithm into the corresponding parallel program, the programmer only has to change a call to function ga by a call to function gaPar indicating an appropriate value for parameter nip. Thus, the only programming effort will be related to selecting a reasonable value for nip, that could depend on the computation cost of each iteration of the genetic algorithm.

5 Conclusions and Future Work

In this paper we have shown that using parallel functional languages like Eden can simplify the task of parallelizing evolutionary algorithms. In particular, we have introduced a new Eden skeleton that allows the programmer to obtain a parallel version of a genetic algorithm without needing to manually deal with the low-level details of its parallelization.

It is important to recall that Eden programs can interact with other programming languages. In particular, C code can be encapsulated inside a Haskell function. Hence, it is not necessary that the whole algorithm is implemented in a functional language. In fact, Eden can be used as a coordination language dealing with the parallel structure of the program, while the core of the parameter functions (crossover, fitnetss, etc.) could be implemented in a computation language like C.

As future work, we plan to extend our library of Eden skeletons to deal with other evolutionary computation methods, like Ant Colony Optimization, Swarm Intelligence, etc. Moreover, we are particularly interested in studying hybrid systems combining two different evolutionary methods.

References

1. Chiong, R. (ed.): Nature-Inspired Algorithms for Optimisation. SCI, vol. 193. Springer, Heidelberg (2009)
2. Cole, M.: Bringing skeletons out of the closet: A pragmatic manifesto for skeletal parallel programming. Parallel Computing 30, 389–406 (2004)
3. de Jong, K.: Evolutionary computation: a unified approach. In: Genetic and Evolutionary Computation Conference, GECCO 2008, pp. 2245–2258. ACM, New York (2008)
4. Eiben, A.E., Smith, J.E.: Introduction to Evolutionary Computing. Springer, Heidelberg (2003)
5. Goldberg, D.E.: Genetic Algorithms in Search, Optimisation and Machine Learning. Addison-Wesley, Reading (1989)
6. Hidalgo-Herrero, M., Ortega-Mallén, Y.: Continuation Semantics for Parallel Haskell Dialects. In: Ohori, A. (ed.) APLAS 2003. LNCS, vol. 2895, pp. 303–321. Springer, Heidelberg (2003)
7. Hidalgo-Herrero, M., Ortega-Mallén, Y., Rubio, F.: Analyzing the influence of mixed evaluation on the performance of Eden skeletons. Parallel Computing 32(7-8), 523–538 (2006)
8. Keller, G., Chakravarty, M.T., Leshchinskiy, R., Peyton Jones, S.L., Lippmeier, B.: Regular, shape-polymorphic, parallel arrays in Haskell. In: International Conference on Functional Programming, ICFP 2010, pp. 261–272. ACM, New York (2010)
9. Klusik, U., Loogen, R., Priebe, S., Rubio, F.: Implementation skeletons in Eden: Low-effort parallel programming. In: Mohnen, M., Koopman, P. (eds.) IFL 2000. LNCS, vol. 2011, pp. 71–88. Springer, Heidelberg (2001)
10. Loogen, R., Ortega-Mallén, Y., Peña, R., Priebe, S., Rubio, F.: Parallelism abstractions in Eden. In: Rabhi, F.A., Gorlatch, S. (eds.) Patterns and Skeletons for Parallel and Distributed Computing, pp. 95–128. Springer, Heidelberg (2002)
11. Marlow, S., Peyton Jones, S.L., Singh, S.: Runtime support for multicore Haskell. In: International Conference on Functional Programming, ICFP 2009, pp. 65–78. ACM Press, New York (2009)
12. Peyton Jones, S.L., Hughes, J.: Report on the programming language Haskell 98. Technical report (February 1999), `http://www.haskell.org`
13. Scaife, N., Horiguchi, S., Michaelson, G., Bristow, P.: A parallel SML compiler based on algorithmic skeletons. Journal of Functional Programming 15(4), 615–650 (2005)
14. Trinder, P.W., Hammond, K., Mattson Jr., J.S., Partridge, A.S., Peyton Jones, S.L.: GUM: a portable parallel implementation of Haskell. In: Programming Language Design and Implementation, PLDI 1996, pp. 79–88. ACM Press, New York (1996)
15. Trinder, P.W., Hammond, K., Loidl, H.-W., Peyton Jones, S.L.: Algorithm + Strategy = Parallelism. Journal of Functional Programming 8(1), 23–60 (1998)

A Case Study on the Use of Genetic Algorithms to Generate Test Cases for Temporal Systems*

Karnig Derderian[1], Mercedes G. Merayo[2], Robert M. Hierons[1], and Manuel Núñez[2]

[1] Department of Information Systems and Computing, Brunel University
Uxbridge, Middlesex, UB8 3PH United Kingdom
derderian@karnig.co.uk, rob.hierons@brunel.ac.uk
[2] Departamento de Sistemas Informáticos y Computación
Universidad Complutense de Madrid, Madrid, Spain
mgmerayo@fdi.ucm.es, mn@sip.ucm.es

Abstract. Generating test data for formal state based specifications is computationally expensive. In previous work we presented a framework that addressed this issue by representing the test data generation problem as an optimisation problem. In this paper we analyze a communications protocol to illustrate how the test case generation problem can be presented as a search problem and automated. Genetic algorithms (GAs) and random search are used to generate test data and evaluate the approach. GAs show to outperform random search and seem to scale well as the problem size increases. We consider a very simple fitness function that can be used with other evolutionary search techniques and automated test case generation suites.

1 Introduction

As computer technology evolves the complexity of current systems increases. Critical parts/aspects of some system are specified using formal specifications in order to better understand and model their behaviour. Communication protocols and control systems, amongst others, have used formal specifications like finite state machines. Unfortunately, in most cases it cannot be guaranteed that system implementations fully comply to the specifications.

Even though testing [1,2] is an important part of the system development process that aims to increase the reliability of the implementation, it can be very expensive. This motivates the research in the combination of formal methods and testing [3,4] since progress in this line of work helps to (partially) automatize the testing process.

In previous work [5] we addressed the issues related to generating test sequences for temporally constrained Extended Finite State Machine (TEFSM) based systems. We focused on generating timed feasible transition paths (TFTPs)

* Research partially supported by the Spanish MEC project TESIS (TIN2009-14312-C02-01).

J. Cabestany, I. Rojas, and G. Joya (Eds.): IWANN 2011, Part II, LNCS 6692, pp. 396–403, 2011.

with specific properties that can in turn be used to generate test input. The problem of generating these paths is represented as a search problem and Genetic Algorithms (GA) can be used to help automate the test data generation process. In short, a GA is a heuristic optimisation technique which derives its behaviour from a metaphor of the processes of evolution in nature [6,7]. GAs have been widely used in search optimisation problems. GAs are known to be particularly useful when searching large, multimodal and unknown search spaces since one of its benefits is their ability to escape local minima in the search for the global minimum. In particular, GAs and other meta-heuristic algorithms have been also used to automate software testing [8,9,10,11,12]. In this paper we present a case study to evaluate our theoretical framework. We consider the Class 2 transport protocol [13] and compare the performance of two GAs and a random algorithm when looking for TFTPs.

The rest of the paper is organized as follows. In Section 2 we introduce the main definitions and concepts that will be used during the presentation of our case study. In Section 3, what constitutes the bulk of the paper, we present our case study. Finally, in Section 4 we present our conclusions and some lines for future work.

2 Preliminaries

In this section we review the concepts that will be used along the paper. In particular, we will introduce the notions of timed extended finite state machine and timed feasible transition path, and discuss on the fitness function that will guide our GAs. This part of the paper was already presented in our previous work [5], where the reader is refered to find further explanations.

We assume that the number of different variables is m. If we assume that each variable x_i belongs to a domain D_i thus the values of all variables at a given point of time can be represented by a tuple belonging to the cartesian product of $D_1 \times D_2 \times ... \times D_m = \Delta$. Regarding the domain to represent time we define that time values belong to a certain domain **Time**.

Definition 1. *A TEFSM M can be defined as* $(S, s_0, V, \sigma_0, P, I, O, T, C)$ *where* S *is the finite set of logical states,* $s_0 \in S$ *is the initial state,* V *is the finite set of internal variables,* σ_0 *denotes the mapping from the variables in* V *to their initial values,* P *is the set of input and output parameters,* I *is the set of input declarations,* O *is the set of output declarations,* T *is the finite set of transitions and* C *is such that* $C \in \Delta$.

A transition $t \in T$ *is defined by* $(s_s, g_I, g_D, g_C, op, s_f)$ *where* s_s *is the start state of* t*;* g_I *is the input guard expressed as* (i, P^i, g_{P^i}) *where* $i \in I \cup \{NIL\}$*;* $P^i \subseteq P$*; and* g_{P^i} *is the input parameter guard that can either be NIL or be a logical expression in terms of variables in* V' *and* P' *where* $V' \subseteq V$*,* $\emptyset \neq P' \subseteq P^i$*;* g_D *is the domain guard and can be either NIL or represented as a logical expression in terms of variables in* V' *where* $V' \subseteq V$*;* $g_C : \Delta \rightarrow$ ***Time*** *is the time the transition needs to take to complete; op is the sequential operation which is made of simple output and assignment statements; and* s_f *is the final state of* t.

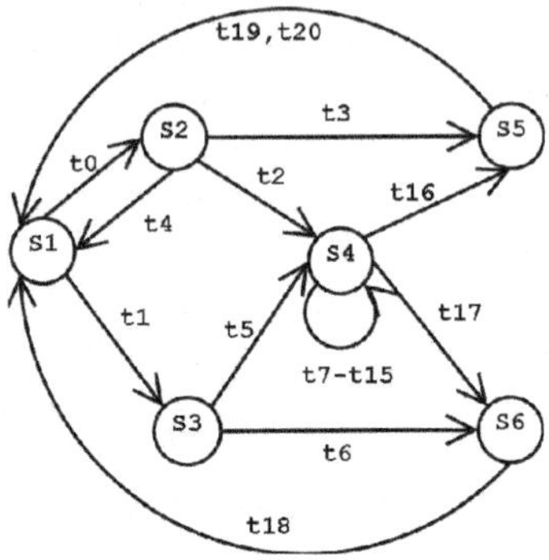

Fig. 1. Class 2 transport protocol TEFSM M_1. The transition table is on Fig. 2.

A TEFSM M is deterministic if any pair of transitions t and t′ initiating from the same state s that share the same input x have mutually exclusive guards. A TEFSM M is strongly connected if for every ordered pair of states (s, s') there is some feasible path from s to s′. A configuration for a TEFSM M is a combination of state and values of the internal variables V of M.

We assume that any TEFSM considered in this paper is deterministic and strongly connected. For example, consider the Class 2 transport protocol [13] represented as a TEFSM in Figure 1.

A timed feasible transition path (TFTP) for state s_i to state s_j of a TEFSM M is a sequence of transitions initiating from s_i that is feasible for at least one combination of values of the finite set of internal variables V (configuration) of M and ends in s_j.

An input sequence (IS) is a sequence of input declarations $i \in I$ with associated input parameters $P^i \subseteq P$ of a TEFSM M.

A predicate branch (PB) is a label that represents a pair of g_{P^i} and g_D for a given state s and input declaration i. A PB identifies a transition within a set of transitions with the same start state and input declaration.

An abstract input sequence (AIS) for M represents an input declaration sequence with associated PBs that triggers a TP in the abstracted M.

We use a very simple fitness function that combines a transition ranking (how likely is to take this transition) and a temporal constrain ranking (how complex the time constraint of a transition is).

Definition 2. *The fitness is a function that given a TP of a TEFSM M, sums the penalty points (assigned through the transition ranking process for M and the temporal constrain ranking for M) for each of the transition of the TP.*

We chose to give equal weight to the rankings following the conclusions of a similar experiment [14] where different weights were used for a similar multi-optimisation problem. However it is possible to consider different ways to combine the two matrices in the fitness function.

t	input	output	feasibility rank	temporal rank
t_0	ICONreq	!CR	0	0
t_1	CC	!ICONconf	0	0
t_2	T_expired	!CR	2	0
t_3	T_expired	!IDISind	1	1
t_4	IDATreq	DT	0	0
t_5	AK		6	1
t_6	AK		6	1
t_7	AK	DT	5	0
t_8	AK	!IDISind	4	0
t_9	T_expired	DT	3	0
t_{10}	T_expired	!IDISind	2	1
t_{11}	DR	!IDISind	0	2
t_{12}	DR	!IDISind	0	2
t_{13}	DR	!IDISind	0	2
t_{14}	DR	!IDISind	0	2

Fig. 2. Temporal constraint ranking and feasibility ranking for all transitions in M_1

3 Case Study

The Class 2 transport protocol M_1 is presented in Figure 1 and the corresponding transition table (excluding the conditions and temporal constraints) is shown in Figure 2. This table also shows the ranked transition table for M_1. For example t_3 and t_{10} share the same temporal constraint classification and therefore are ranked lower than some other transitions however they have different feasibility ranking due to the differently classified guards they have.

The search for a TP that is likely to be feasible and yet have complex temporal constraints is represented as a fitness minimisation problem. The GA is then used to search for appropriate solutions. The same computational effort is also used with a random TP generator using the same fitness function and result verification as the GA. This search problem uses a fitness function that rewards transition sequences with higher ranked transitions and penalises invalid transitions. It produces a numerical value potentially showing how close an input sequence is to defining a valid TFTP. The fitness function represents the search for a TFTP sequence as a function minimisation problem so an AIS with a lower fitness value is considered to be more likely to form a TFTP since it is made up of more highly ranked transitions.

The fitness does not guarantee that a particular transition path can be triggered or that it contains the most complex temporal constraints in M. It makes sure that it is constructed using consecutive transitions that are highly ranked. The verification process then checks if an IS can be generated to trigger such a TP. The verification method consists in evaluating a TP by resetting M to its initial configuration and attempt to trigger the TP in the simulated implementation. The process is repeated several times and the overall result of how many times the TP was correctly triggered are counted and compared to the times it failed. Hence an estimation is derived to measure the feasibility of these TPs.

In our example we looked at a relatively simple temporal constraints range (hence the small range of rankings on Figure 2) and it was easy to manually check the complexity of the temporal constraints for each transition. This was sufficient for our case study, but defining an automated estimation measure for the temporal qualities of a TP remains future work.

In order to compare the performance of the GA and Random algorithms TFTP generation two metrics are used. *State coverage* is the number of cases where at least one TFTP was generated for every TFTP size attempted from each state in M and *success rate* is the number of TFTPs that were generated compared to the total number of attempts it took to generate the results.

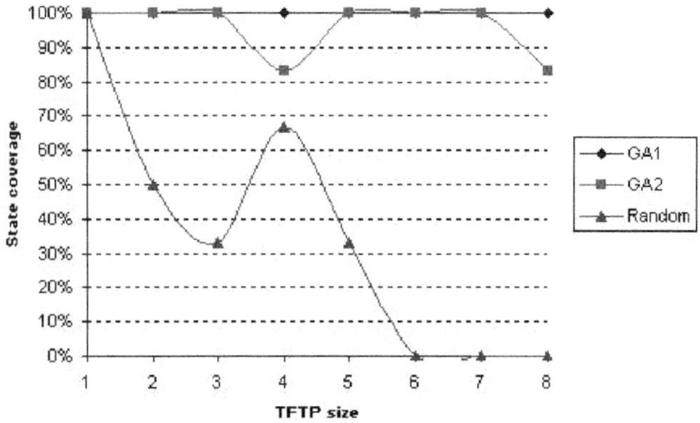

Fig. 3. State coverage for PB notation TFTPs generated using GA and Random generation algorithms for M_1 with 1-8 transitions

In this case study two slightly different GAs were used to compare their performance when applied to this problem. The first GA used a single point crossover and mutation while the second used a complex multiple point crossover and mutation. In general the second GA tended to find a solution slightly faster than the first GA, but they produced the same results.

Figure 5 represents a summary of the result averages. In general the results show that the GAs seem to perform better than the Random generation method according to both metrics. Figure 3 represents the state coverage results for all the different TFTP sizes. GA1 performs well and GA2 fails to find only one TFTP of size 4 and one of size 8, while the random generation algorithm performance peaks when generating TFTPs of size 4 and declines as the TFTP size increases. Clearly the GAs outperform the Random generation method for TFTPs of more than one transition. Figure 4 represents the success rate results for all the different TFTP sizes in our case study. The high fluctuation here can be explained by the different degree of difficulty in generating TFTPs of different sizes for some states. This relates to the guards of transition $t14$, which is one of the core transitions. Although the guard conditions are complex in the context of M_1 they are very easy to satisfy. Hence the random algorithm can easily select $t14$ in its TP search, while the GAs try to avoid it without realising that it should not. GA2 shows the best performance with an average performance of 65%. GA1 performs mostly better than the random generation algorithm, except for TFTPs of size 4 (average performance of 54%). For TFTPs of size 4 the random algorithm performs slightly better than GA1 and GA2. The random generation method did not find any TSTPs for sizes 4 to 6 while the GAs have different success rate for each size. This shows how different states have different properties. Hence future work may focus on more analysis of the guards and the temporal conditions to even out the search performance.

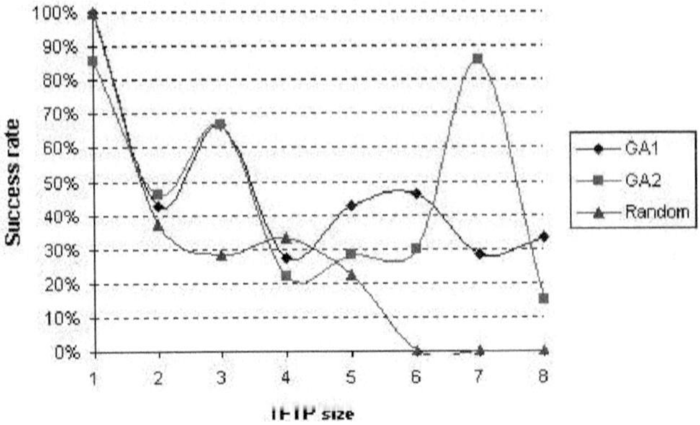

Fig. 4. Success rate for PB notation TFTPs generated using GA and Random generation algorithms for M_1 with 1-8 transitions

For both metrics the two GA search algorithms perform on average better than the random generation algorithm. This suggests that the fitness function here helps guide a heuristic search for TFTPs. In Figure 3 and Figure 4 we observe that as longer TFTPs (and possibly when larger TEFSMs) are considered, the heuristics seems to perform increasingly better than the random generation algorithm when given equal processing effort in terms of fitness evaluations and TFTP verifications. On all occasions the TFTPs generated by the GAs the had equivalent or more complex temporal constraints compared to those generated using the random TP generation method. For a TEFSM of this size, as in our case study, it is expected to have similar performance for the small TFTPs because the search space in those situations is not that big. However as the search space is increased (in or case study by increasing the size of the TFTPs) it becomes clear that a random generation approach finds it hard to generate TPs that feasible and satisfy the temporal constraints.

The state coverage metric is the easier one to satisfy. Not surprisingly the GAs found at least one TFTP for every state in M_1. This measure however discards all the unsuccessful attempts to generate a given TFTP. Hence the success rate metric considers those unsuccessful attempts as well. The success rates results are lower but the GAs seem to outperform the random algorithm.

	State Coverage	Success rate
GA1	100%	48%
GA2	96%	47%
Random	35%	28%

Fig. 5. GA and Random search result averages for the Class 2 protocols for TFTPs with 1-8 transitions

Overall both GAs performed well and generated very similar results. This indicates that the fitness function and the TP representation represent the problem of TFTP generation reasonably well. Hence future work can focus on refining the fitness function and evaluations of larger TEFSMs.

4 Conclusions and Future Work

This paper reported on the application to a communications protocol of a computationally inexpensive method to address the important problem of test data generation for TEFSMs. By taking as initial step our previous work [5], we defined a fitness function that yields some positive results when GA search is applied to the problem. The GA almost fully satisfies the coverage criteria defined and increasingly outperforms random generation as the TFTP size increases. The success rate fluctuated in this case study but the average success rate of the GA generated results was almost double that of the randomly generated results. Overall the limited results suggest that the approach scales well and could possibly be applied to larger TEFSMs. As a conclusion, we claim that our computationally inexpensive fitness function may be used to aid the generation of potentially computationally expensive test input generation sequences in a computationally inexpensive way.

Future work may focus on refining the fitness function to take into account several difficulties to estimate transitions. In addition, it would be interesting to apply our methodology to larger systems. Work on the combination of other related research, like alternative approaches to test sequence generation, with feasibility estimation and temporal constraint satisfaction can also be considered to aid the generation of test input sequences for TEFSMs.

References

1. Myers, G.: The Art of Software Testing, 2nd edn. John Wiley and Sons, Chichester (2004)
2. Ammann, P., Offutt, J.: Introduction to Software Testing. Cambridge University Press, Cambridge (2008)
3. Hierons, R.M., Bowen, J.P., Harman, M. (eds.): FORTEST. LNCS, vol. 4949. Springer, Heidelberg (2008)

4. Hierons, R., Bogdanov, K., Bowen, J., Cleaveland, R., Derrick, J., Dick, J., Gheorghe, M., Harman, M., Kapoor, K., Krause, P., Luettgen, G., Simons, A., Vilkomir, S., Woodward, M., Zedan, H.: Using formal methods to support testing. ACM Computing Surveys 41(2) (2009)
5. Derderian, K., Merayo, M., Hierons, R., Núñez, M.: Aiding test case generation in temporally constrained state based systems using genetic algorithms. In: Cabestany, J., Sandoval, F., Prieto, A., Corchado, J.M. (eds.) IWANN 2009. LNCS, vol. 5517, pp. 327–334. Springer, Heidelberg (2009)
6. Goldberg, D.E.: Genetic Algorithms in search, optimisation and machine learning. Addison-Wesley Publishing Company, Reading (1989)
7. Srinivas, M., Patnaik, L.M.: Genetic algorithms: A survey. IEEE Computer 27, 17–27 (1994)
8. Jones, B.F., Eyres, D.E., Sthamer, H.H.: A strategy for using genetic algorithms to automate branch and fault-based testing. The Computer Journal 41(2), 98–107 (1998)
9. Michael, C.C., McGraw, G., Schatz, M.A.: Generating software test data by evolution. IEEE Transactions on Software Engineering 27(12), 1085–1110 (2001)
10. McMinn, P.: Search-based software test data generation: a survey. Software Testing Verification and Reliability 14(2), 105–156 (2004)
11. Derderian, K., Hierons, R.M., Harman, M., Guo, Q.: Automated Unique Input Output Sequence Generation for Conformance Testing of FSMs. The Computer Journal 49(3), 331–344 (2006)
12. Harman, M., McMinn, P.: A theoretical and empirical study of search-based testing: Local, global, and hybrid search. IEEE Transactions on Software Engineering 36(2), 226–247 (2010)
13. Ramalingom, T., Thulasiraman, K., Das, A.: Context independent unique state identification sequences for testing communication protocols modelled as extended finite state machines. Computer Communications 26(14), 1622–1633 (2003)
14. Derderian, K.: Automated test sequence generation for Finite State Machines using Genetic Algorithms. PhD thesis, Brunel University (2006)

Experimental Comparison of Different Techniques to Generate Adaptive Sequences*

Carlos Molinero[1], Manuel Núñez[1], and Robert M. Hierons[2]

[1] Departamento de Sistemas Informáticos y Computación, Universidad Complutense de Madrid, Madrid, Spain
molinero@fdi.ucm.es, mn@sip.ucm.es

[2] Department of Information Systems and Computing, Brunel University Uxbridge, Middlesex, UB8 3PH United Kingdom
rob.hierons@brunel.ac.uk

Abstract. The focus of this paper is to present the results of a set of experiments regarding the construction of an adaptive sequence by a genetic algorithm and other techniques in order to reach a goal state in a non-deterministic finite state machine.

1 Introduction

Testing ([6,2]) is one of the most important tasks to be undertaken in software engineering. Its development and application covers a high percentage of the total cost of development in any process of software engineering.

Reaching a specific state is a fundamental part of the testing process because it allows the tester to move the implementation to that state and continue the testing of a certain part of a system, such as a specific component of an embedded system. In the case that the tester is confronted with a *non-deterministic finite state machine* (from now on $ndFSM$) this problem belongs to the EXPTIME complete category. Therefore, heuristic methods are used to present a solution.

A non-deterministic finite state machine is, informally, a set of states and labeled transitions with pairs input/output, the characteristic that makes it non-deterministic is that from the same state there can be several transitions labeled with the same input. We restrict our work to observable $ndFSMs$, that is, to machines in which two transitions departing from the same state cannot have the same combination of input/output.

Adaptive sequences [4,3,1] is a method used to reach a state in a non-deterministic setting. An *adaptive sequence* is a tree such that the unique edge that leaves its root will be labeled by an input (to be applied to the $ndFSM$), and it will reach a state such that from this state outgoing edges labeled by outputs (returned from the $ndFSM$) arrives at one state each from where a new input will depart and so on.

We have presented in a previous work [5], the use of a genetic algorithm to create an adaptive sequence to reach deterministically a goal state in a $ndFSM$. The interested

* Research partially supported by the Spanish MICINN project TESIS (TIN2009- 14312-C02-01), the UK EPSRC project Testing of Probabilistic and Stochastic Systems (EP/G032572/1), and the UCM-BSCH programme to fund research groups (GR58/08 - group number 910606).

J. Cabestany, I. Rojas, and G. Joya (Eds.): IWANN 2011, Part II, LNCS 6692, pp. 404–411, 2011.

reader is referred to the aforementioned paper for a more complete understanding of our approach, detailed explanations of the evolution of our GA and a formal definition of the elements present in the system. The goal of this paper is to present a set of experiments regarding the achievement of our genetic algorithm and some other techniques to an extent in which we can assure its validity.

The rest of the paper is organized as follows. In Section 2 we summarize the main aspect of the evolution of our GA. In Section 3 we show the results of our experiments and in Section 4 we present our conclusions.

2 Description of Our GA

In this section we summarize the main concepts behind the evolution of our genetic algorithm. A more detailed description can be found in the aforementioned paper [5].

The inhabitants from the population create, based on their random coefficients, a new adaptive sequence which is their DNA.

This DNA is mutated once every generation, the way this is achieved is by traversing randomly the adaptive sequence and when the algorithm finds a node with no children then it adds a subtree to the adaptive sequence, or deletes the subtree to which the node belongs to (each with a 50% probability). The positive point about using this method to select a node is that it has a similar probability of being chosen as when executing the $ndFSM$. This allows to always modify the nodes that influence in a greater extent the overall functioning of the adaptive sequence.

Crossover is done by selecting the individuals with a higher heuristic value through *roullete wheel selection* and then traversing randomly both instances to try to find a node that represents the same node in the $ndFSM$. If this node is found then the algorithm exchanges the subtree of both adaptive sequences and creates two children that are added to the population. If no node is found following this procedure, then no crossover is performed.

In the beginning of the next generation, all the specimens are judged by a sampling procedure (running 100 times its adaptive sequence), and the algorithm performs a selection of the fittest, maintaining a constant number in the population by eliminating those individuals with the worst heuristic value. The selection of the fittest is an elitist selection, which means that the best individual from the last generation is copied directly into the next one without any mutation or crossover, to make sure that the GA does not lose the best solution found until that moment.

3 Experimental Comparison

The number of experiments that we have conducted was established by taking into account the amplitude of the oscillation in the averaged heuristic values of the runs of different GAs against a series of $ndFSMs$. This value tends to stabilize around 50 experiments. This is one of the main motivations for having extended our experimental setup with respect to our previous work, since, before, we only performed 20 experiments and, as one can see in Figure 1, the value fluctuates at that point greatly.

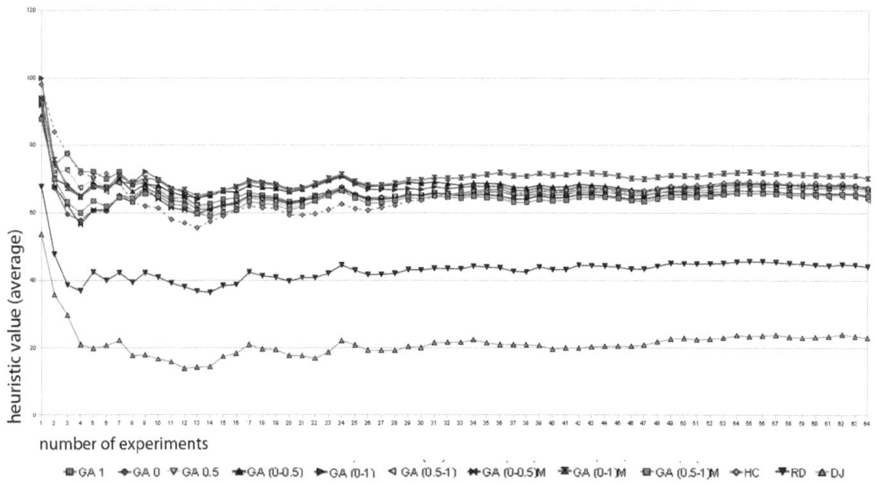

Fig. 1. Evolution of the average heuristic values for several techniques, including various kinds of GAs. The hillclimbing methodology appears in dotted lines.

We were also able to increase the speed of the algorithm, which has led to a modification in the heuristic values and of the total size of the resulting specimens.

3.1 Description of the Experimentation

The experiments were run in a Intel Core2 Duo CPU T7300 at 2.00GHz with 2 GB of RAM.

The different techniques were given separate runs of 200 seconds each to find a solution. The GA was started with a population of 50 individuals, a crossover rate of 25 (half of the population was reproducing and producing new offspring), and a mutation rate of 1 (each individual was mutated once every generation). The highest individual was transferred into the next generation following the normal procedure for elitist evolution.

The hillclimbing specimen mutates as many times as it needs in order to find an specimen with a higher value and then continue its evolution, adding new nodes to its adaptive sequence.

The dijkstra individual is initiated once. In order to do so, first the Dijkstra's shortest path algorithm is ran in the $ndFSM$ in order to calculate the distance from each state to the goal state. The algorithm as is proposed in this paper starts by creating a graph that is an inverted copy of the $ndFSM$, that is, a graph in which for a transition $s_i \xrightarrow{i/o} s_j$ existing in the $ndFSM$ there exist one transition $s_j \rightarrow s_i$ in the inverted graph. Then we use the goal state as the initial state and calculate Dijkstra's shortest path algorithm.

The random individual mutates a random number of times between 0 and the total number of states in the $ndFSM$.

The heuristic that is used is the same for every type of evolution present in the system. The adaptive sequence of each specimen is used a hundred times to run the $ndFSM$, then the $ndFSM$ returns its current state and we apply add of n (where n is the number of states), if it is the goal state, we subtract $n/2$ if it is a node from where the goal is unreachable, or subtract the value of its distance to the goal in any other case. Since the adaptive sequence is applied a hundred times, the total amount of heuristic value that an individual can have is 10000 points, that is considered being 100% fit, which means that every reachable end node of its adaptive sequence is the goal state. A drawback of this heuristic method is that using a sampling rating method, creates a fluctuation in the values for the same adaptive sequence, which makes evolution more complicated.

3.2 Comparison between GAs

The first set of experiments are focused on comparing different GAs, with different random coefficients, and that traverse the $ndFSM$ in a distinctive manner (the results from the experiments are shown in Figure 2). The random coefficient is a number that expresses how likely the GA will mutate using the shortest distance to the goal state. A random coefficient of 0 will behave randomly, and a coefficient of 1 will traverse the $ndFSM$ using the minimum distance, between these values, the specimen will choose some times at random and sometimes the closest node to the goal.

There are three ways of selecting the random coefficient for a new specimen. The first one is that every specimen in the population has a steady coefficient, for example in *GA 0.5* the whole population has 0.5 as its coefficient. The second manner is that it is started randomly from an interval, which is for example the case for *GA (0-0.5)*. And the third approach consist in a hereditary option, in which it is the average of its parents with a small amount of random added, which will be the case for *GA (0-1)m*. Every population labeled with an *m* (*mixed*) behaves in this last manner.

The population that achieved better results was the one started in the range of $(0-1)$ with the hereditary coefficient (*GA (0-1)m*). This population created, in average, adaptive sequences that reached 70.32% of the times the goal state, and obtained the lowest average distance with respect to the maximum achieved by any other method ($\mu = 9.41$, where $\mu = \sqrt{\frac{\sum_{i=0}^{n}(x_i - max\{\cup_{j=0}^{11} technique_j\}_i)^2}{n}}$). This behaves as expected since this population tries every possible random coefficient value and, depending on the configuration of the $ndFSM$ (how much non-determinism contains, how much branching towards the goal following the shortest path) the individuals with highest results pass their configuration to their offspring. The second best population is the one started in the range (0-0.5), this is the population that appeared to behave best in the few experiments that we presented in our previous paper.

The overall values are lower than in Section 3.3 because given the high number of populations, we restricted highly the time that we allowed the populations to evolve.

3.3 Results of the Main Techniques

Next, we will present and comment, the experiments conducted by applying the adaptive sequences, obtained by the main techniques, to $ndFSMs$ of different sizes and connection rates.

Comparison												
HEURISTIC VALUE (%)												
#	1	0	0.5	(0-0.5)	(0-1)	(0.5-1)	(0-0.5)m	(0-1)m	(0.5-1)m	HC	RD	DJ
0	93.86	92.77	93.85	88.6	100.0	92.75	91.75	93.81	87.71	97.92	67.86	53.5
1	54.47	42.05	50.09	51.5	48.94	50.15	43.93	57.67	52.16	70.13	27.23	18.05
2	84.38	43.95	74.3	64.62	56.42	74.95	51.3	50.78	49.46	64.2	21.0	17.34
3	56.35	52.47	51.35	53.57	54.69	52.61	39.0	55.76	50.65	54.63	31.43	-4.56
4	71.88	71.91	80.17	85.38	81.99	69.69	78.01	81.19	77.05	66.68	64.56	15.25
5	58.98	59.41	56.27	56.29	63.6	56.78	61.15	64.33	54.11	76.4	27.57	24.8
6	84.54	92.78	86.11	90.73	79.2	84.37	87.54	93.8	79.29	52.35	55.84	31.36
7	41.81	59.18	60.58	39.64	64.67	30.83	51.9	52.15	61.57	22.07	19.34	-13.26
8	75.16	93.38	70.02	85.5	100.0	85.58	97.92	83.45	77.74	53.04	64.83	19.57
9	45.66	48.21	47.12	65.26	51.02	53.34	37.49	64.03	58.78	56.15	29.27	4.95
10	37.42	42.52	47.91	44.68	41.64	39.89	35.29	39.92	38.08	24.82	21.93	6.25
11	70.06	59.52	46.87	52.12	50.17	49.29	54.47	65.79	58.47	45.29	25.19	-7.81
12	32.8	35.45	42.12	59.73	38.1	37.45	44.29	40.98	31.24	37.73	22.71	18.28
13	68.17	59.0	49.87	71.18	87.7	59.99	81.48	75.41	48.3	82.52	30.83	15.14
14	81.42	79.46	75.17	86.54	83.45	82.45	78.1	82.37	73.88	81.44	65.67	63.95
15	73.17	65.68	81.31	63.52	83.88	75.22	72.05	81.32	76.28	83.52	44.19	31.75
16	91.77	96.93	88.7	100.0	100.0	100.0	98.97	92.85	100.0	82.45	100.0	63.25
17	49.78	54.23	44.37	56.12	58.05	42.61	56.14	61.62	52.7	54.36	23.23	-2.32
18	56.06	59.24	53.38	61.2	59.96	46.76	56.68	58.77	54.78	58.73	32.8	15.07
19	34.66	39.88	31.36	45.75	35.66	29.49	37.51	37.56	33.95	21.91	18.4	-14.39
20	76.68	79.94	76.95	86.34	83.33	83.33	82.1	79.97	82.2	59.36	62.13	17.82
21	96.89	97.94	93.76	94.84	73.08	94.79	72.16	100.0	95.84	67.08	39.94	-1.02
22	75.86	97.94	100.0	100.0	100.0	97.93	98.94	100.0	100.0	87.55	69.59	59.05
23	100.0	100.0	100.0	100.0	100.0	100.0	100.0	100.0	100.0	100.0	100.0	100.0
24	15.22	14.3	19.72	20.84	17.3	16.74	22.49	20.61	12.41	28.03	5.61	-9.54
25	31.68	31.88	34.59	39.65	28.52	31.39	30.63	41.11	23.06	49.36	10.25	-18.26
26	66.05	71.37	73.07	72.77	65.65	78.09	66.09	65.78	70.42	81.97	44.25	15.1
27	80.33	64.73	79.23	74.24	65.77	65.35	69.84	86.1	78.13	79.57	48.22	20.68
28	94.74	91.75	94.72	92.66	74.7	86.89	90.52	94.77	91.64	100.0	75.25	52.0
29	50.88	47.65	44.58	55.89	61.25	55.89	62.37	68.11	51.91	68.44	39.0	12.19
30	92.7	90.64	91.65	85.41	90.68	89.69	88.55	94.86	89.58	100.0	58.54	64.3
31	59.04	61.49	56.16	52.56	54.13	53.13	51.1	63.15	55.32	80.33	42.71	24.85
32	76.79	67.46	86.4	54.64	74.84	55.88	54.49	74.41	53.83	77.3	40.29	21.25
33	87.69	79.42	81.57	89.7	83.14	85.16	80.4	87.65	84.15	95.9	71.28	47.5
34	75.04	80.73	72.69	65.61	62.63	78.16	61.37	85.42	52.11	56.92	34.73	-8.45
35	75.13	56.26	58.42	64.52	71.79	62.66	47.79	93.84	51.51	59.71	33.62	4.5
36	25.43	28.58	30.52	30.72	24.53	23.33	33.51	37.83	26.62	45.9	8.35	21.73
37	63.92	60.53	63.42	59.46	62.39	64.29	67.14	66.67	59.75	76.22	35.57	17.38
38	100.0	86.49	100.0	100.0	100.0	97.94	100.0	100.0	93.75	100.0	100.0	12.78
39	42.03	48.09	37.8	46.55	37.08	34.6	38.32	49.86	49.79	49.1	12.55	-21.52
40	63.9	76.38	80.41	71.12	75.27	78.35	72.15	77.22	72.2	87.68	41.44	34.3
41	100.0	100.0	100.0	100.0	100.0	100.0	100.0	100.0	100.0	74.45	100.0	19.41
42	57.66	66.83	63.96	54.41	57.37	53.5	66.21	63.64	63.74	73.28	37.7	35.52
43	59.53	54.29	59.53	57.81	61.82	53.09	51.02	61.69	66.75	64.78	38.42	27.12
44	50.08	51.91	51.29	44.86	49.93	47.22	45.82	46.78	52.01	56.59	29.83	21.52
45	43.0	39.63	36.09	41.25	40.73	29.9	36.7	34.71	32.36	49.33	18.3	20.2
46	55.69	64.75	63.74	56.68	62.8	64.85	68.89	61.65	48.44	73.31	44.74	36.77
47	100.0	100.0	100.0	100.0	100.0	100.0	100.0	100.0	100.0	100.0	82.53	66.01
48	91.68	93.8	93.8	94.82	92.77	92.83	91.3	92.74	92.7	98.97	85.54	63.6
49	64.2	61.01	67.49	62.89	61.9	64.33	63.78	64.91	67.53	78.71	38.62	33.85
50	71.01	56.26	56.72	75.15	63.64	60.47	55.54	63.88	64.35	75.99	42.08	-0.1
51	75.34	74.22	89.7	92.85	95.91	73.03	73.12	94.87	72.19	76.94	48.37	34.48
52	91.7	76.8	86.55	89.62	86.54	83.42	75.63	92.8	89.63	96.45	52.58	36.76
53	87.52	82.2	80.06	78.14	77.86	76.96	79.58	86.39	89.6	89.6	65.62	64.64
54	76.2	61.8	72.98	59.72	75.09	70.12	73.09	81.37	63.79	73.85	51.35	7.66
55	67.7	65.53	67.71	66.52	68.17	63.55	60.4	54.23	63.48	60.2	42.78	33.74
56	56.52	61.6	57.58	58.02	69.91	53.2	73.93	62.67	61.52	61.43	35.57	38.05
57	54.1	36.59	53.28	59.99	51.13	33.67	62.2	60.22	49.14	44.82	23.62	-9.34
58	66.55	58.43	59.45	60.44	60.46	56.34	55.25	62.25	59.36	70.68	38.33	5.35
59	74.69	64.15	70.35	76.8	74.75	63.11	70.67	66.34	78.87	82.88	26.74	35.1
60	55.95	49.47	55.47	50.53	52.75	50.6	44.29	51.8	45.05	37.38	30.43	35.23
61	86.48	80.08	86.48	85.45	81.15	79.0	83.03	87.17	89.59	92.82	67.44	63.6
62	60.0	43.24	43.35	57.56	60.84	53.73	59.4	64.92	55.56	32.13	36.29	-10.08
63	33.02	20.77	27.09	25.49	22.72	14.16	26.32	24.4	27.98	12.95	10.56	-4.58
$\bar{x}$	67.05	64.76	66.39	67.63	67.4	63.83	65.02	70.32	64.75	67.41	43.98	23.08
μ	12.61	15.25	13.61	12.69	12.74	15.69	15.29	9.41	15.99	16.15	35.95	57.69

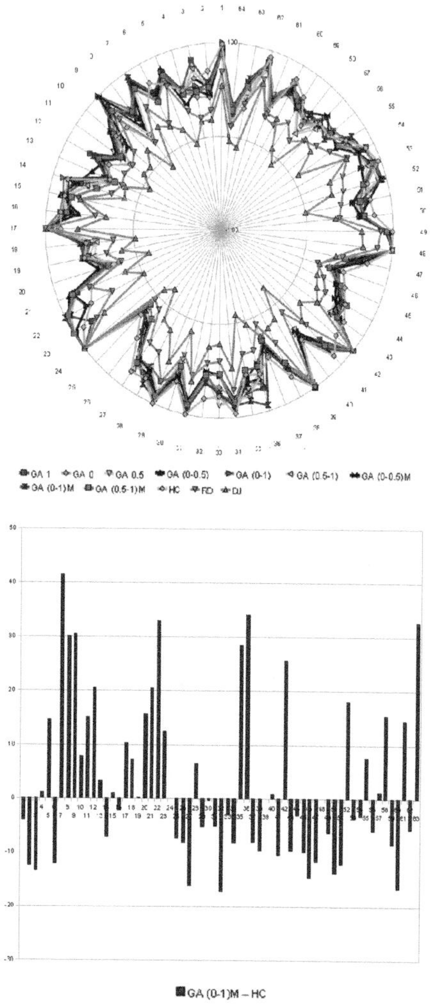

Fig. 2. Heuristic values of the comparisons between different GA methodologies, hillclimbing, random and dijkstra. Graphical representation of the heuristic values (upper right) and difference of the best GA and hillclimbing (GA-HC) (lower right).

Comparison								
	HEURISTIC VALUE (%)				SIZES			
#	GA	HC	RD	DJ	GA	HC	RD	DJ
0	94.85	89.22	79.87	52.0	1158.0	8160.0	261.0	2.0
1	87.44	92.82	59.1	36.16	1696.0	6948.0	265.0	9.0
2	100.0	97.94	81.46	-0.26	12469.0	2642.0	233.0	8.0
3	83.38	74.29	50.4	22.02	33972.0	1214.0	257.0	8.0
4	97.95	96.44	82.92	32.08	789.0	7759.0	263.0	9.0
5	97.93	87.26	49.81	-13.12	2996.0	6723.0	266.0	15.0
6	59.24	51.45	32.88	10.92	2088.0	7016.0	268.0	17.0
7	67.89	71.29	36.51	21.62	4990.0	5223.0	265.0	10.0
8	91.73	98.98	65.39	24.08	1465.0	7309.0	271.0	12.0
9	40.86	31.24	15.76	-11.77	40213.0	1244.0	277.0	14.0
10	23.6	30.72	2.76	-16.14	1771.0	7288.0	272.0	20.0
11	98.97	92.34	64.89	39.03	2295.0	6196.0	265.0	9.0
12	57.85	57.37	40.66	35.69	6214.0	5310.0	266.0	13.0
13	100.0	100.0	73.03	39.54	1205.0	6775.0	260.0	11.0
14	63.37	58.5	19.71	6.82	5249.0	3943.0	266.0	15.0
15	35.25	34.16	12.26	-17.09	4857.0	5221.0	271.0	17.0
16	83.97	73.82	50.8	18.57	29354.0	1972.0	269.0	10.0
17	100.0	85.04	75.24	12.1	931.0	7769.0	276.0	12.0
18	95.82	73.63	41.17	3.25	17124.0	2686.0	265.0	10.0
19	65.73	61.16	43.14	-2.14	24934.0	1902.0	267.0	25.0
20	88.48	89.72	56.17	34.7	20182.0	2150.0	263.0	13.0
21	79.35	77.47	41.91	3.48	6162.0	4523.0	267.0	12.0
22	75.01	77.92	53.88	40.15	5688.0	5844.0	280.0	7.0
23	82.26	75.76	62.67	35.56	54952.0	1003.0	262.0	9.0
24	75.11	47.01	34.81	12.12	4207.0	5519.0	273.0	10.0
25	76.07	70.59	42.27	15.71	4038.0	4839.0	267.0	12.0
26	82.32	68.43	53.03	38.08	3796.0	5571.0	261.0	7.0
27	72.08	51.55	37.13	10.82	59619.0	992.0	266.0	11.0
28	48.14	51.8	20.74	4.94	1331.0	7590.0	271.0	13.0
29	55.1	44.85	25.88	20.33	12127.0	4924.0	272.0	12.0
30	87.51	58.42	23.33	6.67	3637.0	5668.0	278.0	14.0
31	82.21	77.67	52.99	-1.7	1359.0	7236.0	266.0	13.0
32	69.86	62.48	26.7	4.8	23292.0	1848.0	267.0	14.0
33	46.53	46.33	11.16	-13.8	6481.0	4422.0	278.0	16.0
34	80.07	44.61	30.25	20.18	43219.0	1286.0	272.0	12.0
35	73.05	69.37	45.39	-17.26	40626.0	1226.0	248.0	11.0
36	90.75	79.53	49.21	7.35	6071.0	4404.0	269.0	15.0
37	92.76	64.74	43.57	9.16	35659.0	1163.0	271.0	8.0
38	100.0	100.0	90.7	36.4	1129.0	6709.0	257.0	9.0
39	100.0	100.0	76.96	15.51	1328.0	7465.0	268.0	10.0
40	100.0	100.0	100.0	64.64	1428.0	8079.0	263.0	6.0
41	100.0	86.66	54.44	36.41	2182.0	6727.0	263.0	7.0
42	100.0	85.6	75.19	20.21	2433.0	7034.0	269.0	10.0
43	94.83	91.75	70.64	38.45	5342.0	5580.0	259.0	11.0
44	95.88	83.21	55.83	21.61	32788.0	1839.0	264.0	10.0
45	40.5	23.55	7.06	-23.96	3610.0	6644.0	281.0	24.0
46	71.16	60.95	41.44	22.27	12694.0	3254.0	276.0	10.0
47	73.39	82.49	38.43	21.34	1624.0	7011.0	271.0	12.0
48	98.97	95.9	76.42	64.64	4848.0	4201.0	252.0	4.0
49	89.53	64.41	48.12	-13.07	2116.0	6574.0	261.0	14.0
$\bar{x}$	79.34	71.81	48.48	16.58	11994.8	4892.5	266.4	11.7
μ	2.24	12.31	33.74	66.26	-	-	-	-

Comparison								
	HEURISTIC VALUE (%)				SIZES			
#	GA	HC	RD	DJ	GA	HC	RD	DJ
0	57.97	59.38	39.67	39.75	2335.0	5701.0	1320.0	11.0
1	82.1	63.45	40.85	1.69	1135.0	6800.0	1302.0	15.0
2	45.6	40.47	22.74	-13.97	1851.0	6345.0	1306.0	14.0
3	40.58	43.98	16.99	-4.04	2210.0	5490.0	1290.0	16.0
4	73.59	76.19	51.13	-4.71	983.0	6930.0	1312.0	15.0
5	57.77	51.24	30.36	15.48	1202.0	6533.0	1312.0	18.0
6	54.73	42.91	27.95	11.76	1363.0	6358.0	1321.0	15.0
7	45.79	45.04	23.67	-7.88	1219.0	6462.0	1271.0	19.0
8	70.33	80.09	43.12	36.26	1872.0	6265.0	1323.0	15.0
9	48.03	54.49	36.51	37.62	1278.0	7141.0	1335.0	13.0
10	36.16	44.48	19.6	5.86	1315.0	7006.0	1310.0	15.0
11	29.04	33.27	15.69	-15.87	1311.0	6081.0	1315.0	21.0
12	41.36	32.48	28.45	18.27	1356.0	6538.0	1309.0	12.0
13	82.3	83.52	67.31	21.91	867.0	7554.0	1258.0	10.0
14	49.56	41.1	24.14	-8.33	2660.0	5631.0	1284.0	17.0
15	45.54	54.21	31.72	17.2	2360.0	6593.0	1271.0	12.0
16	21.33	19.89	5.21	-17.45	1746.0	6474.0	1348.0	22.0
17	84.3	86.47	66.66	10.6	1133.0	6762.0	1266.0	17.0
18	63.36	70.87	39.84	42.09	1300.0	7057.0	1295.0	7.0
19	49.74	24.62	17.65	-8.0	2418.0	6164.0	1336.0	17.0
20	62.79	36.93	29.67	-12.25	929.0	6811.0	1310.0	16.0
21	55.35	37.5	34.99	11.64	12114.0	2858.0	1281.0	10.0
22	61.44	59.83	37.25	11.56	2288.0	6669.0	1309.0	15.0
23	55.72	49.96	32.73	22.82	1317.0	6467.0	1315.0	12.0
24	81.24	82.93	69.26	68.5	1053.0	7806.0	1317.0	2.0
25	41.72	50.61	25.06	2.82	1130.0	7073.0	1317.0	18.0
26	63.26	72.46	43.89	12.28	784.0	7214.0	1319.0	17.0
27	86.4	85.42	71.96	67.45	707.0	7867.0	1303.0	8.0
28	44.3	41.73	27.69	-0.04	1183.0	7099.0	1319.0	16.0
29	70.44	58.64	32.06	10.43	3888.0	4823.0	1297.0	11.0
30	82.04	66.02	67.91	5.21	955.0	7364.0	1308.0	18.0
31	40.02	27.37	19.3	-9.84	892.0	6564.0	1332.0	17.0
32	17.6	14.39	0.84	-16.62	1028.0	6936.0	1342.0	21.0
33	36.19	22.5	20.29	3.64	1652.0	6045.0	1332.0	18.0
34	32.36	35.66	19.81	19.14	1204.0	7082.0	1341.0	12.0
35	35.85	30.34	16.57	4.96	1159.0	7150.0	1323.0	16.0
36	35.54	35.81	16.48	10.64	1143.0	6423.0	1321.0	15.0
37	74.8	84.39	51.48	36.65	1753.0	6356.0	1279.0	7.0
38	57.39	56.07	35.28	22.2	2470.0	5151.0	1311.0	16.0
39	46.02	32.06	31.15	-12.67	1457.0	6658.0	1318.0	28.0
40	53.61	27.24	28.52	-17.42	12179.0	1992.0	1297.0	17.0
41	53.84	51.13	29.58	-6.26	5989.0	4624.0	1277.0	17.0
42	86.32	87.07	72.36	64.3	708.0	7824.0	1313.0	4.0
43	68.34	73.39	44.11	13.18	1689.0	7223.0	1289.0	13.0
44	67.62	79.35	47.22	39.81	1353.0	6963.0	1277.0	9.0
45	52.77	31.66	33.2	6.14	4991.0	4080.0	1356.0	16.0
46	39.84	41.64	18.11	15.91	1811.0	6771.0	1323.0	10.0
47	42.09	52.96	21.75	2.63	2199.0	5827.0	1316.0	16.0
48	43.21	27.67	14.38	9.8	9047.0	2434.0	1311.0	15.0
49	52.54	25.63	18.86	-21.39	1692.0	5942.0	1303.0	27.0
$\bar{x}$	54.4	50.53	33.22	10.87	2253.6	6279.6	13.1	14.8
μ	4.33	10.46	24.56	49.25	-	-	-	-

Comparison								
	HEURISTIC VALUE (%)				SIZES			
#	GA	HC	RD	DJ	GA	HC	RD	DJ
0	17.78	12.21	0.48	-6.14	2527.0	4578.0	2659.0	67.0
1	42.85	34.23	23.11	-35.1	7276.0	3181.0	2552.0	30.0
2	40.0	33.07	16.18	-14.87	2366.0	5254.0	2571.0	16.0
3	48.31	35.73	20.77	15.5	4518.0	4804.0	2585.0	12.0
4	22.76	17.35	4.45	-0.25	3790.0	4340.0	2636.0	16.0
5	60.18	60.52	49.8	39.58	1719.0	7089.0	2594.0	11.0
6	24.66	22.62	8.01	9.75	2459.0	6113.0	2640.0	13.0
7	53.6	52.46	36.75	19.96	2097.0	5131.0	2589.0	12.0
8	20.9	19.53	8.83	-1.1	2272.0	6802.0	2635.0	20.0
9	54.66	47.64	35.01	12.4	1780.0	7121.0	2627.0	10.0
10	28.53	28.9	15.2	-35.15	2899.0	5528.0	2545.0	23.0
11	80.11	82.65	61.27	65.76	1336.0	7455.0	2644.0	14.0
12	49.99	57.23	32.7	-10.4	2371.0	6047.0	2600.0	28.0
13	37.45	33.78	16.03	19.44	2886.0	4510.0	2596.0	16.0
14	28.55	16.23	11.78	-6.55	2445.0	5443.0	2643.0	21.0
15	42.87	30.32	25.81	1.64	2516.0	5466.0	2589.0	14.0
16	46.39	43.84	23.91	-2.56	7992.0	2765.0	2551.0	14.0
17	65.96	74.06	54.36	24.62	2363.0	6280.0	2593.0	7.0
18	100.0	100.0	100.0	100.0	850.0	7.0	2590.0	7.0
19	57.56	52.16	44.97	40.15	1657.0	6492.0	2567.0	27.0
20	40.59	48.57	23.57	23.12	1267.0	6546.0	2597.0	10.0
21	24.28	20.18	10.75	5.19	2361.0	5588.0	2641.0	16.0
22	46.48	31.53	18.75	-21.27	4118.0	3790.0	2579.0	20.0
23	20.54	23.66	11.76	1.37	2692.0	6065.0	2646.0	20.0
24	17.72	25.46	5.42	4.48	1904.0	6176.0	2625.0	18.0
25	40.59	43.62	19.28	-26.21	2591.0	5170.0	2619.0	34.0
26	50.71	57.1	32.17	-7.79	2372.0	6823.0	2630.0	19.0
27	46.99	35.34	21.26	2.85	2622.0	4882.0	2626.0	12.0
28	32.35	39.64	15.49	4.58	1552.0	5849.0	2582.0	16.0
29	45.77	34.45	31.86	19.36	3352.0	4910.0	2619.0	10.0
30	47.11	32.56	28.73	0.38	2400.0	4924.0	2585.0	19.0
31	29.26	25.05	5.14	-22.12	2102.0	4624.0	2645.0	30.0
32	40.43	35.34	16.96	10.37	14366.0	1863.0	2624.0	17.0
33	35.72	25.86	19.98	-27.36	3748.0	4017.0	2614.0	30.0
34	33.8	27.29	19.61	6.25	18806.0	1407.0	2546.0	17.0
35	45.4	44.15	27.5	-9.46	2287.0	6024.0	2649.0	16.0
36	57.84	44.55	34.53	-2.28	2347.0	6243.0	2573.0	21.0
37	36.11	18.76	8.56	-16.88	2324.0	5069.0	2633.0	21.0
38	70.36	62.78	32.87	14.82	7568.0	2582.0	2581.0	16.0
39	55.79	66.11	29.16	3.88	2196.0	6073.0	2570.0	15.0
40	22.03	32.42	7.7	-14.13	2399.0	6003.0	2591.0	16.0
41	70.04	63.44	45.17	6.79	2163.0	7312.0	2620.0	16.0
42	17.07	15.68	5.43	-4.24	2025.0	5994.0	2632.0	18.0
43	74.72	38.38	65.28	5.06	2367.0	6841.0	2659.0	14.0
44	34.59	28.27	12.84	4.7	2351.0	5389.0	2637.0	16.0
45	28.13	28.58	10.02	5.4	2421.0	5392.0	2639.0	16.0
46	17.75	15.99	9.17	-9.58	2153.0	5757.0	2644.0	21.0
47	56.97	68.18	37.12	-27.05	2179.0	6152.0	2612.0	24.0
48	41.92	47.05	25.98	1.16	2061.0	6552.0	2579.0	21.0
49	32.79	32.66	18.24	-11.45	2387.0	6613.0	2611.0	15.0
$\bar{x}$	42.74	39.34	24.79	1.53	3271	5300.7	2608.3/	18.6
μ	3.82	8.49	20.95	47.97	-	-	-	-

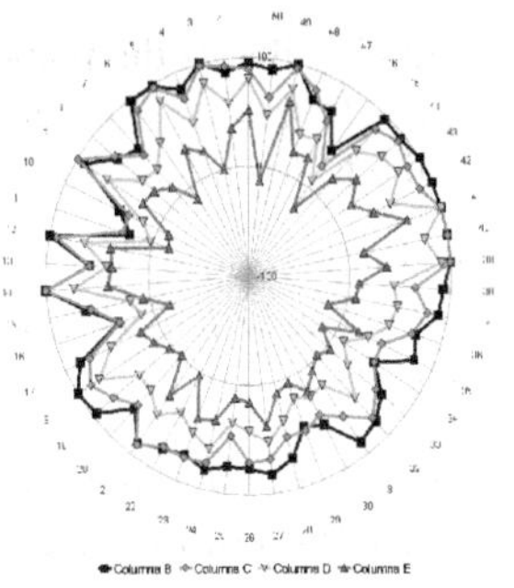

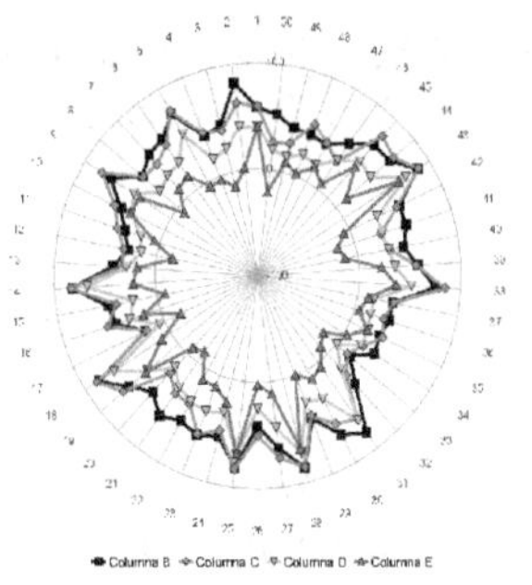

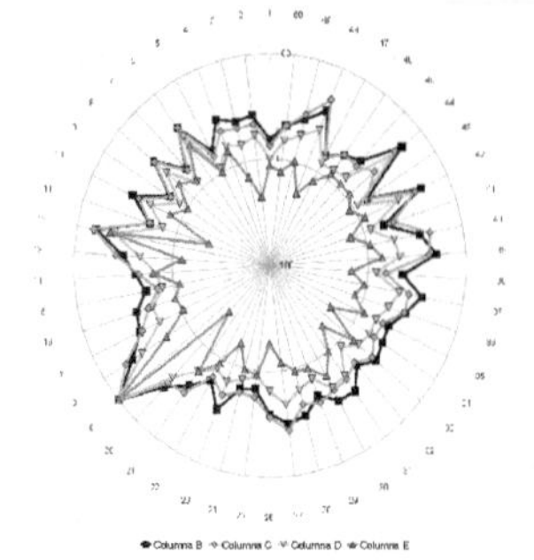

Fig. 3. Overview of the results obtained with $ndFSMs$ with a connection level of 4, including a set of $ndFSMs$ with 100 states (left), 500 states (center) and 1000 states (right)

The connection rate specifies a maximum number of transitions, for example in a *ndFSM* with a hundred states and a connection level of 3 the average number of transitions is 255 and in a *ndFSM* with five hundred states and a connection level of 3 the average number of transitions is 1280.

The experiments were run against *ndFSMs* of 100, 500 and 1000 states, and two connection levels 3 and 4. The results for the connection level 4 are shown in Figure 3 and the results from connection level 3 can be found under `http://www.carlosmolinero.com/GAforAdapSeq.htm`.

As we can see by taking a look at the averages ($\bar{x}$) and the average distances to the maximum (μ), GA outperforms the rest of the methodologies. The value of the average of the heuristic value ($\bar{x}$) depends highly on the number of states, and in a lower percentage in the connection rate. In fact, as expected, the higher the number of states and therefore, of transitions, and the higher connection level (which also influences in the number of transitions) the more difficult it becomes to find a valid adaptive sequence. On the other hand, the distance to the maximum (μ) remains a quasi-constant value for each methodology no matter the *ndFSM* that it is applied to. The lower μ is, the larger the number of times that the technique achieved the maximal heuristic value, it also represents the difference in heuristic value that the technique had when it was not the highest. For our GA $\mu = 3.1$ in average (this average is computed taking into account all the experiments performed) while the hillclimbing method had $\mu = 11.08$.

After performing some scatter plots of the relationships between the heuristic values of the techniques, we realized that GA and hillclimbing has a positive correlation with the random heuristic value. In the case of the GA that uses the random method and the dijkstra method as specimens inside its population this was expected, but the

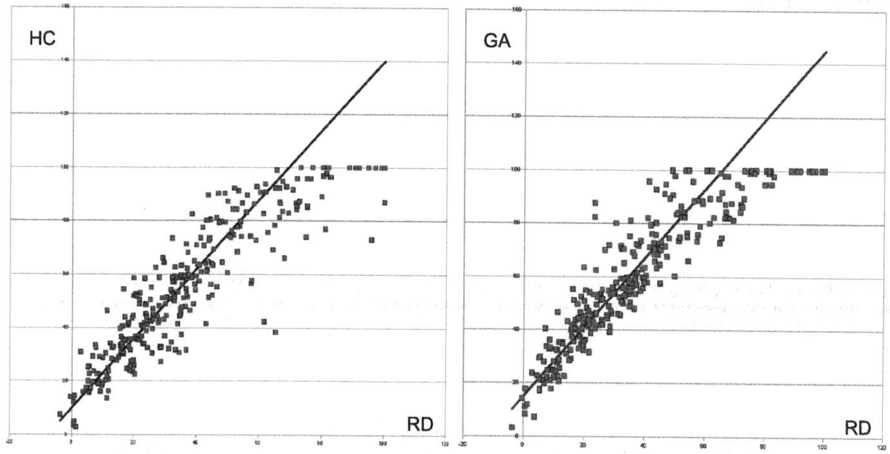

Fig. 4. Representation of the scatter plots of Hillclimbing with respect to Random, and of GA with respect to Random

hillclimbing methodology never uses the random specimen, or a random approach, still they have a correlation. In the case of hillclimbing the trendline is defined by the equation $heurVal(random) * 1.3 + 10$ and in the case of GA the trendline responds to the equation $heurVal(random) * 1.3 + 15$. This scatter plots are presented in Figure 4.

Another thing that we can observe in the scatter plots, is that although hillclimbing and GA behaved proportionally, the samples that are away from the *trendline* behave differently, which is one of the factors that impact on its relative fitness. In the case of hillclimbing, those samples away from the trendline behave worst than expected (they fall mostly in the right lower sector from the trendline) while in the GA they behave better than expected (the locate in the upper left sector from the trendline).

4 Conclusions and Future Work

We have presented in this paper a series of experiments undertaken in the context of our previous work [5]. The purpose of these experiments is to test whether a evolutionary methodology composed of genetic algorithms is able to find *adaptive sequences* that allow to reach a certain goal state in a deterministic fashion in a non-deterministic context.

The comparison with other methodologies such as the hillclimbing or random, was satisfactory in the sense that GA showed a better general performance, and with a higher consistency. In general we can say that the hillclimbing algorithm performs well in most cases, but its results are less consistent than those of the genetic algorithm.

We have experimented with $ndFSMs$ of different sizes and connection levels. As the number of transitions increased there was a decrease in the heuristic values of the adaptive sequences since a higher number of states with a high degree of non-determinism, creates a high level of branching and the existence of a perfect solution becomes more complicated, as well as its discovery.

References

1. Alur, R., Courcoubetis, C., Yannakakis, M.: Distinguishing tests for nondeterministic and probabilistic machines. In: 27th ACM Symp. on Theory of Computing, STOC 1995, pp. 363–372. ACM Press, New York (1995)
2. Ammann, P., Offutt, J.: Introduction to Software Testing. Cambridge University Press, Cambridge (2008)
3. Gromov, M., Yevtushenko, N., Kolomeets, A.: On the synthesis of adaptive tests for nondeterministic finite state machines. Programming and Computer Software 34, 322–329 (2008)
4. Hierons, R.M.: Testing from a non-deterministic finite state machine using adaptive state counting. IEEE Transactions on Computers 53(10), 1330–1342 (2004)
5. Molinero, C., Núñez, M., Hierons, R.M.: Creating adaptive sequences with genetic algorithms to reach a certain state in a non-deterministic FSM. In: IEEE Symposium on Artificial Life, ALIFE 2011. IEEE Computer Society Press, Los Alamitos (to appear 2011)
6. Myers, G.J.: The Art of Software Testing, 2nd edn. John Wiley and Sons, Chichester (2004)

An Efficient Algorithm for Reasoning about Fuzzy Functional Dependencies

P. Cordero, M. Enciso, A. Mora, I. Pérez de Guzmán, and J.M. Rodríguez-Jiménez

Universidad de Málaga, Spain
{pcordero,enciso}@uma.es, {amora,guzman}@ctima.uma.es, jmrodriguezjimenez@alu.uma.es

Abstract. A sound and complete Automated Prover for the Fuzzy Simplification Logic (**FSL** logic) is introduced and based on it a method for efficiently reasoning about fuzzy functional dependencies over domains with similarity relations. The complexity of the algorithm is the same as that of equivalent algorithms for crisp functional dependencies that appear in the literature.

1 Introduction

Constraints are often used to guide the design of relational schema for the sake of consistency of databases and thereby avoiding to avoid the problems of redundancies, anomalies, etc. This statement is valid for any extension of the classical relational model. Different authors have studied fuzzy models and which constraints are more appropriate to extend the well studied relational database theory to their fuzzy database.

Exist many papers that have established the advantages of having a Fuzzy extension of the relational model for databases [1, 2]. Thus, we can affirm that there exists a consensus in the need to have a "good" extension of the classical Codd model to Fuzzy Logic. But it is not only a matter of logicians, several database researchers also demand this extension.

Several approaches for the definition of fuzzy functional dependency [1, 3–6] (FFD) are proposed in the literature. In the same way as the concept of functional dependency (FD) corresponds to the notion of partial function, it should be desired that the concept of FFD would correspond to the notion of fuzzy partial function. The definitions proposed in [1, 5, 6] fit in this idea.

Nevertheless some of these papers preserves the original FD definition and substitutes the equality between values of an attribute by a similarity relation [1, 2, 6, 7]. A proper extension of the concept of functional dependencies requires that we are able to introduce the uncertainty in the FDs that are held by a relation by associating a grade to each FDs [5].

There exists a wide range of dependencies. Each dependency definition is usually followed by its corresponding logic. In [1, 6] the authors propose any generalizations of the well-know Armstrong's Axioms as a useful tool for reasoning with FFDs, but these inferences rules have not been used successfully

J. Cabestany, I. Rojas, and G. Joya (Eds.): IWANN 2011, Part II, LNCS 6692, pp. 412–420, 2011.

in automated deduction. The reason is that this inference system were created to explain dependency semantics more than to design an automated deduction system. In fact, in [6] the authors propose the classical closure algorithm to solve the implication problem and don't directly use the Armstrong's Axioms neither any generalization of them.

Our approach points in this direction. In [8] a novel logic ($\mathbf{SL}_{\mathrm{FD}}$) equivalent to classical Armstrong's axioms was presented. The core of $\mathbf{SL}_{\mathrm{FD}}$ is the *Simplification Rule* that replaces Transitivity Rule (which is the cause of the non applicability of the other logics [8]). The definition of $\mathbf{SL}_{\mathrm{FD}}$ introduces, for the first time, interesting solutions to database problems, which are solved using logic-based automated deduction methods [9, 10]. Our fuzzy FD notion was introduced to have a proper fuzzy definition which allows us to built a Fuzzy extension of the $\mathbf{SL}_{\mathrm{FD}}$ logic [11, 12].

In this work, we illustrate how the *Simplification Rule* can be considered for reasoning with FFDs . We prove that this rule is the key of three equivalence rules which can be considered as efficient tools to manipulate FFDs in a logical way: removing redundancy, solving implication problem etc. We present an automated prover that applies systematically the equivalence rules in order to answer if a FFD can be deduced from a set of FFDs.

This work opens the door to the management of FFD constraints in an efficient and intelligent way. First, we outline the basic notions needed (Section 2) and, in Section 3, we show that the rules of **FSL** logic are equivalence rules and they become adequate tools to remove redundancy. In Section 4 we propose a new automated prover directly based on the equivalences rules of **FSL** logic to solve the FFD implication problem. In Section 5, the soundness and completeness of the algorithm is proved and the complexity is studied and, finally, we establish several conclusions and future works in Section 6.

2 Preliminaries

First, the concept of functional dependency in the relational model of databases is outlined. Let Ω be a finite non-empty set whose elements are named attributes and $\{\mathbb{D}_a \mid a \in \Omega\}$ a family of domains. A database is a relation $R \subseteq \mathbb{D} = \prod_{a\in\Omega} \mathbb{D}_a$ usually represented as a table. The columns are the attributes and the rows of this table are the tuples $t = (t_a \mid a \in \Omega) \in \mathbb{D}$. If $\varnothing \neq X \subseteq \Omega$, D_X denotes $\prod_{a\in X} D_a$ and, for each $t \in R$, $t_{/X}$ denotes the *projection* of t to D_X. That is, $t_{/X} = (t_a \mid a \in X) \in D_X$.

Definition 1. *A* ***functional dependency*** *is an expression* $X{\mapsto}Y$ *where* $X, Y \subseteq \Omega$. *A relation* $R \in \mathbb{D}$ *satisfies* $X{\mapsto}Y$ *if, for all* $t_1, t_2 \in R$, $t_{1/X} = t_{2/X}$ *implies that* $t_{1/Y} = t_{2/Y}$.

The extended method to fuzzify the concept of functional dependency is by using similarity relations instead of the equality. Each domain D_a is endowed with a similarity relation $\rho_a : D_a \times D_a \to [0,1]$, that is, a reflexive ($\rho_a(x,x) = 1$ for all $x \in D_a$) and symmetric ($\rho_a(x,y) = \rho_a(y,x)$ for all $x, y \in D_a$) fuzzy relation.

Given $X \subseteq \Omega$, extensions of these relations to the set $\mathbb{D}$ can be obtained as follows: for all $t, t' \in \mathbb{D}$, $\rho_X(t,t') = \min\{\rho_a(t_a, t'_a) \mid a \in X\}$.

A first step in order to fuzzify, is the following definition of fuzzy functional dependency (FFD) that appears with slight differences in the literature [1, 2, 6].

Remark 1. A relation $R \subseteq \mathbb{D}$ *satisfies* $X \mapsto Y$ *if* $\rho_X(t,t') \leq \rho_Y(t,t')$ *holds, for all* $t, t' \in R$.

However, the functional dependency remains crisp. In [5] the authors add a degree of fuzzyness in the dependency itself and in [12] we generalize this definition of fuzzy functional dependency as follows.

Definition 2. *A* ***fuzzy functional dependency*** *is an expression* $X \xrightarrow{\theta} Y$ *where* $\theta \in [0,1]$ *and* $X, Y \subseteq \Omega$ *with* $X \neq \varnothing$. *A relation* $R \subseteq \mathbb{D}$ *is said to satisfy* $X \xrightarrow{\theta} Y$ *if* $\min\{\theta, \rho_X(t,t')\} \leq \rho_Y(t,t')$, *for all* $t, t' \in R$.

In the literature some authors present a complete axiomatic system defined over FFDs with similarity relations [2, 6, 13] and any axiomatic systems where the dependency is fuzzy [5], and all of them are fuzzy extensions of Armstrong Axiom's having the problem inherent of the transitivity rule in order to apply the axiomatic system in real problems. However, in [12] we introduce **FSL**, a new logic more adequate for the applications, named Simplification Logic for fuzzy functional dependencies. The main novelty of the system is that it is not based on the transitivity rule like all the others, but it is built around a simplification rule which allows the removal of redundancy.

Definition 3. *Given a finite non-empty set of attribute symbols* Ω, *the* ***language*** *of* **FSL** *is* $\mathbf{L} = \{X \xrightarrow{\theta} Y \mid X, Y \in 2^{\Omega},\ X \neq \varnothing \text{ and } \theta \in [0,1]\}$,[1] *the* ***semantics*** *has been introduced in Definition 2 and the* **axiomatic system** *has one axiom scheme:*

Ax: $\vdash X \xrightarrow{1} Y$, *for all* $Y \subseteq X$ **Reflexive Axioms**

and four inferences rules:

InR: $X \xrightarrow{\theta_1} Y \vdash X \xrightarrow{\theta_2} Y$, *if* $\theta_1 \geq \theta_2$ **Inclusion Rule**

DeR: $X \xrightarrow{\theta} Y \vdash X \xrightarrow{\theta} Y'$, *if* $Y' \subseteq Y$ **Decomposition Rule**

CoR: $X \xrightarrow{\theta_1} Y,\ U \xrightarrow{\theta_2} V \vdash XU \xrightarrow{min(\theta_1,\theta_2)} YV$ **Composition Rule**

SiR: $X \xrightarrow{\theta_1} Y,\ U \xrightarrow{\theta_2} V \vdash U\text{-}Y \xrightarrow{min(\theta_1,\theta_2)} V\text{-}Y$, *if* $X \subseteq U$ *and* $X \cap Y = \varnothing$ **Simplification Rule**

The *deduction* ($\vdash$), semantic implication ($\models$) and *equivalence* ($\equiv$) concepts are introduced as usual. Soundness and completeness were proved in [12].

[1] In logic, it is important to distinguish between the language and the metalanguage. So, in a formula, XY denotes $X \cup Y$, $X - Y$ denotes $X \smallsetminus Y$ and $\top$ denotes the empty set.

3 Removing Redundant Information

In database systems redundancy is not desirable in the integrity constraints of a database and finally, in [11] we have outlined that **FSL** logic is adequate for the applications showing its good behavior for removing redundancy. The systematic application of the rules removes redundancy because they can be seen as equivalence rules as the following proposition ensures.

Theorem 1. *If* $X, Y, U, V \subseteq \Omega$ *and* $\theta, \theta_1 \in [0,1]$ *then*

- Decomposition Equivalence (**DeEq**): $\{X \xrightarrow{\theta} Y\} \equiv \{X \xrightarrow{\theta} Y - X\}$.
- Union Equivalence (**UnEq**): $\{X \xrightarrow{\theta} Y, X \xrightarrow{\theta} V\} \equiv \{X \xrightarrow{\theta} YV\}$.
- Simplification Equivalence (**SiEq**): *if* $X \subseteq U$, $X \cap Y = \varnothing$ *and* $\theta \geq \theta_1$, *then* $\{X \xrightarrow{\theta} Y, U \xrightarrow{\theta_1} V\} \equiv \{X \xrightarrow{\theta} Y, U - Y \xrightarrow{\theta_1} V - Y\}$

The proof of this theorem is straightforward and as an immediate consequence of these equivalences there exists other equivalences that are very interesting to remove redundant information.

Corollary 1. *Let* $\theta, \theta_1 \in [0,1]$ *and* $X, Y, U, V \subseteq \Omega$ *with* $X \subseteq U$ *and* $X \cap Y = \varnothing$.

- Simplification+Union Equivalence (**SiUnEq**): $\{X \xrightarrow{\theta} Y, U \xrightarrow{\theta} V\} \equiv \{X \xrightarrow{\theta} YV\}$ *when* $U \setminus Y = X$.
- Simplification+Axiom Equivalence (**SiAxEq**): $\{X \xrightarrow{\theta} Y, U \xrightarrow{\theta_1} V\} \equiv \{X \xrightarrow{\theta} Y\}$ *when* $\theta \geq \theta_1$ *and* $V \setminus Y = \varnothing$.

4 Automated Prover

Given a set of fuzzy functional dependencies Γ, we define the *syntactic closure* of Γ as $\Gamma^+ = \{X \xrightarrow{\theta} Y \mid \Gamma \vdash X \xrightarrow{\theta} Y\}$ that coincides with the semantic closure due to the soundness and completeness of the axiomatic system. So, Γ^+ is the minimum set that contains Γ, all the axioms and is closed for the inference rules. The aim of this section is to give an efficient algorithm to decide if a given FFD belongs to Γ^+.

The input of the algorithm will be a set of fuzzy functional dependencies Γ_0 and a fuzzy functional dependency $\mathrm{A} \xrightarrow{\theta} \mathrm{B}$ and the output will be $\Gamma_0 \vdash \mathrm{A} \xrightarrow{\theta} \mathrm{B}$ or $\Gamma_0 \nvdash \mathrm{A} \xrightarrow{\theta} \mathrm{B}$. We outline the steps of the algorithm in the following:

1. If $\mathrm{A} \xrightarrow{\theta} \mathrm{B}$ is an axiom, the algorithm finishes with the output is $\Gamma_0 \vdash \mathrm{A} \xrightarrow{\theta} \mathrm{B}$.
2. Compute $\Gamma_{\mathrm{A}}^{\theta} = \{\mathrm{A}X \xrightarrow{\theta} Y \mid X \xrightarrow{\theta_1} Y \in \Gamma_0 \text{ with } \theta_1 \geq \theta\}$.
3. If there doesn't exist $X \subseteq \Omega$ such that $\mathrm{A} \xrightarrow{\theta} X \in \Gamma_{\mathrm{A}}^{\theta}$, then the algorithm finishes and the output is $\Gamma_0 \nvdash \mathrm{A} \xrightarrow{\theta} \mathrm{B}$.
4. Otherwise, apply **DeEq** to every formula in $\Gamma_{\mathrm{A}}^{\theta}$ obtaining Γ_1.

5. $\Gamma_1 = \{A\xrightarrow{\theta}C_1\} \cup \Gamma'_1$ such that $A \subseteq X$ for all $X\xrightarrow{\theta}Y \in \Gamma_1$. The FFD $A\xrightarrow{\theta}C_1$ will be named *guide*.
6. Repeat until a fix point is obtained or a *guide* $A\xrightarrow{\theta}C_n$ with $B \subseteq A \cup C_n$.
   - Compute $\Gamma_{i+1} = \{A\xrightarrow{\theta}C_{i+1}\} \cup \Gamma'_{i+1}$ from $\Gamma_i = \{A\xrightarrow{\theta}C_i\} \cup \Gamma'_i$ applying to $A\xrightarrow{\theta}C_i$ and each $X\xrightarrow{\theta}Y \in \Gamma'_i$ the equivalences **SiAxEq**, **SiUnEq** or **SiEq** with this priority ordering.
7. If the *guide* is $A\xrightarrow{\theta}C_n$ and $B \subseteq A \cup C_n$ then the otuput is $\Gamma_0 \vdash A\xrightarrow{\theta}B$. Otherwise, the output is $\Gamma_0 \nvdash A\xrightarrow{\theta}B$.

Example 1. Let $\Gamma = \{\mathtt{ac}\xrightarrow{0.9}\mathtt{def}, \mathtt{fh}\xrightarrow{1}\mathtt{dg}\}$ and the fuzzy functional dependency $\mathtt{cf}\xrightarrow{0.8}\mathtt{beg}$ in order to check if $\Gamma \vdash \mathtt{cf}\xrightarrow{0.8}\mathtt{beg}$. The trace of the execution of the **FSL** *Automated Prover* is the following:

1. $\mathtt{cf} \xrightarrow{0.8} \mathtt{beg}$ is not an axiom, then the algorithm continues.
2. $\Gamma_{\mathtt{cf}}^{0.8} = \{\mathtt{acf}\xrightarrow{0.8}\mathtt{def}, \mathtt{cfh}\xrightarrow{0.8}\mathtt{dg}\}$.
3. Since in $\Gamma_{\mathtt{cf}}$ there is not an FFD $\mathtt{cf}\xrightarrow{0.8}W$ (*guide* is $\varnothing$) then $\Gamma \nvdash \mathtt{cf}\xrightarrow{0.8}\mathtt{beg}$. (**End of FSL *Automated Prover***)

Example 2. Let $\Gamma = \{\mathtt{ac}\xrightarrow{0.9}\mathtt{def}, \mathtt{f}\xrightarrow{1}\mathtt{dg}, \mathtt{de}\xrightarrow{0.9}\mathtt{h}, \mathtt{di}\xrightarrow{0.4}\mathtt{a}, \mathtt{ch}\xrightarrow{0.9}\mathtt{bf}, \mathtt{j}\xrightarrow{0.9}\mathtt{ad}, \mathtt{cd}\xrightarrow{0.8}\mathtt{e}\}$ and the $\mathtt{cf}\xrightarrow{0.8}\mathtt{beg}$ a fuzzy functional dependencies in order to check if $\Gamma \vdash \mathtt{cf}\xrightarrow{0.8}\mathtt{dgh}$. The trace of the execution of the **FSL** *Automated Prover* is the following:

1. $\mathtt{cf} \xrightarrow{0.8} \mathtt{beg}$ is not an axiom and the algorithm continues.
2. $\Gamma_{\mathtt{cf}}^{0.8} = \{\mathtt{acf}\xrightarrow{0.8}\mathtt{def}, \mathtt{cf}\xrightarrow{0.8}\mathtt{dg}, \mathtt{cdef}\xrightarrow{0.8}\mathtt{h}, \mathtt{chf}\xrightarrow{0.8}\mathtt{bf}, \mathtt{cfj}\xrightarrow{0.8}\mathtt{ad}, \mathtt{cdf}\xrightarrow{0.8}\mathtt{e}\}$.
3. There exists $\mathtt{cf}\xrightarrow{0.8}\mathtt{dg} \in \Gamma_{\mathtt{cf}}^{0.8}$ and the algorithms continues.
4. **DeEq** applied to every formula in $\Gamma_{cf}^{0.8}$ gives the set
   $\Gamma_1 = \{\mathtt{acf}\xrightarrow{0.8}\mathtt{de}, \mathtt{cf}\xrightarrow{0.8}\mathtt{dg}, \mathtt{cdef}\xrightarrow{0.8}\mathtt{h}, \mathtt{chf}\xrightarrow{0.8}\mathtt{b}, \mathtt{cfj}\xrightarrow{0.8}\mathtt{ad}, \mathtt{cdf}\xrightarrow{0.8}\mathtt{e}\}$.
5. *guide* $= \{\mathtt{cf}\xrightarrow{0.8}\mathtt{dg}\}$ and
   $\Gamma'_1 = \{\mathtt{acf}\xrightarrow{0.8}\mathtt{de}, \mathtt{cdef}\xrightarrow{0.8}\mathtt{h}, \mathtt{chf}\xrightarrow{0.8}\mathtt{b}, \mathtt{cfj}\xrightarrow{0.8}\mathtt{ad}, \mathtt{cdf}\xrightarrow{0.8}\mathtt{e}\}$.
6. This step can be followed in the following table which shows step by step the application of the equivalence rules.

Equivalence	*guide*	Γ'				
	$\mathtt{cf}\xrightarrow{0.8}\mathtt{dg}$	$\underline{\mathtt{acf}\xrightarrow{0.8}\mathtt{de}}$	$\mathtt{cdef}\xrightarrow{0.8}\mathtt{h}$	$\mathtt{chf}\xrightarrow{0.8}\mathtt{b}$	$\mathtt{cfj}\xrightarrow{0.8}\mathtt{ad}$	$\mathtt{cdf}\xrightarrow{0.8}\mathtt{e}$
SiEq	$\mathtt{cf}\xrightarrow{0.8}\mathtt{dg}$	$\mathtt{acf}\xrightarrow{0.8}\mathtt{e}$	$\underline{\mathtt{cdef}\xrightarrow{0.8}\mathtt{h}}$	$\mathtt{chf}\xrightarrow{0.8}\mathtt{b}$	$\mathtt{cfj}\xrightarrow{0.8}\mathtt{ad}$	$\mathtt{cdf}\xrightarrow{0.8}\mathtt{e}$
SiEq	$\mathtt{cf}\xrightarrow{0.8}\mathtt{dg}$	$\mathtt{acf}\xrightarrow{0.8}\mathtt{e}$	$\mathtt{cef}\xrightarrow{0.8}\mathtt{h}$	$\mathtt{chf}\xrightarrow{0.8}\mathtt{b}$	$\underline{\mathtt{cfj}\xrightarrow{0.8}\mathtt{ad}}$	$\mathtt{cdf}\xrightarrow{0.8}\mathtt{e}$
SiEq	$\mathtt{cf}\xrightarrow{0.8}\mathtt{dg}$	$\mathtt{acf}\xrightarrow{0.8}\mathtt{e}$	$\mathtt{cef}\xrightarrow{0.8}\mathtt{h}$	$\mathtt{chf}\xrightarrow{0.8}\mathtt{b}$	$\mathtt{cfj}\xrightarrow{0.8}\mathtt{a}$	$\underline{\mathtt{cdf}\xrightarrow{0.8}\mathtt{e}}$
SiUnEq	$\mathtt{cf}\xrightarrow{0.8}\mathtt{deg}$	$\underline{\mathtt{acf}\xrightarrow{0.8}\mathtt{e}}$	$\mathtt{cef}\xrightarrow{0.8}\mathtt{h}$	$\mathtt{chf}\xrightarrow{0.8}\mathtt{b}$	$\mathtt{cfj}\xrightarrow{0.8}\mathtt{a}$	×
SiAxEq	$\mathtt{cf}\xrightarrow{0.8}\mathtt{deg}$	×	$\underline{\mathtt{cef}\xrightarrow{0.8}\mathtt{h}}$	$\mathtt{chf}\xrightarrow{0.8}\mathtt{b}$	$\mathtt{cfj}\xrightarrow{0.8}\mathtt{a}$	
SiUnEq	$\mathtt{cf}\xrightarrow{0.8}\mathtt{degh}$		×	$\mathtt{chf}\xrightarrow{0.8}\mathtt{b}$	$\mathtt{cfj}\xrightarrow{0.8}\mathtt{a}$	

In the first column we depicted the equivalence applied between the *guide* and the underlined FFD in each row. The result of each equivalence for the underlined FFD is depicted underneath. The second column depicts the *guide* set that is augmented for **SiUnEq**. The application of **SiAxEq** to an FFD remove this FFD (the symbol × is used). And **SiEq** removes redundancy in an FFD.

7. As the *guide* is $\texttt{cf}\xrightarrow{0.8}\texttt{degh}$ and $\{d, g, h\} \subseteq \{c, f, d, e, g, h\}$ then the output is $\Gamma \vdash \texttt{cf}\xrightarrow{0.8}\texttt{dgh}$. (**End of FSL *Automated Prover***)

5 Soundness, Completeness and Complexity

Tarski's fixed-point theorem ensures that the algorithm finishes because the sequence of the sets C_i is strictly growing in $(2^{\Omega}, \subseteq)$. The following results are oriented to prove that, for all $\Gamma \in \{\Gamma^{\theta}_{A}\} \cup \{\Gamma_i \mid 1 \leq i \leq n\}$,

$$\Gamma_0 \vdash A\xrightarrow{\theta}B \qquad \text{if and only if} \qquad \Gamma \vdash A\xrightarrow{\theta}B$$

Lemma 1. *Let Γ be a set of fuzzy functional dependencies, $X \subseteq \Omega$ and $\theta_1, \theta_2 \in [0, 1]$. If $\theta_1 \leq \theta_2$ then ${\Gamma^{\theta_1}_X}^+ \subseteq {\Gamma^{\theta_2}_X}^+$.*

Proof. From **InR**, $\Gamma^{\theta_1}_X \subseteq {\Gamma^{\theta_2}_X}^+$ and therefore ${\Gamma^{\theta_1}_X}^+ \subseteq {\Gamma^{\theta_2}_X}^+$. □

Lemma 2. *Let Γ be a set of FFDs, $U\xrightarrow{\theta}V$ an FFD and X a set of attributes. If $\Gamma \vdash U\xrightarrow{\theta}V$ then $\Gamma^{\theta}_X \vdash XU\xrightarrow{\theta}V$.*

Proof. From Lemma 1, it is proved by induction that all the elements $U\xrightarrow{\theta}V$ belonging to Γ^+ satisfy that $\Gamma^{\theta}_X \vdash XU\xrightarrow{\theta}V$. □

Theorem 2. *Let Γ be a set of fuzzy functional dependencies, $U\xrightarrow{\theta}V$ an FFD and X a non-empty set of attributes.*

$$\Gamma \vdash X\xrightarrow{\theta}Y \text{ if and only if } \Gamma^{\theta}_X \vdash X\xrightarrow{\theta}Y$$

Proof. The direct implication is an immediate consequence of Lemma 2. Conversely, we prove that $\Gamma^{\theta}_X \subseteq \Gamma^+$. If $U\xrightarrow{\theta}V \in \Gamma^{\theta}_X$ then there exists $U'\xrightarrow{\theta_1}V \in \Gamma$ so $\theta \leq \theta_1$ and $U = X \cup U'$. From $U'\xrightarrow{\theta_1}V$ and the axiom $X\xrightarrow{1}V \cap X$, **CoR**, $U\xrightarrow{\theta_1}V$ is obtained and, by **FrR**, $U\xrightarrow{\theta_1}V \in \Gamma^+$. Finally, $\Gamma^{\theta}_X \subseteq \Gamma^+$ implies that ${\Gamma^{\theta}_X}^+ \subseteq \Gamma^+$. □

With this theorem we have proved that Step 2 in the algorithm is sound and complete. Now, we will prove the soundness of Step 3 and the existence of the *guide* cited in Step 5.

Proposition 1. *If $\Gamma^{\theta}_X \vdash U\xrightarrow{\theta}V$ then one of the following conditions holds:*

A. $U\xrightarrow{\theta}V$ is an axiom.
B. There exists $U'\xrightarrow{\theta}V' \in \Gamma^{\theta}_X$ such that $X \subseteq U' \subseteq U$.

Proof. By induction it is proved that all the fuzzy functional dependencies belonging to ${\Gamma^{\theta}_{X}}^{+}$ satisfy at least one of both conditions. □

Consequently, the existence of $X \subseteq \Omega$ such that $\mathrm{A} \xrightarrow{\theta} X \in \Gamma^{\theta}_{\mathrm{A}}$ is a necessary condition for $\Gamma_0 \vdash \mathrm{A} \xrightarrow{\theta} \mathrm{B}$. This section concludes with the proof of the soundness and completeness of the algorithm.

Theorem 3. *The algorithm is sound and complete.*

Proof. Theorem 2 ensures that $\Gamma \vdash \mathrm{A} \xrightarrow{\theta} \mathrm{B}$ if and only if $\Gamma^{\theta}_{\mathrm{A}} \vdash \mathrm{A} \xrightarrow{\theta} \mathrm{B}$. On the other hand, if the algorithm finishes with the set $\Gamma_n = \{\mathrm{A} \xrightarrow{\theta} \mathrm{C}_n\} \cup \Gamma'_n$ then $\Gamma^{\theta}_{\mathrm{A}} \equiv \Gamma_n$ because equivalence rules have been applied. Moreover, if $\mathrm{B} \not\subseteq \mathrm{A} \cup \mathrm{C}_n$ then $(X \cup Y) \cap \mathrm{C}_n = \varnothing$, for all $X \xrightarrow{\theta} Y \in \Gamma'_n$, and, from Proposition 1, $\Gamma_n \not\vdash \mathrm{A} \xrightarrow{\theta} \mathrm{B}$ because the inference rules can not be applied. However, if $\mathrm{B} \subseteq \mathrm{A} \cup \mathrm{C}_n$ then the following sequence proves that $\Gamma_n \vdash \mathrm{A} \xrightarrow{\theta} \mathrm{B}$.

1. $\mathrm{A} \xrightarrow{\theta} \mathrm{C}_n$ by hypothesis.
2. $\mathrm{A} \xrightarrow{1} \mathrm{A}$ by **AxR**.
3. $\mathrm{A} \xrightarrow{\theta} \mathrm{AC}_n$ by **CoR** to 1. and 2.
4. $\mathrm{A} \xrightarrow{\theta} \mathrm{B}$ by **FrR** to 3. □

Regarding complexity results, the cost of Steps 1 to 5 is $O(|\Gamma|)$ in the worst case. Step 6 has $O(|\Omega|\,|\Gamma|)$ cost because, in the worst case, the prover cross Γ and at least one attribute is added to *guide* and removed in the rest of the set in each iteration. Then the number of operations is lower than $|\Omega|\,|\Gamma|$.

As far as we know, in the literature, the algorithms for automatic reasoning about fuzzy functional dependencies are given for logics with the lowest expressiveness. In [14], the authors give an algorithm for automatic reasoning about (classical) functional dependencies. The complexity of this algorithm is $O(|\Omega|\,|\Gamma|)$ and they say that: "*$O(|\Omega|\,|\Gamma|)$ is usually considered as the order of the input. From this point of view, this is a linear time algorithm*". In the literature there also appears other algorithms for classical functional dependencies with the same cost. However, until now, the unique algorithm for a fuzzy extension of functional dependencies, as far as we know, is the one given in [6]. The complexity of this algorithm is the same. However, the fuzzyfication that they consider is from the first type in which, although they consider fuzzy equalities, the functional dependency remains crisp. So, the expressiveness of our logic is higher [11].

6 Conclusions

In [11] we have introduced the Simplification Logic for the management of fuzzy functional dependencies (**FSL** logic) and we have outlined the advantages of it. Specifically, the absence of transitivity as primitive inference rule that is replaced by the simplification rule. Our logic is conceived thinking for removing redundancies and particular cases of the inference rules are also equivalence rules that automatically remove redundancies.

In this paper, we present a sound and complete Automated Prover to answer the question: $\Gamma \vdash A \xrightarrow{\theta} B$? The basic idea of the algorithm is to replace Γ by Γ_A^θ and remove redundancies in this set by applying systematically the equivalence rules. The complexity of the algorithm is the same as the equivalent ones for crisp functional dependencies that appear in the literature.

Short-term work leads us in the direction of applying our fuzzy model into Formal Concept Analysis using **FSL** logic to manipulate the attribute implications.

References

1. Raju, K.V.S.V.N., Majumdar, A.K.: Fuzzy functional dependencies and lossless join decomposition of fuzzy relational database systems. ACM Trans. Database Syst. 13, 129–166 (1988)
2. Yahia, S.B., Ounalli, H., Jaoua, A.: An extension of classical functional dependency: dynamic fuzzy functional dependency. Information Sciences 119, 219–234 (1999)
3. Cubero, J.C., Vila, M.A.: A new definition of fuzzy functional dependency in fuzzy relational databases. International Journal of Intelligent Systems 9, 441–448 (1994)
4. Prade, H., Testemale, C.: Generalizing database realtional algebra for the treatment of incomplete or uncertain information and vague queries. Information Sciences 34, 115–143 (1984)
5. Sözat, M.I., Yazici, A.: A complete axiomatization for fuzzy functional and multivalued dependencies in fuzzy database relations. Fuzzy Sets and Systems 117, 161–181 (2001)
6. Tyagi, B., Sharfuddin, A., Dutta, R., Tayal, D.K.: A complete axiomatization of fuzzy functional dependencies using fuzzy function. Fuzzy Sets and Systems 151, 363–379 (2005)
7. Cordero, P., Enciso, M., Mora, A., de Guzmán, I.P.: Reasoning about fuzzy functional dependencies. In: XIV Spanish Conference on Fuzzy Logic and Technology, pp. 121–126 (2008)
8. Cordero, P., Enciso, M., Mora, A., de Guzmán, I.P.: SLFD logic: Elimination of data redundancy in knowledge representation. In: Garijo, F.J., Riquelme, J.-C., Toro, M. (eds.) IBERAMIA 2002. LNCS (LNAI), vol. 2527, pp. 141–150. Springer, Heidelberg (2002)
9. Aguilera, G., Cordero, P., Enciso, M., Guzmán, I.P., Mora, A.: A non-explosive treatment of functional dependencies using rewriting logic. In: Bazzan, A.L.C., Labidi, S. (eds.) SBIA 2004. LNCS (LNAI), vol. 3171, pp. 31–40. Springer, Heidelberg (2004)
10. Mora, Á., Enciso, M., Cordero, P., de Guzmán, I.P.: An efficient preprocessing transformation for functional dependencies sets based on the substitution paradigm. In: Conejo, R., Urretavizcaya, M., Pérez-de-la-Cruz, J. (eds.) CAEPIA/TTIA 2003. LNCS, vol. 3040, pp. 136–146. Springer, Heidelberg (2004)
11. Cordero, P., Enciso, M., Mora, A., Guzmán, I.P.: A complete logic for fuzzy functional dependencies over domains with similarity relations. In: Proceedings of the 10th International Work-Conference on Artificial Neural Networks: Part I: Bio-Inspired Systems: Computational and Ambient Intelligence IWANN 2009, pp. 261–269 (2009)

12. Cordero, P., Enciso, M., Mora, A., de Guzmán, I.P.: A complete logic for fuzzy functional dependencies over t-norms. In: XV Spanish Conference on Fuzzy Logic and Technology, pp. 205–210 (2010)
13. Belohlavek, R., Vychodil, V.: Codd's relational model of data and fuzzy logic: Comparisons, observations, and some new results. In: International Conference on Computational Intelligence for Modelling, Control and Automation, 2006 and International Conference on Intelligent Agents, Web Technologies and Internet Commerce, p. 70 (2006)
14. Paredaens, J., De Bra, P., Gyssens, M., Van Gucht, D.: The structure of the relational database model. Springer, New York (1989)

A Sound Semantics for a Similarity-Based Logic Programming Language*

Pascual Julián-Iranzo and Clemente Rubio-Manzano

Department of Information Technologies and Systems,
University of Castilla-La Mancha
{Pascual.Julian,Clemente.Rubio}@uclm.es

Abstract. Bousi∼Prolog is an extension of the standard Prolog language aiming at to make more flexible the query answering process and to deal with vagueness applying declarative techniques. In this paper we precise a model-theoretic semantics for a pure subset of this language. Next, for the first time, we define a notion of correct answer which provides a declarative description of the output of a program and a goal in the context of a similarity relation. Afterwards, we recall both the WSLD-resolution principle and a similarity-based unification algorithm which is the basis of its operational mechanism and then we prove the soundness of WSLD-resolution.

Keywords: Fuzzy Logic Programming, Fuzzy Herbrand Model, Weak Unification, Weak SLD-Resolution, Proximity/Similarity Relations.

1 Introduction

In recent years there has been a renewed interest in amalgamating logic programming with concepts coming from *Fuzzy Logic* or akin to this field. As tokens of this interest we mention the works on *Fuzzy Logic Programming* [3,7,11], *Qualified Logic Programming* [8,1] (which is a derivation of the van Emden's *Quantitative Logic Programming* [10]) or *Similarity-Based Logic Programming* [2,9]. Bousi∼Prolog is a representative of the last class of fuzzy logic programming languages. It replaces the syntactic unification mechanism of the classical **S**election-function driven **L**inear resolution for **D**efinite clauses (SLD–resolution) by a fuzzy unification algorithm based on fuzzy binary relations on a syntactic domain. The result is an operational mechanism, called Weak SLD-resolution, which differs in some aspects w.r.t. the one of [9], based exclusively on similarity relations.

This work can be seen as a continuation of the investigation started in [4]. In this paper after introducing some refinements to the model-theoretic and fix-point semantics of Bousi∼Prolog defined in [4] for definite programs, we introduce for the first time in our framework the concept of a *correct answer*

* This work has been partially supported by FEDER and the Spanish Science and Innovation Ministry under grants TIN2007-65749 and TIN2011-25846 and by the Castilla-La Mancha Regional Administration under grant PII1I09-0117-4481.

J. Cabestany, I. Rojas, and G. Joya (Eds.): IWANN 2011, Part II, LNCS 6692, pp. 421–428, 2011.

providing a declarative description for the output of a program and a goal. It is noteworthy that, although the refinements introduced in the declarative semantics do not dramatically alter the original definitions, given in [4], they are important in order to establish the soundness of our proposal. Afterwards, we recall the operational semantics of Bousi~Prolog and we prove, among other results, its soundness. The soundness theorem is established following a proof strategy comparable with the one appeared in [5]. It is important to remark that, the soundness in our framework will be proven under certain conditions. To be precise, we only consider programs without negation and we restrict ourselves to similarity relations on syntactic domains. It is worthy to say that, along this paper we also clarify some of the existing differences between our framework and the related proposal introduced by [9]. Finally, note that, an extended version is available at the URL `http://dectau.uclm.es/iwann2011/soundness.pdf`, where you can find missed proofs and a more accurate information on the sections of this paper, that had to be dismissed by lack of space.

2 Preliminaries

Fuzzy Relations, Proximity and Similarity Relations. A *binary fuzzy relation* on a set U is a fuzzy subset on $U \times U$ (that is, a mapping $U \times U \longrightarrow [0,1]$). There are some important properties that fuzzy relations may have: i) (Reflexivity) $\mathcal{R}(x,x) = 1$ for any $x \in U$; i) (Symmetry) $\mathcal{R}(x,y) = \mathcal{R}(y,x)$ for any $x, y \in U$; i) (Transitivity) $\mathcal{R}(x,z) \geq \mathcal{R}(x,y) \wedge \mathcal{R}(y,z)$ for any $x, y, z \in U$; where the operator '$\wedge$' is the minimum t-norm. A *proximity relation* is a binary fuzzy relation which is reflexive and symmetric. A proximity relation is characterized by a set $\Lambda = \{\lambda_1, ..., \lambda_n\}$ of approximation levels. We say that a value $\lambda \in \Lambda$ is a *cut value*. A special, and well-known, kind of proximity relations are *similarity* relations, which are nothing but transitive proximity relations.

In classical logic programming different syntactic symbols represent distinct information. Following [9], this restriction can be relaxed by introducing a proximity or similarity relation $\mathcal{R}$ on the alphabet of a first order language. This makes possible to treat as indistinguishable two syntactic symbols which are related by the proximity or similarity relation $\mathcal{R}$ with a certain degree greater than zero. A similarity relation $\mathcal{R}$ on the alphabet of a first order language can be extended to terms and formulas by structural induction in the usual way. See [9] for a precise characterization of this extension.

Interpretations and Truth in the Context of a Proximity Relation. In this section we discuss the notions of interpretation and truth for a first order theory in the context of a proximity relation. A *fuzzy Interpretation* $\mathcal{I}$ of a first order language $\mathcal{L}$ is a pair $\langle \mathcal{D}, \mathcal{J} \rangle$ where $\mathcal{D}$ is the domain of the interpretation and $\mathcal{J}$ is a mapping which assigns meaning to the symbols of $\mathcal{L}$: specifically n-ary relation symbols are interpreted as mappings $\mathcal{D}^n \longrightarrow [0,1]$. In order to evaluate open formulas we need to introduce the notion of variable assignment. A *variable assignment*, ϑ, w.r.t. $\mathcal{I} = \langle \mathcal{D}_\mathcal{I}, \mathcal{J} \rangle$, is a mapping $\vartheta : \mathcal{V} \longrightarrow \mathcal{D}_\mathcal{I}$, which can be extended to the set of the terms of $\mathcal{L}$ by structural induction. Given a fuzzy interpretation $\mathcal{I} = \langle \mathcal{D}, \mathcal{J} \rangle$ and a variable assignment ϑ in $\mathcal{I}$, the *valuation* of a formula w.r.t. $\mathcal{I}$ and ϑ is[1]:

$$\mathcal{I}(p(t_1, \dots, t_n))[\vartheta] = \bar{p}(t_1\vartheta, \dots, t_n\vartheta), \text{ where } \mathcal{J}(p) = \bar{p}$$

$$\mathcal{I}(\mathcal{A} \wedge \mathcal{B}))[\vartheta] = \inf\{\mathcal{I}(\mathcal{A})[\vartheta], \mathcal{I}(\mathcal{B})[\vartheta]\}$$

$$\mathcal{I}(\mathcal{A} \leftarrow \mathcal{B})[\vartheta] = \text{ if } \mathcal{I}(\mathcal{Q})[\vartheta] \leq \mathcal{I}(\mathcal{A})[\vartheta] \text{ then } 1 \text{ else } \mathcal{I}(\mathcal{A})[\vartheta].$$

$$\mathcal{I}((\forall x)\mathcal{A})[\vartheta] = \inf\{\mathcal{I}(\mathcal{A})[\vartheta'] \mid \vartheta' \;\; x\text{–equivalent to } \;\vartheta\}$$

where p is a predicate symbol, and $\mathcal{A}$ and $\mathcal{B}$ are formulas. An assignment ϑ' is x–equivalent to ϑ when $z\vartheta' = z\vartheta$ for all variable $z \neq x$ in $\mathcal{V}$. When the assignment would not be relevant, we shall omit it during the valuation of a formula. In the context of a first order theory equipped with a proximity relation $\mathcal{R}$, characterized by a set $\Lambda = \{\lambda_1, ..., \lambda_n\}$ of approximation levels, it makes sense that the notion of truth be linked to a certain approximation level $\lambda \in \Lambda$. For a fixed value λ and a formula $\mathcal{A}$ of $\mathcal{L}$:

$\mathcal{A}$ is λ-true in $\mathcal{I}$ iff for every assignment ϑ in $\mathcal{I}$, $\mathcal{I}(\mathcal{A})[\vartheta] \geq \lambda$
$\mathcal{A}$ is λ-false in $\mathcal{I}$ iff for every assignment ϑ in $\mathcal{I}$, $\mathcal{I}(\mathcal{A})[\vartheta] < \lambda$
$\mathcal{A}$ is λ-valid iff $\mathcal{A}$ is λ-true for all interpretation $\mathcal{I}$.
$\mathcal{A}$ is λ-unsatisfiable iff $\mathcal{A}$ is λ-false for all $\mathcal{I}$.
$\mathcal{A}$ is λ-satisfiable iff there exists an $\mathcal{I}$ and a ϑ in $\mathcal{I}$ such that $\mathcal{I}(\mathcal{A})[\vartheta] \geq \lambda$.

Intuitively, a cut value λ is delimiting truth degrees equal or greater than λ as true. Since the valuation of a closed formula is completely determined by an interpretation, independently of a variable assignment, we say that an interpretation $\mathcal{I}$ of $\mathcal{L}$ is λ-*model* for $\mathcal{A}$ if and only if $\mathcal{I}(\mathcal{A}) \geq \lambda$.

Closed Conditional Formulas and Models. In this section we elucidate the notion of model for a set of closed conditional formulas in the context of a similarity relation. By conditional formula we mean a formula of the form $\mathcal{C} \equiv A \leftarrow \mathcal{Q}$, where A (called the head) is an atom, $\mathcal{Q}$ a formula (called the body) and all variables are assumed universally quantified. When $\mathcal{Q} \equiv B_1 \wedge \ldots \wedge B_n$ is a conjunction of atoms, the formula $\mathcal{C}$ is called a *Horn clause* or *definite clause.* As it is well known, this kind of formulas play a special role in logic programming where a set of definite clauses is called a *program* and a *goal* is any conjunctive body. A direct naive translation to our context of the classical concept of model for a set of formulas does not work. We need a new definition supported by the notion of what we called an annotated set of formulas of level λ. We want to formalize this concept[2], but before doing that we need some technical definitions introduced to cope with some problems that appear when conditional formulas have non-linear atoms on their heads[3]. Given a non-linear

[1] Note that, $t_i\vartheta$ is equivalent to $\vartheta(t_i)$.
[2] See [4] to obtain more intuitive insides on this idea.
[3] The apparition of this problem in our framework was pointed out by R. Caballero, M. Rodríguez and C. Romero in a private communication. So we want to express them our gratitude.

atom A, the linearization of A (as defined in [1]) is a process by which it is computed the structure $\langle A_l, C_l \rangle$, where: A_l is a linear atom built from A by replacing each one of the n multiple occurrences of the same variable X_i by new fresh variables $Y_k (1 \leq k \leq n_i)$; and S_l is a set of proximity constrains $X_i \sim Y_k$ (with $1 \leq k \leq n_i$). The operator "$s \sim t$" is asserting the proximity of two terms s and t and when interpreted, $\mathcal{I}(s \sim t) = \mathcal{R}(s,t)$, whatever the interpretation $\mathcal{I}$ of $\mathcal{L}$. Now, let $\mathcal{C} \equiv A \leftarrow \mathcal{Q}$ be a conditional formula and $S_l = \{X_1 \sim Y_1, \ldots, X_n \sim Y_n\}$, $lin(\mathcal{C}) = A_l \leftarrow X_1 \sim Y_1 \wedge \ldots \wedge X_n \sim Y_n \wedge \mathcal{Q}$. For a set Γ of conditional formulas, $lin(\Gamma) = \{lin(\mathcal{C}) \mid \mathcal{C} \in \Gamma\}$. The following algorithm, which is a reformulation of the one that appears in [4] to cope with the linearization process, gives a precise procedure for the construction of the set of annotated formulas of level λ.

Algorithm 1

Input: A set of conditional formulas Γ and a proximity relation $\mathcal{R}$ with a set of levels Λ and a cut value $\lambda \in \Lambda$.
Output: A set Γ^λ of annotated formulas of level λ.
Initialization: $\Gamma_l := lin(\Gamma)$ and $\Gamma^\lambda := \{\langle \mathcal{C}, 1 \rangle \mid \mathcal{C} \in \Gamma_l\}$
For each conditional formula $\mathcal{C} \equiv A \leftarrow \mathcal{Q} \in \Gamma_l$ do
- $\mathcal{K}_\lambda(\mathcal{C}) = \{\langle \mathcal{C}' \equiv A' \leftarrow \mathcal{Q}, \alpha \rangle \mid \mathcal{R}(A, A') = \alpha \geq \lambda\}$
- *For each element $\langle \mathcal{C}', \alpha \rangle$ in $\mathcal{K}_\lambda(\mathcal{C})$ do:*
  If $\langle \mathcal{C}', L \rangle \in \Gamma^\lambda$ then $\Gamma^\lambda = (\Gamma^\lambda \setminus \{\langle \mathcal{C}', L \rangle\}) \cup \langle \mathcal{C}', L \wedge \alpha \rangle$
  else $\Gamma^\lambda = (\Gamma^\lambda \cup \{\langle \mathcal{C}', \alpha \rangle\})$

Return Γ^λ

The general idea behind this algorithm is to start annotating each formula in the set Γ_l with a truth degree equal to 1. On the other hand, the rest of the formulas generated by proximity, starting from formulas of the original set Γ_l, are annotated with its corresponding approximation degree (regarding the original formula). Afterward, if several formulas of the set generate the same approximate formula, with different approximations degrees, we take the least degree as annotation. Now we are ready to define the core concepts of a model for a set of closed conditional formulas and logical consequence of level λ w.r.t. a proximity relation. Let Γ be a set of closed conditional formulas of a first order language $\mathcal{L}$, $\mathcal{R}$ be a proximity relation which is characterised by a set Λ of approximation levels with cut value $\lambda \in \Lambda$ and $\mathcal{I}$ be a fuzzy interpretation of $\mathcal{L}$.

1) $\mathcal{I}$ is λ-model for $\{\Gamma, \mathcal{R}\}$ iff for all annotated formula $\langle \mathcal{A}, \lambda' \rangle \in \Gamma^\lambda$, $\mathcal{I}(\mathcal{A}) \geq \lambda'$;

2) $\mathcal{A}$ is a λ-logical consequence of $\{\Gamma, \mathcal{R}\}$ if and only if for each fuzzy interpretation $\mathcal{I}$ of $\mathcal{L}$, $\mathcal{I}$ is a λ-model for $\{\Gamma, \mathcal{R}\}$ implies that $\mathcal{I}$ is a λ-model for $\mathcal{A}$.

3 Declarative Semantics

In this section we recall the declarative semantics of Bousi~Prolog. Roughly speaking, Bousi~Prolog programs are sequences of (normal) clauses plus a proximity relation. However, in this and the following sections we restrict ourselves to definite clauses.

Fuzzy Herbrand Interpretations and Models. Herbrand interpretations are defined on a syntactic domain, called the Herbrand universe. For a first order language $\mathcal{L}$, the *Herbrand universe* $\mathcal{U}_{\mathcal{L}}$ for $\mathcal{L}$, is the set of all ground terms in $\mathcal{L}$. On the order hand, the *Herbrand base* $\mathcal{B}_{\mathcal{L}}$ for $\mathcal{L}$ is the set of all ground atoms which can be formed by using the predicate symbols of $\mathcal{L}$ jointly with the ground terms from the Herbrand universe taken as arguments. As in the classical case, it is possible to identify a Herbrand interpretation with a fuzzy subset of the Herbrand base. That is, a *fuzzy Herbrand interpretation* for $\mathcal{L}$ can be considered as a mapping $\mathcal{I} : \mathcal{B}_{\mathcal{L}} \longrightarrow [0,1]$. The ordering $\leq$ in the lattice $[0,1]$ can be easily extended to the set of Herbrand interpretations $\mathcal{H}$ (see, for instance [7]). It is important to note that the pair $\langle \mathcal{H}, \sqsubseteq \rangle$ is a complete lattice.

In the following, we focus our attention on Herbrand λ-models. For this special kind of λ-models we proved in [4] an analogous property to the *model intersection property* and we defined the *least Herbrand model of level* λ, for a program Π and a proximity relation $\mathcal{R}$, as the mapping $\mathcal{M}_{\Pi}^{\lambda} : \mathcal{B}_{\mathcal{L}} \longrightarrow [0,1]$ such that, $\mathcal{M}_{\Pi}^{\lambda}(A) = \inf\{\mathcal{I}(A) \mid \mathcal{I}$ is a λ-model for Π and $\mathcal{R}\}$, for each $A \in \mathcal{B}_{\mathcal{L}}$. The interpretation $\mathcal{M}_{\Pi}^{\lambda}$ is the natural interpretation for a program Π and a proximity relation $\mathcal{R}$, since, as it was proved in [4], for each $A \in \mathcal{B}_{\mathcal{L}}$ such that $\mathcal{M}_{\Pi}^{\lambda}(A) \neq 0$, A is a logical consequence of level λ for Π and $\mathcal{R}$.

Fixpoint Semantics. Let Π be a definite program and $\mathcal{R}$ be a proximity relation. We define the immediate consequences operator of level λ, $\mathcal{T}_{\Pi}^{\lambda}$, as a mapping $\mathcal{T}_{\Pi}^{\lambda} : \mathcal{H} \longrightarrow \mathcal{H}$ such that, for all $A \in \mathcal{B}_{\mathcal{L}}$, $\mathcal{T}_{\Pi}^{\lambda}(\mathcal{I})(A) = \inf\{\mathcal{PT}_{\Pi}^{\lambda}(\mathcal{I})(A)\}$, where $\mathcal{PT}_{\Pi}^{\lambda}$ is a non deterministic operator such that $\mathcal{PT}_{\Pi}^{\lambda}(\mathcal{I}) : \mathcal{B}_{\mathcal{L}} \longrightarrow \wp([0,1])$ and it is defined as follows: Let $\Pi_l = lin(\Pi)$,

1. For each fact $H \in \Pi_l$, let $K_{\lambda}(H) = \{\langle H', \lambda' \rangle \mid \mathcal{R}(H, H') = \lambda' \geq \lambda\}$ be the set of approximate atoms of level λ for H. Then $\mathcal{PT}_{\Pi}^{\lambda}(\mathcal{I})(H'\vartheta) \ni \lambda'$, for all H' and assignment ϑ.
2. For each clause $\mathcal{C} \equiv (A \leftarrow \mathcal{Q}) \in \Pi_l$. Let $K_{\lambda}(\mathcal{C}) = \{\langle \mathcal{C}' \equiv A' \leftarrow \mathcal{Q}, \lambda' \rangle \mid \mathcal{R}(A, A') = \lambda' \geq \lambda\}$ be the set of approximate clauses of level λ for $\mathcal{C}$. Then $\mathcal{PT}_{\Pi}^{\lambda}(\mathcal{I})(A'\vartheta) \ni \lambda' \wedge \mathcal{I}(\mathcal{Q}'\vartheta)$, for all $\mathcal{C}'$ and assignment ϑ.

In [4], we proved that the immediate consequences operator (of level λ) is monotonous and continuous and the least fuzzy Herbrand model (of level λ) coincides with its least fixpoint.

Correct Answer. In this section we define for the first time the concept of a correct answer, which provides a declarative description of the desired output from a program, a proximity relation, and a goal. This is a central concept for the later theoretical developments.

Definition 1 (Correct Answer of level λ). *Let Π be a definite program and $\mathcal{R}$ be a proximity relation, which is characterised by a set Λ of approximation levels with cut value $\lambda \in \Lambda$. Let $\mathcal{G} \equiv \leftarrow A_1, ..., A_k$ be a goal. We say that $\langle \theta, \beta \rangle$ is a correct answer of level λ for $\{\Pi, \mathcal{R}\}$ and $\mathcal{G}$ if: i) $\forall(A_1, ..., A_k)\theta$ is a λ-logical consequence of $\{\Pi, \mathcal{R}\}$;ii) $\mathcal{M}_{\Pi}^{\lambda}(\forall(A_1, ..., A_k)\theta) \leq \beta$.*

4 Operational Semantics

Weak Unification Based on Similarity Relations. Bousi∼Prolog uses a weak unification algorithm that, when we work with similarity relations, coincides with the one defined by M. Sessa [9]. However, there exists some remarkable differences between our proposal and Sessa's proposal that we shall try to put in evidence along this section. In presence of similarity relations on syntactic domains, it is possible to define an extended notion of a unifier and a more general unifier of two expressions[4].

Definition 2. *Let $\mathcal{R}$ be a similarity relation, λ be a cut value and $\mathcal{E}_1$ and $\mathcal{E}_2$ be two expressions. The substitution θ is a* weak unifier *of level λ for $\mathcal{E}_1$ and $\mathcal{E}_2$ w.r.t $\mathcal{R}$ (or λ-unifier) if its* unification degree, *$\mathcal{D}eg_{\mathcal{R}}(\mathcal{E}_1\theta, \mathcal{E}_2\theta)$, defined as $\mathcal{D}eg_{\mathcal{R}}(\mathcal{E}_1\theta, \mathcal{E}_2\theta) = \mathcal{R}(\mathcal{E}_1\theta, \mathcal{E}_2\theta)$, is greater than λ.*

Note that in Sessa's proposal the idea of "cut value" is missed. Also in order that a substitution θ be a weak unifier for $\mathcal{E}_1$ and $\mathcal{E}_2$ she put a strong constrain: the unification degree of $\mathcal{E}_1$ and $\mathcal{E}_2$ w.r.t. θ must be the maximum of the unification degrees of $\mathcal{D}eg_{\mathcal{R}}(\mathcal{E}_1\varphi, \mathcal{E}_2\varphi)$ for whatever substitution φ. Therefore, some substitution that we consider as a weak unifier, are disregarded by her proposal.

Definition 3. *Let $\mathcal{R}$ be a similarity relation and λ be a cut value. The substitution θ is* more general *than the substitution σ* with level λ, *denoted by $\theta \leq_{\mathcal{R},\lambda} \sigma$, if there exists a substitution δ such that, for any variable x in the domain of θ or σ, $\mathcal{R}(x\sigma, x\theta\delta) \geq \lambda$.*

Definition 4. *Let $\mathcal{R}$ be a similarity relation and $\mathcal{E}_1$ and $\mathcal{E}_2$ be two expressions. The substitution θ is a* weak most general unifier *(w.m.g.u.) of $\mathcal{E}_1$ and $\mathcal{E}_2$ w.r.t $\mathcal{R}$, denoted by $wmgu(\mathcal{E}_1, \mathcal{E}_2)$, if:*

1. *θ is a λ-unifier of $\mathcal{E}_1$ and $\mathcal{E}_2$; and*
2. *$\theta \leq_{\mathcal{R},\lambda} \sigma$, for any λ-unifier σ of $\mathcal{E}_1$ and $\mathcal{E}_2$.*

The weak unification algorithm we are using is a reformulation of the one appeared in [9], which, in turn, is an extension of Martelli and Montanari's unification algorithm for syntactic unification [6]. The main difference is regarding the so called *decomposition rule*[5]: Given the unification problem $\langle\{f(t_1, \ldots, t_n) \approx g(s_1, \ldots, s_n)\} \cup E, \sigma, \alpha\rangle$, if $\mathcal{R}(f, g) = \beta > \lambda$, it is not a failure but it is equivalent to solve the new configuration $\langle\{t_1 \approx s_1, \ldots, t_n \approx s_n\} \cup E, \sigma, \alpha \wedge \beta\rangle$, where the approximation degree α has been compounded with the degree β. It is important to note that, differently to [9], the resulting approximation degree is casted by a cut value λ. The weak unification algorithm allows us to check if a set of expressions $S = \{\mathcal{E}_1 \approx \mathcal{E}'_1, \ldots, \mathcal{E}_n \approx \mathcal{E}'_n\}$ is weakly unifiable. The w.m.g.u. of the set S is denoted by $wmgu(S)$. In general, a w.m.g.u. of two expressions $\mathcal{E}_1$ and $\mathcal{E}_2$ is not unique [9]. Therefore, the weak unification algorithm computes a representative of a w.m.g.u. class.

---

[4] We mean by "expression" a first order term or an atomic formula.

[5] Here, the symbol "$\mathcal{E}_1 \approx \mathcal{E}_2$" represents the potential possibility that two expressions $\mathcal{E}_1$ and $\mathcal{E}_2$ be close.

Weak SLD-Resolution. Let Π be a definite program, $\mathcal{R}$ be a similarity relation and λ be a cut value. A *Weak SLD* (WSLD) *resolution step of level* λ is dedined by the inference rule:

$$\frac{\mathcal{C} = (\mathcal{A} \leftarrow \mathcal{Q}) \ll \Pi, \sigma = wmgu(\mathcal{A},\mathcal{A}') \neq fail,\ \beta = \mathcal{R}(\mathcal{A}\sigma,\mathcal{A}'\sigma)) \geq \lambda}{\leftarrow \mathcal{A}',\mathcal{Q}' \overset{[\mathcal{C},\sigma,\beta]}{\Longrightarrow}_{\text{WSLD}} \leftarrow (\mathcal{Q}, \mathcal{Q}')\sigma}$$

where $\mathcal{Q}$, $\mathcal{Q}'$ are conjunctions of atoms, the notation "$\mathcal{C} \ll \Pi$" is representing that $\mathcal{C}$ is a standardized apart clause in Π. A WSLD *derivation of level* λ for $\Pi \cup \{\mathcal{G}_0\}$ and $\mathcal{R}$ is a sequence of steps of level λ: $\mathcal{G}_0 \overset{[\mathcal{C}_1,\theta_1,\beta_1]}{\Longrightarrow}_{\text{WSLD}} \cdots \overset{[\mathcal{C}_n,\theta_n,\beta_n]}{\Longrightarrow}_{\text{WSLD}} \mathcal{G}_n$. That is, each $\beta_i \geq \lambda$. And a WSLD *refutation of level* λ for $\Pi \cup \{\mathcal{G}_0\}$ and $\mathcal{R}$ is a WSLD derivation of level λ for $\Pi \cup \{\mathcal{G}_0\}$ and $\mathcal{R}$: $\mathcal{G}_0 \overset{\theta,\beta}{\Longrightarrow}_{\text{WSLD}}{}^* \Box$, where the symbol "$\Box$" stands for the empty clause, $\theta = \theta_1\theta_2\ldots\theta_n$ is the *computed substitution* and $\beta = \bigwedge_{i=1}^{n} \beta_i$ is its *approximation degree*. The output of a WSLD refutation is the pair $\langle \theta_{\restriction(\mathcal{V}ar(\mathcal{G}))}, \beta \rangle$, which is said to be the *computed answer*. Certainly, a WSLD refutation computes a family of answers, in the sense that, if $\theta = \{x_1/t_1, \ldots, x_n/t_n\}$ then, by definition, whatever substitution $\theta' = \{x_1/s_1, \ldots, x_n/s_n\}$, holding that $\mathcal{R}(s_i, t_i) \geq \lambda$, for any $1 \leq i \leq n$, is also a computed substitution with approximation degree $\beta \wedge (\bigwedge_1^n \mathcal{R}(s_i, t_i))$. Observe that our definition of proximity based SLD resolution is parameterized by a cut value $\lambda \in \Lambda$. This introduces an important conceptual distinction between our approach and the similarity based SLD resolution presented in [9] (see [4] for details).

5 Soundness of WSLD-Resolution

In this section we establish the soundness of WSLD-Resolution, but before proving the main result of the paper we need to introduce some important intermediate lemmas.

Lemma 1. *Let Π be a definite program, $\mathcal{R}$ be a proximity relation and λ be a cut value. Given $(\mathcal{A} \leftarrow \mathcal{Q}) \in \Pi$ and $\mathcal{A}'$ an atom such that $\mathcal{R}(\mathcal{A}, \mathcal{A}') = \alpha \geq \lambda$. If $(\forall \mathcal{Q})$ is a λ-logical consequence of $\{\Pi, \mathcal{R}\}$ then $(\forall \mathcal{A}')$ is a λ-logical consequence of $\{\Pi, \mathcal{R}\}$.*

Lemma 2. *Let $\mathcal{A}$ and $\mathcal{B}$ be two atoms such that $\mathcal{A} \leq \mathcal{B}$. Then, $\mathcal{I}(\forall \mathcal{A}) \leq \mathcal{I}(\forall \mathcal{B})$.*

Lemma 3. *Let $\mathcal{A}$ and $\mathcal{B}$ be two atoms, $\mathcal{R}$ be a proximity relation, with cut level λ, and θ be a λ-unifier for $\mathcal{A}$ and $\mathcal{B}$ with degree α. Then, there exists an atom $\mathcal{A}'$ such that, $\mathcal{R}(\mathcal{A}, \mathcal{A}') = \alpha$ and $\mathcal{A}'\theta = \mathcal{B}\theta$ (That is there exists $\mathcal{A}'$ which is close to $\mathcal{A}$, with degree α which unifies syntactically with $\mathcal{B}$, through the unifier θ).*

Theorem 1 (Soundness of the WSLD–Resolution). *Let Π be a definite program, $\mathcal{R}$ a similarity relation, λ a cut value and $\mathcal{G}$ a definite goal. Then every computed answer $\langle \theta, \beta \rangle$ of level λ for $\{\Pi, \mathcal{R}\}$ and $\mathcal{G}$ is a correct answer of level λ for $\{\Pi, \mathcal{R}\}$ and $\mathcal{G}$.*

6 Conclusions and Future Work

In this paper we revisited the declarative semantics of Bousi∼Prolog which were defined for a pure subset of this language and presented in [4]. We have given more accurate definitions for the semantic concepts and thereby solved some problems that may arise when we work with non-linear programs. Moreover, we introduce for the first time a notion of correct answer inside our framework. Then, after recalling both the WSLD-resolution principle and a similarity-based unification algorithm, which is the basis of the Bousi∼Prolog operational mechanism for definite programs, we prove the soundness of WSLD-resolution as well as other auxiliary results. Finally, it is worthy to say that, along this paper we have clarified some of the existing differences between our framework and the related proposal introduced by [9].

As a matter of future work we want to continue proving the completeness theorem for this restricted subset of Bousi∼Prolog. On the other hand, at the present time we know that a naive extension of Sessa's unification algorithm to proximity relations does not work, because correctness problems may arise. Therefore, it is necessary to define a complete new algorithm able to deal with proximity relations and to lift some of the current results to the new framework.

References

1. Caballero, R., Rodríguez, M., Romero, C.A.: Similarity-based reasoning in qualified logic programming. In: Proc. PPDP 2008, pp. 185–194. ACM, New York (2008)
2. Arcelli, F., Formato, F.: A similarity-based resolution rule. Int. J. Intell. Syst. 17(9), 853–872 (2002)
3. Guadarrama, S., Muñoz, S., Vaucheret, C.: Fuzzy Prolog: A new approach using soft constraints propagation. Fuzzy Sets and Systems 144(1), 127–150 (2004)
4. Julián, P., Rubio, C.: A declarative semantics for Bousi∼Prolog. In: PPDP, pp. 149–160. ACM, New York (2009)
5. Lloyd, J.W.: Foundations of Logic Programming. Springer, Berlin (1987)
6. Martelli, A., Montanari, U.: An Efficient Unification Algorithm. ACM Transactions on Programming Languages and Systems 4, 258–282 (1982)
7. Medina, J., Ojeda, M., Vojtáš, P.: Similarity-based unification: a multi-adjoint approach. Fuzzy Sets and Systems 146(1), 43–62 (2004)
8. Rodríguez, M., Romero, C.A.: Quantitative logic programming revisited. In: Garrigue, J., Hermenegildo, M.V. (eds.) FLOPS 2008. LNCS, vol. 4989, pp. 272–288. Springer, Heidelberg (2008)
9. Sessa, M.I.: Approximate reasoning by similarity-based sld resolution. Theoretical Computer Science 275(1-2), 389–426 (2002)
10. van Emden, M.H.: Quantitative deduction and its fixpoint theory. Journal of Logic Programming 3(1), 37–53 (1986)
11. Vojtáš, P.: Fuzzy Logic Programming. Fuzzy Sets and Systems 124(1), 361–370 (2001)

A Static Preprocess for Improving Fuzzy Thresholded Tabulation*

P. Julián[1], J. Medina[2], P.J. Morcillo[3], G. Moreno[3], and M. Ojeda-Aciego[4]

[1] Dept. of Information Technologies and Systems, University of Castilla-La Mancha
pascual.julian@uclm.es
[2] Department of Mathematics, University of Cadiz
jesus.medina@uca.es
[3] Department of Computing Systems, University of Castilla-La Mancha
gines.moreno@uclm.es, pmorcillo@dsi.uclm.es
[4] Department of Applied Mathematics, University of Málaga
aciego@uma.es

Abstract. Tabulation has been widely used in most (crisp) declarative paradigms for efficiently running programs without the redundant evaluation of goals. More recently, we have reinforced the original method in a fuzzy setting, by the dynamic generation of thresholds which avoid many useless computations leading to insignificant solutions. In this paper, we draw a static technique for generating such filters without requiring the consumption of extra computational resources at execution time.

Keywords: Fuzzy Logic Programming, Tabulation, Thresholding, Unfolding.

1 Introduction

Fuzzy logic programming represents a flexible and powerful declarative paradigm amalgamating fuzzy logic and logic programming, for which there exists different promising approaches described in the literature [5,9,2,11]. One step beyond of [6], in this work we refine an improved fuzzy query answering procedure for the so-called MALP (*Multi-Adjoint Logic Programming*) approach [10,11], which avoids the re-evaluation of goals and the generation of useless computations thanks to the combined use of tabulation [13,4] with thresholding techniques [8], respectively. As shown in Section 2, the general idea is that, when trying to perform a computation step by using a given program rule R, we firstly analyze if such step might contribute to reach further significant solutions (not tabulated - saved, stored- yet). When it is the case, it is possible to avoid useless computation steps via rule R by using thresholds/filters based on the truth degree of R, as well as a safe, accurate and dynamic estimation of the maximum truth degree

* Work supported by the Spanish MICINN projects TIN2009-14562-C05-01, TIN2009-14562-C05-03, TIN 2007-65749 and TIN2011-25846, and by the Andalucía and Castilla-La Mancha Administrations under grants P09-FQM-5233 and PII1I09-0117-4481.

J. Cabestany, I. Rojas, and G. Joya (Eds.): IWANN 2011, Part II, LNCS 6692, pp. 429–436, 2011.

associated to its body. Moreover, in Section 3, we propose too a static preprocess with links to well-known unfolding techniques [3,14,1,7] in order to build and manage a powerful kind of filters which largely enhances the benefits achieved by thresholding when combined with fuzzy tabulation.

The MALP approach (see the original formal definition in [10,11] a real implementation in [12]) considers a language, $\mathcal{L}$, containing propositional variables, constants, and a set of logical connectives. In our fuzzy setting, we use implication connectives ($\leftarrow_1, \leftarrow_2, \ldots, \leftarrow_m$) together with a number of aggregators. They will be used to combine/propagate truth values through the rules. The general definition of aggregation operators subsumes conjunctive operators (denoted by $\&_1, \&_2, \ldots, \&_k$), disjunctive operators ($\vee_1, \vee_2, \ldots, \vee_l$), and average and hybrid operators (usually denoted by $@_1, @_2, \ldots, @_n$). Aggregators are useful to describe/specify user preferences: when interpreted as a truth function they may be considered, for instance, as an arithmetic mean or a weighted sum. The language $\mathcal{L}$ will be interpreted on a *multi-adjoint lattice*, $\langle L, \preceq, \leftarrow_1, \&_1, \ldots, \leftarrow_n, \&_n \rangle$, which is a complete lattice equipped with a collection of adjoint pairs $\langle \leftarrow_i, \&_i \rangle$, where each $\&_i$ is a conjunctor[1] intended to provide a *modus ponens*-rule w.r.t. $\leftarrow_i$. In general, the set of truth values L may be the carrier of any complete bounded lattice but, for simplicity, in the examples of this work we shall select L as the set of real numbers in the interval $[0, 1]$.

A *rule* is a formula $A \leftarrow_i \mathcal{B}$, where the *head* A is an propositional symbol and the *body* $\mathcal{B}$ is a formula built from propositional symbols $B_1, \ldots, B_n$ ($n \geq 0$), truth values of L and conjunctions, disjunctions and aggregations. Rules with an empty body are called *facts*. A *goal* is a body submitted as a query to the system. Roughly speaking, a MALP program is a set of pairs $\langle R; \alpha \rangle$, where R is a rule and α is a value of L, which might express the confidence which the user of the system has in the truth of the rule R (note that the truth degrees in a given program are expected to be assigned by an expert). In contrast with the fuzzy extension of SLD-resolution described for MALP programs in [10,11], in what follows we recast from [6] the much more efficient procedural principle based on thresholded tabulation for efficiently executing MALP programs.

2 The Fuzzy Thresholded Tabulation Procedure

Tabulation arises as a technique to solve two important problems in deductive databases and logic programming: termination and efficiency. The datatype we will use for the description of the proposed method is that of a *forest*, i.e., a finite set of trees. Each one of these trees has a root labeled with a propositional symbol together with a truth-value from the underlying lattice (called the *current value* for the *tabulated* symbol); the rest of the nodes of each of these trees are labeled with an "extended" formula in which some of the propositional symbols have been substituted by its corresponding value.

[1] An increasing operator satisfying boundary conditions with the top element.

The following descriptions is considered, in order to prune some useless branches or, more exactly, by avoiding the use (during unfolding) of those program rules whose weights do not surpass a given "threshold" value

- Let $\mathcal{R} = \langle A \leftarrow_i \mathcal{B}; \vartheta \rangle$ be a program rule.
- Let $\mathcal{B}'$ be an expression with no atoms, obtained from body $\mathcal{B}$ by replacing each occurrence of a propositional symbol by $\top$.
- Let $v \in L$ be the result of interpreting $\mathcal{B}'$ under a given lattice.
- Then, $Up\_body(\mathcal{R}) = v$.

Apart from the truth degree ϑ of a program rule $\mathcal{R} = \langle A \leftarrow_i \mathcal{B}; \vartheta \rangle$ and the maximum truth degree of its body $Up\_body(\mathcal{R})$, in the multi-adjoint logic setting, we can consider a third kind of filter for reinforcing thresholding. The idea is to combine the two previous measures by means of the adjoint conjunction $\&_i$ of the implication $\leftarrow_i$ in rule $\mathcal{R}$. Now, we define the *maximum truth degree of a program rule*, symbolized by function Up_rule, as: $Up\_rule(\mathcal{R}) = \vartheta \&_i (Up\_body(\mathcal{R}))$.

Operations for Tabulation with Thresholding

The tabulation procedure requires four basic operations: Root Expansion, New Subgoal/Tree, Value Update, and Answer Return.

In the first operation, the *filters* for thresholding argued previously are implemented, from which the number of nodes in trees can be drastically diminished. Note that by avoiding the generation of a single node, the method implicitly avoids the construction of all its possible descendants as well. On the other hand, the time required to properly evaluate the filters is largely compensated. Anyway, in order to perform an efficient evaluation of filters, it must be taken into account that a condition only is checked if none of the previous ones fails. In particular, the only situation in which the three filters are completely evaluated appears only when the first two ones do not fail.

New Subgoal is applied whenever a propositional variable is found without a corresponding tree in the forest. Value update is used to propagate the truth-values of answers to the root of the corresponding tree. Finally, answer return substitutes a propositional variable by the current truth-value in the corresponding tree. We now describe formally the operations:

Rule 1: Root Expansion. Given a tree with root $A: r$ in the forest, if there is at least a program rule $\mathcal{R} = \langle A \leftarrow_i \mathcal{B}; \vartheta \rangle$ not consumed before and verifying the three conditions below, append the new child $\vartheta \&_i \mathcal{B}$ to the root of the tree.

- Condition 1. $\vartheta \not\leq r$.
- Condition 2. $Up\_body(\mathcal{R}) \not\leq r$.
- Condition 3. $Up\_rule(\mathcal{R}) \not\leq r$.

Rule 2: New Subgoal/Tree. Select a non-tabulated propositional symbol C occurring in a leaf of some tree (this means that there is no tree in the forest with the root node labeled with C), then create a new tree with a single node, the root $C: \bot$, and append it to the forest.

Rule 3: Value Update. If a tree, rooted at $C\colon r$, has a leaf $\mathcal{B}$ with no propositional symbols, and $\mathcal{B} \rightarrow_{IS}{}^{*} s$, where $s \in L$, then update the current value of the propositional symbol C by the value of $\sup_L\{r, s\}$.

Furthermore, once the tabulated truth-value of the tree rooted by C has been modified, for all the occurrences of C in a non-leaf node $\mathcal{B}[C]$ such as the one in the left of the figure below then, update the whole branch substituting the constant u by $\sup_L\{u, t\}$ (where t is the last tabulated truth-value for C, i.e. $\sup_L\{r, s\}$) as in the right of the figure.

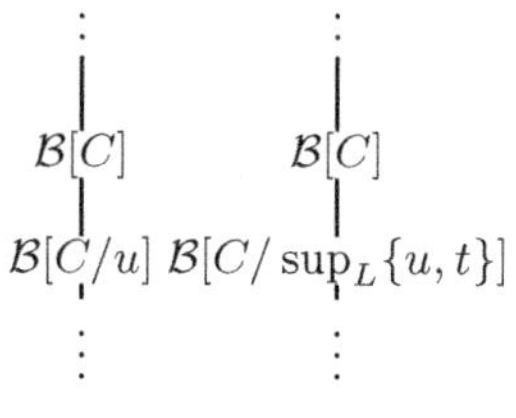

Rule 4: Answer Return. Select in any leaf a propositional symbol C which is tabulated, and assume that its current value is r; then add a new successor node as shown below:

The non-deterministic thresholded tabulation procedure and the correctness was proved in [6]. Furthermore, a deterministic procedure was presented using the four basic operations above.

3 Improvements Based on Static Preprocess

Before illustrating the fast execution method explained in the previous section, we would like to enhance a particularity of the first "**Root Expansion Rule**". Note that its application requires (in the worst case) the dynamic generation of three filters aiming to brake, when possible, the expansion of degenerated branches on trees. Such filters can be safely compiled on program rules after applying an easy static preprocess whose benefits will be largely be redeemed on further executions of the program. The following example will be considered as motivation, where the labels P, G and L stand for *Product*, *Gödel* and *Łukasiewicz* connectives.

Example 1. Let $\mathbb{P}$ be a program where there are only two rules with head p:

$$\mathcal{R}_1 : \langle\, p \;\leftarrow_{\mathtt{P}} \quad ; 0.4 \,\rangle$$
$$\mathcal{R}_2 : \langle\, p \;\leftarrow_{\mathtt{P}} \; q \,; 0.5 \,\rangle$$

When p is selected as non-tabulated propositional symbol, then *New Subgoal/Tree* creates a new tree with a single node, $p\colon \bot$. The following step is to apply *Root Expansion*, which selects a rule, for example $\mathcal{R}_1 : \langle p \leftarrow_{\mathtt{P}} ; 0.4\rangle$, and the following three conditions are checked:

- Condition 1. $\vartheta \nleq r$.
- Condition 2. *Up_body*($\mathcal{R}$) $\nleq r$.
- Condition 3. *Up_rule*($\mathcal{R}$) $\nleq r$.

therefore the root is updated to p: 0.4. In the next step, *Root Expansion* is applied to $\mathcal{R}_2$ and the three conditions must be checked too, obtaining the final root p: 0.5.

Consider now a modified version of the program where a new rule with head p is considered:

$$\begin{array}{l} \mathcal{R}_1 : \langle\, p \leftarrow_{\mathtt{P}} \qquad\qquad ; 0.4 \,\rangle \\ \mathcal{R}_2 : \langle\, p \leftarrow_{\mathtt{P}} q \qquad\quad ; 0.5 \,\rangle \\ \mathcal{R}_3 : \langle\, p \leftarrow_{\mathtt{L}} r \wedge_{\mathtt{P}} 0.8 \,; 0.9 \,\rangle \end{array}$$

In this case, the same procedure must be applied to rules $\mathcal{R}_1$ and $\mathcal{R}_2$, checking once again the three conditions for both rules. Finally, when rule $\mathcal{R}_3$ is considered by *Root Expansion*, the conditions are satisfied and the root is changed to p: 0.7.

The actualization of the program can be usual in a lot of applications and the verification of the three conditions is worse than directly calculating the value *Up_rule*($\mathcal{R}$).

Therefore, for any MALP program $\mathbb{P}$, we can obtain its extended version $\mathbb{P}+$ (for being used during the "query answering" process) by adding to its program rules their *proper threshold* Up_rule($\mathcal{R}$) as follows:

$$\mathbb{P}+ = \{\langle A \leftarrow_i \mathcal{B}; \vartheta;\, Up\_rule(\mathcal{R})\rangle \;\mid\; \mathcal{R} = \langle A \leftarrow_i \mathcal{B}; \vartheta\rangle \in \mathbb{P}\}.$$

Assuming the extended program $\mathbb{P}+$, we consider the new Rule 1:

Rule 1: Root Expansion. Given a tree with root A: r in the forest, if there is at least a program rule $\mathcal{R} = \langle A \leftarrow_i \mathcal{B}; \vartheta;\, Up\_rule(\mathcal{R})\rangle$ not consumed before and verifying $Up\_rule(\mathcal{R}) \not\leq r$, append the new child $\vartheta \&_i \mathcal{B}$ to the root of the tree.

Example 2. Continuing with the example above, note that calculating the values *Up_rule*($\mathcal{R}$) in the first step is better than checking the three conditions.

$$\begin{array}{l} \mathcal{R}_1 : \langle\, p \leftarrow_{\mathtt{P}} \quad ; 0.4 \,; 0.4\rangle \\ \mathcal{R}_2 : \langle\, p \leftarrow_{\mathtt{P}} q \,; 0.5 \,; 0.5\rangle \end{array}$$

Moreover, when the program is updated, the extended program is

$$\begin{array}{l} \mathcal{R}_1 : \langle\, p \leftarrow_{\mathtt{P}} \qquad\qquad ; 0.4 \,; 0.4\rangle \\ \mathcal{R}_2 : \langle\, p \leftarrow_{\mathtt{P}} q \qquad\quad ; 0.5 \,; 0.5\rangle \\ \mathcal{R}_3 : \langle\, p \leftarrow_{\mathtt{L}} r \wedge_{\mathtt{P}} 0.8 \,; 0.9 \,; 0.7\rangle \end{array}$$

and only one condition is required to be checked for each rule, thus reducing considerably the number of computations.

Now, two examples presented in [6] will be adapted to this new framework. Let $\mathbb{P}+$ be the following extended program with mutual recursion and query $?p$, on the unit interval of real numbers $([0,1], \leq)$:

$$\begin{array}{l} \mathcal{R}_1 : \langle\, p \leftarrow_{\mathtt{P}} q \,; 0.6 \,; 0.6 \,\rangle \\ \mathcal{R}_2 : \langle\, p \leftarrow_{\mathtt{P}} r \,; 0.5 \,; 0.5 \,\rangle \\ \mathcal{R}_3 : \langle\, q \leftarrow \quad ; 0.9 \,; 0.9 \,\rangle \\ \mathcal{R}_4 : \langle\, r \leftarrow \quad ; 0.8 \,; 0.8 \,\rangle \\ \mathcal{R}_5 : \langle\, r \leftarrow_{\mathtt{L}} p \,; 0.9 \,; 0.9 \,\rangle \end{array}$$

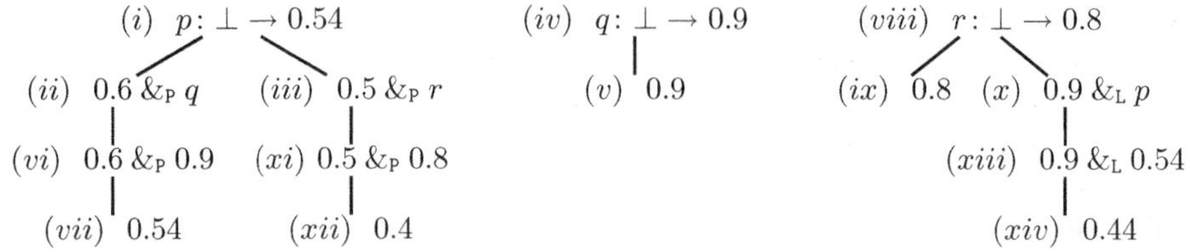

Fig. 1. Example forest for query $?p$

Firstly, the initial tree consisting of nodes $(i), (ii), (iii)$ is generated, see Figure 1. Then *New Subgoal* is applied on q, a new tree is generated with nodes (iv) and (v), and its current value is directly updated to 0.9.

By using this value, *Answer Return* extends the initial tree with node (vi). Now *Value Update* generates node (vii) and updates the current value of p to 0.54.

Then, *New Subgoal* is applied on r, and a new tree is generated with nodes $(viii), (ix)$ and (x). *Value Update* increases the current value to 0.8.

By using this value, *Answer Return* extends the initial tree with node (xi). Now *Value Update* generates node (xii). The current value is not updated since its value is greater than the newly computed one.

Finally, *Answer Return* can be applied again on propositional symbol p in node (x), generating node $(xiii)$. A further application of *Value Update* generates node (xiv) and the forest is terminated, as no rule performs any modification.

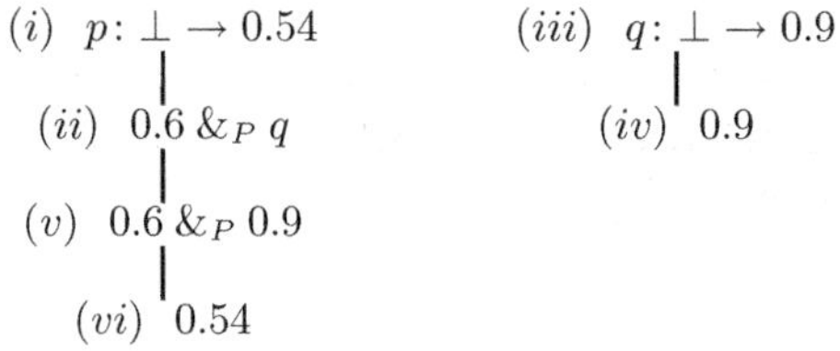

Fig. 2. Example threshold forest for p

In order to illustrate the advantages of our improved method, consider that in our extended program we replace the second program rule by:

$$\mathcal{R}'_2 : \langle p \ \leftarrow_{\text{P}} \ (r \&_{\text{P}} 0.9) \ \ ; \ \ 0.55; \ \ 0.495 \ \rangle$$

It is important to note now that even when the truth degree of the rule is 0.55, its threshold decreases to $Up\_rule(\mathcal{R}'_2) = 0.55 * 0.9 = 0.495 < 0.54$, which avoids extra expansions of the tree as Figure 2 shows.

As revealed in the previous examples, the presence of truth degrees on the body of program rules, is always desirable for optimizing the power of thresholding at tabulation time. In [7], we show that it is possible to transform a program rule into a semantically equivalent set of rules with the intended shape. The key point is the use of classical unfolding techniques initially described for crisp (i.e. not fuzzy) settings in [3,14,1], in order to optimize programs. The underlying idea is to "apply computational steps" on program rules, whose benefits remain compiled in their bodies. Now, we can give a new practical taste to these techniques in order to empower the benefits of thresholding when executing fuzzy programs in a tabulated way. For instance, given a MALP program like:

$$\begin{array}{llllll} \mathcal{R}_1: & \langle\ p & \leftarrow & & ;0.4 & \rangle \\ \mathcal{R}_2: & \langle\ p & \leftarrow_{\mathrm{P}} & q_1 & ;0.9 & \rangle \\ \mathcal{R}_3: & \langle\ q_1 & \leftarrow_{\mathrm{P}} & q_2 & ;0.9 & \rangle \\ \mathcal{R}_4: & \langle\ q_2 & \leftarrow_{\mathrm{P}} & q_3 & ;0.9 & \rangle \\ \mathcal{R}_5: & \langle\ q_3 & \leftarrow_{\mathrm{P}} & q_4 & ;0.9 & \rangle \\ & \ \vdots & & & & \end{array}$$

... for which the tabulation procedure would never end if the program be infinite (regarding goal p), the simple application of 9 unfolding steps on the second rule could produce the following extended program:

$$\begin{array}{lllllll} \mathcal{R}'_1: & \langle\ p & \leftarrow & & ;0.4 & ;0.4 & \rangle \\ \mathcal{R}'_2: & \langle\ p & \leftarrow_{\mathrm{P}} & 0.9\&_{\mathrm{P}}0.9\ldots\&_{\mathrm{P}}0.9\ \&_{\mathrm{P}}\ q_{10} & ;0.9 & ;0.3874204890 & \rangle \\ & \ \vdots & & & & & \end{array}$$

The reader may easily check that following our improved thresholded tabulation technique, the unique solution (0.4) for our initial query (p) could be easily found by simply applying a very few number of computation steps.

Another interesting example where the powerful of the static preprocess is showed arises if we consider a program with the following rules:

$$\begin{array}{llllll} \mathcal{R}_1: & \langle\ p & \leftarrow & & ;0.4 & \rangle \\ \mathcal{R}_2: & \langle\ p & \leftarrow_{\mathrm{L}} & (r\&_{\mathrm{L}}\ 0.5) & ;0.8 & \rangle \\ \mathcal{R}_3: & \langle\ p & \leftarrow_{\mathrm{L}} & (r\&_{\mathrm{G}}\ 0.8) & ;0.6 & \rangle \\ \mathcal{R}_4: & \langle\ p & \leftarrow_{\mathrm{P}} & (r\&_{\mathrm{G}}\ 0.7) & ;0.6 & \rangle \\ \mathcal{R}_5: & \langle\ p & \leftarrow_{\mathrm{P}} & (r\&_{\mathrm{P}}\ 0.9) & ;0.55 & \rangle \end{array}$$

If the extended program is not assumed, then, surely, all the rules will be considered in the tabulation procedure. However, if we calculate the proper threshold, it is possible to reorder the rules in order to improve the efficiency of the procedure.

$$\begin{array}{llllll} \mathcal{R}_5: & \langle\ p & \leftarrow_{\mathrm{P}} & (r\&_{\mathrm{P}}\ 0.9) & ;0.55;\ 0.495 & \rangle \\ \mathcal{R}_4: & \langle\ p & \leftarrow_{\mathrm{P}} & (r\&_{\mathrm{G}}\ 0.7) & ;0.6;\ 0.42 & \rangle \\ \mathcal{R}_1: & \langle\ p & \leftarrow & & ;0.4;\ 0.4 & \rangle \\ \mathcal{R}_3: & \langle\ p & \leftarrow_{\mathrm{L}} & (r\&_{\mathrm{G}}\ 0.8) & ;0.6;\ 0.4 & \rangle \\ \mathcal{R}_2: & \langle\ p & \leftarrow_{\mathrm{L}} & (r\&_{\mathrm{L}}\ 0.5) & ;0.8;\ 0.3 & \rangle \end{array}$$

In this last case, when fact $\mathcal{R}_6 : \langle r\leftarrow\ ; 1\rangle$ is considered, only one of the five rules above is applied by the thresholded tabulation procedure.

4 Conclusions and Future Work

In this paper, we were concerned with same static improvements that can be easily achieved on the thresholded tabulation procedure we have recently designed in [6] for the fast execution of MALP programs. Before lifting our results to the first order case and implementing it into our FLOPER platform [12], for the near future we plan to formally define the unfolding process of the method drawn here, providing stopping criteria and guides for applying the unfolding operation to program rules in a satisfiable way.

References

1. Alpuente, M., Falaschi, M., Moreno, G., Vidal, G.: Rules + Strategies for Transforming Lazy Functional Logic Programs. Theoretical Computer Science 311(1-3), 479–525 (2004)
2. Baldwin, J.F., Martin, T.P., Pilsworth, B.W.: Fril- Fuzzy and Evidential Reasoning in Artificial Intelligence. John Wiley & Sons, Inc., Chichester (1995)
3. Burstall, R.M., Darlington, J.: A Transformation System for Developing Recursive Programs. Journal of the ACM 24(1), 44–67 (1977)
4. Damásio, C.V., Medina, J., Ojeda-Aciego, M.: A tabulation proof procedure for residuated logic programming. In: Proc. of the European Conf. on Artificial Intelligence, Frontiers in Artificial Intelligence and Applications, vol. 110, pp. 808–812 (2004)
5. Ishizuka, M., Kanai, N.: Prolog-ELF Incorporating Fuzzy Logic. In: Joshi, A.K. (ed.) Proceedings of the 9th International Joint Conference on Artificial Intelligence (IJCAI 1985), pp. 701–703. Morgan Kaufmann, San Francisco (1985)
6. Julián, P., Medina, J., Moreno, G., Ojeda, M.: Efficient thresholded tabulation for fuzzy query answering. Studies in Fuzziness and Soft Computing (Foundations of Reasoning under Uncertainty) 249, 125–141 (2010)
7. Julián, P., Moreno, G., Penabad, J.: On Fuzzy Unfolding. A Multi-adjoint Approach. Fuzzy Sets and Systems 154, 16–33 (2005)
8. Julián, P., Moreno, G., Penabad, J.: Efficient reductants calculi using partial evaluation techniques with thresholding. Electronic Notes in Theoretical Computer Science 188, 77–90 (2007)
9. Kifer, M., Subrahmanian, V.S.: Theory of generalized annotated logic programming and its applications. Journal of Logic Programming 12, 335–367 (1992)
10. Medina, J., Ojeda-Aciego, M., Vojtáš, P.: Multi-adjoint logic programming with continuous semantics. In: Eiter, T., Faber, W., Truszczyński, M. (eds.) LPNMR 2001. LNCS (LNAI), vol. 2173, pp. 351–364. Springer, Heidelberg (2001)
11. Medina, J., Ojeda-Aciego, M., Vojtáš, P.: Similarity-based Unification: a multi-adjoint approach. Fuzzy Sets and Systems 146, 43–62 (2004)
12. Morcillo, P.J., Moreno, G.: Programming with fuzzy logic rules by using the FLOPER tool. In: Bassiliades, N., Governatori, G., Paschke, A. (eds.) RuleML 2008. LNCS, vol. 5321, pp. 119–126. Springer, Heidelberg (2008)
13. Swift, T.: Tabling for non-monotonic programming. Annals of Mathematics and Artificial Intelligence 25(3-4), 201–240 (1999)
14. Tamaki, H., Sato, T.: Unfold/Fold Transformations of Logic Programs. In: Tärnlund, S. (ed.) Proc. of Second Int'l Conf. on Logic Programming, pp. 127–139 (1984)

Non-deterministic Algebraic Structures for Soft Computing*

I.P. Cabrera, P. Cordero, and M. Ojeda-Aciego

Dept. Matemática Aplicada, Universidad de Málaga, Spain
{ipcabrera,pcordero,aciego}@uma.es

Abstract. The need of considering non-determinism in theoretical computer science has been claimed by several authors in the literature. The notion of non-deterministic automata as a formal model of computation is widely used, but the specific study of non-determinism is useful, for instance, for natural language processing, in describing interactive systems, for characterizing the flexibility allowed in the design of a circuit or a network, etc. The most suitable structures for constituting the foundation of this theoretical model of computation are non-deterministic algebras. The interest on these generalized algebras has been growing in recent years, both from a crisp and a fuzzy standpoint. This paper presents a survey of these structures in order to foster its applicability for the development of new soft computing techniques.

Keywords: Non-determinism, multialgebras, hyperalgebras, non-deterministic algebras.

1 Hyperstructures, Multistructures and nd-structures

The difficulty of handling non-determinism has been sometimes avoided by simulating it using specific algorithms on deterministic automata. Nonetheless, the need of developing a formal theory which considers non-determinism as an inherent aspect of computation, instead of merely simulating it, is widely accepted.

A usual direct approach to non-determinism is the use of multialgebras [25, 41, 48], also called multivalued algebras or hyperalgebras, in which the arguments of the operations are individuals and the result is a set of possible outcomes.

Hyperalgebra, or hyperstructure theory, was introduced in [36] when Marty defined hypergroups, began to analyze their properties, and applied them to groups, rational fractions and algebraic functions. Nowadays, a number of different hyperstructures have been widely studied from both the theoretical and applicative point of view, and for their use in applied mathematics and artificial intelligence; note, however, that an important obstacle for the study of these structures is the lack of consensus in the terminology. These hyperstructures can be roughly classified as follows:

* Partially supported by Spanish Ministry of Science project TIN09-14562-C05-01 and Junta de Andalucía project P09-FQM-5233.

J. Cabestany, I. Rojas, and G. Joya (Eds.): IWANN 2011, Part II, LNCS 6692, pp. 437–444, 2011.

- *Generalizations of group theory.* The most general hyperstructure, the hypergroupoid, is just a nonempty set H endowed with a binary operation $H \times H \to \mathcal{P}(H) \setminus \{\varnothing\}$. *Semihypergroups, quasihypergroups, hypergroups, join spaces*, etc, are different classes of hypergroupoids with different sets of additional requirements. A theoretical study of these and other structures and a wide survey of its applications can be found in [14], which describes applications to geometry, graph theory, fuzzy set theory, rough set theories, cryptography, codes, etc. Recently, several results relating hypergroups and fuzzy set theory have been obtained, see [16, 18, 19, 49].
- *Extensions of ring theory.* In this topic, the most referenced structures are hyperrings and hyperfields, which were defined by Krasner in [30, 31] and have been applied to geometry and number theory [11]. A weakening of these structures (multiring and multifield) was introduced in [33].
- *Lattice-related structures.* A number of structures are inspired in lattice theory although not all of them are proper extensions of the structure of lattice, for instance, *nearlattices* [10], *near lattices* [43], *hyperlattices* [29], or *superlattices* [38].

  Specially interesting, in this context, is the structure of multilattice (see Section 2), which provides a convenient generalization of lattices both from the algebraic and the coalgebraic points of view.

It is remarkable that most of the structures above consider that the codomain of the operations is always a nonempty set. This restriction does not suit certain applications, and that is why we introduced the non-deterministic algebras (briefly, *nd-algebras*) [34] by considering operations of type $A_1 \times \cdots \times A_n \to \mathcal{P}(A)$. Thus, a non-deterministic groupoid (or *nd-groupoid*) is just a hypergroupoid in which the restriction of the images being nonempty is dropped.

Among the applications which demand nd-operations we can find a number of them requiring partial ordered sets (*posets*) which are not lattices, but have similar properties. The notion of partially ordered set has proven to be very relevant in modern mathematics, perhaps being lattice theory one of the best examples. Note, however, that it is not difficult to find situations in which posets arise that are not lattices as, for example, in *divisibility* theory, *special relativity* theory, ... These posets, although lacking a proper lattice structure, share some of their properties.

2 Multilattices: Algebraically and Coalgebraically

It was Benado [3] who firstly proposed an approach to generalizing the notion of lattice in which the supremum and the infimum are replaced by the set of minimal upper bounds, named *multisupremum*, and the set of maximal lower bounds, named *multiinfimum*, respectively. This structure is called *multilattice*. Notice that the operators which compute the multi-suprema and multi-infima in a poset provide precisely nd-groupoids or, if we have for granted that at least a multi-supremum always exists, a hypergroupoid. Although other generalizations of the notion of lattice have been developed so far, see above, we are focusing our attention on multilattices because of their computational properties.

The idea underlying the algebraic study of multilattices is the development of a new theory involving *non-deterministic operators* as a framework for formalizing key notions in computer science and artificial intelligence. For instance, non-determinism has been considered under the combination of modal and temporal logics to be used in communication systems; new results have been recently obtained in database theory as well. A lot of effort is being put in this area, as one can still see recent works dealing with non-determinism both from the theoretical and from the practical point of view [27, 47].

Although Benado's original motivation was purely theoretical (he used multilattices to work with *Dedekind connections*, *Schreier's refinement* theorem and *evaluation* theory) multilattices (and relatives such as multisemilattices) have been identified in several disparate research areas: (1) in the field of automated deduction, specifically when devising a theory about implicates and implicants for certain temporal logics during the development of automated theorem provers for those logics [13]; (2) unification for logical systems, whose starting point was the existence of a most general unifier for any unifiable formula in Boolean logic: Ghilardi [22] proved that there are no most general unifiers in intuitionistic propositional calculus but a finite set of maximal general unifiers instead; and (3) multilattices play important roles in computation, for instance the set of words builded from an alphabet by considering the *"be a subword"* ordering.

As stated above, the notions of ordered and algebraic multilattice were introduced by Benado in [3]. An alternative algebraic characterization was introduced by Hansen in [24] and, later, Johnston studies ideals and distributivity on this algebras [26]. However, the first applicable algebraic characterization is relatively recent, Martínez *et al.* [34], and it reflects much better the corresponding classical theory about lattices than those given previously. Moreover, this algebraic characterization allows natural definitions of related structures such as *multisemilattices* and, in addition, is better suited for applications. For instance, [46] shows several examples in process semantics where the carrier set has the structure of multilattice, and Medina *et al.* [37] developed a general approach to fuzzy logic programming based on a multilattice as underlying set of truth-values for the logic.

Certain abstract structures can be thought of both algebraically and coalgebraically. The context and the aims of the work usually indicates which framework one should consider; for instance, when non-deterministic behavior is assumed, the coalgebraic framework is generally preferred because it appears to fit more naturally, since coalgebras are becoming an ideal framework for formalization in diverse branches of computer science (Kripke structures, labeled transition systems, various types of non-deterministic automata, etc).

Following this trend, we started a research line consisting in the development of a coalgebraic view of several mathematical structures of interest for the handling of non-determinism, in particular, for multilattices. In [8], we have defined a suitable class of coalgebras, the ND-coalgebras, and developed a thorough analysis of the required properties in order to achieve a convenient coalgebraic characterization of multilattices which complements the algebraic one given in [35]. The class of ND-coalgebras can be regarded as a collection of coalgebras underlying non-deterministic situations, and creates a setting in which many other structures could be suitably described.

3 Congruences, Homomorphisms and Ideals on Non-deterministic Structures

In traditional mathematics, congruences, homomorphisms and ideals are usually considered as different views of the same phenomenon, as stated by the so-called isomorphism theorems. Note, however, that in the realm of nd-structures there are several plausible generalizations of these notions which do not necessarily preserve the existing relationships in the classical case.

The study of congruences is important both from a theoretical standpoint and for its applications in the field of logic-based approaches to uncertainty. Regarding applications, the notion of congruence is intimately related to the foundations of fuzzy reasoning and its relationships with other logics of uncertainty [21]. More focused on the theoretical aspects of computer science, some authors [2, 40] have pointed out the relation between congruences, fuzzy automata and determinism. There have also been studies on qualitative reasoning about the morphological relation of congruence. A spatial congruence relation is introduced in [15] which, moreover, provides an algebraic structure to host relations based on it.

3.1 Crisp Approaches

To begin with, a discussion on the most suitable extension of the notions of congruence and homomorphism on a given nd-structure is needed. In [6], we consider the notion of homomorphism on nd-groupoids and how it preserves the different subhyperstructures. Likewise, in this general framework, the relation between nd-homomorphisms and crisp congruences on a hyper-structure is investigated. In [4], we dealt with congruences on a hypergroupoid or nd-groupoid. Specifically, the set of congruences on an nd-groupoid need not be a lattice unless we assume some extra properties. This problem led us to review some related literature and, as a result, we found one counter-example even in the context of crisp congruences on a hypergroupoid. The previous example motivated the search for a sufficient condition which granted the structure of complete lattice for the set of congruences on a hypergroupoid and, by extension, on an nd-groupoid; this property turned out to be that the underlying nd-structure should be a certain sort of multisemilattice.

The next step in this context is to study congruence relations in the more general structure of multilattices, together with a suitable definition of homomorphism. In [12] the classical relationship between homomorphisms and congruences was suitably adapted, as well as a proof that the set of congruences of a certain class of multilattices is a complete lattice.

In a subsequent work, the focus was put on the notion of ideal. This is not a trivial matter since several definitions have been proposed for the notion of ideal of a multilattice: for instance, one can find the notion of s-ideals introduced by Rachůnek, or the l-ideals of Burgess, or the m-ideals given by Johnston [26, 42]. In [7], we introduced an alternative definition more suitable for extending the classical results about congruences and homomorphisms. This approach led to generalize the result about the lattice structure of the set of congruences to be applied to *any* multilattice.

3.2 Fuzzy Approaches

The systematic generalization of crisp concepts to the fuzzy case has proven to be an important theoretical tool for the development of new methods of reasoning under uncertainty, imprecision and lack of information. One can find disparate extensions of classical algebraic structures to a fuzzy framework in the literature; moreover, recently, hyperstructures and fuzzy theory are being studied jointly, giving rise to the so-called fuzzy hyperalgebra and, consequently, several areas within artificial intelligence and soft computing have been benefitted from the new results obtained [1, 32, 44, 50, 51].

Regarding the generalization level, since the inception of fuzzy sets and fuzzy logic, there have been approaches to consider underlying sets of truth-values more general than the unit interval; for instance, consider the L-fuzzy sets introduced in [23], where L is a complete lattice. Furthermore, one can even consider the study of M-fuzzy sets where M has the structure of a multilattice.

Several papers have been published about the lattice of fuzzy congruences on different algebraic structures[17, 20, 39, 45]. A previous step before studying the fuzzy congruences on multilattices and the suitable generalizations of the concept of L-fuzzy and M-fuzzy congruence, is to define fuzzy congruence relations on nd-groupoids.

Our generalization to the context of nd-groupoids is introduced in [5], following the trend initiated in [4]. Concerning the study of the lattice structure of fuzzy congruence relations, the main result obtained is a set of conditions guaranteeing that the set of fuzzy congruences on an nd-groupoid is a complete lattice, since in general this is not always the case.

Unlike the development of the fuzzy versions of other crisp concepts in mathematics like congruence relation, the fuzzy extension of the notion of function has been studied from several standpoints, and this fact complicates the choice of the most suitable definition of fuzzy homomorphism: the most convenient definition seems to depend on particular details of the underlying algebraic structure under consideration. The definition of fuzzy function introduced in [28] is used in [6] in order to establish the relation between fuzzy congruences and perfect fuzzy homomorphisms, leading to a fuzzy version of the canonical decomposition theorem for certain class of fuzzy homomorphisms. Specifically, a given $\varphi\colon A \to B$ in this class can be decomposed[1] as $\varphi = \iota \circ \bar{\varphi} \circ \pi$ where $\pi\colon A \to A/\rho_\varphi$ is the

[1] Note that all the notions involved in the decomposition are fuzzy.

fuzzy projection from A to its quotient set over the kernel congruence relation ρ_φ induced by φ, $\bar{\varphi}: A/\rho_\varphi \to \operatorname{Im}\varphi$ is the induced isomorphism, and $\iota: \operatorname{Im}\varphi \to B$ is the inclusion.

The previous approaches are extended to the general theory of hyperrings in [9], where the theory of hyperrings and fuzzy homomorphisms between them is studied. Specifically, isomorphism theorems are established which relate fuzzy homomorphisms between hyperrings, fuzzy congruences and fuzzy hyperideals.

4 Conclusions

(Multi, hyper, nd)-algebras provide a suitable theory for the foundation of non-determinism. Although this theory was originated in 1934, currently a lot of effort has been put on them, mostly due to its applicability, especially in computer science: the current trend being the fuzzy extension of hyperalgebra and its relation to soft computing.

In this work, we have reviewed a class of these structures in order to foster its applicability for the development of new soft computing techniques. Specifically, a brief survey of the most cited hyperalgebras in the literature has been presented. Then, the notion of non-deterministic algebra (nd-algebra) is introduced; this is a general notion which includes, in a common framework, algebras, partial algebras and hyperalgebras. Later, the focus is put on two important classes of nd-algebras: multisemilattices and multilattices. The importance of these structures is due to the fact that they extend the classical results about lattice theory to a wide range of partially ordered sets and they appear in several areas in theoretical computer science. The final section is devoted to the recent advances related to congruences (and its relatives, homomorphisms and ideals) on non-deterministic structures, due to its intrinsic interest both from a theoretical standpoint and for its applications in the field of logic-based approaches to uncertainty.

References

1. Ameri, R., Nozari, T.: Fuzzy hyperalgebras. Computers and Mathematics with Applications 61(2), 149–154 (2011)
2. Bělohlávek, R.: Determinism and fuzzy automata. Information Sciences 143, 205–209 (2002)
3. Benado, M.: Les ensembles partiellement ordonnés et le théorème de raffinement de Schreier. I. Čehoslovack. Mat. Ž 4(79), 105–129 (1954)
4. Cabrera, I.P., Cordero, P., Gutiérrez, G., Martínez, J., Ojeda-Aciego, M.: Congruence relations on some hyperstructures. Annals of Mathematics and Artificial Intelligence 56(3-4), 361–370 (2009)
5. Cabrera, I.P., Cordero, P., Gutiérrez, G., Martínez, J., Ojeda-Aciego, M.: Fuzzy congruence relations on nd-groupoids. International Journal on Computer Mathematics 86, 1684–1695 (2009)
6. Cabrera, I.P., Cordero, P., Gutiérrez, G., Martínez, J., Ojeda-Aciego, M.: On congruences and homomorphisms on some non-deterministic algebras. In: Proc. of Intl. Conf. on Fuzzy Computation, pp. 59–67 (2009)

7. Cabrera, I.P., Cordero, P., Gutiérrez, G., Martínez, J., Ojeda-Aciego, M.: On congruences, ideals and homomorphisms over multilattices. In: EUROFUSE Workshop Preference Modelling and Decision Analysis, pp. 299–304 (2009)
8. Cabrera, I.P., Cordero, P., Gutiérrez, G., Martínez, J., Ojeda-Aciego, M.: A coalgebraic approach to non-determinism: applications to multilattices. In: Information Sciences, vol. 180, pp. 4323–4335 (2010)
9. Cabrera, I.P., Cordero, P., Gutiérrez, G., Martínez, J., Ojeda-Aciego, M.: On fuzzy homomorphisms between hyperrings. XV Congreso Español Sobre Tecnologías Y Lógica Fuzzy – ESTYLF 2010 (2010)
10. Chajda, I., Kolařík, M.: Nearlattices. Discrete Math. 308(21), 4906–4913 (2008)
11. Connes, A., Consani, C.: The hyperring of adèle classes. J. Number Theory 131(2), 159–194 (2011)
12. Cordero, P., Gutiérrez, G., Martínez, J., Ojeda-Aciego, M., Cabrera, I.P.: Congruence relations on multilattices. In: Intl FLINS Conference on Computational Intelligence in Decision and Control, FLINS 2008, pp. 139–144 (2008)
13. Cordero, P., Gutiérrez, G., Martínez, J., de Guzmán, I.P.: A new algebraic tool for automatic theorem provers. Annals of Mathematics and Artificial Intelligence 42(4), 369–398 (2004)
14. Corsini, P., Leoreanu, V.: Applications of hyperstructure theory. Kluwer, Dordrecht (2003)
15. Cristani, M.: The complexity of reasoning about spatial congruence. Journal of Artificial Intelligence Research 11, 361–390 (1999)
16. Cristea, I., Davvaz, B.: Atanassov's intuitionistic fuzzy grade of hypergroups. Information Sciences 180(8), 1506–1517 (2010)
17. Das, P.: Lattice of fuzzy congruences in inverse semigroups. Fuzzy Sets and Systems 91(3), 399–408 (1997)
18. Davvaz, B., Corsini, P., Leoreanu-Fotea, V.: Fuzzy n-ary subpolygroups. Computers & Mathematics with Applications 57(1), 141–152 (2009)
19. Davvaz, B., Leoreanu-Fotea, V.: Applications of interval valued fuzzy n-ary polygroups with respect to t-norms (t-conorms). Computers & Mathematics with Applications 57(8), 1413–1424 (2009)
20. Dutta, T.K., Biswas, B.: On fuzzy congruence of a near-ring module. Fuzzy Sets and Systems 112(2), 399–408 (2000)
21. Gaines, B.R.: Fuzzy reasoning and the logics of uncertainty. In: Proc. of ISMVL 1976, pp. 179–188 (1976)
22. Ghilardi, S.: Unification in intuitionistic logic. The Journal of Symbolic Logic 64(2), 859–880 (1999)
23. Goguen, J.A.: L-fuzzy sets. J. Math. Anal. Appl. 18, 145–174 (1967)
24. Hansen, D.J.: An axiomatic characterization of multilattices. Discrete Math. 33(1), 99–101 (1981)
25. Hesselink, W.H.: A mathematical approach to nondeterminism in data types. ACM Trans. Program. Lang. Syst. 10, 87–117 (1988)
26. Johnston, I.J.: Some results involving multilattice ideals and distributivity. Discrete Math. 83(1), 27–35 (1990)
27. Khan, J., Haque, A.: Computing with data non-determinism: Wait time management for peer-to-peer systems. Computer Communications 31(3), 629–642 (2008)
28. Klawonn, F.: Fuzzy points, fuzzy relations and fuzzy function. In: Novák, V., Perfilieva, I. (eds.) Discovering World with Fuzzy Logic, pp. 431–453. Physica-Verlag, Heidelberg (2000)
29. Konstantinidou, M., Mittas, J.: An introduction to the theory of hyperlattices. Math. Balkanica 7, 187–193 (1977)

30. Krasner, M.: Approximation des corps values complets de caracteristique p $\neq 0$ par ceux de caracteristique 0. In: Colloque d'algebre superieure, Centre Belge de Recherches Mathematiques Etablissements, pp. 129–206 (1957)
31. Krasner, M.: A class of hyperrings and hyperfields. Internat. J. Math. & Math. Sci. 6(2), 307–312 (1983)
32. Ma, X., Zhan, J., Leoreanu-Fotea, V.: On (fuzzy) isomorphism theorems of Γ-hyperrings. Computers and Mathematics with Applications 60(9), 2594–2600 (2010)
33. Marshall, M.: Real reduced multirings and multifields. Journal of Pure and Applied Algebra 205(2), 452–468 (2006)
34. Martínez, J., Gutiérrez, G., de Guzmán, I.P., Cordero, P.: Generalizations of lattices via non-deterministic operators. Discrete Math. 295(1-3), 107–141 (2005)
35. Martínez, J., Gutiérrez, G., Pérez de Guzmán, I., Cordero, P.: Multilattices via multisemilattices. In: Topics in applied and theoretical mathematics and computer science, pp. 238–248. WSEAS (2001)
36. Marty, F.: Sur une généralisation de la notion de groupe. In: Proceedings of 8th Congress Math. Scandinaves, pp. 45–49 (1934)
37. Medina, J., Ojeda-Aciego, M., Ruiz-Calviño, J.: Fuzzy logic programming via multilattices. Fuzzy Sets and Systems 158(6), 674–688 (2007)
38. Mittas, J., Konstantinidou, M.: Sur une nouvelle généralisation de la notion de treillis: les supertreillis et certaines de leurs propriétés générales. Ann. Sci. Univ. Clermont-Ferrand II Math. 25, 61–83 (1989)
39. Murali, V.: Fuzzy congruence relations. Fuzzy Sets and Systems 41(3), 359–369 (1991)
40. Petković, T.: Congruences and homomorphisms of fuzzy automata. Fuzzy Sets and Systems 157, 444–458 (2006)
41. Pickett, H.E.: Homomorphisms and subalgebras of multialgebras. Pacific Journal of Mathematics 21(2), 327–342 (1967)
42. Rachůnek, J.: 0-idéaux des ensembles ordonnés. Acta Univ. Palack. Fac. Rer. Natur. 45, 77–81 (1974)
43. Schweigert, D.: Near lattices. Math. Slovaca 32(3), 313–317 (1982)
44. Sun, K., Yuan, X., Li, H.: Fuzzy hypergroups based on fuzzy relations. Computers and Mathematics with Applications 60(3), 610–622 (2010)
45. Tan, Y.: Fuzzy congruences on a regular semigroup. Fuzzy Sets and Systems 117(3), 399–408 (2001)
46. Vaida, D.: Note on some order properties related to processes semantics. I. Fund. Inform. 73(1-2), 307–319 (2006)
47. Varacca, D., Winskel, G.: Distributing probability over non-determinism. Mathematical Structures in Computer Science 16(1), 87–113 (2006)
48. Walicki, M., Meldal, S.: A complete calculus for the multialgebraic and functional semantics of nondeterminism. ACM Trans. Program. Lang. Syst. 17, 366–393 (1995)
49. Yamak, S., Kazancı, O., Davvaz, B.: Applications of interval valued t-norms (t-conorms) to fuzzy n-ary sub-hypergroups. Information Sciences 178(20), 3957–3972 (2008)
50. Yamak, S., Kazancı, O., Davvaz, B.: Normal fuzzy hyperideals in hypernear-rings. Neural Computing and Applications 20(1), 25–30 (2011)
51. Yin, Y., Zhan, J., Xu, D., Wang, J.: The L-fuzzy hypermodules. Computers and Mathematics with Applications 59(2), 953–963 (2010)

Fuzzy Computed Answers Collecting Proof Information*

Pedro J. Morcillo, Ginés Moreno, Jaime Penabad, and Carlos Vázquez

University of Castilla-La Mancha
Faculty of Computer Science Engineering
02071, Albacete, Spain
{pmorcillo,cvazquez}@dsi.uclm.es,
{Gines.Moreno,Jaime.Penabad}@uclm.es

Abstract. MALP (i.e., the so-called *Multi-Adjoint Logic Programming approach*) can be seen as a promising fuzzy extension of the popular, pure logic language Prolog, including too a wide repertoire of constructs based on fuzzy logic in order to support uncertainty and approximated reasoning in a natural way. Moreover, the *Fuzzy LOgic Programming Environment for Research*, FLOPER in brief, that we have implemented in our research group, is intended to assists the development of real-world applications written with MALP syntax. Among other capabilities, the system is able to safely translate fuzzy code into Prolog clauses which can be directly executed inside any standard Prolog interpreter in a completely transparent way for the final user. In this fuzzy setting, it is mandatory the use of lattices modeling truth degrees beyond $\{true; false\}$. As described in this paper, FLOPER is able to successfully deal (in a very easy way) with sophisticated lattices modeling truth degrees in the real interval $[0, 1]$, also documenting -via declarative traces- the proof procedures followed when solving queries, without extra computational cost.

Keywords: Fuzzy Logic Programming, Logic Proofs, Declarative Debugging.

1 Introduction

Logic Programming (LP) [8] has been widely used for problem solving and knowledge representation in the past, with recognized applications in AI and related areas. Nevertheless, traditional LP languages do not incorporate techniques or constructs to treat explicitly with uncertainty and approximated reasoning. To overcome this situation, during the last years, several fuzzy logic programming systems have been developed where the classical inference mechanism of SLD–Resolution has been replaced by a fuzzy variant able to handle partial truth and

* This work was supported by the EU (FEDER), and the Spanish Science and Innovation Ministry (MICINN) under grants TIN 2007-65749 and TIN2011-25846, and by the Castilla-La Mancha Administration under grant PII1I09-0117-4481.

J. Cabestany, I. Rojas, and G. Joya (Eds.): IWANN 2011, Part II, LNCS 6692, pp. 445–452, 2011.

to reason with uncertainty [3,1,10], with promising applications in the fields of Computational Intelligence, Soft-Computing, Semantic Web, etc.

Informally speaking, in the MALP framework of [10,9], a program can be seen as a set of rules each one annotated by a truth degree, and a goal is a query to the system, i.e., a set of atoms linked with connectives called *aggregators*. A *state* is a pair $\langle \mathcal{Q}, \sigma \rangle$ where $\mathcal{Q}$ is a goal and σ a substitution (initially, the identity substitution). States are evaluated in two separate computational phases. Firstly, *admissible steps* (a generalization of the classical *modus ponens* inference rule) are systematically applied by a backward reasoning procedure in a similar way to classical resolution steps in pure logic programming, thus returning a computed substitution together with an expression where all atoms have been exploited. This last expression is then interpreted under a given lattice, hence returning a pair $\langle$*truth degree*; *substitution*$\rangle$ which is the fuzzy counterpart of the classical notion of computed answer traditionally used in LP.

In the present paper, we draw the last developments performed on the FLOPER system (see [11,12] and visit `http://www.dsi.uclm.es/investigacion/dect/FLOPERpage.htm`), which currently provides facilities for compiling, executing and manipulating such kind of fuzzy programs, by means of two main representation (high/low-level, Prolog-based) ways which are somehow antagonistics regarding simplicity and accuracy features. The main purpose of the present paper is to highlight a collateral effect of the last feature implemented into the tool, regarding the possibility of introducing different notions of multi-adjoint lattices which can be easily defined with a Prolog taste. Only a few number of clauses suffices for modeling rich notions of truth degrees incorporating augmented information about the program rules used in a derivation sequence as well as the set of fuzzy connectives evaluated at execution time when reaching the whole set of solutions for a given program and goal. The most surprising fact reported here is that this kind of "*extra proof information*" can be freely collected on fuzzy computed answers without requiring any additional computational resource.

The outline of this work is as follows. In Section 2 we detail the main features of multi-adjoint logic programming, both syntax and procedural semantics. Section 3 explains the current menu of programming resources implemented into the FLOPER tool, which nowadays is being equipped with new options for performing advanced program manipulation tasks (transformation, specialization, optimization) with a clear fuzzy taste. The benefits of our present approach regarding how to obtain fuzzy computed answers containing debugging information on execution proofs, are highlighted in Section 4. Finally, in Section 5 we present our conclusions and propose some lines of future work.

2 Multi-adjoint Logic Programs

In what follows, we present a short summary of the main features of our language (we refer the reader to [10] for a complete formulation). We work with a first order language, $\mathcal{L}$, containing variables, function symbols, predicate symbols, constants, quantifiers ($\forall$ and $\exists$), and several (arbitrary) connectives to increase language expressiveness. In our fuzzy setting, we use implication connectives ($\leftarrow_1, \leftarrow_2, \ldots, \leftarrow_m$) and also other connectives which are grouped under the name of "aggregators" or "aggregation operators". They are used to combine/propagate truth values through the rules. The general definition of aggregation operators subsumes conjunctive operators (denoted by $\&_1, \&_2, \ldots, \&_k$), disjunctive operators ($\vee_1, \vee_2, \ldots, \vee_l$), and average and hybrid operators (usually denoted by $@_1, @_2, \ldots, @_n$). Although the connectives $\&_i$, $\vee_i$ and $@_i$ are binary operators, we usually generalize them as functions with an arbitrary number of arguments. By definition, the truth function for an n-ary aggregation operator $[\![@]\!] : L^n \rightarrow L$ is required to be monotone and fulfills $[\![@]\!](\top, \ldots, \top) = \top$, $[\![@]\!](\bot, \ldots, \bot) = \bot$. Additionally, our language $\mathcal{L}$ contains the values of a multi-adjoint lattice, $\langle L, \preceq, \leftarrow_1, \&_1, \ldots, \leftarrow_n, \&_n \rangle$, equipped with a collection of adjoint pairs $\langle \leftarrow_i, \&_i \rangle$, where each $\&_i$ is a conjunctor intended to the evaluation of *modus ponens*. In general, the set of truth values L may be the carrier of any complete bounded lattice but, for simplicity, in this paper we shall select L as the set of real numbers in the interval $[0, 1]$.

A *rule* is a formula $A \leftarrow_i \mathcal{B}$, where A is an atomic formula (usually called the *head*) and $\mathcal{B}$ (which is called the *body*) is a formula built from atomic formulas $B_1, \ldots, B_n$ ($n \geq 0$), truth values of L and conjunctions, disjunctions and aggregations. Rules with an empty body are called *facts*. A *goal* is a body submitted as a query to the system. Variables in a rule are assumed to be governed by universal quantifiers. Roughly speaking, a multi-adjoint logic program is a set of pairs $\langle \mathcal{R}; v \rangle$, where $\mathcal{R}$ is a rule and v is a *truth degree* (a value of L) expressing the confidence which the user of the system has in the truth of the rule $\mathcal{R}$. Often, we will write "$\mathcal{R}$ with v" instead of $\langle \mathcal{R}; v \rangle$.

In order to describe the procedural semantics of the multi–adjoint logic language, in the following we denote by $\mathcal{C}[A]$ a formula where A is a sub-expression (usually an atom) which occurs in the –possibly empty– context $\mathcal{C}[]$ whereas $\mathcal{C}[A/A']$ means the replacement of A by A' in context $\mathcal{C}[]$. Moreover, $\mathcal{V}ar(s)$ denotes the set of distinct variables occurring in the syntactic object s, $\theta[\mathcal{V}ar(s)]$ refers to the substitution obtained from θ by restricting its domain to $\mathcal{V}ar(s)$ and $mgu(E)$ denotes the *most general unifier* of an equation set E. In the following definition, we always consider that A is the selected atom in goal $\mathcal{Q}$.

Definition 1 (Admissible Step). *Let $\mathcal{Q}$ be a goal and let σ be a substitution. The pair $\langle \mathcal{Q}; \sigma \rangle$ is a* state. *Given a program $\mathcal{P}$, an* admissible computation *is formalized as a state transition system, whose transition relation $\rightarrow_{AS}$ is the smallest relation satisfying the following* admissible rules*:*

1) $\langle \mathcal{Q}[A]; \sigma \rangle \rightarrow_{AS} \langle (\mathcal{Q}[A/v\&_i\mathcal{B}])\theta; \sigma\theta \rangle$ *if* $\theta = mgu(\{A' = A\})$, $\langle A' \leftarrow_i \mathcal{B}; v \rangle$ *in* $\mathcal{P}$ *and $\mathcal{B}$ is not empty.*
2) $\langle \mathcal{Q}[A]; \sigma \rangle \rightarrow_{AS} \langle (\mathcal{Q}[A/v])\theta; \sigma\theta \rangle$ *if* $\theta = mgu(\{A' = A\})$, *and* $\langle A' \leftarrow_i; v \rangle$ *in* $\mathcal{P}$.

Apart for exploiting atoms by using program rules, in this setting we can also evaluate expressions composed by truth degrees and fuzzy connectives by directly interpreting them w.r.t. lattice L following our definition recasted from [6]:

Definition 2 (Interpretive Step). *Let $\mathcal{P}$ be a program, $\mathcal{Q}$ a goal and σ a substitution. Assume that $[\![@]\!]$ is the truth function of connective @ in the lattice $\langle L, \preceq\rangle$ associated to $\mathcal{P}$, such that, for values $r_1, \ldots, r_n, r_{n+1} \in L$, we have that $[\![@]\!](r_1, \ldots, r_n) = r_{n+1}$. Then, we formalize the notion of* interpretive computation *as a state transition system, whose transition relation $\rightarrow_{IS}$ is defined as the least one satisfying:* $\langle \mathcal{Q}[@(r_1, \ldots, r_n)]; \sigma\rangle \rightarrow_{IS} \langle \mathcal{Q}[@(r_1, \ldots, r_n)/r_{n+1}]; \sigma\rangle$

Example 1. In order to illustrate our definitions, consider now the following program $\mathcal{P}$ and lattice $([0, 1], \leq)$, where $\leq$ is the usual order on real numbers.

$\mathcal{R}_1 : p(X) \leftarrow_{\mathtt{P}} q(X, Y) \&_{\mathtt{G}} r(Y)$ *with* 0.8 $\qquad \mathcal{R}_2 : q(a, Y) \leftarrow_{\mathtt{P}} s(Y)$ *with* 0.7
$\mathcal{R}_3 : q(b, Y) \leftarrow_{\mathtt{L}} r(Y)$ *with* 0.8 $\qquad \mathcal{R}_4 : r(Y) \leftarrow$ *with* 0.7
$\mathcal{R}_5 : s(b) \leftarrow$ *with* 0.9

The labels P, G and L mean for *Product logic*, *Gödel intuitionistic logic* and *Łukasiewicz logic*, respectively. That is, $[\![\&_{\mathtt{P}}]\!](x, y) = x \cdot y$, $[\![\&_{\mathtt{G}}]\!](x, y) = min(x, y)$, and $[\![\&_{\mathtt{L}}]\!](x, y) = max(0, x+y-1)$. In the following derivation for the program $\mathcal{P}$ and goal $\leftarrow p(X)$, we underline the selected expression in each computation step, also indicating as a superscript the rule/connective exploited/evaluated in each admissible/interpretive step (as usual, variables of program rules are renamed after being used):

$$\begin{aligned}
\langle \underline{p(X)}; \{\}\rangle &\rightarrow_{AS1}{}^{\mathcal{R}_1} \langle 0.8 \&_{\mathtt{P}} (\underline{q(X_1, Y_1)} \&_{\mathtt{G}} r(Y_1)); \{X/X_1\}\rangle \\
&\rightarrow_{AS1}{}^{\mathcal{R}_2} \langle 0.8 \&_{\mathtt{P}} ((0.7 \&_{\mathtt{P}} \underline{s(Y_2)}) \&_{\mathtt{G}} r(Y_2)); \{X/a, X_1/a, Y_1/Y_2\}\rangle \\
&\rightarrow_{AS2}{}^{\mathcal{R}_5} \langle 0.8 \&_{\mathtt{P}} (\underline{(0.7 \&_{\mathtt{P}} 0.9)} \&_{\mathtt{G}} r(b)); \{X/a, X_1/a, Y_1/b, Y_2/b\}\rangle \\
&\rightarrow_{IS}{}^{\&_{\mathtt{P}}} \langle 0.8 \&_{\mathtt{P}} (0.63 \&_{\mathtt{G}} \underline{r(b)}); \{X/a, X_1/a, Y_1/b, Y_2/b\}\rangle \\
&\rightarrow_{AS2}{}^{\mathcal{R}_4} \langle 0.8 \&_{\mathtt{P}} \underline{(0.63 \&_{\mathtt{G}} 0.7)}; \{X/a, X_1/a, Y_1/b, Y_2/b, Y_3/b\}\rangle \\
&\rightarrow_{IS}{}^{\&_{\mathtt{G}}} \langle \underline{0.8 \&_{\mathtt{P}} 0.63}; \{X/a, X_1/a, Y_1/b, Y_2/b, Y_3/b\}\rangle \\
&\rightarrow_{IS}{}^{\&_{\mathtt{P}}} \langle 0.504; \{X/a, X_1/a, Y_1/b, Y_2/b, Y_3/b\}\rangle
\end{aligned}$$

So, after focusing our interest in variables belonging to the original goal, the final fuzzy computed answer (f.c.a., in brief) is $\langle 0.504; \{X/a\}\rangle$, with the obvious meaning that the original goal is true at a 50.4% when X be a.

3 The FLOPER System

As detailed in [11,12], our parser has been implemented by using the classical DCG's (*Definite Clause Grammars*) resource of the Prolog language, since it is a convenient notation for expressing grammar rules. Once the application is loaded inside a Prolog interpreter (in our case, Sicstus Prolog v.3.12.5), it shows a menu which includes options for loading, parsing, listing and saving fuzzy programs, as well as for executing fuzzy goals. All these actions are based in the translation of the fuzzy code into standard Prolog code. The key point is to extend each atom with an extra argument, called *truth variable* of the form `_TVi`, which is intended to contain the truth degree obtained after the subsequent evaluation of the atom. For instance, the first clause in our target program is translated into: "`p(X, _TV0) : -q(X, Y, _TV1), r(Y, _TV2), and_godel(_TV1, _TV2, _TV3),`

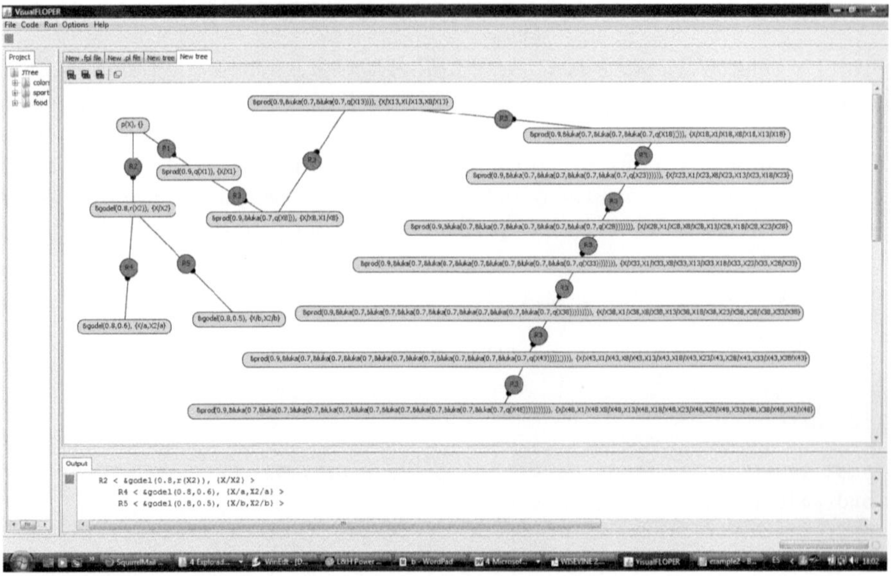

Fig. 1. Building a graphical interface for FLOPER

and_prod(0.8, _TV3, _TV0). ", where the definition of the "aggregator predicates" are: "and_prod(X, Y, Z) : −Z is X ∗ Y." and "and_godel(X, Y, Z) : −(X =< Y, Z = X; X > Y, Z = Y).". The last clause in the program, becomes the pure Prolog fact "s(b, 0.9)." while a fuzzy goal like "p(X)", is translated into the pure Prolog goal: "p(X, Truth_degree)" (note that the last truth degree variable is not anonymous now) for which the Prolog interpreter returns the two desired fuzzy computed answers [Truth_degree=0.504,X=a] and [Truth_degree=0.4,X=b].

The previous set of options suffices for running fuzzy programs: all internal computations (including compiling and executing) are pure Prolog derivations whereas inputs (fuzzy programs and goals) and outputs (fuzzy computed answers) have always a fuzzy taste, which produces the illusion on the final user of being working with a purely fuzzy logic programming tool.

Moreover, it is also possible to select into the FLOPER's goal menu, options "tree" and "depth", which are useful for tracing execution trees and fixing the maximum length allowed for their branches (initially 3), respectively. Working with these options is crucial when the "run" choice fails: remember that this last option is based on the generation of pure logic SLD-derivations which might fall in loop or directly fail in some cases as the experiments of [11] show, in contrast with the traces (based on finite, non-failed, admissible derivations) that the "tree" option displays. By using the graphical interface we are implementing for FLOPER, Figure 1 shows a tree evidencing an infinite branch where states are colored in yellow and program rules exploited in admissible steps are enclosed in circles.

4 Fuzzy Computed Answers with Extended Information

Strongly related with the last paragraph of the previous section and also connecting with the results we plan to explain in what follows, the "`ismode`" choice is useful for deciding among three levels of detail when visualizing the interpretive computations performed during the generation of "evaluation trees". This last option, together with the possibility of loading new lattices into the system, represents our last developments performed on FLOPER, as reported in [12].

```
member(X) :- number(X),0=<X,X=<1.

bot(0).                    top(1).

leq(X,Y) :- X=<Y.

and\_luka(X,Y,Z):- pri_add(X,Y,U1),pri_sub(U1,1,U2),pri_max(0,U2,Z).
and_godel(X,Y,Z):- pri_min(X,Y,Z).
and_prod(X,Y,Z) :- pri_prod(X,Y,Z).

or_luka(X,Y,Z)  :- pri_add(X,Y,U1),pri_min(U1,1,Z).
or_godel(X,Y,Z) :- pri_max(X,Y,Z).
or_prod(X,Y,Z)  :- pri_prod(X,Y,U1),pri_add(X,Y,U2),pri_sub(U2,U1,Z).

agr_aver(X,Y,Z) :- pri_add(X,Y,U),pri_div(U,2,Z).

pri_add(X,Y,Z)  :- Z is X+Y.    pri_min(X,Y,Z) :- (X=<Y,Z=X;X>Y,Z=Y).
pri_sub(X,Y,Z)  :- Z is X-Y.    pri_max(X,Y,Z) :- (X=<Y,Z=Y;X>Y,Z=X).
pri_prod(X,Y,Z) :- Z is X * Y.  pri_div(X,Y,Z) :- Z is X/Y.
```

Fig. 2. Multi-adjoint lattice modeling truth degrees in the real interval [0,1]

We have recently conceived a very easy way to model truth-degree lattices for being included into the FLOPER tool by using the "`lat/show`" options. All relevant components of each lattice can be encapsulated inside a Prolog file which must necessarily contain the definitions of a minimal set of predicates defining the set of valid elements (predicate `member`), including special mentions to the "`top`" and "`bottom`" ones, the full or partial ordering established among them (predicate `leq`), as well as the repertoire of fuzzy connectives which can be used for their subsequent manipulation.

For instance, in Figure 2 we have modeled the lattice that we used in our examples, which enables the possibility of working with truth degrees in the infinite space of the real numbers between 0 and 1, allowing too the possibility of using conjunction and disjunction operators recasted from the three typical fuzzy logics proposals described before (i.e., the *Łukasiewicz*, *Gödel* and *product* logics), as well as a useful description for the hybrid aggregator *average*. Note also that we have included definitions for auxiliary predicates, whose names always begin with the prefix "pri_". All of them are intended to describe primitive/arithmetic operators (in our case +, −, ∗, /, *min* and *max*) in a Prolog style, for being appropriately called from the bodies of clauses defining predicates with higher levels of expressivity (this is the case for instance, of the three kinds of fuzzy connectives we are considering: conjuntions, disjunctions and agreggations).

One step beyond, we can also conceive a more complex lattice whose elements could have two components, coping with truth degrees and "labels" collecting information about the program rules and fuzzy connectives used when executing programs. In order to be loaded into FLOPER, we must define in Prolog the new lattice, whose elements could be expressed, for instance, as data terms of the form "`info(Fuzzy_Truth_Degree, Label)`". Moreover, the clauses defining some predicates required for managing them are:

```
member(info(X,_)):-number(X),0=<X,X=<1.

bot(info(0,_)).            top(info(1,_)).

leq(info(X1,_),info(X2,_)) :- X1 =< X2.

and_prod(info(X1,X2),info(Y1,Y2),info(Z1,Z2)):-
            pri_prod(X1,Y1,Z1,DatPROD),pri_app(X2,Y2,Dat1),
            pri_app(Dat1,'&PROD.',Dat2),pri_app(Dat2,DatPROD,Z2).

pri_app(X,Y,Z):-name(X,L1),name(Y,L2),append(L1,L2,L3),name(Z,L3).

pri_append([],X,X).            append([A|B],C,[A|D]):-append(B,C,D).
```

Here, we have seen that when implementing for instance the conjunction operator of the Product Logic, in the second component of our extended notion of "truth degree", we have *appended* the labels of its arguments with the label `'&PROD.'` (see clauses defining `and_prod`, `pri_app` and `append`). Of course, in the fuzzy program to be run, we must also take into account the use of labels associated to the program rules. For instance, in our example the first rule must have the form:

```
p(X) <prod q(X,Y) &godel r(Y) with info(0.8,'RULE1.').
```

And now, after executing goal `p(X)` we obtain the two desired computed answers (including the sequence of program rules exploited and connective definitions evaluated till finding each solution):

```
[ Truth_degree=info(0.504,   RULE1.RULE2.RULE5.&PROD.
                             RULE4.&GODEL.&PROD.}),          X=a]

[ Truth_degree=info(0.4,     RULE1.RULE3.RULE4.&LUKA.
                             RULE4.&GODEL.&PROD.),           X=b]
```

5 Conclusions and Future Work

The experience acquired in our research group regarding the design of techniques and methods based on fuzzy logic in close relationship with the so-called multi-adjoint logic programming approach ([2,5,7,4]), has motivated our interest for putting in practice all our developments around the design of the FLOPER environment [11,12]. Our philosophy is to friendly connect this fuzzy framework with Prolog programmers: our system, apart for being implemented in Prolog, also translates the fuzzy code to classical clauses and, what is more, in this paper we have also shown that a wide range of lattices modeling powerful and flexible notions of truth degrees can be easily used into FLOPER for augmenting fuzzy computed answers with proof traces without requiring additional cost.

Apart for our ongoing efforts devoted to providing FLOPER with a graphical interface as illustrated in Figure 1, nowadays we are especially interested in implementing all the manipulation tasks developed in our group on fold/unfold transformations [2,5], partial evaluation [7] and thresholded tabulation [4].

References

1. Baldwin, J.F., Martin, T.P., Pilsworth, B.W.: Fril- Fuzzy and Evidential Reasoning in Artificial Intelligence. John Wiley & Sons, Inc., Chichester (1995)
2. Guerrero, J.A., Moreno, G.: Optimizing fuzzy logic programs by unfolding, aggregation and folding. Electronic Notes in Theoretical Computer Science 219, 19–34 (2008)
3. Ishizuka, M., Kanai, N.: Prolog-ELF Incorporating Fuzzy Logic. In: Joshi, A.K. (ed.) Proceedings of the 9th International Joint Conference on Artificial Intelligence (IJCAI 1985), pp. 701–703. Morgan Kaufmann, San Francisco (1985)
4. Julián, P., Medina, J., Moreno, G., Ojeda, M.: Efficient thresholded tabulation for fuzzy query answering. Studies in Fuzziness and Soft Computing (Foundations of Reasoning under Uncertainty) 249, 125–141 (2010)
5. Julián, P., Moreno, G., Penabad, J.: On Fuzzy Unfolding. A Multi-adjoint Approach. Fuzzy Sets and Systems 154, 16–33 (2005)
6. Julián, P., Moreno, G., Penabad, J.: Operational/Interpretive Unfolding of Multi-adjoint Logic Programs. Journal of Universal Computer Science 12(11), 1679–1699 (2006)
7. Julián, P., Moreno, G., Penabad, J.: An Improved Reductant Calculus using Fuzzy Partial Evaluation Techniques. Fuzzy Sets and Systems 160, 162–181 (2009)
8. Lloyd, J.W.: Foundations of Logic Programming. Springer, Berlin (1987)
9. Medina, J., Ojeda-Aciego, M., Vojtáš, P.: A procedural semantics for multi-adjoint logic programing. In: Brazdil, P.B., Jorge, A.M. (eds.) EPIA 2001. LNCS (LNAI), vol. 2258, pp. 290–297. Springer, Heidelberg (2001)
10. Medina, J., Ojeda-Aciego, M., Vojtáš, P.: Similarity-based Unification: a multi-adjoint approach. Fuzzy Sets and Systems 146, 43–62 (2004)
11. Morcillo, P.J., Moreno, G.: Programming with fuzzy logic rules by using the FLOPER tool. In: Bassiliades, N., Governatori, G., Paschke, A. (eds.) RuleML 2008. LNCS, vol. 5321, pp. 119–126. Springer, Heidelberg (2008)
12. Morcillo, P.J., Moreno, G., Penabad, J., Vázquez, C.: A Practical Management of Fuzzy Truth-Degrees Using FLOPER. In: Dean, M., Hall, J., Rotolo, A., Tabet, S. (eds.) RuleML 2010. LNCS, vol. 6403, pp. 20–34. Springer, Heidelberg (2010)

Implication Triples versus Adjoint Triples*

M[a] Eugenia Cornejo, Jesús Medina**, and Eloisa Ramírez

Department of Mathematics, University of Cádiz
jesus.medina@uca.es

Abstract. Implication triples and adjoint triples are two of the more general residuated operators which have been applied independently in manifold important fields. This paper presents diverse properties of adjoint triples in order to relate them to implication triples. As a consequence of this relation, we obtain, for example, that a multi-adjoint lattice in multi-adjoint logic programming is a particular case of a complete adjointness lattice.

1 Introduction

Several different types of algebraic structures form a background for many domains in mathematics and information sciences, such as many-valued logics, generalized measure and integral theory, quantum logics, quantum computing, etc.

Usually, the operators used are the t-norms which are defined in the unit interval $[0, 1]$. However, these operators are very restrictive and manifold applications need more general structures, e.g., that the carrier will be a lattice, as well as operators on these structures which provide a more flexible environment.

Adjoint triples are used as basic operators to carry out the calculus in several frameworks. For example, adjoint triples play an important role in two important environments: multi-adjoint logic programming and multi-adjoint concept lattices. Although the adjoint triples are defined in a general environment, more properties must be assumed in order to ensure the mechanism for the calculus needed to resolve the problems in, e.g., these frameworks.

On the other hand, implication triples and adjointness algebras were introduced in [15] and many properties were studied in [16,1]. In fact, an extension - the logic of tied implications - was presented in [13,14,12]. Implication triples are operators defined in a general structure that follow the same motivation of the adjoint triples in order to reduce the mathematical requirements of the basic operators used in, e.g., logic programming.

This paper presents diverse properties of adjoint triples and, as a result, the relation between implication triples and adjoint triples is fixed. This relation provides that an multi-adjoint lattice in multi-adjoint logic programming is a particular case of a complete adjointness lattice.

---

* Partially supported by the Spanish Science Ministry under grant TIN2009-14562-C05-03 and by Junta de Andalucía under grant P09-FQM-5233.

** Corresponding author.

J. Cabestany, I. Rojas, and G. Joya (Eds.): IWANN 2011, Part II, LNCS 6692, pp. 453–460, 2011.

The plan of this paper is the following: Section 2 recalls adjoint triples, multi-adjoint logic programming and multi-adjoint concept lattices; Section 3 introduces implication triples and adjointness algebras, presents several properties of adjoint triples and the relation of these operators to implication triples; the paper ends with a number of conclusions and prospects for future work.

2 Adjoint Triples and Applications

Assuming non-commutativity on the conjunctor, directly provides two different ways of generalizing the well-known adjoint property between a t-norm and its residuated implication, depending on which argument is fixed.

Definition 1. *Let* $(P_1, \leq_1)$, $(P_2, \leq_2)$, $(P_3, \leq_3)$ *be posets and* $\&\colon P_1 \times P_2 \longrightarrow P_3$, $\swarrow\colon P_3 \times P_2 \longrightarrow P_1$, $\nwarrow\colon P_3 \times P_1 \longrightarrow P_2$ *be mappings, then* $(\&, \swarrow, \nwarrow)$ *is an* adjoint triple *with respect to* P_1, P_2, P_3 *if:*

1. $\&$ *is order-preserving on both arguments, i.e. if* $x_1, x_2, x \in P_1$, $y_1, y_2, y \in P_2$ *and* $x_1 \leq_1 x_2$, $y_1 \leq_2 y_2$, *then* $(x_1 \& y) \leq_3 (x_2 \& y)$ *and* $(x \& y_1) \leq_3 (x \& y_2)$*; and*
2. $\swarrow, \nwarrow$ *are order-preserving on the first argument and order-reversing on the second argument, i.e. if* $x_1, x_2, x \in P_1$, $y_1, y_2, y \in P_2$, $z_1, z_2, z \in P_3$ *and* $x_1 \leq_1 x_2$, $y_1 \leq_2 y_2$, $z_1 \leq_3 z_2$, *then* $(z_1 \swarrow y) \leq_1 (z_2 \swarrow y)$, $(z \swarrow y_2) \leq_1 (z \swarrow y_1)$, $(z_1 \nwarrow x) \leq_2 (z_2 \nwarrow x)$ *and* $(z \nwarrow x_2) \leq_2 (z \nwarrow x_1)$*; and*
3. $x \leq_1 z \swarrow y$ *iff* $x \& y \leq_3 z$ *iff* $y \leq_2 z \nwarrow x$, *where* $x \in P_1$, $y \in P_2$ *and* $z \in P_3$.

Note that in the domain and codomain of the considered conjunctor we have three (in principle) different sorts, thus providing a more flexible language to a potential user. Furthermore, notice that no boundary condition is required, in difference to the usual definition of multi-adjoint lattice [11] or implication triple [15].

Adjoint triples are used as basic operators to make the calculus in manifold frameworks. Adjoint triples play important roles in two important environments: multi-adjoint logic programming and multi-adjoint concept lattices.

Although adjoint triples are defined in a general environment, more properties must be assumed in order to ensure the mechanism for the calculus needed to resolve the problems in, e.g., these frameworks. The following subsections introduce a summary of multi-adjoint logic programming and multi-adjoint concept lattices and several comments about the need for new properties required by the adjoint triples.

2.1 Multi-adjoint Logic Programing

Multi-adjoint logic programming was introduced in [10] as a refinement of both initial work in [17] and residuated logic programming [5]. The main concept in the extension to logic programming to the fuzzy case is that of *adjoint pair*.

Definition 2 (Adjoint pair). *Let* $\langle P, \preceq\rangle$ *be a partially ordered set and a pair* $(\&, \leftarrow)$, *of binary operations in* P, *such that:*

(a1) *Operation & is order-preserving in both arguments.*
(a2) *Operation $\leftarrow$ is order-preserving in the first argument (the consequent) and order-reversing in the second argument (the antecedent).*
(a3) *For any $x, y, z \in P$, we have that $x \preceq (y \leftarrow z)$ holds if and only if $(x \,\&\, z) \preceq y$ holds.*

Then we say that $(\&, \leftarrow)$ *forms an* adjoint pair *in* $\langle P, \preceq \rangle$.

Note that, if $(\&, \swarrow, \nwarrow)$ is an adjoint triple with respect to P_1, P_2 and P_3, where $P_1 = P_2 = P_3$, then $(\&, \swarrow)$ is an adjoint pair.

The need of the monotonicity of operators $\leftarrow$ and & is clear, if they are going to be interpreted as generalised implications and conjunctions. The third property in the definition corresponds to the categorical adjointness, and can be adequately interpreted in terms of multiple-valued inference as asserting both that the truth-value of $y \leftarrow z$ is the maximal x satisfying $x \,\&\, z \preceq y$, and also the validity of the following generalised modus ponens rule [6].

In addition to $(a1)$–$(a3)$ it will be necessary to assume the existence of bottom and top elements in the poset of truth-values (the zero and one elements), and the existence of joins (suprema) for every directed subset; that is, we will assume a structure of complete lattice, but nothing about associativity, commutativity and general boundary conditions of & .

Extending the results in [5,17] to a more general setting, in which different implications (Łukasiewicz, Gödel, product) and thus, several modus ponens-like inference rules are used, naturally leads to considering different *adjoint pairs* in the lattice:

Definition 3 (Multi-Adjoint Lattice). *Let $\langle L, \preceq \rangle$ be a lattice. A* multi-adjoint lattice $\mathcal{L}$ *is a tuple $(L, \preceq, \leftarrow_1, \&_1, \ldots, \leftarrow_n, \&_n)$ satisfying the following items:*

(l1) *$\langle L, \preceq \rangle$ is bounded, i.e. it has bottom ($\bot$) and top ($\top$) elements;*
(l2) *$(\&_i, \leftarrow_i)$ is an adjoint pair in $\langle L, \preceq \rangle$ for $i = 1, \ldots, n$;*
(l3) *$\top \,\&_i\, \vartheta = \vartheta \,\&_i\, \top = \vartheta$ for all $\vartheta \in L$ for $i = 1, \ldots, n$.*

The multi-adjoint programs are defined on a language $\mathfrak{F}$ constructed on a set of propositional symbol Π and the set of operators Ω and using a multi-adjoint lattice $\mathcal{L}$ as the set of truth-values [10].

A general multi-adjoint programs, where P_1, P_2 and P_3, can be different, is presented in [3]. Moreover, in [9] adjoint triples are considered in the multi-adjoint programs instead of adjoint pairs.

2.2 Multi-adjoint Concept Lattices

Multi-adjoint concept lattices considers at the beginning the adjoint triples as in Definition 2. Furthermore, in order to provide more flexibility into the language, the existence of different adjoint triples for a given triplet of posets is considered. Notice, however, that since these triplets will be used as the underlying structures of the multi-adjoint concept lattice, it is reasonable to require the lattice structure on some of the posets in the definition of adjoint triple.

Definition 4. *A multi-adjoint frame $\mathcal{L}$ is a tuple $(L_1, L_2, P, \&_1, \ldots, \&_n)$, where $(P, \leq)$ is a poset, $(L_1, \preceq_1)$ and $(L_2, \preceq_2)$ are complete lattices and, for all $i \in \{1, \ldots, n\}$, $(\&_i, \swarrow^i, \nwarrow_i)$ is an adjoint triple with respect to L_1, L_2, P.*

The definition of multi-adjoint context follows similarly as in the classical case.

Definition 5. *Given a multi-adjoint frame $(L_1, L_2, P, \&_1, \ldots, \&_n)$, a* context *is a tuple (A, B, R, σ) such that A and B are non-empty sets (usually interpreted as attributes and objects, respectively), R is a P-fuzzy relation $R\colon A \times B \to P$ and $\sigma\colon B \to \{1, \ldots, n\}$ is a mapping which associates any element in B with some particular adjoint triple in the frame.*

Once we have fixed a multi-adjoint frame and a context for that frame, we can define the following mappings ${}^{\uparrow_\sigma}\colon L_2^B \longrightarrow L_1^A$ and ${}^{\downarrow^\sigma}\colon L_1^A \longrightarrow L_2^B$ which can be seen as generalisations of those given in [2,7]:

$$g^{\uparrow_\sigma}(a) = \inf\{R(a, b) \swarrow^{\sigma(b)} g(b) \mid b \in B\} \tag{1}$$

$$f^{\downarrow^\sigma}(b) = \inf\{R(a, b) \nwarrow_{\sigma(b)} f(a) \mid a \in A\} \tag{2}$$

It is not difficult to show that these two arrows generate a Galois connection [8]. Moreover, if A and B are finite, then we can consider in the definition of context that the lattices $(L_1, \preceq_1)$ and $(L_2, \preceq_2)$ do not need to be complete.

As usual in the different frameworks of formal concept analysis, a *multi-adjoint concept* is a pair $\langle g, f\rangle$ satisfying that $g \in L_2^B$, $f \in L_1^A$ and that $g^{\uparrow_\sigma} = f$ and $f^{\downarrow^\sigma} = g$; with $({}^{\uparrow_\sigma}, {}^{\downarrow^\sigma})$ being the Galois connection defined above.

Definition 6. *The* multi-adjoint concept lattice *associated to a multi-adjoint frame $(L_1, L_2, P, \&_1, \ldots, \&_n)$ and a context (A, B, R, σ) is the set*

$$\mathcal{M} = \{\langle g, f\rangle \mid g \in L_2^B, f \in L_1^A \text{ and } g^{\uparrow_\sigma} = f, f^{\downarrow^\sigma} = g\}$$

in which the ordering is defined by $\langle g_1, f_1\rangle \preceq \langle g_2, f_2\rangle$ if and only if $g_1(b) \preceq_2 g_2(b)$, for all $b \in B$ (equivalently $f_2(a) \preceq_1 f_1(a)$, for all $a \in A$).

The ordering just defined above actually provides $\mathcal{M}$ with the structure of a complete lattice.

3 Properties of Adjoint Triples to Be Related to Implication Triples

This section introduces several properties of adjoint triples. Note that the goal of this paper is to relate adjoint triples to implication triples, hence these operators are developed in order to reach this goal and, first of all, implication triples and adjointness algebras must be introduced.

Implication triples and adjointness algebras were introduced in [15] and a lot of properties were studied in [16,1], indeed an extension, the logic of tied implications, was presented in [13,14,12].

Definition 7. *An* adjointness algebra *is an 8-tuple* $(L, \leq_L, P, \leq_P, \top_P, A, K, H)$, *in which* $(L, \leq_L)$, $(P, \leq_P)$ *are two posets with a top element* $\top_P$ *in* $(P, \leq_P)$, *and the following four conditions are satisfied:*

1. *The operation* $A\colon P \times L \to L$ *is antitone in the left argument and monotone in the right argument, and it has* $\top_P$ *as a left identity element, that is* $A(\top_P, \gamma) = \gamma$, *for all* $\gamma \in L$. *We call A an* implication on (L, P).
2. *The operation* $K\colon P \times L \to L$ *is monotone in each argument and has* $\top_P$ *as a left identity element, that is* $K(\top_P, \beta) = \beta$, *for all* $\beta \in L$. *We call K a* conjunction on (L, P).
3. *The operation* $H\colon L \times L \to P$ *is antitone in the left argument and monotone in the right argument, and it satisfies, for all* $\beta, \gamma \in L$, *that:*

$$H(\beta, \gamma) = \top_P \quad \textit{if and only if} \quad \beta \leq_L \gamma$$

   We call H a forcing implication on L.
4. *The three operations* A, K *and* H, *are mutually related by the following conditions, for all* $\alpha \in P$ *and* $\beta, \gamma \in L$:
   Adjointness : $\beta \leq_L A(\alpha, \gamma)$ *iff* $K(\alpha, \beta) \leq_L \gamma$ *iff* $\alpha \leq_P H(\beta, \gamma)$

   We call the ordered triple (A, K, H) *an* implication triple *on* (L, P).

An adjointness lattice *is an adjointness algebra whose two underlying posets are lattices. A* complete adjointness lattice *is an adjointness algebra over two complete lattices. An* adjointness chain *is an adjointness lattice whose two orders are linear.*

Now, diverse properties of adjoint triples will be presented directed to related both definitions adjoint triples and implication triples. From now on, the posets $(P_1, \leq_1)$, $(P_2, \leq_2)$ and $(P_3, \leq_3)$, and the adjoint triple $(\&, \swarrow, \nwarrow)$ will be fixed.

The first statements show what the posets and conjunctor $\&$ must satisfy in order to assure that $\swarrow$ and $\nwarrow$ are the expected generalizations of the classical two-valued material implication connective. The proof of these results follows directly by the hypothesis and the adjoint property and it was commented in previous papers [4].

Lemma 1. *If* $P_2 \subseteq P_3$ *and* P_1 *has a maximum* $\top_1$ *as a left identity element for* $\&$, *that is* $\top_1 \& y = y$, *for all* $y \in P_2$, *then we obtain that:*

$$\top_1 = z \swarrow y \quad \textit{if and only if} \quad y \leq_3 z$$

Lemma 2. *If* $P_1 \subseteq P_3$, *and* P_2 *has a maximum* $\top_2$ *as a right identity element for* $\&$, *that is* $x \& \top_1 = x$, *for all* $x \in P_1$, *then we obtain that:*

$$\top_2 = z \nwarrow x \quad \textit{if and only if} \quad x \leq_3 z$$

The properties above are used to prove the following technical results, which are a first step to relating adjoint triples to implication triples. First of all, the forcing implication property is related to the boundary conditions.

Proposition 1. *If* $P_2 \subseteq P_3$ *and* P_1 *has a maximum* $\top_1$, *then we obtain that*

$$\swarrow \textit{ is a forcing-implication} \quad \textit{if and only} \quad \top_1 \& y = y, \textit{ for all } y \in P_2$$

Proposition 2. *If $P_1 \subseteq P_3$ and P_2 has a maximum $\top_2$, then we obtain that*

$$\nwarrow \text{ is a forcing-implication} \quad \text{if and only} \quad x \,\&\, \top_2 = x, \text{ for all } x \in P_1$$

The second step is given by the next results which relate the implication property, in implication triples, to the boundary conditions.

Proposition 3. *If $P_2 \subseteq P_3$ and P_1 has a maximum $\top_1$, then the following equivalence is provided*

$$z \nwarrow \top_1 = z \quad \text{if and only if} \quad \top_1 \,\&\, y = y, \text{ for all } y \in P_2$$

Proposition 4. *If $P_1 \subseteq P_3$ and P_2 has a maximum $\top_2$, then the following equivalence is provided*

$$z \swarrow \top_2 = z \quad \text{if and only if} \quad x \,\&\, \top_2 = x, \text{ for all } x \in P_1$$

Assuming the oposite mappings of the adjoint implications $\swarrow$ and $\nwarrow$, that is the mappings $\swarrow^{op} \colon P_2 \times P_3 \to P_3$, $\nwarrow_{op} \colon P_1 \times P_3 \to P_2$, defined as $y \swarrow^{op} z = z \swarrow y$ and $x \nwarrow_{op} z = z \nwarrow x$, for all $x \in P_1$, $y \in P_2$ and $z \in P_3$, we have the following result:

Theorem 1. *Given an adjoint triple $(\&, \swarrow, \nwarrow)$, with respect to the posets $(P_1, \leq_1)$, $(P_2, \leq_2)$, $(P_3, \leq_3)$, where $P_2 = P_3$, and P_1 has a maximum $\top_1$ as a left identity element for $\&$, we obtain that $(\&, \swarrow^{op}, \nwarrow_{op})$ is an implication triple.*

Proof. If the adjoint triple $(\&, \swarrow, \nwarrow)$ satisfies $\top_1 \,\&\, y = y$, for all $y \in P_2$, then by Proposition 1, the mapping $\swarrow^{op}$ is a forcing implication and by Proposition 3, the mapping $\nwarrow_{op}$ is an implication. Consequently, $(\&, \swarrow^{op}, \nwarrow_{op})$ is an implication triple.

Therefore, the definition of adjoint triple is more general than implication triple. Indeed, the definition of implication triple can be given as follows.

Definition 8. *Let $(L, \leq_L)$ and $(P, \leq_P)$ be two posets with a maximum element $\top_P$ in $(P, \leq_P)$. An implication triple (A, K, H) is an ordered triple in which $A \colon P \times L \to L$ and $H \colon L \times L \to P$ are antitone in the left argument and monotone in the right argument, $K \colon P \times L \to L$ is monotone in each argument and has $\top_P \in P$ as a left identity element, and the adjointness conditions is verified, for all $\alpha \in P$, $\beta, \gamma \in L$:*
Adjointness: $\beta \leq A(\alpha, \gamma)$ *iff* $K(\alpha, \beta) \leq \gamma$ *iff* $\alpha \leq H(\beta, \gamma)$.

Therefore, instead of saying that (A, K, H) is an implication triple, we would say that it is a left implication triple, since K satisfies the identity on the left. Likewise, the implication on the right is defined as follows:

Definition 9. *Let $(L, \leq_L)$ and $(P, \leq_P)$ be two posets with a maximum element $\top_P$ in $(P, \leq_P)$. An implication triple (A, K, H) is an ordered triple in which $A \colon L \times L \to P$ and $H \colon P \times L \to L$ are antitone in the left argument and monotone in the right argument, $K \colon L \times P \to L$ is monotone in each argument and has $\top_P \in P$ as a right identity element, and the adjointness conditions is verified, for all $\beta \in P$, $\alpha, \gamma \in L$:*
Adjointness: $\beta \leq A(\alpha, \gamma)$ *iff* $K(\alpha, \beta) \leq \gamma$ *iff* $\alpha \leq H(\beta, \gamma)$.

As a result, if $P = L$ and (A, K, H) is a left implication triple (resp. right implication triple) which satisfies that $\top_P \in P$ is a right identity element for K (resp. $\top_P \in P$ is a left identity element for K), then (A, K, H) is a right implication triple (resp. (A, K, H) is a left implication triple).

Furthermore, a multi-adjoint lattice in multi-adjoint logic programming is a particular case of complete adjointness lattice. The relation is not so strong with respect to the multi-adjoint concept lattice environment, since an adjointness algebra cannot consider three different kinds as a support and moreover, needs K to satisfy the identity on the left, a property now required in this environment. However, the posets do need to be (complete) lattices.

Now we will present an example of adjoint triple, which will be compared with the definition of implication triple. The following adjoint triple was used in [11] to obtain information from a relational data base obtained form journal citation reports.

In this paper the authors consider a multi-adjoint frame with three different lattices: one for handling the information taken from the JCR, which is rounded to the second decimal digit; a second one to handle information about the attributes, in which we estimate steps of 0.05 in order to appreciate a qualitative difference; and a third one, which is used to set the different levels of preference of the journal, which is considered to be of 0.125 (hence the unit interval is divided into eight equal pieces)

Hence, they assume the multi-adjoint frame

$$([0,1]_{20}, [0,1]_8, [0,1]_{100}, \leq, \leq, \leq, \&_P^*, \&_L^*)$$

where $[0,1]_m$ denotes a regular partition of $[0,1]$ into m pieces, $\&_P^* \colon [0,1]_{20} \times [0,1]_8 \to [0,1]_{100}$ and $\&_L^* \colon [0,1]_{20} \times [0,1]_8 \to [0,1]_{100}$ are the discretisations of the product and Łukasiewicz conjunctors respectively.

For example, the operator $\&_P^*$, is defined for each $x \in [0,1]_{20}$ and $y \in [0,1]_8$ as:

$$\&_P^*(x, y) = \frac{\lceil 100 \cdot x \cdot y \rceil}{100}$$

where $\lceil \_ \rceil$ is the ceiling function.

For this operator, the corresponding residuated implications $\swarrow_P^* \colon [0,1]_{100} \times [0,1]_8 \to [0,1]_{20}$ and $\nwarrow_P^* \colon [0,1]_{100} \times [0,1]_{20} \to [0,1]_8$ can be defined as in [11], obtaining that $(\&_P^*, \swarrow_P^*, \nwarrow_P^*)$ is an adjoint triple, although it is not an implication triple.

Effectively, $\&_P^*$ is non-commutative and non-associative as well, although it does not satisfy that 1 is a left identity for $\&_P^*$ ($\&_P^*(1, y) = y$, for all $y \in [0,1]_8$). For instance, if $y = 0.625$, then $\&_P^*(1, 0.625) = 0.63$. Furthermore, $[0,1]_8 \neq [0,1]_{20}$.

4 Conclusions and Future Work

Several properties of adjoint triples have been presented in order to relate these operators to implication triples. Consequently, we have shown that it is possible to shorten the definition of implication triple. Moreover, this relation provides that a multi-adjoint lattice in multi-adjoint logic programming is a particular case of a complete adjointness lattice; however, the relation is not so strong with respect to the multi-adjoint concept lattice environment.

In the future, more properties of adjoint triples will be proved and compared with those given for implication triples. Moreover, practical examples will be studied in which these general operators and their properties will be useful.

References

1. Abdel-Hamid, A., Morsi, N.: Associatively tied implicacions. Fuzzy Sets and Systems 136(3), 291–311 (2003)
2. Bělohlávek, R.: Concept lattices and order in fuzzy logic. Annals of Pure and Applied Logic 128, 277–298 (2004)
3. Damásio, C., Medina, J., Ojeda-Aciego, M.: Sorted multi-adjoint logic programs: Termination results and aplications. In: Alferes, J.J., Leite, J. (eds.) JELIA 2004. LNCS (LNAI), vol. 3229, pp. 252–265. Springer, Heidelberg (2004)
4. Damásio, C., Medina, J., Ojeda-Aciego, M.: Termination of logic programs with imperfect information: applications and query procedure. Journal of Applied Logic 5, 435–458 (2007)
5. Damásio, C.V., Pereira, L.M.: Monotonic and residuated logic programs. In: Benferhat, S., Besnard, P. (eds.) ECSQARU 2001. LNCS (LNAI), vol. 2143, pp. 748–759. Springer, Heidelberg (2001)
6. Hájek, P.: Trends in Logic. In: Metamathematics of Fuzzy Logic. Kluwer Academic, Dordrecht (1998)
7. Krajči, S.: A generalized concept lattice. Logic Journal of IGPL 13(5), 543–550 (2005)
8. Medina, J., Ojeda-Aciego, M., Ruiz-Calviño, J.: Formal concept analysis via multi-adjoint concept lattices. Fuzzy Sets and Systems 160(2), 130–144 (2009)
9. Medina, J., Ojeda-Aciego, M., Valverde, A., Vojtáš, P.: Towards biresiduated multi-adjoint logic programming. In: Conejo, R., Urretavizcaya, M., Pérez-de-la-Cruz, J.-L. (eds.) CAEPIA/TTIA 2003. LNCS (LNAI), vol. 3040, pp. 608–617. Springer, Heidelberg (2004)
10. Medina, J., Ojeda-Aciego, M., Vojtáš, P.: Multi-adjoint Logic Programming with Continuous Semantics. In: Eiter, T., Faber, W., Truszczyński, M. (eds.) LPNMR 2001. LNCS (LNAI), vol. 2173, pp. 351–364. Springer, Heidelberg (2001)
11. Medina, J., Ojeda-Aciego, M., Vojtáš, P.: Similarity-based unification: a multi-adjoint approach. Fuzzy Sets and Systems 146, 43–62 (2004)
12. Morsi, N., El-Zekey, M.: Applications of tied implications to approximate reasoning and fuzzy control. In: 70 Years of FCE STU (2008)
13. Morsi, N., Lotfallah, W., El-Zekey, M.: The logic of tied implications, part 1: Properties, applications and representation. Fuzzy Sets and Systems 157(15), 647–669 (2006)
14. Morsi, N., Lotfallah, W., El-Zekey, M.: The logic of tied implications, part 2: Syntax, fuzzy sets and systems. Fuzzy Sets and Systems 157(17), 2030–2057 (2006); Corrigendum in: Fuzzy Sets and Systems 157 (17), 2416-2417
15. Morsi, N.N.: Propositional calculus under adjointness. Fuzzy Sets and Systems 132(1), 91–106 (2002)
16. Morsi, N.N., Roshdy, E.M.: Issues on adjointness in multiple-valued logics. Information Sciences 176, 2886–2909 (2005)
17. Vojtáš, P.: Fuzzy logic programming. Fuzzy sets and systems 124(3), 361–370 (2001)

Confidence-Based Reasoning with Local Temporal Formal Contexts*

Gonzalo A. Aranda-Corral[1], Joaquín Borrego Díaz[2], and Juan Galán Páez[2]

[1] Universidad de Huelva, Department of Information Technology, Crta. Palos de La Frontera s/n. 21819 Palos de La Frontera, Spain
[2] Universidad de Sevilla, Department of Computer Science and Artificial Intelligence, Avda. Reina Mercedes s/n. 41012 Sevilla, Spain

Abstract. Formal Concept Analysis (FCA) is a theory whose goal is to discover and to extract Knowledge from qualitative data. It provides tools for reasoning with implication basis (and association rules). In this paper we analyse how to apply FCA reasoning to increase confidence in sports betting, by means of detecting temporal regularities from data. It is applied to build a Knowledge based system for confidence reasoning.

1 Introduction

Context modelling and reasoning represents a major paradigm in Artificial Intelligence (AI). It is a useful approach for pragmatic and realistic reasoning in AI. An interesting issue in some types of context reasoning problems is Knowledge's temporal dimension. Knowledge Bases (KB), or databases, may contain information from temporal stamps, bounds or duration. The correctness of reasoning with them depends on a sound selection for time-dependent data (among other features) that will be used in each context. It represents a problem in data mining, particularly for reasoning with association rules, thinking on them as implications with no exact confidence.

Formal Concept Analysis (FCA) [8] is a mathematical theory for data analysis using formal contexts and concept lattices as key tools. Domains can be formally modelled according to the extent and the intent of each formal concept. In FCA, the basic data structure is a formal context (with a qualitative nature) which represents a set of objects and their properties and it is useful both to detect and to describe regularities and structures of concepts. It also provides a sound formalism for reasoning with such structures, mainly Stem Basis and association rules. Therefore, it is interesting to consider its application for reasoning with temporal qualitative data (see e.g. [14]) in order to discover temporal trends.

In this paper, FCA application scope is the challenge of sports betting, specifically, the forecasting of soccer league's results. Forecasting sport results is a fast

* Partially supported by TIN2009-09492 project (Spanish Ministry of Science and Innovation), cofinanced with FEDER founds and *Proyecto de excelencia* TIC-6064 of *Junta de Andalucía*.

J. Cabestany, I. Rojas, and G. Joya (Eds.): IWANN 2011, Part II, LNCS 6692, pp. 461–468, 2011.

growing research area, because of its economic impact in betting markets as well as for its potential application to problems with similar behaviour (markets) [1].

Roughly speaking, three dimensions have been considered for analysing/synthesizing prediction systems: 1)Those which analyse information on teams (endogenous) versus those which analyse results (exogenous); 2)Those which exploit quantitative data versus those which exploit qualitative knowledge, and finally, 3)Statistic-based ones versus other methods. Usually, one can work with hybrid models, and rarely with pure qualitative and exogenous reasoning systems appear in literature, although their use is considered for experiments (for example, frugal methods [3] and based on the recognition heuristic [10]) or as part of hybrid systems (see e.g. [13]). There are two reasons that may justify this point.

On the one hand, transformation from a large quantitative dataset to a qualitative problem is faced with the selection of an acceptable threshold and the discovery of better relations (see e.g. [12]). On the other hand, a qualitative dataset must be accomplished with some amount of information based on confidence, trust or probability of these data sets.

The aim of this paper is to present a method for FCA reasoning on contexts with temporal dimensions that allows the detection of some kind of regularity in data focusing on results from the Spanish soccer league as the source of temporal qualitative information. The method is bet-oriented and its performance is evaluated within a confidence-based reasoning system that increases the number of hits in soccer matches forecasting by using the discovery of temporal trends on data mining and association rules reasoning.

The structure of the paper is as follows. The next section reviews the main features of FCA and association rules on formal contexts. Temporal formal contexts are defined in Sect. 3. The confidence-based reasoning system is described in Sect. 4, and some comments on experimentation are discussed in Sect. 5. Section 6 is devoted to describe future work.

2 Formal Concept Analysis

According R. Wille, FCA mathematizes the philosophical understanding of a concept as a unit of thoughts composed of two parts: the extent and the intent [8]. The extent covers all objects belonging to this concept, while the intent comprises of all common attributes valid for all the objects under consideration. It also allows the computation of concept hierarchies from data tables. In this section, we succinctly present basic FCA elements, although it is assumed that the reader is familiar with this theory (the fundamental reference is [8]).

We represent a formal context as $M = (O, A, I)$, which consists of two sets, O (objects) and A (attributes) and a relation $I \subseteq O \times A$. Finite contexts can be represented by a 1-0-table (representing I as a Boolean function on $O \times A$). The FCA main goal is the computation of the concept lattice associated to the context. In this paper it works with logical relations on attributes which are valid in the context. For $X \subseteq O$ and $Y \subseteq A$ we can define

$$X' := \{a \in A \mid oIa \text{ for all } o \in X\} \quad Y' := \{o \in O \mid oIa \text{ for all } a \in Y\}$$

Logical expressions in FCA are *implications between attributes*, pair of sets of attributes, written as $Y_1 \to Y_2$, which is true with respect to $M = (O, A, I)$ according to the following definition. A subset $T \subseteq A$ *respects* $Y_1 \to Y_2$ if $Y_1 \not\subseteq T$ or $Y_2 \subseteq T$. It says that $Y_1 \to Y_2$ holds in M ($M \models Y_1 \to Y_2$) if for all $o \in O$, the set $\{o\}'$ respects $Y_1 \to Y_2$. In that case, $Y_1 \to Y_2$ is *an implication* of M.

Definition 1. *Let $\mathcal{L}$ be a set of implications and L an implication of M.*
1. *L follows from $\mathcal{L}$ ($\mathcal{L} \models L$) if each subset of A respecting $\mathcal{L}$ also respects L.*
2. *$\mathcal{L}$ is complete if every implication of the context follows from $\mathcal{L}$.*
3. *$\mathcal{L}$ is non-redundant if for each $L \in \mathcal{L}$, $\mathcal{L} \setminus \{L\} \not\models L$.*
4. *If $\mathcal{L}$ is a basis for M is complete and non-redundant.*

It can obtain a basis from the *pseudo-intents* [11] called *Stem basis*:
$\mathcal{L} = \{Y \to Y'' \; : \; Y \text{ is a pseudointent}\}$
The so-called *Armstrong rules* provides an implicational reasoning:

$$R1 : \frac{}{X \to X} \qquad R2 : \frac{X \to Y}{X \cup Z \to Y} \qquad R3 : \frac{X \to Y,\; Y \cup Z \to W}{X \cup Z \to W}$$

Let $\vdash_A$ be the proof relation by Armstrong rules. It holds that implicational bases are $\vdash_A$-complete [6]: If $\mathcal{L}$ is a implicational basis for M, and L an implication, then $M \models L$ if and only if $\mathcal{L} \vdash_A L$.

In order to work with formal contexts, stem basis and association rules, the Conexp[1] software has been selected. It has been used as a library to build the module which provides implications (and association rules) to the reasoning module. This module is a production system based on which was designed for [4]. It works with Stem Basis, and entailment is based on the following result.

Theorem 1. *Let $\mathcal{S}$ be a stem basis associated with the context M, o a new document tagged with $A_1, \ldots, A_n$. The following conditions are equivalent:*

1. *$\mathcal{S} \cup \{A_1, \ldots A_n\} \vdash_p Y$ ($\vdash_p$ is the entailment from the production system).*
2. *$S \vdash_A A_1, \ldots A_n \to Y$*
3. *$M \models \{A_1, \ldots A_n\} \to Y$.*

We can consider a Stem Basis as an adequate production system in order to reason and predict results. However, Stem Basis is designed for entailing true implications only, without any exceptions into the object set nor implications with a low number of counterexamples in the context. Another more important question is about predictions, we are interested in obtaining some methods for selecting a result among all obtained results (even if they are mutually incoherent), and theorem 1 does not provide such a method. Therefore, it is better to consider rules with confidence instead of true implications and the initial production system must be revised for working with confidence.

Researching on sound logical reasoning methods with association rules is a relatively recent research line with promising applications [7]. In FCA, association rules are implications among sets of attributes. Confidence and support are

[1] `http://sourceforge.net/projects/conexp/`

defined as usual. Recall that the *support* of X, $supp(X)$, of a set of attributes X is defined as the proportion of objects which satisfy every attribute of X, and the *confidence* of an association rule is $conf(X \Rightarrow Y) = supp(X \cup Y)/supp(X)$. Confidence can be interpreted as an estimate of the probability $P(Y|X)$, the probability of an object satisfying every attribute of Y under the condition that it also satisfies every one of X. Conexp software provides the association rules, as well as, their confidence for contexts.

3 Data and Temporal Contexts

A temporal context on a set of objects is defined as follows:

Definition 2. *Let O be a set of objects.*

1. *A* **temporal context** *on O is a context $M = (O_1, A, I)$ where $O_1 \subseteq O \times \mathbb{N}$*
2. *A* **contextual selection** *is a map $s : O \to \mathcal{P}(O_1) \times \mathcal{P}(A)$*
3. *A* **contextual KB for an object** *o w.r.t. a selection s with confidence γ is a subset of association rules with confidence greater or equal that γ of the formal context associated to $s(o) = (s_1(o), s_2(o))$, that is, to the context*

$$M(s(o)) := (s_1(o), s_2(o), I_{\restriction s_1(o) \times s_2(o)})$$

In this paper only set of association rules extracted by Conexp with confidence greater than a threshold γ for a contextual selection are used as contextual KB.

3.1 Temporal Contexts for Soccer League

For both selecting data and building contexts, some assumptions on forecasting in soccer league matches have been considered. Reconsiderations of such decisions can be easily computed in the system. First, we consider that the regularity of team's behaviour only depends on the contextual selection that has been considered. This contextual selection is obtained by taking matches from the last X weeks backwards, starting from the week just before the one we want to forecast. Second, since FCA methods are used to discover regularity features, thus it does not consider forecasting exceptions (unexpected results). Therefore, the model can be considered as a starting point for betting expert who would adjust attributes, in order to more personalised criteria.

These attributes have to be computed and used to entail the forecasting. This analysis is assisted by Conexp. ConExp software is used to compute and analyze the concept laticces associated to the temporal contexts. In this way, the expert can evaluate the goodness of the attributes (and the thresholds defining them) (See Fig. 1). The attribute $ID\_1\_T\_16$ is defined by: 'the budget of $team_2$ is greater than γ_1 times the budget of $team_1$', where γ_1 is the threshold the expert must estimate. In the concept lattice we can observe that the biggest concept containing the attributes $team_2\_wins$ and $ID\_1\_T\_16$ covers the about the 10% of the objects owned by the first attribute, therefore it is suggested to use the second attribute for reasoning with association rules to get a prediction.

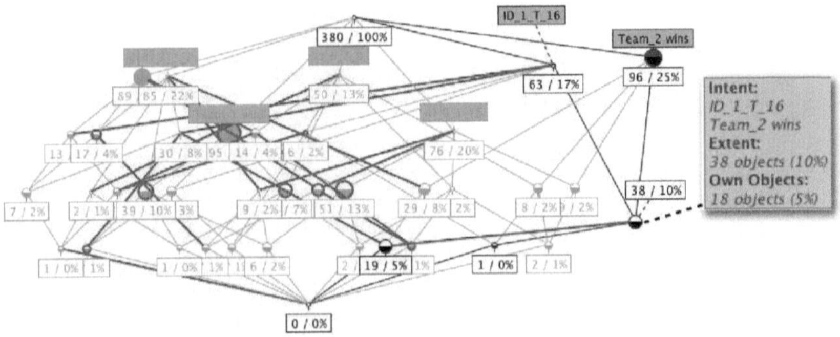

Fig. 1. Concept Lattice for the match *Málaga-Sevilla* (week 31, season 2009-10)

The system computes the value of an amount of attributes on objects. Experimentally a boolean combination of attributes is possible. Once the temporal context has been computed, the system can build contextual selections by selecting the match and the attribute set. The selection of attributes was made by considering four kinds of factors: those related with the classification, the history of teams' matches in the recent past, results of direct matches and other non related results, as for example the difference between team budgets. Seventeen relevant attributes were selected.The attribute set has three special attributes, $Team_1$ wins (1), $Team_2$ wins (2) and draws (X).

4 Confidence-Based Reasoning System

The reasoning system works on facts of the type (a, c), where a is an attribute and c is the estimated probability of the trueness of a, which we also call confidence (by similarity with the same term for association rules).

The system has a module for a confidence-based reasoning system (Fig. 2). Its entries for a match $Team_1$ - $Team_2$ are: the contextual Knowledge basis for a threshold given as rule set and attribute values for the current match

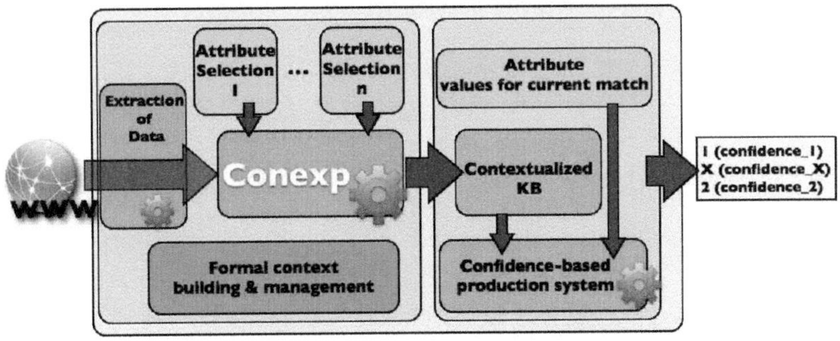

Fig. 2. Context based reasoning system

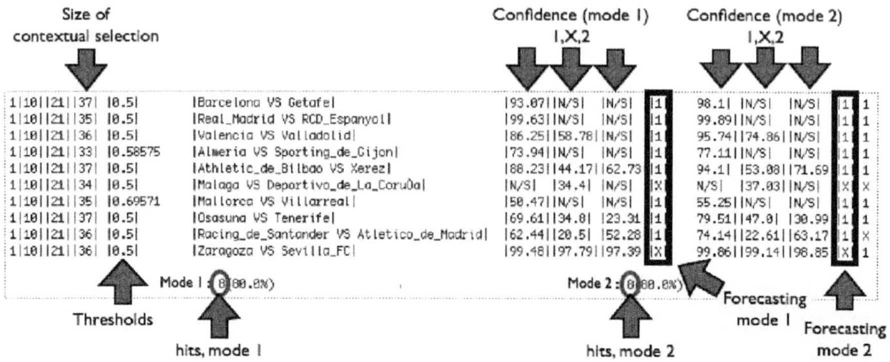

Fig. 3. Forecasting results screenshot

(except 1,X,2) as facts, all of them with a confidence (whose value depends on the reasoning mode, see below). The production system is executed and the output is a triple $< (1, c_1), (X, c_x), (2, c_2) >$ of attribute, confidence for this match. The attribute with greater confidence is selected as the prediction.

The execution of the production system is as regular. There exist several modes of confidence computing of facts, which are based in uncertain reasoning in Expert Systems [9]. Any attribute/fact a is initialized with confidence

$$conf(a) := \frac{|\{o \ : \ oIa\}| + 1}{|O| + 1}$$

The most promising computation modes used are:

Mode 1: As usual in Expert systems: If rule $r : \{a_1, \ldots, a_k\} \to \{c_1, \ldots, c_n\}$ with confidence rule $conf(r)$ is fired on the facts $(a_1, conf(a_1)), \ldots, (a_k, conf(a_k))$, the confidence estimated for each c_i by the rule is

$$conf_n(c_i) = conf(r) \cdot \min(conf(a_1), \ldots, conf(a_k))$$

and it updates $conf(c_i)$ as $conf(c) := conf(c) + conf_n(c) \cdot (1 - conf(c)$.

Mode 2: If c is obtained by firing the rule r, define $conf(c) = P(c) \cdot Q(c)$ where

$$P(c) := f_p(c, r) = conf(r), \ \ Q(c) := f_q(c, r) = \min(conf(a_1), \ldots, conf(a_k))$$

If it entails c by firing of other rule r' produces the update of $conf(c)$ by updating

$$P(c) := P(c) + f_p(c, r') - P(c) \cdot f_p(c, r'), \quad Q(c) := Q(c) + f_q(c, r') - Q(c) \cdot f_q(c, r')$$

With respect to the thresholds for confidence, currently the system allows the user to select them by hand or by using the automatic selection mode which is:

$$\gamma = \max(\{conf(a) \ : \ \text{exists } \{\emptyset\} \to Y \text{ rule of the KB s.t. } a \in Y\} \cup \{0.5\})$$

Fig. 3 shows forecasting for week 21 of the Spanish premier league (2009-10).

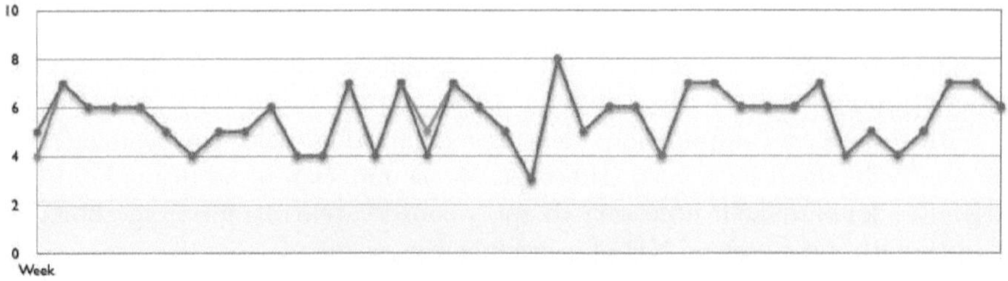

Fig. 4. Hits on 2009-10 league

5 Experiments

It ran an experiment for the Spanish premier and second division soccer leagues from 2009-10. In Fig. 4 hits for premier league are graphically depicted.

About data source, temporal contexts for forecasting results were built by data extracted from the RSSSF Archive (http://www.rsssf.com). Objects are matches (with temporal stamp (week, year)) and attributes are computed. Data was collected for the past four years. The size of the temporal context is about 300 objects and 17 attributes (although several of them are parametrized, i.e. , ranking difference above a threshold). Thus, $|I|$ is about 5,100 pairs.

Experiments with the system show forecasts of about 57.37% in mode 2 and 56.32% in mode 1 by a selection of ten qualitative attributes and contextual selection based on the previous 38 matches of each team (Fig. 4). Such an percentage for a qualitative reasoning system may be considered as an acceptable result comparable with expectable results of experts [3].

It is interesting to comment that the contextual selection for the premier league is not the best for the second league that we have found. For the second league is better to consider complete sets of results. The results, under the conditions of official spanish bet system are: three awards are achieved: 583.42 eur. of earning, with a cost of 38 eur. corresponding to 76 bets (two bets by week).

6 Concluding Remarks and Future Work

A confidence-based reasoning system that works on sub-contexts extracted from a temporal formal context, built for soccer bets, is presented. The system has some similarities with [13], although the reasoning system based on FCA is qualitative while the cited system is hybrid (bayesian reasoning). Pure qualitative reasoning was selected based on the aim of discovering trends (under a contextual selection) represented in the form of association rules with high confidence. It is worth noting that due to the proprietary nature of prediction models, it is difficult to compare them with our system.

Part of our ongoing work includes three research lines. Firstly, it is more interesting to apply methods for the automated definition of new attributes [2]. Secondly, since Attribute logic based on implications does not suffer from inconsistencies (two mutually results can be derived), it was necessary to select the attribute with higher confidence. However, it seems more sound to decide this by using more sophisticated methods. And, finally, the selection of thresholds can be refined to achieve a better dependence among attributes, for this, methods in data mining could be used (see e.g. [12]).

With respect to computational features, computing tasks are feasible (due to the relatively small data size). However, the summation of additional data and attributes could make it necessary to apply conservative retraction methods [5,2] to work with a contextual KB of a feasible size.

References

1. Why Spain will win..., Engineering & Technology (June 5–18, 2010)
2. Alonso-Jiménez, J.A., Aranda-Corral, G.A., Borrego-Díaz, J., Fernández-Lebrón, M.M., Hidalgo-Doblado, M.J.: Extending Attribute Exploration by Means of Boolean Derivatives. In: Proc. 6th Int. Conf. Concept Lattices and Their Applications (CLA 2008), pp. 121–132 (2008)
3. Andersson, P., Ekman, M., Edman, J.: Forecasting the fast and frugal way: A study of performance and information-processing strategies of experts and non-experts when predicting the World Cup 2002 in soccer, Working Paper Series in Business Administration 2003:9, Stockholm School of Economics (2003)
4. Aranda-Corral, G.A., Borrego-Díaz, J.: Reconciling Knowledge in Social Tagging Web Services. In: Corchado, E., Graña Romay, M., Manhaes Savio, A. (eds.) HAIS 2010. LNCS(LNAI), vol. 6077, pp. 383–390. Springer, Heidelberg (2010)
5. Aranda-Corral, G.A., Borrego-Díaz, J., Fernández-Lebrón, M.M.: Conservative Retractions of Propositional Logic Theories by Means of Boolean Derivatives: Theoretical Foundations. In: Carette, J., Dixon, L., Coen, C.S., Watt, S.M. (eds.) MKM 2009, Held as Part of CICM 2009. LNCS, vol. 5625, pp. 45–58. Springer, Heidelberg (2009)
6. Armstrong, W.: Dependency structures of data base relationships. In: Proc. of IFIP Congress, Geneva, pp. 580–583 (1974)
7. Balcázar, J.L.: Redundancy, Deduction Schemes, and Minimum-Size Bases for Association Rules. Logical Methods in Computer Science 6(2), 1–23 (2010)
8. Ganter, B., Wille, R.: Formal Concept Analysis - Mathematical Foundations. Springer, Heidelberg (1999)
9. Giarratano, J.C., Riley, G.D.: Expert Systems: Principles and Programming. Brooks/Cole Publishing Co., Pacific Grove (2005)
10. Goldstein, D.G., Gigerenzer, G.: Models of ecological rationality: the recognition heuristic. Psychological review 109(1), 75–90 (2002)
11. Guigues, J.-L., Duquenne, V.: Familles minimales d' implications informatives resultant d'un tableau de donnees binaires. Math. Sci. Humaines 95, 5–18 (1986)
12. Imberman, S.P., Domanski, B., Orchard, R.A.: Using Booleanized Data To Discover Better Relationships Between Metrics. In: Int. CMG Conference, pp. 530–539 (1999)
13. Min, B., Kim, J., Choe, C., Eom, H., McKay, R.I.: A compound framework for sports results prediction: A football case study. Know.-Based Syst. 21(7), 551–562 (2008)
14. Neouchi, R., Tawfik, A.Y., Frost, R.A.: Towards a Temporal Extension of Formal Concept Analysis. In: Proc.14th Conf. Canadian Soc. on Comp. Studies of Intell. LNCS, vol. 2056, pp. 335–344. Springer, Heidelberg (2001)

Application of Independent Component Analysis for Evaluation of Ashlar Masonry Walls

Addisson Salazar, Gonzalo Safont, and Luis Vergara

[1] Instituto de Telecomunicaciones y Aplicaciones Multimedia, Universidad Politécnica de Valencia, Camino de Vera s/n, 46022, Valencia, Spain
{asalazar,lvergara@dcom}@dcom.upv.es
gonsaar@upvnet.upv.es

Abstract. This paper presents a novel application of Independent Component Analysis (ICA) to the evaluation of ashlar masonry walls inspected with Ground Penetrating Radar (GPR). ICA is used as preprocessor to eliminate the background from the backscattered signals. Thus, signal-to-noise ratio of the GPR signals is enhanced. Several experiments were made on scale models of historic ashlar masonry walls. These models were loaded with different weights, and the corresponding B-Scans were obtained. ICA shows the best performance to enhance the quality of the B-Scans compared with classical methods used in GPR signal processing.

Keywords: ICA, GPR, Clutter, NDT.

1 Introduction

Non-destructive testing (NDT) has been supported by computational intelligent methods such as neural networks [1] and independent component analysis (ICA) [2]. The use of ICA for Ground Penetrating Radar (GPR) signal processing has been recently studied in [3], [4], [5], [6]. Most of these works deal with the use of stepped-frequency GPR for the detection of non-metallic land mines. In this paper, a novel application of ICA to Non-Destructive Testing (NDT) of historical walls using GPR is presented. The overall objective is to reduce the clutter in the captured GPR signal in order to enhance the radargrams of the wall internal structure. A radargram is an image that represents values of the measured signals at different depths of the material through the points of a trajectory used to examine the material. Each of the signals collected is called an "A-Scan" and the map formed by the collection of all the A-Scans in a trajectory is called a "B-Scan" (i.e., the radargram); see examples of B-Scan in Fig. 4.

Two scale models of historical ashlar masonry walls were analyzed: the first one was homogeneous and the second one was previously mechanized to create imperfections in specific locations. We will show the detection of these inhomogeneities inside the wall and the characterization of the propagation of electromagnetic waves inside the masonry under different loads. Fig. 1 shows an outline of the walls that depicts the following: ashlars, mortar interfaces between

J. Cabestany, I. Rojas, and G. Joya (Eds.): IWANN 2011, Part II, LNCS 6692, pp. 469–476, 2011.

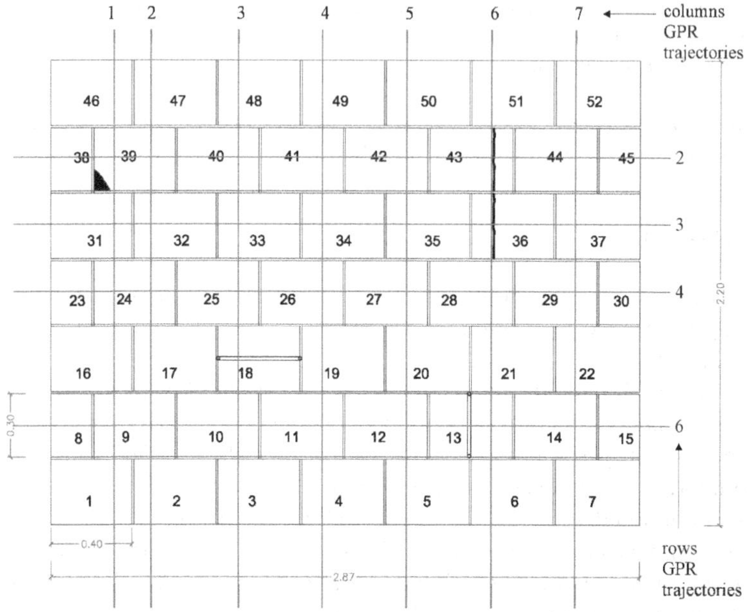

Fig. 1. Geometry of the wall under test. Ashlars, mortar interfaces and imperfections are depicted.

ashlars, imperfections, and trajectories of the GPR data acquisition (7 columns and 4 rows). The walls were 2.87 m. x 2.2 m. x 0.204 m. (width, height and length).

An ICA model for the received GPR signals that allows the background component from the backscattered components to be separated is proposed. The background component was used to assess the deformation of the wall under a load applied at the centre of the top of the wall using hydraulic pressure. The backscattered components were analyzed to detect the imperfections inside the wall. The algorithms employed for background estimation at the first step of the GPR signal processing were the following: (i) ICA algorithms - Mixca [7], JADE [8], TDSEP [9]; (ii) polynomial estimation; and (iii) spatial mean [10].

2 Statement of the ICA Problem

The final objective of this application is to obtain clear maps (B-Scans) of a wall inspected by GPR that allow the imperfections inside the wall to be detected. Thus, we propose ICA for a preprocessing step to separate the so-called background (reflections from the air-wall interfaces) from the rest of the backscattered measured signal. The background is a kind of interference that has to be removed in order to enhance the signal-to-noise ratio (SNR) of the GPR signals [11]. Thus, we considered the measured signal as a mixture obtained from the backscattering of the material microstructure plus the background.

The signal of backscattering measured by GPR, under assumptions on the wavelength and the scattering size, can be modelled as a stochastic process. This model is composed of a homogeneous non-dispersive media, and randomly distributed punctual scatters depicting the composite nature of the received grain noise signal instead of a rigorous description of the material microstructure. Thus, the backscattering model can be written as [12],

$$\left\{\tilde{Z}(p,t)\right\}=\sum_{n=1}^{N(p)}\tilde{A}_{n,p}f\left(t-\tilde{\tau}_n(p)\right) \tag{1}$$

where p is the location of the GPR antenna in the B-Scan. The random variable (r.v.) $\tilde{A}_{n,p}$ is the scattering cross-section of the n th scatter in the location p. The r.v. $\tilde{\tau}_n$ is the delay of the signal backscattered by the n th scatter and $N(p)$ is the number of scatters contributing from this position. The function $f(t)$ is the pulse emitted by the GPR, which is defined as

$$f(t)=b\cdot\left(1-(\omega_0 t)^2\right)e^{\frac{-(\omega_0 t)^2}{2}} \tag{2}$$

where w_0 is the central frequency of the GPR antenna (in this application it was 1.6 GHz) and b is a normalization factor.

The recorded signals can be modelled as the superposition of the backscattered signal plus sinusoidal phenomena representing the background. The extraction of the background was made using a sliding window of size 100 A-Scans with an overlap of 99 A-Scans. For each window, a background was estimated using ICA. This can be written as,

$$x_{k,l}(t)=s_{k,l}(t)+\sum_{i=1}^{N-1}\alpha_{ikl}e^{j(\omega_i t+\theta_{ik})} \qquad k=1...M \quad l=1...L \tag{3}$$

where M is the number of A-Scans, L is the number of windows, $x_{k,l}(t)$ is the received signal from the material at position k of the B-Scan estimated in the window l; $s_{k,l}(t)$ is the backscattering signal that depends on the material microstructure; and $\alpha_{ikl}e^{j(\omega_i t+\theta_{ik})}$ $i=1\ldots N-1, k=1\ldots M, l=1\ldots L$ are the sinusoidal sources to be analyzed with α amplitude, ω angular frequency, and θ initial phase.

In the proposed application, the wall was scanned for 7 columns (vertical scan) and 4 rows (horizontal scan), see Fig. 1. Due to constraints for the movement of the antenna survey wheel, actual length of the acquisition trajectories was 1.7 m. and 2.1 m. for columns and rows respectively. The scan density was 200 A-Scans per metre and 1024 samples were acquired for each A-Scan. Thus, $M=340$ for vertical scanning and $M=420$ for horizontal scanning. This number of scanning was adequate to include the anomalies of the material and background. Background estimation is performed for each window, thus its performance is not affected by the sampling configuration. On the contrary, B-Scan resolution depends on the number of data available to build the wall image, even though this resolution can be enhanced by using interpolated data. The suitability of this sampling configuration for the present application is demonstrated in Section 4 (Results and Discussion).

Obviously, the estimated sinusoidal components have the same frequencies along the B-Scan, with possibly changing amplitude and phase. From a statistical point of view, considering the background as a sinusoid with deterministic but unknown amplitude and uniform random phase, it is clearly guaranteed that the backscattering signal and the background are statistically independent. Therefore, ICA algorithms can separate the background (characterized for one sinusoidal source estimated by ICA) from the backscattering (the rest of sources estimated by ICA) contribution.

3 Performance Analysis of Background Estimation Methods

Historic walls suffer a degradation of its physical properties with the pass of time. This can generate a strong clutter in the measured signals using GPR. For this work, two walls were only available with particular SNR. Thus, we generated more cases of historic walls in different conditions by adding K-distributed noise to the measured signals. K-distribution has been demonstrated to be a good model for radar clutter [13]. The SNRs were 0, 2, 4, 6, 8, 10, 15, 20, 25, 30, 40, 50, 60 dB. For each different SNR, the background was estimated using the following methods: ICA algorithms (JADE, Mixca, and TDSEP), polynomials, and spatial mean, see Fig. 2.

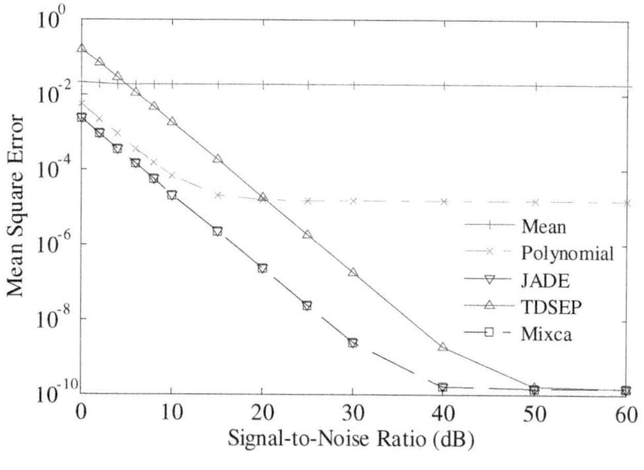

Fig. 2. Performance analysis of different methods for background estimation

Fig. 2 shows the Mean Square Error (MSE) between the background of one of the original B-scans and the background estimated with added K-distributed noise. JADE and Mixca achieve the best results, i.e. the background is well-estimated even for low SNRs. ICA methods show similar behaviours, improving the estimation with higher SNR, while classical methods (polynomial, mean) do not improve significantly. The separation of the source corresponding to the background with any small energy level is possible using ICA since it is based on statistical independence of the components and not on the energy associated to each frequency component. We selected the Mixca algorithm for the preprocessing step. This algorithm implements non-parametric source density estimation, which has allowed its application to different applications such as NDT [2][14] and biosignal processing [15]. Fig. 3 shows one of the sources (ICA component) obtained for the background.

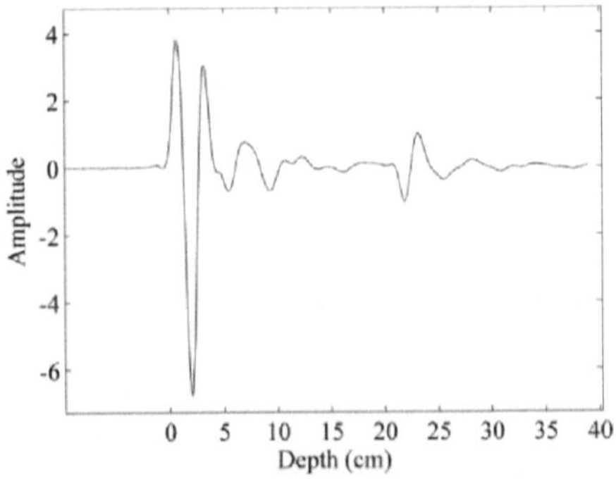

Fig. 3. Source signal corresponding to the background estimated in a single window

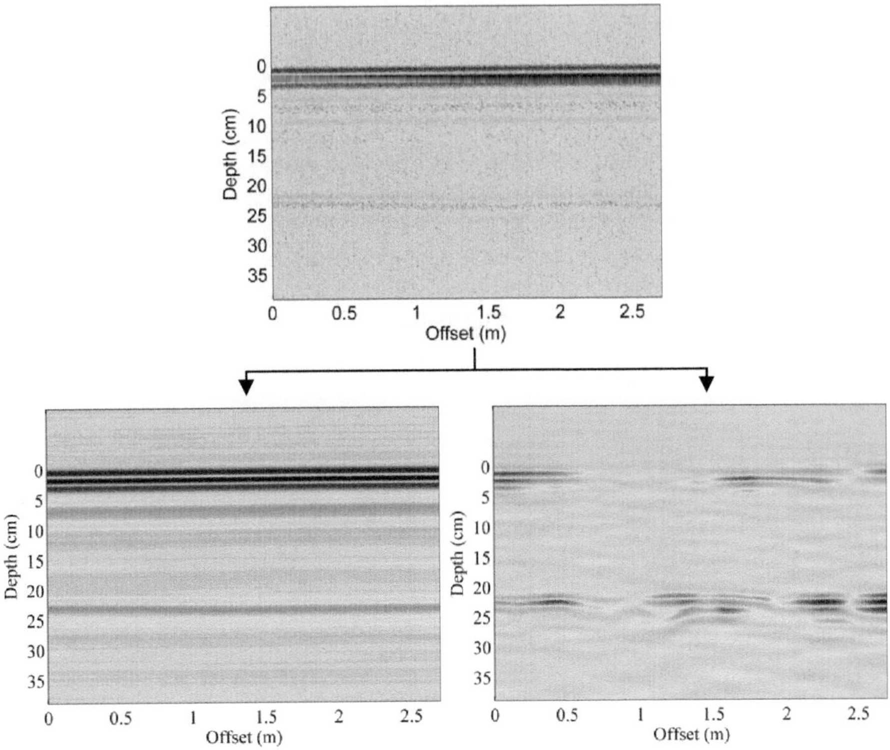

Fig. 4. B-Scan from GPR signals added with K-distributed noise for 0 dB SNR (upper graph); estimated background component (lower left graph); backscattered components (lower right graph)

Fig. 4 shows the separation of the background component and the backscattered components from the GPR measured signals with 0 dB SNR. The pulse generated by the GPR equipment propagates from the antenna to the wall through an air interface. After this, the pulse travels through the wall until the opposite side of the wall. Finally, it keeps propagating through free air. The signal reflected from these two interfaces forms the background component. Grain noise and imperfections in the wall's structure form the backscattered components. Note that it is very difficult to obtain any information about the wall's condition from the graph in the upper part of Fig. 4. However, some locations of interest are clearly depicted in the lower right graph of Fig. 4. In the next section, we will show that there were imperfections in the wall located in these locations.

4 Results and Discussion

The equipment employed consisted of a GPR system, SIR 3000 from Geophysical Survey Systems Inc. We used a 1.6 GHz mounted on an encoder. The receiving antenna had a size of 3.8 cm. x 10 cm. x 16.5 cm that was adequate for both vertical and horizontal measures. The configuration parameters used for data acquisition were: distance mode (200 scans per meter), range of 10 ns and 1024 samples per scan. In order to obtain enhanced B-Scans of the wall.

4.1 Analysis of the Homogeneous Wall (Background Component)

Fig. 5 shows the background signals of captured radargrams for the homogeneous ashlar masonry wall, for two different loads. A variation of the propagation conditions of the wall is shown. This is noticeable the way the opposite side of the wall seems to move away (see Fig. 5.b) that indicates a loss in velocity of propagation inside the material. Thus, worsening of transmission properties in the wall was produced by the weight load. This behaviour seemed consistent with *in situ* measurements of the wall's distortions, i.e. the wall suffered a strong buckling that was emphasized on its uppermost part.

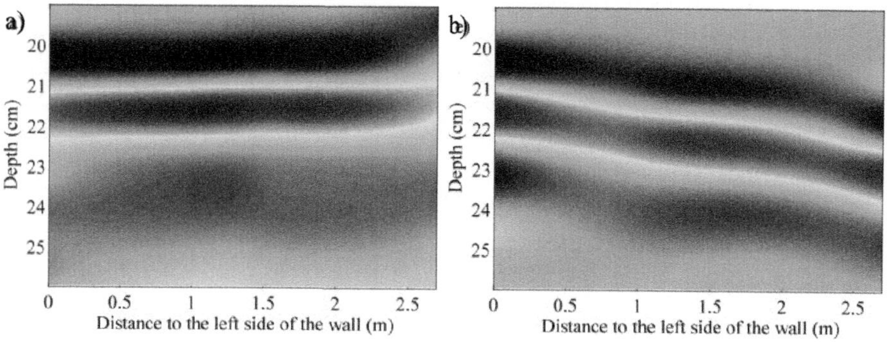

Fig. 5. Background at the opposite side of the wall for different values of the load. Wall 1, row 6; a) no load; b) 80 metric tons (mt.) load.

4.2 Analysis of the Wall with Imperfections (Backscattered Components)

For the purpose of flaw detection, some algorithms were implemented in order to highlight the discontinuities (typically due to changes in the material) in the radargrams [10]. These methods, as seen on previous sections, were: background removal, depth resolution enhancing, Kirchhoff migration and improvement of the contrast in the B-scan. Fig. 6 shows the radargram obtained after this processing.

Fig. 6 corresponds to the radargram of row 2 with a load of 80 mt. A nook and a crack in the wall can be visually detected. Note the difference in perceived amplitude between both flaws. This can be explained by analysing the geometry of the flaws. The straight geometry of the crack accentuates the reflection of the waves producing higher amplitude values in the received signal. On the contrary, the nook is irregularly-shaped and rough-edged, which attenuates the reflections of the incident waves in the flaw scatters. Thus, received signal values for nook location are lower than those measured at the location of the crack.

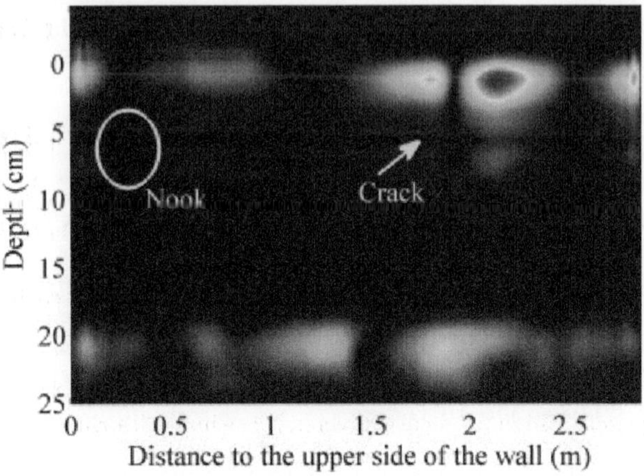

Fig. 6. Processed radargram with the imperfections highlighted

The detection results were better for high load values. The reflection of incident waves from the flaws is strengthened with the increase in compression, which allows a better definition of the shapes of the imperfections in the radargrams.

5 Conclusion

The proposed method for the separation of background and backscattered components using ICA has demonstrated accurate separation in GPR signals from historic ashlar masonry walls. This has allowed enhanced images of the wall were obtained. The auscultation of historical masonry walls with ground-penetrating radar (GPR) has proved to be effective for the detection of imperfections and the characterization of walls under load. It was possible to detect flaws with millimetre sizes and variations in the interfaces between ashlars and mortar caused by the effect of the compression suffered under load.

Acknowledgments. This work has been supported by the *Generalitat Valenciana* under grant PROMETEO/2010/040, and the Spanish Administration and the FEDER Programme of the European Union under grant TEC 2008-02975/TEC.

References

1. Salazar, A., Unió, J.M., Serrano, A., Gosalbez, J.: Neural networks for defect detection in non-destructive evaluation by sonic signals. In: Sandoval, F., Prieto, A.G., Cabestany, J., Graña, M. (eds.) IWANN 2007. LNCS, vol. 4507, pp. 638–645. Springer, Heidelberg (2007)
2. Salazar, A., Vergara, L., Llinares, R.: Learning material defect patterns by separating mixtures of independent component analyzers from NDT sonic signals. Mechanical Systems and Signal processing 24(6), 1870–1886 (2010)
3. Zhao, A., Jiang, Y., Wang, W.: Exploring Independent Component Analysis for GPR Signal Processing. In: Progress In Electromagnetics Research Symposium 2005, pp. 750–753. The Electromagnetics Academy, Cambridge (2005)
4. Abujarad, F., Omar, A.: Comparison of Independent-Component Analysis (ICA) Algorithms for GPR Detection of Non-Metallic Land Mines. In: Bruzzone, L. (ed.) Proceedings of SPIE Image and Signal Processing for Remote Sensing XII, vol. 6365, pp. 636516.1–636516.12. SPIE, Bellingham (2006)
5. Liu, J.X., Zhang, B., Wu, R.B.: GPR Ground Bounce Removal Methods Based on Blind Source Separation. In: Progress In Electromagnetics Research Symposium 2006, pp. 256–259. The Electromagnetics Academy, Cambridge (2006)
6. Verma, P.K., Gaikwad, A.N., Sigh, D., Nigam, M.J.: Analysis of Clutter Reduction Techniques for Through Wall Imaging in UWB Range. In: Progres. Electromagnetics Research B 2009, vol. 17, pp. 29–48. The Electromagnetics Academy, Cambridge (2009)
7. Salazar, A., Vergara, L., Serrano, A., Igual, J.: A General Procedure for Learning Mixtures of Independent Component Analyzers. Pattern Recognition 43(1), 69–85 (2010)
8. Cardoso, J.F., Souloumiac, A.: Blind beamforming for non Gaussian signals. IEE Proceedings-F 140(6), 362–370 (1993)
9. Ziehe, A., Muller, K.R.: TDSEP - An Efficient Algorithm for Blind Separation Using Time Structure. In: Proceedings of the Eighth International Conference on Artificial Neural Networks ICANN 1998, Perspectives in Neural Computing, pp. 675–680 (1998)
10. Reynolds, J.M.: An Introduction to Applied and Environmental Geophysics. Wiley, Chichester (1997)
11. Igual, J., Camacho, A., Vergara, L.: A blind source separation technique for extracting sinusoidal interferences in ultrasonic non-destructive testing. Journal of VLSI Signal Processing 38, 25–34 (2004)
12. Salazar, A., Gosálbez, J., Igual, J., Llinares, R., Vergara, L.: Two applications of independent component analysis for non-destructive evaluation by ultrasounds. In: Rosca, J.P., Erdogmus, D., Príncipe, J.C., Haykin, S. (eds.) ICA 2006. LNCS, vol. 3889, pp. 406–413. Springer, Heidelberg (2006)
13. Raghavan, R.S.: A Model for Spatially Correlated Radar Clutter. IEEE Trans. on Aerospace and Electronic Systems 27, 268–275 (1991)
14. Salazar, A., Vergara, L.: ICA mixtures applied to ultrasonic nondestructive classification of archaeological ceramics. EURASIP Journal on Advances in Signal Processing, Article ID 125201, 11 (2010), doi:10.1155/2010/125201
15. Salazar, A., Vergara, L., Miralles, R.: On including sequential dependence in ICA mixture models. Signal Processing 90(7), 2314–2318 (2010)

Fast Independent Component Analysis Using a New Property

Rubén Martín-Clemente*, Susana Hornillo-Mellado, and José Luis Camargo-Olivares

Departamento de Teoría de la Señal y Comunicaciones, Escuela S. de Ingenieros, Avda. de los Descubrimientos, s/n., 41092 Seville, Spain
University of Seville, Spain
ruben@us.es, susanah@us.es, jlcamargo@yahoo.es

Abstract. In this paper we present a new theoretical characterization of the solutions to the Independent Component Analysis (ICA) problem. As an application, we also propose an algorithm that is directly based on that theoretical characterization. The algorithm has a very low computational complexity, and the experiments show that it is fast and reliable.

Keywords: Independent Component Analysis, Blind Signal Separation.

1 Introduction

In Independent Component Analysis (ICA), we deal with a model of the form

$$\mathbf{x} = \mathbf{A}\,\mathbf{s} \tag{1}$$

where $\mathbf{s}$ is a random vector whose components s_i are statistically independent, and matrix $\mathbf{A}$ is deterministic and unknown. The goal is to estimate the elements s_i of $\mathbf{s}$ from the sole observation of $\mathbf{x}$. These elements are often called *independent components* in the literature.

ICA has been applied to problems in fields as diverse as speech processing, brain imaging, electrical brain signals (e.g., EEG signals), telecommunications, and stock market prediction [1–5]. For example, it has been suggested that ICA mimics the biological processing in the simple cells of the mammalian primary visual cortex [1, 2]. This fact has been used to develop advanced models of the human visual system and the statistical structure of natural images.

In this paper we propose a new theoretical characterization of the solutions to ICA. It is based in performing ICA by taking into account prior information about the positive support of the signal of interest. The resulting method is based on a conditional mean, an is thus cost-efficient. The work may be reminiscent of the conditional second order statistical approaches developed in [6, 7]. As an application, we also propose an iterative new method for determining

* Corresponding author.

J. Cabestany, I. Rojas, and G. Joya (Eds.): IWANN 2011, Part II, LNCS 6692, pp. 477–483, 2011.

the independent components. The paper is organized as follows: in Section 2, the pre-processing known as 'whitening' is reviewed. Section 3 is devoted to the description of the proposed method. Section 4 illustrates the performance of the method via experiments. Finally, Section 5 is devoted to the conclusions.

2 Preprocessing

Given an ensemble of T data points $\mathbf{x}_1, \ldots, \mathbf{x}_T$ of zero sample mean, i.e.,

$$\bar{\mathbf{x}} = \frac{1}{T} \sum_{n=1}^{T} \mathbf{x}_n = \mathbf{0}. \tag{2}$$

The first step in most ICA approaches is the so-called whitening transformation [2]:

$$\mathbf{z}_n = \mathbf{C}_x^{-1/2} \mathbf{x}_n, \quad n = 1, \ldots, T, \tag{3}$$

where $\mathbf{C}_x^{1/2}$ is any square root of the sample covariance matrix

$$\mathbf{C}_x = \frac{1}{T} \sum_{n=1}^{T} \mathbf{x}_n \, \mathbf{x}_n'. \tag{4}$$

It holds that the components of $\mathbf{z}_n$ are of *zero mean*, *unit variance*, and are also *uncorrelated* among themselves, i.e.

$$\bar{\mathbf{z}} = \frac{1}{T} \sum_{n=1}^{T} \mathbf{z}_n = \mathbf{0} \tag{5}$$

$$\mathbf{C}_z = \frac{1}{T} \sum_{n=1}^{T} \mathbf{z}_n \, \mathbf{z}_n' = \mathbf{I}. \tag{6}$$

It can be shown that $\mathbf{Q} = \mathbf{C}_x^{-1/2} \mathbf{A}$ is an orthogonal matrix. Thus, thanks to the preprocessing, the problem of finding vector $\mathbf{s}$ simply reduces to that of determining the inverse of $\mathbf{Q}$.

3 Proposed Approach

To fix ideas, and without loss of generality, consider that the goal is to estimate the first independent component s_1. Let $s_{11}, \ldots, s_{1T}$ be the samples of s_1, and let

$$\mathbb{S} = \{n : s_{1n} \geq 0\}$$

be the set of indexes corresponding to s_{1n} greater than zero. Now, define:

$$y_n = \mathbf{w}' \, \mathbf{z}_n, \tag{7}$$

where

$$\mathbf{w} = \frac{1}{Z} \sum_{i \in \mathbb{S}} \mathbf{z}'_i \tag{8}$$

and

$$Z = \left\| \sum_{i \in \mathbb{S}} \mathbf{z}_i \right\| . \tag{9}$$

Consider the following result:

Lemma 1. *Under mild hypotheses, $y_n \approx s_{1n}$ for all n. In other words, (7) is an estimate of the first independent component.*

Sketch of proof: after some algebra, (7) can be rewritten as:

$$y_n = \frac{1}{Z} \sum_{k \in \mathbb{S}} \mathbf{s}'_k \, \mathbf{s}_n . \tag{10}$$

We can always assume, without loss of generality, that

$$\sum_{k=1}^{T} \mathbf{s}_k = \mathbf{0} \tag{11}$$

(i.e. the independent components are assumed to be zero-mean). Therefore,

$$\sum_{k \in \mathbb{S}} \mathbf{s}_k = -\sum_{k \notin \mathbb{S}} \mathbf{s}_k . \tag{12}$$

Observe also that

$$sign\{s_{1k}\} = 1 \tag{13}$$

if $k \in \mathbb{S}$, and

$$sign\{s_{1k}\} = -1 \tag{14}$$

otherwise. Therefore,

$$\sum_{k \in \mathbb{S}} \mathbf{s}_k = \sum_{k \in \mathbb{S}} sign\{s_{1k}\} \, \mathbf{s}_k \tag{15}$$

and

$$-\sum_{k \notin \mathbb{S}} \mathbf{s}_k = \sum_{k \notin \mathbb{S}} sign\{s_{1k}\} \, \mathbf{s}_k . \tag{16}$$

It readily follows that

$$\sum_{k \in \mathbb{S}} \mathbf{s}_k = \frac{1}{2} \sum_{k=1}^{T} sign\{s_{1k}\} \, \mathbf{s}_k . \tag{17}$$

Substituting in (10) we get:

$$y_n = \frac{1}{2\,Z} \sum_{k=1}^{T} sign\{s_{1k}\} \, \mathbf{s}'_k \, \mathbf{s}_n =$$

$$= \frac{1}{2\,Z} \left[\sum_{k=1}^{T} sign\{s_{1k}\} \, s_{1k} \, s_{1n} + \sum_{k=1}^{T} sign\{s_{1k}\} \, s_{2k} \, s_{2n} + \ldots \right] . \tag{18}$$

Since the independent components are statistically independent, we have that:

$$\frac{1}{T}\sum_{k=1}^{T} sign\{s_{1k}\}\, s_{2k}\, s_{2n} \approx \frac{1}{T^2}\sum_{k=1}^{T} sign\{s_{1k}\} \sum_{k=1}^{T} s_{2k}\, s_{2n}, \tag{19}$$

i.e.

$$\sum_{k=1}^{T} sign\{s_{1k}\}\, s_{2k}\, s_{2n} \approx \left(\frac{1}{T}\sum_{k=1}^{T} sign\{s_{1k}\}\right) \sum_{k=1}^{T} s_{2k}\, s_{2n}, \tag{20}$$

which is, under very general conditions, negligible in comparison with

$$\sum_{k=1}^{T} sign\{s_{1k}\}\, s_{1k}\, s_{1n} \tag{21}$$

Then, it readily follows that:

$$\boxed{y_n \approx s_{1n}} \tag{22}$$

maybe up to an irrelevant constant or a sign[1]. □

The Lemma has the potential to be useful for estimating an independent component, but, first, we have to 'guess' the set $\mathbb{S}$. The following algorithm refines that guess iteratively until an acceptable solution is found:

Given $\mathbf{z}_1, \ldots, \mathbf{z}_T$:

1. Define $\mathbb{S}$ at random.
2. Generate y_n using (7).
3. Replace $\mathbb{S}$ with $\{n : y_n \geq 0\}$.
4. Return to step 2. and repeat until convergence.

Observe that, to speed up the process, $\mathbb{S}$ should be determined during step 2. Note also that the algorithm has a low computational complexity since, unlike in other ICA methods, it does not need to compute higher-order statistics. To estimate more than one independent component, we can use the procedure described in Chapter 4 of book [2]. Basically, this procedure consists of the following steps:

1. Remove y_n from $\mathbf{z}_n$ by $\hat{\mathbf{z}}_n = \mathbf{z}_n - \mathbf{w}\, y_n$.
2. Substitute $\mathbf{x}_1, \ldots, \mathbf{x}_T$ with $\hat{\mathbf{z}}_1, \ldots, \hat{\mathbf{z}}_T$.
3. Apply whitening again and reduce the dimensionality of the data in one unit.
4. Apply the algorithm to estimate a new independent component.

This procedure is repeated until all the independent components are recovered.

[1] We are grateful to one of the anonymous reviewers, who provided an alternative proof. This alternative proof is based on the computation of the conditional expectation $E[\mathbf{s}|s_1]$, whose elements $E[s_i|s_1]$ are zero for $i \neq 1$ by virtue of the source independence and zero-mean assumptions. Noting that $\frac{1}{Z}\sum_{k\in\mathbb{S}} \mathbf{s}_k \propto E[\mathbf{s}|s_1]$, the proof readily follows.

4 Experiments

To evaluate the performance of the proposed algorithm, we have carried out a comparison with FastICA [8], which is a celebrated algorithm for ICA. FastICA is also based on kurtosis maximization, and it is commonly regarded as one of the faster and more reliable methods. We performed 100 independent experiments. In each experiment, the independent components were samples drawn from an uniform distribution, and matrix **A** was generated at random. We performed the experiments with $N = 2$ independent components, $T = 1000$ samples each. The independent components were estimated using both the proposed method and FastICA. Table 1 reports the signal to noise ratio obtained by means of both methods. It is shown that the performance of the proposed method is acceptable (the mean SNR of the estimated sources equals 33.76 dB), while, at the same time, the speed of the algorithm is high (the method is 1.5 times faster than FastICA). Similar results are obtained when the sample size is increased. Experiments were carried out using an Intel Core Duo processor (2.26 GHz) and Matlab.

As a further example, in Figure 1 we show the observed data in a particular case in which the signals s_i are, respectively, a sinus, an square signal, and a

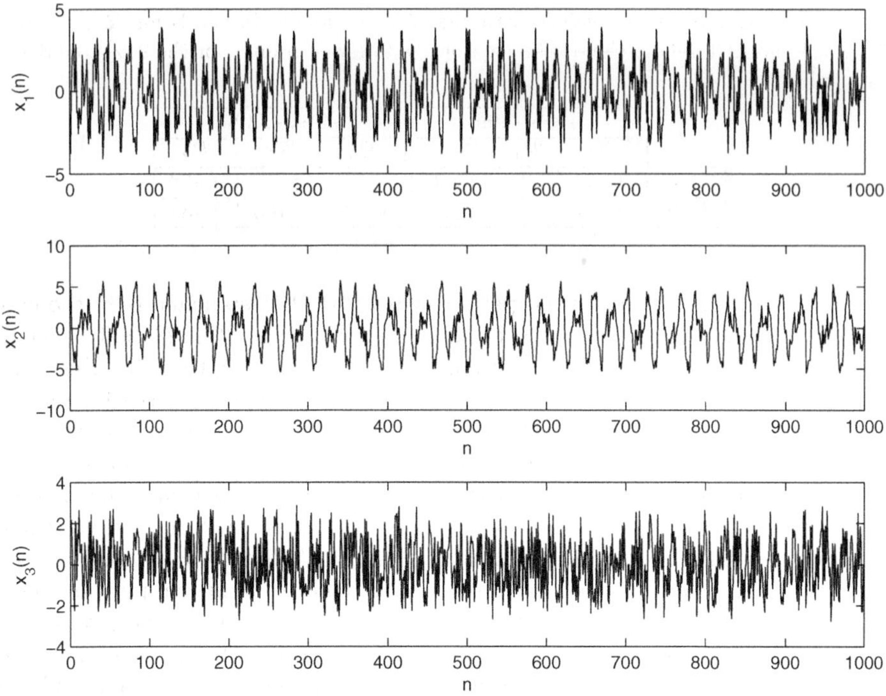

Fig. 1. Observed data

Table 1. Comparison between the proposed method and FastICA. SNR and execution time are averaged over 100 independent experiments. The numbers between brackets are the standard deviations.

	SNR (dB)	execution time (s)
Proposed method	33.76 (2.6)	0.0029 (0.012e-4)
`FastICA`	31.46 (8.8)	0.0400 (7.66e-4)

random noise. Figure 2 shows the estimated source signals s_i. The mean SNR of the estimated sources is equal to 35.11 dB.

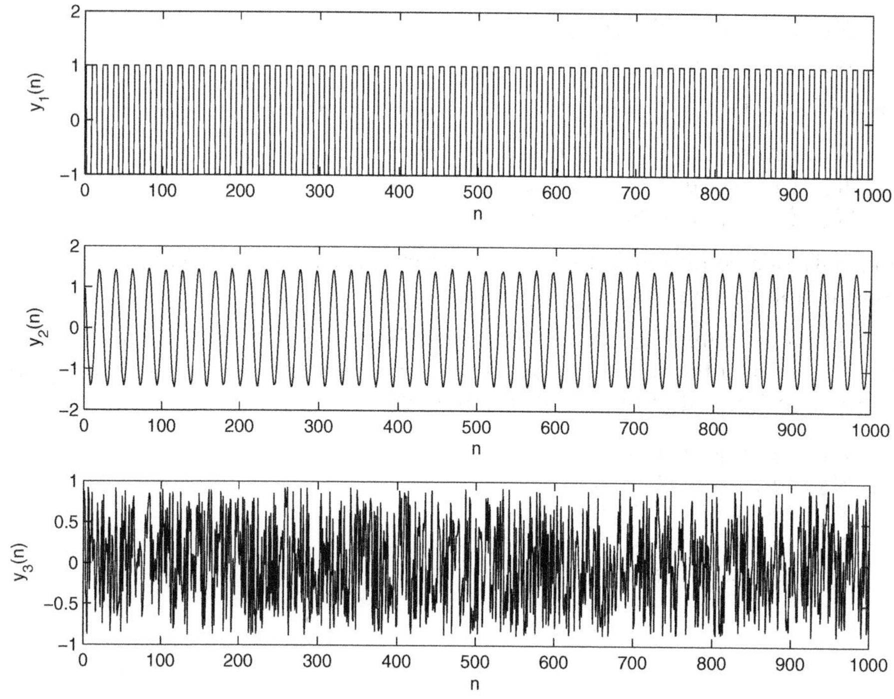

Fig. 2. Estimated signals s_i

5 Conclusions

We have presented a new solution for ICA that is used to develop an iterative algorithm. The algorithm has a very low complexity since, unlike what it is usual in ICA, it does not to compute higher-order statistics. Even though the research is still in its early stages, and much work has to be done, experiments show that the proposed approach is promising.

Acknowledgements. This work was supported by a grant from the 'Junta de Andalucía' (Spain) with reference P07-TIC-02865.

References

1. Comon, P., Jutten, C. (eds.): Handbook of blind source separation, independent componet analysis and applications. Elsevier, Amsterdam (2010)
2. Cichocki, A., Amari, S.I.: Adaptive blind signal and image processing: learning algorithms and applications. Wiley, Chichester (2002)
3. Zarzoso, V., Nandi, A.K.: Noninvasive fetal electrocardiogram extraction: blind separation versus adaptive noise cancelation. IEEE Tr. on Biomedical Engineering 48(1), 12–18 (2001)
4. Camargo-Olivares, J.L.R., Hornillo, S., Roman, I.: The maternal a abdominal ecg as input to mica in the fetal ecg extraction problem. IEEE Signal Processing Letters 18(3), 161–164 (2011)
5. Martín-Clemente, R., Camargo-Olivares, J., Hornillo-Mellado, S., Roman, I.: Fast a technique for noninvasive fetal ecg extraction. IEEE Tr. on Biomedical Engineering 58(2), 227–230 (2011)
6. Xerri, B., Borloz, B.: An iterative method using conditional second-order statistics applied to the blind source separation problem. IEEE Trans. on Signal Processing 52(2), 313–328 (2004)
7. Phlypo, R., Zarzoso, V., Lemahieu, I.: Source extraction by maximizing the variance in the conditional distribution tails. IEEE Trans. on Signal Processing 58(1), 305–316 (2010)
8. Hyvärinen, A.: Fast and robust fixed-point algorithms for independent component a analysis. IEEE Transactions on Neural Networks 10(3), 626–634 (1999)

Using Particle Swarm Optimization for Minimizing Mutual Information in Independent Component Analysis

Jorge Igual[1], Jehad Ababneh[2], Raul Llinares[1], and Carmen Igual[1]

[1] Universitat Politecnica de Valencia, Camino de Vera s/n,
46022 Valencia, Spain
[2] Jordan University of Science and Technology
22110 Irbid, Jordan
jigual@dcom.upv.es

Abstract. Independent Component Analysis (ICA) aims to recover a set of independent random variables starting from observations that are a mixture of them. Since the prior knowledge of the marginal distributions is unknown with the only restriction of at most one Gaussian component, the problem is usually formulated as an optimization one, where the goal is the maximization (minimization) of a cost function that in the optimal value approximates the statistical independence hypothesis. In this paper, we consider the ICA contrast function based on the mutual information. The stochastic global Particle Swarm Optimization (PSO) algorithm is used to solve the optimization problem. PSO is an evolutionary algorithm where the potential solutions, called particles, fly through the problem space by following the current optimum particles. It has the advantage that it works for non-differentiable functions and when no gradient information is available, providing a simple implementation with few parameters to adjust. We apply successfully PSO to separate some selected benchmarks signals.

Keywords: Independent Component Analysis, Particle Swarm Optimization.

1 Independent Component Analysis

The seminal paper [1] established the theoretical basis of Independent Component Analysis (ICA). Some books give a full explanation of the problem and present some applications and extensions, e.g., [2], [3]. In this paper we will concentrate in the linear noiseless instantaneous model:

$$\mathbf{x} = \mathbf{A}\mathbf{s} \tag{1}$$

where $\mathbf{x}$ is the observed vector that is a linear transformation (matrix $\mathbf{A}$) of a random vector $\mathbf{s}$ whose components are statistically independent, i.e., the joint probability is the product of the marginal densities $p(\mathbf{s}) = \prod_i p(s_i)$. In Blind Source Separation (BSS) terminology, $\mathbf{x}$ is the mixture vector, $\mathbf{A}$ is the mixing matrix and $\mathbf{s}$

J. Cabestany, I. Rojas, and G. Joya (Eds.): IWANN 2011, Part II, LNCS 6692, pp. 484–491, 2011.

is the source vector. For the sake of simplicity, we will assume the square problem, i.e., the number of sources and mixtures are equal. As it is well known, if the number of sources is not known and the signal to noise ratio is high enough, the number of sources can be estimated using the Principal Component Analysis of the observations. In fact, many ICA algorithms can be decomposed into two steps; in the first one, the second order statistics are exploited, imposing the decorrelation of the signals (whitening step). Because of the inherent indeterminacies of ICA (sign, order and amplitude), it is common to assume that the sources are unit variance; in addition we will assume that they are zero mean. In this case, the first step is usually called sphering. The second step consists on the estimation of an orthogonal matrix that imposes the independence, being necessary the use of higher order statistics. Note that the orthogonal restriction is mandatory in order to preserve the uncorrelated outputs. In matrix notation,

$$\mathbf{y} = \mathbf{Bx} = \mathbf{UWx} \tag{2}$$

where $\mathbf{W}$ is the whitening matrix and $\mathbf{U}$ is the orthogonal one. The whitened vector is expressed as $\mathbf{z} = \mathbf{WAs}$, with $E[\mathbf{zz}^T] = \mathbf{I}$, and the uncorrelated unit variance constraint $E[\mathbf{yy}^T] = \mathbf{I}$, being $\mathbf{y}$ the recovered or estimated sources.

Remember that in ICA the only assumption is the independence of the source components. It implies that to solve the problem without other assumptions, the distribution of the sources $p(\mathbf{s})$ must be known. In this case, it is straightforward to obtain the distribution of the observed vector:

$$p(\mathbf{x}) = |\det \mathbf{A}|^{-1} \prod_i p(s_i) \tag{3}$$

and consequently the Maximum Likelihood estimate $\mathbf{B} = \mathbf{A}^{-1}$ and the recovered sources $\mathbf{y} = \mathbf{A}^{-1}\mathbf{x}$. But in most of real applications, the marginal distributions are not available and the source distributions must be approximated. As a consequence, the independence condition $p(\mathbf{s}) = \prod_i p(s_i)$ can not be enforced. Therefore, the estimation of the sources or the demixing matrix is transformed in an optimization problem defined by a contrast (cost) function that is minimum when the estimated sources are as independent as possible, i.e., the matrix $\mathbf{BA}$ equals the product of a permutation (order indeterminacy) and a diagonal matrix (scale indeterminacy). A natural way to measure this independence is using the mutual information. We present in this paper the Particle Swarm Optimization (PSO) approach to solve the ICA problem for the case of contrast functions based on mutual information. For a detailed statistical study of the ICA problem, see for example [4].

2 Contrast Functions in ICA Based on Mutual Information

The Mutual Information (MI) is defined as the Kullback-Leibler divergence between the joint density and the product of the marginal distributions; it is non negative and equals to zero only if the distributions are the same, i.e., MI is a contrast function for ICA:

$$MI(\mathbf{y}) = KL(\mathbf{y}; \prod_i p(y_i)) = \int p(\mathbf{y}) \log \frac{p(\mathbf{y})}{\prod_i p(y_i)} d\mathbf{y} \tag{4}$$

It is related to the differential entropy, $MI(\mathbf{y}) = \sum_i H(y_i) - H(\mathbf{y})$, where the differential entropy of a random variable u is defined as $H(u) = -\int p(u) \log p(u) du$, that can be seen as a measure of the randomness of the variable u. Using $H(\mathbf{y}) = H(\mathbf{x}) + \log|\det \mathbf{B}|$, the contrast, up to an additive constant term, can be expressed as:

$$MI(\mathbf{y}) = \sum_i H(y_i) - \log|\det \mathbf{B}| \tag{5}$$

with the advantage that only one dimensional distributions are involved instead of multidimensional densities. In the case where the observations are first whitened as we explained before, we only have to estimate the remaining orthogonal matrix, and the contrast is reduced to the sum of the marginal entropies of $\mathbf{y}$:

$$MI(\mathbf{y}) = \sum_i H(y_i) \quad , E[\mathbf{y}\mathbf{y}^T] = \mathbf{I} \tag{6}$$

As a conclusion, ICA can be interpreted as a minimum entropy method under the whitening assumption that is reflected in the second term in (6). It must also be pointed out that $-H(u)$, up to a constant, is the KL divergence between the random variable u and the zero mean unit variance Gaussian density. Hence, the MI contrast is equivalent to find the marginal distributions as far as possible from Gaussianity. In addition, this contrast is closely related to the Maximum Likelihood (ML) solution. As it has been mentioned before, the ML approach requires a prior source distribution $p(\mathbf{s})$. Hence, the Kullback-Leibler divergence between the output vector and the assumed source distribution $KL(\mathbf{y};\mathbf{s})$, up to a constant, can be seen as a contrast, i.e., find the matrix $\mathbf{B}$ that makes the distribution of $\mathbf{y} = \mathbf{B}\mathbf{x}$ as close as possible to the hypothesized distribution of the sources [5]. One of the most popular ICA algorithms, the infomax [6], can be understood from the ML point of view; see for instance [7]. This approach obviously has problems if the prior source distribution is far from the real one. On the other hand, the optimization step admits an easy gradient based implementation (usually the so called relative [8] or natural gradient [9]). In front of the ML solution, the MI minimizes the divergence with respect not only to the matrix $\mathbf{B}$ but to the source model distribution. The cost we must pay in the MI solution is that it is not possible to obtain a feasible gradient based algorithm. And if we want to obtain an adaptive algorithm, some approximations must be introduced in the estimation of the MI. The most famous empirical contrast functions are based on cumulant approximations. Another solution is to explore non gradient based algorithms, such as PSO.

In conclusion, we have seen that contrasts based on non Gaussianity, mutual information or higher order statistics (cumulants) are closely connected and they constitute the basis for many of the contrasts found in the literature.

3 Particle Swarm Optimization

PSO is a global optimization algorithm with many similarities with evolutionary computation techniques. The system is initialized with a population of random solutions (particles) and searches for optima by updating generations simulating the swarming behavior of birds, bees, fish, etc [10].

It aims at increasing the probability of finding the global minimum (maximum), without performing an exhaustive search within the entire solution space. It has been shown in [11] that the PSO outperformed the GA in certain instances. A major advantage of the PSO is its ease of implementation in both the context of coding and parameter selection. The algorithm is much simpler and intuitive to implement than complex, probability based selection and mutation operators required for evolutionary algorithms such as the GA.

Like other evolutionary algorithms, the PSO starts with an initial population of individuals (to be termed swarm of particles). Each individual in the swarm is randomly assigned an initial position and velocity within the solution space. The position of the particle is an N-dimensional vector that represents a possible set of the unknown parameters to be optimized. Each particle in the swarm starts from its initial position at its initial velocity in order to find the position with global minimum (or maximum). During the algorithm search, the velocity and position of each particle is updated based on the individual and the swarm experience according to:

$$\begin{gathered} v_{mn}^{t} = v_{mn}^{t-1} + U_{n1}^{t}(0,\varphi_1)(pbest_{mn}^{t} - x_{mn}^{t-1}) + U_{n2}^{t}(0,\varphi_2)(gbest_{mn}^{t} - x_{mn}^{t-1}) \\ x_{mn}^{t} = x_{mn}^{t-1} + \Delta t\, v_{mn}^{t} \end{gathered} \tag{7}$$

where v_{mn} and x_{mn} represents the velocity and position of the *m-th* particle in the nth dimension, respectively. The superscripts *t* and *t-1* denote the time index of the current and the previous iterations, $U_{n1}^{t}(0,\varphi_1)$ and $U_{n2}^{t}(0,\varphi_2)$ are two different, uniformly distributed random numbers in the intervals [0, φ_1] and [0, φ_2], respectively. These random numbers are generated at each iteration and for each particle. The first term in (7) indicates that particle's current velocity depends on its previous velocity while the other two terms represent the effect of the individual (particle's best position (pbest)) and the swarm experience (neighborhood best position (gbest)) on the behavior of the particle. In (7), Δt represents a given time step (usually chosen to be one). The goodness of the new particle position (possible solution) is measured by evaluating a suitable contrast or fitness function. The process of updating velocities and positions continues and results in that one of the particles in the swarm finds the position with best fitness (global minimum or maximum). Eventually, all the particles will be drawn to this position since they will not be able to find a better one. The algorithm may be terminated if the number of iteration reached a pre-specified maximum number of iteration or the value corresponding to gbest is close enough to a desired number.

3.1 ICA Contrasts and PSO

The optimal selection of the contrast in ICA mainly depends on two factors: the prior knowledge about the distribution of the sources and the statistical properties of the estimate. Not considering the estimate of the contrast function, the best contrast would be the mutual information, since it minimizes the divergence with respect to the demixing matrix and the source distribution. But it is difficult to obtain a good estimate of it, because to use the definition of entropy we need an estimate of the density. Therefore, in practice, some approximations of mutual information can work better. One example is the cumulant-based approximation: it has the advantage of simplifying the use of mutual information, but the problem is that we do not consider all the higher order statistics (usually up to the fourth order cumulant due again to the problem of obtaining a good estimate for higher than fourth order). The same can be said with respect to the kurtosis; it is focused in obtaining components as far as possible from Gaussianity, like the negentropy. An additional problem in the case of the kurtosis is that it is very sensitive to outliers; this is because smooth versions of it are preferred, since the finite-sample statistical properties of the estimates (asymptotic variance and robustness) are better. Of course, a prior knowledge about the source distributions will help, but the problem will be no more called blind; in this case, the ML estimate will be better, since we do not have to optimize with respect the source distribution. Since we are using an algorithm that does not require mathematical derivations but only the estimation of the cost function in the adapting process, we will use the mutual information; its calculation is reduced to the estimation of the numerical marginal entropies for whitened signals.

With respect to the implementation of PSO, there have been many modifications that mainly reformulate the velocity equation. One of them, multiply the first term in (6) by a parameter w (called the "inertia weight"; it is a number in the range [0-1] that specifies the weight by which the particle's current velocity depends on its previous velocity). It has been shown in [12] that the PSO algorithm converges faster if w is linearly damped with iterations, for example starting at 0.9 at the first iteration and finishing at 0.4 in the last iteration. In another modification, Clerc and Kennedy [13] proposed a strategy for the placement of a constriction coefficient (χ) by which the first term in (7) is multiplied. This coefficient is used to control the convergence of the particle, prevent explosion and ensure convergence. In Clerc's constriction method,

$$\chi = \frac{2}{\varphi - 2 + \sqrt{\varphi^2 - 4\varphi}} \tag{8}$$

where $\varphi = \varphi_1 + \varphi_2 > 4$. The parameter φ is commonly set to 4.1 and $\varphi_1 = \varphi_2$ which result in an approximate value of 0.7298 for χ.

4 Results and Discussion

We test the performance of the algorithm with some signals included in the ICALAB package [14]. To measure the performance of the PSO, we calculate the distorsion of the estimated sources.

The signal to interference value for the *i-th* recovered component y_i is defined as:

$$SIR(y_i) = 10\log_{10}\frac{|\langle y_i, s_i\rangle|}{\langle y_i, y_i\rangle - |\langle y_i, s_i\rangle|} \tag{9}$$

where $\langle f, g\rangle$ is the standard inner product of two vectors.

We use Clerc's constriction PSO method. A swarm size of 25 particles, 500 iterations, $\varphi_1 = \varphi_2 = 2.05$, $\chi = 0.7298$ and circular population topology were used. The results obtained are compared with ICA algorithm FastICA [15]. The elements of the PSO output vector are reshaped to form the demixing matrix **B**, which is then normalized and made orthogonal before used in the contrast function in order to impose these constraints on the demixing matrix. The PSO output is an N-dimensional vector (position of the particle with best fitness) that represents a possible set of all the elements of the matrix **B**.

Two cases were analyzed. In the first experiment, 10 sparse (smooth bell-shape) sources that are approximately independent are randomly mixed. The original sources are shown in Figure 1. The experiment is repeated 10 times in order to test computationally the robustness and stability. In Table I (Example 1 rows) we show, for every source (S1 up to S5), the maximum (Max.), average (Avg.) and standard deviation (Std.) SIR for PSO (left columns) and FastICA (right columns).

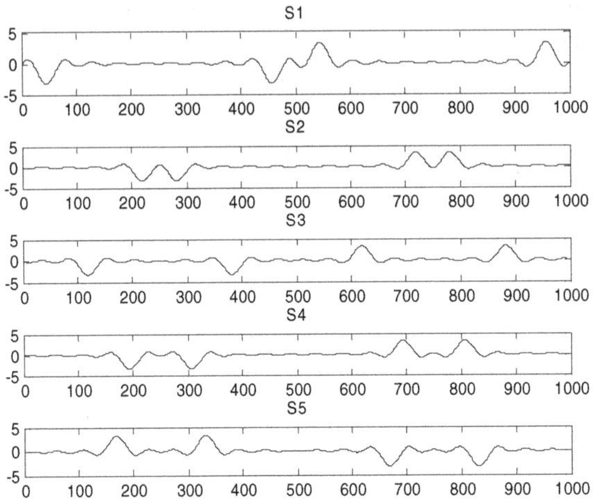

Fig. 1. The original signals for the first experiment

The second experiment involves five fourth order colored sources with a distribution close to Gaussian. Since the signals are close to Gaussian and ICA algorithms can estimate at most one Gaussian signal, this is a very challenging experiment. Figure 2 shows the original signals.

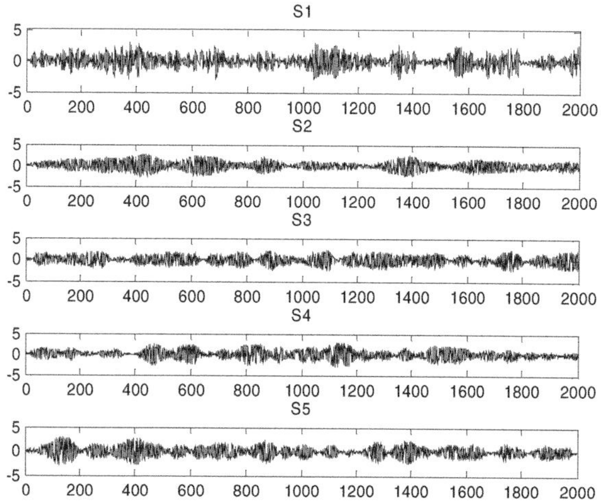

Fig. 2. The original signals for the second experiment

Table 1. The SIR in decibels for the simulated examples

		PSO			Fast ICA		
Example		Max	Avg	Std	Max	Avg	Std
1	S1	57.2	56.93	0.37	33.2	31.68	1.1
	S2	36.24	35.96	0.09	27.17	24.85	2.4
	S3	53.44	52.92	0.18	34.33	33.67	0.57
	S4	35.74	35.35	0.13	21.87	20.39	1.76
	S5	40.55	40.48	0.08	24.68	22.29	2.09
2	S1	14.37	14.18	0.58	Algorithm fails		
	S2	18.25	17.59	0.23	Algorithm fails		
	S3	21.04	19.92	0.39	Algorithm fails		
	S4	0.52	0.48	0.22	Algorithm fails		
	S5	1.12	1.06	0.19	Algorithm fails		

The quality of the extraction can be analyzed in Table I (Example 2 rows). The example was run also 10 times, showing the Max., Avg. and Std. *SIR*. As we can see, PSO was successful in recovering some of the original signals with acceptable quality (S1, S2 and S3). It failed in the recovering of the other two sources. On the other hand, the FastICA algorithm was not able to recover any original source.

Acknowledgments. This work has been partially funded by the Valencia Regional Government (Generalitat Valenciana) through project GV/2010/002 (Conselleria d'Educacio) and by the Universidad Politecnica de Valencia under grant no. PAID-06-09-003-382.

References

[1] Comon, P.: Independent Component Analysis; a new concept? Signal Processing 36(3), 287–314 (1994)
[2] Hyvarinen, A., Karhunen, J., Oja, E.: Independent Component Analysis. Wiley, New York (2001)
[3] Cichocki, A., Amari, S.I.: Adaptive Blind Signal and Image Processing. Wiley, New York (2002)
[4] Cardoso, J.F.: Blind Signal Separation: statistical principles. Proceedings of the IEEE 86(10), 2009–2025 (1998)
[5] Cardoso, J.F.: Higher-order contrasts for Independent Component Analysis. Neural Computation 11, 157–192 (1999)
[6] Bell, A.J., Sejnowski, T.J.: An information-maximization approach to blind separation and blind deconvolution. Neural Computation 7(6), 1004–1034 (1995)
[7] Cardoso, J.F.: Infomax and maximum likelihood for blind source separation. IEEE Signal Processing Letters 4(4), 112–114 (1997)
[8] Cardoso, J.F., Laheld, B.: Equivariant adaptive source separation. IEEE Trans. Signal Processing 44, 3017–3030 (1996)
[9] Amari, S.I.: Natural gradient works efficiently in learning. Neural Computation 10, 251–276 (1998)
[10] Kennedy, J., Eberhart, R.C.: Particle swarm optimization. In: Proceedings of IEEE International Conference on Neural Networks, Piscataway, NJ, pp. 1942–1948 (1995)
[11] Kennedy, J., Spears, W.M.: Matching algorithm to problems: an experimental test of the particle swarm and some genetic algorithms on multi modal problem generator. In: Proc. IEEE Int. Conf. Evolutionary Computation (1998)
[12] Trelea, I.C.: The particle swarm optimization algorithm: convergence analysis and parameter selection. Information Processing Letters 85, 317–325 (2003)
[13] Clerc, M., Kennedy, J.: The particle swarm—Explosion, stability, and convergence in a multidimensional complex space. IEEE Trans. Evol. Comput. 6, 58–73 (2002)
[14] Cichocki, A., Amari, S., Siwek, K., Tanaka, T., Phan, A.H., Zdunek, R.: ICALAB - MATLAB Toolbox Ver. 3 for signal processing
[15] Hyvarinen, A., Oja, E.: A fast fixed point algorithm for independent component analysis. Neural Computation 9(7), 1483–1492 (1997)

Regularized Active Set Least Squares Algorithm for Nonnegative Matrix Factorization in Application to Raman Spectra Separation

Rafal Zdunek

Institute of Telecommunications, Teleinformatics and Acoustics,
Wroclaw University of Technology, Wybrzeze Wyspianskiego 27,
50-370 Wroclaw, Poland
rafal.zdunek@pwr.wroc.pl

Abstract. Nonnegative Matrix Factorization (NMF) is an important tool in spectral data analysis. Various types of numerical optimization algorithms have been proposed for NMF, including multiplicative, projected gradient descent, alternating least squares and active-set ones. In this paper, we discuss the Tikhonov regularized version of the FC-NNLS algorithm (proposed by Benthem and Keenan in 2004) that belongs to a class of active-set methods in the context of its application to spectroscopy data. We noticed that starting iterative updates from a large value of a regularization parameter, and then decreasing it gradually to a very small value considerably reduces the risk of getting stuck into unfavorable local minima of a cost function. Moreover, our experiments demonstrate that this algorithm outperforms the well-known NMF algorithms in terms of Peak Signal-to-Noise Ratio (PSNR).

1 Introduction

NMF decomposes an input matrix into lower-rank factors that have nonnegative values and usually some physical meaning or interpretation. Hence, NMF has already found diverse applications in spectral data analysis [1–8].

Raman spectroscopy is commonly used for identification and measurement of abundance/concentration of the chemical compounds in a given specimen. Each chemical compound has unique Raman spectra that can be regarded as a fingerprint by which the compound can be identified. Due to many reasons such as measurements errors, impure specimen, and background fluorescence, the observed spectra represent a linear mixture of spectra underlying consistent compounds, and the aim of NMF is to extract the desired spectra and provide the information on their abundance.

Probably the most popular algorithms for NMF are based on multiplicative updates [9] that assure a monotonic convergence but their convergence is usually very slow. To tackle this problem, several algorithms that are based on additive updates have been proposed, including Projected Gradient (PG) descent, Alternating Least Squares (ALS) and active-set algorithms. A survey of PG and ALS

J. Cabestany, I. Rojas, and G. Joya (Eds.): IWANN 2011, Part II, LNCS 6692, pp. 492–499, 2011.

algorithms for NMF can be found in [8]. Since the estimated factors in NMF-based spectroscopy are expected to be large and very sparse, a good choice seems to be active-set algorithms.

In this paper, we discuss the selected active-set algorithms that are inspired by the Non-Negative Least Squares (NNLS) algorithm, proposed by Lawson and Hanson [10] in 1974. Bro and de Jong [11] considerably accelerate the NNLS algorithm by rearranging computations for cross-product matrices. The solution given by the NNLS algorithms is proved to be optimal according to the Karush-Kuhn-Tucker (KKT) conditions. Kim and Park [12] applied the modified NNLS algorithm to the l_1- and l_2-norm regularized LS problems in NMF, and showed that the regularized NNLS algorithms work very efficiently for gene expression microarrays. Their approach assumes constant regularization parameters to enforce the desired degree of sparsity.

Unfortunately, the basic NNLS algorithms are not very efficient for solving nonnegativity constrained linear systems with multiple Right-Hand Side (RHS) vectors since they compute a complete pseudoinverse once for each RHS. To tackle this problem, Benthem and Keenan [13] devised the Fast Combinatorial NNLS (FC-NNLS) algorithm and experimentally demonstrated it works efficiently for energy-dispersive X-ray spectroscopy data.

Here we apply the regularized version of the FC-NNLS algorithm to Raman spectra recovering, assuming the regularization parameter decreases gradually with iterations to trigger a given character of iterative updates.

The paper is organized in the following way. The next section discusses the concept of NMF for Raman spectra recovering. Section 3 is concerned with the NNLS algorithms. The experiments for spectra separation are presented in Section 4. Finally, the conclusions are given in Section 5.

2 NMF for Raman Spectra Recovering

The aim of NMF is to find such lower-rank nonnegative matrices $\boldsymbol{A} \in \mathbb{R}^{I \times J}$ and $\boldsymbol{X} \in \mathbb{R}^{J \times T}$ that $\boldsymbol{Y} \cong \boldsymbol{A}\boldsymbol{X} \in \mathbb{R}^{I \times T}$, given the matrix $\boldsymbol{Y}$, the lower rank J (the number of pure spectra), and possibly a prior knowledge on the matrices $\boldsymbol{A}$ and $\boldsymbol{X}$. Assuming each row vector of $\boldsymbol{Y}$ represents an observed mixed spectrum, and J is *a priori* known number of pure spectra, we can interpret each row vector of $\boldsymbol{X}$ as the unknown constituent Raman spectrum, and the corresponding column vector of $\boldsymbol{A} = [\boldsymbol{a}_1, \ldots, \boldsymbol{a}_J]$ as abundance/concentration of the constituent material.

To estimate the matrices $\boldsymbol{A}$ and $\boldsymbol{X}$ from $\boldsymbol{Y}$, we assume the Tikhonov regularized Euclidean function:

$$D(\boldsymbol{Y}||\boldsymbol{A}\boldsymbol{X}) = \frac{1}{2}||\boldsymbol{Y} - \boldsymbol{A}\boldsymbol{X}||_F^2 + \frac{\lambda}{2}||\boldsymbol{A}||_F^2, \tag{1}$$

where λ is a regularization parameter. The function is then alternatingly minimized with the modified NNLS algorithm.

3 NNLS Algorithms

The NNLS algorithm was originally proposed by Lawson and Hanson [10], and currently it is used by the command `lsqnonneg(.)` in Matlab. This algorithm iteratively partitions the unknown variables into the passive set P that contains basic variables and the active set R that contains active constraints, and updates only the basic variables until the complementary slackness condition of KKT conditions is met. Let $P = \{j : x_{jt} > 0\}$ and $R = \{1, \ldots, J\}\backslash P$, and the partition: $\forall t : \boldsymbol{x}_t = [\boldsymbol{x}_t^{(P)}; \boldsymbol{x}_t^{(R)}]^T \in \mathbb{R}^J$, and $\boldsymbol{g}_t = \nabla_{x_t} D(\boldsymbol{y}_t || \boldsymbol{A}\boldsymbol{x}_t) = [\boldsymbol{g}_t^{(P)}; \boldsymbol{g}_t^{(R)}]^T \in \mathbb{R}^J$. The columns of $\boldsymbol{A}$ can be also partitioned in the similar way: $\boldsymbol{A} = [\boldsymbol{A}_P \ \boldsymbol{A}_R]$, where $\boldsymbol{A}_P = [a_{*,P}]$ and $\boldsymbol{A}_R = [a_{*,R}]$. The basic variables can be estimated by solving the unconstrained LS problem:

$$\min_{x_t^{(P)}} \left\{ ||\boldsymbol{A}_P \boldsymbol{x}_t^{(P)} - \boldsymbol{y}_t||_2 \right\}, \tag{2}$$

where $\boldsymbol{A}_P$ has full column rank. The nonbasic variables in the KKT optimality point should be equal zero. In contrary to the projected ALS, the NNLS algorithm does not replace negative entries with zero-values, which is equivalent to determine nonbasic variables from unconstrained LS updates, but it starts from all nonbasic variables and tries to iteratively update a set of basic variables.

Bro and de Jong [11] considerably speed up this algorithm for $I >> J$ by precomputing the normal matrix $\boldsymbol{A}^T\boldsymbol{A}$ and the vector $\boldsymbol{A}^T\boldsymbol{y}_t$, and then solving the problem (2) as follows:

$$\bar{\boldsymbol{x}}_t^{(P)} = \left((\boldsymbol{A}^T\boldsymbol{A})_{P,P} \right)^{-1} (\boldsymbol{A}^T\boldsymbol{y}_t)_P. \tag{3}$$

Unfortunately, the inverse of $(\boldsymbol{A}^T\boldsymbol{A})_{P,P}$ must be computed for each t, which is very expensive if the number of RHS is very large.

Van Benthem and Keenan [13] tackled the problem of a high computational cost for multiple RHS. They noticed that for a sparse solution with multiple column vectors, a probability of finding columns vectors that have the same layout of the zero-entries (active constraints) is high. Hence, after detecting such vectors in $\boldsymbol{X}$, their passive entries can be updated by computing the inverse of $(\boldsymbol{A}^T\boldsymbol{A})_{P,P}$ only once. The NNLS algorithm proposed by Van Benthem and Keenan is referred to as FC-NNLS. The CSSLS algorithm (Algorithm 2) is a part of the FC-NNLS algorithm.

Applying the FC-NNLS algorithm to the objective function $D(\boldsymbol{Y}||\boldsymbol{A}\boldsymbol{X}) = \frac{1}{2}||\boldsymbol{Y} - \boldsymbol{A}\boldsymbol{X}||_F^2 + \frac{\lambda}{2}||\boldsymbol{X}||_F^2$ we obtained the Regularized FC-NNLS (RFC-NNLS) algorithm given by Algorithm 1.

To minimize (1) with respect to $\boldsymbol{A}$, the RFC-NNLS algorithm should be applied to the transposed system: $\boldsymbol{X}^T\boldsymbol{A}^T = \boldsymbol{Y}^T$. Finally, the complete algorithm (RNNLS) for updating the matrices $\boldsymbol{A}$ and $\boldsymbol{X}$ is given in Algorithm 3.

The aim of using regularization for updating $\boldsymbol{A}$ is rather to trigger the character of iterations than to stabilize ill-posed problems since the matrix $\boldsymbol{X}$ is not expected to be ill-conditioned. Motivated by [14], we propose to gradually

Algorithm 1. RFC-NNLS Algorithm

Input : $\boldsymbol{A} \in \mathbb{R}^{I \times J}$, $\boldsymbol{Y} = \mathbb{R}^{I \times T}$, λ

Output: $\boldsymbol{X}^* \geq 0$ such that $\boldsymbol{X}^* = \arg\min_X ||\boldsymbol{AX} - \boldsymbol{Y}||_F + \lambda||\boldsymbol{X}||_F$

Initialization: $M = \{1, \ldots, T\}$, $N = \{1, \ldots, J\}$;

Precompute: $\boldsymbol{B} = [b_{ij}] = \boldsymbol{A}^T\boldsymbol{A} + \lambda \boldsymbol{I}_J$ and $\boldsymbol{C} = [c_{it}] = \boldsymbol{A}^T\boldsymbol{Y}$;

$\boldsymbol{X} = \boldsymbol{B}^{-1}\boldsymbol{C}$; `// unconstrained minimizer`

$\boldsymbol{P} = [p_{jt}]$, where $p_{jt} = \begin{cases} 1 \text{ if } x_{jt} > 0, \\ 0 \text{ otherwise} \end{cases}$; `// passive entries`

$F = \{t \in M : \sum_j p_{jt} \neq I\}$; `// set of columns to be optimized`

$x_{jt} \leftarrow \begin{cases} x_{jt} \text{ if } p_{jt} = 1, \\ 0 \text{ otherwise} \end{cases}$;

while $F \neq \emptyset$ **do**

- $\boldsymbol{P}_F = [p_{*,F}] \in \mathbb{R}^{J \times |F|}$, $\boldsymbol{C}_F = [c_{*,F}] \in \mathbb{R}^{J \times |F|}$;
- $[x_{*,F}] = \texttt{cssls}(\boldsymbol{B}, \boldsymbol{C}_F, \boldsymbol{P}_F)$; `// Solved with the CSSLS algorithm`
- $H = \{t \in F : \min_{j \in N}\{x_{jt}\} < 0\}$; `// Columns with negative vars.`
- **while** $H \neq \emptyset$ **do**
  - $\forall s \in H$, select the variables to move out of the passive set $\boldsymbol{P}$;
  - $\boldsymbol{P}_H = [p_{*,H}] \in \mathbb{R}^{J \times |H|}$, $\boldsymbol{C}_H = [c_{*,H}] \in \mathbb{R}^{J \times |J|}$;
  - $[x_{*,H}] = \texttt{cssls}(\boldsymbol{B}, \boldsymbol{C}_H, \boldsymbol{P}_H)$;
  - $H = \{t \in F : \min_{j \in N}\{x_{jt}\} < 0\}$; `// Columns with negative vars.`
- $\boldsymbol{W} = [w_{jt}] = \boldsymbol{C}_F - \boldsymbol{BX}_F$, where $\boldsymbol{X}_F = [x_{*,F}]$; `// negative gradient`
- $J = \{t \in F : \sum_j w_{jt}(1 - \boldsymbol{P}_F)_{jt} = 0\}$; `// set of optimized columns`
- $F \leftarrow F \backslash J$; `// set of columns to be optimized`
- $p_{jt} = \begin{cases} 1 \text{ if } j = \arg\max_j \{w_{jt}(1 - \boldsymbol{P}_F)_{jt}, \forall t \in F\}, \\ p_{jt} \text{ otherwise} \end{cases}$; `// updating`

decrease the regularization parameter λ with alternating iterations for updating the matrix $\boldsymbol{A}$, starting from a large value λ_0. The regularization parameter for updating $\boldsymbol{X}$ can be neglected or set up for the minimal value. Considering updates for $\boldsymbol{A}$ in terms of Singular Value Decomposition (SVD) of $\boldsymbol{X}$, it is obvious that if λ is large, only the left singular vectors that correspond to the largest singular values take part in the updating process. When the number of iterations is large (in practice about 100), $\lambda \to \bar{\lambda}$, where we take $\bar{\lambda} = 10^{-12}$, and the singular vectors corresponding to the smallest singular values participate in the updates.

Furthermore, when $\lambda >> \sigma_{max}(\boldsymbol{X})$, where $\sigma_{max}(\boldsymbol{X})$ is the largest singular value of $\boldsymbol{X}$, the LS update rule for $\boldsymbol{A}$ can be approximated by

$$\boldsymbol{A} \leftarrow \boldsymbol{A} - \frac{1}{\lambda}\boldsymbol{G}_A, \tag{4}$$

where $\boldsymbol{G}_A = (\boldsymbol{AX} - \boldsymbol{Y})\boldsymbol{X}^T$ is the gradient of (1) with respect to $\boldsymbol{A}$. Thus, starting the iterative updates for NMF from a large value of λ, the initial iterations have the nature of gradient descent updates with a small step size. This kind of iterations is beneficial when the updates are far away from the desired minimum.

Algorithm 2. CSSLS Algorithm

Input : $\boldsymbol{B} \in \mathbb{R}^{J \times J}$, $\boldsymbol{C} = \mathbb{R}^{J \times K}$, $\boldsymbol{P} \in \mathbb{R}^{J \times K}$
Output: $\boldsymbol{X}$

$M = \{1, \ldots, K\}$, $N = \{1, \ldots, J\}$, $\boldsymbol{P} = [\boldsymbol{p}_1, \ldots, \boldsymbol{p}_K]$;
Find the set of L unique columns in $\boldsymbol{P}$: $\boldsymbol{U} = [\boldsymbol{u}_1, \ldots, \boldsymbol{u}_L] = \text{unique}\{\boldsymbol{P}\}$;
$\boldsymbol{d}_j = \{t \in M : \boldsymbol{p}_t = \boldsymbol{u}_j\}$; `// columns with identical passive sets`
for $j = 1, \ldots, L$ **do**
  $x_{\boldsymbol{u}_j, \boldsymbol{d}_j} = \left([\boldsymbol{B}]_{\boldsymbol{u}_j, \boldsymbol{u}_j}\right)^{-1} [\boldsymbol{C}]_{\boldsymbol{u}_j, \boldsymbol{d}_j}$

Algorithm 3. RNNLS Algorithm

Input : $\boldsymbol{Y} \in \mathbb{R}^{I \times T}$, J - lower rank, λ_0 - initial regularization parameter, $\bar{\lambda}$ - minimal value of regularization parameter,
Output: Factors $\boldsymbol{A}$ and $\boldsymbol{X}$

Initialize (randomly) $\boldsymbol{A}$ and $\boldsymbol{X}$;
repeat
  $k \leftarrow k + 1$;
  $\lambda = \max\left\{\bar{\lambda}, 2^{-k}\lambda_0\right\}$; `// Regularization parameter schedule`
  $\boldsymbol{X} \leftarrow$ `RFCNNLS`$(\boldsymbol{A}, \boldsymbol{Y}, 10^{-12})$; `// Update for X`
  $\bar{\boldsymbol{A}} \leftarrow$ `RFCNNLS`$(\boldsymbol{X}^T, \boldsymbol{Y}^T, \lambda)$; `// Update for A^T`
  $\boldsymbol{A} = \bar{\boldsymbol{A}}^T$;
  $a_{ij} \leftarrow \frac{a_{ij}}{\sum_{i=1}^{I} a_{ij}}$; `// Normalization of the columns to unit l_1-norm`
until `Stop criterion` is satisfied ;

When the alternating steps proceed, the nature of the updates for $\boldsymbol{A}$ gradually changes to the Newton steps which explore the local minimum deeply.

4 Experiments

The numerical experiments are carried out for two different benchmarks of spectral signals. Benchmark A contains 40 random positive spiky signals where the sparsity of each signal amounts to be about 80 %. The total number of samples is 1000, i.e. $T = 1000$. Benchmark B is created from 5 real Raman spectra taken from RRUFF™ Project[1]. We selected the spectra for the following minerals: Spurrite from Okayama in Japan, Zaratite from Tasmania in Australia, Epsomite from Death Valley in California, and Demesmaekerite from Katanga in Congo. Fig. 1 illustrates these spectra in the following order from the top: Spurrite measured at the wavelength 532nm, Zaratite at 514nm, Zaratite at 532nm, Epsomite at 785nm, and Demesmaekerite at 532nm. All the signals in benchmark B are resampled to 2000 samples ($T = 2000$).

For both benchmarks the observed spectra are obtained by mixing the source spectra with an uniformly distributed random matrix $\boldsymbol{A}$. We used $\boldsymbol{A} \in \mathbb{R}^{160 \times 40}$

[1] `http://rruff.info`

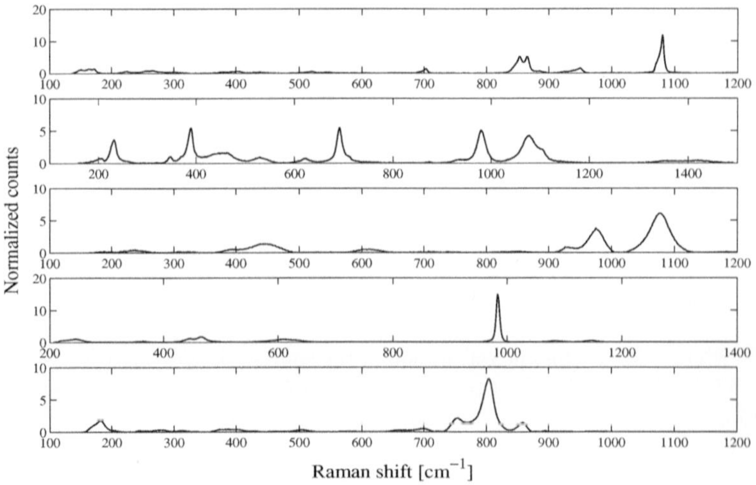

Fig. 1. Normalized Raman spectra for the following minerals (from the top): Spurrite (532 nm), Zaratite (514 nm), Zaratite (532 nm), Epsomite (785 nm), Demesmaekerite (532 nm)

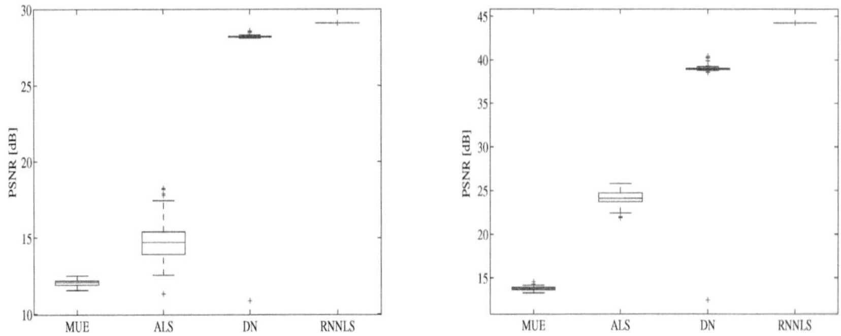

Fig. 2. PSNR statistics for the estimation of the sources in $\boldsymbol{X}$ using various algorithms (MUE, ALS, DN, RNNLS) and the benchmark A ($I = 160$, $J = 40$): (left) worst source; (right) mean over the sources

and $\boldsymbol{A} \in \mathbb{R}^{15\times5}$ for benchmarks A and B, respectively. Moreover, the mixed signals for benchmark B are additionally corrupted with a zero-mean Gaussian noise with $SNR = 20$[dB].

We tested the following algorithms: standard Lee-Seung NMF for the Euclidean distance (referred to as MUE) [9], Projected ALS [7, 8], Damped Newton (DN) [14], and RNNLS. All the algorithms are terminated after 500 iterations for the benchmark A and 50 iterations for benchmark B. For the RNNLS algorithm, we set the following parameter: $\bar{\lambda} = 10^{-12}$, $\lambda_0 = 10^8$.

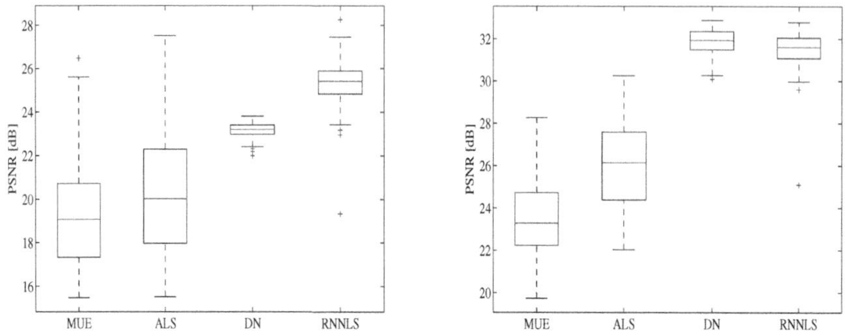

Fig. 3. PSNR statistics for the estimation of the sources in $\boldsymbol{X}$ using various algorithms (MUE, ALS, DN, RNNLS) and the benchmark B ($I = 15$, $J = 5$): (left) worst source; (right) mean over the sources

Each tested algorithm is run for 100 Monte Carlo (MC) trials with a random initialization. The algorithms are evaluated in terms of the Peak Signal-to-Noise Ratio (PSNR) averaged over 100 trials. This measure, defined in Chapter 3 [8], is particular useful for evaluating spiky signals. The statistics of PSNR samples for benchmarks A and B are shown in Figs. 2 and 3, respectively. The computational complexity can be roughly estimated with the elapsed time that for the benchmark A and 500 iterations is 3.1, 4.3, 92, 178 seconds for the MUE, ALS, DN, and RNNLS algorithms, respectively.

5 Conclusions

We applied the Tikhonov regularized version of the FC-NNLS algorithm to NMF and demonstrated that this approach is very useful for spectral data recovering. The regularization parameter should decrease gradually with alternating steps, starting from a large initial value that can be set up in a large range (e.g. 10^4 – 10^{15}). Obviously, larger value of λ_0 needs more alternating steps to run but the risk of getting stuck into unfavorable local minima is lower. We used the rule: $\lambda \leftarrow \lambda/2$, however, other rules are also possible. For example, the exponential rule: $\lambda = \lambda_0 \exp\{-\tau k\}$ works even better but it needs two parameters to set up in advance. The experiments demonstrate that the RNNLS algorithm outperforms all the tested algorithms in terms of PSNR and resistance to initialization for sparse data (the variance of PSNR samples in Fig. 2 is the smallest for the RNNLS).

Acknowledgment

This work was partially supported by the habilitation grant N N515 603139 (2010-2012) from the Ministry of Science and Higher Education, Poland.

References

[1] Sajda, P., Du, S., Brown, T., Parra, L., Stoyanova, R.: Recovery of constituent spectra in 3D chemical shift imaging using nonnegative matrix factorization. In: Proc. of 4th International Symposium on Independent Component Analysis and Blind Signal Separation, Nara, Japan, pp. 71–76 (2003)

[2] Sajda, P., Du, S., Brown, T.R., Stoyanova, R., Shungu, D.C., Mao, X., Parra, L.C.: Nonnegative matrix factorization for rapid recovery of constituent spectra in magnetic resonance chemical shift imaging of the brain. IEEE Transaction on Medical Imaging 23(12), 1453–1465 (2004)

[3] Pauca, V.P., Pipera, J., Plemmons, R.J.: Nonnegative matrix factorization for spectral data analysis. Linear Algebra and its Applications 416(1), 29–47 (2006)

[4] Li, H., Adali, T., Wang, W., Emge, D., Cichocki, A.: Non-negative matrix factorization with orthogonality constraints and its application to Raman spectroscopy. The Journal of VLSI Signal Processing 48(1-2), 83–97 (2007)

[5] Gobinet, C., Perrin, E., Huez, R.: Application of nonnegative matrix factorization to fluorescence spectroscopy. In: Proc. European Signal Processing Conference (EUSIPCO 2004), Vienna, Austria, September 6–10 (2004)

[6] Jia, S., Qian, Y.: Constrained nonnegative matrix factorization for hyperspectral unmixing. IEEE Transactions on Geoscience and Remote Sensing 47(1), 161–173 (2009)

[7] Berry, M., Browne, M., Langville, A., Pauca, P., Plemmons, R.: Algorithms and applications for approximate nonnegative matrix factorization. Computational Statistics and Data Analysis 52(1), 155–173 (2007)

[8] Cichocki, A., Zdunek, R., Phan, A.H., Amari, S.I.: Nonnegative Matrix and Tensor Factorizations: Applications to Exploratory Multi-way Data Analysis and Blind Source Separation. Wiley and Sons, Chichester (2009)

[9] Lee, D.D., Seung, H.S.: Learning of the parts of objects by non-negative matrix factorization. Nature 401, 788–791 (1999)

[10] Lawson, C.L., Hanson, R.J.: Solving Least Squares Problems. Prentice-Hall, Englewood Cliffs (1974)

[11] Bro, R., Jong, S.D.: A fast non-negativity-constrained least squares algorithm. Journal of Chemometrics 11, 393–401 (1997)

[12] Kim, H., Park, H.: Non-negative matrix factorization based on alternating non-negativity constrained least squares and active set method. SIAM Journal in Matrix Analysis and Applications 30(2), 713–730 (2008)

[13] Benthem, M.H.V., Keenan, M.R.: Fast algorithm for the solution of large-scale non-negativity-constrained least squares problems. Journal of Chemometrics 18, 441–450 (2004)

[14] Zdunek, R., Phan, A., Cichocki, A.: Damped Newton iterations for nonnegative matrix factorization. Australian Journal of Intelligent Information Processing Systems 12(1), 16–22 (2010)

A Decision-Aided Strategy for Enhancing Transmissions in Wireless OSTBC-Based Systems

Tiago M. Fernández-Caramés, Adriana Dapena, José A. García-Naya, and Miguel González-López

Departamento de Electrónica e Sistemas
Universidade da Coruña
Campus de Elviña s/n, 15071, A Coruña, Spain
{tmfernandez,adriana,jagarcia,mgonzalezlopez}@udc.es

Abstract. Orthogonal Space-Time Block Coding (OSTBC) allows multiple antenna wireless systems to exploit spatial and temporal diversity. When transmitting multimedia content, OSTBC-based system's performance can be greatly enhanced through the optimization of different parameters. For instance, data rates can be improved if the overhead related to sending side information (pilot symbols) is minimized. Also, processing speed can be increased if the decoding algorithms computational load is reduced. In this paper, it is presented a decision-aided channel estimation method aimed specifically at addressing the aforementioned two issues: the method incorporates unsupervised strategies to estimate the channel without using pilot symbols and a simple decision rule to determine if channel estimation is required for each frame, what allows transceiver designers to implement techniques to avoid channel estimation.

1 Introduction

In the last decade, a large number of Space-Time Coding (STC) techniques have been proposed in the literature to exploit spatial diversity in systems with multiple elements at both transmission and reception (see, for instance, [1, 2] and references therein). Orthogonal Space Time Block Coding (OSTBC) is remarkable in that it is able to provide full diversity gain with linear decoding complexity. The basic premise of OSTBC is the encoding of the transmit symbols into a unitary matrix so as to spatially decouple their Maximum Likelihood (ML) detection, which can be seen as a matched filter followed by a symbol-by-symbol detector.

To address the issue of decoding complexity, S. M. Alamouti proposed a popular OSTBC scheme for transmitting in systems with two antennas at the transmitter and only one at the receiver [3]. This scheme is the only OSTBC capable of achieving full spatial rate for complex constellations. Other OSTBCs have been proposed for more than two transmit antennas, but they suffer from important

J. Cabestany, I. Rojas, and G. Joya (Eds.): IWANN 2011, Part II, LNCS 6692, pp. 500–507, 2011.

losses in spatial rate [4, 5]. Due to the advantages mentioned, the Alamouti code has been incorporated in some of the latest wireless communication standards, like IEEE 802.11n [6] or IEEE 802.16 [7].

In such recent wireless communication standards, to perform coherent detection when using OSTBC systems, it is required the identification of a unitary channel matrix. For this purpose, the standards define the inclusion of pilot symbols into the data frame. Pilot symbols help the receiver to invert the channel, but they reduce the maximum achievable data rate and increase the computational load due to the channel inversion procedure. This paper focuses on proposing an alternative to solve these two important issues, which are critical in the field of digital transmissions, specially in systems with real-time constraints.

The proposed method uses unsupervised (also called blind) algorithms to decrease the overhead associated with the pilot symbols. Such algorithms consider the transmitted symbol substreams to be unknown sources that can be recovered from their mixtures, which are obtained at each receiving antenna. The channel matrix can be seen as the mixing transformation between the sources and the observations [8]. The term unsupervised refers to the fact that little or nothing is known or assumed about the sources and the mixing matrix structure.

In order to reduce the computational load of the decoding algorithms, our method exploits the information obtained during time synchronization, which is the first step required to receive each frame. We use such information to determine the instant when the channel suffers a considerable variation and, as a consequence, to determine the instant when the channel has to be estimated. The resulting strategy has a very low computational load and can be combined with supervised and unsupervised methods.

This paper is organized as follows. Section 2 describes the signal model of the Alamouti coding scheme. Section 3 presents unsupervised algorithms that allow eliminating the transmission of pilot symbols. Section 4 proposes the decision-aided method to reduce the computational load at the receiver side. Finally, Section 6 is devoted to the conclusions.

2 Alamouti Coding Systems

Figure 1 shows the baseband representation of an Alamouti-based system with two antennas at the transmitter and one antenna at the receiver. A digital source in the form of a binary data stream, b_i, is mapped into symbols which are split into two substreams, s_1 and s_2. We assume that s_1 and s_2 are independent equiprobable discrete random variables that take values from a finite set of symbols belonging to a real or complex modulation (PAM, PSK, QAM...). The vector $\mathbf{x} = [x_1\ x_2]^T$ of the received signals (observations) can be written as $\mathbf{x} = \mathbf{H}\mathbf{s} + \mathbf{v}$, where $\mathbf{s} = [s_1\ s_2]^T$ is the source vector, $\mathbf{v} = [v_1\ v_2]^T$ is the additive white Gaussian noise vector and the 2×2 channel matrix has the form

$$\mathbf{H} = \begin{bmatrix} h_1 & h_2 \\ h_2^* & -h_1^* \end{bmatrix} \tag{1}$$

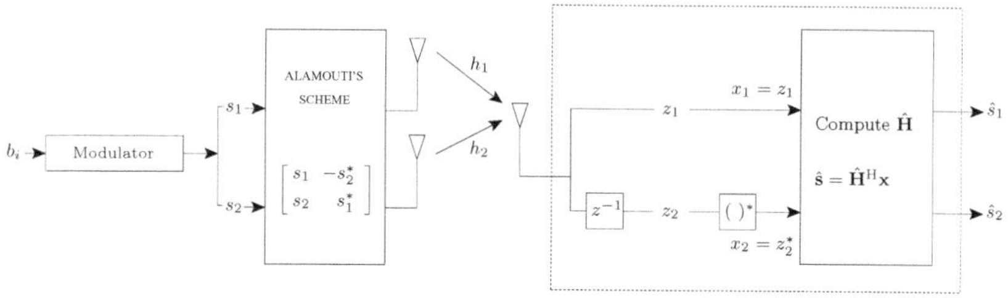

Fig. 1. Alamouti coding scheme

Matrix $\mathbf{H}$ is unitary up to a scalar factor, i.e., $\mathbf{H}\mathbf{H}^{\mathrm{H}} = \mathbf{H}^{\mathrm{H}}\mathbf{H} = \|\mathbf{h}\|^2\mathbf{I}_2$ where $\|\mathbf{h}\|^2 = |h_1|^2 + |h_2|^2$ is the squared Euclidean norm of the channel vector, $\mathbf{I}_2$ is the 2×2 identity matrix and $(\cdot)^{\mathrm{H}}$ is the Hermitian operator. It follows that the transmitted symbols can be recovered, up to scale, as $\hat{\mathbf{s}} = \hat{\mathbf{H}}^{\mathrm{H}}\mathbf{x}$, where $\hat{\mathbf{H}}$ is a suitable estimate of the channel matrix. As a result, this scheme supports maximum likelihood detection based only on linear processing at the receiver.

The decoding procedure requires to estimate the channel matrix. For such purpose, current digital communication standards insert pilot symbols into the transmitted frames and a supervised method is used to obtain the estimate at the receiver side. A simple estimation technique consists in using the Least Mean Squares (LMS) method [9] or the so-called Widrow-Hoff solution, given by

$$\mathbf{W} = \mathbf{C}_{\mathbf{x}}^{-1}\mathbf{C}_{\mathbf{xp}}, \tag{2}$$

where $\mathbf{C}_{\mathbf{x}} = E[\mathbf{x}\mathbf{x}^{\mathbf{H}}]$ is the autocorrelation matrix of the observations and $\mathbf{C}_{\mathbf{xp}} = E[\mathbf{x}\mathbf{p}^{\mathrm{H}}]$ is the matrix of the cross–correlation between the observations and the pilot symbols $\mathbf{p}$. In the case of the Alamouti code, the weight matrix $\mathbf{W}$ is equal to $\mathbf{H}^{\mathrm{H}}$ and, therefore, the transmitted signals can be estimated using $\hat{\mathbf{s}} = \mathbf{W}^{\mathrm{H}}\mathbf{x}$.

3 Pilot Symbols Elimination

The transmission of pilot symbols can be avoided by using unsupervised approaches. Unsupervised algorithms can estimate the mixing matrix $\mathbf{H}$ and the realizations of the source vector $\mathbf{s}$ directly from the corresponding realizations of the observed vector $\mathbf{x}$. This lack of prior knowledge may limit the achievable performance, but makes unsupervised approaches more robust to calibration errors (i.e., deviations from the model assumptions) than conventional array processing techniques [10]. A property commonly exploited is the statistical independence of the sources. Depending on the degree of independence considered, two main group of techniques can be distinguished: principal component analysis (PCA), which are based on Second-Order Statistics (SOS), and independent component analysis (ICA), which exploits Higher-Order Statistics (HOS). In the literature, it can be found a large number of unsupervised approaches (see, for instance, the book [11]). In particular, in the case of the Alamouti scheme, eigen-based approaches have received a great deal of attention [12–14].

Research on higher-order eigen-based approaches began with Cardoso's early work on the so-called quadricovariance, a folded version of the fourth-order moment array, and culminated in the popular ICA method known as joint approximate diagonalization of eigenmatrices (JADE) [10], which jointly diagonalizes a set of matrices formed by fourth-order cross-cumulants of the observations. The description of the JADE algorithm is beyond the scope of this paper, but we encourage the interested reader to see [10].

In Alamouti coding scheme, the channel matrix is essentially unitary and, therefore, the JADE algorithm can be simplified by considering the diagonalization of a reduced set of fourth-order cross cumulants matrices. In particular, in [14], Dapena et al. proposed a method called Blind Channel Estimation based on Eigenvalue Spread (BCEES) where the matrix to be diagonalized is selected taking into account the absolute difference between the eigenvalues (eigenvalue spread).

4 Reduction of the Computational Load

In this section, we propose a simple method to reduce the computational load of the above-mentioned channel estimation methods. In static environments, it is common to assume that the channel remains constant during the transmission of several frames. On the contrary, in mobile environments the channel variations happen faster (each frame or during the transmission of each frame). In both cases, it is needed to perform a time synchronization that is usually carried out in two steps: a coarse synchronization, which considers the received symbols, and a fine synchronization, that works at a sample level.

A coarse synchronization method consists basically in computing complex cross–correlations of the received signals and a preamble that is known at reception. Since in the Alamouti OSTBC all symbols are received by the same antenna, the frame synchronization in time can be done by correlating the received signal with a linear combination of the two transmitted, orthogonal preambles (a single preamble is transmitted by each of the two antennas). For instance, we consider that the two preambles consists of Pseudo-Noise (PN) sequences mapped to BPSK symbols and that the preamble transmitted by the first antenna contains zeros at even time instants whereas the one transmitted by the second antenna contains zeros at odd time instants. The peak values of the cross-correlation between the linear combination of the two preambles and the acquired signal demarcate the position of the first symbols in the acquired frame. After performing the time synchronization, we can split the received signal in two streams corresponding to each transmit antenna.

Using the model described in Section 2, the received signals corresponding to the preambles, denoted by r_1 and r_2, have the form

$$x_1 = h_1\, r_1 + h_2\, r_2 + v_1, \quad x_2 = h_2^*\, r_1 - h_1^* r_2 + v_2 \tag{3}$$

The correlation between the preamble and the received signal is computed by using a sliding window. The synchronization is performed by finding the peak value. Note that, the peak value is given by

$$c_{11} = \frac{E[x_1 r_1^*]}{E[|r_1|^2]} = h_1, \quad c_{21} = \frac{E[x_2 r_1^*]}{E[|r_1|^2]} = h_2^* \tag{4}$$

It is interesting to note that the correlations in Equation (4) produce an estimation of the two channel coefficients h_1 and h_2. This fact allows us to use the synchronization procedure to determine the instants when the channel suffers a considerable variation and, therefore, when the channel has to be estimated. Our idea consists in comparing the cross–correlation values obtained in two consecutive frames. Denoting by $c_{11}[k]$ and $c_{21}[k]$ the peak correlation values obtained for the k-th frame, the proposed procedure is the following

Step 1: Compute the error $Error_1[k] = |c_{11}[k] - c_{11}[k-1]|$ and $Error_2[k] = |c_{21}[k] - c_{21}[k-1]|$
Step 2: Use the decision criterion

$$(Error_1[k] > \beta) \text{ OR } (Error_2[k] > \beta) \rightarrow \text{Estimate the channel}$$

where β is a real-valued threshold.

Note that this decision rule can be combined with supervised and unsupervised channel estimation approaches.

5 Simulation Results

This section presents the performance evaluation of the channel estimation strategies studied in this paper. The experiments have been performed using QPSK source symbols coded with the Alamouti scheme. In total, 20 frames of 200 symbols have been transmitted by each antenna. Additionaly, each frame contains a preamble of 10 BPSK per antenna. The channel coefficients h_1 and h_2 were generated randomly according to a spatially-white Rayleigh distribution. The channels remain constant during the transmission of 5 frames.

We have evaluated the following methods:

- Supervised approach, where the coefficient matrix is computed for each frame using Equation (2). A 5% of pilot symbols (10 pilot symbols) was included in each frame.
- Unsupervised approaches (JADE and BCEES), where the channel is estimated for each frame. The fourth-order cross-cumulants computed in these approaches have been calculated by sample averaging over the symbols of each frame (200 symbols per frame).
- Decision-Aided approaches (DA-Supervised, DA-JADE and DA-BCEES).

The results have been obtained by averaging 1 000 independent realizations.

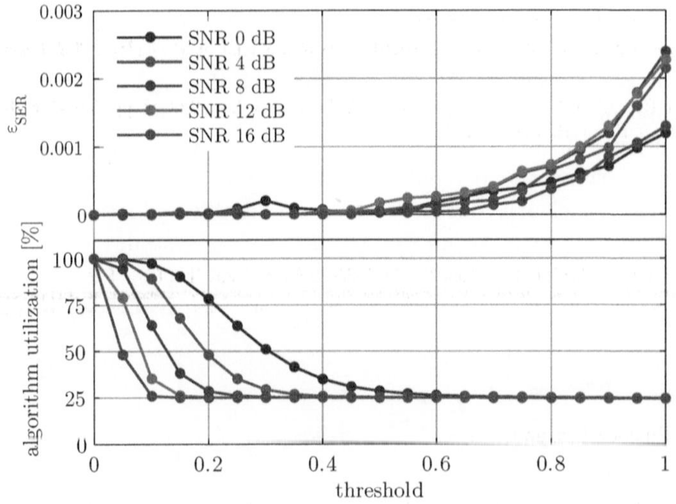

Fig. 2. Selection of parameter β: SER versus SNR and percentage of algorithm utilization

5.1 Determination of the Threshold for the Simulations

Before beginning the averaging process, it is necessary to consider the problem of selecting the threshold value used in the decision criterion. In order to obtain a good estimation of the cross–correlations, the simulations performed in these tests contained a preamble with 100 symbols per antenna. To quantify the difference —in terms of SER (Symbol Error Rate) versus SNR (Signal-to-Noise Ratio)— between the DA-Supervised and the Supervised approaches, the following expression is introduced:

$$\epsilon_{\mathrm{SER}} = 1 - \frac{1 + \mathrm{SER}_{\mathrm{Supervised}}}{1 + \mathrm{SER}_{\mathrm{DA\text{-}Supervised}}}. \tag{5}$$

Thus, Figure 2 plots ϵ_{SER} as well as a percentage that indicates the number of frames that require to estimate the channel (i.e. the number of frames where the channel was estimated divided by the total number of frames). We can see that a value of $\beta = 0.6$ gives a good tradeoff between SER and channel estimation, since SER is almost zero and the channel estimation is equal to 25 %, which corresponds to estimate the channel only 5 times for the 20 transmitted frames (for the first frame and for each channel variation).

5.2 Algorithm Performance Comparison

Figure 3 presents the SER for the different approaches considered. We have used a threshold $\beta = 0.6$ for the decision-aided approaches. It can be observed that JADE and DA-JADE provide the same performance as the supervised approach, but with the advantage of not requiring the transmission of pilot symbols. BCEES has a loss due to the reduced number of symbols used to estimate the fourth-order cross-cumulants. Note also that the decision-aided strategies provide the same performance as when the channel estimation is carried out for all frames. Figure 3 also gives a good idea on the processing savings that can be achieved, showing the percentage of the frames that need to estimate the channel. It can be clearly observed the considerable reduction obtained when the decision criterion is incorporated to the receiver system.

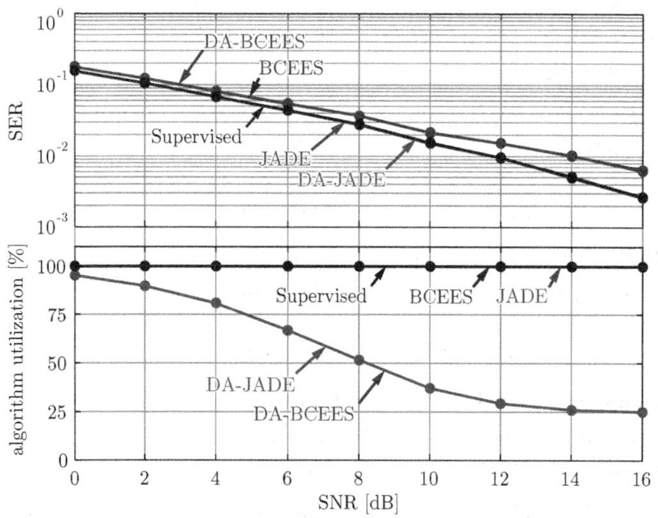

Fig. 3. Performance comparison for spatially-white Rayleigh distributed channels: SER versus SNR and percentage of algorithm utilization

6 Conclusions

We have proposed a simple scheme to detect channel variations in Alamouti coded systems. This novel approach uses information obtained during the synchronization procedure to determine such channel variations. When channel variations are significant, the system estimates the channel matrix using a supervised or unsupervised method. In other case, decoding is performed using a previous estimate. The main advantage of the proposed method is that it does not require to perform additional operations and, selecting an adequate threshold value, it is possible to obtain the same performance as when the channel is estimated for each frame. In addition, the incorporation of unsupervised approaches avoids the transmission of pilot symbols. Future work will have to deal with the determination of an analytical expression for the threshold parameter used by the decision criterion.

Acknowledgements

This work was supported by Xunta de Galicia through contracts 10TIC105003PR and 09TIC008105PR, and by the Spanish Ministerio de Ciencia e Innovación under grants TEC2010-19545-C04-01 and CSD2008-00010.

References

1. Gesbert, D., Shafi, M., Shan-Shiu, D., Smith, P.J., Naguib, A.: From theory to practice: an overview of MIMO space-time coded wireless systems. IEEE Journal on Selected Areas in Communications 21, 281–302 (2003)
2. Paulraj, A.J., Papadias, C.B.: Space-time processing for wireless communications. IEEE Signal Processing Magazine 14(6), 49–83 (1997)
3. Alamouti, S.M.: A simple transmit diversity technique for wireless communications. IEEE Journal on Selected Areas in Communications 16, 1451–1458 (1998)
4. Tarokh, V., Jafarkhani, H., Calderbank, A.R.: Space-time block codes from orthogonal designs. IEEE Transactions on Information Theory 45(5), 1456–1467 (1999)
5. Larsson, E.G., Stoica, P.: Space-Time Block Coding for Wireless Communications. Cambridge University Press, Cambridge (2003)
6. IEEE, IEEE Standard for Information technology–Telecommunications and information exchange between systems–Local and metropolitan area networks–Specific requirements Part 11: Wireless LAN Medium Access Control (MAC) and Physical Layer (PHY) Specifications Amendment 5: Enhancements for Higher Throughput (October 2009)
7. IEEE, IEEE 802.16-2009: Air interface for fixed broadband wireless access systems (May 2009)
8. Zarzoso, V., Nandi, A.K.: Blind source separation, ch. 4, pp. 167–252. Kluwer Academic Publishers, Dordrecht (1999)
9. Haykin, S.: Neural Networks A Comprehensive Foundation. Macmillan College Publishing Company, New York (1994)
10. Cardoso, J.-F., Souloumiac, A.: Blind beamforming for non-Gaussian signals. IEE Proceedings F 140(46), 362–370 (1993)
11. Comon, P., Jutten, C.: Handbook of Blind Source Separation, Independent Component Analysis and Applications. Academic Press, London (2010)
12. Beres, E., Adve, R.: Blind channel estimation for orthogonal STBC in MISO systems. In: Proc. of Global Telecommunications Conference, vol. 4, pp. 2323–2328 (November 2004)
13. Vía, J., Santamaría, I., Pérez, J., Ramírez, D.: Blind decoding of MISO-OSTBC systems based on principal component analysis. In: Proc. of International Conference on Acoustic, Speech and Signal Processing, vol. IV, pp. 545–549 (2006)
14. Dapena, A., Pérez-Iglesias, H., Zarzoso, V.: Blind Channel Estimation Based on Maximizing the Eigenvalue Spread of Cumulant Matrices in (2 x 1) Alamouti's Coding Schemes. In: Wireless Communications and Mobile Computing, Article published online (October 5, 2010) (accepted) doi:10.1002/wcm.992

Nonlinear Prediction Based on Independent Component Analysis Mixture Modelling

Gonzalo Safont, Addisson Salazar, and Luis Vergara

Instituto de Telecomunicaciones y Aplicaciones Multimedia, Universidad Politécnica de Valencia, Camino de Vera s/n, 46022, Valencia, Spain
gonsaar@upvnet.upv.es,
{asalazar,lvergara}@dcom.upv.es

Abstract. This paper presents a new algorithm for nonlinear prediction based on independent component analysis mixture modelling (ICAMM). The data are considered from several mutually-exclusive classes which are generated by different ICA models. This strategy allows linear local projections that can be adapted to partial segments of a data set while maintaining generalization (capability for nonlinear modelling) given the mixture of several ICAs. The resulting algorithm is a general purpose technique that could be applied to time series prediction, to recover missing data in images, etc. The performance of the proposed method is demonstrated by simulations in comparison with several classical linear and nonlinear methods.

Keywords: ICA, ICAMM, nonlinear prediction, Kriging, Wiener structure.

1 Introduction

Independent component analysis (ICA) is an intensive area of research that is progressively finding more applications for both blind source separation (BSS) and for feature extraction/modelling. The goal of ICA is to perform a linear transformation of the observed sensor signals, such that the resulting transformed signals (the sources or prior generators of the observed data) are as statistically independent of each other as possible [1]. The linear ICA method is extended in independent component analysis mixture modelling (ICAMM) to a kind of nonlinear ICA model, i.e., multiple ICA models are learned and weighted in a probabilistic manner. Thus, the ICA mixture model is a conditional independence model, i.e., the independence assumption holds only within each class and there may be dependencies among the classes [2]. ICAMM has recently emerged as a flexible approach to model arbitrary data densities with non-gaussian distributions for the independent components (i.e., relaxing the restriction of modelling every component by a multivariate Gaussian probability density function).

ICA has been shown to improve the prediction of time series by considering certain realistic assumptions [3]. There is also a close relationship between ICA and the minimization of algorithmic complexity [4]. This has been used to improve the prediction of financial time series, see for instance [5]. A procedure that includes an

J. Cabestany, I. Rojas, and G. Joya (Eds.): IWANN 2011, Part II, LNCS 6692, pp. 508–515, 2011.

ICA-based preprocessing step followed by prediction using neural networks has been applied for data prediction in several fields; see for instance [6]. Other applications benefit by using ICA for preprocessing since the independence among the variables is required [7].

This paper presents a novel procedure based on independent component analysis mixture modelling (ICAMM). The novelty of the procedure consists of the mixture of ICAs is not only proposed as a preprocessor to be applied before prediction, but the prediction itself is made by estimating the ICAMM parameters. The data are considered from several mutually-exclusive classes which are generated by different ICA models. This strategy allows linear local projections that can be adapted to partial segments of a data set while maintaining generalization (capability for nonlinear modelling) given the mixture of several ICAs.

The following sections of the paper include the following: the theoretical foundations of ICA and ICAMM; the proposed procedure for nonlinear prediction based on ICAMM; the results obtained for several simulations in comparison with two classical prediction methods (Kriging and a Wiener structure-based method); and the conclusions of the paper.

2 Independent Component Analysis Mixture Modelling

The standard noiseless instantaneous ICA formulates a $M \times 1$ random vector $\mathbf{x}$ by linear mixtures of M random variables that are mutually independent $(s_1,\ldots,s_M)$ whose distributions are totally unknown. That is, for $\mathbf{s}=(s_1,\ldots,s_M)^T$ and some matrix $\mathbf{A}$:

$$\mathbf{x} = \mathbf{A}\mathbf{s} \ . \tag{1}$$

The essential principle is to estimate the so-called mixing matrix $\mathbf{A}$, or equivalently $\mathbf{B} = \mathbf{A}^{-1}$ (the demixing matrix). The matrix $\mathbf{A}$ contains the coefficients of the linear transformation that represents the transfer function from sources to observations. Thus, given N i.i.d. observations $(\mathbf{x}_1,\ldots,\mathbf{x}_N)$ from the distribution of $\mathbf{x}$, $\mathbf{A}^{-1}$ can be applied to separate each of the sources $\mathbf{s}_i = \mathbf{B}_i\,\mathbf{x}$, where $\mathbf{B}_i$ is the ith row of $\mathbf{B}$. This can be seen as a projection pursuit density estimation problem to find M directions such that the corresponding projections are the most mutually independent. For the sake of simplicity, we will assume the square problem (the same number of sources as mixtures, and thus the order of $\mathbf{A}$ is $M \times M$).

ICAMM is proposed in the framework of pattern recognition, considering that the observed data come from a mixture model and they can be categorized into several mutually exclusive classes. ICAMM assumes the underlying process that generated observed data is composed by multiple ICA models (data of each class are modelled as an ICA, i.e., linear combinations of independent non-gaussian sources).

The general formulation of ICAMM is:

$$\mathbf{x}_t = \mathbf{A}_k\,\mathbf{s}_k + \mathbf{b}_k,\ \ k=1\ldots K \ . \tag{2}$$

where C_k denotes the class k, and each class is described by an ICA model with a mixing matrix $\mathbf{A}_k$ and a bias vector $\mathbf{b}_k$. Essentially, $\mathbf{b}_k$ determines the location of the cluster and $\mathbf{A}_k\,\mathbf{s}_k$ its shape. The goal of an ICA mixture model algorithm is to determine the parameters for each class.

ICAMM was introduced in [2] considering a source model switching between Laplacian and bimodal densities. Recently, a generalization of the ICAMM framework called Mixca was proposed in [8], which includes non-parametric density estimation, semi-supervised learning, using any ICA algorithm for parameter updating, and correction of residual dependencies.

3 ICAMM-Based Nonlinear Predictor

Let us consider data vector $\mathbf{x}$, of size $(N \times 1)$, which can be modelled through an ICAMM model with $k = 1...K$ classes. The parameters $\mathbf{W}_k, \mathbf{s}_k, \mathbf{b}_k$ for this model have to be estimated from a training data set. Assuming that the last N_2 values of vector $\mathbf{x}$ are unknown, we can group known and unknown values of $\mathbf{x}$ into two smaller vectors, $\mathbf{y}$ (known values) and $\mathbf{z}$ (unknown values). Thus, we can write

$$\mathbf{x} = \begin{pmatrix} \mathbf{y} \\ \mathbf{z} \end{pmatrix} . \tag{3}$$

We would like to predict $\mathbf{z}$ using the known values $\mathbf{y}$. Following the ICAMM model, the probability density function of data vector $\mathbf{x}$ can be expressed as

$$p(\mathbf{x}) = p(\mathbf{y},\mathbf{z}) = \sum_{k=1}^{K} p(\mathbf{y},\mathbf{z}/C_k)\, p(C_k) = \sum_{k=1}^{K} \left|\det \mathbf{W}_k\right| p(\mathbf{s}_k) p(C_k) , \tag{4}$$

where

$$\mathbf{s}_k = \mathbf{A}_k^{-1} \begin{pmatrix} \mathbf{y} \\ \mathbf{z} \end{pmatrix} - \mathbf{b}_k = \mathbf{W}_k \begin{pmatrix} \mathbf{y} \\ \mathbf{0} \end{pmatrix} + \mathbf{W}_k \begin{pmatrix} \mathbf{0} \\ \mathbf{z} \end{pmatrix} - \mathbf{b}_k . \tag{5}$$

In order to maximize the joint probability density function, we propose a maximum a posteriori (MAP) estimator:

$$\mathbf{z}_{MAP} = \underbrace{\max}_{z} p(\mathbf{y},\mathbf{z}) . \tag{6}$$

Thus, we have to solve the following equations:

$$\frac{\delta p(\mathbf{y},\mathbf{z})}{\delta \mathbf{z}} = \sum_{k=1}^{K} \left|\mathbf{W}_k\right| \frac{\delta p(\mathbf{s}_k)}{\delta \mathbf{z}} p(C_k) = 0 \tag{7}$$

$$\frac{\delta p(\mathbf{s}_k)}{\delta \mathbf{z}} = \sum_{n=1}^{N} \frac{\delta p(\mathbf{s}_k)}{\delta s_{kn}} \frac{\delta s_{kn}}{\delta \mathbf{z}} . \tag{8}$$

From (5), we can obtain the contribution of the sources corresponding to the variables $\mathbf{z}$:

$$\mathbf{W}_k \cdot \begin{pmatrix} \mathbf{0} \\ \mathbf{z} \end{pmatrix} = \begin{pmatrix} \mathbf{r}_{k1}^{\mathrm{T}} \mathbf{z} \\ \vdots \\ \mathbf{r}_{kN}^{\mathrm{T}} \mathbf{z} \end{pmatrix} , \tag{9}$$

where $\mathbf{W}_k = \begin{pmatrix} \cdots & \mathbf{r}_{k1}^{\mathrm{T}} \\ \cdots & \vdots \\ \cdots & \mathbf{r}_{kN}^{\mathrm{T}} \end{pmatrix}$ (columns N_1, N_2). Thus, the sources can be expressed as

$$s_{kn} = \text{constant} + \mathbf{r}_{kn}^{T} \mathbf{z} . \tag{10}$$

The partial derivative of the sources with respect to the elements of vector $\mathbf{z}$ is

$$\frac{\delta \mathrm{s}_{\mathrm{kn}}}{\delta \mathbf{z}} = \begin{bmatrix} \dfrac{\delta \mathrm{s}_{\mathrm{kn}}}{\delta z_1} \\ \vdots \\ \dfrac{\delta \mathrm{s}_{\mathrm{kn}}}{\delta z_{N_2}} \end{bmatrix} = \mathbf{r}_{kn} . \tag{11}$$

Assuming independence among the sources, and replacing (11) in (7), we obtain the objective function,

$$\frac{\delta p(\mathbf{y},\mathbf{z})}{\delta \mathbf{z}} = \sum_{k=1}^{K} \left|\det \mathbf{W}_k\right| p(C_k) \sum_{n=1}^{N} c_{kn} \mathbf{r}_{kn} = \sum_{k=1}^{K} \left|\det \mathbf{W}_k\right| p(C_k) \mathbf{R}_k \mathbf{c}_k , \tag{12}$$

where $\mathbf{R}_k = [\mathbf{r}_{k1} \ldots \mathbf{r}_{kN}]$ of dimension $N_2 \times 1$ and $\mathbf{c}_k = [c_{k1} \ldots c_{kN}]^T$ of dimension $N \times 1$. We have called the proposed algorithm PREDICAMM (prediction based on ICAMM).

The optimization of the value of $\mathbf{z}$ is done using a gradient method. Applying a steepest ascent optimization technique, we can write

$$\mathbf{z}_{(i+1)} = \mathbf{z}_{(i)} + \alpha_{(i)} \cdot \nabla f\left(\mathbf{z}_{(i)}\right) , \tag{13}$$

where $\alpha_{(i)}$ is the stepsize and $\nabla f\left(\mathbf{z}_{(i)}\right)$ is the gradient of the cost function. The stepsize α and the ascent direction $\nabla f(\mathbf{z})$ both set the convergence rate of the algorithm. There are several different combinations of stepsize and direction for gradient methods [9]. For simplicity, we used a steepest ascent method combined with a constant stepsize,

$$\mathbf{z}_{(i+1)} = \mathbf{z}_{(i)} + \alpha \cdot \left. \frac{\delta p(\mathbf{y},\mathbf{z})}{\delta \mathbf{z}} \right|_{(i)} . \tag{14}$$

We selected a constant greater than 1 for α. This is due to the probability density values are lower in magnitude, and so its derivative has a much lower absolute value than $\mathbf{z}$.

4 Results

In order to test the proposed algorithm, a total of 13 different data sets were simulated, as shown in Table 1. The data sets consisted of ICA mixtures with up to three classes, which were obtained by using the Mixca algorithm [8] embedding the so-called JADE ICA algorithm [10] for parameter updating. This procedure has demonstrated flexibility for data modelling in several fields such as non-destructive testing [11][12] and biomedical problem diagnosis [13].

Table 1. Data sets used in simulations

Dataset number	Known variables	Unknown variables	Number of classes	Density functions
01	1	1	1	Uniform, Laplacian
02	2	1	1	Uniform, Laplacian, K1
03	2	2	1	Uniform, Laplacian, K1, K10
04	2	2	1	Uniform
05	2	2	1	Laplacian
06	2	1	2	Uniform, Laplacian
07	2	2	2	Uniform, Laplacian
08	2	2	2	Laplacian, K1
09	2	2	2	K1, K10
10	2	1	3	Uniform, Laplacian, K1
11	2	2	3	Uniform, Laplacian, K1
12	2	1	3	Uniform, Laplacian, K1, K10
13	2	2	3	Uniform, Laplacian, K1, K10

From 2 to 4 variables were considered in simulations, defining 1 or 2 of them as unknowns for prediction purposes. The following source data densities were simulated: uniform (between ±1), Laplacian distribution with a sharp peak at the bias and heavy tails, and two K distributions [14] with shape parameters $\upsilon = 1$ (named K1) and $\upsilon = 10$ (named K10). Examples of these densities are shown in Table 2. All densities had zero mean and unit variance. A total of 100 Montecarlo simulations were obtained for each of the data sets of Table 1.

Table 2. Probability density functions used in the simulations. All densities had zero mean and unit variance.

Density	Skewness	Kurtosis	Sample
Uniform	0	-1.2	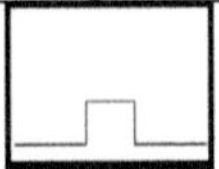
Laplacian	0	3	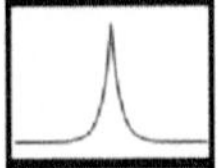
K1	1.7	5.3	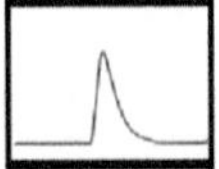
K10	0.8	0.7	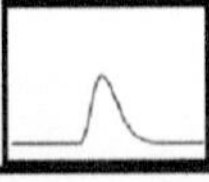

Fig. 1 shows the results of a Montecarlo experiment to test the stability of the prediction algorithm. Both graphs in Fig. 1 show three different regions separated by different shadowing. The first region, named NC for Non-Convergence, is not shadowed. This region is composed by input values that do not converge at all and their end values are approximately equal to their starting values.

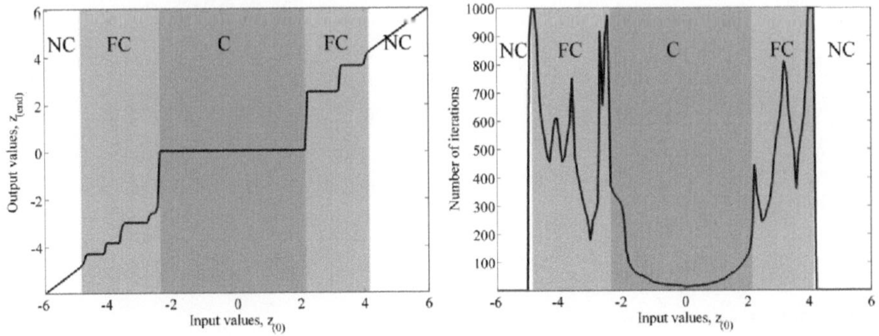

Fig. 1. Stability study for case number 1. Abbreviations (NC: nonconvergence, FC: failed convergence, convergence).

The second region, named FC for Failed Convergence, is shadowed in light gray. It is composed of input values that converge partially, but their value does not match the right prediction. The last region, named C for Convergence, is shadowed in darker gray. The values in this region converge to the right prediction and do so in the least amount of iterations. Note that the axis values in Fig. 1 (both for $\mathbf{z}_{(end)}$ and for $\mathbf{z}_{(0)}$) are normalized in such a way that 0 is the value of the right prediction. The size of the convergence zone (C) spans approximately from -2 to +2, which is about four times the standard deviation of the mixed signals. Thus, Fig. 1 shows that the algorithm was able to converge to the right value starting from relatively far values. We compared PREDICAMM with two classical predictors. The first predictor was ordinary kriging, a linear unbiased predictor. The second predictor was a Wiener structure composed by

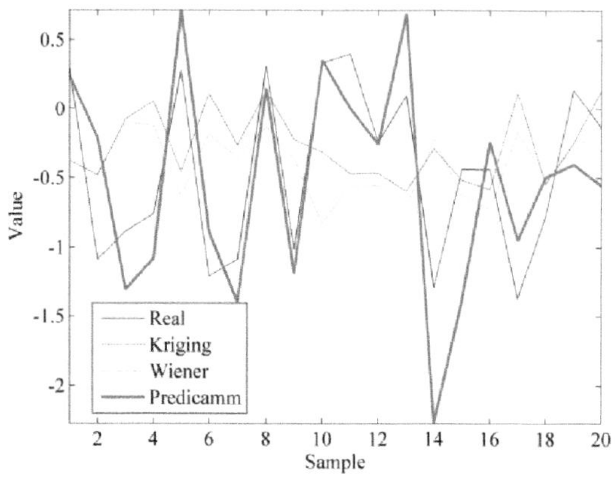

Fig. 2. Comparison of some predicted samples estimated by different methods

a linear step performed by ordinary kriging followed by a non-linear step performed by the conditional expectation of predicted to real data.

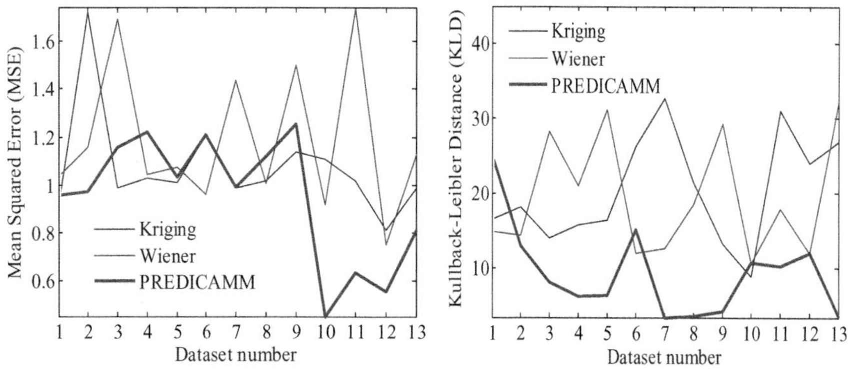

Fig. 3. Performance comparison of the proposed method with classical prediction methods

Fig. 2 shows an example of 20 samples estimated by different prediction methods. The best results are obtained by the proposed method PREDICAMM. The curve estimated by the proposed method follows the changes in slope and magnitude of the real values. The values estimated by Kriging and the Wiener-based predictor are smoothed version of the real values.

Two figures of merit were defined in order to obtain a general evaluation of the method performance: (1) the mean squared error (MSE) and (2) the symmetric Kullback-Leibler divergence (KLD) between the probability density of the predicted data and the probability density of the real data. Fig. 3 shows the figures of merit estimated for the different simulated data sets. The PREDICAMM method obtained the lowest values for MSE and KLD demonstrating the best quality of the prediction. The difference of PREDICAMM with classical methods is higher for the datasets generated from models with multiple ICAs since the prediction is made using a method that is based on the generative model of the data.

5 Conclusion

A novel method for prediction has been presented. The method is based on ICAMM which allows complex data densities to be dealt with. The method is a general purpose technique that could be applied to several fields such as time series prediction, recovering of missing data in images, etc. The performance of the method has been demonstrated in several simulations outperforming the results obtained by classical prediction methods such as kriging and Wiener structures.

Acknowledgments. This work has been supported by the *Generalitat Valenciana* under grant PROMETEO/2010/040, and the Spanish Administration and the FEDER Programme of the European Union under grant TEC 2008-02975/TEC.

References

1. Hyvärinen, A., Karhunen, J., Oja, E.: Independent Component Analysis. John Wiley & Sons, New York (2001)
2. Lee, T.W., Lewicki, M.S., Sejnowski, T.J.: ICA mixture models for unsupervised classification of non-gaussian classes and automatic context switching in blind signal separation. IEEE Trans. on Patt. Analysis and Mach. Intellig. 22(10), 1078–1089 (2000)
3. Malaroiu, S., Kiviluoto, K., Oja, E.: ICA Preprocessing for Time Series Prediction. In: 2nd International Workshop on ICA and BSS (ICA 2000), pp. 453–457 (2000)
4. Pajunen, P.: Extensions of Linear Independent Component Analysis: Neural and Information-Theoretic Methods. Ph.D. Thesis, Helsinki University of Technology (1998)
5. Gorriz, J.M., Puntonet, C.G., Salmeron, G., Lang, E.W.: Time Series Prediction using ICA Algorithms. In: Proceedings of the 2nd IEEE International Workshop on Intelligent Data Acquisit. and Advanc. Comput. Systems: Technology and Applications, pp. 226–230 (2003)
6. Wang, C.Z., Tan, X.F., Chen, Y.W., Han, X.H., Ito, M., Nishikawa, I.: Independent component analysis-based prediction of O-Linked glycosylation sites in protein using multi-layered neural networks. In: IEEE 10th Internat. Conf. on Signal Processing, pp. 1–4 (2010)
7. Zhang, Y., Teng, Y., Zhang, Y.: Complex process quality prediction using modified kernel partial least squares. Chemical Engineering Science 65, 2153–2158 (2010)
8. Salazar, A., Vergara, L., Serrano, A., Igual, J.: A general procedure for learning mixtures of independent component analyzers. Pattern Recognition 43(1), 69–85 (2010)
9. Bersektas, D.: Nonlinear programming. Athena Scientific, Massachusetts (1999)
10. Cardoso, J.F., Souloumiac, A.: Blind beamforming for non gaussian signals. IEE Proceedings-F 140(6), 362–370 (1993)
11. Salazar, A., Vergara, L., Llinares, R.: Learning material defect patterns by separating mixtures of independent component analyzers from NDT sonic signals. Mechanical Systems and Signal processing 24(6), 1870–1886 (2010)
12. Salazar, A., Vergara, L.: ICA mixtures applied to ultrasonic nondestructive classification of archaeological ceramics. EURASIP Journal on Advances in Signal Processing, vol. 2010, p.11, Article ID 12520111 (2010), doi:10.1155/2010/125201
13. Salazar, A., Vergara, L., Miralles, R.: On including sequential dependence in ICA mixture models. Signal Processing 90(7), 2314–2318 (2010)
14. Raghavan, R.S.: A Model for Spatially Correlated Radar Clutter. IEEE Trans. on Aerospace and Electronic Systems 27, 268–275 (1991)

Robustness of the "Hopfield Estimator" for Identification of Dynamical Systems*

Miguel Atencia[1], Gonzalo Joya[2], and Francisco Sandoval[2]

[1] Departamento de Matemática Aplicada, Universidad de Málaga, Spain
[2] Departamento de Tecnología Electrónica, Universidad de Málaga, Spain
Campus de Teatinos, 29071 Málaga, Spain
matencia@ctima.uma.es

Abstract. In previous work, a method for estimating the parameters of dynamical systems was proposed, based upon the stability properties of Hopfield networks. The resulting estimation is a dynamical system itself, and the analysis of its properties showed, under mild conditions, the convergence of the estimates towards the actual values of parameters. Also, it was proved that in the presence of noise in the measured signals, the estimation error remains asymptotically bounded. In this work, we aim at advancing in this robustness analysis, by considering deterministic disturbances, which do not fulfill the usual statistical hypothesis such as normality and uncorrelatedness. Simulations show that the estimation error asymptotically vanishes when the disturbances are additive. Thus the form of the perturbation affects critically the dynamical behaviour and magnitude of the estimation, which is a significant finding. The results suggest a promising robustness of the proposed method, in comparison to conventional techniques.

Keywords: Parameter Estimation, Dynamical Systems, Hopfield Neural Networks.

1 Introduction

System identification [12] is a multidisciplinary field that aims at building models of dynamical systems from measured data. In this context, a system is defined by an Ordinary Differential Equation (ODE) although the techniques implied can be extended to more general models, such as Stochastic Differential Equations and Delay Differential Equations. It is often the case that the structure of a model results from basic principles, whereas the numeric value of the parameters comprised in the model remain uncertain and, possibly, time-varying. The algorithms that are designed to compute values of such parameters are referred to as parameter estimation methods. A significant trend within

* This work has been partially supported by the Spanish Ministerio de Ciencia e Innovación (project no. TIN2008-04985), the Junta de Andalucía (project no. P08-TIC-04026), and the Agencia Española de Cooperación Internacional para el Desarrollo (project no. D/030223/10).

J. Cabestany, I. Rojas, and G. Joya (Eds.): IWANN 2011, Part II, LNCS 6692, pp. 516–523, 2011.

parameter estimation stems from the concept of statistical regression, leading to the development of least-squares techniques. Such statistically oriented methods perform reasonably well in most cases, but require the addition of some adjustment mechanism in order to deal with time-varying parameters. In previous work, a method for parameter estimation was defined [5] with the same rationale as classical techniques: estimation can be formulated as an optimization problem. The underlying optimization algorithm is the continuous Hopfield network [9], which is defined in a natural way as a dynamical system itself. Hence, the defined method is suitable for handling time-varying parameters, and a rigourous theoretical analysis proved that the estimation error remains bounded, at least asymptotically, under mild assumptions. Further research dealt with the estimator robustness [2], proving that the property of bounded estimation error holds even when the system states are affected by disturbances. In Section 2, these results are stated. The main goal of this paper is to deepen this robustness analysis.

The usual analysis of parameter estimation techniques has often a statistical approach: the disturbance of the system states is defined as a probability distribution, and the corresponding distribution of the estimates is studied. However, the assumptions on the characteristics of the disturbances are often too restrictive to be considered realistic, and they cannot be proved in practical settings. This is the case, in particular, with the independence of the noise at different time instants. In contrast, in this work we consider a deterministic disturbance, which could result from a systematic measurement error. The aim is to determine whether the proposed method is able to perform an accurate estimation even in the presence of disturbances or, contrarily, the noise destroys the qualitative, dynamical properties of the method, i.e. the asymptotical convergence towards the correct parameter values. The estimation performance is empirically studied by applying the proposed estimator to a system of ODEs. Such system was previously defined as a model of HIV-AIDS epidemics [3], which focuses on the different strategies to detect unknown infected individuals, in order to assess the health policies in Cuba. This model was numerically integrated and the obtained system states were perturbed with two different classes of disturbances, to ascertain the effect of each class of noise on the estimator dynamical behaviour. The most significant finding is that the effect of additive noise asymptotically vanishes when the system states grow. These results, which are presented in Section 3, open the way for further analysis and comparisons to other conventional techniques, so Section 4 comprises these directions of future research.

2 Continuous Hopfield Networks for Parameter Estimation

The main tool of the proposed parameter estimation method is the capability of Hopfield neural networks [8,9] to solve optimization problems [10]. Hopfield networks are recurrent systems that, in the Abe formulation [1], are defined by the following system of ODEs:

$$\frac{d\,u_i}{d\,t} = \sum_j w_{i\,j} s_j - I_i \; ; \quad s_i = \tanh \frac{u_i}{\beta} \tag{1}$$

where s_i is the state of neuron i, $w_{i\,j}$ and I_i are the network weights and biases, respectively, and β is a design variable than can be, in principle, fixed arbitrarily. The rationale to apply Hopfield networks to optimization stems from the stability theory of dynamical systems [11]. From the dynamical point of view, Hopfield networks are stable systems because a Lyapunov function can be defined:

$$V(\boldsymbol{s}) = -\frac{1}{2} \sum_i \sum_j w_{ij}\, s_i\, s_j + \sum_i I_i\, s_i \tag{2}$$

The main condition for a function to be a Lyapunov function is to be decreasing over time: $\frac{d\,V}{d\,t} \leq 0$, thus the evolution of the states of the network heads towards a stable equilibrium, where the Lyapunov function presents a (possibly local) minimum. Therefore, an optimization problem where the target function has the same structure of the Lyapunov function given by Equation (2) is solved by matching the target and the Lyapunov function, so that the weights and biases are obtained. Finally, the network is "constructed", which usually means that some numerical method is used to integrate Equation (1) until the states reach an equilibrium, which provides the minimum.

The formalization of the parameter estimation problem starts by considering a dynamical system that is modelled by a—possibly vectorial—first order ODE, which contains some unknown or uncertain parameters. A usual assumption is that the parameters appear linearly into the model, so the system model, which is said to be in the *Linear In the Parameters* (LIP) form, can thus be written as:

$$\frac{d\,\boldsymbol{x}(t)}{d\,t} = \boldsymbol{A}\,(\boldsymbol{x}(t))\;\boldsymbol{\theta}\,(t) + \boldsymbol{b}\,(\boldsymbol{x}(t)) \tag{3}$$

where $\boldsymbol{\theta}$ is a vector of parameters. The notation can be simplified by defining the vector $\boldsymbol{y} = \frac{d\,\boldsymbol{x}}{d\,t} - \boldsymbol{b}$, leading to the equation:

$$\boldsymbol{y} = \boldsymbol{A}\,\boldsymbol{\theta} \tag{4}$$

where the explicit dependence on time has been dropped for brevity. Parametric identification is accomplished by producing an estimation $\hat{\boldsymbol{\theta}}\,(t)$ that is intended to minimize the estimation error $\tilde{\boldsymbol{\theta}} = \boldsymbol{\theta} - \hat{\boldsymbol{\theta}}$ at every instant t. Consider the optimization problem resulting from minimizing the squared norm of the prediction error $\boldsymbol{e} = \boldsymbol{y} - \boldsymbol{A}\,\hat{\boldsymbol{\theta}} = \boldsymbol{A}\,\tilde{\boldsymbol{\theta}}$:

$$V = \frac{1}{2}\,\|\mathbf{e}\|^2 = \frac{1}{2}\,\mathbf{e}^\top \boldsymbol{e} = \frac{1}{2}\,\left(\boldsymbol{A}\,\tilde{\boldsymbol{\theta}}\right)^\top \left(\boldsymbol{A}\,\tilde{\boldsymbol{\theta}}\right) = \frac{1}{2}\,\tilde{\boldsymbol{\theta}}^\top\,\boldsymbol{A}^\top\,\boldsymbol{A}\,\tilde{\boldsymbol{\theta}} \tag{5}$$

This target function given by Equation (5) can be matched to the standard Lyapunov function of Hopfield networks, Equation (2). Then, some straightforward algebraic manipulations yield the appropriate weights and biases:

$$\boldsymbol{W} = -\boldsymbol{A}^\top \boldsymbol{A}\,; \qquad \boldsymbol{I} = -\boldsymbol{A}^\top \boldsymbol{y} \tag{6}$$

It is noticeable that these weights and biases are time-varying, whereas conventional Hopfield networks applied to combinatorial optimization problems possess constant weights [4].

The fact that weights are time-varying makes the theoretical analysis of such 'Hopfield estimator' considerably more complicated than conventional Hopfield networks. This analysis was undertaken in previous work [5], proving that the estimations converges towards a bounded region around the actual values of parameters, even when parameters are time-varying. Later, the robustness of the estimator was studied [2], stating that asymptotical boundedness of the estimation error is preserved when the measurements $\boldsymbol{y}$, $\boldsymbol{A}$ of the dynamical model are perturbed by a bounded disturbance. However, the procedure is fundamentally non-constructive, so there is no practical way to relate the size of such bounded region to the magnitude of the signal disturbance. Thus, the rest of this contribution aims at exploring the dynamical behaviour of the estimation error in different conditions of signal disturbance.

3 Empirical Results on the Estimator Robustness

3.1 Experimental Setting

In order to perform an estimation experiment on a system extracted from a real-world situation, a model of HIV epidemics in Cuba has been defined (see [6] and references therein):

$$\begin{aligned}\frac{dx}{dt} &= (\lambda - k_1)\,x + \lambda'\,(y+z) - k_2\frac{x(y+z)}{x+y+z} \\ \frac{dy}{dt} &= k_1\,x - \mu\,y \\ \frac{dz}{dt} &= -\,\mu\,z + k_2\frac{x(y+z)}{x+y+z}\end{aligned} \tag{7}$$

In Equation (7), the state variables x, y and z represent three different infected populations, according to whether the infection is known and, in this case, how it has been detected. The variable x comprises those individuals whose HIV infection is unknown. Some infected individuals are detected through the *contact tracing* program: individuals that are found to be seropositive, are encouraged to identify all their sexual partners, who are analysed in order to find out whether they are infected. Then, those individuals whose HIV infection has been detected through the contact tracing program form the population z. Finally, those individuals whose infection has been detected by chance (i.e., unrelated to the contact tracing program) belong to the population y. The parameters λ, λ' and μ are assumed to be known, whereas k_1 and k_2 are parameters that must be estimated. In Equation (7), it can be observed that the parameters k_1, k_2 measure the rate of transition from hidden infected population x to known infection (y, z), but they model different situations. The rate of detection by means of the contact tracing programs is modelled by a nonlinear term that depends both on x and (y, z) and the corresponding coefficient is the parameter k_2. All other forms of infection detection are assumed to depend only on the size of the population x, multiplied by the parameter k_1. Consequently, the estimation of these parameters is critical to provide an assessment of health policies. In order to test the estimation in several situations, k_1 is assumed to be time-varying, whereas k_2 is constant. The model is numerically integrated from the initial value $(x_0, y_0, z_0, t_0) = (200, 0, 0, 1986)$, and system states are recorded at discrete weekly intervals $\Delta t = \frac{7}{365.25}$ years. All these settings have been chosen in order to provide an approximate, but realistic, model of HIV epidemics in Cuba.

3.2 Experimental Results

The aim of the experiments was to determine the influence of signal disturbance on the estimation error $\hat{\boldsymbol{\theta}}$, thus first of all the estimator was applied to the noiseless data set, obtained by the procedure above described. The estimator "runs" by numerically integrating Equation (1), with the weights given by Equation (6), where the "measures" $\boldsymbol{A}$, $\boldsymbol{y}$ are computed from the simulation of the system defined by Equation (7), once rewritten in the LIP form of Equation (4). The numerical method uses a variable time step, which is much finer than the sampling interval of recorded data, therefore discrete data are extended to continuous signals by means of an interpolation algorithm. Results show that the Mean Squared Error (MSE) is negligible: $MSE = \frac{1}{n} \sum_t \left(\hat{\boldsymbol{\theta}} - \boldsymbol{\theta}\right)^2 < 10^{-3}$, where n is the number of discrete data points.

Next the analysis of the influence of disturbances is undertaken. In order to provide further insight on the class of results that are being pursued, the classical theory of statistical estimation [12] has been reviewed, paying particular attention to the results on the variance of estimators. In this regard, the so-called *maximum likelihood estimator* is optimal, because it provides the minimal variance among all unbiased estimators, i.e. those whose mean is the actual parameter value. Such minimal variance V_{MLE}, which is called the Cramér-Rao bound, can be computed when data is a sequence of N independent, identically distributed random variables, yielding $V_{MLE} = \frac{1}{\sqrt{N}} \boldsymbol{M}^{-1}$, where $\boldsymbol{M}$ is called the *Fisher information matrix*. To summarize, a "good" estimator is expected to have a small variance, as close to the Cramér-Rao bound as possible, and in particular variance should asymptotically vanish as the sample size grows. Although these statistical results are out of the scope of this work, they constitute a source of inspiration for the dynamical analysis of the estimator. For instance, the variance of a sample is simply the MSE, which can thus be used to assess the quality of the proposed estimator. The translation of the Cramér-Rao bound to a deterministic setting, from the viewpoint of dynamical systems, suggests that the estimation error of an optimal estimator should converge towards zero. Certainly, we cannot expect to prove such strong result when the statistical requirements on signal disturbances do not hold. Rather, we model disturbance just the other way round: a systematic measurement error necessarily presents autocorrelation and it is thus not a sequence of independent variables. But with the Cramér-Rao bound in mind, we aim to analyse which kind of errors is the estimator more sensitive to, as a first step towards enhancing the robustness of the proposed method.

All experiments proceed by introducing a disturbance in the value of the variable x of the dynamical system defined by Equation (7). This choice is coherent with the real situation, where the number of undetected HIV-positive individuals is obviously the most difficult to measure accurately. Notice that this disturbance affects nontrivially the estimator computations, because of the nonlinear dependance of the "signals" $\boldsymbol{A}$, $\boldsymbol{y}$ on the original variable x.

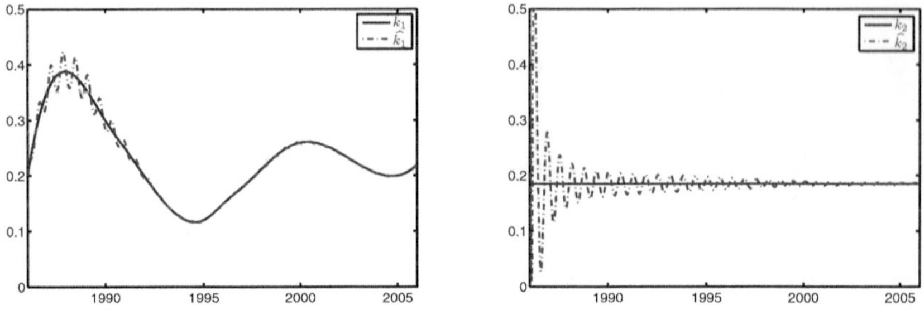

Fig. 1. Estimation of parameters under absolute disturbance

Two classes of disturbances were considered, absolute and relative. Firstly, an absolute disturbance is defined, yielding a noisy variable x^*:

$$x^* = x + 30 \sin(10\,t)$$

Then, the estimation algorithm to these perturbed data, producing the estimates that are shown in Figure 1. A significant finding is the fact that the estimation errors tend to vanish as the number of samples grow, even though the noise is strongly correlated.

A second set of experiments were carried out by introducing a relative noise so that the perturbed variable x^{**} is defined by:

$$x^{**} = x\ (1 + 0.07 \sin(10\,t))$$

The resulting parameter estimates when signals are perturbed by relative noise are shown in Figure 2. It is noticeable that, in this case, the estimation error does not fade, although it is still bounded. It can be stated that the estimator presents a relative robustness, since the disturbance does not destroy the fundamental dynamical behaviour of the estimator, i.e. the asymptotical convergence towards a bounded neighbourhood of the correct parameter value. Incidentally, it can be mentioned that a similar situation arises in the statistical framework, where the analysis of estimation methods is considerably more complicated when multiplicative, rather than additive, noise is considered (see e.g. [7] and references therein).

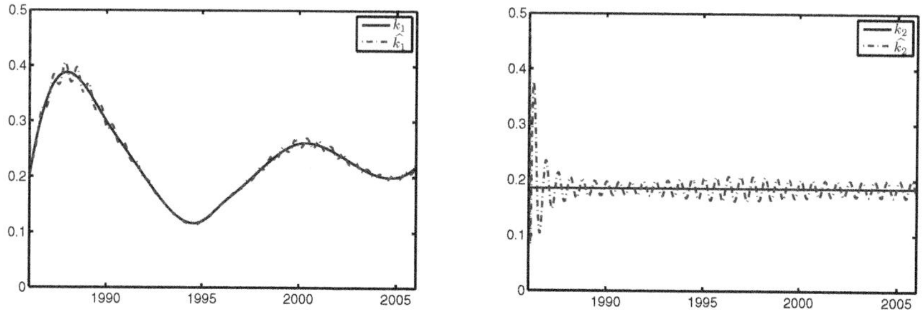

Fig. 2. Estimation of parameters under relative disturbance

4 Conclusions and Future Directions

We have analysed the dynamical behaviour of an algorithm based upon the optimization capability of Hopfield networks, previously proposed for parameter estimation of dynamical systems. Empirical results suggest that the estimation error remains bounded, even in the presence of systematic signal disturbances. Indeed, when the disturbances are additive, the estimate converges towards the correct parameter value, just as in the statistical framework, though no specifical assumption has been required on the disturbance distribution. A battery of experiments, with different disturbance levels, have been performed, though they are not here presented for the sake of brevity. The presented results correspond to the maximal admissible perturbation; for stronger noise amplitude, the estimation error is still bounded, as predicted in [2], but the qualitative average convergence is lost, thus rendering this theoretical bound useless.

We are currently developing further experiments in order to determine the influence of diverse factors on the estimator robustness. Firstly, we are performing comparisons with conventional techniques, such as least squares. Preliminary results suggest a favourable performance of the "Hopfield estimator", in terms of asymptotical estimation error. Secondly, we are assessing the estimator accuracy when applied to different systems. In this regard, it is noticeable that the estimation error is strongly determined by the sensitivity of the system model to the estimated parameters. Once again, there is a parallelism with the statistical framework, since the mentioned Fisher information matrix is the stochastic equivalent to the sensitivity equations. Finally, some experiments are devoted to find an optimal choice of the design variable β, possibly by assigning different values for each parameter. The rationale for this procedure is the observation that some parameters are more difficult to estimate because the sensitivity of the system is larger, thus β could act as a regularization parameter. The overall aim of this programme is to find an expression, either empirical or theoretical, of the error bounds, related to the magnitude of the signal disturbances, and to develop adjustments to the algorithm in order to increase the estimator robustness.

References

1. Abe, S.: Theories on the Hopfield Neural Networks. In: Proc. IEE International Joint Conference on Neural Networks, vol. I, pp. 557–564 (1989)
2. Alonso, H., Mendonça, T., Rocha, P.: Hopfield neural networks for on-line parameter estimation. Neural Networks 22(4), 450–462 (2009)
3. Atencia, M., Joya, G., Sandoval, F.: Modelling the HIV-AIDS Cuban Epidemics with Hopfield Neural Networks. In: Mira, J., Álvarez, J. (eds.) IWANN 2003. LNCS, vol. 2687, pp. 1053–1053. Springer, Heidelberg (2003)
4. Atencia, M., Joya, G., Sandoval, F.: Dynamical Analysis of Continuous Higher Order Hopfield Networks for Combinatorial Optimization. Neural Computation 17(8), 1802–1819 (2005)
5. Atencia, M., Joya, G., Sandoval, F.: Hopfield Neural Networks for Parametric Identification of Dynamical Systems. Neural Processing Letters 21(2), 143–152 (2005)
6. Atencia, M.A., Joya, G., García-Garaluz, E., de Arazoza, H., Sandoval, F.: Estimation of the Rate of Detection of Infected Individuals in an Epidemiological Model. In: Sandoval, F., Prieto, A.G., Cabestany, J., Graña, M. (eds.) IWANN 2007. LNCS, vol. 4507, pp. 948–955. Springer, Heidelberg (2007)
7. Ghogho, M.: Maximum likelihood estimation of amplitude-modulated time series. Signal Processing 75(2), 99–116 (1999)
8. Hopfield, J.J.: Neural networks and physical systems with emergent collective computational abilities. Proceedings of the National Academy of Sciences of the United States of America 79(8), 2554–2558 (1982)
9. Hopfield, J.J.: Neurons with graded response have collective computational properties like those of two-state neurons. Proceedings of the National Academy of Sciences of the United States of America 81(10), 3088–3092 (1984)
10. Hopfield, J.J., Tank, D.W.: "Neural" computation of decisions in optimization problems. Biological Cybernetics 52, 141–152 (1985)
11. Khalil, H.: Nonlinear systems, 2nd edn. Prentice Hall, Upper Saddle River (1996)
12. Ljung, L.: System identification: theory for the user. Prentice-Hall, Englewood Cliffs (1999)

Modeling Detection of HIV in Cuba

Héctor de Arazoza[1,2,4], Rachid Lounes[2], Andres Sánchez[1], Jorge Barrios[1], and Ying-Hen Hsieh[3]

[1] Facultad de Matemática y Computación, Universidad de la Habana, Cuba
arazoza@matcom.uh.cu
[2] LaboratoireMAP5, UMR-CNRS 8145, University Paris Descartes, 45 rue des Saints-Pères, 75270 Paris CEDEX 06, France
rachid.lounes@parisdescartes.fr
[3] Department of Public Health and Center for Infectious Disease Education and Research, China Medical University Taichung, Taichung, Taiwan
hsieh@mail.cmu.edu.tw
[4] Laboratoire Paul Painlevé, U. Lille 1, 59655 Villeneuve d'Ascq Cedex, France

Abstract. A nonlinear compartmental model is developed for the HIV detection system in Cuba with different types of detections, some random and others non-random. We analyze the dynamics of this system, compute the reproduction numbers, and use the data from the Cuban HIV/AIDS epidemic between 1986-2008 to fit the model. We obtain estimates for the detection-related parameters during two separate time periods to reflect the timeline of the implementation of various types of searches. The reproduction numbers for each time period are also computed from the sets of values of the parameters. We found that random screening is most important as a mean of surveillance. Moreover, local asymptotic stability for the Disease Free Equilibrium can be achieved if (i) random screening is sufficiently effective and (ii) infection by detected HIV-positive individuals is minimal. Our results highlight the importance of education for the known infectious for the purpose of preventing further infection. Fitting the 1986-2008 HIV data to obtain the model parameter estimates indicates that the HIV epidemic in Cuba is currently approaching an endemic equilibrium. A Genetic Algorithm is used.

1 Introduction

The Cuban HIV/AIDS program was established in 1983 and the detection of the first HIV-positive person in Cuba took place in December 1985. The first AIDS case was diagnosed in Cuba in April 1986, this signaled the official start of the AIDS epidemic in the country.

The Cuban HIV/AIDS epidemic has the lowest prevalence rate in the Caribbean region [10]. The UNAIDS Epidemiological Fact Sheet on HIV and AIDS for Cuba reports an HIV prevalence of less than 0.1 % for adults [11].

The Cuban HIV/AIDS program includes a detection system that allows for detection of HIV-positive cases from several sources. Some of these sources were started at the beginning of the program, while others were introduced later and

J. Cabestany, I. Rojas, and G. Joya (Eds.): IWANN 2011, Part II, LNCS 6692, pp. 524–531, 2011.

some have been discontinued in time. Since 1993, this detection system is composed, among others, of 6 major sources. These were screenings of all blood donors, persons that were treated for other sexually transmitted infections, persons admitted to a hospital with a suspicion of HIV infection or subject to specific procedures like dialysis, persons that volunteered to be tested, persons having received a recommendation for HIV testing from his/her general practitioner (family doctor) and through sexual partner tracing. Other minor sources include testing of all pregnant women and prison inmates [3].

From 1986 up to 2008 more than 34 million tests were performed, in recent years since 2002, the total number of tests performed has stabilized to 1.6-1.7 million test every year [12]. From 1986, in keeping with the "partner notification program", a person tested to be HIV-positive is invited to give names and contact details of his/her sexual partners during the past two years. These partners are then traced and a recommendation for voluntary HIV testing is made.

The detection system has changed over the years. For example, during the period 1986-1999, the partner notification and contact tracing detected 30.7% of the new HIV infections while the Family Doctors Program detected only 8.9% of new infections reported. From 2000 and up to September 2008, the family doctors were responsible for the detection of 31.7% of the new cases while contact tracing detected 20.4% of the new cases. From this information we can deduce that the detection system is not static in terms of the contribution to the detection of new HIV cases. Before 1999 the most important contribution was from the contact tracing and partner notification program, while after 1999 the family doctors play a dominant role in the detection system. We will focus on modeling this change in the detection system by dividing the time period into two periods and introducing model parameters that will differentiate the detection by the family doctors from contact tracing and other (random) searches.

2 The Model

As noted earlier, the primary objective of the Cuban Program to control the HIV/AIDS epidemic is the active search of persons infected with HIV long before they show any signs of AIDS. Our focus is not to model how new infections by HIV are generated, but how the HIV-infected persons are detected. We will consider the following variables and parameters.

Model Variables: $X(t)$ the number of HIV infected persons that do not know they are infected (at time t), $Y(t)$ the number of HIV infected persons that know they are infected, $Z(t)$ the number of persons with AIDS.

Model Parameters:

1. λ the rate of recruitment of new HIV infected persons infected by X,
2. λ' the rate of recruitment of new HIV infected persons infected by Y,
3. k_1, k_2 and k_3 rates at which the unknown HIV infected persons are detected by the system through different methods of detection.
4. β the rate at which the undetected HIV-positive persons develop AIDS, reciprocal of the mean incubation

5. β' the rate at which the detected HIV-positive persons develop AIDS, the reciprocal of the mean time it takes to go from Y to Z,
6. μ the mortality rate of the sexually active population,
7. μ' the mortality rate of the population with AIDS.

The dynamics is described by the following system:

$$\begin{aligned} \frac{dX}{dt} &= \lambda X + \lambda' Y - k_1 X - k_2 XY - k_3 X^2 - \beta X - \mu X, \\ \frac{dY}{dt} &= k_1 X + k_2 XY + k_3 X^2 - \beta' Y - \mu Y, \\ \frac{dZ}{dt} &= \beta X + \beta' Y - \mu' Z. \end{aligned} \tag{1}$$

We denote $\gamma = \beta + \mu$, $\sigma = \lambda - k_1 - \gamma$, $\gamma' = \beta' + \mu$ and we consider the system only in the region $\mathcal{D} = \{X \geq 0,\ Y \geq 0,\ Z \geq 0\}$. It is clear that $\mathcal{D}$ is positively invariant under the flow induced by (1).

We give the following remarks regarding model (1):

1. There are three ways for individuals to go from "unknown HIV infected" (X) to "known HIV infected" (Y). One is through the nonlinear term $k_2 XY$ for contact tracing, where the individual is found through his contacts with persons that are known to live with HIV. The term $k_3 X^2$ models the detection through family doctors. The third way they can be detected is through the term $k_1 X$ which models all the other "random" ways of searching for seropositives. It is important to note that $1/k_1$ can be viewed as the mean time from infection to detection for the persons found through a random screening. We can look at all the ways that an "unknown" can go from X to Y as a term of the form "$F(X,Y)\,X$", where $F(X,Y)$ is a recruitment function from the class X into the class Y. For our study we take $F(X,Y)$ as a polynomial of degree 1: $k_1 + k_2 Y + k_3 X$.
2. We assume that the known HIV-infected persons are infectious, but at a much lower rate than those that do not know they are infected due to education or change of behavior.
3. We assume that once a person develops AIDS it is no longer infectious.

We consider the following two cases:
Case 1. If $\lambda' - \gamma' = 0$, the system has a disease-free equilibrium $P_0 = (0,0,0)$ if $\sigma + k_1 = \lambda - \gamma \neq 0$ and a set of endemic equilibria $P^* = (X^*, Y^*, Z^*)$ of the form $(\lambda' - k_2 X^*)Y^* + (\sigma + k_3 X^*)X^* = 0$ and $Z^* = \frac{\beta X^* + \beta' Y^*}{\mu'}$ if $\sigma + k_1 = \lambda - \gamma = 0$.
Case 2. If $\lambda' - \gamma' \neq 0$, the system has two equilibria, $P_0 = (0,0,0)$ is the disease-free equilibrium and $P^* = (X^*, Y^*, Z^*)$ is the endemic equilibrium.
Set

$$\mathcal{R}_1 = \frac{\lambda'}{\gamma'}, \ \mathcal{R}_2 = \frac{\lambda}{\gamma} \text{ and } \mathcal{R}_0 = \frac{k_1}{k_1 + \gamma}\mathcal{R}_1 + \frac{\gamma}{k_1 + \gamma}\mathcal{R}_2. \tag{2}$$

Then

$$X^* = \frac{\gamma'(k_1 + \gamma)(\mathcal{R}_0 - 1)}{\gamma k_2(\mathcal{R}_2 - 1) + \gamma' k_3(1 - \mathcal{R}_1)}, \quad Y^* = \frac{X^*}{\gamma\gamma'}\frac{\mathcal{R}_2 - 1}{1 - \mathcal{R}_1}, \quad Z^* = \frac{\beta X^* + \beta' Y^*}{\mu'}.$$

The endemic equilibrium is feasible if and only if $\mathcal{R}_0 - 1$, $\mathcal{R}_2 - 1$ and $1 - \mathcal{R}_1$ have the same sign.

The parameters R_0, R_1 and R_2 play a significant role in the analysis of the behaviour of trajectories for (1).

Since Case 1 requires specific values of the parameters and hence is of little practical importance, we will suppose that $\lambda' - \gamma' \neq 0$.

The analysis of the stability of the equilibria points and the asymptotic behaviour of the trajectories is given in Table 1.

Table 1. Asymptotic states for the model. "GAS" denotes equilibrium is globally asymptotically stable, "LAS" denotes it is locally asymptotically stable and "NE" denotes equilibrium does not exist in the domain

$\mathcal{R}_0$	$\mathcal{R}_1$	$\mathcal{R}_2$	P_0	P^*	$(X,Y,Z) \rightarrow$
< 1	< 1	< 1	GAS	NE	P_0
< 1	< 1	> 1	GAS	NE	P_0
< 1	> 1	< 1	LAS	unstable	P_0 or unbounded
> 1	< 1	> 1	unstable	GAS	P^*
> 1	> 1	< 1	unstable	NE	unbounded
> 1	> 1	> 1	unstable	NE	unbounded

3 Conclusions

In this work, we have identified three threshold parameters of epidemiological importance, namely $\mathcal{R}_0$, $\mathcal{R}_1$, and $\mathcal{R}_2$, for the dynamic behavior of the system in question. A summary of their respective roles in the asymptotic state of the system is given in Table 1. Previous modeling studies of HIV with secondary reproduction numbers includes [5], [6] and [8]. The biological interpretations of these parameters is the following. $\mathcal{R}_0$ gives the number of infections by an infective who is detected via random screening; $\mathcal{R}_1$ is the number of infections caused by an infective after he/she has been tested positive for HIV; and $\mathcal{R}_2$ is the number of infections caused by an infective that is not detected during asymptomatic period, i.e., if an infective either develops AIDS-related illness or pass away before progression to AIDS. We note that $\mathcal{R}_0$ can be considered as the basic reproduction number, if we do not consider the detection of HIV-positive individuals in Cuba as part of disease surveillance instead of an intervention measure, especially in light of the realistic model assumption that those detected to be HIV-positive can still infected others, albeit at a lower level, before the onset of AIDS-defined illnesses [2]. Moreover, $\mathcal{R}_0$ determines whether the DFE (P_0) is locally asymptotically stable, as is the typical role of a basic reproduction number. $\mathcal{R}_1$ and $\mathcal{R}_2$ are secondary reproduction numbers, which help us determine the asymptotic behavior of the system, often relating to the endemic equilibrium. In the present model, whether $\mathcal{R}_1$ is larger than unity determines the existence of an unstable endemic equilibrium P^* when $\mathcal{R}_0 < 1$. When $\mathcal{R}_0 > 1$, $\mathcal{R}_1 < 1$ in combination with $\mathcal{R}_2 > 1$ ensure the existence of a globally asymptotically stable equilibrium P^*. Serving as threshold parameters for the model system, these reproduction numbers have an obvious and epidemiologically meaningful interpretation. For the proposed model of detection and surveillance of HIV epidemic in Cuba, our results indicate that random screening is most important as a mean of surveillance, since the number of infections due to an infective detected through random determines whether the DFE is (locally) asymptotically stable. In other words, if the averaged total number of infections by an infective detected through random screening exceeds one, then there will always be an epidemic. On the other hand, local asymptotic stability for the DFE can be achieved (i.e., R_0 can be brought down to less than one) if: (i) random screening is sufficiently effective (k_1 large), and (ii) new infections by detected HIV-positive individuals is minimal (λ' small). Our results further highlight the importance of the education for the known infectious, in light of the second threshold parameter $\mathcal{R}_1$. If the known infectious do not change, through education, their behavior and continue to practice risky sexual habits and their infection rate is still high (i.e., λ' sufficiently large) so that the average number of infections by a known infective ($\mathcal{R}_1$) exceeds unity, then the endemic equilibrium is always unstable and there is always a possibility for the total number of infectious (i.e., $X + Y$) to increase without bound. This is true even when $\mathcal{R}_0 < 1$, provided that the initial population sizes are outside the domain of attraction of the DFE. This scenario of adverse impact of public health measures, which had been shown previously to be theoretically possible in [1], [9], is only possible if k_1 is sufficiently small compared to γ, knowing that $\mathcal{R}_0$ is a convex combination of $\mathcal{R}_1$ and $\mathcal{R}_2$ (see 2). In other words, an ill-designed detection system might adversely lead to the epidemic increasing without bound if (i) random screening is not comprehensive enough (k_1 too small); (ii) lack of an education program to change behavior (λ' too high); and (iii) the prevalence is too high when the system is first implemented (initial infective populations outside of the domain of attraction of the DFE). This result further highlights the importance of universal testing in high-prevalence regions [4], [7]. On the other hand, if through an adequate education to change the behavior of the known infectious (so that λ' is sufficiently low) the average number of infections by a known infective is less than one, then either the DFE P_0 or the endemic equilibrium P^* is globally asymptotically stable, leading to a more manageable epidemic for the public health purposes, even if the disease is not eradicated.

3.1 Application to the Cuban HIV/AIDS Data

The modeling results are clearly relevant to our understanding of the current state of the HIV epidemic in Cuba. We will use the model (1) to fit the data of the known HIV positives and AIDS cases in Cuba. We have divided the period 1986-2008 into two different time periods, namely 1986-1999 and 1999-2008, to take into account the introduction of the family doctors in the detection system. In each period we have parameters that can be estimated from the data. These are β, β', μ, μ' and $X(0)$. $Y(0)$ and $Z(0)$ are known. In previous work [2], [8] estimates for λ and λ' have been obtained.

The family doctors program had actually started after 1990, but only as a pilot project where the family doctors typically did not prescribe HIV testing. It is only after 1999 that detection through the family doctors started to take on an important role in the yearly detection figures, arriving at more than 30% of new detections in a year. Another significant difference between the two periods chosen is λ'. In the first years (1986-1999) the sanatorial system played an important role in preventing HIV transmission from persons that had been detected. Hence we suppose that there is practically no transmission from the persons living with HIV (Y) and assume $\lambda' = 0$ for the first period.

We fit the model to the data to obtain values for k_1, k_2 and k_3 by minimizing an error function. As traditional optimization methods failed to work properly we used a genetic algorithm approach to find an initial point for starting the optimization method using a gradient method. To compute standard errors for the parameters, 200 fitting runs were made using different values of the known parameters, taken randomly from their confidence interval.

Using PET a software written on MATLAB, we obtain the least-square estimates for the unknown model parameters (k_1, k_2 and k_3), for each of the two periods of the Cuban HIV epidemic from 1986 to 2008, by fitting the Cuban HIV data of the persons known to live with HIV to the model as described previously. Here the resulting numbers for 1999 obtained from the first stage of estimation using data from 1986-1999 were used as initial values for the second stage of estimation using the 1999-2008 data. The estimated mean values of the parameters k_1, k_2, and k_3 for each of the two periods with 95% confidence intervals, obtained from the 100 best fits, are given in Table 2.

Table 2. Estimated mean values with the 95% confidence intervals for parameters k_1, k_2 and k_3 for both periods. UCI and LCI denote the respective upper and lower bounds for the 95% confidence intervals

	1986-1999 ($k_3 = 0$)		1999-2008		
	k_1	k_2	k_1	k_2	k_3
Mean	0.1347	2.381×10^{-5}	0.2195	2.728×10^{-5}	6.801×10^{-5}
LCI	0.1345	2.366×10^{-5}	0.2192	2.725×10^{-5}	6.783×10^{-5}
UCI	0.1349	2396×10^{-5}	0.2198	2.732×10^{-5}	6.818×10^{-5}

We can also compute the theoretical values of the number of the unknown persons living with HIV, $X(t)$, from the estimation results. For the unknown X we have between 2300 and 2400 at the end of 2008. By comparing the estimated results for the two periods, we conclude that detection by random screening (k_1) improved significantly after 1999, perhaps reflecting the steeper increase in reported cases after 2000 [3], while detection via contact tracing (k_2) was at a similar level throughout the whole course of the epidemic. Detection by family doctors (k_3) was slightly higher than that of contact tracing after 1999 but of similar magnitude. Both the analytical result (of the dynamics) and the data-fitting parameter estimates indicate that random screening was the most effective route of detection, while contact tracing and family doctors played mainly secondary roles, as had been previously proposed in [7]. The estimated values of the parameters in the model also allow us to calculate the three reproduction numbers, $\mathcal{R}_0$, $\mathcal{R}_1$, and $\mathcal{R}_2$, with the 95% confidence intervals for each of the two time periods which are given in Table 3.

Table 3. Estimated mean values with the 95% confidence intervals for $\mathcal{R}_0$, $\mathcal{R}_1$, and $\mathcal{R}_2$. UCI and LCI denote the respective upper and lower bounds for the 95% confidence intervals

	1986-1999			1999-2008		
	$\mathcal{R}_0$	$\mathcal{R}_1$	$\mathcal{R}_2$	$\mathcal{R}_0$	$\mathcal{R}_1$	$\mathcal{R}_2$
LCI	1.824	0	4.446	1.858	0.762	4.428
UCI	1.837	0	4.665	1.859	0.765	4.434

As we can see, for both time periods the Cuba HIV epidemic is in the case of $\mathcal{R}_0 > 1$, $\mathcal{R}_1 < 1$ and $\mathcal{R}_2 > 1$. P_0 is unstable and P^* is globally asymptotically stable with the trajectories approaching P^* asymptotically. Hence we can conclude that the HIV epidemic in Cuba is approaching (in the long term) an endemic steady state which we can estimate from our parameters using the expression obtained from our modeling for $P^* = (X^*, Y^*, Z^*)$ given in Equation (2). That is, assuming no drastic changes in the prevention, transmission, detection, or treatment of HIV in Cuba in the long term future, there could be, eventually, around 2700 persons living with HIV that do not know they are infected. Further noting that the theoretical number of unknown persons living with HIV from the model is a little under 2400, we speculate with optimism that, at the endemic steady state, the number of persons living with HIV that represent the main core for the transmission of the epidemic in Cuba will not increase drastically, in the long term future.

The assumption that the parameters do not change in the long term is not a very real one, taking into account that every year new information is gained on the virus and that there is ample research on treatment, vaccines and other aspects that affect the dynamics of the epidemic. For example existing therapy reduces the probability of transmission for HIV, in terms of our model this means that coefficient λ' is reduced and this will make $\mathcal{R}_1$ smaller, changing the value of the asymptotic point. But therapy delays the onset of AIDS, this means that β' gets smaller, also changing $\mathcal{R}_1$ but making it bigger. So long term predictions (as in the case of asymptotic behavior) make sense only as an indicator of how is the epidemic going to behave. In this sense in table 1, we see that if $\mathcal{R}_1 > 1$ trajectories could become unbounded, and this would mean that the epidemic is out of control. It is very important to manage the value of $\mathcal{R}_1$.

In this work we chose to use parameters that are constant or step functions (with 2 steps for now) that also produce a change in the model itself, changing the recruitment function from X to Y, $F(X, Y)$, from a linear polynomial in Y for the period 86-99, to a linear polynomial in X and Y for the period 99-2008.

Acknowledgments. This work was carried out during visits to the University of Paris Descartes by YHH and H de A. H de A, YHH and RL received support from the French "Agence National pour la Recherche" project "Viroscopy". H de A also received support as visiting professor from the Laboratoire Paul Painlevé of the University Lille 1, France. H. de A., A.S. and J.B. received support from the Spanish AECID, from their projects PCI D/023835/09 and D/030223/10. For all the support we have received from all the different sources we are grateful.

References

1. Anderson, R.M., Gupta, S., May, R.M.: Potential of community-wide chemotherapy or immunotherapy to control the spread of HIV-1. Nature 350, 356–359 (1991)
2. de Arazoza, H., Lounes, R.: A non linear model for a sexually transmitted disease with contact tracing. IMA. J. Math. Appl. Med. Biol. 19, 221–234 (2002)
3. de Arazoza, H., Joanes, J., Lounes, R., Legeai, C., Clémençon, S., Pérez, J., Auvert, B.: The HIV/AIDS epidemic in Cuba: Description and tentative explanation of its low HIV prevalence. BMC Infectious Diseases 7:130, 1–6 (2007)
4. Granich, R.M., Gilks, C.F., Dye, C., De Cock, K.M., Williams, B.G.: Universal voluntary HIV testing with immediate antiretroviral therapy as a strategy for elimination of HIV transmission: a mathematical model. Lancet 373, 48–57 (2009)
5. Hsieh, Y.H., Cooke, K.: Behavior Change and Treatment of Core Group and Bridge Population: Its Effect on the Spread of HIV/AIDS. IMA J. of Math. Appl. Biol. Med. 17(3), 213–241 (2000)
6. Hsieh, Y.H., de Arazoza, H., Lounes, R., Joanes, J.: A Class of Models for HIV Contact Tracing in Cuba: Implications for Intervention and Treatment. In: Tan, W.Y. (ed.) Deterministic and Stochastic Models for AIDS Epidemics and HIV Infection with Interventions. World Scientific, Singapore (2005)
7. Hsieh, Y.H., de Arazoza, H.: Correspondence to "Universal voluntary HIV testing and immediate antiretroviral therapy". Lancet 373, 1079–1080 (2009)
8. Hsieh, Y.H., Wang, Y.S., de Arazoza, H., Lounes, R.: HIV Model with Secondary Contact Tracing: Impact of the Partner Notification Program in Cuba. BMC Infectious Diseases 10:194, 1–9 (2010)
9. Hsu, S.B., Hsieh, Y.H.: Modeling intervention measures and public response during SARS outbreak. SIAM J. Appl Math. 66(2), 627–647 (2006)
10. Inciardi, J.A., Syvertsen, J.L., Surratt, H.L.: HIV/AIDS in the Caribbean Basin. AIDS Care 17(suppl. 1), S9–S25 (2005)
11. WHO. World Health Organisation: Cuba, `http://www.who.int/countries/cub/en/` (accessed June 20, 2006)
12. MINSAP. CUBA:PLAN ESTRATEGICO NACIONAL ITS/VIH/SIDA 2007-2011 (2006), `http://www.sld.cu/galerias/pdf/servicios/sida/anexo_2,_plan_estrategico__2007-2011.pdf`

Flexible Entrainment in a Bio-inspired Modular Oscillator for Modular Robot Locomotion

Fernando Herrero-Carrón, Francisco B. Rodríguez, and Pablo Varona*

Grupo de Neurocomputación Biológica (GNB),
Dpto. de Ingeniería Informática, Escuela Politécnica Superior,
Universidad Autónoma de Madrid,
Calle Francisco Tomás y Valiente, 11
28049 Madrid, Spain
{fernando.herrero,pablo.varona,f.rodriguez}@uam.es

Abstract. The ability of a Central Pattern Generator to adapt its activity to the mechanical response of the robot is essential for robust autonomous locomotion in unknown environments. In previous works we have introduced a new oscillator model for locomotion in modular robots. In this paper, we study the ability of our oscillator model to entrain a servo. For a given configuration of the oscillator, we simulate different servos with different responsiveness, ranging from very slow to very fast servos. The result is that our oscillator adapts its frequency of oscillation to the responsiveness of the servo up to several orders of magnitude, without changing the parameters of the oscillator itself.

1 Introduction

Central Pattern Generators (CPGs) are neural networks responsible for rhythmic motion in animals. CPGs can autonomously generate and coordinate rhythmic signals in a robust yet flexible manner modulated by sensory feedback [9, 13]. The study of living CPGs has a long tradition and these circuits are probably the best known neural networks in neuroscience research. While the idea of using CPGs to control locomotion in robotics has been extensively exploited (for a review see [7]), recent results in CPG research provide new elements of bio-inspiration for robotics and, in particular, modular approaches.

Living CPGs have been shown to have single neurons with rich dynamics [13], non-open topologies [6] for network connections, neural signatures in the form of cell-specific inter-spike intervals in their bursting activity [8, 14], dynamical invariants between period and phase lags [11] and homeostatic mechanisms for self regulation [9]. To our knowledge, none of these features have been used for the design of CPGs for locomotion control in robots. Each of them can potentially add different benefits for autonomous locomotion.

Artificial CPG circuits are particularly suitable for the design of autonomous modular robots, as CPG control fully fits the idea of having variable number of

* Work supported by MICINN BFU2009-08473 and TIN 2010-19607.

J. Cabestany, I. Rojas, and G. Joya (Eds.): IWANN 2011, Part II, LNCS 6692, pp. 532–539, 2011.

modules that are organized by the same scalable principles. In previous works [2–5] we introduced a new bio-inspired oscillator model for modular locomotion, based on some of this recent CPG research. We successfully used it in open loop to control a real robot with eight connected modules. The work presented in this paper is the beginning of our work on adaptation and integration of sensory information.

In the next section we review different modular oscillator models, from low biological inspiration to higher biological plausibility. Then, we provide the definition of our novel bio-inspired CPG for modular robots. Finally, we show how one single oscillator is able to adapt its frequency to the responsiveness of a simulated servo using position error as feedback.

2 Entrainment between a Simulated Servo and the Bio-Inspired Modular Oscillator

The true power of a CPG controller lies in its ability to maintain a certain rhythmic activity while adapting itself to external conditions. We have performed several experiments to show how a novel modular oscillator will react when coupled to different servos in different working conditions, without modifying the configuration of the CPG itself.

Our goal is to design a modular oscillator that should drive a servomotor simulated using the following equation:

$$e_n = u_n - s_n \tag{1}$$

$$s_{n+1} = s_n + \mu e_n \tag{2}$$

where s_n is the position of the servo at time n, μ represents the "responsiveness" of the servo, ranging from infinitely slow ($\mu = 0$) to infinitely fast ($\mu = 1$), and u_n is the reference position at time u_n. The error term e_n will be used as feedback for the CPG.

2.1 Multiple Time Scales Neuron Model

CPG neuron models for robot locomotion typically use oscillator models with one single time scale. The negotiation capacity of these units to produce robust yet flexible rhythms within the network is more limited than that of neurons with multiple time scales. Multiple time scales can account for a wider variety of bifurcations and transient dynamics to provide autonomous coordinated responses [10]. Indeed, multiple time scales behaviors are ubiquitous in real living CPG neurons [13].

In our work, we use a neuron model that mimics the activity of real, multiple time scale neurons [12]. Several characteristics have played in its favor: its mathematical simplicity and the possibility to easily control the possible set of behaviors depending on the selection of a few parameters. Three stable regimes may be selected by combination of its parameters: silent, in which the potential of the neuron (variable x_n in (4)) remains in a constant resting state; tonic spiking, in which the neuron emits spikes at a constant rate; and tonic bursting, in which bursts of spikes are emitted at a constant rate, with a silent interval in between. Of these behaviors, tonic bursting is the one of greater interest for us. See figure 1 for an overall idea of the model working in tonic bursting regime.

The mathematical description of the neuron model used in this work is as follows:

$$f(x,y) = \begin{cases} \frac{\alpha}{1-x} + y, & \text{if } x \leq 0 \\ \alpha + y, & \text{if } 0 \leq x < \alpha + y \\ -1, & \text{otherwise} \end{cases} \tag{3}$$

$$x_{n+1} = f(x_n, y_n) \tag{4}$$

$$y_{n+1} = y_n - \mu(x_n + 1) + \mu\sigma + \mu I_n \pm Ae_n \tag{5}$$

with $\mu = 0.001$ in all experiments.

This is a bi-dimensional model, where variable x_n represents a neuron's membrane voltage and y_n is a slow dynamics variable with no direct biological meaning, but with similar meaning as *gating* variables in biological models, that represents the fraction of open ion-channels in the cell. While x_n oscillates on a fast time scale, representing individual spikes of the neuron, y_n keeps track of the bursting cycle, a sort of context memory. Units are dimensionless, and can be rescaled to match the requirements of the robot. The combination of σ and α selects the working regime of the model: silent, tonic spiking or tonic bursting. In the bursting regime, these parameters also control several properties of neural activity like period or number of spikes per cycle.

Finally, I_n is synaptic current flowing from the reciprocal cell to achieve synchronization, and e_n is the error term between desired and actual position of the servo, scaled by a factor A to modulate the importance of this input. The promotor neuron will receive positive feedback of the error ($+Ae_n$) while the remotor neuron will receive negative feedback ($-Ae_n$).

2.2 Kinetic Synapse Model for Inter-neuron Communication

A key property of CPGs is that they are autonomous, i.e. the different units in the circuit talk to each other to negotiate the overall function. Here we present the model we have chosen to implement synapses, the communication channel of neurons. In this work we use a chemical synapse model [1].

Chemical synapses are unidirectional. When a potential spike arrives from the presynaptic neuron, the synapse releases a certain amount of neurotransmitter molecules that bind to the postsynaptic neuron's receptors. With time, neurotransmitter molecules begin to *unbind*. If a succession of spikes arrives within a short time, the synaptic response to each of them may overlap. Therefore the state of the synapse is dependent upon past events, a mechanism of context memory.

The additional time-scale provided by kinetic synapses in a CPG enriches synchronization between bursting neurons. For instance, we may choose to synchronize two bursting neurons upon the spike (fast) time scale or the burst (slow) time scale. We have selected the kinetics of the binding and unbinding processes such that synapses act as filters of the fast time scale and synchronization occurs at the slow time scale. That is, the basic unit of synchronization will be the burst as a whole, not every individual spike. Beyond this, synapses may introduce delays for finer control of phase difference between neurons.

The mathematical description of the model follows:

$$\dot{r} = \begin{cases} \lambda[T](1-r) - \beta r, & \text{if } t_{\mathrm{f}} < t < t_{\mathrm{f}} + t_{\mathrm{r}} \\ -\beta r, & \textit{otherwise} \end{cases} \tag{6}$$

This equation defines the ratio of bound chemical receptors in the postsynaptic neuron, where r is the fraction of bound receptors, λ and β are the forward and backward rate constants for transmitter binding and $[T]$ is neurotransmitter concentration. The equation is defined piecewise, depending on the specific times when the presynaptic neuron fires (t_{f}): during t_{r} units of time, the synapse is considered to be releasing neurotransmitters that bind to the postsynaptic neuron. After the release period, no more neurotransmitter is released and the only active process is that of unbinding, as described by the second part of the equation. Times t_{f} are determined as the times when the presynaptic neuron's membrane potential crosses a given threshold θ.

Synaptic current is then calculated as follows:

$$I(t) = g \cdot r(t) \cdot (X_{\mathrm{post}}(t) - E_{\mathrm{syn}}) \tag{7}$$

where $I(t)$ is postsynaptic current at time t, g is synaptic conductance, $r(t)$ is the fraction of bound receptors at time t, $X_{\mathrm{post}}(t)$ is the postsynaptic neuron's membrane potential and E_{syn} its reversal potential, the potential at which the net ionic flow through the membrane is zero.

In order to couple the continuous synapse equations, we can integrate r using a simple Euler method with time step h, so that:

$$r_{n+1} = r_n + \begin{cases} \lambda h([T](1-r) - \beta r), & \text{if } t_{\mathrm{f}} < nh < t_{\mathrm{f}} + t_{\mathrm{r}} \\ -\beta h r, & \textit{otherwise} \end{cases} \tag{8}$$

$$I_n = g \cdot r_n \cdot (x_n^{\mathrm{post}} - E_{\mathrm{syn}}) \tag{9}$$

2.3 Translating from Neural Code to Motor Actuator Commands: Motoneurons

The oscillator we propose consists of two mutually inhibiting neurons (promotor and remotor), each of them working in tonic bursting regime, and thus capable of rhythmic activity in isolation. Through inhibition, anti-phase synchronization of the promotor (P) and remotor (R) neurons is achieved.

Movement information is robustly encoded in each neuron's bursting episodes. A neuron called *motoneuron* is then responsible of *decoding* this information and translating it into the signal that will finally be sent to the servo controller. This signal tells the angle at which the servo should be positioned, in degrees. Figure 1

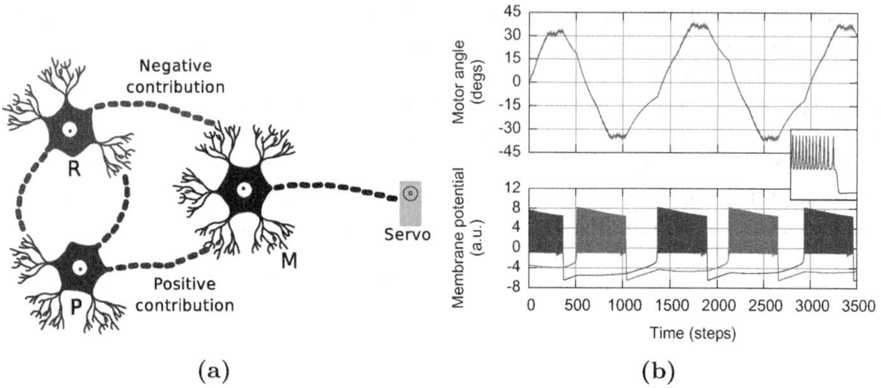

Fig. 1. (a) Organization of the CPG within one module. The promotor (P) and remotor (R) neurons are interconnected with inhibitory synapses so that they synchronize in anti-phase. The motoneuron M sends a command signal to the servomotor specifying the angle at which the servo should position itself. The signal generated by R is directly input to M, and the opposite of the signal generated by P is input to M. M integrates its input according to (12). (b) Activity sample (variable x_n in (4)). Neuron P contributes positively and raises M to 30°; neuron 'R' does exactly the opposite and drives M towards -30°. P and R are synchronized in anti-phase. With no input, M tends to 0°. Parameters for P and R: $\alpha = 15$, $\mu = 0.001$, $\sigma = -0.33$, $\beta_{\mathrm{e}} = 0$, $\sigma_{\mathrm{e}} = 1$; parameters for the inhibitory synapses between P and R: $\lambda = 0.5$, $\beta = 10$, $\tau = 0$, $E_{syn} = 9$, $g_{\mathrm{syn}} = 1.5$, $T = 1$, $t_{\mathrm{r}} = 0.01$; Parameters for M: $\gamma = 30$, $\tau = 0.5$, $\nu = -1.5$.

shows the organization and an example pattern of activity of a module in its steady state, after an initial transient period of self-adjustment.

Motoneurons *read* the activity of the modular oscillator through a pair of synapses. These synapses connect R and P neurons to the motoneuron, and are governed by a very simple threshold equation:

$$s(x, \nu) = \begin{cases} 1, \text{ if } x > \nu \\ 0, \text{ otherwise} \end{cases} \tag{10}$$

The role of this function is to detect individual spikes of neurons. By setting the threshold to, for example, $\nu = -1.5$ a.u., this function applied to the potential trace of one neuron will have value 1 during individual spikes and 0 otherwise. In this way, communication between neurons is event-based, i.e., the actual shape of neural activity is not important, only the timing. We have argued that this is a useful mechanism to lower the impact of noise [3].

The role of motoneuron M is now to integrate the individual events emitted by each one of the neurons. If neuron P emits a spike, motoneuron M will move the servo a little bit in a positive angle. If it emits a second spike close enough to the first one, the servo will be positioned a little bit further. Analogously, the R neuron will make the motoneuron move the servo towards negative angle positions. If both neurons are silent, motoneuron M will slowly drive the servo to a resting position of angle 0. This is accomplished through the following equation governing motoneurons in our CPGs:

$$C(t) = \gamma[s(x_{\mathrm{p}}(t), \nu) - s(x_{\mathrm{r}}(t), \nu)] \tag{11}$$

$$\tau\dot{m} = C(t) - m(t) \tag{12}$$

where, $m(t)$ is the output of neuron M (in degrees), the two $s(.)$ terms are the threshold function (10) applied to input from R and P, τ is a time constant that controls how quick the output signal $m(t)$ will change, and γ defines the maximum amplitude of signal $m(t)$.

In this equation, P contributes positively and R negatively. Given the fact that P and R oscillate in anti-phase, the solution $m(t)$ is an oscillatory function bounded between $-\gamma$ and γ. When the motoneuron receives no input because P and R are silent, it will go back to zero due to the leak term $(-m(t))$ in (12) (see decay between bursts in figure 1).

3 Results and Discussion

We build a CPG with the parameters specified in figure 1, coupled to a servo that follows (2). We perform a simulation of 100.000 time steps and record the trajectory generated by the CPG (see (12)), and the trajectory followed by the servo. For each simulation, we calculate amplitude (in degrees) and period (in simulation time steps) of the servo. In total, we perform ten simulations for each combination of parameters with different initial conditions and average the results.

Figure 2 shows three example simulations for different servo responses. For a fast servo (figure 2a), i.e. a high value of μ, the CPG oscillates at its nominal frequency, since the positions error is negligible, and so feedback to the neurons will be close to zero. A noticeable adaptation in frequency occurs when the servo has a slower response (figure 2b). In fact, a positive neuronal feedback has as a consequence prolonged bursts. This in turns, has also as a consequence that neuron M achieves higher values, up to a saturation value. Finally, for very slow servos (figure 2c), neural bursts are prolonged well over orders of magnitude further (notice the change in range of the abscissas). The result of all simulations is shown in figure 3.

There are several results from recent research on living CPGs that have contributed to a better understanding of how these systems generate robust and, at the same time, flexible and adaptive rhythms to control motor activity. These in principle contradictory features from a classical engineering perspective can be achieved by the right combination of individual dynamical adaptability in single neurons and the overall network coordination mechanisms. In this paper we have applied recent knowledge from living CPG function to design a servo controller for a modular robot. Our study shows that the proposed CPG oscillator adapts its frequency to the responsiveness of the servo remarkably up to several orders of magnitude. The resulting adaptability in a single module can lead to an increased overall adaptability and to a more autonomous locomotion for modular robots.

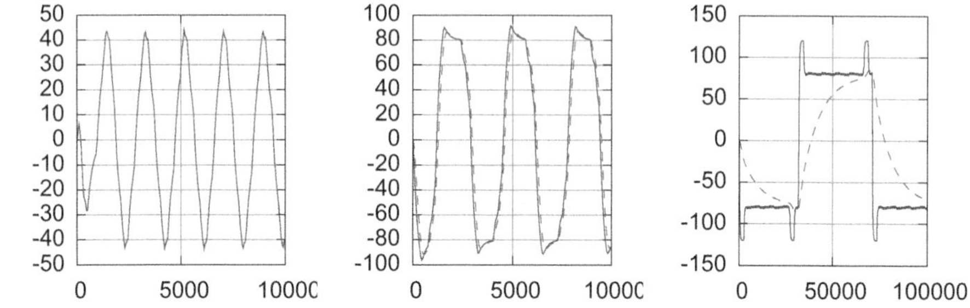

(a) $\mu = 0.1083$, $A = 1$ (b) $\mu = 0.0117$, $A = 1$ (c) $\mu = 0.0001$, $A = 1$

Fig. 2. Output of our oscillator and entrained servo. Output of the CPG is plotted in red, solid lines; servo response is plotted in green, dashed lines. (a) If the servo is fast enough to follow the oscillator, the CPG works at its nominal regime, with an amplitude of 40 degrees and a period of approximately 1600 units of time. (b) Our CPG gracefully adapts, without changing any of its parameters, to a slightly slower servo. As a side effect of sustained neural activity the amplitude raises above the amplitude in the previous case. Period is almost double as in the previous case. (c) Our CPG is able to sustain its oscillations well over orders of magnitude (notice the change of scale of the abscissa) in case the servo cannot follow it on time.

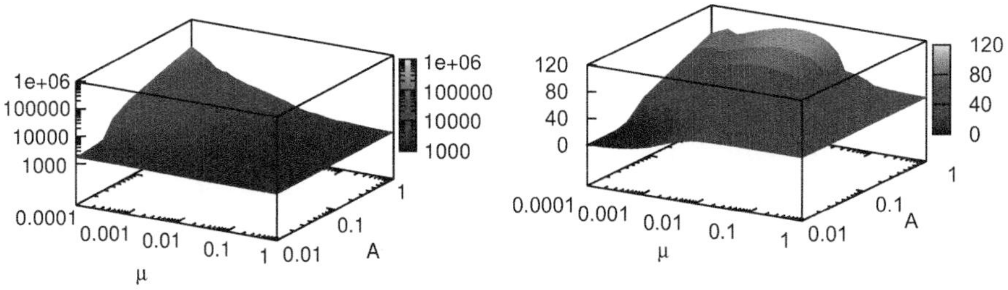

(a) Servo period (simulation steps). (b) Servo amplitude (degrees).

Fig. 3. Oscillation parameters of a servo controlled and entrained by our oscillator. Ten different simulations were carried out for each pair of servo response 'μ' and feedback scaling 'A'. For each of them, the period and the amplitude of the oscillations performed by the servo were analyzed, and then averaged. Notice logarithmic scales on 'μ' and 'A' axes in both figures, and also 'z' axis in figure (a). Period of the servo effectively adapts over several orders of magnitude if feedback strength is enough. As a consequence of prolonged bursts, amplitude also increases for low values of 'μ'.

References

1. Destexhe, A., Mainen, Z.F., Sejnowski, T.J.: An efficient method for computing synaptic conductances based on a kinetic model of receptor binding. Neural Computation 6(1), 14–18 (1994)

2. Herrero-Carrón, F., Rodríguez, F.B., Varona, P.: Dynamical invariants for CPG control in autonomous robots. In: Filipe, J., Cetto, J.A., Ferrier, J.L. (eds.) 7th International Conference on Informatics in Control, Automation and Robotics, Funchal, Portugal, June 2010, vol. 2, pp. 441–445 (2010)
3. Herrero-Carrón, F., Rodríguez, F.B., Varona, P.: Novel modular CPG topologies for modular robotics. In: IEEE 2010 International Conference on Robotics and Automation workshop "Modular Robots: The State of the Art", pp. 89–93. IEEE, Los Alamitos (2010)
4. Herrero-Carrón, F., Rodríguez, F.B., Varona, P.: Studying robustness against noise in oscillators for robot control. In: International conference on automation, robotics and control systems, Orlando, USA, pp. 58–63 (July 2010)
5. Herrero-Carrón, F., Rodríguez, F.B., Varona, P.: Bio-inspired design strategies for central pattern generator control in modular robotics. Bioinspiration & Biomimetics 6(1), 016006+ (2011)
6. Huerta, R., Varona, P., Rabinovich, M.I., Abarbanel, H.D.: Topology selection by chaotic neurons of a pyloric central pattern generator. Biological Cybernetics 84(1), L1–L8 (2001)
7. Ijspeert, A.: Central pattern generators for locomotion control in animals and robots: A review. Neural Networks 21(4), 642–653 (2008)
8. Latorre, R., Rodríguez, F.B., Varona, P.: Neural signatures: multiple coding in spiking-bursting cells. Biological Cybernetics 95(2), 169–183 (2006)
9. Marder, E., Bucher, D.: Understanding circuit dynamics using the stomatogastric nervous system of lobsters and crabs. Annual Review of Physiology 69(1), 291–316 (2007)
10. Rabinovich, M.I., Varona, P., Selverston, A.I., Abarbanel, H.D.I.: Dynamical principles in neuroscience. Reviews of Modern Physics 78(4), 1213–1265 (2006)
11. Reyes, M., Huerta, R., Rabinovich, M., Selverston, A.: Artificial synaptic modification reveals a dynamical invariant in the pyloric CPG. European Journal of Applied Physiology 102(6), 667–675 (2008)
12. Rulkov, N.F.: Modeling of spiking-bursting neural behavior using two-dimensional map. Physical Review E 65(4), 041922+ (2002)
13. Selverston, A.I., Rabinovich, M.I., Abarbanel, H.D.I., Elson, R., Szucs, A., Pinto, R.D., Huerta, R., Varona, P.: Reliable circuits from irregular neurons: A dynamical approach to understanding central pattern generators. Journal of Physiology-Paris 94(5-6), 357–374 (2000)
14. Szucs, A., Pinto, R.D., Rabinovich, M.I., Abarbanel, H.D., Selverston, A.I.: Synaptic modulation of the interspike interval signatures of bursting pyloric neurons. Journal of Neurophysiology 89(3), 1363–1377 (2003)

Dengue Model Described by Differential Inclusions

Jorge Barrios[1,2,⋆], Alain Piétrus[1], Aymée Marrero[2], Héctor de Arazoza[2], and Gonzalo Joya[3]

[1] Université des Antilles et de la Guyane, LAMIA(EA4540)
apietrus@univ-ag.fr
[2] Universidad de La Habana, Facultad de Matemática y Computación
{jbarrios,aymee,arazoza}@matcom.uh.cu
[3] Universidad de Málaga, Dpto. Tecnología Electrónica, E.T.S. Ing. Telecomunicación
gjoya@uma.es

Abstract. In this paper we formulate a differential inclusion to model an epidemic outbreak of Dengue fever in the Cuban conditions. The model takes into account interaction of human and mosquito populations as well as vertical transmission in the mosquito population. Finally, we propose a mathematical framework allowing us to make suitable predictions about the populations of humans, mosquitoes and eggs infected during the epidemic time.

Keywords: Reachable Sets, Differential Inclusion, Dengue.

1 Introduction

Dengue is classified by the World Health Organization (WHO) as the most common and rapidly spreading mosquito-borne viral disease of humans. The risk factors for transmission of dengue are associated with an epidemiological triad - the vector, the dengue virus and susceptible hosts. Aedes aegypti (Linnaeus, 1762) (Diptera: Culicidae) is the primary vector for the four closely related, but antigenically distinct, dengue virus serotypes designated as DEN-1, DEN-2, DEN-3, and DEN-4 of the genus Flavivirus (family: Flaviviridae). A person infected by one of the four serotypes will never be infected again by the same serotype.

The Aedes aegypti mosquito is characterized by its biting pattern, which consists of multiple bloodmeals during each egg-laying cycle, and the ability of its immature stages (Egg, Larva and Pupa) to grow in clean water. These features make it an ideal vector for dengue virus transmission, especially in large urban areas, where there are high human population densities and numerous artificial containers where the aquatic stages of Aedes aegypti flourish. Aedes aegypti is infected by sucking infected human blood. Humans are infected with dengue

⋆ This work has been partially supported by the AECID, Projects PCI D/023835/09 and D/030223/10.

J. Cabestany, I. Rojas, and G. Joya (Eds.): IWANN 2011, Part II, LNCS 6692, pp. 540–547, 2011.

viruses by the bite of an infective mosquito. Reduction of vector populations - both adult mosquitoes and immature states- is currently the only way to prevent dengue because there is no commercially available vaccine or clinical cure.

Dengue is endemic in many areas in Central America and the Caribbean. In Cuba, the Health System implemented a strong program to eradicate mosquitoes and breeding sites, to eliminate the epidemic and to prevent the disease from becoming endemic in Cuba.

The development of mathematical models to simulate the dynamics of Dengue epidemic is considered to be one of the basic and applied research that can have positive implications for dengue control. The main problem in these models is the lack of a precise knowledge of parameters. To deal with the uncertainty in parameters the stochastic approach is commonly used. Also fuzzy differential equations (DEs) have been suggested as a way of modeling uncertain and incompletely specified systems. However, formulations of fuzzy DEs can be interpreted in terms of differential inclusions. This reinterpretation is more significant than the classical definition, because differential inclusions have much richer properties. For example, one can speak of stability, attractors, periodicity and bifurcation for inclusions whereas these are barren concepts for the theory of fuzzy DEs because of possibilistic irreversibility. Consequently, the inclusion approach is much better suited to modeling practical situations under uncertainty and imprecision [5]. The approach proposed in this paper is based on differential inclusions and it is deterministic, in the sense that uncertain parameter it is not assumed to have any probabilistic structure whatsoever. In this case, the only assumption on the uncertain parameter is that it belongs to some finite interval.

A great number of papers and books are available on the differential inclusions theory (e.g., [1]) and applications (e.g., [2]). In this paper we present a model of dengue disease described by a differential inclusion to express the uncertainty in one parameter in order to analyze the behavior of an epidemic outbreak of dengue in Cuba. In our model, the uncertainty is consider in the daily mortality rate of mosquitoes (which we denote by μ) -we justify this choice below-. Also, we establish a mathematical framework, based on the computation of the reachable set of the differential inclusion, that will allow us to make a suitable prediction of the number of Dengue infected humans, mosquitoes and its eggs during the epidemic time.

The paper is organized as follows. Section 2 presents some assumptions on dynamics of Dengue. Section 3 is devoted to a Dengue model described by a differential inclusion. Section 4 is concerned with an approach based in a discrete approximation of the reachable set of the model to make predictions. Section 5 concludes the paper.

2 Biological Assumptions

In nature, only female mosquitoes require a blood meal, and therefore, only they can transmit the infection by biting. We assume the vertical transmission in mosquito population -transmission of the infection from a female mosquito to its offspring - as another via of infection, but we suppose that born infected males do not transmit the virus during mating. Therefore our model only considered female mosquitoes to be involved in the dynamics of the epidemic. Once a female mosquito is infected through vertical transmission or by bitting an infected person, they remain infected for life. We assume vector and human density to be spatially homogeneous. Also, all infective humans/mosquitoes have the same capacity of transmission of the infection. And, a single serotype is considered during an epidemic, which mostly occur in Cuba, where dengue is not an endemic disease.

The Uncertainty. Among the life history parameters of Aedes aegypti, the natural mortality rate of adult females is one of the most important. A small decrease in the mortality rate may substantially increase the vectorial capacity of mosquitoes. As a rule, vectors must survive longer than the extrinsic incubation period to be able to infect another human. Low mortality rates increase the chance of a mosquito to blood feed in a infectious human host, become infective, and transmit the virus during subsequent feeding attempts. Aedes aegypti females usually takes multiple blood feedings during a single gonotrophic cycle and dengue transmission could be enhanced there.

In most previous studies it was assumed that natural mosquito mortality remains constant with age [8]. This assumption has been used to assess the role of mosquitoes in pathogen transmission and predict public health consequences of vector control strategies. This hypothesis supposes that natural mortality factors (e.g., ambient temperature and humidity conditions, predation and disease) would kill mosquitoes before they had an opportunity to die of old age. However, others studies (e.g. [3]) suggest that with advancing age the mosquito mortality rate increases and that mosquitoes are no different from most other organisms in experiencing deleterious functional and structural changes with age. Despite the growing evidence of age-dependent mortality recent disease transmissions models assume constant mortality [7]. This disinclination to change can be attributed to two factors. First, there are no published reports that provide the detailed mortality trajectory necessary to properly test the assumption. Second, the assumption of age-independent mortality greatly simplifies the mathematics of models and reduces the number and complexity of variables that need to be considered. The major limitation of assuming age-independent mortality is that it leads to the unrealistic, simplified view that all mosquitoes have the same potential as vector, regardless of their age. In the last 50 years, only a few publications have considered the effect of mosquito age on transmission factors. Moreover, because vector mortality is a particularly sensitive component of pathogen transmission, quantitative models that assume age-independent mortality produce results with substantial errors [10].

Given the previous discussion, we propose to consider a new point of view for the behavior of the natural mortality of mosquitoes, which we will denote by μ_N. Our assumption is that μ_N can be seen as an uncertain variable where we only assume an interval where it belongs and nothing more. This view is not associated to any of the above hypothesis, although it can include them. It also allows us to consider a wider range of possibilities to incorporate a more diverse behavior and to reflect an non-exactly known parameter.

Daily mortality rate of mosquitoes μ is defined as the sum of the natural mortality μ_N and the mortality due to intervention. Intervention mortality μ_I (caused by chemical control) is assumed as a constant parameter in this work. This is justified in the case of Cuba due to the intensity and frequency of chemical control programs, through the application of fumigation inside and outside homes in the affected and adjacent areas where the epidemic develops. One objective of this intensive campaign is to reduce and control the vector population. Therefore, μ is an uncertainty obtained from the sum of an uncertain variable and a constant. This uncertainty and variation can be incorporated into the model by associating a range of possible values to this parameter.

3 The Model

Let t_0 and T $(T > t_0)$ be the initial and final time, respectively, where the model is defined. Let μ be an uncertain parameter which belongs to the known real interval $\mathcal{K} = [a, b]$. The dynamical system that describes the transmission dynamics of dengue in the three components of its transmission - humans, adult and immature stages of mosquitoes (we group the immature stages in a single state which we call **eggs**)- is given by the differential inclusion:

$$\underbrace{\begin{bmatrix} \frac{dH_S}{dt} \\ \frac{dH_L}{dt} \\ \frac{dH_I}{dt} \\ \frac{dH_R}{dt} \\ \frac{dM_S}{dt} \\ \frac{dM_L}{dt} \\ \frac{dM_I}{dt} \\ \frac{dE_S}{dt} \\ \frac{dE_I}{dt} \end{bmatrix}}_{=\dot{x}} \in G(x) = \left\{ \underbrace{\begin{bmatrix} -dp_M M_I \frac{H_S}{H} - \mu_H H_S + b_H H \left(1 - \frac{H}{c_H}\right) \\ dp_M M_I \frac{H_S}{H} - (\mu_H + \gamma_H) H_L \\ \gamma_H H_L - (\mu_H + m_H + r_H) H_I \\ r_H H_I - \mu_H H_R \\ \epsilon E_S - \mu M_S - dp_H M_S \frac{H_I}{H} \\ dp_H M_S \frac{H_I}{H} - (\mu + \gamma_M) M_L \\ \gamma_M M_L + \epsilon E_I - \mu M_I \\ \alpha (M_S + M_L + (1 - \sigma) M_I) \left[1 - \frac{E_S + E_I}{c_E}\right] - \epsilon E_S - \mu_E E_S \\ \sigma \alpha M_I \left[1 - \frac{E_S + E_I}{c_E}\right] - \epsilon E_I - \mu_E E_I \end{bmatrix}}_{=g(x,\mu)} : \mu \in \mathcal{K} \right\}, \quad (1)$$

where $x : [t_0, T] \to \mathbb{R}^9$, $g : \mathbb{R}^9 \times \mathbb{R} \to \mathbb{R}^9$ is a vector valued function, and $G : \mathbb{R}^9 \to 2^{\mathbb{R}^9}$ is a set valued map or multifunction (see the definition in [1]).

The model classifies hosts in four epidemiological states:

- $H_S(t)$: is the number of susceptible humans at time t.
- $H_L(t)$: is the number of infected but not infectious humans at time t.

- $H_I(t)$: is the number of infected and infectious humans at time t.
- $H_R(t)$: is the number of recovered (and immune) humans at time t.

Female mosquitoes are classified in three states:

- $M_S(t)$: is the number of susceptible mosquitoes at time t.
- $M_L(t)$: is the number of infected but not infectious mosquitoes at time t.
- $M_I(t)$: is the number of infected and infectious mosquitoes at time t.

And, two states for eggs:

- $E_S(t)$: is the number of non-infected eggs at time t.
- $E_I(t)$: is the number of infected eggs at time t.

The total number of humans, mosquitoes and eggs at time t are given by $H(t) = H_S(t) + H_L(t) + H_I(t) + H_R(t)$, $M(t) = M_S(t) + M_L(t) + M_I(t)$ and $E(t) = E_S(t) + E_I(t)$, respectively. Also, the model combines the above defined state variables with the following constant coefficients:

- d: Daily biting rate of female mosquitoes.
- p_H: Probability of virus transmissions from human to mosquitoes.
- b_H: Birth rate of humans.
- μ_H: Humans natural mortality rate.
- m_H: Dengue induced mortality rate in humans.
- c_H: Humans carrying capacity.
- γ_H: Humans incubation rate.
- r_H: Humans recovery rate.
- ϵ: Eggs hatching rate.
- p_M: Probability of virus transmissions from mosquitoes to human.
- γ_M: Mosquitoes incubation rate (inverse of extrinsic incubation period).
- α: Oviposition rate.
- c_E: Eggs carrying capacity.
- μ_E: Natural mortality rate of eggs.
- σ: Proportion of infected eggs.

We assume as known, the initial value vector $x(t_0) = x_0$ of the system (1). Also, since the model (1) monitors human, mosquitoes and egg population, all variables and parameter of the model are nonnegative. Then, we consider the system (1) only in the biologically feasible region

$$\mathscr{S} = \{(H_S, H_L, H_I, H_R, M_S, M_L, M_I, E_S, E_I)^T \in \mathbb{R}^9_+ : 0 < H \leq c_H,\ 0 \leq M \leq c_M,\ 0 \leq E \leq c_E\}. \quad (2)$$

It is natural to assume the existence of an upper bound c_M for the mosquito population from the fact that there is a carrying capacity for their eggs. In addition, we assume in the model that $b_H > \mu_H$, thus the human population $H(t)$ is always bounded since $H(t)$ follows a logistic equation.

4 Approximating the Reachable Set of the Model

Consider the Initial Value Problem (IVP) formed by (1) and the known initial valued vector $x(t_0) = x_0$. By definition, a solution of this IVP is any absolutely continuous function $x : [t_0, T] \to \mathbb{R}^9$ that satisfies the IVP for almost all $t \in [t_0, T]$.

Let us denote by $\mathcal{X}$ the set of all solutions or trajectories $x(\cdot)$ of the IVP on $[t_0, T]$ and let us denote by $\mathcal{X}(b)$ the set of elements x of $\mathcal{X}$ such that $x(t_0) = b$.

Several existence results for the IVP are available under various continuity assumptions on the set-valued map G (see [1]). In this paper we assume existence of solutions for our IVP and we will not enter in theoretical aspects on this topic. Nevertheless, we can give a result on the stability of the set of solutions $\mathcal{X}(b)$.

Let B be an origin-centered closed ball in $\mathbb{R}^9$ and $CC(B)$ the family of nonempty closed convex subsets of B equipped with Hausdorff distance $d_H(\cdot, \cdot)$ (see the definition in [1]). Assuming that G in (1) is a continuous map of B into $CC(B)$ satisfying

$$d_H(G(x), G(y)) \leq k\|x - y\| \tag{3}$$

for some $k > 0$. Then, for $b \in B$, $\mathcal{X}(b)$ is nonempty [6, Th. 2.1] and compact [4].

Proposition 1. *If $kT < 1$, then the set $\mathcal{X}(b)$ of all trajectories of (1) beginning in b is stable.*

Proof. Let $C[0, T]$ be the set of continuous maps of $[0, T]$ into $\mathbb{R}^9$ with the sup norm. By [11, Th. 4] we have the continuity of the set $\mathcal{X}(b)$ from B into the family of nonempty compact subsets of $C[0, T]$ equipped with the Hausdorff distance. Thus, this gives the conclusion of the proof.

We defined the reachable set $R(\tau)$ at the time τ of the IVP as

$$R(\tau) \doteq \{x(\tau) : x(\cdot) \in \mathcal{X}\} . \tag{4}$$

In order to compute an approximation $R_N(N)$ of the reachable set $R(T)$, a set-valued version of Heun's method [9] is revisited.

HEUN's METHOD. *For $N \in \mathbb{N}' \subset \mathbb{N}$ choose a grid $t_0 = t_0 < t_1 < ... < t_N = T$, with stepsize $h = \frac{T - t_0}{N} = t_j^N - t_{j-1}^N \quad (j = 1, ..., N)$. Let the starting approximation y_0 equal to a know initial value vector and for $j = 0, ..., N - 1$ compute $R_N(j + 1)$ from*

$$R_N(j + 1) = \bigcup_{y \in R_N(j)} \{y + \frac{h}{2}(g_{\mathcal{K}}(y, \mu) + g_{\mathcal{K}}(y + hg_{\mathcal{K}}(y, \mu), \mu)) : \mu \in \mathcal{K}\}. \tag{5}$$

Here, as usual $\mathbb{N}'$ denotes a subsequence of $\mathbb{N}$ converging to infinity.

Now, rewrite the differential inclusion (1) as follows:

$$\underbrace{\begin{bmatrix} \frac{dH_S}{dt} \\ \frac{dH_L}{dt} \\ \frac{dH_I}{dt} \\ \frac{dH_R}{dt} \\ \frac{dM_S}{dt} \\ \frac{dM_L}{dt} \\ \frac{dM_I}{dt} \\ \frac{dE_S}{dt} \\ \frac{dE_I}{dt} \end{bmatrix}}_{=\dot{x}} \in \left\{ \left[\underbrace{\begin{bmatrix} -dp_M M_I \frac{H_S}{H} - \mu_H H_S + b_H H \left(1 - \frac{H}{c_H}\right) \\ dp_M M_I \frac{H_S}{H} - (\mu_H + \gamma_H) H_L \\ \gamma_H H_L - (\mu_H + m_H + r_H) H_I \\ r_H H_I - \mu_H H_R \\ \epsilon E_S - dp_H M_S \frac{H_I}{H} \\ dp_H M_S \frac{H_I}{H} - \gamma_M M_L \\ \gamma_M M_L + \epsilon E_I \\ \alpha (M_S + M_L + (1-\sigma) M_I) \left[1 - \frac{E_S + E_I}{c_E}\right] - \epsilon E_S - \mu_E E_S \\ \sigma \alpha M_I \left[1 - \frac{E_S + E_I}{c_E}\right] - \epsilon E_I - \mu_E E_I \end{bmatrix}}_{=f(x)} + \underbrace{\begin{bmatrix} 0 \\ 0 \\ 0 \\ 0 \\ -M_S \\ -M_L \\ -M_I \\ 0 \\ 0 \end{bmatrix}}_{=F(x)} \mu \right] : \mu \in \mathcal{K} \right\} \quad (6)$$

with the same fixed initial condition, where $f : \mathbb{R}^9 \to \mathbb{R}^9$ and $F : \mathbb{R}^9 \to \mathbb{R}^9$.

The following theorem gives the possibility of obtaining convergence results and estimates for the reachable set of the IVP using the Heun's method (5).

Theorem 1. *Assuming that all trajectories of the model (6) are contained in* $\mathscr{S}$, *and* μ *belongs to* $\mathcal{K}$. *Given* N, *let the approximating set-valued dynamical system (5) where* $g_{\mathcal{K}}(x, \mu) = f(x) + F(x)\mu$. *Then, there exists a constant* C *such that*

$$d_H(R_N(N), R(T)) \leq Ch^{1.5} .$$

In addition, for those N *for which the sets* $R_N(N)$ *are convex*

$$d_H(R_N(N), R(T)) \leq Ch^2 .$$

Proof. It is easy to see that the components of f and F are differentiable with respect to t and x, and all the first derivatives are Lipschitz continuous with respect to $(t, x) \in [t_0, T] \times \mathscr{S}$ since the constants parameters c_H and c_E are always greater than zero and there exists a strictly positive lower bound for $H(t)$ for all t. Moreover the Lie bracket structural condition assumption of Theorem 4.1 of [9] is always fulfilled since the dimension of μ is equal to one. Then all assumptions of Theorem 4.1 [9] are satisfied (or Theorems 4.1 and 4.2 of [12]), its application gives the conclusion and both estimations follow.

5 Conclusions and Future Work

In this paper, we formulate a dengue model as a Differential Inclusion and we prove that it is possible to calculate an approximation of the reachable set of our model (1) using the set valued version of Heun's method (5). However, the practical viability of the theorem is limited by the computational complexity and the computational representation of non-finite sets in (5). Inspired by this limitation, in a working paper we are developing a practical version for (5) that handle the memory in a controlled fashion during all iterations, and thus it suits to the hardware features. The practical method is based on an approximation of $\mathcal{K}$ by discretization, and successive approximations of the resulting discrete sets via projections. Both the discretization of $\mathcal{K}$ and the projection of each discrete set are performed on at most a fixed number of points belonging to a regular mesh. This mesh is build in each case from the set to approximate. Therefore, the algorithm perform on a variable mesh. Theoretically, when the meshes density goes to infinity, our algorithm is the same that (5), and on the assumptions of the Theorem 1 estimations $O(h^2)$ or $O(h^{1.5})$ for the reachable set, still holds.

The approach described in this paper allows us to make predictions in time for all state variables of the model (1) using projections of the corresponding reachable set in each of the state variable axes. Of particular interest is the resulting intervals for the dengue latent (H_L) and infectious (H_I) humans, respectively, at time t. These variables would significantly influence the course of the epidemic and can be used to assist in successful health policy decisions.

References

1. Aubin, J.-P., Cellina, A.: Differential Inclusions, Set-Valued Maps and Viability Theory. Springer, Heidelberg (1984)
2. Barrios, J., Piétrus, A., Marrero, A., De Arazoza, H.: Hiv model described by differential inclusions. In: Cabestany, J., Sandoval, F., Prieto, A., Corchado, J.M. (eds.) IWANN 2009. LNCS, vol. 5517, pp. 909–916. Springer, Heidelberg (2009)
3. Briegel, H., Kaiser, C.: Life-span of mosquitoes (culicidae, diptera) under laboratory conditions. Gerontologia 19, 240–249 (1973)
4. Castaing, C.: Sur les équations différentielles multivoques. C.R. Acad. Sci. Paris 263, 63–66 (1966)
5. Diamond, P.: Time-dependent differential inclusions, cocycle attractors and fuzzy differential equations. IEEE Transactions on Fuzzy Systems 7(6) (December 1999)
6. Hermes, H.: The generalized differential equation $\dot{x} \in r(t, x)$. Adv. in Math. 4, 149–169 (1970)
7. Ishikawa, H., Ishii, A., et al.: A mathematical model for the transmission of plasmodium vivax malaria. Parasitol. Int. 52, 81–93 (2003)
8. Keener, G.: Detailed observations on the life history of anopheles quadrimaculatus. J. Natl. Malaria. Soc. 4, 263–270 (1945)
9. Lempio, F., Veliov, V.: Discrete approximation of differential inclusions. Bayreuther Mathematische Schriften 54, 149–232 (1998)
10. Styer, L.M., Carey, J.R., et al.: Mosquitoes do Senesce: Departure from the Paradigm of Constant Mortality. Am. J. Trop. Med. Hyg. 76(1), 111–117 (2007)
11. Teck-Cheong, L.: On fixed point stability for set-valued contractive mappings with application to generalized differential equations. Journal of Mathematical Analysis and Applications 110, 436–441 (1985)
12. Veliov, V.: On the time-discretization of control systems. SIAM J. Control Optim. 35(5), 1470–1486 (1997)

Simulating Building Blocks for Spikes Signals Processing

A. Jimenez-Fernandez, M. Domínguez-Morales, E. Cerezuela-Escudero,
R. Paz-Vicente, A. Linares-Barranco, and G. Jimenez*

Departamento de Arquitectura y Tecnología de Computadores
ETS Ingeniería Informática - Universidad de Sevilla
Av. Reina Mercedes s/n, 41012-Sevilla, Spain
angel@us.es

Abstract. In this paper we will explain in depth how we have used Simulink with the addition of Xilinx System Generation to design a simulation framework for testing and analyzing neuro-inspired elements for spikes rate coded signals processing. Those elements have been designed as building blocks, which represent spikes processing primitives, combining them we have designed more complex blocks, which behaves like analog frequency filter using digital circuits. This kind of computation performs a massively parallel processing without complex hardware units. Spikes processing building blocks have been written in VHDL to be implemented for FPGA. Xilinx System Generator allows co-simulating VHDL entities together with Simulink components, providing an easy interface for presented building block simulations and analysis.

1 Introduction

Living beings brains allow them to interact dynamically with their environment, unlike robotics systems that need in best cases a controlled environment, being a great current challenge to improve robots adaptability and cognitive skills. Neuromorphic systems provide a high level of parallelism, interconnectivity, and scalability; doing complex processing in real time, with a good relation between quality, speed and resource consumption. Neuromorphic engineers work in the study, design and development of neuro-inspired systems developed, like aVLSI chips for sensors [1][2], neuro-inspired processing, filtering or learning [3][4][5][6], neuro-inspired control pattern generators (CPG), neuro-inspired robotics [8][17] and so on. Neuromorphic engineering community grows every year as it demonstrate the success of the Telluride and Capocaccia Cognitive Neuromorphic workshops [9][10]. Spiking systems are neural models that mimic the neurons layers of the brain for processing purposes. Signals in spikes-domain are composed by short pulses in time, called spikes. Information is carried by spikes frequency or rate [11], following a Pulse Frequency Modulation (PFM) scheme, and also from other point of view, in the inter-spike-time (ISI) [5]. Spikes rate coded information can be processed in a continuous way, without codifying information into discrete samples. Because of the simplicity of these models, spikes processing do not need a complex hardware to perform

* This work has been supported by the Spanish project VULCANO (TEC2009-10639-C04-02).

J. Cabestany, I. Rojas, and G. Joya (Eds.): IWANN 2011, Part II, LNCS 6692, pp. 548–556, 2011.

information processing, in consequence, this hardware can be easily replicated performing a massively parallel information processing [13][14]. Previous work presented a set of components, designed as building blocks, to process spikes coded signals in real time, mimicking elementary analog components to design spikes filters equivalent to classical analog filters [18], written in VHDL to be stored in a digital devices like a FPGA. This way of processing spikes was previously called Spikes Signals Processing (SSP). This work is focused on presenting in a detailed way a simulation framework to test, measure and analyze SSP building blocks behavior. MATLAB with Simulink provides a good set of tools for simulating almost everything, so they are ideal for simulating our SSP blocks. However we need to simulate together VHDL components with Simulink blocks, for this purpose Xilinx provides a MATLAB toolkit (Xilinx System Generator) that allows the desired kind of simulation and many other things.

2 Simulation FrameWork

Simulation framework has two levels:

a) Lowest level is composed by a Simulink simulation model, which will contain SSP VHDL components and also Simulink ideal elements.
b) A MATLAB script is located in the highest level, this script will set up the component that will be simulated with its parameters. This script will also launch the simulation within, and after simulations it will reconstruct and analyze the spikes from the SSP VHDL components.

Figure 1 a) shows Simulink model, this model contains source stimulus, VHDL components. Stimulus signal is applied to a Synthetic Spikes Generator (this component will be the first studied in next section), which is used to convert Simulink signals to spikes. Those spikes will be processed by a SSP component (figure center). Each SSP component has its own set of parameters, which will be set up using constant blocks. Processed spikes (simulation outputs) will be monitored and analyzed by two Simulink components (Scope and a link to MATLAB workspace respectively). At figure bottom, it has been placed a set of Simulink elements that will perform the same processing to stimulus signal, but with continuous elements (without spikes), as S domain transfer functions. Ideal components will allow us to compare each SSP responses with their ideal equivalents. So for simulating any designed SSP component we only have to change the component under test, its parameters and its ideal transfer function, and we will be ready to simulate and to analyze exhaustively its behavior. Simulations of the Simulink model are managed by a high level MATLAB script. Figure 1 b) shows the flow chart of this script. First, it will set up simulation elements and parameters, then, it will launch the simulation, next, it will reconstruct output simulation spikes, and finally output analysis will be performed. Simulation output should be spikes fired by the simulated SSP blocks in time, which are a stream of short pulses. To analyze them it is necessary to transform output spikes to discrete numbers. For this task we have written in MATLAB a function that takes spikes output from simulation and returns a vector with the spikes average frequency for a fixed period of time (sample time). Sample time represents the number of clock cycles to measure average spikes frequency. The strategy followed is to count the number of spikes fired during the sample time, being the average spikes rate the number of counted spikes divided by the sample time and the clock frequency used in simulations for VHDL components. Once spikes are converted to discrete numbers, we are ready to analyze simulation results.

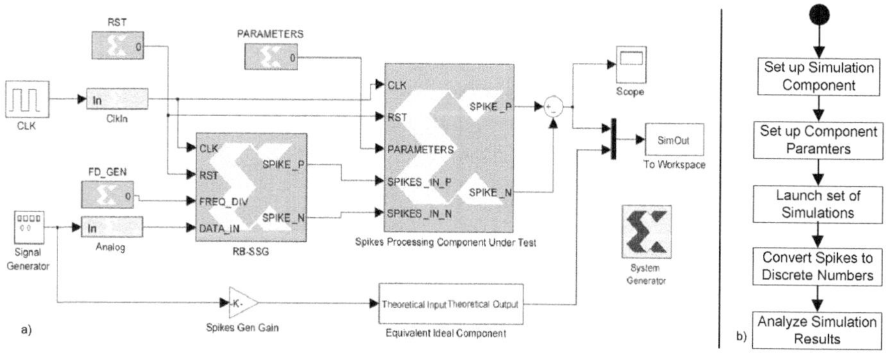

Fig. 1. A) Simulation scenario designed for spikes processing components testing. B) MATLAB script for simulation managing.

3 Building Blocks for SSP

In this section we make a brief introduction to SSP building blocks already presented in [18]. Then we show and comment simulations made for those SSP components in a detailed way. SSP components have been designed using four basic building blocks, and finally all of them can be combined for implementing, for example, a spike-based low pass filters.

3.1 Synthetic Spikes Generator (RB-SSG)

A Synthetic Spikes Generator (SSG) will transform a digital word (SSG input) into a frequency rate of spikes (SSG output). This element is necessary in those scenarios where is needed to transform digital signals to spikes, as commented before about Simulink source signals. There are several ways to design SSGs as presented in [15], selected SSG implements the reverse bitwise method found detailed in [16] for synthetic AER images generation. Figure 2 shows the internal components of the Reverse Bitwise SSG (RB-SSG) implemented. Since a reference or an analog signal can be negative, it is necessary to generate positive and negatives spikes. RB-SSG gain ($k_{BWSpikesGen}$) can be calculated as follows:

$$k_{BWSpikesGen} = \frac{F_{CLK}}{2^{N-1}(genFD + 1)} \tag{1}$$

Where Fclk represents system clock frequency, N the RB-SSG bits length, and genFD clock frequency divider value. RB-SSG has been the first element simulated, to verify (1) and its right behavior. Figure 2 b) bottom shows RB-SSG output spikes in time, and at top we can find spikes reconstructed frequency and theoretical value. Input stimulus signal from Simulink is a 2kHz sine, in consequence, RB-SSG output spikes rate starts to increase, until input signal reach its maximum value. From that moment, input signal starts to decrease, as output spikes rate, crossing 0 and changing of sign. Output spikes also changes its sing, observing negatives spikes, although spikes if different signs are fired by different signals, from the point of view of MATLAB workspace, positive spikes are represented by '1', negative ones by '-1', and the absence of spikes by a '0'.

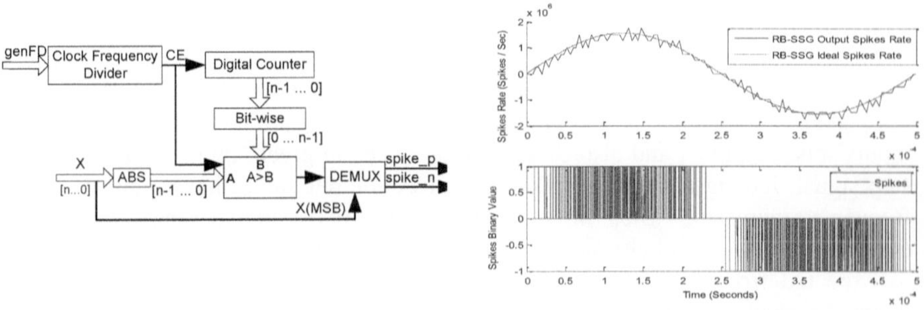

Fig. 2. A) Reverse Bitwise Synthetic Spikes Generator block diagram. B) RB-SSG output and reconstructed spikes rate, for sinusoidal input RB-SSG Simulation Parameters.

3.2 Spikes Integrate and Generate (SI&G)

Spikes Integrate & Generate (SI&G) is composed by a spikes counter, and by a RB-SSG, as showed in Figure 3 a). Spikes counter is a digital counter that is increased by one when a positive spike is received, and its value is decreased by one with a negative spike. Counter output is the RB-SSG input. Therefore, new spikes generated have a frequency proportional to spikes integration. The SI&G gain is set by RB-SSG parameters. With these considerations, SI&G spikes output frequency, $f_{I\&G}$, and SI&G gain, $k_{SI\&G}$, can be expressed as:

$$f_{SI\&G} = k_{SI\&G} * \int f_{inputSpikes} dt = \frac{F_{CLK}}{2^{N-1}(genFD+1)} \int f_{inputSpikes} dt \qquad (2)$$

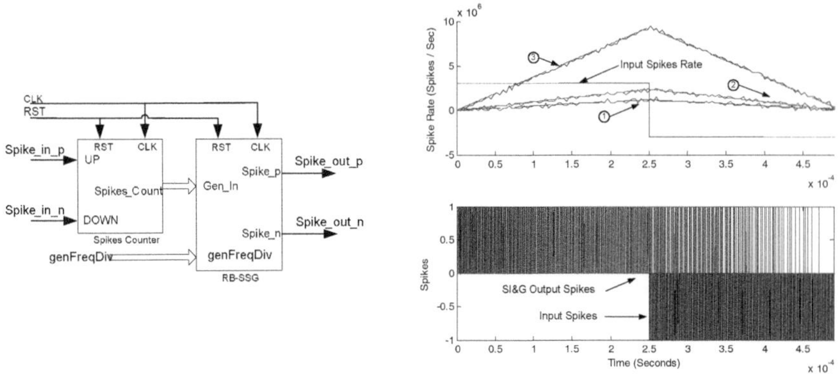

Fig. 3. A) Spikes Integrate & Generate block diagram. B) SI&G simulation output spikes rate reconstructed, top, and input/output spikes, bottom.

Similarly to analog systems, we can calculate equivalent SI&G transfer function in "spikes-domain" using Laplace transform. SI&G transfer function is presented in (2), being equivalent to an ideal integrator with a gain of $k_{I\&G}$.

$$SI\&G(s) = \frac{F_{SI\ \&G}(s)}{F_{inputSpikes}(s)} = \frac{k_{SI\&G}}{s} = \frac{F_{CLK}}{2^{N-1}(genFD+1)*s} \tag{3}$$

For SI&G testing purpose we have executed a set of simulations for different pairs of parameters, in Table 1, and a constant rate of input spikes. Figure 4 b) shows SI&G temporal simulation results, at figure top is represented the frequency of input spikes in blue (a step signal), and three SI&G frequency output spikes couples for different parameters. These couples are composed by simulated reconstructed output (solid line) and theoretical response (discontinuous line). At figure bottom we can see SI&G inputs spikes in blue, and also SI&G output spikes in green. Input spikes represent a step signal, having a constant spike rate, SI&G output spikes have a constant linear frequency increasing when input spikes are positives, like a ramp, with a slope equivalent to $k_{SI\&G}$, and also decreasing output spikes frequency with negative spikes, as expected from an ideal analog integrator with same features. Simulations also denoted a good accuracy with theoretical responses.

Table 1. SI&G Simulation Parameters

Sim. Case	N (Bits) - genFD	$k_{SI\&G}$
1	13 - 0	12.207*10^3
2	14 – 1	3.052*10^3
3	16 – 0	1.526*10^3

3.3 Spikes Hold&Fire (SH&F)

This block performs the subtraction between two spikes stream. SH&F will allow us to implement feedback in the SI&G block, obtaining new transfer functions. Subtracts a spikes input signals (f_U) to another (f_Y), means to get a new spikes signal which spike rate ($f_{SH\&F}$) will be the difference between both inputs spikes rate:

$$f_{SH\&F} = f_U - f_Y \tag{4}$$

The procedure of the SH&F is to hold the incoming spikes a fixed period of time while monitoring the input evolution to decide output spikes: holding, canceling, or firing spikes according to input spikes ports and sign. This block has been successfully used previously to design spike-based closed-loop control systems in mobile robots by authors [17][18]. And will be used later to feedback the SI&G to design spikes filters [18]. Figure 4 contains simulation results, SH&F has been excited with two saw tooth signals with different amplitude and sign. At simulation begin input spikes has opposed sign, in consequence output spikes frequency is the addition of inputs spikes frequency. When input spikes frequency is the same, there is no output spikes, with a frequency of 0. Finally, in the case that both inputs spikes have the same sign, SH&F output spikes are the difference between both signals, performing perfectly the subtraction operation.

3.4 Spikes Frequency Divider (SFD)

This block will divide the spike rate of an input signal by a constant. To ensure spikes distribution we have implemented this block inspired in the way that RB-SSG works. Figure 5 shows SFD internal components. In SFD *spikesDiv* behaves like the constant to divide. Output buffers only enable output spikes according *spikesDiv* homogenously in time. SFD transfer function can be calculated using (5), where N represents the SFD number of bits, and *spikesDiv* the signal presented before. SFD transfer function is equivalent to a gain block with a value in the range of [0, 1], with 2^N possible steps. SFD accuracy can also be adjusted with N (N=16 bits in Fig.5) getting an accuracy of 2^{-N}.

$$k_{SFD} = \frac{F_{outSpikes}}{F_{inputSpikes}} = \frac{spikesDiv}{2^N} \tag{5}$$

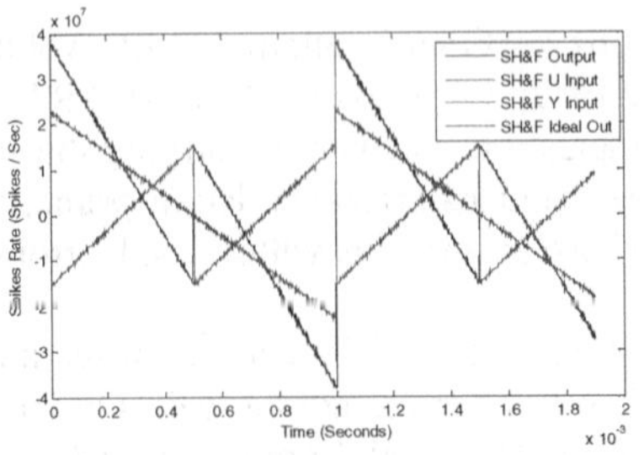

Fig. 4. Spike Hold&Fire temporal simulations

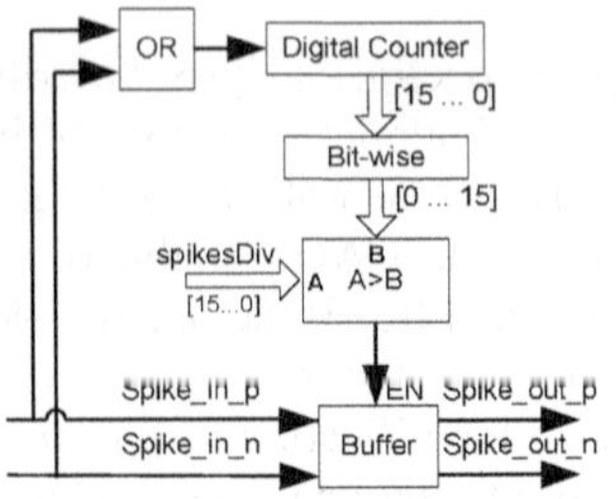

Fig. 5. Spikes Frequency Divider internal components

3.5 Spikes Low-Pass Filter (SLPF)

As analog frequency filters modifies signal frequency components from amplitude changes, spikes filters will work on spikes rate changes frequency components. For example, a spikes low pass filter will attenuate spikes rate high frequency components, if constant spikes rate signal is applied to spikes filter input, as an input step, spikes output frequency will start to increase exponentially, as expected from an analog signal amplitude, until spikes rate reach some frequency. To build a SLPF a SI&G have been feedback using a SH&F and two SFD, as it is shown in Figure 6.

Next equation shows SLPF ideal transfer function, where SI&G(s) can be obtained from (3), k_{SDOut} represents output SFD gain and k_{SDFB} the gain of the SFD placed in the feedback loop, both detailed in (5).

$$F_{SLPF}(s) = \frac{k_{SFDOut} * SI\&G(s)}{1 + k_{SFDFB} * SI\&G(s)} = \frac{k_{SFDOut} * k_{I\&G}}{s + k_{SFDFB} * k_{I\&G}} \tag{6}$$

Finally we are going to simulate SLPF with different parameters. So different features spikes filters have been simulated, fixing different parameters sets, getting equivalent filters with various cut-off frequencies and gain. Simulations results are presented in Figure 7, this figure is composed by two different kinds of simulations, in a) there are spikes filters temporal response, and in b) the frequency response. As first simulations we present a set of temporal simulations, Figure 7 a), simulating spikes filters for a fixed time, being X axis simulation time, and Y axis instantaneously spikes rate. For temporal simulation these spikes filters have been excited by a constant spike rate input, as an analog voltage step. Filter output spikes have been analyzed and their instantaneous frequencies have been reconstructed. Theoretical responses of analog ideal equivalent filters have been also added to the figure, to compare simulation response respect to ideal response. In the case of SLPF (1-5), output spikes frequency start to increase exponentially, like expected from an ideal low pass-filter, with a rise time near to 4 times $\grave{u}_{cut-off}$. SLPF output spikes rate reach a steady state value proportional to SLPF gain, reaching input spikes frequency when is 1 (1-3), higher output spikes rates in the case that SLPF gains is higher than 1 (5), and opposed effect happens when SLPF gains is lower than 1 (4).

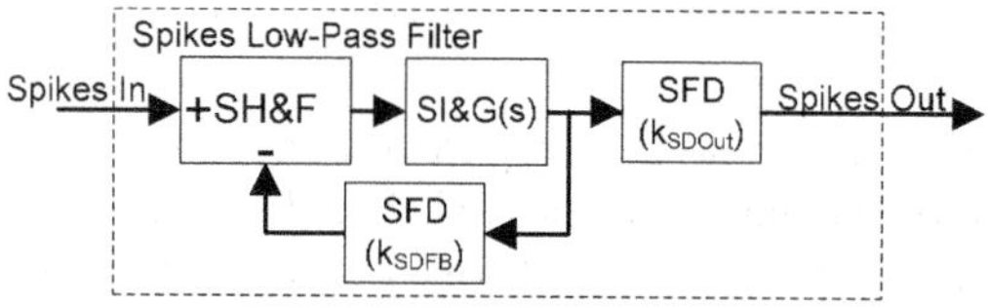

Fig. 6. Spikes Low Pass Filter Architecture

Table 2. SLPF Simulation Parameters

Sim. Case	N (Bits) - genFD	Spikes Div. Out	Spikes Div. Feedback	$\grave{u}_{cut-off}$	Filter Gain (abs)	Rise Time (mSec)
1	13 – 0	0.5147	0.5147	1kHz	1	0.6366mSec
2	11 – 0	0.634	0.634	5kHz	1	0.1273mSec
3	9 – 0	0.634	0.634	20kHz	1	0.0318mSec
4	13 – 0	0.6691	0.5147	1kHz	0.5	0.6366mSec
5	13 – 0	0.6691	0.5147	1kHz	1.3	0.6366mSec

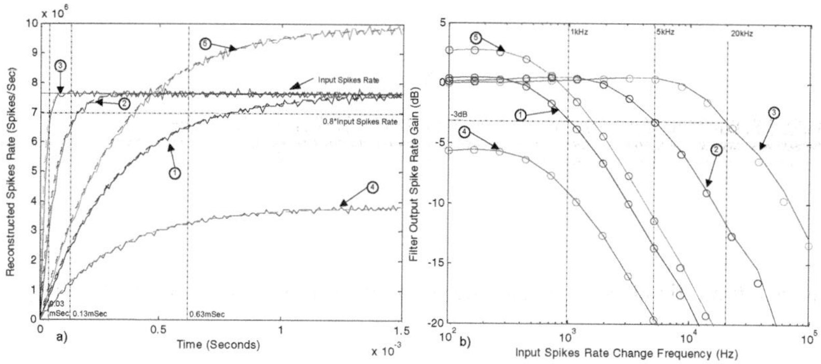

Fig. 7. Simulation Results: a) Spikes Filters time response b) Spikes Filters Bode diagram

After simulating the temporal behavior of designed spikes filters, we have studied their frequency responses. If our aim were to characterize an analog filter in frequency domain, one simple way to perform this task could be to excite the analog filter with pure frequency tones, and annotate analog filter output power for each tone. Translating this experiment to spikes domain, we are going to stimulate RB-SSG with an input sinusoidal signal, getting a spikes output which frequency changes according the sinusoidal signal, codifying a pure tone as spikes rate changing. Generated spikes will stimulate the spikes filters, whose output will be a spikes stream with same input frequency tone, but with its amplitude and phase modified by the spikes filters. So we have exited every spikes filter with a set of spikes coded tones, and recording all the spikes filter output power at that tone, obtaining finally the spikes filter Bode diagram. Results of these simulations are presented in Figure 7 b), X axis represents the frequency of input tones in Hz, and Y axis contains the spikes filter gain in dB. Figure contains SLPF simulated and theoretical responses marked with a cross, where theoretical responses are marked in these cases with circle. SLPF with a gain of 1 (1-3) have predicted gain of 0dB in the pass band, and then gain starts to decrease when frequency is near to cut-off frequencies, cutting them with a gain of 3dB, providing an attenuation of 20dB by decade as expected of analog filters. SLPF with different gains (4-5), having a gain in the band pass of -6dB and +2.6dB respectively.

4 Conclusions

This paper presents a framework for the simulation of building blocks to process signals using a spike rate coded representation. Simulations have been possible thanks to the power, flexibility and scalability of Simulink and Xilinx System Generator, providing an excellent scenario to co-simulate VHDL files with other kind of components. We have also presented the SSP building blocks that had been simulated, analyzing, and comparing with ideal components, in an exhaustive way for each presented SSP block. Simulations have denoted an accurate behavior of designed SSP blocks compared with ideal results, validating theoretical equations. Presented framework is nowadays being used in our lab for designing and testing new SSP building blocks that will improve current blocks features, and they will also allow the development of more complex SSP components.

References

1. Lichtsteiner, P., et al.: A 128×128 120dB 15 us Asynchronous Temporal Contrast Vision Sensor. IEEE Journal on Solid-State Circuits 43(2) (February 2008)
2. Chan, V., et al.: AER EAR: A Matched Silicon Cochlea Pair With Address Event Representation Interface. IEEE TCAS I 54(1) (January 2007)
3. Serrano-Gotarredona, R., et al.: On Real-Time AER 2-D Convolutions Hardware for Neuromorphic Spike-Based Cortical Processing. IEEE TNN 19(7) (July 2008)
4. Oster, M., et al.: Quantifying Input and Output Spike Statistics of a Winner-Take-All Network in a Vision System. In: IEEE International Symposium on Circuits and Systems, ISCAS 2007 (2007)

5. Hafliger, P.: Adaptive WTA with an Analog VLSI Neuromorphic Learning Chip. IEEE Transactions on Neural Networks 18(2) (March 2007)
6. Indiveri, G., et al.: A VLSI Array of Low-Power Spiking Neurons and Bistables Synapses with Spike-Timig Dependant Plasticity. IEEE Transactions on Neural Networks 17(1) (January 2006)
7. Gomez-Rodríguez, F., et al.: AER Auditory Filtering and CPG for Robot Control. In: IEEE International Symposium on Circuits and Systems, ISCAS 2007 (2007)
8. Linares-Barranco, A., et al.: Using FPGA for visuo-motor control with a silicon retina and a humanoid robot. In: IEEE International Symposium on Circuits and Systems, ISCAS 2007 (2007)
9. Telluride Cognitive Neuromorphic workshop, `https://neuromorphs.net/`
10. Capo Caccia Cognitive Neuromorphic workshop, `http://capocaccia.ethz.ch`
11. Shepherd, G.: The Synaptic Organization of the Brain. Oxford University Press, Oxford (1990)
12. Chicca, E., et al.: An event based VLSI network of integrate-and-fire neurons. In: IEEE International Symposium on Circuits and Systems, ISCAS 2004 Misha Mahowald. VLSI Analogs of Neuronal Visual Processing: A Synthesis of Form and Function. PhD. Thesis, California Institute of Technology Pasadena, California (1992)
13. Serrano-Gotarredona, R., et al.: CAVIAR: A 45k-neuron, 5M-synapse AER Hardware Sensory-Processing-Learning-Actuating System for High-Speed Visual Object Recognition and Tracking. IEEE Trans. on Neural Networks 20(9), 1417–1438 (2009)
14. Gomez-Rodriguez, F., et al.: Two Hardware Implementation of the Exhaustive Synthetic Aer Generation Method. In: Cabestany, J., Prieto, A.G., Sandoval, F. (eds.) IWANN 2005. LNCS, vol. 3512, pp. 534–540. Springer, Heidelberg (2005)
15. Paz-Vicente, R., et al.: Synthetic retina for AER systems development. In: International Conference on Computer Systems and Applications, AICCSA 2009 (2009)
16. Jiménez-Fernández, A., et al.: AER-based robotic closed-loop control system. In: IEEE International Symposium on Circuits and Systems, ISCAS 2008 (2008)
17. Jiménez-Fernández, A., et al.: AER and dynamic systems co-simulation over Simulink with Xilinx System Generator. In: IEEE International Conference on Electronics, Circuits and Systems, ICECS 2008 (2008)
18. Jimenez-Fernandez, A., et al.: Building Blocks for Spike-based Signal Processing. In: IEEE International Joint Conference on Neural Networks, IJCNN 2010 (2010)

Description of a Fault Tolerance System Implemented in a Hardware Architecture with Self-adaptive Capabilities

Javier Soto, Juan Manuel Moreno, and Joan Cabestany

Technical University of Catalunya (UPC), Barcelona - Spain
Department of Electronic Engineering, Advanced Hardware Architectures (AHA)
{javier.soto.vargas,joan.manuel.moreno,joan.cabestany}@upc.edu

Abstract. This paper describes a Fault Tolerance System (FTS) implemented in a new self adaptive hardware architecture. This architecture is based on an array of cells that implements in a distributed way self-adaptive capabilities. The cell includes a configurable multiprocessor, so it can have between one and four processors working in parallel, with a programmable configuration mode that allows selecting the size of program and data memories. The self-elimination and self-replication capabilities of cell(s) are performed when the FTS detects a failure in any of the processors that include it, so that this cell(s) will be self-discarded for future implementations. Other self-adaptive capabilities of the system are self-routing, self-placement and runtime self-configuration.

Keywords: Self-adaptive, self-placement, self-routing, self-elimination, self-replication, Multicomputer, MIMD, dynamic fault tolerance, fault tolerance.

1 Introduction

In coming years the self-adaptive systems promise to change radically our experience, in order to provide high performance computation power to serve the increasing need of large, complex and parallel applications. Self-adaptation is defined as the ability of a system to react to its environment in order to optimize its performance. The system could also change its behavior to adapt its functionality to a new mission, with respect to the environment and the user needs [1]. Adaptive computing system is a promising technology with respect to classical computing architectures.

Self-healing is a special feature of an adaptive system, where hardware failures should be detected, handled and corrected by the system automatically. A fault tolerance system in an adaptive system together with other self-adaptive capabilities could provide this functionality.

In this paper we aim to show a new hardware architecture, whose principal characteristics are the self-adaptive capabilities implemented, that are executed autonomously and in distributed way by the system members (cells). Basically,

J. Cabestany, I. Rojas, and G. Joya (Eds.): IWANN 2011, Part II, LNCS 6692, pp. 557–564, 2011.

this is a novel unconventional MIMD hardware architecture with self-adaptive capabilities like self-routing, self-placement, self-configuration, self-elimination and self-replication, which includes a fault tolerance system that permits a given subsystem to modify autonomously its structure in order to achieve fault detection and fault recovery.

The architecture proposed includes two mechanisms of fault tolerance (FT). One of these is the Dynamic Fault Tolerance Scaling Technique [2], that has the ability to create and eliminate the redundant copies of the *functional* section of a specific application. The other mechanism of fault tolerance (presented in this paper) is a dedicated or static Fault Tolerance System (FTS), it provides redundant processing capabilities that are working continuously. When a failure in the execution of a program is detected, the processors of the cell are stopped and the self-elimination and self-replication processes starts for the cell (or cells) involved in the failure. This cell(s) will be self-discarded for future self-placement processes.

Other projects that propose fault tolerance on self-adaptive architectures have been developed, like ReCoNet-platform [3] that presents a framework for increasing fault tolerance and flexibility by solving the problem of hardware/software codesign online. It is based on field-programmable gate arrays (FPGAs) in combination with CPUs that allow migrating tasks implemented in hardware or software from one node to another. The principal difference with the platform presented in this paper is its high scalability and that it could provide a lot of processing units, with less processing capacities, but where any processor could implement a fault tolerance system. Additionally it provides self-adaptive capabilities like self-placement and self-routing in a large array of processing units.

Reference [4] proposes a framework under which different fault tolerant schemes can be incorporated in applications using an adaptive method. Under this framework, applications are able to choose near optimal fault tolerance schemes at run time according to the specific characteristics of the platform on which the application is executing. This framework is presented for high performance parallel and distributed computing. Compared with this framework the architecture proposed in this paper provides adaptive capabilities at hardware level different from application level.

2 Description of the Architecture

Any application scheduled to the system has to be organized in components, where each component is composed by one or more interconnected cells. The interconnection of cells inside of a component is made at cell level, while the physical interconnections of components are made in switch matrix level.

In the initial state, all cells are free, i.e., they do not belong to any component. Then, the predefined components have to be placed and connected for data processing and information exchange. This is a sequential process where each cell has to be placed and routed in the system. For this purpose, the cells execute in a distributed way the self-placement and self-routing algorithms. All cells have

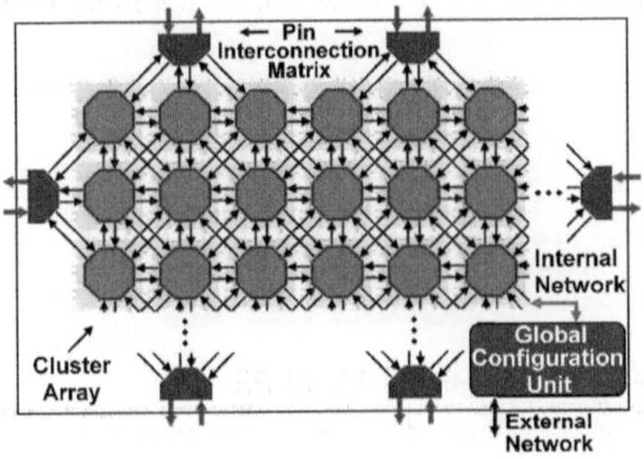

(a) Architecture overview.

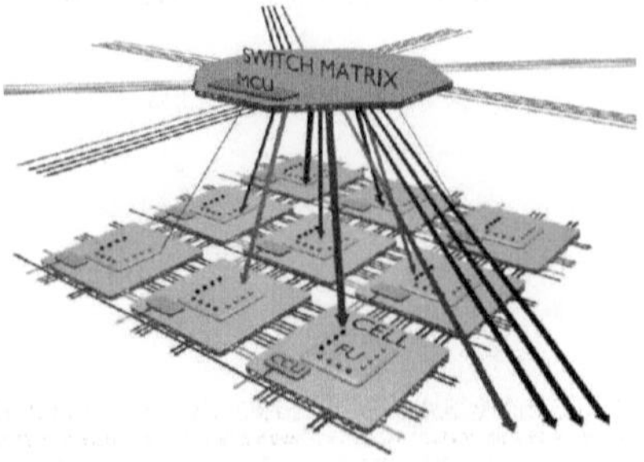

(b) 3D view of a cluster.

Fig. 1. System Architecture. Composed by a cluster array, pin interconnections matrices and GCU. Cluster: 3x3 cell array and switch matrix.

got a 32-bit unique identifier called *address*. This field is divided into two 16-bit words, called *id_component* and *id_cell*. The *id_component* is the component unique identifier, where the value FFFFh is reserved for special features and the value 0000h to indicate that the cell is free and does not belong to any component. Therefore, it is possible to instantiate up to 65534 different components. The *id_cell* is the cell unique identifier in a component, so there may be up to 65536 cells in a component and a maximum close to 2^{32} cells in the system.

The physical implementation of this architecture is depicted in Figure 1, which shows the representation of a chip that includes an array of clusters, pin interconnection matrices and a Global Configuration Unit (GCU). Figure 1b shows a cluster, that is composed by a 3x3 cell array and a switch matrix.

This two-layer implementation is composed by interconnected cells in the first level and interconnected switch and pin matrices in the second level, which are interconnected to the GCU by means of an Internal Network. Several chips can be interconnected by means an External Network connected to the GCUs. These networks have been designed to support the system with the necessary functionality to carry out the self-adaptive capabilities.

Cell Architecture. The cell is the basic element of the proposed self-adaptive architecture. Therefore, the cell has to include the necessary hardware to carry out the basic principles of self-adaptation; dynamic and distributed self-routing [5] [6] [7], dynamic and distributed self-placement, self-elimination, self-replication, scalability and distributed control.

The cell consists of the Functional Unit (FU), the Cell Configuration Unit (CCU) and multiplexers that allow the interconnection between FU ports of two cells. The cell is interconnected with its four direct neighbors by means of local, remote and expansion ports. The cell architecture and port distribution is depicted in Figure 2a. The local and remote ports are 9-bit wide bus (8-bit for data and 1 bit for read enable). The read enable (RE) is set to logic '1' for a clock pulse when a processor performs a write operation over the corresponding output port.

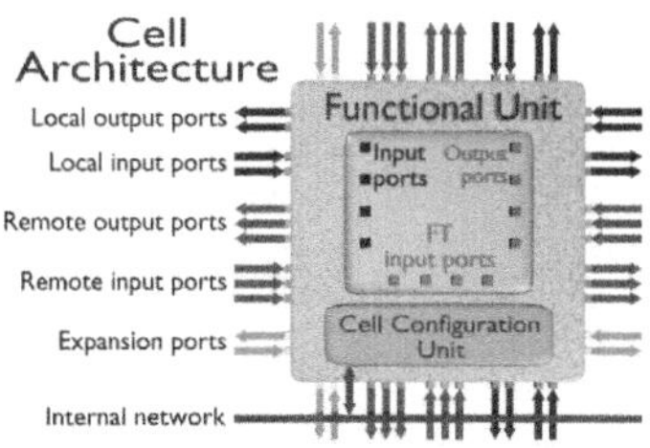

(a) Cell Block Diagram.

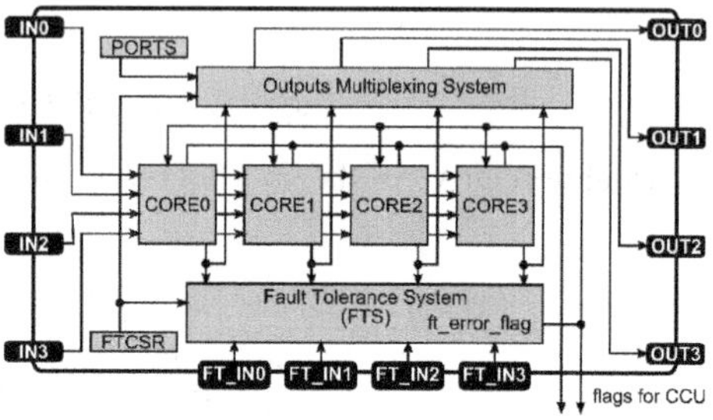

(b) Block diagram of FU.

Fig. 2. Cell Architecture

Functional Unit (FU). The FU is in charge of executing the processes scheduled to the cell. The FU can be described as a configurable multicomputer with four cores (Figure 2b). The FU has four 9-bit input ports and four 9-bit output ports. Additionally the FU includes four 9-bit FT input ports exclusively for FTS, if it is enabled. The FU includes an Outputs Multiplexing System that allows the cores to write data in the output ports, as well as direct the data flow of cores to outputs when the FTS is enabled and any of its cores is working as a redundant processor. The FTS is explained in detail in section 3.

Each core has program memory, data memory and additional hardware necessary for its functionality. The instruction set is composed of 44 instructions, which includes arithmetic, logic, shift, branch, conditional branch and special instruction for the execution of microthreads [8]. All instructions can be executed in a single clock pulse. The program memory can store up to 64 instructions. The data memory is 8-bit and it is composed of 8 general-purpose registers and 14 Configuration and Status Registers.

The *configuration modes* of the FU consists basically in grouping cores, where the expansion of data and program memory describes the specific configuration mode. The data memory can be combined in width and length, achieving combinations for data processing of 8, 16, 24 and 32 bits. The program memory can only be joined in length, making possible to have programs of 64, 128, 192 or 256 instructions. There are twelve different configuration modes, ranging between one and four processor working in parallel.

Cell Configuration Unit (CCU). Using a distributed working principle, the CCUs of the cells in the array are responsible for the execution of the required algorithms for the implementation of the self-adaptive capabilities of the system, specifically the self-placement and self-routing algorithms. These algorithms are executed by the CCU using the Internal Network and the expansion ports.

The *self-placement algorithm* is responsible for finding out the most suitable position in the cell array to insert the new cell of a component. The *self-routing algorithm* allows interconnecting the ports of the functional unit of two cells. The *self-elimination* and *self-replication* of cells allows the replication and elimination of cells that are suspect of having a hardware failure, these are self-adaptive capabilities that are closely tied with the FTS (see section 3).

3 Fault Tolerance System (FTS)

The FTS enables the system to continue operating properly in the event of the failure of some of its processors. The FTS consists of a specific hardware that allows the comparison of two identical processors each time that a instruction is executed, thats means each clock pulse. This involves the comparison of 2, 4, 6 or 8 cores, depending of the configuration mode of the FTS (FT_mode).

These processors that will be compared are called primary and redundant. These processors must share the same inputs, but on the other hand, the primary processor takes over writing the output ports, because two outputs can not be routed to the same location.

The redundant processor may or may not be included in the same cell where the primary processor is located, this depends on the processor on which the FTS is to be implemented (configuration mode). If the redundant processor is located in the same cell that includes the primary processor (FT_modes 0, 1 , 2, 3 or 4), the FT_input ports are not used, and the FTS only performs comparison between cores of the cell.

Otherwise, when the redundant processor is located in another cell (FT_modes 5, 6, 7 or 8), the FTS of the primary cell must perform a comparison between cores of different cells, in this case the Output Multiplexing System of the cell that includes the redundant processor must drive the output of the data flow of the core(s) to the output of the cell (RE is set to logic 1), which in turn should be connected to the FT_input of the FU of the cell that includes the primary processor.

When a hardware failure is detected by the FTS, the damaged cell(s) is self-eliminated and self-replicated to another location. The routing resources of these cells remain working. The next steps describe the procedure for self-elimination and self-replication of a single damaged cell, the procedure is repeated for a second damaged cell if the redundant processor is located there:

- The cores of damaged cell are blocked, no more instructions will be executed. This cell requests to the GCU to start the process.
- The GCU sends a command through the Internal Network requesting the cell array to perform a self-derouting process for all connections with the damaged cell.
- The damaged cell is self-eliminated, this cell is configured as *busy cell* and its *address* will be fixed to FFFF0001 to avoid future use.
- The GCU starts the self-replication process. The GCU sends the appropriate information to start the execution of the processes for self-placement and self-routing the cell again. The program memories and the appropriate configuration registers of the new cell are configured again.
- The GCU asks to the system to start the process of self-routing at component level to route the missing connections in the system. The GCU generates a reset to start the functionality of the system again.

Fault Tolerance Configuration Modes. The FTS could be implemented in any processors available in the system. Table 1 shows the 9 FT_modes available

Table 1. Fault Tolerances Modes (FT_modes)

FT_mode	Core Comparison
0	C0⇔C1
1	C0⇔C1 & C2⇔C3
2	C2⇔C3
3	C0⇔C2
4	C0⇔C2 & C1⇔C3
5	C0⇔C0*
6	C0⇔C0* & C1⇔C1*
7	C0⇔C0* & C1⇔C1* & C2⇔C2*
8	C0⇔C0* & C1⇔C1* & C2⇔C2* & C3⇔C3*

⇔ denotes a comparion between cores. & denotes a logic AND. * denotes a core in the redundant cell. A connection between FUs of primary and redundant cell is assumed.

and the comparison performed. This table indicates the cores that are compared in each FT_mode, which could be combined with any of the 12 configuration modes available for the system, obtaining 108 possible combinations. It is responsibility of the developer to configure an appropriate combination between configuration mode of cells (always necessary) and FT_mode (if FTS is enabled).

When a specific application needs to implement a processor with fault tolerance capacity, the developer has to set the configuration mode and the *FT_mode* by means of the FTCSR register, which also includes the next fields: *FT_enable* that enable the FTS. If it is not enabled, the other bits of FTS are not taken into account. The *FT_error_flag* indicates when the FTS has found an error while performing a comparison between two processors, this bit stops the execution of the programs in the cores and alerts the CCU to start the self-elimination and self-replication processes of damaged cells. The FT_redundant_cell bit indicates if the cell is a redundant cell, it must be set in the redundant cell for the FT_modes 5, 6, 7 and 8 only.

Example of an Application that Implements a FTS. Lets suppose a 32-bit sequence of data that has to be generated by a processor with a capacity for 250 instructions, which has to include a FTS to protect the reliability of the sequence. As a condition, the sequence has to be generated at a low speed, very much lower than the clock available, so it is necessary to implement a delay in the sequence, therefore this should be implemented in the processor of another cell with a capacity for at least 50 instructions.

The problem can be solved in many ways, the figure 3a shows a specific solution that includes two components (AAAA and BBBB). The cell identified as AAAA0001 includes the primary processor, which generates the sequence, and the cell AAAA0002 includes the redundant processor that generates the same sequence. These cells are configured in configuration mode 11 and in FT_mode 8, that allows the FTS of the primary cell making the comparison of the 8 cores (primary and redundant cells). The cell BBBB0000 contains a processor that performs the delay. This will be in mode 4, and its FTS will be disabled.

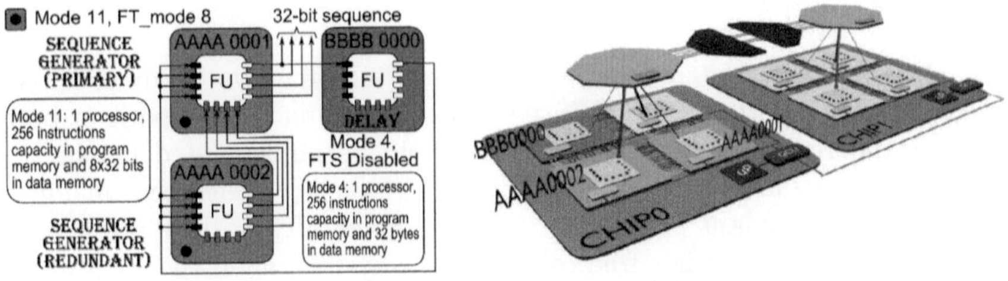

(a) Components configuration. (b) Prototype.

Fig. 3. FT application example

When the primary cell generates one data of the sequence, the four output ports of the cells are written, producing an RE flag in all of them. One of them is conducted to the cell BBBB0000, which waits this RE pulse from cell AAAA0001 to start the generation of a delay (this RE is received by mean of the special instruction called BLMOV "Blocked Move", which reads the port, saves the data in a memory location and follows with the next instruction when a RE pulse is produced.) The data read is not important for the cell that produces the delay (BBBB0000). This cell executes the delay algorithm, and then writes any data in its OUT0 port, which produces a RE pulse that has to be conducted to the four inputs (due to the 32-bit mode implemented) of the primary and redundant cell. When these cells receive the RE pulse they generate the next data of the special sequence, and so on. While the delay is performed, the primary and redundant processors could calculate the next data of the sequence in a parallel way.

If the primary or redundant processor have a hardware failure, the cells AAAA0001 and AAAA0002 are self-eliminated, the address FFFF0001 will be fixed in both cells, and they will not be used in subsequent self-placement operations. Its routing resources continue available. Next, the self-replication starts for the cells AAAA0001 and AAAA0002, it involves the self-placement and self-routing of this cells in another location inside the cell array. Once this process ends, the sequence starts the generation again.

At this point of the explanation, the reader could be thinking "and what happens if...?",turning the sample in a more elaborated and complex situation. Therefore, it is important to take into account that the system provides more cells available, therefore this application could be extended, even another entities that read the sequence could be created, so that these will be working in a parallel way.

Prototype Architecture. For demonstration purposes, the original architecture previously described has been modified for the construction of a prototype, due principally to the physical limitations in the FPGAs used for the implementation of the system (Figure 3b shows the location of cells in a prototype if placed in the order BBBB0000, AAAA0001 and AAAA0002.). The prototype has been developed in two chips, each one is a Virtex4 Xilinx FPGA (XC4VLX60), with an utilization rate close to 80% of their capacity. The system can have between 8 and 32 processors working in parallel, it depends of the configuration mode implemented.

4 Conclusion and Future Work

A Fault Tolerance System (FTS) for a novel self-adaptive hardware architecture has been reported. The FTS performs a continuous comparison between primary and redundant processors, that could be implemented in one or two cells, depending on the configured *FT_mode*. The cell (or cells) that includes the primary and redundant processors involved in the FTS will be self-eliminated and self-replicated to other free cell(s) by the system when the FTS detects a failure. This cell(s) will be self-discarded for future implementations.

As future work, we consider the implementation of a development system that provides a C-compiler to facilitate the implementation of applications.

References

1. AETHER Project Home, `http://www.aether-ist.org`
2. Soto, J., Moreno, J., Madrenas, J., Cabestany, J.: Implementation of a Dynamic Fault-Tolerance Scaling Technique on a Self-adaptive Hardware Architecture. In: Proceedings of the International Conference on ReConFigurable Computing and FPGAs, pp. 445–450 (2009)
3. Streichert, T., Koch, D., Haubelt, C., Teich, J.: Modeling and Design of Fault-Tolerant and Self-Adaptive Reconfigurable Networked Embedded Systems. Hindawi Publishing Corp., New York (2006)
4. Chen, Z., Yang, M., Francia, G., Dongarra, J.: Self Adaptive Application Level Fault Tolerance for Parallel and Distributed Computing. In: IEEE International Parallel and Distributed Processing Symposium, IPDPS 2007, pp. 1–8 (2007)
5. Macias, N., Durbeck, L.: Self-Assembling Circuits with Autonomous Fault Handling. In: Proceedings of the 2002 NASA/DoD Conference on Evolvable Hardware (EH 2002), pp. 46–55. IEEE Computer Society Press, Los Alamitos (2002)
6. Moreno, J., Thoma, Y., Sanchez, E.: POEtic: A Prototyping Platform for Bio-inspired Hardware. In: Proceedings of the 6th International Conference on Evolvable Systems (ICES), pp. 180–182
7. Moreno, J., Sanchez, E., Cabestany, J.: An In-System Routing Strategy for Evolvable Hardware Programmable Platforms. In: Proceedings of the Third NASA/DoD Workshop on Evolvable Hardware, pp. 157–166. IEEE Computer Society Press, Los Alamitos (2001)
8. Vu, T., Jesshope, C.: Formalizing SANE virtual processor in thread algebra. In: Butler, M., Hinchey, M.G., Larrondo-Petrie, M.M. (eds.) ICFEM 2007. LNCS, vol. 4789, pp. 345–365. Springer, Heidelberg (2007)

Systems with Slope Restricted Nonlinearities and Neural Networks Dynamics

Daniela Danciu and Vladimir Răsvan

Department of Automatic Control, University of Craiova
13, A.I. Cuza str., 200585 - Craiova, Romania
{daniela.danciu,vrasvan}@automation.ucv.ro

Abstract. The quite large standard class of additive neural networks is considered from the point of view of the qualitative theory of differential equations. Connections with the theory of absolute stability are pointed out and a new class of Liapunov functions is introduced, starting from the positiveness theory (Yakubovich-Kalman-Popov lemma). The results are valid for a quite large class of dynamical systems and they are tested on some neural network structures. In the concluding part some perspective research is mentioned, including synchronization and time-delay effects.

Keywords: Hopfield neural networks, KWTA networks, cellular neural networks, Liapunov function, absolute stability.

1 Introduction. State of the Art

A. In the last 10-15 years it has been understood finally that all neural networks, both natural and artificial, are characterized by two kinds of dynamics: the "learning dynamics" and the "intrinsic" dynamics. The first one is the sequential (discrete-time) dynamics of the choice of the synaptic weights. The intrinsic dynamics is the dynamics of the neural network (or of a technology system displaying a neural network structure) viewed as a dynamical system *after the weights have been established by learning*. The two dynamics are never considered simultaneously. The starting point should be, according to our point of view, [6,4,7] the fact that emergent computational capabilities of a neural network can be achieved provided it has many equilibria. Also the network task *is achieved* provided it approaches these equilibria. This assertion is valid e.g. for such networks as classifiers [19], content addressable networks [3], KWTA networks [2].

On the other hand, the best suited neural network structures for such tasks are the so-called recurrent neural networks containing feedback connections that may induce instabilities: this is valid for Hopfield neural networks and for cellular neural networks, too. This shows the importance of the second, intrinsic dynamics of the neural networks viewed as dynamical systems. These dynamical systems have their structure induced *a posteriori* by the learning process (the first dynamics) that has established the synaptic weights but, however, it is not compulsory that this *a posteriori* dynamics should have the required properties; therefore it appears that they have to be checked separately.

J. Cabestany, I. Rojas, and G. Joya (Eds.): IWANN 2011, Part II, LNCS 6692, pp. 565–572, 2011.

B. The basic concepts for qualitative behavior of dynamical systems arise from the theory of Liapunov and its subsequent development. The standard stability properties (Liapunov, asymptotic, exponential) are properties of a single equilibrium. Their counterpart for several equilibria are: mutability, global asymptotics, gradient behavior [11,17,21,4,7]. For the class of recurrent neural networks, the "best" qualitative behavior is the so-called *gradient behavior* whose significance is as follows: each state trajectory approaches asymptotically some equilibrium. This does not mean that all equilibria are attractive: on the contrary, the stationary (equilibria) set contains repellers but it is important for a neural network to have enough attractors.

C. The main analysis tool for these properties is the Liapunov function but, however, finding a suitable Liapunov function is still a matter of art. Fortunately, many standard recurrent neural networks e.g. bidirectional associative memories, Hopfield, cellular, Cohen-Grossberg, possess some "natural" (i.e. associated in a natural way) Liapunov functions which allow to obtain the required properties (for some of the applications of such natural Liapunov functions [16,21]). While the "natural" Liapunov functions – e.g. the Hopfield "energy function" and the Cohen-Grossberg function – are such that they can provide gradient-like behavior under sufficiently large assumptions on the synaptic weights, they have nevertheless some drawbacks. A quite restrictive assumption is that of a finite number of equilibria, since this is not always true, especially in degenerate cases or high-gain structures when standard (natural) Liapunov functions fail in giving the required answer [9]. These aspects together with the quasi-permanent task of the improvement for the analysis tools in order to obtain less restrictive conditions on the neural networks' parameters speak for a unique approach to be followed: to discover new Liapunov functions for the neural networks dynamics.

A specific feature of the neural network is represented by the sigmoidal functions that describe the neuron activation functions. The common feature of various sigmoids is that they are $\mathscr{C}^1$-functions, sector restricted, monotonically increasing, slope restricted, bounded and equal to zero at the origin. Such functions are those for which the standard theory of the absolute stability is constructed. This fact combined with the standard structure of the so-called *additive neural networks* allows application of all achievements of the absolute stability theory.

The standard sense of the notion of *absolute stability* is "global asymptotic stability of the equilibrium for all nonlinear (and linear) elements belonging to some class" (i.e. satisfying some sector and slope/monotonicity requirements). This (informal) definition clearly speaks about some robustness of this property with respect to some uncertainty (the lack of information about a specific nonlinear characteristic: all we know is that the nonlinearity is subject to some restrictions defining a class of nonlinear elements).

Due to the similarity with the classical absolute stability problem, our approach will be to use the achievements of this theory, mainly the positiveness (Yakubovich-Kalman-Popov) lemma in order to obtain new and general Liapunov functions which take into account monotonicity and slope restrictions of the nonlinearities. Such attempts have already been made in [19,5], where some existing Liapunov functions, obtained for nonlinear systems with slope restricted nonlinearities, have been adapted to systems with several equilibria. It was later discovered that for such rather sharp analysis tools, the properties of the Liapunov functions are rather dependent of the proof approach.

Moreover, older techniques (developed however for single nonlinearity systems) showed better than more recent ones. The research on this topics is in progress [22].

In this paper we want to show the application of these most recent results of [22] to neural networks in order to obtain the improvement of the gradient-like behavior conditions. Consequently, what remains of the paper is organized as follows: first a Liapunov function result is presented, containing only slope information about the nonlinear functions. This result has already been applied to neural networks [19,5] but the basics had their own drawbacks due to the technique of proof [15]. Those drawbacks have been up to now "covered" by the fact that the applications avoided "critical cases" (see [14,13]). The obtained Liapunov function together with the frequency domain inequality that generated it are applied to some neural networks and the paper ends with a review of possible extensions and future applications.

2 The Basic (Liapunov Like) Results

In this section we shall give the theoretical background from the stability theory in order to deduce the corresponding properties for the dynamics of the neural networks. The first instrument is a lemma of Liapunov type for systems with more than one equilibrium.

Lemma 1. *[11,17] Consider the system*

$$\dot{x} = f(x) \ , \ \dim x = \dim f = n \tag{1}$$

and its equilibria set $\mathscr{E} = \{c \in \mathbb{R}^n : f(c) = 0\}$. *Suppose there exists* $V : \mathbb{R}^n \mapsto \mathbb{R}$ *continuous with the following properties: i)* $V^*(t) = V(x(t))$ *is non-increasing along the solutions of (1); ii) if* $V(x(t)) \equiv const.$ *for some bounded on* $\mathbb{R}$ *solution of (1), then* $x(t) \equiv c$*; iii)* $\lim_{|x| \to \infty} V(x) = +\infty$. *Then system (1) has global asymptotics i.e. each solution approaches asymptotically (for* $t \to \infty$*) the stationary set* $\mathscr{E}$.

Let us remark, as in [11,17], that if $\mathscr{E}$ consists of isolated equilibria, then (1) is *gradient like* i.e. each solution approaches asymptotically some equilibrium that is $\lim_{t \to \infty} x(t, x_0) = c \in \mathscr{E}$.

We shall give now a stability theorem for systems with several slope restricted nonlinearities and several equilibria. These systems have a rather general dynamics that incorporates the dynamics of several types of neural networks [5,19,18,9]:

$$\dot{x} = Ax - \sum_1^m b_j \phi_j(c_j^* x) + h \tag{2}$$

where $h \equiv const.$

This model has already been used in [5] where the stability result was relying upon another result of [15] incorporating only slope information about the nonlinear elements. In [5] a certain static decoupling assumption had been relaxed in order to cope with the standard assumptions of neural networks. Here we shall give an improved version which is based on a more recent result [22]; this result (theorem) will be now formulated for systems with several equilibria.

Theorem 1. *Consider system (2) under the following basic assumptions: i)* $\det A \neq 0$*; ii)* (A,B) *is a controllable pair and* (C^*,A) *is an observable pair, where* B *is the* $n \times m$ *matrix with* b_j, $j=\overline{1,m}$ *as columns and* C^* *is the* $m \times n$ *matrix with* c_j^*, $j=\overline{1,m}$ *as rows; iii)* $\det C^*A^{-1}B \neq 0$*; iv) there exists* $\widetilde{\varphi}_i \in [\underline{\nu}^i, \overline{\nu}^i]$ *such that* $A - \sum_1^m b_i \widetilde{\varphi}_i c_i^*$ *is a Hurwitz matrix. Let the* $\mathscr{C}^1$*-functions* $\phi_j : \mathbb{R} \to \mathbb{R}$ *be such that satisfy the slope restrictions* $\underline{\nu}^j \leq \phi_j'(\nu) \leq \overline{\nu}^j$, $j=\overline{1,m}$*. Suppose there exist a set of nonnegative parameters* $\tau_j \geq 0$ *and a set of real parameters* $\theta_j \in \mathbb{R}$, $j=\overline{1,m}$ *such that the following matrix frequency domain inequality holds*

$$D_1 + Re\left(D_1(\underline{G}+\overline{G}) - \frac{D_2}{\imath\omega}\right)\widetilde{K}(\imath\omega) + \widetilde{K}^*(-\imath\omega)\underline{G}D_1\overline{G}\widetilde{K}(\imath\omega) \geq 0 \tag{3}$$

where D_1 *and* D_2 *are the diagonal matrices having as diagonal elements* $\tau_j \geq 0$ *and* θ_j, $j=\overline{1,m}$*, respectively and* $\widetilde{K}(\sigma) = C^*(\sigma I - A)^{-1}B$ *is the* $m \times m$ *matrix transfer function of the linear part of (2) i.e. the linear system*

$$\dot{x} = Ax + Bu(t)\ ,\ y = C^*x \tag{4}$$

Also, D_2 *is such that* $D_2(C^*A^{-1}B)^{-1}$ *is Hermitian. Assume also the following alternative is valid: either (3) is strict* (>0) *including for* $\omega \to \infty$ *and the slope restrictions are non-strict or (3) is non-strict* (≥ 0) *but then all* $\tau_j > 0$ *and the slope conditions are strict. If the nonlinear functions satisfy*

$$\liminf_{\lambda \to \infty} \frac{\theta_j}{\lambda^2}\left(\frac{1}{2}\lambda\phi_j(\lambda) - \int_0^\lambda \phi_j(\lambda)\mathrm{d}\lambda\right) \leq 0 \tag{5}$$

then system (2) has global asymptotics. If, additionally, all its equilibria are isolated, then it has a gradient-like behavior.

The proof of this theorem is very involved, clearly outside the aims and largely the size of the present paper. It may be done in several steps, following the hints of [5] and the line of [15] with the improvement of [22].

We give below some comments concerning the result. The main tool of the proof is the Liapunov function resulted from the frequency domain inequality *via* Yakubovich-Kalman-Popov lemma in the multivariable case [20]. Some additional explanation appears as useful and necessary. Using the information on the nonlinearities (sector and slope restrictions), an integral quadratic index is constructed and associated to (4). If (3) holds, the application of the Yakubovich-Kalman-Popov lemma ensures existence of some matrices allowing a certain writing of the integral index along the solutions of (4). Then the same integral index is viewed along the solutions of (2). Equating the two forms of the integral index, a state function is "revealed" ("discovered") which is non-increasing along the trajectories of (2). Consequently this state function is most suitable for a *candidate Liapunov function*. Since in the case of the neural networks the Liapunov function is useful for estimates of the attraction domains for stable equilibria, we reproduce it here:

$$\mathscr{V}(x) = (Ax - Bf(C^*x) + h)^* H (Ax - Bf(C^*x) + h) - \\ -\frac{1}{2}x^*CD_2(C^*A^{-1}B)^{-1}C^*x + \sum_1^m \theta_i \int_0^{c_i^*x} \phi_i(\lambda)\mathrm{d}\lambda \tag{6}$$

where $\theta_i \in \mathbb{R}$ are those of the frequency domain inequality (3) and the Hermitian (symmetric in the real case) matrix H is prescribed by the Lurie type matrix inequalities of the Yakubovich-Kalman-Popov lemma in the multivariable case. During the proof of the theorem the following inequalities are obtained for function's derivative:

$$\frac{\mathrm{d}}{\mathrm{d}t}\mathscr{V}(x(t)) \leq -\varepsilon|Ax(t) - Bf(C^*x(t)) + h|^2 \leq 0 \tag{7}$$

if (3) is strict and

$$\frac{\mathrm{d}}{\mathrm{d}t}\mathscr{V}(x(t)) \leq -\sum_1^m \tau_i \left(-\phi_i'(c_i^*x(t)) + \underline{\nu}^i\right)\left(-\phi_i'(c_i^*x(t)) + \overline{\nu}^i\right) \cdot \\ \cdot \left(c_i^*\left(Ax(t) - Bf(C^*x(t)) + h\right)\right)^2 \leq 0 \tag{8}$$

if (3) is non-strict. At its turn $\mathscr{V} : \mathbb{R}^n \mapsto \mathbb{R}$ is always subject to the following estimate from below

$$\mathscr{V}(x) \geq \delta_2|x|^2 - \delta_1|x| - \delta_0 + \sum_1^m \theta_i \left(\int_0^{c_i^*x} \phi_i(\lambda)\mathrm{d}\lambda - \frac{1}{2}(c_i^*x)\phi_i(c_i^*x)\right) \tag{9}$$

Combining (7) or (8) with (9) shows that (6) is satisfying the assumptions of Lemma 1. Application of this lemma gives the global asymptotics (also called quasi-gradient like behavior [17]) and, if the equilibria are isolated, the gradient like behavior.

A special comment is due to the assumption for the matrix $D_2(C^*A^{-1}B)^{-1}$ to be Hermitian. This is not a technical assumption but a necessary consequence of the special structure of the Lurie-like matrix equalities in this case. Several authors did not even remark it e.g. [14] but it is mentioned by [19,18]. In an earlier paper [15] this condition is replaced by a stronger one – the static decoupling of the linear part i.e. $C^*A^{-1}B$ is a diagonal matrix.

The Hermitian character of $D_2(C^*A^{-1}B)^{-1}$ with D_2 diagonal appears to follow as a special case of the matrix equation $A^*X = XA$, at its turn a special case of $AX = XB$ where A and B are square of different dimensions, hence X is rectangular (see [10], Chapter VIII).

The fact that this equation may be solved in the most general case (note that the solution is not unique) will appear as most useful in the case of the neural networks.

3 Applications to the Case of Neural Networks Dynamics

Following [19,18], we shall consider the case of the continuous time Hopfield type classification networks described by

$$\frac{dv_i}{dt} = -\frac{1}{C_i}\left(\frac{1}{R_i} + \sum_{j=1}^{n}\frac{1}{R_{ij}}\right)v_i + \frac{1}{C_i}\sum_{j=1}^{n}\frac{1}{R_{ij}}\phi_j(v_j)\ i = \overline{1,n} \tag{10}$$

This system has the form (2) but in a rather particular case with A a diagonal Hurwitz matrix, $C^* = I$, $B = -B_0\Lambda$ with B_0 diagonal and Λ – the synaptic matrix of the neural network. Since A is Hurwitz we may take $\widetilde{\varphi}_i = 0$ and if we take into account the sigmoidal character of the nonlinear functions, then clearly $\underline{v}^j = 0$. We have also

$$H(\sigma) = -(\sigma I - A)^{-1}B_0\Lambda \tag{11}$$

where $(\sigma I - A)^{-1}B_0$ is the diagonal matrix with the diagonal elements $(\sigma C_i + \frac{1}{R_i} + \sum_{j=1}^{n}\frac{1}{R_{ij}})^{-1}$. For sigmoidal functions we have $0 < \overline{v}^i < +\infty$, hence (3) can be multiplied by $\overline{G}^{-1}$. With the choice of $\tau_i \geq 0$, $\theta_i \geq 0$ (which is slightly different of that from [19,5]) namely

$$\tau_i = \delta_i C_i\ ,\ \theta_i = -\delta_i\left(\frac{1}{R_i} + \sum_{j=1}^{n}\frac{1}{R_{ij}}\right)\ ,\ \delta_i > 0 \tag{12}$$

and denoting by D_3 the diagonal matrix with the entries δ_i we obtain

$$\overline{G}^{-1}D_3 - \frac{1}{i\omega}(D_3\Lambda - \Lambda^* D_3) > 0. \tag{13}$$

Obviously, the inequality (13) is fulfilled if Λ – the synaptic matrix – is symmetric. This assumption is quite known in the theory of the neural networks. But we may use D_3 for the choice $D_3\Lambda = \Lambda^* D_3$ as previously mentioned. This may provide some relaxation of the stability conditions. The result reads as follow

Theorem 2. *Consider the Hopfield classification network described by (10) with $\phi_i : \mathbb{R} \mapsto \mathbb{R}$ being $\mathscr{C}^1$ sigmoidal functions verifying the slope restrictions $0 \leq \phi_j'(v) \leq \overline{v}^j < +\infty$, $j = \overline{1,n}$. If there exists some positive numbers $\delta_i > 0$, $i = \overline{1,n}$ such that the matrix with the entries δ_i/R_{ij}, $i, j = \overline{1,n}$ is symmetric and the equilibria are isolated, then the network is gradient like.*

In order to display the advances and the drawbacks of the criterion, we shall refer to some neural networks model and associated Liapunov functions of [12]. The Hopfield model above seems to be the most general used there. The so called Pineda model given by

$$\frac{\mathrm{d}x_i}{\mathrm{d}t} = -\alpha_i x_i + \beta_i g_i\left(\sum_{1}^{n} w_{ij}x_j + s_i\right) \tag{14}$$

may be reduced to a Hopfield one by the change of variables

$$u_i = \sum_{1}^{n} w_{ij}x_j + s_i \tag{15}$$

Note however that the standard Hopfield structure is obtained provided the synaptic weight matrix is symmetric. For the case of KWTA networks the structure of the neural network is the same but some specific elements connected with equilibria set properties [8] lead to a very particular structure of Λ which in any case is symmetric. More precisely the synaptic matrix is defined by $w_{ii} = a < 0, \forall i, w_{ij} = b > 0, \forall i \neq j$. The cellular neural networks may contain non-symmetric synaptic matrices, hence in this cases the improved criterion may be useful.

An additional problem is that the "natural" Hopfield energy function, may ensure the quasi-gradient like behavior without the symmetry assumption. For the gradient like behavior, however, the assumption of isolated equilibria is again required.

At the same time there is still room for criteria improving e.g. by making use of the multivariable extension of the criterion of Yakubovich from [23,24], see [22].

4 Some Conclusions

The development of the absolute stability criteria for systems with slope restricted nonlinearities allowed some relaxation of the stability conditions which turned to be useful for the case of neural networks dynamics. As shown in the applications part, there exist some possibilities to relax the symmetry conditions for the synaptic matrix.

In order to emphasize a possible field of research we refer here to the relatively recent paper [9] where there is considered the problem of modeling the neuron activation functions by the *piecewise linear functions*. These functions are also chosen to be monotonically increasing and they are already sector restricted, piecewise $\mathscr{C}^1$ and slope restricted, also zero at the origin. As pointed there [9] the use of such functions confers to the additive neural networks several advantages "among which we mention the capability to exactly locate the neural network's equilibrium points". This in turn has permitted to derive effective techniques to design neural networks for solving specific signal processing tasks. Unfortunately, as pointed also in [9] "there are fundamental and somewhat unexpected difficulties to analyze their complete stability". This speaks again for the importance of an adequate analysis of the second dynamics mentioned in the Introduction.

With respect to this, the application of the Yakubovich type dissipativeness criterion in the multivariable case [1] might eliminate the symmetry assumption for the matrix $D_2(C^*A^{-1}B)^{-1}$ hence, another significant relaxation of the stability criteria. This aspect, as well as the problem of the time delays and the synchronization, are also left for some future research.

References

1. Barbălat, I., Halanay, A.: Conditions de comportement "presque linéaire" dans la théorie des oscillations. Rev. Roum. Sci. Techn.-Electrotechn. et Energ. 19(2), 321–341 (1974)
2. Calvert, B.D., Marinov, C.A.: Another K-Winners-Take-All analog neural network. IEEE Trans. Neural Networks 11(4), 829–838 (2000)
3. Cohen, M.A.: The construction of arbitrary stable dynamics in nonlinear neural networks. Neural Networks 5, 83–103 (1992)

4. Danciu, D.: Dynamics of neural networks as nonlinear systems with several equilibria. In: Pazos, A., Sierra, A., Buceta, W. (eds.) Advancing Artificial Intelligence through Biological Process Applications, pp. 331–357. Medical Information Science Reference, IGI Global (2009)
5. Danciu, D., Răsvan, V.: On Popov-type stability criteria for neural networks. In: Elec. J. Qual. Theory Diff. Equ (Proc. 6th Coll. Qual. Theory Diff. Equ. QTDE), pp. 1–10 (1999), http://www.math.u-szeged.hu/ejqtde/6/623.pdf
6. Danciu, D., Răsvan, V.: Dynamics of neural networks - some qualitative properties. In: Sandoval, F., Prieto, A.G., Cabestany, J., Graña, M. (eds.) IWANN 2007. LNCS, vol. 4507, pp. 8–15. Springer, Heidelberg (2007)
7. Danciu, D., Răsvan, V.: Neural networks. equilibria, synchronization, delays. In: Dopico, J.R., Dorado, J., Pazos, A. (eds.) Encyclopedia of Artificial Intelligence, pp. 1219–1225. Information Science Reference, IGI Global (2009)
8. Danciu, D., Răsvan, V.: Gradient like behavior and high gain design of KWTA neural networks. In: Cabestany, J., Sandoval, F., Prieto, A., Corchado, J.M. (eds.) IWANN 2009. LNCS, vol. 5517, pp. 24–32. Springer, Heidelberg (2009)
9. Forti, M.: Some extensions of a new method to analyze complete stability of neural networks. IEEE Trans. Neural Networks 13(5), 1230–1238 (2002)
10. Gantmakher, F.R.: Theory of matrices, fourth augmented edn."Nauka" Publishing House, Moscow (1988) (in Russian)
11. Ghelig, A.K., Leonov, G.A., Yakubovich, V.A.: Stability of Stationary Sets in Control Systems with Discontinuous Nonlinearities. World Scientific, Singapore (2004)
12. Gupta, M.M., Jin, L., Homma, N.: Dynamic Neural Networks. IEEE Press & J. Wiley, N. Y (2003)
13. Haddad, W.M.: Correction to "absolute stability criteria for multiple slope-restricted monotonic nonlinearities". IEEE Trans. on Autom. Control 42(4), 591 (1997)
14. Haddad, W.M., Kapila, V.: Absolute stability criteria for multiple slope-restricted monotonic nonlinearities. IEEE Trans. on Autom. Control 40(2), 361–365 (1995)
15. Halanay, A., Răsvan, V.: Absolute stability of feedback systems with several differentiable non-linearities. Int. J. Systems Sci. 22(10), 1911–1927 (1991)
16. Halanay, A., Răsvan, V.: Applications of Liapunov Methods to Stability, Mathematics and its Applications, vol. 245. Kluwer Academic Publishers, Dordrecht (1993)
17. Leonov, G.A., Reitmann, V., Smirnova, V.B.: Non-local methods for pendulum-like feedback systems, Teubner Texte zur Mathematik, vol. 132. Teubner Verlag, Stuttgart-Leipzig (1992)
18. Noldus, E., Loccufier, M.: An application of Liapunov's method for the analysis of neural networks. Journ. of Comp.and Appl. Math. 50, 425–432 (1995)
19. Noldus, E., Vingerhoeds, R., Loccufier, M.: Stability of analogue neural classification networks. Int. Journ. Systems. Sci. 25(1), 19–31 (1994)
20. Popov, V.M.: Hyperstability of Control Systems. Editura Academiei & Springer-Verlag, Bucharest & Heidelberg (1973)
21. Răsvan, V.: Dynamical systems with several equilibria and natural Liapunov functions. Archivum mathematicum 34(1), 207–215 (1998)
22. Răsvan, V., Danciu, D., Popescu, D.: On absolute (robust) stability: slope restrictions and stability multipliers. Int. J. Rob. Nonlin. Contr. (2011) (submitted)
23. Yakubovich, V.A.: Frequency domain conditions of absolute stability and dissipativeness of control systems with a single differentiable element (in Russian). Doklady Akad. Nauk SSSR 160(2), 298–301 (1965)
24. Yakubovich, V.A.: The method of the matrix inequalities in the theory of stability for nonlinear controlled systems II. Absolute stability in the class of slope restricted nonlinearities (in Russian). Avtom. i telemekh. 29(4), 588–590 (1965)

Bio-inspired Systems. Several Equilibria. Qualitative Behavior

Daniela Danciu

Department of Automatic Control, University of Craiova
13, A.I. Cuza str., 200585 - Craiova, Romania
daniela.danciu@automation.ucv.ro

Abstract. This paper is a survey on the approaches and results concerning the encountered behaviors and the qualitative properties of a class of bio-inspired dynamical learning machines. An overview and some results of several qualitative behaviors encountered both in natural and artificial dynamical systems – including synchronization and the dynamics affected by time-delays are presented in the main part of the paper. Some conclusions and open problems will end the survey.

Keywords: neural networks, multiple equilibria, Liapunov function(al), Popov-like frequency domain inequality, synchronization, time-delays.

1 Introduction

There are two main categories of approaches for the issue of *bio-inspired systems*. One includes the research involving the understanding, modelling and/or extracting the basic features and properties of the mechanisms behind some specific biological behaviors such as movement coordination, learning, pattern recognition, associative memory, but also the emergent systems, swarm intelligence and so on. The other includes the research aimed to give some algorithms, methods, conditions to be fulfilled in order to finally implement the artificial bio-inspired systems for some specific tasks to be used for unfriendly environments and/or heavy/difficult/long-lasting actions, jobs, activities including also the research ones.

From the dynamical point of view, a specific feature of an important class of bio-inspired dynamical learning machines – the recurrent neural networks (RNN) – is that their state space consists of multiple equilibria, not necessary all stable. Thus the usual local concepts of stability are not sufficient for an adequate description. Accordingly, the analysis have to be done within both the frameworks of the *stability theory* and the *qualitative theory of systems with several equilibria*. The results give conditions to be fulfilled by the network parameters in order the NN to have the desirable dynamical properties. These conditions have to be checked after the functional design of the neural structure – stage which firstly aims the "global pattern formation" and not the qualitative properties as stability and a "good" global behavior of the AI device. There is a large amount of research contributions concerning the stability and global behavior of RNN without time-delays – see for instance the references of the surveys [5,11].

J. Cabestany, I. Rojas, and G. Joya (Eds.): IWANN 2011, Part II, LNCS 6692, pp. 573–580, 2011.

Another important behavior of the biological systems involves oscillations. For instance, the work [15] refers to the *synchronization* of the oscillatory responses with the time-varying stimuli, while in [16] there are discussed the rhythmic activities in the nervous system. A section of this paper will present a summary of the results of [3,6,9] for such behaviors in the case of the neural bio-inspired dynamical systems.

Time-delays occur both in natural and artificial neural networks and they have undesirable effects on the dynamics of systems leading to oscillations or instabilities. In the *Neuroscience* literature this topic has been studied since '90 from different points of view and frameworks: stability [18,13,4,6,7,8,9]; autooscillations [1] and forced oscillations [24].

The paper aims to be an overview on the approaches and results concerning the qualitative properties of the dynamical neural systems. The rest of the paper is organized as follows: next section introduces the basic notions and results of the qualitative theory of systems with several equilibria; the main part will give some results for RNN in what concerns both the stability and the global qualitative behavior and also the problems of both the synchronization and the time-delays; some conclusions and open problems will end the paper.

2 The Theoretical Background: Notions and Basic Results

We shall introduce in the sequel the basic concepts of the framework of the *qualitative theory of systems with several equilibria* [17]. Consider the system

$$\dot{x} = f(t,x) \tag{1}$$

with $f : \mathbb{R}_+ \times \mathbb{R}^n \mapsto \mathbb{R}^n$ continuous, at least in the first argument.

Definition 1. *a) Any constant solution of (1) is called equilibrium. The set of equilibria $\mathscr{E}$ is called stationary set. b) A solution of (1) is called convergent if it approaches asymptotically some equilibrium:* $\lim_{t\to\infty} x(t) = c \in \mathscr{E}$ *c) A solution is called quasi-convergent if it approaches asymptotically the stationary set $\mathscr{E}$:* $\lim_{t\to\infty} d(x(t),\mathscr{E}) = 0 \in \mathscr{E}$*, where $d(\cdot,\mathscr{E})$ denotes the distance from a point to the set $\mathscr{E}$.*

Definition 2. *The system (1) is called: a) monostable, if every bounded solution is convergent; b) quasi-monostable, if every bounded solution is quasi-convergent; c) gradient-like, if every solution is convergent; d) quasi-gradient-like, if every solution is quasi-convergent.*

Since there are also other terms designating the above qualitative behaviors (see [22] for comments) in the rest of the paper we shall use the following notions: a) *dichotomy* – all bounded solutions tend to the equilibrium set; b) *global asymptotics* – all solutions tend to the equilibrium set; c) *gradient-like behavior* – the set of equilibria is stable in the sense of Liapunov and any solution tends asymptotically to some equilibrium point.

The Liapunov-like results of [17] for systems with multiple equilibria are:

Lemma 1. *Consider the nonlinear system*

$$\dot{x} = f(x)\ ,\ x \in \mathbb{R}^n \tag{2}$$

and its equilibria set $\mathscr{E} = \{c \in \mathbb{R}^n : f(c) = 0\}$. *Suppose there exists* $V : \mathbb{R}^n \mapsto \mathbb{R}$ *continuous with the following properties: i)* $V^*(t) = V(x(t))$ *is non-increasing along the solutions of (2); ii) if* $V(x(t)) \equiv const.$ *for some bounded on* $\mathbb{R}$ *solution of (2), then* $x(t) \equiv c$. *Then the system (2) is dichotomic.*

Lemma 2. *If the assumptions of Lemma 1 hold and, additionally,* $\lim_{|x| \to \infty} V(x) = +\infty$, *then system (2) has global asymptotics i.e. each solution approaches asymptotically (for* $t \to \infty$*) the stationary set* $\mathscr{E}$.

Lemma 3. *If the assumptions of Lemma 2 hold and the set* $\mathscr{E}$ *is discrete (i.e. it consists of isolated equilibria only) then the system (2) is gradient-like (i.e. each solution approaches asymptotically some equilibrium that is* $\lim_{t \to \infty} x(t, x_0) = c \in \mathscr{E}$*).*

Remark 1. *(Moser [19]) Consider the rather general nonlinear autonomous system*

$$\dot{x} = -f(x) \ , \ x \in \mathbb{R}^n \tag{3}$$

where $f(x) = gradG(x)$ *and* $G : \mathbb{R}^n \mapsto \mathbb{R}$ *is such that: i)*$\lim_{|x| \to \infty} G(x) = +\infty$ *and ii) the number of its critical points is finite. Under these assumptions any solution of (3) approaches asymptotically one of the equilibria (which is also a critical point of G – where its gradient, i.e. f vanishes) and thus the system's behavior is gradient-like.*

As a conclusion of this section, embedding the issue of the dynamics of NN within the framework of the *theory of systems with multiple equilibria* give us a general overview concerning the requirements of a well-designed neural network: a) it has to have several (but finite number) fixed-point equilibrium states; b) the network has to be convergent, i.e. each solution of the dynamics has to converge to an equilibrium state. If the first condition is fulfilled, the second property is equivalent to a gradient-like behavior.

3 Dynamical Neural Networks with Multiple Equilibria

3.1 Cohen-Grossberg Competitive Networks

Consider the general case of the Cohen-Grossberg competitive neural network [2]

$$\dot{x}_i = a_i(x_i) \left[b_i(x_i) - \sum_1^n c_{ij} d_j(x_j) \right] \ , \ i = \overline{1,n} \tag{4}$$

with $c_{ij} = c_{ji}$. To system (4) it is associated the "natural" Lyapunov function [2,22]

$$V(x_1, \ldots, x_n) = \frac{1}{2} \sum_1^n \sum_1^n c_{ij} d_i(x_i) d_j(x_j) - \sum_1^n \int_0^{x_i} b_i(\lambda) d_i'(\lambda) d\lambda \tag{5}$$

The condition on its derivative along the solutions of (4)

$$W(x_1, \ldots, x_n) = -\sum_1^n a_i(x_i) d_i'(x_i) \left[b_i(x_i) - \sum_1^n c_{ij} d_j(x_j) \right]^2 \leq 0 \tag{6}$$

is fulfilled provided $a_i(\lambda) > 0$ and $d_i(\lambda)$ are monotone non-decreasing. If additionally, $d_i(x_i)$ are strictly increasing, then the set where $W = 0$ consists of equilibria only; according to Lemma 2 in this case the system has *global asymptotics*. Moreover, if $d_i(x_i)$ are strictly increasing functions, denoting $A_{ij}(x) = \frac{a_i(x_i)}{d_i'(x_i)}\delta_{ij}$, the system (4) may be written as $\dot{x} = -A(x)gradV(x)$ which is a *pseudo-gradient* system comparing to (3). Since the model of system (4) is more general and the stationary set consists of a large number of equilibria, the study of *gradient-like behavior* have to be done on particular cases.

3.2 K-Winners-Take-All – KWTA Neural Networks

The KWTA (K-Winners-Take-All) artificial NN are a special type of analog electrical circuit that selects K largest elements from a given set of N real numbers. Consider de KWTA networks structure with the form

$$T_i\dot{u}_i(t) = -u_i(t) - \sum_1^N v_{ij}g_j(\lambda u_j) + b_i \ , \ i = \overline{1,N} \tag{7}$$

The result of Danciu and Răsvan [10] concerning the global behavior of the system is based on the associated "natural" Liapunov function of the Cohen-Grossberg type

$$V(u_1,\dots,u_N) = \frac{1}{2}\sum_1^N\sum_1^N v_{ij}g_i(\lambda u_i)g_j(\lambda u_j) + \sum_1^N \int_0^{\lambda u_i}(\theta - b_i\lambda)g_i'(\theta)\mathrm{d}\theta \tag{8}$$

Theorem 1. *[10] Consider system (7) with $T_i > 0$, $v_{ij} \in \mathbb{R}$, $\lambda > 0$ and $g_i : \mathbb{R} \mapsto (-1,1)$ sigmoidal functions having the following properties: i) strictly monotonic and continuous; ii) $g_i(0) = 0$, $\lim_{\sigma\to\pm\infty} g_i(\sigma) = \pm 1$; iii) Lipschitz hence differentiable a.e. with integrable derivative. If the matrix of the weights $\Gamma = [v_{ij}]_{i,j=1}^N$ is symmetric, then the system is quasi-gradient like. If, additionally, the equilibria set $\mathscr{E}$ is discrete, then it is gradient like. Moreover, if the matrix of the weights Γ is nonnegative definite ($\Gamma \geq 0$) all equilibria are Liapunov stable and if $\mathscr{E}$ is discrete their stability is also asymptotic.*

3.3 Hopfield-Type Neural Networks

Consider the continuous time Hopfield-type classification networks described by

$$\frac{dv_i}{dt} = -\frac{1}{C_i}\left(\frac{1}{R_i} + \sum_{j=1}^n \frac{1}{R_{ij}}\right)v_i + \frac{1}{C_i}\sum_{j=1}^n \frac{1}{R_{ij}}\phi_j(v_j) \ , \ i = \overline{1,n} \tag{9}$$

The result of [12] written for the particular case of the Hopfield-type neural networks is

Theorem 2. *Consider the Hopfield classification network described by (9) with $\phi_i : \mathbb{R} \mapsto \mathbb{R}$ being $\mathscr{C}^1$ sigmoidal functions verifying the slope restrictions $0 \leq \phi_j'(v) \leq \overline{v}^j < +\infty$, $j = \overline{1,n}$. If there exists some positive numbers $\delta_i > 0$, $i = \overline{1,n}$ such that the matrix with the entries δ_i/R_{ij}, $i,j = \overline{1,n}$ is symmetric and the equilibria are isolated, then the network is gradient like.*

4 Dynamical Neural Networks Affected by Time-Delays

4.1 Synchronization for RNN with Time-Delays

Synchronization refers to a periodic or almost periodic external stimulus which has to be tracked. From the mathematical point of view, "the synchronization problem refers to the finding of some conditions ensuring existence and exponential stability of a unique global solution that has the features of a limit regime, i.e. it is not defined by the initial conditions and is of the same type as the stimulus is – periodic, or almost-periodic, respectively." [11]. With the usual notations, the mathematical model of Hopfield-type neural networks with time-delays and time-varying external stimuli has the form

$$\dot{x}(t) = -a_i x_i(t) - \sum_{j=1}^{n} w_{ij} f_j(x_j(t-\tau_{ij})) + s_i(t)\ ,\ i=\overline{1,n} \tag{10}$$

Due to the properties of the nonlinear sigmoidal functions, there are two possible approaches: the Liapunov method and Popov-like frequency domain inequalities.

A. The Liapunov-Like Approach results of the synchronization problem given in [9,6] are based mainly on the Liapunov-Krasovskii functional suggested by [20] and restricted to be only quadratic in the state variables: $V : \mathbb{R}^n \times \mathscr{L}^2(-\tau,0;\mathbb{R}^n) \to \mathbb{R}_+$

$$V(x) = \sum_{i=1}^{n} \left[\frac{1}{2}\pi_i x_i^2(0) + \sum_{j=1}^{n} \rho_{ij} \int_{-\tau_{ij}}^{0} x_j^2(\theta)d\theta \right] \tag{11}$$

with $\tau = \max_{i,j}\tau_{ij}$ and $\pi_i > 0$, $\rho_{ij} \geq 0$ some free parameters which have to be chosen.

Theorem 3. *[6] Consider system (10) under the following assumptions: i) $a_i > 0, \forall i = \overline{1,n}$; ii) $|s_i(t)| < M_i, \forall i = \overline{1,n}$; iii) the nonlinearities $f_i(\cdot)$ verify the Lipschitz condition*

$$0 \leq \frac{f_i(\sigma_1)-f_i(\sigma_2)}{\sigma_1-\sigma_2} \leq L_i\ ,\ \forall \sigma_1 \neq \sigma_2\ ,\ with f_i(0)=0; \tag{12}$$

If the synaptic weights w_{ij} are such that it is possible to choose $\pi_i > 0$ and $\rho_{ij} > 0$ for $\forall i,j = \overline{1,n}$ in order to satisfy

$$\begin{gathered}\alpha \in (0,a_i), \qquad \Delta_i = (a_i-\alpha)^2 - \sum_{j=1}^{n} \frac{(w_{ij}L_j e^{\alpha\tau_{ij}})^2}{\rho_{ij}} \sum_{j=1}^{n}\rho_{ji} > 0 \\ 2\frac{(a_i-\alpha)-\sqrt{\Delta_i}}{\sum_{j=1}^{n}\frac{(w_{ij}L_j e^{\alpha\tau_{ij}})^2}{\rho_{ij}}} < \pi_i < 2\frac{(a_i-\alpha)+\sqrt{\Delta_i}}{\sum_{j=1}^{n}\frac{(w_{ij}L_j e^{\alpha\tau_{ij}})^2}{\rho_{ij}}}\end{gathered} \tag{13}$$

then the system (10) has a unique global solution $\tilde{x}_i(t), i=\overline{1,n}$ which is bounded on $\mathbb{R}$ and exponentially stable. Moreover, this solution is periodic or almost periodic according to the character of $s_i(t)$ – periodic or almost periodic, respectively.

B. Popov-Like Frequency Domain Inequalities Approach

Theorem 4. *[3] Consider system (10) under the following assumptions: i) $a_i > 0, \forall i = \overline{1,n}$; ii) the nonlinearities $f_i(\cdot)$ verify the Lipschitz global condition (12); iii) $|s_i(t)| < M$, $i = \overline{1,n}$, $t \in \mathbb{R}$. If there exist $\theta_i \geq 0$, $i = \overline{1,n}$ such that the frequency domain condition*

$$\Theta L^{-1} + \frac{1}{2}[\Theta K(i\omega) + K^*(i\omega)\Theta] > 0 \; , \; \omega > 0 \tag{14}$$

holds (the star denoting the transpose and complex conjugation), then there exists a solution of (10) which is bounded on the whole real axis and it is periodic or almost periodic according to the character of the stimuli $s_i(t)$ – periodic or almost periodic, respectively. Moreover, this solution is exponentially stable.

The notations within the Theorem 4 are: $K(\sigma) = diag\left[(\sigma + a_i)^{-1}\right]_{i=1}^{n} \cdot \left[w_{ij}\mathrm{e}^{-\sigma\tau_{ij}}\right]_{i,j=1}^{n}$; $L^{-1} = diag\left[L_i^{-1}\right]_{i=1}^{n}$; $\Theta = diag[\theta_i]_{i=1}^{n}$.

4.2 RNN with Multiple Equilibria and Time-Delays

A. The Popov-like Results Using Comparison which are of interest here are concerned with dichotomy and global asymptotics. In order to apply the Popov comparison techniques and results [21], we have used the following time-delays neural system [23]

$$\dot{u}_i(t) = -u_i(t) - \sum_1^n w_{ij}h_j(u_j(t-\tau_j)) \; , \; i = \overline{1,n} \tag{15}$$

with the "general assumptions" that the nonlinearities h_j are uniformly Lipschitz and strictly increasing with constants $0 < \varepsilon < L_j < 1$ and $h_j(0) = 0$. It is shown that system (15) can be written as a "convolution system" of the form considered by Popov

$$u + f + W\kappa * h(u) = 0 \tag{16}$$

where the matrix kernel κ is diagonal with the diagonal entries

$$\kappa_j = \begin{cases} \mathrm{e}^{-(t-\tau_j)}, & t \geq \tau_j \\ 0 & , \; elsewhere \end{cases} \tag{17}$$

thus verifying the "regularly decreasing" condition (see the definition in [21]).

Theorem 5. *[23] Consider the neural system (15) with $W = [w_{ij}]_{i,j=1}^n$ symmetric and nonsingular and $h_i(\cdot)$ satisfying the "general assumption" from above. Then the system (15) is dichotomic. If there exists $\varepsilon \in (0,1)$ such that $1 - n\varepsilon|W| > 0$ and the set of equilibria is finite then the system (15) has global asymptotics.*

B. Admissible Time-Delays for Preserving the Gradient Behavior

Theorem 6. *[7] Consider the Hopfield neural network*

$$\dot{x}(t) = -a_i x_i(t) - \sum_{j=1}^n w_{ij} f_j(x_j(t)) + S_i \; , \; i = \overline{1,n} \tag{18}$$

be gradient-like with the nonlinear functions $f_i(\cdot)$ verifying the Lipschitz condition (12). If $W = [w_{ij}]_{i,j=1}^n$ is a symmetric doubly dominant matrix, then the time-delay network (10) with $s_i(t) = S_i = const.$ has a gradient-like behavior, as well as the network (18), provided that the delays are sufficiently small satisfying the following inequality

$$\max_i \tau_i \leq \frac{\min_i a_i}{(1+\sum_1^n L_i)(\sum_{i=1}^n \max_j |w_{ij}|)} \cdot \frac{1}{\sum_{j=1}^n L_j \left(a_j + \sum_{k=1}^n |w_{jk}|\right)}. \qquad (19)$$

5 Some Conclusions and Open Problems

The present paper emphasizes our point of view that for the intrinsic dynamics and goal achievement the best approach is to consider the neural networks (both natural and artificial) as dynamical systems with several equilibria. Moreover, the best qualitative behavior to aim at is the gradient behavior (or at least the quasi-gradient behavior).

Since the NN dynamics is described by nonlinear systems with sigmoidal nonlinearities, the methods for analysis are the equivalent approaches of the either Liapunov function(al) or Popov-like frequency domain inequalities. For systems described by ODE the equivalence takes place at the finite dimensional level. In the time-delay case the equivalence of the frequency domain inequalities and the Liapunov functionals is valid on a properly chosen Hilbert space. For this reason there are used *ad hoc* Liapunov functionals, more or less inspired by the theory of functional differential equations.

With respect to this problem, we maintain our opinion from [9] – the section entitled "The extension of the LaSalle like theory" – that an open problem is to obtain the counterparts of the Liapunov like lemmas given for systems with multiple equilibria (see Section 2 of this paper) within the LaSalle theory for the systems with time-delays. Section 4 of [9] gives an short overview and some basic results regarding this issue, recalling two facts: 1) a theorem of Barbašin-Krasovskii-LaSalle type already exists for time-delay systems and 2) Liapunov-like lemmas of the type of Lemma 1 in [9] are easy to obtain within the framework of the general LaSalle invariance principle. All this speaks for the advantages of the Liapunov functional approach. We are thus "pushed back" to the art of "guessing" a suitable Liapunov functional. With respect to this we like to draw the attention to the approach of [14].

The technique introduced in [14] starts from a prescribed derivative of the Liapunov functional which may be thus taken in order to vanish on the equilibria set only. The result is a quadratic functional satisfying a quadratic estimate from below. We may then modify it – in order to make it suitable for nonlinear systems with sector restricted nonlinearities – by making use of an approach that goes back to I. G. Malkin and was already used in our research – see [12] and some of its references.

References

1. Bélair, J., Campbell, S., van den Driessche, P.: Frustration, stability and delay induced oscillations in a neural network model. SIAM J. Appl. Math. 56, 254–265 (1996)
2. Cohen, M.A., Grossberg, S.: Absolute stability of pattern formation and parallel storage by competitive neural networks. IEEE Trans. of Syst., Man, Cyber. 13, 815–826 (1983)

3. Danciu, D.: Qualitative behavior of the time delay Hopfield type neural networks with time varying stimulus. Annals University of Craiova, Series: El. Eng (Automatics, Comp., Electronics) 26(1), 72–82 (2002)
4. Danciu, D.: Systems with several equilibria. Applications to the neural networks. Control Engineering, Universitaria Publ. House, Craiova, Romania (2006) (in Romanian)
5. Danciu, D.: Dynamics of neural networks as nonlinear systems with several equilibria. In: Pazos, A., Sierra, A., Buceta, W. (eds.) Advancing Artificial Intelligence through Biological Process Applications, pp. 331–357. Medical Information Science Reference, IGI Global (2009)
6. Danciu, D., Ionete, C.: Synchronization problem for time-delay recurrent neural networks. In: Proc. 8th IFAC Workshop on Time Delays Systems TDS 2009 (2009)
7. Danciu, D., Răsvan, V.: Gradient-like behaviour for Hopfield-type neural networks with delay. In: Proc. 3rd International Workshop on Intelligent Control Systems ICS 2001, Bucharest, Romania, pp. 20–24. Printech (2001)
8. Danciu, D., Răsvan, V.: Stability results for cellular neural networks with time delays. In: Cabestany, J., Prieto, A.G., Sandoval, F. (eds.) IWANN 2005. LNCS, vol. 3512, pp. 366–373. Springer, Heidelberg (2005)
9. Danciu, D., Răsvan, V.: Dynamics of neural networks - some qualitative properties. In: Sandoval, F., et al. (eds.) IWANN 2007. LNCS, vol. 4507, pp. 8–15. Springer, Heidelberg (2007)
10. Danciu, D., Răsvan, V.: Gradient like behavior and high gain design of KWTA neural networks. In: Cabestany, J., et al. (eds.) IWANN 2009. LNCS, vol. 5517, pp. 24–32. Springer, Heidelberg (2009)
11. Danciu, D., Răsvan, V.: Neural networks. equilibria, synchronization, delays. In: Dopico, J.R., Dorado, J., Pazos, A. (eds.) Encyclopedia of Artificial Intelligence, pp. 1219–1225. Information Science Reference, IGI Global (2009)
12. Danciu, D., Răsvan, V.: Systems with slope restricted nonlinearities and neural networks dynamics. In: Cabestany, J., Rojas, I., Joya, G. (eds.) IWANN 2011, Part II. LNCS, vol. 6692, pp. 565–572. Springer, Heidelberg (2011)
13. Gopalsamy, K., He, X.: Stability in asymmetric Hopfield nets with transmission delays. Physica D (76), 344–358 (1994)
14. Kharitonov, V., Zhabko, A.: Lyapunov-Krasovskii approach to the robust stability analysis of time-delay systems. Automatica 39, 15–20 (2003)
15. König, P., Schillen, J.: Stimulus dependent assembly formation of oscillatory responses: I. Synchronization. Neural Computation (3), 155–166 (1991)
16. Koppell, N.: We got the rhythm: dynamical systems of the nervous system. Notices AMS (47), 6–16 (2000)
17. Leonov, G.A., Reitmann, V., Smirnova, V.B.: Non-local methods for pendulum-like feedback systems, Teubner Texte zur Mathematik, vol. 132. Teubner Verlag, Stuttgart-Leipzig (1992)
18. Marcus, C., Westervelt, R.: Stability of analog neural networks with delay. Physical Review A 39, 347–359 (1989)
19. Moser, J.: On nonoscillating networks. Quarterly Applied Mathematics 25, 1–9 (1967)
20. Nishimura, M., Kitamura, S., Hirai, K.: A Lyapunov functional for systems with multiple non-linearities and time lags. Technology Repts. Osaka Univ. 19, 83–88 (1969)
21. Popov, V.: Monotonicity and mutability. J. Diff. Eqs. 31, 337–358 (1979)
22. Răsvan, V.: Dynamical systems with several equilibria and natural Liapunov functions. Archivum mathematicum 34(1), 207–215 (1998)
23. Răsvan, V., Danciu, D.: Neural networks - global behavior versus delay. Sci. Bulletin Politehnica Univ. of Timisoara, Trans. Autom. Contr. and Computer Sci. 49(63), 11–14 (2004)
24. Yi, Z.: Global exponential stability and periodic solutions of delay Hopfield neural networks. Int. J. Syst. Sci. 27(2), 227–231 (1996)

Biologically Inspired Path Execution Using SURF Flow in Robot Navigation

Xavier Perez-Sala[1], Cecilio Angulo[1], and Sergio Escalera[2]

[1] CETpD-UPC. Technical Research Centre for Dependency Care and Autonomous Living, Universitat Politècnica de Catalunya,
Neàpolis, Rambla de l'Exposició, 59-69, 08800 Vilanova i la Geltrú, Spain
xavier.perez-sala@upc.edu, cecilio.angulo@upc.edu

[2] MAiA-UB. Dept. Matemàtica Aplicada i Anàlisi, Universitat de Barcelona,
Gran Via de les Corts Catalanes 585, 08007 Barcelona, Spain
sergio@maia.ub.es

Abstract. An exportable and robust system using only camera images is proposed for path execution in robot navigation. Motion information is extracted in the form of optical flow from SURF robust descriptors of consecutive frames, so the method is called SURF flow. This information is used to correct robot displacement when a straight forward path command is sent to the robot, but it is not really executed due to several robot and environmental concerns. The proposed system has been successfully tested on the legged robot Aibo.

Keywords: Robot navigation, Path execution, Optical flow, SURF.

1 Introduction

Navigation for autonomous mobile robots, for any kind of platform and independently to its task, implies to solve two related problems: path planning and path execution. Path planning can be defined as a high level robot guidance from a place to another place, while path execution refers to low level processes needed to fulfill path planning decisions [16]. This work is about, given a certain path plan, how to ensure path execution when the only available information for the robot is data extracted from its on-board camera. Especially, no landmarks in the environment will be considered.

Unexpected robot behaviours can be observed during path execution when a system is asked for reaching a place or set point, though it acted properly in simulated or ideal conditions. Failures in path execution, even for simple path executions like a 'go straight forward' path command, are due to several reasons: noise in the sensors, damages in the actuators, perturbations, model errors or shocks. Consequently, a feedback control would be interesting to be implemented to correct the robot from possible motion deviations.

A common approach for obtaining feedback is to consider some landmarks in the environment that help the robot to be localized in [15,16]. However, for a general solution, no landmark should be considered, and no exact final place in

J. Cabestany, I. Rojas, and G. Joya (Eds.): IWANN 2011, Part II, LNCS 6692, pp. 581–588, 2011.

the path where to arrive exist, which could act like a landmark. Another solutions focus on constrain robot motion and camera localization on the robot in order to obtain robot egomotion [2,4,5]. Since nor robot configuration, neither camera localization will be constrained, but be placed in the front direction, egomotion can not be considered. The general problem at hands is to ensure the execution of a 'go straight forward' path command by a general mobile robot, when frames from the on-board frontal camera is the only available information.

Our proposed approach, like those based on optical flow [2], will use consecutive frames from the on-board robot camera to extract an approximation of the displacement direction by observing 2-D displacements of brightness patterns in the image. However, unlike standard solutions, the robot direction will be computed online by extracting the so-called *SURF flow*, i.e. motion information from SURF robust descriptors of consecutive frames of image sequences provided by the robot camera. This knowledge will be the only one needed to close the control loop, and to achieve the desired straight forward movement.

Otical flow is a measure closely related with motion field [1], i.e. the projection of 3-D relative velocity vectors of the scene points onto the 2-D image plane. During a frontal displacement, motion field shows a radial configuration: vectors radiate from a common origin, the *Vanishing Point* (VP) of the translation direction. In particular, forward displacements generate vectors point away from this point, named *Focus Of Expansion* (FOE), else it is named *Focus Of Contraction* (FOC). It is proposed in this work to achieve straight forward control for mobile robots by maintaining the FOE in the center of the SURF flow.

The remaining work is organized as follows: the state of the art about robot navigation using optical flow is introduced in Section 2. Section 3 describes the solution proposed for the straight forward robot motion. In Section 4, experiments are described and results are discussed. Finally, possible improvements and further reserarch lines are listed in Section 5.

2 Related Work

Biological principles of insect vision [7,11] have inspired vision-based solutions in robot navigation for obstacle avoidance. Insects extract qualitative 3-D information using image motion to avoid obstacles. Vision-based control techniques try to balance the optical flow divergences betwen eyes/sides of the image. In [8], an approach from ecological psychology was presented to avoid obstacles based on the visual field with the lowest time to contact. As indicated in [6], qualitative measures of flow field divergence are a reliable indicator of the presence of obstacles. In the same way, it has been proposed [10] and demonstrated [9] that humans use optical flow to perceive translational direction of self-motion: radial patterns generated by the optical flow during frontal movement guide human locomotion.

Besides qualitative information, motion field can provide more accurate measurements. It is possible to estimate the relative motion between camera and scene, i.e. egomotion, by considering some hard assumptions. In [2], constraints are met and optical flow is used as an approximation of motion field to compute translational and angular velocities of a car. Egomotion can also be used to localize the robot in the environment. In [4,5], the navigation task is divided in three phases: localization, path finding, and path execution. Optical flow is used to correct localization. In [4], odometry computed from wheel encoders is improved with an inaccurate egomotion, computing vehicle speed from optical flow. In [5], better results are presented from visual odometry, and localization is made only using egomotion. However, for path execution, our goal, global localization is a hard task to be avoided. Hence, a system is described in [3] allowing a wheeled robot to drive through the center of a corridor by controlling the steering angle. Robot navigates aligning the camera to the wall, at a certain distance, only using a rigidly mounted camera.

Using steering angle as control signal, a novel method will be proposed to detect translational direction without global localization (egomotion) or relative references (landmarks or a wall). Mimicking the human use of optical flow, steering angle will be calculated from radial patterns around the vanishing point (FOE in our case) that optical flow generates during translational movements. Several works exist where FOE is located from optical flow, but none of them use it as a feedback signal to correct robot navigation. For pure translation displacements, FOE calculation is completely described in [1]. Else, when the rotational component is non-zero, optical flow vectors will not intersect on FOE. However, it is the most trivial method to compute FOE, as it was pointed out in [14], where FOE is computed for locomotion control using an Artificial Neural Network, but it was never implemented for this goal. A simple method to solve rotations was introduced in [13] by discounting arbitrary rotations and applying the method for pure translation. However, it is claimed in [12] that navigation methods using optical flow are usually based on unrealistic assumptions about the scene, and unrealistic expectations about the capabilities of motion estimation techniques. Better results could be obtained by directly determining general qualitative properties of the motion structure (FOE computation), instead of a precise analysis of rotational parameters.

3 Robot Navigation Control

A method to control the path execution during the navigation of mobile robots is introduced. A closed loop is implemented to control straight forward displacements, with feedback signal extracted from robot camera images. Proposed procedure is composed by three steps: firstly, motion information is extracted from consecutive frames through SURF flow computation. Next, instantaneous direction of translation is computed by finding the Focus Of Expansion (FOE) from SURF flow vectors. Finally, control loop is closed, maintaining constant the direction of translation. Hence, straight forward displacements are ensured without the use of egomotion, odometry information is omitted, robot localizationt is avoided, and computational resources are dedicated to achieve reliable orientation measurements for the control module.

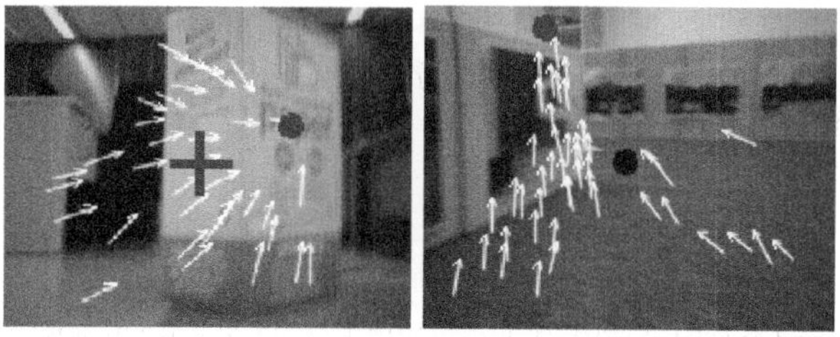

(a) (b)

Fig. 1. (a) Error signal depends on distance from the center of the image to the VP (b) Average Vanishing Point computation. Red-top point represents the current vanishing point (VP_k), and blue-centred point is the averaged one (AVP_k).

Procedure 1. Vision-based navigation control at instant k

Input: Current image I_k from the camera (Fig. 1(b)), number of frames taken during a robot step h, horizontal camera resolution res_x, horizontal opening angle oa_x and set point in pixels $sp_x^p = res_x/2$

Output: Steering angle: $e_{x_k}^o$

loop
- Compute SURF descriptors and keypoint locations: P_k
- Find correspondences between P_k and P_{k-1}: M_k
- Compute intersections of motion vectors M_k: C_k
- Estimate Vanishing Point from highest density region in C_k: (VP_{x_k}, VP_{y_k})
- Apply temporal filter using h last VP: $\mathrm{AVP}_k = \frac{1}{h+1}\sum_{i=k-h}^{k} \mathrm{VP}_i$
- Compute horizontal error in pixels: $e_{x_k}^p = AVP_{x_k} - sp_x^p$
- Transform error $e_{x_k}^p$ to angles: $e_{x_k}^o = e_{x_k}^p\,(oa_x/res_x)$

end loop

3.1 Feedback Control

To achieving a straight forward displacement, the robot motion target will be to hold the same orientation during all the path execution. From the camera point of view, this target is similar to hold the vanishing point in the center of the image (Fig. 1(a)). The error signal to close the loop will be calculated from video signal feedback, by computing distance between VP and the actual center of the image. Since the control variable will be the steering angle, only horizontal component of distance will be used to define it.

3.2 Vanishing Point

During frontal displacements, motion field displays a radial vector configuration around a common origin, the vanishing point of the translation direction. Motion field is not a directly accessible measure, but it is closely related with optical flow, under certain circumstances [2]: (1) robot moves on a flat ground, with (2) on-board camera translating in parallel to the ground and (3) its angular velocity is perpendicular to the ground plane. For general robots like that used in this work, nevertheless, constraints do not meet. The Sony Aibo robot is a quadruped robot with a camera on its "nose". Thus, image sequence are more instable than those provided by a wheeled vehicle, with a camera mounted rigidly on its structure. Image instability is due to neck joints, causing head vibrations transmitted to the camera, and specially, for robot walking. Legged robot steps produce very different movements than wheeled robot displacements, usually smoother than Sony Aibo gait. Walk behaviour in our experiments generates vertical and left-right pendular movements, i.e. camera suffers simultaneous roll and pitch rotations. Only the first assumption could be fulfilled in this case.

The hardest assumption of our approach is made at this point. Since Aibo robot gait is symmetric and periodic, restrictions two and three can be assumed as satisfied 'in average' and they will be extrapoled, during robot displacements, for instantaneous translation. Therefore, Sony Aibo gait deviations will be considered like shocks and vibrations which the controller will correct. As shown in Section 4, our qualitative approach is enough to control the desired legged robot navigation. A temporal filter is performed to compute VPs as averaged during robot gait. The Averaged Vanishing Point (AVP), described in Algorithm 1, is the point from which is computed the steering control.

As it was pointed out, calculated optical flow vectors do not converge to an unique point (FOE), even when assumptions are met. Hence, VP has been extracted by clustering intersections, since they form a cloud around VP.

3.3 SURF Flow

SURF flow is defined as 2-D displacements of SURF patterns in the image, where SURF is referred to Speeded Up Robust Features [17]. It is the field resulting from correspondences between SURF keypoints from consecutive frames in a video sequence. Unlike optical flow or the more similar SIFT flow [19], SURF flow is not a dense flow. It is only performed between high confidence keypoints in the image, selected by using a multi-scale Hessian detector to find image corners. SURF flow computation is faster than SIFT flow, since correspondences are only searched for a few hundreds of keypoints in each image (depending on the image texture), and corner detection and SURF description are computed using Haar wavelets on the integral image representation. Result of this correspondence is shown in Fig. 2(a) and Fig. 2(b).

Moreover, an image correspondence post-processing is applied in order to achieve better VP computation. This refinement, shown in Fig. 2(c), takes place once SURF flow is extracted and an estimation of VP is computed (see Section 3.2). It consists on search for better correspondences for each keypoint in current image, looking for similar SURF descriptors in a restricted area of previous image. This search area is defined by the triangle ABC, where vertex A is the keypoint in current image, the middle point of edge BC is the estimated VP and angle $\widehat{BAC}$ defines the search range. Once correspondences are refined, VP is computed again, using the same process described above.

Method effectiveness depends, as usual, on assuming that keypoints are found in images, i.e. a textured environment exists. In fact, typical human-hand scenes

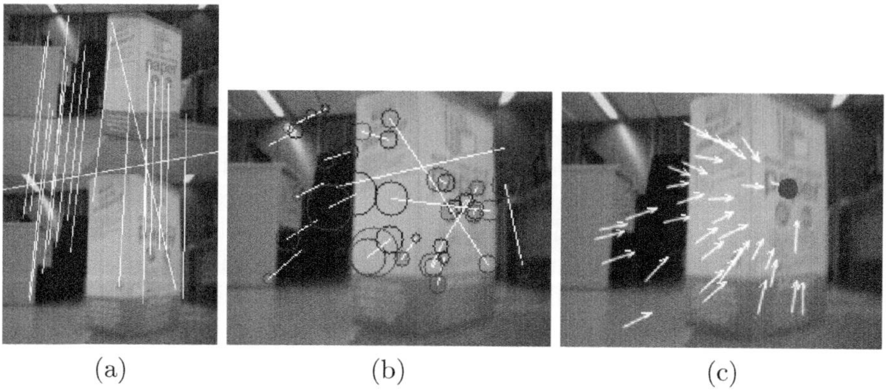

Fig. 2. (a) Keypoint correspondences between consecutive images (b) Motion vectors in the newest image (c) Refined motion vectors with the correspondent vanishing point

have enough corners for achieve SURF flow performance. On the other hand, SURF flow is robust to optical flow methods' limitations [20]: brightness constancy, temporal persistence or "small movements", and spatial coherence.

4 Results and Discussion

Results presented in this work are obtained using a Sony Aibo ERS-7 robot wirelessly communicated with a standard dual-core PC. Experiments are performed using the robot for environment interaction and the computer for hard computation processing. Path execution has been divided in reactive collision avoidance and straight forward control. Obstacle avoidance procedure is performed on-board, as a reactive behaviour using the robot infrared sensor, and computation to go straight forward is executed in the external computer. Sony Aibo camera captures the image and it is sent to the PC every 100ms, through wireless connection. Application running on the computer, first of all, extracts SURF flow from consecutive frames; then, the VP of the translation direction and the steering angle are computed; and finally, walking direction is sent to Sony Aibo. Gait behaviour for the robot is based on the Tekkotsu software[1].

Experiments are performed in an artificial grass surface of about $4m^2$, containing two crossing corridors. It is a natural scenario without artificial landmarks and small variability of the light level. To allow a future development in unstructured environments, corridor walls are wallpapered with pictures of real halls and corridor walls; providing enough textures to the system to ensure the correct performance of image processing algorithms. Used image resolution is 208×159 pixels.

In order to achieve qualitative results of the system performance in different relative positions between the robot and walls, 8 representative starting positions and orientations are chosen around the scenario, equally distributed, and 5 trials

[1] `http://www.tekkotsu.org/`

Fig. 3. Navigation sequence in open loop control

Fig. 4. Navigation sequence, with straight forward control

are launched for each one. Results show the difference between non-controlled straight forward behaviour and the controlled one. In open loop control, due to their mechanical characteristics, robot walks drawing a curve (Fig. 3). When feedback control is applied, Sony Aibo robot goes successfully straight forward (Fig. 4), correcting faulty displacements and performing the desired behaviour.

Some problems with wireless connection are observed, and sometimes image is not sent at time from robot to computer. When it occurs in consecutive images, it produces large oscillations, which can be corrected or not, depending on the number of frames lost. If problem persist, it can produce uncontrolled behaviours. A precise study about the maximum number of lost images supported should be completed depending on the last order sent by the computer, which will be repeated during all the non-informed period.

5 Conclusions and Future Work

In this work it is proposed a biological inspired vision-based navigation control to walk straight forward in a reliable way. Moreover, implementation is exportable to other robotic platforms with different configurations. Results shown that objectives introduced in this work have been accomplished without the use of artificial landmarks, taking into account some assumptions about the robot movement. Since Aibo's camera suffers simultaneous roll and pitch rotations during the robot gait, future work will avoid the hardest assumption proposed. The robot will correct its trajectory using motor information.

Moreover, shocks and vibrations suffered by the camera will be compensated by tacking in account robot configuration. Future work will be an improving of the system presented in this work, to be used in legged robots. In [2], motion field is formulated supposing an error component due to shocks and vibrations. Nevertheless, motion field error in x and y axis are roughly estimated. At this point, we are in an advantageous position, because it is assumed that our shocks and vibrations are movements resulting to the quadruped robot gait, and these movements are possible to be modelled through direct kinematics. Other improvements include decreassing sampling rate and the duration of actions.

References

1. Trucco, E., Verri, A.: Introductory Techniques for 3-D Computer Vision. Prentice Hall PTR, Upper Saddle River (1998)
2. Giachetti, A., Campani, M., Torre, V.: The Use of Optical Flow for Road Navigation. IEEE Trans. on Robotics and Automation 14, 34–48 (1998)
3. Dev, A., Krose, B., Groen, F.: Navigation of a mobile robot on the temporal development of the optic flow. In: IEEE/RSJ International Conference on Intelligent Robots and Systems, vol. 2, pp. 558–563 (1997)
4. Nagatani, K., Tachibana, S., Sofne, M., Tanaka, Y.: Improvement of odometry for omnidirectional vehicle using optical flow information. In: IEEE/RSJ International Conference on Intelligent Robots and Systems, vol. 1, pp. 468–473 (2000)
5. Sorensen, D.K., Smukala, V., Ovinis, M., Lee, S.: On-line optical flow feedback for mobile robot localization/navigation. In: IEEE/RSJ International Conference on Intelligent Robots and Systems, vol. 2, pp. 1246–1251 (2003)
6. Nelson, R.C., Aloimonos, J.: Obstacle avoidance using flow field divergence. Trans. on Pattern Analysis and Machine Intelligence 11, 1102–1106 (2002)
7. Srinivasan, M., Chahl, J., Weber, K., Venkatesh, S., Nagle, M., Zhang, S.: Robot navigation inspired by principles of insect vision. Robotics and Autonomous Systems 26, 203–216 (1999)
8. Duchon, A.P., Warren, W.H.: Robot navigation from a Gibsonian viewpoint. In: IEEE International Conference on Systems, Man, and Cybernetics, 'Humans, Information and Technology', vol. 3, pp. 2272–2277 (1994)
9. Warren, W.H., Kay, B.A., Zosh, W.D., Duchon, A.P., Sahuc, S.: Optic flow is used to control human walking. Nature Neuroscience 4, 213–216 (2001)
10. Warren, W.H., Hannon, D.J.: Direction of self-motion is perceived from optical flow. Nature 336, 162–163 (1988)
11. Santos-Victor, J., Sandini, G., Curotto, F., Garibaldi, S.: Divergent stereo in autonomous navigation: From bees to robots. International Journal of Computer Vision 14, 159–177 (1995)
12. Thompson, W.B., Kearney, J.K.: Inexact Vision. In: Proc. Workshop on Motion: Representation and Analysis, pp. 15–22 (1986)
13. Negahdaripour, S., Horn, B.K.P.: A Direct Method for Locating the Focus of Expansion. Computer Vision, Graphics, and Image Processing 46, 303–326 (1989)
14. Branca, A., Stella, E., Attolico, G., Distante, A.: Focus of Expansion estimation by an error backpropagation neural network. Neural Computing & Applications 6, 142–147 (1997)
15. Yoon, K., Jang, G., Kim, S., Kweon, I.: Color landmark based self-localization for indoor mobile robots. J. Control Autom. Syst. Eng. 7(9), 749–757 (2001)
16. Deng, X., Milios, E., Mirzaian, A.: Landmark selection strategies for path execution. Robotics and Autonomous Systems 17, 171–185 (1996)
17. Herbert, B., Tinne, T., Luc, V.: Surf: Speeded up robust features. In: Leonardis, A., Bischof, H., Pinz, A. (eds.) ECCV 2006. LNCS, vol. 3951, pp. 404–417. Springer, Heidelberg (2006)
18. Lowe, D.G.: Object recognition from local scale-invariant features. In: IEEE International Conference on Computer Vision, USA, vol. 2, p. 1150 (1999)
19. Liu, C., Yuen, J., Torralba, A., Sivic, J., Freeman, W.: SIFT flow: dense correspondence across different scenes. In: Forsyth, D., Torr, P., Zisserman, A. (eds.) ECCV 2008, Part III. LNCS, vol. 5304, pp. 28–42. Springer, Heidelberg (2008)
20. Lucas, B.D., Kanade, T.: An iterative image registration technique with an application to stereo vision. In: IJCAI, vol. 3, pp. 674–679. Citeseer (1981)

Equilibrium-Driven Adaptive Behavior Design

Paul Olivier and Juan Manuel Moreno Arostegui

Technical University of Catalunya (UPC), Department of Electronic Engineering,
Campus Nord, Building C4, c/Jordi Girona 1-3, 08034, Barcelona, Spain
paul.olivier@upc.edu

Abstract. In autonomous robotics, scalability is a primary discriminator for evaluating a behavior design methodology. Such a proposed methodology must also allow efficient and effective conversion from desired to implemented behavior. From the concepts of equilibrium and homeostasis, it follows that behavior could be seen as driven rather than controlled. Homeostatic variables allow the development of need elements to completely implement drive and processing elements in a synthetic nervous system. Furthermore, an autonomous robot or system must act with a sense of meaning as opposed to being a human-command executor. Learning is fundamental in adding adaptability, and its efficient implementation will directly improve scalability. It is shown how using classical conditioning to learn obstacle avoidance can be implemented with need elements instead of an existing artificial neural network (ANN) solution.

Keywords: homeostasis synthetic nervous system need element autonomous robotics classical conditioning behavior design methodology

1 Introduction

Developing autonomous robotic systems has many challenges. For one, it is not unrealistic to foresee that such systems will exhibit a large set of behaviors with a significant capacity for adaptation. Therefore, when evaluating technologies and methods for behavior design, a fundamental aspect is scalability. Reaching a significant level of scalability requires that the fundamental elements used in behavior design be simple yet powerful in meeting our demands for complex behavior. (Note that "elements" refer here to the building blocks of a synthetic nervous system, similar to neurons in a biological nervous system.) But will such a design approach based on simple elements be effective in realizing the desired behavior? The trend over the last few decades in favor of artificial evolution and ANN's for nervous system design, suggests that such elements either do not exist or did not stand the test of time. Methods of behavior design not based on artificial evolution and ANN's tend to be only reactive, that is, learning is absent. Yet, where the intention is to operate in unpredictable environments, autonomy requires learning. However, which parameters must be learned? An additional aspect of scalability is the computing resources required to implement the synthetic nervous system and the learning processes that it employs. More

J. Cabestany, I. Rojas, and G. Joya (Eds.): IWANN 2011, Part II, LNCS 6692, pp. 589–596, 2011.

processing-intensive designs require larger power-hungry microprocessors, which in turn either requires that the robot be fitted with larger batteries or causes reduced period of operation between battery charging events.

In addition, the design methodology applied must allow construction and integration of behavior mechanisms in a systematic way such that design expansions and changes can be made in a controlled manner. In the end, the objective is to first establish a desired set of behaviors to be exhibited by the robot, followed by construction of the synthetic nervous system with as few steps as possible in between. What is the shortest path to go from desired to implemented behavior? One approach is to say that the terminology used to describe the behavior must match the terminology of the behavior elements and mechanisms used for the implementation (elaborated further below). In such a case the design never leaves the behavior realm. The complete design is defined, developed and evaluated in terms of behavior terminology. In addition, the design methodology must take into account the current issue of getting robots to act with a sense of meaning. That is, an autonomous robot is best not seen as a human-command executor such as found in tele-operated robotics. To solve this, one approach is to focus on the reason that the robot performs any behavior.

This paper is structured as follows: Section 2 describes the need element to serve as the basic building block for a behavior design methodology described in section 3. Furthermore, in section 3 an example design showing a mobile robot that avoids touching objects is used to elucidate how this methodology converts desired behavior into a synthetic nervous system. Section 4 shows how learning can, in principle, be integrated into need elements in a way compliant with existing learning theories such as classical conditioning.

2 The Need Element

It is imperative to understand that the starting point for the development of the behavior design methodology discussed in this paper was to give the robot a reason to perform any behavior. Having a reason to do anything infers that behavior is driven. This reason is here linked with internal significance, that is, a behavior is driven for the reason of satisfying some need internal to the robot. The concept of "behavior being driven for the reason of internal significance" is easily understood when looking at homeostatic variables. These variables are physiological variables, such as blood glucose and body fluid levels, that the body attempts to maintain stable, that is, within an operational range. The corrective action taken by the body when the variable's value falls outside this range is aimed at regaining stability in the variable. The concept of equilibrium is naturally linked to homeostasis in that equilibrium and stability are often used to imply the same phenomenon [1]. In terms of psychology, homeostasis is defined as "a state of psychological equilibrium obtained when tension or a drive has been reduced or eliminated" [2]. This is essentially what a need element is.

The need element model is shown in Fig. 1 in its two variations: drive and processing need elements. When driving behavior, the drive need element is used to serve as the origin of the behavior mechanism. In addition, there must also be elements that perform processing, given that a neuron in essentially an input-output element. (Note that neurons serve only as inspiration in the development of the need element model, such that a need element must not be seen as a neuron.) Therefore, input and output processing around the drive need element are done using processing need elements. Each need element contains a value, which is updated according to the signals entering the need via four types of Need Input Ports (NIP): DR (drive), INH (inhibition), RI (Rate of Increase) and RD (Rate of Decrease). There can be a variable number of inputs of the same type as long as there are always at least one DR, RI and RD input. Each input has an associated weight, much like a synaptic weight. When any DR input is nonzero, the value increases nonlinearly according to the RI and INH inputs; else, the value decreases nonlinearly according to the RD inputs (more detail is given in Fig. 1).

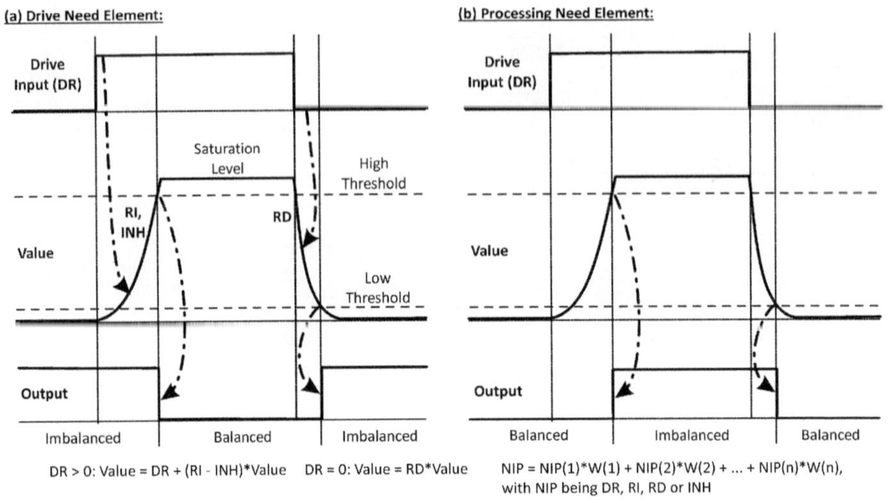

Fig. 1. The drive need element model is shown on the left and the processing need element model on the right. The only difference is in terms of the effect that the drive input has on the state of the need element.

Each need element is always in any of two states: equilibrium (balance), or disequilibrium (imbalance). The current state is calculated using the current value and a pair of high and low thresholds (see Fig. 1). In a drive need element, the DR input drives the need element into a state of equilibrium. In a processing need element, the DR input drives the need element into a state of imbalance. Whenever the need element is imbalanced, it asserts its output to a nonzero value.

3 Behavior Design Methodology

Before describing the design methodology based on need elements, let us describe the desired behavior. The robot used is the e-puck [3]: it has a circular shape of 7 cm diameter, two differentially driven wheels and (amongst other sensors) an array of eight infrared proximity sensors located such that objects can be detected to the front, sides and rear. For the current design only one of the front-facing proximity sensors is used for distance measurement. In addition, all eight proximity sensors are used to emulate a single touch sensor. The desired behavior is for the robot to move forward (first behavior) while avoiding touching any object (second behavior). However, since the detection area of the beam emitted from the distance sensor is narrower than its body, the robot could potentially drive into objects located in front and slightly off-centre to its forward movement direction, or when approaching a wall at a narrow angle. Therefore, the robot cannot simply move forward until it detects an object; it must scan for objects. Since the sensor cannot move, the robot itself must do right and left turns to perform the scanning motion. Therefore, even if there is no object nearby, the robot will have to alternate between moving forward for a short distance, then scanning, then moving forward again, then scanning again, and so forth.

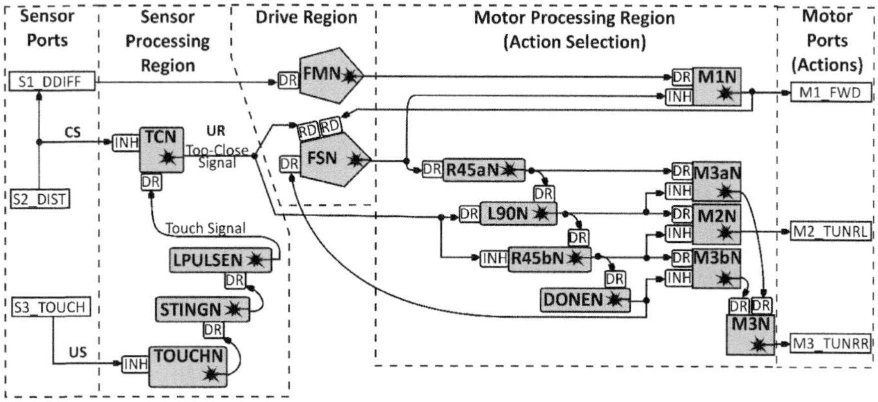

Fig. 2. The complete synthetic nervous system. Elements *TCN*, *FSN* and *FMN* are shown in full detail in Fig. 3. All need element inputs with constant values are omitted (see Fig. 3 for details). *US*, *CS*, and *UR* are discussed in section 4. Note that drive need elements have a different shape than processing need elements.

How is the desired behavior converted into implemented behavior? First, it must be described how need elements constitute the basic behavior mechanism. Up to now, the need element has been described as a generic input-output element. To generate behavior a drive need element's input must be connected to processed or unprocessed sensory input, and its output must be connected to motor output via an action generation stage, which might include processing or not. Designing these connections is guided by the following rule: "While a drive need element receives appropriate stimulus (input), it maintains a state of equilibrium. When there is a lack in appropriate stimulus, the drive need element becomes imbalanced, upon which its output is asserted to a nonzero value (the drive signal). This signal indicates its imbalanced state to the rest of the nervous system, expecting that the rest of the nervous system take appropriate action that will generate the appropriate stimulus to return to equilibrium." Thus, interaction with the environment is required to reestablish equilibrium. The observed interaction is the behavior. Note that the rule can also be applied to processing need elements where it is not a lack of an appropriate stimulus that causes imbalance but the presence of an excitatory stimulus.

The design therefore starts by asking "Which need can be added to the robot such that, when imbalanced, will drive the desired behavior? " For the desired behavior of robot moving forward, let us define a Forward Movement Need (FMN). Thus, the robot has a need for forward movement, and a lack in an appropriate stimulus to indicate forward movement will lead to imbalance and the assertion of the FMN output. The expected consequence of this output is that the robot will move forward. Regarding an appropriate stimulus, let us assume the environment contains only static objects. This means that any change in the value measured by the distance sensor would indicate movement by the robot. Note the cyclic nature: a lack in movement leads to FMN imbalance, which leads to FMN output assertion, which leads to movement, which leads to appropriate input, which leads to FMN equilibrium, which leads to FMN output deassertion, which leads to stopping and therefore once again a lack in movement. The complete synthetic nervous system in Fig. 2 shows that this behavior mechanism is constituted by the elements S1_DDIFF (Distance Difference), FMN, M1N (described further below) and M1_FWD (forward movement action).

For the second desired behavior of not touching any object, the desired phenomenon is that the robot must stop moving forward and initiate object avoidance actions when an object is detected closer than a certain distance (called the too-close distance). When applying the notion of danger to the event of touching objects (for example, an object on fire) one can therefore think in terms of forward safety. It is safe to move forward? For this reason the drive need element is called the forward safety need (FSN). A state of imbalance implies "not safe", thereby driving object avoidance actions. The appropriate input is simply a signal (called the too-close signal) that is asserted whenever an object is measured within the too-close distance. The too-close signal is generated by the Too-Close Need (TCN).

The object avoidance actions are simply the scanning motion (mentioned above) which is a sequence of Turn Right 45deg, Turn Left 90deg, Turn Right 45deg (implemented via the processing need elements R45aN, L90N, R45bN and DONEN). Thus, the scanning motion is designed to find a 90deg zone that is safe for forward movement, that is, it contains no too-close objects. At the end of this sequence, the DONEN drives the FSN back into equilibrium to indicate that forward movement is safe once more. As mentioned above, the scanning motion must be regularly preformed even if no object is detected due to the detection area limitations of the proximity sensor. As a result, the scanning motion is triggered either by sufficient forward movement (M1N to FSN RD) or the too-close signal (TCN to FSN RD). Depending on what triggered the scanning motion and the presence of too-close objects, the scanning motion will complete when a safe direction for forward movement has been determined that is either the same than or to the left of the previous direction. Note that all behavior exhibited by the robot comes from the nervous system attempting to keep all its drive need elements in a state of equilibrium.

The output stage (M1N to M3N) consists of processing need elements (called motor needs) that perform arbitration to select which of the three winner-takes-all actions (move forward, turn right, turn left) is to be selected. Complete design detail for the TCN, FMN and FSN is shown in Fig. 3.

From this section it can be seen that the behavior design methodology enhances scalability by direct transcription from desired to implemented behavior. Fig. 2 is the full design workspace; it is neither a conceptual nor a architectural block diagram. Every parameter (weights and thresholds) has a direct and predictable effect on the resulting behavior. Thus, the design is condensed into the necessary parameters. Additional behaviors are added and integrated systematically. Coordinated perception and action (an important aspect in autonomous robotic design [4]) are easily integrated since their definition is required to complete the drive need element description.

4 Learning

The synthetic nervous system behavior mechanisms described so far are designed a priori, that is, the design parameters are set before robot operation. Essentially this nervous system can be divided into two parts: its organization (interconnections, sensor inputs, motor outputs, need element configuration) and a set of design parameters. These parameters consist of the weight value at each need input and the need element thresholds. As mentioned above, any weight change has a direct and predictable effect on the behavior displayed. One such parameter is the weight that determines the too-close distance. As seen from Fig. 3, this weight (W1) is located at the set point input (DR) to the TCN. If the weight value is increased, the too-close distance increases thereby increasing the sensitivity of the robot to too-close objects.

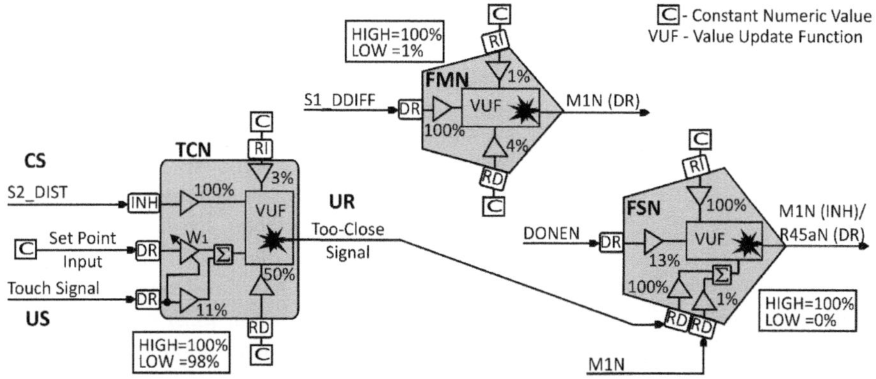

Fig. 3. The design details of the *TCN*, *FMN* and *FSN* elements. The *touch signal* and *W1* are described in the next section. The *Value Update Function* is given in Fig. 1.

Thus, it is possible a priori to set the weight value via an iterative design process (or even to calculate it) such that the robot will for a particular controlled environment avoid touching any object. However, many factors influence the weight value. For example, as the amount of ambient light changes, the infrared sensor will give different readings for the same distance. Thus, it is simply practical to rather enable the robot itself to determine what the weight value must be, given its current environmental conditions.

To learn the weight value, the touch modality is added. Fig. 2 shows how the touch signal is generated by the processing need elements TOUCHN, STINGN and LPULSEN which, as a brief description, converts the unprocessed touch input into a learning pulse of sufficient duration that effectively emulates typical too-close signal generation, which will then trigger the scanning motion. The learning solution implemented here corresponds with previous work on classical conditioning [5][6]. That is, using classical conditioning terminology, a touch signal as unconditioned stimulus (US) is used for the learning of a conditioned stimulus (CS) to generate the unconditioned response (UR) which is the too-close signal. Different from classical conditioning, the CS is not neutral but the stimulus that is supposed to (that is, by design) implement too-close detection. Rather than punishment or reinforcement, the touch signal is helping (strengthening) the TCN to perform its intended function. Learning of W1 is activated by the touch signal input (TSI) in that, while present, W1 is increased according to the update rule

$$W1(i+1) = W1(i) + TSI(i)*\delta \,, \qquad (1)$$

where i is the need network update iteration and δ is the learning rate.

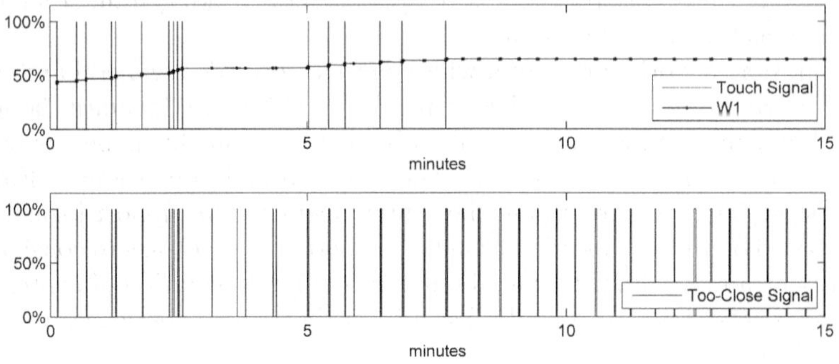

Fig. 4. Learning occurs as the value of the weight W1 increases until the robot ceases to touch any object. All magnitudes are shown as a percentage of the maximum possible value. The learning rate was 0.03 % of the weight value.

The need network (the set of need elements that constitute the synthetic nervous system) and motor outputs are updated at the same time instance in order to emulate complete parallelism across the nervous system. Fig. 4 shows a graph of how W1 increases until the touch signal is not asserted anymore since before touching any object the too-close signal is asserted via the too-close distance. For this test the robot was placed in a walled rectangular space of 20 cm x 30 cm.

In terms of scalability, different from an ANN, here only a single parameter needs to be learned, which is a result of the need element's encoding of behavior. Therefore, learning (which is normally costly in terms of processing and operational time) is principally kept to the minimum. In addition, the current need network update period is a mere 2 ms. Performance-wise, an iteration of the complete application running on the e-puck requires less than 750 us of the 2 ms period, without any compiler optimization. Thus, ample room is left for adding additional need elements.

5 Conclusion

The objective of this article was to show one possible approach to adaptive behavior design that is based on homeostasis rather than an ANN or artificial evolution. Similar to homeostatic variables, it was shown how drive and processing need elements, alternating between states of equilibrium and imbalance, can be used to completely design object avoidance behavior for the e-puck robot using a single proximity sensor. For a drive need element to remain in equilibrium it must receive an appropriate stimulus, else its output will be asserted, thereby expecting the rest of the nervous system to generate appropriate action that will once again generate the appropriate stimulus. Processing need elements implement functions of sensor and motor processing as well as action coordination and arbitration. All behavior exhibited by the robot is a result of the synthetic nervous system attempting to maintain its drive need elements in equilibrium.

The nervous system design can be divided into an organizational part and a set of design parameters that directly influence the exhibited behavior. Learning mechanisms can be added, using processing need elements, to enable the learning of parameters. The implemented learning mechanism coincides with previous work based on classical conditioning that uses touch perception as the unconditioned stimulus. Only the necessary design parameters need to be learned, that is, those that are best not set a priori. Overall, the synthetic nervous system shows efficient and effective conversion from desired to implemented behavior, while maintaining a low demand for processing resources.

Regarding future work there are the following issues: The learned weight value, W1, cannot only increase; an additional method or mechanism must allow decreasing W1. In addition, using infrared sensors to emulate touch is not the best option since their performance is susceptible to ambient light changes. One possible solution is to use proximity sensing in combination with proprioceptive sensing of the motors [6].

References

1. Ashby, W.R.: Design for a brain: The origin of adaptive behaviour, 2nd edn. Chapman and Hall, London (1960)
2. homeostasis, Dictionary.com, http://dictionary.reference.com/browse/homeostasis
3. E-puck education robot, http://www.e-puck.org/
4. Pfeifer, R., Bongard, J., Iida, F.: New Robotics: Design Principles for Intelligent Systems. In: Artificial Life (2005)
5. Verschure, P., Kröse, B., Pfeifer, R.: Distributed Adaptive Control: The Self-Organization of Behavior. Robotics and Autonomous Systems 9, 181–196 (1992)
6. Salomon, R.: Improving the DAC Architecture by Using Proprioceptive Sensors. In: Pfeifer, R., et al. (eds.) SAB 1998. LNCS, vol. 6226, pp. 232–241. Springer, Heidelberg (1998)

Gait Identification by Using Spectrum Analysis on State Space Reconstruction

Albert Samà, Francisco J. Ruiz, Carlos Pérez, and Andreu Català

CETpD - Technical Research Center for Dependency Care and Autonomous Living
Neàpolis Building, Rambla de l'Exposició, 59-69, 08800 Vilanova i la Geltrú, Barcelona, Spain
{albert.sama, francisco.javier.ruiz, carlos.perez_lopez, andreu.catala}@upc.edu

Abstract. This paper describes a method for identifying a person while walking by means of a triaxial accelerometer attached to the waist. Human gait is considered as a dynamical system whose attractor is reconstructed by time delay vectors. A Spectral Analysis on the state space reconstruction is used to characterize the attractor. Parameters involved in the reconstruction and characterization process are evaluated to examine the effect in gait identification. The method is tested in five volunteers, obtaining an overall accuracy of 92%.

Keywords: Gait identification, spectral methods, inertial sensors.

1 Introduction

Human movement analysis is a research field with clinical and biometrics application. It has been shown useful in the objective measurement of gait, balance, falls risk assessment and mobility monitoring [1]. Biometric identification is also a field of great interest whose research covers security and access control applications. Typical identification systems analyze fingerprints, speech or iris. Some studies try to perform it by more complex patterns like those obtained by gait [2]. The existent gait identification methods can be grouped into three categories: vision based, floor sensor based and inertial sensor based. In this work, we focus our study in the third category.

Accelerometers, gyroscopes and magnetometers are the most common sensors applied for movement analysis. It can be distinguished two main approaches of its signal treatment. Firstly, direct approaches are those which integrate sensor measures directly into a mathematical model [3]. For instance, gait kinematics may be characterized as an inverted pendulum model and angular velocity provided by gyroscopes may be integrated to extract gait properties [4]. Secondly, indirect approaches characterize movement in an indirect way by using features extracted from the signal (mean, variance, kurtosis, etc.). Those models may provide qualitative information by using a classification approach, for instance daily life activities can be detected by Support Vector Machines (SVM) [5] or Hidden Markov Models (HMM) [6], or may provide quantitative characteristics by using a regression model, such that step length and velocity [7].

J. Cabestany, I. Rojas, and G. Joya (Eds.): IWANN 2011, Part II, LNCS 6692, pp. 597–604, 2011.

Dynamical systems give us a different approach to analyze human movement. This approach may be considered between direct and indirect models. It is based on Taken's theorem; thus sensor measures are treated as time series to reconstruct the attractor of the dynamical system being sensed, similarly to direct models. Then, reconstructed space is characterized by some features, as indirect models do. Such a combined approach is followed by this work, and it has been tested previously in order to extract step length and velocity by a triaxial accelerometer [8].

This paper analyzes the reconstruction of attractors in dynamical systems in the context of gait identification. The objective is to identify a person while walking by the measures of an accelerometer located at the waist. The effect of different parameters that affects the reconstruction space in the accuracy of the method is being analyzed. Gait identification may be useful in medical applications since normal and pathological gait may be then recognized.

The methodology used is based on spectral analysis on state space reconstruction. Similar methods have been previously used for human full-body pose tracking by means of six inertial sensors [9]. A similar approach to the presented in this work was used in [10], where activity classification was performed by using a Spectral method to the state space reconstruction. However, the effect of the different parameters that affects the reconstruction space was not evaluated.

The paper is organized as follows: in the next Section, a brief introduction to the theory of state space reconstruction and spectrum analysis and some remarks on practical implementation is presented. Section 3 is devoted to describe the approach used in this work, which is based on applying the described spectral method to state space reconstruction in order to perform gait identification. Experiments description and the analysis of the results are described in section 4. Finally, section 5 includes the conclusion and future research issues

2 State Space Reconstruction

This section presents a brief introduction to the theory of state space reconstruction and some remarks on practical implementation. State space reconstruction methods have been developed as a mean to obtain a topologically equivalent representation of the state space from one or more observed signals of a dynamical system.

2.1 Delay Coordinates

A scalar time series can be considered as a one-dimensional observed measures obtained from a smooth d-dimensional dynamical system. The original d-dimensional state space of the dynamical system cannot be directly observed from the time series. However, it is possible to reconstruct this original state space or, at least, a topologically equivalent embedded space from the called *delay coordinates* [11].

Considering a single time series measured every time step Δt $\{s_{t_0}, s_{t_0+\Delta t}, \ldots\}$ (where Δt is the inverse of the sampling frequency), the *delay coordinates set* with dimension m and time lag τ is formed by the time delayed values of the scalar measurements $\mathbf{r}_t=\{s_{t-\tau(m-1)\Delta t}, \ldots, s_{t-\tau\Delta t}, s_t\} \in R^m$. For notation simplicity, henceforth, time step Δt is avoided. Takens proved in 1980 the well known *Takens' embedding theorem* [12], which states that if the time series comes from a noiseless observation of a smooth dynamical system, the attractor recovered by delay coordinates is topologically equivalent to the original attractor in the state space. Even though Takens' theorem does not give guarantees of the success of the embedding procedure in the noisy case, the method has been found useful in practice.

There is a large literature of the "optimal" choice of the embedding parameters m and τ. It turns out, however, that what constitutes the optimal choice largely depends on the application [13]. In terms of the time lag τ, one of the most extended method to determine the optimal delay time was suggested by Fraser and Swinney [14]. They suggest using the first minimum in delayed average mutual information function. On the other hand, a method to determine the minimal sufficient embedding dimension m was proposed by Kennel et al. [15] [16]. The idea is related to topological properties of the embedding and consists of computing the percentage of false neighbors, i.e. closer points that are no longer neighbors if the embedding dimension increases, which allows the sufficient embedding dimension to be determined.

2.2 Singular Spectrum Analysis

If Taken's theorem requirements are accomplished, the time delay coordinates leads to an embedding of the original state's space. Then, every linear transformation of sufficient rank from the time delay coordinates also leads to an embedding. A good choice of linear transformation is known as *principal component analysis* (PCA). This technique is widely used, for example to reduce multivariate data to a few dimensional data. The idea is to introduce a new set of orthonormal basis vectors in embedding space such that projections onto a given number of these directions preserve the maximal fraction of the variance of the original vectors. Solving this problem leads to an eigenvalue problem. The orthogonal eigenvectors obtained from the autocovariance matrix determine the *principal directions*. By considering only a few of this directions (those with largest eigenvalues) is sufficient to represent most part of the embedded attractor.

Singular Spectrum Analysis (SSA) [17] consists of applying a PCA, or other similar methods of spectra decomposition, to the set of delay coordinates, hereafter to be called *reconstructed attractor*. This analysis is applied in this case as follows: given a time delayed vector $\mathbf{r}_t=(s_{t-\tau(m-1)},\ldots,s_t)$, which reconstructs the attractor for the actual state $\mathbf{x}_t$ at time t, a matrix which reconstructs the trajectory from time t to time $t+w$ is:

$$\mathbf{M}_t=[\ \mathbf{r}_t\ \mathbf{r}_{t+\tau} \ldots \mathbf{r}_{t+k\,\tau}\ ]^{\mathrm{T}} \tag{1}$$

where $k=w/\tau$.

Such matrix is first set to have zero mean (that leads to matrix $\mathbf{M}^0{}_t$) and then analyzed by applying a PCA process, so $\mathbf{M}^0{}_t$ is decomposed such that:

$$\mathbf{M}^0{}_t=\mathbf{U\Sigma V}^* \tag{2}$$

$\mathbf{V}$ represents a change of basis between the reconstruction and the called latent space. Then, $\mathbf{VM}^0{}_t$ identifies the trajectory of the reconstruction of the states $\mathbf{x}_t$, $\mathbf{x}_{t+1},\ldots,\mathbf{x}_{t+w}$ in the latent space.

For high embedding dimension it is usually considered that those directions with smallest eigenvalues correspond to noisy components of the signal and do not contain relevant information about the trajectories.

SSA has been successfully applied in mid-term prediction of geoclimatic time series [18] and extracting information from human movement measures provided by accelerometers [8]. Moreover, this approach has been shown able to detect low-variance oscillations in short and noisy data even if the data is non-stationary [13].

3 Gait Identification Approach

This section describes the approach used in this work, which is based on applying the described spectral method to state space reconstruction in order to perform gait identification.

The accelerometer signal is measured at each time t as a triplet composed of the three accelerations from the three axes. Its magnitude is used as the scalar measure to reconstruct the state space:

$$s_t = (x_t^2 + y_t^2 + z_t^2)^{1/2} \tag{3}$$

Thus, only magnitudes measures are going to be used, providing a method independent of the orientation.

Gait is a process of cyclic nature and its sequence of states is expected to be essentially periodic. Thus, trajectories in the state space reconstruction should be more o less complex but closed orbits, as it will be observed in the generated matrix $\mathbf{M}_t$ through recurrence plots. Different gaits are expected to provide different orbits, so the characterization of those orbits may allow us to identify which person belongs to. The orbit characterization from PCA is considered by two ways. Firstly, the directions where maximum variance is achieved are expected to characterize the dynamical system, as each trajectory should take a different form. Secondly, the eigenvalues are assumed to describe the transformation between latent and reconstruction space, so each transformation would be particular for each gait.

Embedding dimension m, time-lag τ and window size w are set through the characteristics of the dynamical system. Different values are tested for m and w considering both the attractor dimension and the number of states that a cycle takes. The time lag will be fixed by the results obtained by Average Mutual Information. This approach is different from classical spectral methods where an arbitrarily large value for m parameter is fixed without evaluating its effect [13].

4 Experiments

Five healthy volunteers walked 20 m at normal speed twice. A device containing a triaxial accelerometer developed at CETpD [8] and located at the lateral side of the waist logged accelerations in a sampling frequency of 200 Hz. Figure 1 shows an example of the signal obtained while volunteers walked.

Figure 2 shows average mutual information (AMI) results for all five signals obtained. Time lag influences the attractor reconstruction by its order of magnitude but not by its specific value. Thus, a time lag of 10 times the time step is selected as a mean value of the local minima among all volunteers.

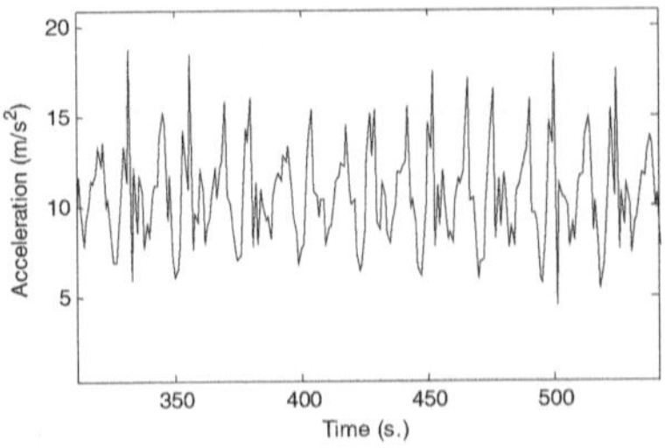

Fig. 1. Example of signal obtained while walking. Acceleration norm is showed

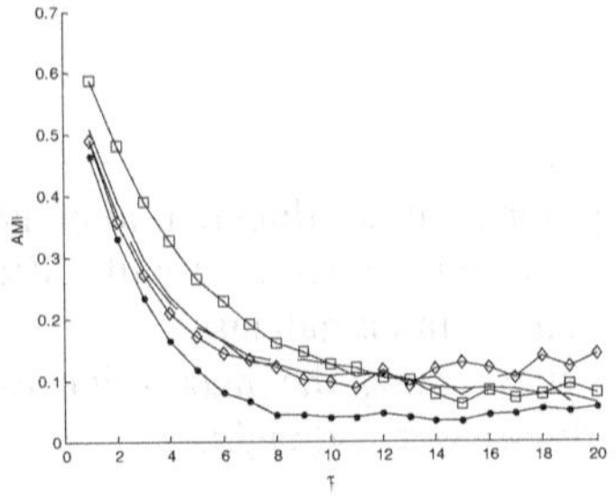

Fig. 2. AMI results for all five signals. From this analysis, a time lag of 10 is suitable for all time series

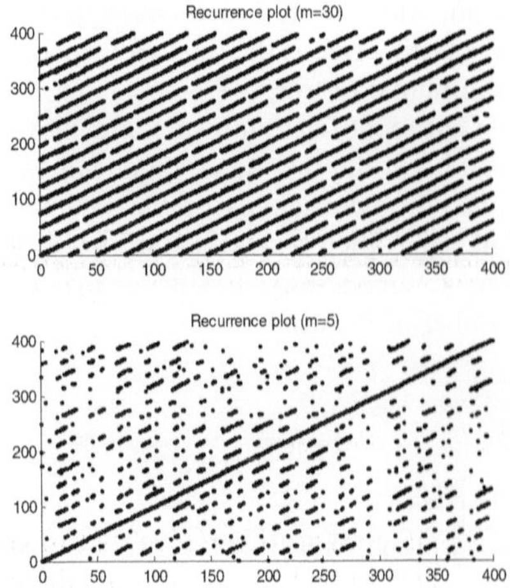

Fig. 3. Recurrence plot when using embedding dimension 5 (up) and 30 (down)

Recurrence plots are a common technique helpful to visualize the recurrences of dynamical systems. The essential periodic motions are reflected by long and non-interrupted diagonals. The vertical distance between these lines corresponds to the period of the oscillation. Figure 3 shows the recurrence plot for a volunteer when using embedding dimension 5 and 30. For the lower dimension, the periodic motion is not as clear as in the higher dimension, where the cyclic motion appears obvious.

4.1 Results

From Figure 3 it is shown that the period of the orbit in the state space is the same for both embedding dimensions, and is reckoned to be ~30 reconstructed states. The rest of volunteers provide a similar period. FNN algorithm gives 5 as the minimum embedding dimension for all volunteers. Taking into account results from recurrence plots and FNN, *m* parameter is tested with values from 5 to 30.

Since orbits comprise 30 samples and sampling frequency was 200 Hz, a whole orbit takes $30 \cdot \tau/200=1.5$ sec. approximately. In order to test whether half a period, one or two periods enable to recognize the system, window size values used are: w=0.75 s., w=1.5 s. and w=3 s.

The spectral analysis described in previous sections is applied while volunteers walked. Training is performed by Classification and Regression Trees (CART) using the first 20 m. walked by volunteers. Accuracies are obtained classifying the second 20 m. walk. CART methodology used is the standard cross-validation prune, where the optimal tree is the one with least nodes whose accuracy is within 1 std. error of the minimum cost tree.

Gait identification results are shown in Figure 4. Two different sets of features are tested. On the one hand, eigenvalues are taken to characterize the gait. On the other hand, the coefficients of the latent variables or Principal Components (PC) which determine their directions are used. Accuracies are shown as a function of window size *w*, embedding dimension *m* and either the number of eigenvalues used or the number of PC used.

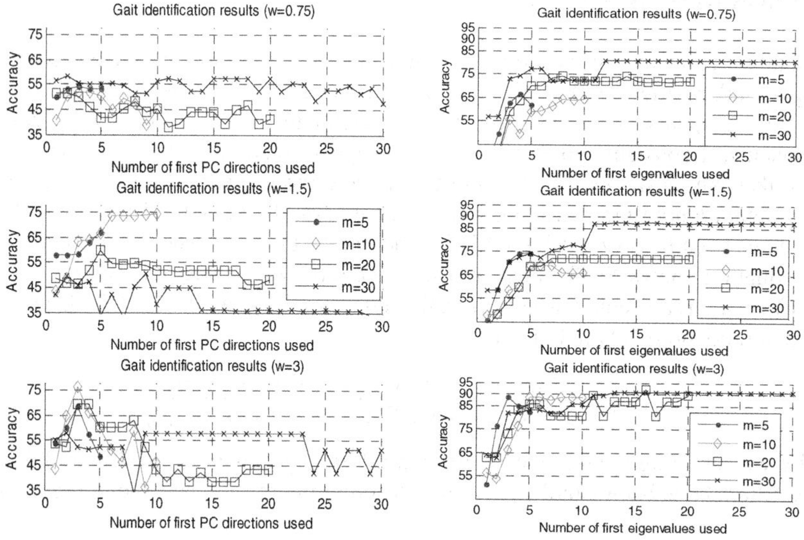

Fig. 4. Gait identification results when using PC directions are shown on the left and those obtained by eigenvalues are on the right

4.2 Discussion

Results show that it is possible to identify a person from a gait classifier with an overall accuracy of 90%. Latent variables direction does not allow the dynamical system to be recognized as accurately as eigenvalues does: eigenvalues achieve 92.3% while PC directions 76.2%.

The direction of the latent variables is not enough to identify the dynamical system when half an orbit is considered (*w*=0.75), but provides reasonable good results (~75%) for larger window sizes. Best results are obtained for *m*=10 and *m*=20, so highest accuracies are obtained for lowest embedding dimensions. The number of PC that provides best results is a fraction of total, e.g. for *m*=10 and *w*=3 it should be used

3 PC to obtain highest accuracy. This agrees with the consideration that last PC's contains only noise, since including them does not increment identification accuracy.

When gait identification is performed by means of eigenvalues, a better identification is also obtained for larger window sizes. A window size containing half a period (w=0.75) provides a maximum accuracy of 82%, for a whole period (w=1.5) achieves 88% of accuracy, and for two-periods size (w=3) obtains the highest accuracy: 92.3%. These results suggest that it is possible to identify the dynamical system with a reasonable accuracy without taking a whole orbit, but it is advisable to use it fully in order to obtain a better identification. Furthermore, higher accuracies are used when using two cycles instead of a single one, which means that considering a unique oscillation in the reconstruction space is not enough to obtain best results.

Best classification results are obtained for m=30 when window sizes are w=0.75 and w=1.5. For w=3 such value of m also provides the highest accuracy, but m=20 does as well. Thus, it can be concluded that in general unfolding whole orbits provide good results for gait identification. However, it may be used lower embedding dimensions for some window sizes, which would save computational costs.

Observing the number of eigenvalues needed to achieve the maximum accuracy for m=30, it is seen that less than a half are needed. Thus, the embedding approach used is able to characterize the dynamical system with a number of parameters which is lower than the half of the dimensions used to reconstruct it with the highest accuracy.

5 Conclusions

A methodology to identify people by gait is proposed and tested. It considers human gait as a dynamical system, whose attractor is reconstructed and characterized in order to recognize it. The characterization is performed by a spectral analysis of the reconstruction space, and it is tested into 5 people achieving an overall accuracy of 92.3%. A triaxial accelerometer located in the waist is needed to perform it.

Attractor's reconstruction is based on Taken's theorem. The different parameters involved in the reconstruction and the characterization are tested in order to evaluate the effect in gait identification. It is concluded from results that unfolding a whole orbit seems to provide the best identification, though in some cases unfolding a part of the orbit may be enough. Suitable window sizes for identification are those equal or larger than the orbit duration. The approach used provides its best results by using a number of parameters less than a half the embedding dimension. Thus, the method characterizes the dynamical system by relatively few parameters when comparing to its reconstruction space dimension.

Further research is needed in order to use the same methodology for rehabilitation applications trying to evaluate, for example, the gait progress in a patient after a clinical intervention measuring the attractor changes.

Acknowledgments

This work is supported by the Spanish project SENSORIAL (TIN2010-20966-C02-02) Spanish Ministry of Education and Science.

References

1. Culhane, K.M., O'Connor, M., Lyons, D., Lyons, G.M.: Accelerometers in rehabilitation medicine for older adults. Age and Ageing 34(6), 556–560 (2005)
2. Yazdanpanah, A.P., Faez, K., Amirfattahi, R.: Multimodal biometric system using face, ear and gait biometrics. In: 10th International Conference on Information Sciences Signal Processing and their Applications, pp. 251–254 (2010)
3. Sabatini, A.M., Martelloni, C., Scapellato, S., Cavallo, F.: Assessment of walking features from foot inertial sensing. IEEE Trans. Biomed. Eng. 52(3), 486–494 (2005)
4. Salarian, A., Russmann, H., Vingerhoets, F.J.G., Dehollain, C., Blanc, Y., Burkhard, R.P., Aminian, K.: Gait Assessment in Parkinson's Disease: Toward an Ambulatory System for Long-Term Monitoring. IEEE Trans. Biomed. Eng. 51(8), 1434–1443 (2004)
5. Parera, J., Angulo, C., Rodríguez-Molinero, A., Cabestany, J.: User daily activity classification from accelerometry using feature selection and SVM. In: Cabestany, J., Sandoval, F., Prieto, A., Corchado, J.M. (eds.) IWANN 2009. LNCS, vol. 5517, pp. 1137–1144. Springer, Heidelberg (2009)
6. Mannini, A., Sabatini, A.M.: Computational methods for the automatic classification of postures and movements from acceleration data. Gait & Posture 30(S.1), S68-S69 (2009)
7. Perrin, O., Terrier, P., Ladetto, Q., Merminod, B., Schutz, Y.: Improvement of walking speed prediction by accelerometry and altimetry, validated by satellite positioning. Med. Biol. Eng. Comput. 38, 164–168 (2000)
8. Samà, A., Pardo, D., Cabestany, J., Rodríguez-Molinero, A.: Time Series Analysis of inertial-body signals for the extraction of dynamic properties from human gait. In: 2010 International Joint Conference on Neural Networks (IJCNN), pp. 1–5 (2010)
9. Schwarz, L.A., Mateus, D., Navab, N.: Multiple-Activity Human Body Tracking in Unconstrained Environments. In: Perales, F.J., Fisher, R.B. (eds.) AMDO 2010. LNCS, vol. 6169, pp. 192–202. Springer, Heidelberg (2010)
10. Frank, J.: Learning state space models from time series data. In: Multidisciplinary Symposium on Reinforcement Learning (2009)
11. Sauer, T., Yorke, J.A., Casdagli, M.: Embedology. J. Stat. Phys. 65(3/4), 579–616 (1991)
12. Takens, F.: Detecting strange attractors in turbulence. Lecture Notes in Math. vol. 898, pp. 366–381 (1981)
13. Kantz, H., Schreiber, T.: Nonlinear Time Series Analysis. Cambridge University Press, Cambridge (2004)
14. Fraser, A.M., Swinney, H.L.: Independent coordinates for strange attractors from mutual information. Phys. Rev. A 33, 1134–1140 (1986)
15. Kennel, M.B., Brown, R., Abarbanel, H.D.: Determining embedding dimension for phase-space reconstruction using a geometrical construction. Phys. Rev. A 45, 3403–3411 (1992)
16. Fredkin, D.R., Rice, J.A.: Method of false nearest neighbors: A cautionary note. Physical Review E 51(4), 2950–2954 (1995)
17. Vautard, R., Yiou, P., Ghil, M.: Singular Spectrum Analysis: A toolkit for short, noisy chaotic signals. Physica D 58, 95–126 (1992)
18. Ghil, M., Allen, R.M., Dettinger, M.D., Ide, K., Kondrashov, D.: Advanced spectral methods for climatic time series. Rev. Geophys. 40(1), 3.1–3.41 (2002)

Aibo JukeBox – A Robot Dance Interactive Experience

Cecilio Angulo, Joan Comas, and Diego Pardo

CETpD - Technical Research Centre for Dependency Care and Autonomous Living
UPC - Technical University of Catalonia,
Neàpolis Building. Rambla de l'Exposició 59-69, 08800 Vilanova i la Geltrú, Spain
{cecilio.angulo,joan.comas-fernandez,diego.pardo}@upc.edu
http://www.upc.edu/cetpd/

Abstract. This paper presents a human-robot interaction system based on the Aibo platform. This robot is both, complex and empathetic enough to generate a high level of interest from the user. The complete system is an interactive JukeBox intending to generate affective participation, i.e., empathy, from the user towards the robot and its behavior. This application is based on a robotic dance control system that generates movements adequate to the music rhythm using a stochastic controller. The user can interact with the system selecting or providing the songs to be danced by the robot. The application has been successfully presented in different non-scientific scenarios.

Keywords: Human Robot Interaction, dancing robots, interactive environment.

1 Introduction

Social robotics is a main research area in the Technical Research Centre for Dependency Care and Autonomous Living (CETpD), a research centre associated to the Technical University of Catalonia (UPC). One of its main objectives is user acceptability when integrating robots in domestic environments. The Aibo robot has been employed as robotic platform for this kind of experiences for several years. It was originally launched by Sony as an entertainment robotic pet, nevertheless it quickly became an appropriate platform for research due to its flexibility and technical features.

Some of the most important out-of-the-box features of Aibo are those concerning dancing movements. Dance is a very important behavior demanded by the users when interacting with it. Using dancing behaviors the user-friendly features of this robot are exposed. Moreover, Sony realized that this friendly behavior motivates the human robot interaction, thus, the First Aibo Dance Contest (2005) was proposed[1]. Diverse robot dances routines were developed for exploiting the capacities of the robot demonstrating imagination and creativity.

[1] "Aibo Does Daft-Punk" programming contest, Sony Entertainment Robot Europe.

J. Cabestany, I. Rojas, and G. Joya (Eds.): IWANN 2011, Part II, LNCS 6692, pp. 605–612, 2011.

Lately, entertainment robots have been focused towards mechanically simple platforms, mainly rollers i.e., reproducing music systems that rolls on the ground following the perceived music rhythm. This approach fits commercial purposes, however it could be improved for user interactivity. Hence, this paper introduces a human-robot interaction system for the Aibo platform that uses dance as a form of social communication. This platform is both, complex and empathetic enough to obtain a high level of user interest. The complete system is an interactive JukeBox intending to generate affective participation, i.e., empathy, from the user towards the robot and its behavior.

The paper is structured as follows. Next section presents related work on music/dancing robots. Section 3 describes the Aibo JukeBox, while Section 4 describes in detail the diverse modules of the application. Finally, Section 5 enumerates three experiences of real human robot interaction in different environments. Main conclusions and future works are presented in Section 6.

2 Background and Related Work

Music robots developed in Tohoku University are conceived for entertainment purposes, therapy or research. For instance, the "Partner Ballroom Dance Robot" [1,2], features a woman's face and a sensor around its waist detecting movements. When interacting with a human, the robot analyzes his/her movements and figures out how to accompany him/her with its shoulders, elbows, waist and neck. Another well-known example is the Toyota "Partner Robot" [3]. Toyota announced that they developed artificial lips that move with the same finesse as human lips, allowing the robot to play musical instruments, e.g., a trumpet, the same way humans do.

The most promising dancing robot for therapy is Keepon [4]. Its mechanical shape remains very simple. It successfully interacts with children based on environmental sounds. In [5] it is defended that human social behavior is rhythmic, so synchrony plays an important role in coordinating and regulating our interactions. They presented two experiments in which Keepon dances with children listening music (see also [6]), and in which the effects on engagement and rhythmic synchrony are examined.

Entertainment robotics is an area of interest for the growing area of commercial robots. Regarding music robots, ZMP Inc, a Japanese robotic company based in Tokyo, develops a platform named *miuro* for music innovation based on utility robot technology. Miuro is a music player that dances while activates its LEDs whereas its two-wheeled twist movements synchronizes with the music. A second example is Sega Toys and its Music Robot *ODO*, which bears a resemblance an affordable alternative to miuro. Sony's music robot *Rolly* is a third example. It plays music and dances around while colored lights flash. These commercial efforts demonstrate a high interest on music robots, nevertheless, the user interaction offered by these platform is limited. A natural extension is to allow the users to interact with this type of robots and let them feed the system with their expectations and feedback.

Some studies already exist trying to incorporate user interactivity to the robot behaviors. An early attempt is presented in [7], where a Human-Robot Dance Interaction Challenge using Sony's QRIO was proposed with a simple goal: to keep human's interest as long as possible. Robot dancing movements were based on imitation of a human. However, for this goal, the robotic physical capabilities are still too far from required. Recently, inspired by experiences like RoboDance contests that take place in RoboCup competitions, a robotic system has been developed in the form of a humanoid based on the Lego Mindstorms NXT [8], which tries to simulate the human rhythmic perception from audio signals. Unfortunately, no real experience has been reported yet, and authors seem more interested on developing a didactic application framework for the competition.

3 The JukeBox System

The Aibo JukeBox application is a control system for the AIBO dancing behavior. The system interacts with the user and generates a random sequence of movements for the robot dancing. As a result, Aibo dances the music chosen/proposed by the user with adequate rhythmic motions.

Inheriting from a work presented in the 2005 World's First Aibo Dance Contest, an early version of the application (Aibo JukeBox 1.0) was developed without taking into account user interaction. In this primitive version songs danced by the robot were selected from a list, then, Aibo moved using random dance steps following the rhythm of the song. An external detection software was used to extract the BPM (beats per minute) of the song. The output of this process was preprogrammed in the application and related with the list of songs. A database of 19 programmed steps was available in the robot memory. The dancing steps were selected depending on the posture state of the robot (laying, sitting or standing), and transitions between these states were also available.

The purpose of Aibo Jukebox 2.0 is to reduce the distance between the technological application and the user. Using a touch screen, users will select a song, either from a preset list or adding their own songs from media devices (smart phone, usb, etc.). The robot dancing behavior synchronizes its motions with the music rhythm. For the development of this new application, a modular approach was followed tackling individual goals independently. Figure 1 shows a descriptive scheme of the application. The following modules were developed,

- Stochastic dance control algorithm (Director).
- Dancing steps data base (Dancer).
- Robot Communication Protocol (Distributed System).
- Music files treatment (BPM Extraction).
- Music files capture and reproduction.
- Graphical User Interface (GUI).

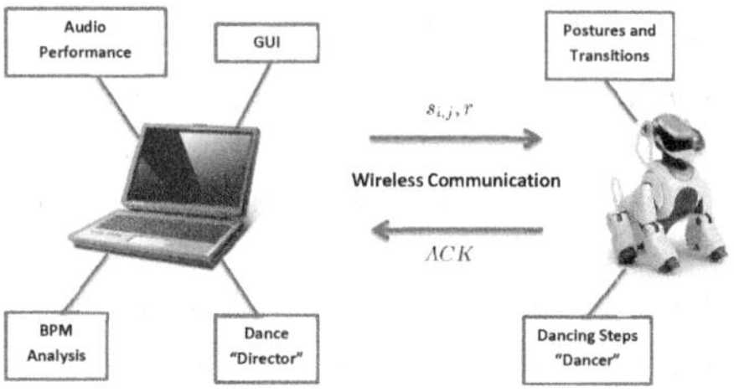

Fig. 1. System modules

4 Application Architecture

As shown in Fig.1, the Aibo Jukebox is a distributed system. The software selecting the sequence of dancing steps is programmed in the computer side. This decision algorithm acts as the dancing "Director" which connects with the robot controller to command the dancing steps. Motions are programmed inside the robot memory (Dancer), then the execution of the dancing steps is independent from the Director. A communication protocol is required for synchronization purposes. In the application side, a GUI was developed to interact with the user, who is allowed to introduce their own themes to the song list. Besides, the GUI also informs the user about the state of the system. Finally, modules for BPM extraction and Audio functionality were also developed.

4.1 BPM Extraction

Audio files (MP3 format) are stored in a repository together with the output of an online BPM analysis. The BPM is a measurement unit denoting the beats of quarter notes in a piece. This index is a quantity proportional to the speed of a given song, therefore, this parameter is required by the robot to complete its motions adequately. The Adion's BPM Detection Library[2] was used to process the MP3 files and extract the BPM index.

4.2 Dancing Steps Database

A total of basic dancing steps (fifteen) were created in the robot memory using the Urbi-script language. These are simple motions of the robot limbs that were manually designed and programmed. The velocity of execution for the motions was parameterized so they can couple with diverse types of music. Moreover, since steps are cyclical motions, the number of repetitions is also parameterized. Three starting postures are considered for the robot: Standing, Sitting and Laying. Several steps were created for each posture as well as transition motions between them. Every posture has associated a series of parameterized dancing steps, so not all the dancing steps can be executed for every posture. Figure 2 shows the postural alternatives and the available transitions between them.

[2] http://adionsoft.net/bpm/

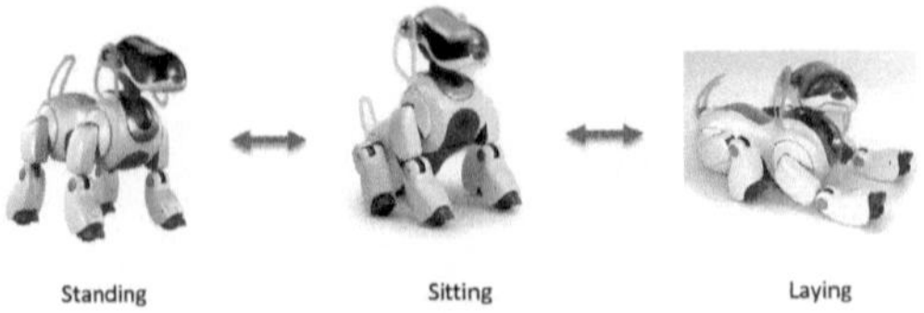

Fig. 2. Aibo postures

To formalize the dancing behavior, let $p = \{standing, sitting, laying\}$ denote the Aibo postures, while $s_{i,j}(b, r)$ represents the transition to the j-th dancing step of the i-th posture, with parameters b and r standing for the motion velocity (or rhythm) and the number of repetitions, respectively. Moreover, transitions $t_{i,j}$ indicate the motion between correspondent postures. Therefore, for a certain song m, a dance is the sequence of steps denoted as $d(m) = \{s_{0,j}, ..., t_{0,j}, ...\}$.

4.3 Stochastic Dance Control Algorithm

Once the song has been selected and its corresponding BPM index extracted, the Director conducts a stochastic dance control. The purpose of this module is to decide among the steps available in the database in order to create the dance. The Director should create a natural dancing behavior avoiding to generate a "machine dancing-style" from the user perspective. The dancing, i.e., the series of steps, cannot be pre-established, whereas a completely random system may generate weirdness due to repetitive transitions between postures. For the case of a completely random dance, transitions would indiscriminately interfere the step sequence and the dancing coherence.

The state machine shown in Fig. 3 models the dancing behavior. Links between states represent steps and postures transitions. Assuming that the motion of the robot is a random variable, the probability of a step or posture transition is given by $P_{s_{i,j}}$ and $P_{t_{i,j}}$, respectively. The sum of the possible transitions in a given posture must add up to one

$$\sum_i P_{s_{i,j}} + \sum_i P_{t_{i,j}} = 1. \tag{1}$$

The algorithm changes individual probabilities using Eq. 1 as restriction. The probability of a given transition is updated every time a step (and its repetitions) is completed. New values depend on the number of steps the robot has performed in the corresponding posture, this is, the probabilities of the m transitions associated to a given posture are updated using,

$$P_{t_{i,j}}^{h+1} = P_{t_{i,j}}^{h} + \eta \tag{2}$$

$$P_{s_k}^{h+1} = P_{s_k}^{h} + \gamma \tag{3}$$

where $0 < \eta < 0.5$ and $\gamma = -\frac{\eta}{2 \times m}$. A higher probability is given to the posture transitions than to that of a step change. Using this update rule, restriction in

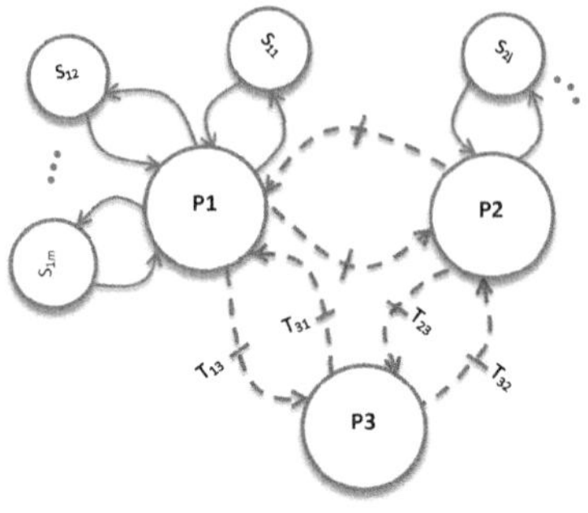

Fig. 3. Aibo states and transitions model

Eq.1 is met. The outcome of this strategy is that the robot performs random steps in a given posture leaving that posture for sure after a certain number of steps, creating artlessness effect from the user perspective.

4.4 Robot Communication Protocol

In order to couple the stochastic controller in the computer with the local process controlling the dancing steps a simple communication protocol is established. When starting a song, the Director moduel sends an initialization message to the robot which stops any ongoing process and changes to the laying posture. After any transition completed the Dancer module sends an "ACK" signal to the Director informing that the process may continue. The complete connection and data transmission protocol is presented in Fig. 4.

4.5 GUI Design

Finally, a GUI module has been designed to interact with the user. Its main screen contains the list of available songs. A button to start/stop the dancing is present in the main screen. The playing time is also displayed in order to let the user decide wether to wait for the end of the dancing routine or to try another song. In an auxiliary screen users can incorporate their own music themes. It is simple and intuitive. An administrator window is also launched in background, where advance functions are available such communication settings, battery level supervision, etc.

4.6 Auxiliary Modules

The application is based on Microsoft C#[3]. The FMOD[4] library was used to play music files in several formats and provide audio functionality. The robot was controlled using an URBI Server[5], which allows a remote connection from a C-based library client (liburbi).

[3] `http://msdn.microsoft.com/en-us/vcsharp/aa336809.aspx`
[4] `http://www.fmod.org/`
[5] Universal Robot Body Interface. `http://www.urbiforge.org/`

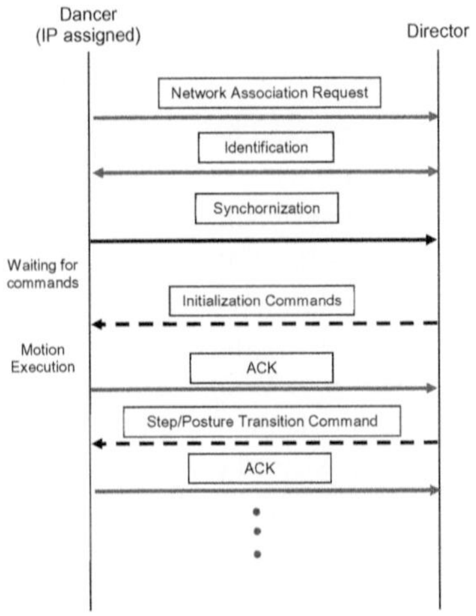

Fig. 4. Director-Dancer communication protocol

5 Experiences with Users

The Aibo JukeBox experience has evolved using feedback obtained from user interaction. User feedback was helping to test diverse implementations until its nowadays format. First experience with the early version of Aibo JukeBox was in CosmoNit, in CosmoCaixa (Scientific Museum in Barcelona, Spain) in June 2007. The activity, entitled "How can a robot dance?". A couple of Aibo robots performed a synchronized dance according to the rhythm of a music theme chosen from a list for an user. Needs for user-provided music themes, beats per minute analysis, and user-friendly screen were reported from this experience. Surprisingly, spontaneity in the robot dance was recognized by the public, as well as diversity in the movements. Only those users standing for more than four songs were able to recognize pre-programmed basic dance movements.

Empathy and socialization, achieved in the first experience, were tested in a more general, not so scientific-technological, long term scenario (June 5th-7th 2008), the 'Avante 2008' Exhibition on Personal Independence and Quality of Life, in Barcelona, Spain. Avante exhibits solutions for people affected by dependency and disability. The Aibo JukeBox was running only under user demand, for battery saving and empathy measurement. Although no behavioral results were obtained, robot empathy was enough to create interactivity and comments for the improvement of the system were accepted.

The third experimentation with the system was performed on demand of an Architecture Studio from Barcelona (Cloud-)), for their presentation in the Collective Exhibition "Out There: Architecture Beyond Building", in the Biennale di Venezia, Mostra Internazionale di Archittectura, from September, 14th to November, 23rd 2008. The main goal in this very long term exhibition was to show how robotics can interact with humans in the usual human environment.

6 Conclusion and Future Work

The system fulfill the expectations of creating a dancing behaviors that have been rated as artlessness by the users. The dancing steps are relative motions, no absolute movements (e.g., walking) were considered which which also empowered the naturalness effect. The stochastic director generates random but consisted steps sequences avoiding indiscriminate connections of steps.

Intervention of the user received positive feedback. Users perceive their participation as important being able to decide the song that the robot dances, moreover, the possibility of incorporate new (and known) songs to the list encourages the user to engage with the application. Adaptation of the robot motions to the music rhythm was also valued as a fundamental feature of the application.

Acknowledgments. This work is partly supported by Grant TSI-020301-2009-27 (ACROSS project), by the Spanish Government and the FEDER funds.

References

1. Aucouturier, J.J.: Cheek to chip: Dancing robots and AI's future. IEEE Intelligent Systems 23(2), 74–84 (2008)
2. Liu, Z., Koike, Y., Takeda, T., Hirata, Y., Chen, K., Kosuge, K.: Development of a passive type dance partner robot. In: IEEE/ASME International Conference on Advanced Intelligent Mechatronics, AIM 2008, pp. 1070–1075 (2008)
3. Toyota. Toyota partner robot, `http://www.toyota.co.jp/en/special/robot/`
4. Kozima, H., Michalowski, M.P., Nakagawa, C.: Keepon. International Journal of Social Robotics 1(1), 3–18 (2009)
5. Michalowski, M.P., Simmons, R., Kozima, H.: Rhythmic attention in child-robot dance play. In: Proceedings of the IEEE International Workshop on Robot and Human Interactive Communication (ROMAN-2009), Toyama, Japan (2009)
6. Hattori, Y., Kozima, H., Komatani, K., Ogata, T., Okuno, H.G.: Robot gesture generation from environmental sounds using inter-modality mapping. In: Proc. of the 5th Int Workshop on Epigenetic Robotics: Modeling Cognitive Development in Robotic Systems. Lund University Cognitive Studies, vol. 123, pp. 139–140 (2005)
7. Tanaka, F., Suzuki, H.: Dance interaction with qrio: a case study for non-boring interaction by using an entrainment ensemble model. In: 13th IEEE Int. Workshop on Robot and Human Interactive Communication, September 2004, pp. 419–424 (2004)
8. Oliveira, J., Gouyon, F., Reis, L.: Towards an interactive framework for robot dancing applications. In: Barbosa, A., ed.: Artech 2008 Proc. of the 4th Int. Conf. on Digital Arts, Porto, Portugal, Universidade Católica Portuguesa (November 2008)

On Planning in Multi-agent Environment: Algorithm of Scene Reasoning from Incomplete Information

Tomasz Grzejszczak and Adam Galuszka

Silesian University of Technology,
Akademicka 16, 44-100 Gliwice, Poland
{Tomasz.Grzejszczak,Adam.Galuszka}@polsl.pl

Abstract. Planning belongs to fundamental AI domains. Examples of planning applications are manufacturing, production planning, logistics and agentics. In real world applications knowledge about environment is incomplete, uncertain and approximate. It implies that planning in the presence of different kind of uncertainty is more complex than classical planning. Aim of this paper is to show the way of reasoning basing on the incomplete information about the initial state of planning problem. The proper reasoning about the state of the problem can reduce such understood uncertainty and then increase efficiency of planning. The article presents an algorithm created in order to reason the state of scene from block world basing on incomplete information from two cameras observing the scene from top and side. The algorithm is explained using an example. Additionally, possible types of uncertainties are presented.

Keywords: AI planning, Block World, reasoning algorithm, objects detection, semantic representation, uncertainty.

1 Introduction

This paper deals with semantic representation of the current state of planning problem in the presence of uncertainty. Planning problem is a problem of finding a set of actions (often also called operators) which transform an initial state into a desired goal situation. It should be distinguished from scheduling - well-known and frequently used technique of improving the cost of a plan. Planning is understood as causal relations between actions, while scheduling is concerned with metric constraints on actions [11]. When all states of a planning problem (including an initial and a goal) are described by a given set of predicates, then the problem is called STRIPS planning problem ([5],[11]). There are many applications of the planning problems in industrial processes, production planning, logistics and robotics ([3],[4],[6],[8],[9],[11],[12]). The STRIPS system has been successfully applied in planning modules of Deep Space One Spacecraft and for elevators control in Rockefeler Center in New York [10].

The presented algorithm was created in order to recognize the scene from the Block World, where 2 manipulators are operating. In typical Elementary Block World problem there is a finite number of blocks, each on the working space. Each block can be either on the table or on another block. The whole scene contains towers of blocks placed on the table [11]. In given task there were a number of blocks on the scene that could be on the table in form of a line or on each other. Each block was individual and

J. Cabestany, I. Rojas, and G. Joya (Eds.): IWANN 2011, Part II, LNCS 6692, pp. 613–620, 2011.

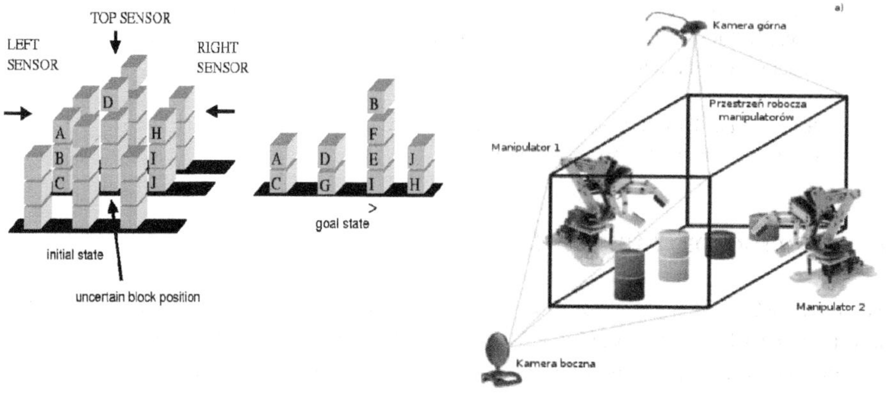

Fig. 1. Block World with incomplete information and corresponding scene construction

could be distinguished by color. There were two manipulators that were moving the blocks. They were operating on the blocks from left and right.

In order to recognize the blocks, two cameras were placed from top and side. Each of those cameras was able to recognize the blocks that was seen either from top or side of blocks piles. In that case some of the blocks was not seen and needed to be concluded. The problem is that this conclusion can lead to a set of possible initial states of planning problem. In this case, the algorithm must seek for a robust plan by evaluating all eventualities. This approach is called Conformant planning [1],[2] and usually increases computational complexity of searching for the plan. The proper reasoning about the state of the problem can reduce such understood uncertainty and increase efficiency of planning.

The main aim of the presented algorithm is to process the knowledge obtained from side and top recognition and conclude the whole scene construction in form of a matrix of data containing the number of block. Basing on this matrix it is easy to generate set of STRIPS predicates describing the scene.

The algorithm is working by means of semantic representation. The term is used when one tries to write down the rules describing a way of obtaining conclusions from input data [2]. In application to Block World and scene recognition, the semantic representation is understood as the way of transforming the scene into the set of predicates describing current state of planning problem with STRIPS representation. The input of semantics are two frames from cameras and output is an array of blocks positions. In this case semantic representation is an algorithm in form of rules, that can recognize the scene with a given level of uncertainty.

2 STRIPS System

A STRIPS system is represented by four lists (C, O, I, G) ([3], [11]):

- a finite set of ground atomic formulas (C), called predicates;
- a finite set of operators (O);
- a finite set of predicates that denotes initial state (I);
- a finite set of predicates that denotes goal situation (G).

The initial state describes the physical configuration of the blocks. This description should be complete i.e. it should deal with every true predicate corresponding to this state. The goal situation describes what should be true. Each goal consists of subgoals and has a form of conjunction of predicates. In multi-agent environment each agent defines its own goal. This description does not need to be complete, i.e. does not need to describe a *state* of the problem. The algorithm results in an ordered set of operators which transforms the initial state I into a state with true predicates mentioned in the goal situation G. Operators in STRIPS representation consist of three sublists: a precondition list (*pre(o)*), a delete list (*del(o)*) and an add list (*add(o)*). The precondition list is a set of predicates that must be satisfied to apply this operator. The delete list is a set of predicates that will be false after applying the operator and the add list is a set of predicates that are true after the operator is applied. The two last lists show the effects of applying the operator into a current problem state ($S \subset C$). Let an operator $o \in O$ takes the form $pre(o) \rightarrow add(o), del(o)$. Following (Koehler and Hoffmann 2000), the set of operators $<o_1, o_2,\ldots, o_n>$ in a plan is denoted by P^O.

If an operator is applied to the current state of the problem then the state is modified. This modification is described by function *Result*:

$Result(S, <o>) = (S \cup add(o)) \setminus del(o)$ if $pre(o) \in S$, S in the opposite case,
$Result(S, <o_1, o_2,\ldots, o_n>) = Result(Result(S, <o_1>), <o_2,\ldots, o_n>)$.

3 Preparation of Data Necessary for Reasoning

The main information that is obtained is a identification number of block. The goal is to create a matrix of all positions and fill it with the number of block that is in a given position.

Two cameras can observe the scene from top and side and can create a vector of seen block. First camera can produce a vector of all blocks seen from top, that is all top blocks from each pile of blocks and also can determine there are those piles. The second can observe the scene from side, calculating the vector of blocks seen from side of some piles. This information can determine the count of blocks from the highest pile. During the detection, algorithm searches for circles and rectangles, which are a 2d projections of 3d cylinder blocks [7].

3.1 Ways of Obtaining the Top/Side Vector

In order to obtain the vectors of information a vision system needs to be constructed. In the simplest case the blocks can differ by color and the system can work on HSV color space.

HSV is an model based on human eye perception. It contains of three channels: hue, saturation, and value, also called brightness. Hue is a color indicated as an angle. Colors can have different saturation. Saturation is indicated by the second component. For example pure color have maximal saturation. Lower the saturation, color becomes more faded. If the brightness is changed, denoted as value, the color becomes more dark. Value can be treated as gray scale image recognizable for human. The detailed algorithm has been described in [7].

4 Reasoning Algorithm Used to Recognize the Scene

If the both vectors are calculated properly, the next step is to pass them to reasoning algorithm, recognizing the state of the matrix. The size of the matrix is determined by the number of blocks. If a scene contains of *n* blocks, there are two extremal cases. Each block is on the table, creating n long top vector, or each block is in one pile, creating *n* long side vector. This indicates that the vectors needs to be n long and a matrix needs to be *n* x *n*. In other cases simply most of the records in matrix would be empty, however it needs to be prepared to hold the extremal cases.

The algorithm of reasoning the final state matrix from vectors is:

1. Assign the blocks detected on side camera. First block from side vector should be also detected on top camera. If it has been detected, then its position is known. All positions above it and in front of are for sure empty. All positions below it and after are unknown.
2. If next block has not been detected on top camera, it is below previous. If it has been detected, it is assigned to new position. Under the previous block there are some unknown blocks, because they can not be hold in air.
3. Repeat step 2 until the end of side vector.
4. Put rest of blocks from top camera on the scene. Calculate the number of detected blocks, and calculate how many blocks has not been detected.
5. Put the undetected blocks in left positions.

In order to fully understand the algorithm it is advised to read the example.

4.1 Example

In order to illustrate the algorithm an example has been created. This example shows the case with 6 blocks.

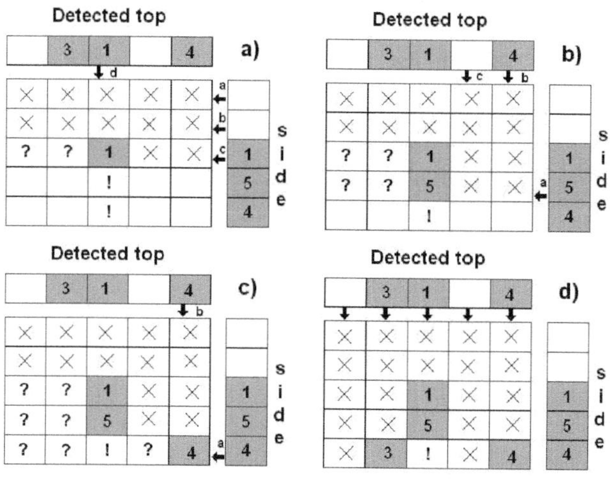

Fig. 2. Example of scene recognition

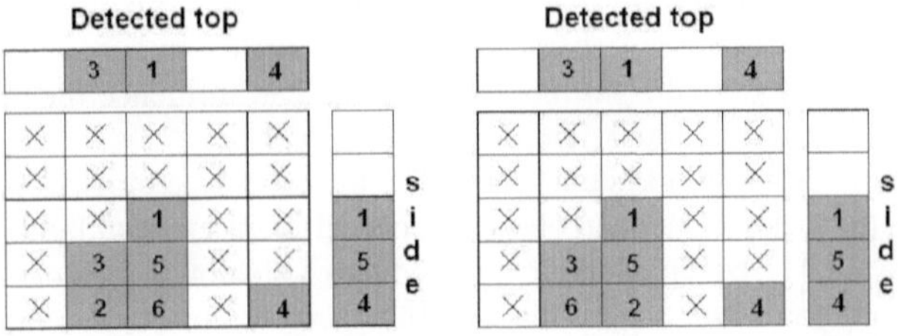

Fig. 3. Two possible results of the recognition

First of all it is important to notice that scene consists of 6 blocks, while blocks 2 and 6 has not been detected either on side nor top camera. Those blocks position needs to be concluded.

a) The algorithm analyzes the side vector from top to bottom. The first two empty values indicates that there are no blocks on this level. First found block is number 1. Top vector should also consist the number 1. The position of the highest block has been found. All positions from the left of this blocks are unknown, all positions below for sure consist a block.

b) Next block from side vector is block 5. This time it does not occur in top vector. This means that in needs to be under block 1.

c) Last block is number 4. It occurs in top vector. This means that we have no more information about previous pile, and another lower pile has been detected. In this case the new pile consist of only one block with number 4. All blocks on the left of it are unknown.

d) The bottom of the side vector has been reached. Now the top vector is being analyzed. All empty records means that there is no piles in this position and the used information about block 1 and 4 remains unchanged. The only new information is that there is a pile with block 3 on top. This block is put on the table for a moment.

Until this moment all known information has been processed. 4 blocks has been detected and there are two more. There is one sure position under block 5. This remains the only possibility that one of unknown blocks is under block 3, causing the unknown pile to be two blocks high. It is impossible to determine which block from those unknown are on which position. This cause the two possible results. In order to determine the correct result, one block needs to be removed and the algorithm needs to be done once again searching for 5 blocks.

4.2 Types of Uncertainty during Scene Recognition Process

The problem of the scene construction is that vision system does not have full knowledge about the environment. In this case system need to work under uncertainty. This is the case where uncertainty is caused by the incompleteness of reasoning. On the other hand uncertainty can be also caused by the errors, while obtaining facts. If the knowledge is mistaken, the conclusions would also be wrong.

Uncertainty is a widely handled problem in artificial intelligence. The uncertainty occurs when a obtained information differs in some percentage from the actual real state or can not be concluded at all [2].

4.2.1 Uncertainty Caused by the Recognition Errors in Vision System

Dependent on construction of vision system, the errors in recognition process can vary. Changing the external lightning of the scene can influence the detected color of the blocks, causing wrong block identification.

Let us assume that vision system would like to find a blue block denoted as 100 hue value. During the tests, it has been observed that different external lightning can slightly change this value. Moreover, each blocks produces a shadow which can also disturb the results. All those parameters cause that this block contain pixels in range about 96-104. This leads to following problem. Let us assume that we have two blocks: blue (110) and cyan (90). Due to influence of previously described phenomenons, a block with dominant value 100 has been detected. The problem is: is it blue or is it cyan.

This problem can not be solved. The only possibility is to reduce the error influences or calibrate the vision system. However, in less extremal cases, the detected pixels should be assigned to the closest defined color.

4.2.2 Uncertainty Caused by the Incompleteness of Information

This type of uncertainty occurs only in some special cases. If there are only up to two blocks, there is no uncertainty. If there are more blocks, the uncertainty is more and more often and complicated. There are two types of this uncertainty. The case when, it is sure that block is in certain position, but system do not know which color, and a case when system knows that there is one block that was not detected and it do not know where this block is.

Following examples shows the types of uncertainty for growing number of blocks.

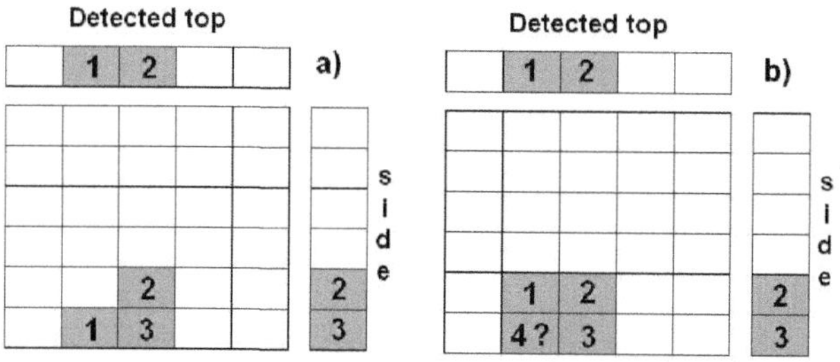

Fig. 4. Recognition of scene (a)with three blocks, (b) with four blocks

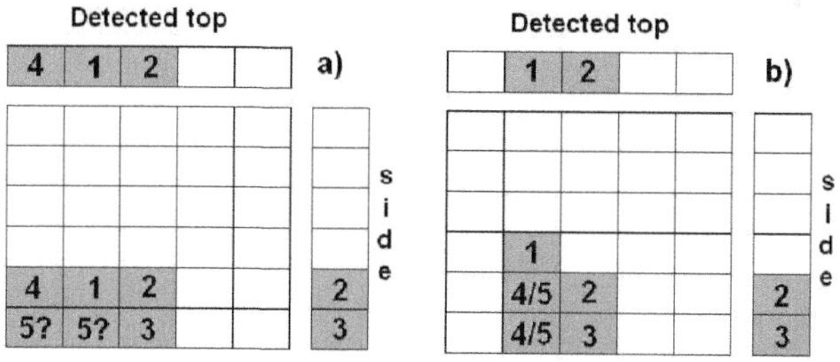

Fig. 5. Recognition with unknown (a) where to place fifth block, (b) which blocks put on detected positions

The situation with one block is too simple to consider. If there are two blocks, no matter of combination, they would always be detected with no uncertainty. If there are three blocks, there can be one situation where an uncertainty occurs. This is the case when it is sure that a block is in this certain place, but system do not know which block. If system has all information about blocks, knowing that two blocks has been placed, the missing third block can be placed only in one position. All other cases with three blocks are recognized without uncertainty. When the recognition state is as shown on Fig. 4, it can be recognized in two ways. The first recognition is correct if scene contains three blocks. This is the case without uncertainty. However if system knows that it should look for four blocks, it finds the place where the block can be placed. In case of four blocks, all detected uncertainty can be recognized completely. This example shows that it is crucial to know how many blocks there are on the scene in order to properly recognize it.

In case of five blocks the uncertainty level becomes more complicated. Two more uncertainties are introduced on Fig. 5. First case shows the situation when system do not know where to put remaining fifth block. There are two possibilities, and no way of determining the true solution. In second situation, it has been detected that two positions contains unseen blocks. In this case both blocks can be on either positions. Those cases shows the incomplete recognition. In both cases all recognition possibilities are passed to final recognition state.

5 Conclusion

Proper problem state recognition and representation is crucial for efficient planning. In our case some of problem environment components was not seen by sensors and needed to be concluded. The problem was that this conclusion could lead to a set of possible initial states of planning problem. The proposed reasoning algorithm about the state of the problem can reduce such understood uncertainty and increase efficiency of planning. Presented algorithm works correctly if all information from vision system is with no errors. For small number of blocks the output data can be useful, however if the number of blocks grow, the uncertainty in detection becomes more frequent, giving more uncertainty then facts on the output.

Acknowledgments. This work has been supported by Ministry of Science and Higher Education In the years 2010 – 2012 as development project O R00 0113 12 for the second author.

References

1. Baral, C., Kreinovich, V., Trejo, R.: Computational complexity of planning and approximate planning in presence of incompleteness. Artificial Intelligence 122, 241–267 (2000)
2. Blythe, J.: An Overview of Planning Under Uncertainty. Pre-print from AI Magazine 20(2), 37–54 (1999)
3. Bylander, T.: The computational complexity of propositional STRIPS planning. Artificial Intelligence 69, 165–204 (1994)
4. Bylander, T.: A linear programming heuristic for optimal planning. In: Proceedings of the 14th National Conference on Artificial Intelligence, pp. 694–699 (1997)

5. Cocosco C.A.: A review of STRIPS: A new approach to the application of theorem proving to problem solving by R.E. Fikes, N.J. Nillson, 1971. For 304-526B Artificial Intelligence (1998)
6. Galuszka, A., Swierniak, A.: Planning in Multi-agent Environment Using Strips Representation and Non-cooperative Equilibrium Strategy. Journal of Intelligent and Robotic Systems 58(3), 239–251 (2010)
7. Grzejszczak, T.: Semantic representation of Block World Environment: algorithm of scene reasoning from incomplete information. Electrical Review, R. 87 NR 2/2011 (2011) (to be published)
8. Gupta, N., Nau, D.S.: On the complexity of Blocks World planning. Artificial Intelligence 56(2-3), 223–254 (1992)
9. Kim, K.H., Hong, G.-P.: A heuristic rule for relocating blocks. Computers & Operations Research 33, 940–954 (2006)
10. Koehler, J., Schuster, K.: Elevator Control as a Planning Problem. In: The Fifth International Conference on Artificial Intelligence Planning and Scheduling Systems Breckenridge, CO, April 15-19, pp. 331–338 (2000)
11. Nillson N.J., R.E. Fikes: STRIPS: A new approach to the application of theorem proving to problem solving. Technical Note 43, SRI Project 8259, Artificial Intelligence Group, Stanford Research Institute (1970)
12. Slaney, J., Thiebaux, S.: Block World revisited. Artificial Intelligence 125, 119–153 (2001)

Research Opportunities in Contextualized Fusion Systems. The Harbor Surveillance Case

Jesus Garcia[1], José M. Molina[1], Tarunraj Singh[2], John Crassidis[2], and James Llinas[2]

[1] GIAA, Universidad Carlos III de Madrid. Av. Univ. Carlos III, 22, Colmenarejo, Spain
{jesus.garcia, josemanuel.molina}@uc3m.es
[2] CMIF, State University of New York at Buffalo, Bell Hall 313. Buffalo, NY 14260, USA
{llinas,tsingh,sudit,crassidis}@buffalo.edu

Abstract. The design of modern Information Fusion (IF) systems involves a complex process to achieve the requirements in the selected applications, especially in domains with a high degree of customization. In general, an advanced fusion system is required to show robust, context-sensitive behavior and efficient performance in real time. It is necessary to exploit all potentially relevant sensor and contextual information in the most appropriate way. Among modern applications for IF technology is the case of surveillance of complex harbor environments that are comprised of large numbers of surface vessels, high-value and dangerous facilities, and many people. The particular conditions and open needs in the harbor scenario are reviewed in this paper, highlighting research opportunities to explore in the development of fusion systems in this area.

Keywords: Harbor surveillance, Sensor Fusion, Context Representation, Situation Analysis.

1 Introduction

The continuous development of new applications of Information Fusion (IF) has increased the research interest in applicable techniques to extend capabilities of current ones. Nowadays, fusion systems are oriented to the integration of all types of sensor data and available information in databases, knowledge experience, contextual information, user mission, etc, in order to generate value-adding fused estimates of the conditions of interest in dynamically-changing situations. In particular, among the modern research challenges for IF process and algorithm design, has been the design and development of techniques to exploit contextual information, which is any kind of information which helps better characterize the situation/state. Contextual information can aid both the formulation of an improved state estimate and the interpretation of a computed estimate. However, the incorporation of contextual information adds complexity and demands new, hybrid techniques be developed in order to use different sources of information in an integrated reasoning process.

A current need is the study of novel methods to enhance current IF systems, applying adaptive paradigms to the management of processes, sensors and other available knowledge. Usually fusion systems are described according to JDL fusion model. Recent articles suggesting revisions and extensions of this model point to key aspects

J. Cabestany, I. Rojas, and G. Joya (Eds.): IWANN 2011, Part II, LNCS 6692, pp. 621–628, 2011.

in real applications [1],[2] such as evaluation and quality-control processes to increase reliability, the adaptation mechanisms to maximize the output value in the application, the need for and exploitation of an ontologically-based approach, or the role of distributed IF.

The two general objectives highlighted in this work are (i) exploring ways of modeling contextual information in the fusion process, looking for possible general formalizations, and (ii) design and development methods for adapting the data processing systems based on the fusion quality, context and requirements imposed by the application. The research in these areas will be focused in studying IF technology applications to maritime surveillance, with a particular emphasis on the harbor/port surveillance problem.

2 The Maritime Scenario. Requirements and Research Needs

The selected application, maritime surveillance, is a high-priority aspect for national and international security programs, especially in the coastal waters, borders and ports. The surveillance in these zones faces challenging problems to analyze such as terrorist threats, maritime and ecological accidents, illegal immigration, illegal fishing, drug trafficking, etc. Therefore, the research in maritime surveillance is mainly promoted by state agencies such as NATO, who supports to research on Maritime Surveillance, and national programs like Hawkeye or Centurion 0 projects in the United States to prevent threats in ports. In the European Maritime Policy "Blue Paper" [4], the European Commission states a general target of European surveillance network, composed of "interoperable surveillance system" for maritime safety, protection of the marine environment, fisheries control, control of external borders and other law enforcement activities.

2.1 Sensor Fusion for Maritime Surveillance

In order to achieve the required level of quality in maritime surveillance, it is necessary to use a heterogeneous network of sensors and a global multi-sensor tracking and fusion infrastructure capable of processing the data. The system must determine potential threatening objects within a scene containing a complex, moving background, minimizing the errors (both false positives and negatives), and exploiting scene-related knowledge to exploit the sensor data with maximum accuracy.

There are varied technologies for detection and location (coastal radar, video cameras, IR, automatic identification system, etc), but none of them alone are able to ensure reliable surveillance for handling complex scenarios. For example, high resolution coastal radar technology is effective with high accuracy and availability, but usually presents difficulties which make it necessary to supplement with cooperative location technologies. Radar can have problems such as occlusions, shadows, clutter, etc., and difficulty detecting small boats, because they are very small with low detectablity (for instance small inflatable boats in trafficking activities or skiffs in piracy, both with poor radar returns). Automatic Identification System (AIS) technology can provide situational awareness with positive identification of approaching vessels, but they are obviously insufficient on their own, because of needed cooperation, and occasional presence of anomalous data, losses in coverage etc. Therefore it is a usual situation to seek help of additional sensor sources such as computer vision systems to improve the detectability of all type of targets. The fusion system must take into account the characteristics of all data sources. Research of appropriate architectures and algorithms for multi-sensor fusion in this environment is needed, especially with large and heterogeneous areas and high density spaces with large numbers of very diverse tracked objects (tankers, ferries, sailboats, inflatable boats, etc.).

2.2 The Harbor Scenario as a Highly Contextualized Case

The harbor is one of the most complex maritime scenarios. The surveillance system in this area must analyze the situation to monitor and control entry to ports and land waterways using the available sensors and other sources with very specific information. The concerns of surveillance system are safety and efficiency in traffic operations, with additional considerations regarding the operation of oil and gas stations. Besides, the representation of situation will allow provide the pilots with aids such as navigation in port and docking assistance (which can be mandatory depending on vessels type and harbor configuration).

Fig. 1. Configuration of harbor La Spezia, Italy

An example of port configuration is shown in Fig. 1. We have selected Port of La Spezia [3] for illustration because it clearly illustrates the complex portfolio of activities that can take place in large ports. We can appreciate the very diverse nature of operations carried out in a big harbor like La Spezia, in Italy. In a reduced area there is coexistence of very diverse operations: cargo containers traffic, routes for related with Liquid Natural Gas re-gasification terminal (LNG), passenger operations, recreational boats in marinas, defense, etc.

There are clear indications for pre-planned ship mooring arrangements, approach-speed monitors and mooring strain sensors to facilitate vessel arrival and support safe mooring operations to eliminate damage to facilities. Some examples of mandatory and suggested routes and traffic indications in this specific site are indicated below (depicted in Fig. 2):

Outer channels:

- Merchant ships must follow a "safe speed", taking care of traffic conditions and circumstances at any moment, and avoiding the production of waves that can cause troubles to small boats/vessels which, in particular during summertime, are used to sail in the area and along the coast.
- Some zones are considered dangerous for maritime traffic due to military exercises. Three compulsory tracks for landing at la Spezia Port are defined

Inner channels:

- Towed ships coming in or out of the inner channels have precedence over all other ships or boats inside inner channels; merchant ships cannot exceed six knots in speed, except extraordinary circumstances imposed by rigging demands.
- When the vessel meets vessels sailing in the opposite direction and/or when are closed to berthed vessels, they have to decrease their speed if possible.
- The connection between Punta San Bartolomeo and the white light of Molo Lagora it is possible the transit of vessel with at most 30' deep draft.
- In the southern the connection between the top of Enel wharf and the root of Molo Lagora it is possible the transit of vessel with at most 27' deep draft.

Fig. 2. Routes for traffic of the harbor in La Spezia, Italy

So there is much information related with regulations and predefined behavior in the harbor area, information that should be used to characterize the situation, interpret the sensor data and focus the analysis in the expected normal operations. The process of assessing situations and threats requires monitoring not only critical harbor facilities but also linked coastal areas, sea surface, underwater, etc. Some of the issues defining this scenario are:

- Large number of vessels ranging from small recreational sailboats, tug boats, jet skis, to commercial vessels.
- Detection difficulties such as clutter and low resolution
- Tracking requires good data association for a multi-sensor tracking scenario
- Threats and potential conflicts for traffic operation (such as lack of adequate separation) must be detected.

3 General Approach for Context-Based Adaptive Fusion System

Complexity in maritime and harbor scenarios calls for designing advanced fusion systems capable of processing all available sensor data and information to support the multiple decision makers at different levels. Figure 3 shows a general approach to context-based adaptive fusion. It follows the structure proposed by Llinas [2] for a general framework of robust information fusion systems, where multiple processes work in parallel to achieve context-sensitive and adaptive behavior.

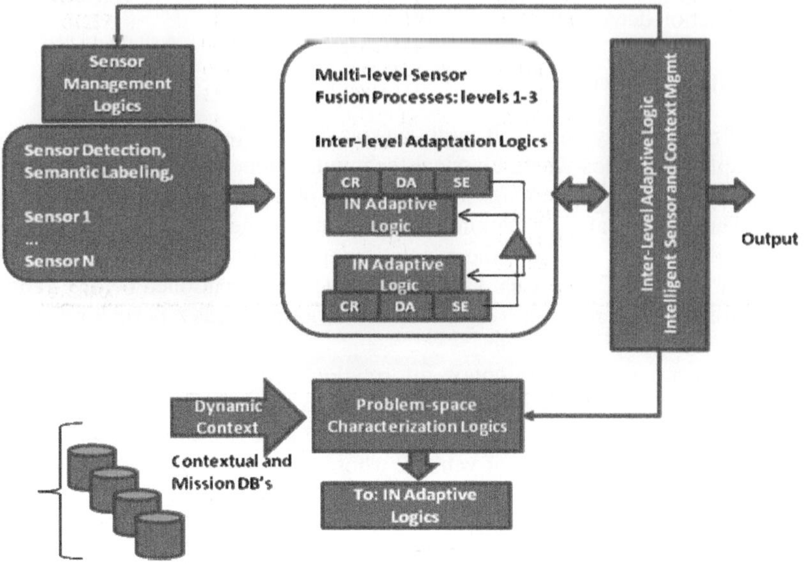

Fig. 3. Overall architecture of the general context reasoning environment [2]

Here, the core with fusion processes contains levels 1-3, and the external adaptation process belongs to level 4. Every IF process in JDL is abstracted in the three functions of Common Referencing (CR), Data Association (DA, and State Estimation (SE). Several aspects are considered with respect to a non-adaptive conventional fusion system:

- A function module called Problem Space Characterization Logic is in charge of adaptively managing the system. Making use of contextual knowledge and a library of alternative algorithms, an intelligent algorithm manager can terminate and invoke the best algorithm for the current problem-space condition. Although advanced fusion algorithms work close to optimality, they usually operate under limited conditions (assumed models, sensor performance, information consistency, etc.). The combination of different intelligent techniques is a possibility to overcome limitations of individual techniques and available implementation of fusion algorithms, with the aim of adapting the fusion system performance to the requirements, and to different situations.
- Contextual information has an essential role, which feeds this knowledge base and adaptation logic. The adaptation of data processing algorithms (extensible to sensor management when they have control parameters, such as coverage, refresh cycle, etc.) needs a model of interrelation between all aspects of fusion.

4 Research Areas in Harbor Scenario Applications

Finally, among the open challenges to develop advanced fusion systems in the context-dependent situation as a harbor case, we indicate three specific research challenges as immediate steps: context-aided tracking systems, situation analysis and coordinated action detection.

4.1 Contextualized Sensor Fusion

The area of contextually-enhanced multi-sensor, multi-object tracking is very important in complex scenarios where classical techniques alone (typically based solely on observational data) are not enough to model the objects behavior and dynamics of situation. The contextual information about the environment allows a more accurate interpretation of sensor data and the adaptation/optimization of system performance. Examples of sources of contextual knowledge have been mentioned before: geometrical configuration of harbor, moving lines and mooring areas, speed limit, types of vessel and associated priorities, etc.

Depending on the nature of context information, different strategies are appropriate. The a priori known static context can be used to determine the representation and constraints to be taken into account in the inference processes. For example, assign a detected object with its category of vehicle, whose constraints are known, can be used to refine its dynamic model. In this way, in [5], [6], fusion parameters are adapted depending on regions (sea, ocean, regions of high clutter density, etc.). In recent works related in port areas [7], the limitations derived from channel depth, limited areas, etc. are used to characterize the "trafficability" and refine the prediction models. Knowledge of dynamic context also enables algorithms to interpret the data received [8]. A dynamic representation of contextual information, inferred with the help of other fusion levels and their contextual information can be used to enhance the sensor data processing. An example can be the description of multi-object situations such as vessels in coordinated motion according to traffic movement protocols, expected maneuvers, knowledge of relationships between entities, active pilotage/towage operations among vessels and tug boats, etc.

4.2 Situation Analysis

The inference of suspicious object behavior is one of the main objectives of fusion systems, in order to focus the operator attention. Many factors can be used to characterize the situation: object classes, speeds, relative positions (with respect to allowed zones according to its class, etc.). The surveillance system must decide which events are anomalous, to recognize the situation of mobile elements with respect to the static elements in the scene. Contextual information can clearly help understand whether certain behaviors are anomalous or not.

Rule-based systems can be configured to identify situations that are out of the norm and provide alarms, such as the presence of ships unidentifiable or vessels approaching unsafe waters. The definition of normal behavior accordingly to maritime/harbor rules should be formalized, with the possibility of ontology formalisms to represent and include reasoning mechanisms at different levels [9].

4.3 Collective Behavior Analysis

An important challenge for situation analysis in the harbor and maritime scenarios is the capability to detect or anticipate collective behavior that may represent threats. An example is a coordinated attack of several boats against a certain target (big ship, land facility, etc.). Contextual information can help in such IF-based estimation techniques by providing inputs that define known coordinated vessel activities such as planned tugboat-ship operations, coordinated derding operations, etc.

Group behavior recognition would be based on features referred to several objects. A possibility is a trajectory-based approach in which multi-agent action recognition involves the compact representation and modeling of actions and interactions, as well as their logical and temporal relations. This type of approach has been used in works related with robotics and sports analysis [10]; most of them divide the collective activity into individual actions. The application of these approaches to this problem would need in the first step to have available the individual trajectories of all objects, which depends on the sensor resolution and capability of tracking systems.

5 Conclusion

We have presented some important open issues for modern information fusion systems. The case of harbor surveillance is a representative example since it opens challenging research problems, requiring processing information at different levels, from multi-sensor data to domain knowledge. Contextual information is essential for building a coherent and accurate surveillance concerned with security and safety events, to interpret sensed observations about a high number of different types of vessels appearing in diverse operations. The future work will focus in the formalization of the domain knowledge, extension of fusion framework with adaptation mechanisms, and multi-level fusion strategies to integrate sensor data and structured knowledge.

Acknowledgements

This work was supported in part by Projects CICYT TIN2008-06742-C02-02/TSI, CICYT TEC2008-06732-C02-02/TEC and CAM CONTEXTS S2009/TIC-1485. A funded stay of Jesus Garcia in CMIF allowed the collaboration to define research strategies. The authors would like especially thank to Dr. Kessel, at NATO Undersea Research Centre, for helpful discussions on port traffic in the presented case of study.

References

1. Steinberg, A.N., Bowman, C.L.: Revisions to the JDL data fusion model. In: Liggins, M.E., Hall, D.L., Llinas, J. (eds.) Handbook of Multisensor Data Fusion. CRC Press, Boca Raton (2009)
2. Llinas, J.: A survey and analysis of frameworks and framework issues for information fusion applications. In: Graña Romay, M., Corchado, E., Garcia Sebastian, M.T. (eds.) HAIS 2010. LNCS, vol. 6076, pp. 14–23. Springer, Heidelberg (2010)
3. The Port La Spezia. Port Authority (2010), `http://www.porto.laspezia.it` (accessed November 2010)
4. On a Draft Roadmap towards establishing the Common Information Sharing Environment for the surveillance of the EU maritime domain. Commission of the European Communities, An Integrated Maritime, `http://eur-lex.europa.eu/` (accesed February 2011)
5. Benavoi, A., Chisci, L., Farina, A., Immediata, S., Timmoneri, L.: Knowledge-Based System for Multi-Target Tracking in a Littoral Environment. IEEE Trans. on Aerospace and Electronic Systems 42(N3), 1100–1119 (2006)
6. García, J., Guerrero, J.L., Luís, A., Molina, J.M.: Robust Sensor Fusion in Real Maritime Surveillance Scenarios. In: Proceedings of the 13th International Conference on Information Fusion (Fusion 2010), Edinburgh, UK (2010)
7. George, J., Crassidis, J.L., Singh, T.: Threat assessment using context-based tracking in a maritime environment Information Fusion. In: 12th International Conference on Issue Date: July 6-9, seattle, WA (2009)
8. Rhodes, B.J.: Knowledge Structure Discovery and Exploitation from Multi-Target Classifier Output. In: 7th International Conference on Information Fusion, Stockholm, Sweden (2004)
9. Gómez-Romero, J., Patricio, M.A., García, J., Molina, J.M.: Ontological representation of context knowledge for visual data fusion. In: Ontological representation of context knowledge for visual data fusion. 12th International Conference on Information Fusion, pp. 2136 – 2143 (2009)
10. Perse, Matej, Ljubljana: A trajectory-based analysis of coordinated team activity in a basketball game. Journal Computer Vision and Image Understanding 113(5), 612–621 (2009)

Multiagent-Based Middleware for the Agents' Behavior Simulation

Elena García, Sara Rodríguez, Juan F. De Paz, and Juan M. Corchado

Computers and Automation Department, University of Salamanca, Salamanca, Spain
{elegar,srg,fcofds,corchado}@usal.es

Abstract. Nowadays, simulation is used for several purposes ranging from work flow to system's procedures representation. The main goal of this work is the design and development of a middleware to communicate the current technology in charge of the development of the multiagent systems (MAS) and the technology in charge of the simulation, visualization and analysis of the behavior of the agents. It is a key element when considering that MAS are autonomous, adaptive and complex systems and provides advances abilities for visualization. The adaptation of technology in charge of the development of MAS to support the notion of time is the most important and complex feature of our proposal The proposed middleware infrastructure makes it possible to visualize the emergent agent behaviour and the entity agent.

Keywords: Multiagent systems, Simulation, JADE, Repast.

1 Introduction

Agents and multiagent systems (MAS) are adequate for developing applications in dynamic and flexible environments. Autonomy, learning and reasoning are especially important aspects for an agent. These capabilities can be modelled in different ways and with different tools [11]. The contribution from agent based computing to the field of computer simulation mediated by ABS (Agent Based Simulation) is a new paradigm for the simulation of complex systems that require a high level of interaction between the entities of the system. Possible benefits of agent based computing for computer simulation include methods for evaluation of multi agent systems or for training future users of the system [3]. The properties of ABS makes it especially suitable for simulating this kind of systems. The idea is to model the behaviour of the human users in terms of software agents.

Mainly there are two ways for visualizing multiagent systems simulation: the agents interaction protocol and the agent entity. In the former, it is visualized a sequence of messages between agents and the constraints on the content of those messages. On the other hand, the latter method visualizes the entity agent and its iteration with the environment. Most software programs, such as JADE platform [1][8] or Zeus toolkit [2], provide graphical tools that allow the visualization of the messages exchanged between agents. The toolkits MASON [5], Repast (Recursive Porous Agent Simulation Toolkit) [6][9] and Swarm [10] provide the visualization of the entity agent and its interaction with the environment. Repast seeks to support the development of

J. Cabestany, I. Rojas, and G. Joya (Eds.): IWANN 2011, Part II, LNCS 6692, pp. 629–636, 2011.

extremely flexible models of living social agents, but is not limited to modelling living social entities alone. Repast is differentiated from other systems since it has multiple pure implementations in several languages and built-in adaptive features such as genetic algorithms and regression [7].

The most well-known agent platforms (like JADE [8]) offer basic functionalities for the agents, such as AMS (Agent Management System) and DF (Directory. Facilitator) services; but designers must implement nearly all organizational features by themselves, like simulation constraints imposed by the MAS topology. In order to model open and adaptive simulated systems, it becomes necessary to have an infrastructure than can use agent technology in the development of simulation environments. The presented middleware makes use of JADE [8] and Repast [9], and combines them so that it is possible to use their capabilities to build highly complex and dynamic systems.

The main contribution of this paper is the reformulation of the FIPA protocol used in JADE [8], the most widely used platform for based software agents middleware, achieving several advantages: (i) development of a new middleware that provides independence between the model and visualization components; (ii) improvement on the visualization component that makes it possible to use the concept of "time", essential for simulation and analysis of the behavior of agents; (iii) improvements to the user capabilities to which several tools were added, such as message visualization, analysis behavioral, statistics, etc.

The article is structured as follows: Section 1 makes a review of agent-modeling toolkits and presents the challenges for simulated multiagent systems. Sections 2 introduces a description of the middleware specifically adapted to the simulation of multiagent systems within dynamic environments. Finally, some results and conclusions are given in Sections 3.

2 Middleware for Behavior Simulation

MISIA (*Middleware Infrastructure to Simulate Intelligent Agents)* is a middleware infrastructure that allows to model JADE multiagent systems with the possibility of being represented in Repast. The main concept introduced in this environment is the notion of time in JADE, which means it is possible to render in real time the events into Repast. One of the main differences between JADE and Repast is that in JADE, there not exists the concept of time as such, and the agents interact each other based on changes or events that occur in the execution environment. However, Repast has a time unit: the tick, which is what sets the pace and allows simulations. Agents in the JADE context are implemented based on FIPA standards. This allows to create multi-agent systems in open environments, which is not possible within Repast. These differences are what MISIA solved, integrating these two environments and achieving a working environment for creation and simulation of multiagent systems more powerful and versatile.

It is necessary to synchronize JADE to work simultaneously to Repast. This is achieved by keeping the JADE agents informed about the tick of the simulation they are involved. Moreover, agents are informed when a tick is elapsed. To obtain versatile simulations, it is necessary that all events occurring in JADE are rendered

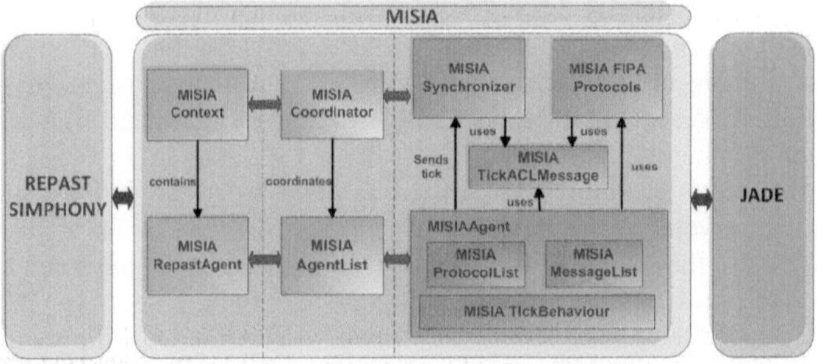

Fig. 1. Functional structure of MISIA

instantly on Repast. The minimum unit of time is the tick, thus, the idea is that every JADE agent can perform functions in a tick (must be simple actions, such as sending a message, receiving or re-establishment of their state) and once finished, they can be updated in Repast. This must occur during the course of all ticks, which are getting updated in real time all events.

The bottom layer of the framework is which connects JADE, and is divided into four functional blocks: (i) *MISIAAgent***,** is the extension of JADE agent. Performs the same functions, but adapting them to the presence of ticks. It consists of a number of features to manage the time in JADE. (ii) *MISIATickACLMessage***.** JADE messages are used for communication between agents. MISIAAgent agents communicate between them with MISIATickACLMessage messages. MISIATickACLMessage is the extension of JADE ACL message that incorporates the concept of time. It includes aspects such as the tick where to send the message, and the delay that the message has when achieves its destination. In JADE, the messages exchanged between agents are sent and arrive instantly, but in real life, that is not the case. It aims to simulate and view the evaluation of the system as time passes, and to achieve this, it is necessary that messages are not instant, but must have a shipping time and a different reception time. (iii) *MISIAFIPAProtocols***.** JADE implements FIPA standards, which, among other things, specify multiple communication protocols. These define a series of patterns that respond to different types of communication that two or more agents can perform. The objective is to adapt FIPA protocols defined in JADE with Repast ticks. (iv) *MISIASynchronizer* is a JADE agent that acts of notificator. It is responsible for notifying the MISIAAgent when a tick goes by. Is the system clock synchronization. When a tick goes by, MISIASynchronizer is notified in order to notify MISIAAgents. It is made through MISIATickACLMessage messages with a special performative.

The top layer is the contact with Repast. Contains two functional blocks, which are: (i) *MISIARepastAgent*. Each MISIAAgent existing in the system will be represented by a MISIARepastAgent in the context of Repast. This means that for every agent that we want to have on the system actually have to create two: a MISIAAgent agent running on JADE, and its respective MISIARepastAgent released on Repast. It can be seen as follows: a logical agent, and two physical agents. MISIARepastAgents have an important role: they cannot updated their status until their respective MISIAAgents does not end with all the work they need to perform during that tick. This is a very important aspect, since it is the characteristic of the framework as a system in real time. (ii) *MISIAContext* has two important goals. One is to establish the synchronism in the execution. When a tick goes by, lets know MISIASynchronizer agent that it is necessary to notify MISIAAgent agents that following tick happened. The other goal is to incorporate new agents MISIARepastAgent that entry in the context of the Repast simulation. For each new MISIAAgent that appears in the system, MISIAContext will create their respective MISIARepastAgent and will added it to the simulation environment.

Finally, the intermediate layer is divided into two functional blocks, and its goal is to join adjacent layers. These modules are: (i) *MISIAAgentList*, as its name implies, stores all agents in the system at a given time. It plays an important role because it enables communication between a MISIAAgent and their respective MISIARepastAgent, and vice versa. The diagram shows two-way information flows ranging from MISIARepastAgent to MISIAAgentList and MISIAAgentList to MISIAAgent. These flows are representing that communication, that union between the two physical agents, to confine a logical agent. (ii) *MISIACoordinator* coordinates communication between the two adjacent layers. It is necessary the presence of a coordinator to maintain synchronism between both layers. Thanks to MISIACoordinator, MISIAContext can notify the occurrence of a tick to MISIASynchronizer, and MISIASynchronizer can assure that its purpose is served to MISIAContext, reporting that all MISIAAgent received tick. This kind of communication is necessary to maintain full synchronization between the two platforms.

2.1 Redefinition of FIPA Protocols

JADE has a number of implemented FIPA protocols, which help the programmer. With these protocols, it abstracts the developer from having to prepare messages to be sent, sending, or to of manage the reception them, among other things. In this framework has been re-implemented FIPA protocols defined in JADE to support the notion of time.

In the FIPA protocols implemented in Jade [4], it is possible to observe the presence of two roles: *Initiator* and *Responder* o *Participant.* Jade provides a predefined class for each role and each type of FIPA interaction protocol, or rather, for a certain group of FIPA protocols. In the *jade.proto* package are all the classes that, in the form of behaviors , facilitate the implementation of the FIPA communication protocols. Each pair of classes is indicated to implement a series of protocols. MISIA aims to adapt all these classes to their environment, so that an end user can use them as in Jade, without worrying about the presence of time. For example, with the first pair adapted (*AchieveREInitiator* and *AchieveREResponder*), it is possible to implement *FIPA-Request, FIPA-Query, FIPA-Recruiting, FIPA-Request-When* y *FIPA-Brokering* protocols.

To implement any of these protocols in MISIA, it is necessary to use *AchieveREInitiator* (Jade class) and *MisiaAchieveREResponder*, the adapted class of the *Responder* role. *MisiaAchieveREResponder* intends to replace *AchieveREResponder* (Jade class). It's provides two handling methods, such as Jade: *manejarPeticionRequest,* to send the first message in response, and *manejarResultadoPeticionRequest,* to send a second message to the agent with the Initiator role. In addition, it implements the exceptions, to try to provide the same Jade interface (*MisiaNotUnderstoodException, MisiaRefuseException* y *MisiaFailureException).* The exceptions are important because Jade uses them to send messages of rejection or no understanding of a task (i.e. if *Responder* role sends a message of acceptation for a task, the execution flow does not diverge in exception).

Table 1. *jade.proto* package

Behaviors	FIPA protocols
AchieveREInitiator **AchieveREResponder** **SimpleAchieveREInitiator** **SimpleAchieveREResponder** **IteratedAchieveREInitiator** **SSIteratedAchieveREResponder**	FIPA-Request FIPA-Query FIPA-Recruiting FIPA-Request-When FIPA-Brokering
ContractNetInitiator **ContractNetResponder** **SSContractNetResponder**	FIPA-Contract-Net
SubscriptionInitiator **SubscriptionResponder**	FIPA-Subscribe FIPA-Request-Whenever
ProposeInitiator **ProposeResponder**	FIPA-Propose

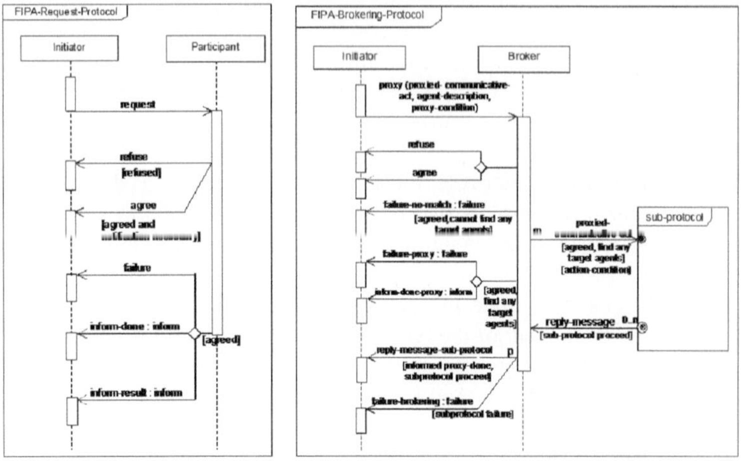

Fig. 2. FIPA-Request Protocol, FIPA-Brokering Interaction Protocol

The messages of *refuse, failure* and *notUnderstood* (Fig. 2) will diverge in exceptions, that also be adapted to the notion of time to send these messages in the desired tick. Thus, the equivalence between Jade classes (and methods) and MISIA is as shown in the table below.

Table 2. Relation AchieveREResponder (Jade) - MisiaAchieveREResponder (MISIA)

Jade	MISIA
`AchieveREResponder` (class)	`MisiaAchieveREResponder` (class)
`protected ACLMessage handleRequest(ACLMessage request)` (AchieveREResponder method)	`protected MisiaTickACLMessage manejarPeticionRequest (ACLMessage requestMessage)` (MisiaAchieveREResponder method)
`protected ACLMessage prepareResultNotification(ACLMessage request, ACLMessage response)` (AchieveREResponder method)	`protected MisiaTickACLMessage manejarResultadoPeticionRequest (ACLMessage requestMessage, ACLMessage responseMessage)` (MisiaAchieveREResponder method)
`NotUnderstoodException` (class)	`MisiaNotUnderstoodException` (class)
`RefuseException` (class)	`MisiaRefuseException` (class)
`FailureException` (class)	`MisiaFailureException` (class)

The communication protocols JADE defines two roles, which starts the conversation (Initiator role) and which is evolved in the conversation (Responder role). The Initiator agent role will begin by the conversation by sending a message to the recipient. Therefore, it follows the logic developed with the message queue. When a MISIAAgent agent wishes to follow a communication protocol in a given tick, just add the protocol of communication to the agent in the tick established. Therefore, one of the functions of MISIAAgent agent after receiving a tick is to add communication protocols. The rest of communication for sending and receiving messages is reimplementing, recording different behaviors that make the different functions of the protocols. The novelty is that these new behaviors support MISIA modules redefined for JADE, such as support MISIA-TickACLMessage messages or the ability to respond to a message in a certain tick, without being immediately.

An example reimplemented is the FIPA-Request protocol, which is like follows: the agent with Initiator role sends a request to agent with Responder role. Responder replies, accepting or rejecting the request, and immediately returns to answer the agent with Initiator role informing the result (if the request was made correctly, or there was a problem). With the new definition by MISIA of this protocol, it is possible to send messages during the tick chosen. In this case, MISIA only redefines the role Responder. The Initiator is not necessary because it only sends a message to the beginning.

In the case of the Responder role, must send two messages, as discussed above. So, MISIA provides to programmers two handles, like JADE; one to send the first message, and another to send the second one, abstracting from all the system logic that is to managing ticks.

Below is a fragment of code in Java where it shown how a behavior is reimplemented to manage the arrival of the request by the agent with Initiator role. In this example, *handleMISIARequest* is the procedure that the final developer overwrites to provide the message he want to send in response.

```
registerPrepareResponse(new OneShotBehaviour(){
   public void action() {
        //Get DataStore to obtain the request message
     DataStore ds = getDataStore();
     ACLMessage requestMessage = (ACLMessage) ds.get(REQUEST_KEY);
     TickACLMessage agreeMessage = null;
     try {  agreeMessage = handleMISIARequest(requestMessage);
      } catch (Exception e) {}
    //If the message isn't null, send
    if (agreeMessage != null)    jadeAgent.MISIASend(agreeMessage);    }
```

3 Experimental Results and Conclusions

It has been developed a case study using this middleware to create a multiagent system aimed at facilitating the employment of people with disabilities, so it is possible to simulate the behavior of the agents in the work environment and observe the agents actions graphically in Repast. This is a simple example that defines four jobs, which are occupied by four people with certain disabilities. Every job is composed of a series of tasks. Agents representing the workers have to do them, and according to their capabilities, carry out the assignment with varying degrees of success. Performing various simulations, and seeing the evolution in time, the results can be assessed to

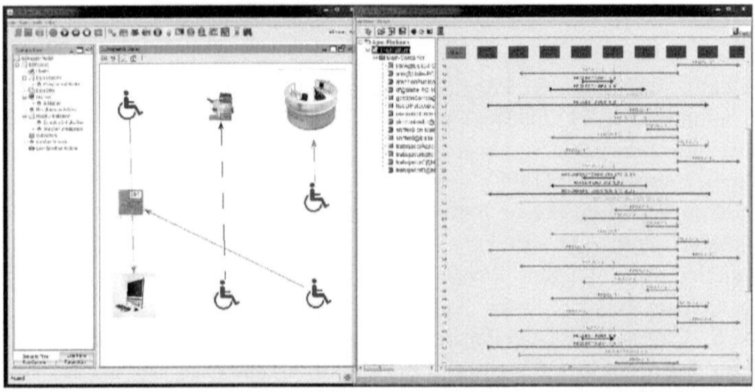

Fig. 3. Case Study MISIA

determine what would be the most suitable job for each employee. Below is an example of the execution of this case study. There are two ways for visualizing multiagent systems simulation: the agents interaction protocol and the agent entity.

MISIA provides the capabilities visualize the sequence of messages between agents and the entity agent and its iteration with the environment. The union of these two platforms involves having a highly efficient environment for the creation of multiagent systems, getting the benefits of JADE to create the systems, as is the use of FIPA standards, and also the visual representation and extraction of simulation data to different applications provided by Repast.

Simulation is a helpful tool for understanding complex problems. Therefore, the simulation of multiagent systems in several levels of details and the emergent behavior s fundamental for analyzing the systems processes. In this study, a list of basic concepts and advances is presented for the development of simulated multiagent systems. MISIA allows simulation, visualization and analysis of the behavior of agents. With the MAS behavior simulator it is possible to visualize the emergent phenomenon that arises from the agents' interactions.

The proposed visualization system also suggests further developments. One of them is make the agent representation more realistic. A 3D agent visualization in more levels of details showing the interaction them would make the system complete and realistic. Another future work is to improve interactivity with the user. The goal is to improve the interactivity by means of allowing the interaction of the specialists with the live execution besides the basic functionalities such as play, pause, stop and increase/decrease the speed, by means of putting some substances in the position and observing the emergent behavior. It would allow the self-organization optimization and the proposal of new hypotheses. Even more: generation of reports about the information visualized during the simulation process in several levels of detail, which could increase the comprehension about the process. MISIA is the ideal framework for this purpose.

Acknowledgments. This work has been supported by the Spanish Ministry of Science and Innovation, Project T-Sensitive, TRA2009_0096.

References

[1] Bellifemine, F., Caire, G., Poggi, A., Rimassa, G.: Jade a white paper. EXP in search of innovation 3(3), 6–19 (2003)

[2] Collis, J.C., Ndumu, D.T., Nwana, H.S., Lee, L.C.: The zeus agent building tool-kit. BT Technol. Journal 16(3) (1998)

[3] Davidsson, P.: Multi Agent Based Simulation: Beyond social simulation. In: Moss, S., Davidsson, P. (eds.) MABS 2000. LNCS (LNAI), vol. 1979, Springer, Heidelberg (2001)

[4] Foundation for Inteligent Physical Agents."FIPA Agent Management Specification". Disponible en, http://www.fipa.org/specs/fipa00001/SC00001L.html

[5] Luke, S., Cioffi-Revilla, C., Panait, L., Mason, S.K.: A new multiagent simulation toolkit. In: Proceedings of the 2004 SwarmFest Workshop (2004)

[6] North, M.J., Howe, T.R., Collier, N.T., Vos, J.R.: The repast symphony runtime system. In: Proceedings of the Agent Conference on Generative Social Processes, Models, and Mechanisms (2005)

[7] North, M.J., Collier Nicholson, T., Vos Jerry, R.: Experiences Creating Three Implementations of the Repast Agent Modeling Toolkit. ACM Transactions on Modeling and Computer Simulation 16(1), 1–25 (2006)

[8] JADE, Java Agent Development Platform, http://JADE.tilab.com

[9] Repast, http://repast.sourceforge.net/repast_3/index.html

[10] Swarm, http://www.swarm.org

[11] Wooldridge, M., Jennings, N.R.: Agent Theories, Architectures, and Languages: a Survey. In: Wooldridge, M., Jennings, N.R. (eds.) Intelligent Agents, pp. 1–22. Springer, Heidelberg (1995)

A Dynamic Context-Aware Architecture for Ambient Intelligence

José M. Fernández, Rubén Fuentes-Fernández, and Juan Pavón

Facultad de Informática de la Universidad Complutense de Madrid
Avda. Complutense, s/n. 28040 Madrid, Spain
{jmfernandezdealba,ruben,jpavon}@fdi.ucm.es
http://grasia.fdi.ucm.es/main/

Abstract. Ambient Intelligence (AmI) deals with scenarios where users receive services according to their state. This is possible thanks to environments populated with multiple sensors and actuators. The *context-aware* features focus on considering a rich knowledge about users, including the current events but also preferences, activities or social context. The effective availability of this information demands mechanisms that seamlessly gather and propagate it between different settings, and react dynamically to changes in the needs and the environment. Their design is one of the key difficulties in AmI. Our work addresses this problem with an architecture for the context subsystem of AmI applications. It specifies several layers of abstraction for information, the relevant components to provide their storage, management and distribution, and the automated lifecycle and binding of components to support the previous functionality. A case study on path guidance illustrates the architecture and is the basis to discuss the related work.

Keywords: context-awareness, architecture, ambient intelligence, component.

1 Introduction

Ambient-Intelligence (AmI) is a paradigm for the design of computational systems that presents the concept of *intelligent environment* [8]. An intelligent environment uses a set of embedded and unobtrusive sensors and actuators to determine the identity, features, state, and current activities of the users within its domain, anticipating their needs and offering adequate assistance [5]. A common use of these systems is people assistance in their daily life, which is known as Ambient Assisted Living (AAL), as in [6].

In this field, context-awareness is defined as the ability to use context to provide relevant services depending on the user's task. Context is any information relative to the different participants in the interaction that the system could need to achieve their goals [1]. It includes knowledge about the physical and human environment and the user, both current and got in the past, and both observed and inferred. In order to provide services in a comfortable and unobtrusive way

J. Cabestany, I. Rojas, and G. Joya (Eds.): IWANN 2011, Part II, LNCS 6692, pp. 637–644, 2011.

for users, the conversational bandwidth with the users must be high, i.e. systems should be able to obtain as much information of this context as possible without user involvement. This implies building and keeping updated the context, and making it available where suitable and required in order to minimize the need of explicit interaction with the user.

The construction of the context information consist on a process of data abstraction from the low-level information of the different sensors to the high-level information of the application components and services. Its main challenge is to orchestrate the system components for this information flow in a flexible way. There exists several frameworks that facilitate this task [2], but they present relevant limitations. The main one is that they do not usually specify how the lifecycle of the components is managed or how their bindings are solved in order to obtain a dynamic system. This makes difficult that the resulting systems take advantage of its experience to improve their context management and behaviour, or react to unexpected changes in its configuration or that of the environment.

Our work has developed a framework to address this problem with a component model that includes information management issues. It proposes splitting the information in abstraction layers that constitute the context, and gives guidelines to determine what information corresponds to each layer. The framework also includes an architecture with functionality for lifecycle management and automatic service discovering in order to automatically communicate the different abstraction layers of information in a transparent way for the developer. Each component only needs to declare what context information want to observe, and then use it relieving the binding and lifecycle details to the core components of the architecture. The discussion of this paper focuses on this component model.

The case study of a system that guides a user following a path within a room illustrates this model. The system knows the room map and the features of the available sensors. The guidance is based on a target path given by the activity in which the user is engaged, and the user's current position and past path. The position is inferred using the low-level information from distance sensors.

The rest of the paper is structured as follows. Section 2 presents the architecture, which section 3 applies to the case study. The case study drives the discussion on section 4 about alternative approaches to deal with the context subsystem in AmI. Finally, section 5 presents some conclusions and future work on the approach.

2 An Architecture for Context Management Subsystems

The architecture presented in this work is summarized in Figure 1. It is built on top of existing state-of-the-art component-oriented middleware. The only requirement for these middlewares is that they have to provide support for the dynamic boot of new components, the management of their dependencies and bindings, and service discovering. CORBA and OSGi are examples of platforms that support these functionalities.

Over the previous infrastructure, our context-aware architecture provides the following services:

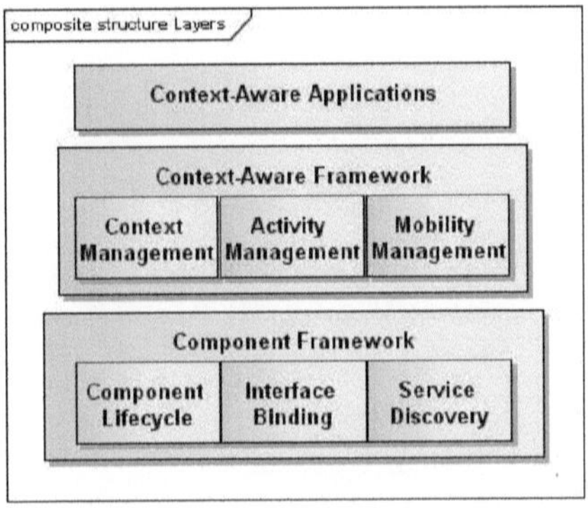

Fig. 1. Architecture layers of a context-aware AmI application

- **Context Management.** This service allows that application components request the desired context elements by using a predefined *ContextContainer* component. This component performs the required coordination in the framework to fulfill the request.
- **Activity Management.** This service provides activity detection and monitoring in the environment given a description of each activity in the form of state machines.
- **Mobility Management.** It manages the propagation of relevant information among different settings of the system.

Due to space reasons, the remaining discussion focuses on the first service. The use of the *context management* in a system requires designers to provide a description of the context and its related components in terms of the abstractions used in our architecture, that is, that they indicate for each component its type and relationships with other context elements. Figure 2 shows the main abstractions for this purpose available in our work:

- **ContextContainer.** This element is responsible of storing and retrieving context elements. The context elements are instantiated on-demand when requested for the first time, and destroyed when they are not being used.
- **InitializationDocument.** It contains an initial set of context elements that the context container reads and stores when booting.
- **ContextElement.** It represents the information items that constitute the context elements used by the system.
- **Entity.** It is either a *Person*, *Object*, *Place* or *Service.* These are the participants in the system interactions. The type differentiation its necessary to know which properties are applicable in each case.
- **Property.** It is any information that characterizes an entity and is necessary for some component in the system.
- **Person.** This element represents a person who interacts with the system.
- **Object.** A physical object that exists in the system or its environment, e.g. a physical sensor, a robot or a piece of furniture.

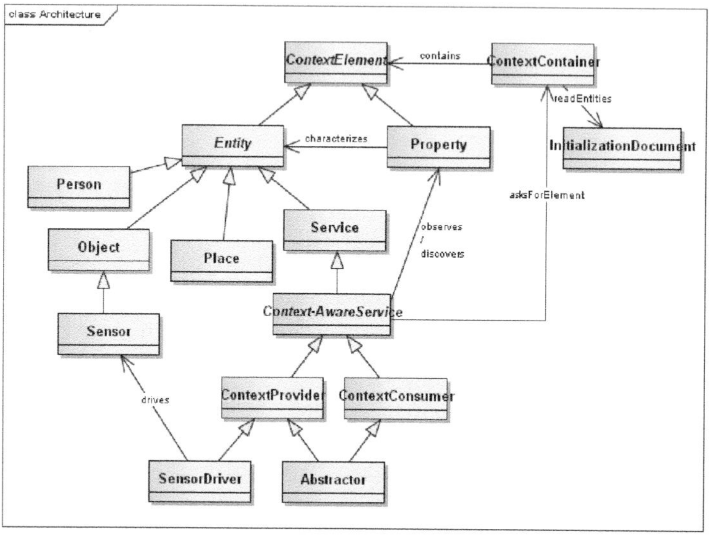

Fig. 2. Main classes of the architecture and its relationships

- **Sensor.** An object that corresponds to a sensor or peripheral that collects data from outside the system boundaries.
- **Place.** A location in the environment, e.g. a room or a spot within a room.
- **Service.** It is an abstract (software) service that runs in the system. A database or a web server are examples of services.
- **Context-AwareService.** Either a context provider or consumer.
- **ContextProvider.** It is a type of service able to calculate a property of an entity in the context, e.g. a component that can obtain the user's preferences.
- **SensorDriver.** A component that communicates directly with sensors and provides their state to other components in the system as a context property.
- **ContextConsumer.** It is any service using a context property for its function, e.g. a component whose processing takes the user's preferences as input.
- **Abstractor.** This represents a kind of service that observes a set of properties and produces or inferres from them the value of a new property.

With these elements, the proposed framework is able to manage the context of AmI applications in a general way. Their actual use for a system is illustrated in next section.

3 Case Study: Path Guidance

To illustrate the previous architecture, this section considers an application that helps a user to find her path in a room in the context of a given activity. The development of the context components is divided into two tasks: identifying and structuring the relevant information and initial components of the system

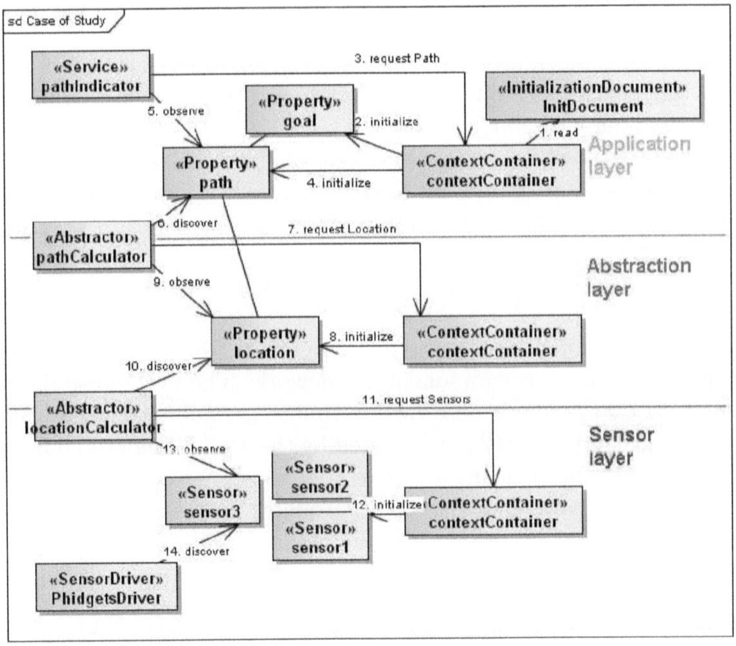

Fig. 3. Collaboration diagram for the path guidance system

(i.e. sensors and processors of information in the environment and domain), and determining the components for their management.

As part of the methodology for the development of applications within the proposed framework, a first recommended step is to to divide the initial components in layers, as shown in Figure 3. The main elements in the figure are:

- **Sensor layer.** This layer contains the lowest abstraction-level context information, i.e. that information relative to the state of the sensors and peripherals. Components usually collect this raw information to infer new and more abstract information.
- **Abstraction layer.** It contains abstract information inferred from the sensor layer or from the aggregation of information from this same layer. This information is used by the high-level abstractors in order to obtain the business-logic information necessary for the application.
- **Application layer.** This layer contains the business-logic information, i.e. the information that is directly referenced by the system requirements or analysis. For example, if the system needs to show a path, then the path is the referenced information, and its calculation procedure is transparent at this level.

The path guidance system uses the previous elements of the architecture in the following process:

1. At the system boot, one of the *ContextContainers* reads its *InitializationDocument.*
2. Following the information in the initialization document, the context container initializes the property "goal", which contains the information of the target that the user should reach.
3. Also at the system boot, the *PathIndicator* requests the property "path", since the definition of this component contains a "read" dependency with that property. This means that it needs to be notified about the value changes of this property.
4. The *ContextContainer* searches the property it in its repository. It does not found the property, so it creates a new instance in the system.
5. The *PathIndicator* discovers the property "path" in the system. This is done automatically by the component framework. Then, it subscribes itself to changes to that property in order to update its state accordingly.
6. The new "path" property is discovered by the *PathCalculator.* The definition of this component contains a "write" dependency with that property. This means that it makes changes to the property instance that will be notified to its readers.
7. The *PathCalculator* needs the "location" property in order to calculate the path and so it requests it to the *ContextContainer*, proceeding as in step 3
8. ... (The same for the lower layers).

Once all the bindings have been established, the context representation is changed dynamically to reflect the progress of the situation. In this process, the lower layers hide the specific details to the upper layers. This way, as the user walks through the room, the sensors produce a lot of information, which is processed by the location calculator, but it only updates the Location property when the sensors context reflect a change of location. The same may be done in the upper layers, but in this case, the path is modified along with the location, as it depends directly on it and the defined goal. When the goal is reached, the path property is updated accordingly, and the observer components will change its behavior in order to reflect this circumstance.

As shown in the example, this way of working has two main advantages. First, when a component able to calculate a, property is found in the system, it is automatically bound to the property and it updates its value. This way, it is possible to change the calculator components during runtime with other implementations and the rebinding is automatically done. Second, components are not requested to calculate properties unless it is completely necessary. Property instances are created on demand and consequently a component can stop if there is no interest in that particular property from any other component. The same way, once a property is instantiated, any number of consumers may read its state. The component container maintains a list of property readers, so that it can delete the instance if there are no interested components.

4 Related Work

The problem of the support to manage the context in AmI applications has been already considered in some works in the literature. Most of them provide some infrastructure or middleware to support the storage and management of information, but few consider how to organize this information to facilitate its use or make up a suitable design for the related components.

Examples of works focused on the infrastructure aspects of the problem are [3,4]. The first one proposes a three-layered architecture based on Java libraries to develop context-aware applications. The architecture is simple and easy to understand, and is well-documented. Moreover, it has been designed with the goal of flexibility regarding communication protocols. Regarding its limitations, it does not deal automatically with the management of dependencies among the components. The work in [4] also proposes a three-layer architecture, in this case service-oriented. A relevant feature of this work is the use of ontologies to facilitate the management of the information and its use. It adopts the RDF (Resource Description Framework) language to describe meta-knowledge about the information and its management, and the OWL (Ontology Web Language) for ontologies with its related logics to represent the context and reason about it. The key advantage of this approach is the use of well-established technologies to represents and use the information. However, it does not guide designers in the development of the specific components of their systems, so they need to rely on their own knowledge to determine the proper use of the infrastructure in their project.

Finally, there are some works that offer advice on the design of the system-specific components for the context. In [7], the authors show an architecture that differentiates between context consumers and providers to better characterize the nature of components. This distinction is in line with our architecture, but that work does not specify the actual protocols these components use to function together.

5 Conclusions and Future Work

This paper has introduced a general architecture for the context-aware subsystem of AmI applications. It includes support for the definition of the elements of information in the context and their management and use. The definition of the context information is made through a hierarchy of components commonly present or required in this kind of applications and their dependencies (provider or consumer). The infrastructure to work with that information automatically manages and coordinates the different components to obtain and propagate the context. It supports the dynamic reconfiguration of the system when available components change by taking care of new bindings between components. The generation of information elements is made on demand, which saves resources if there is no application component interested in a certain property (e.g. a peripheral can turn into stand by if there is no component interested in its state).

All this functionality can be built adapting the services of existing component frameworks (e.g. component containers and lifecycle management). With these features, the infrastructure relieves designers from working from scratch these aspects in each new AmI system.

The presented work is part of a wider effort to provide a general architecture and infrastructure for AmI applications. This architecture includes the activity and mobility management services already mentioned in this paper. Specifically related to the context management there are two main open issues. First, our work is studying how to represent the context elements to support logical propositions based on their properties, as in [7]. Second, it is necessary to consider ways to represent and use information about the temporal evolution of context. This would allow systems that actually learn from their past experiences to improve their performance.

Acknowledgments

The authors acknowledge support from the project *Agent-based Modelling and Simulation of Complex Social Systems (SiCoSSys)*, supported by Spanish Council for Science and Innovation, with grant TIN2008-06464-C03-01.

References

1. Abowd, G.D., Dey, A.K., Brown, P.J., Davies, N., Smith, M., Steggles, P.: Towards a Better Understanding of Context and Context-Awareness. In: Gellersen, H.-W. (ed.) HUC 1999. LNCS, vol. 1707, pp. 304–307. Springer, Heidelberg (1999)
2. Baldauf, M., Dustdar, S., Rosenberg, F.: A Survey on Context-Aware Systems. International Journal of Ad Hoc and Ubiquitous Computing 2(4), 263–277 (2007)
3. Bardram, J.E.: The Java Context Awareness Framework (JCAF) – A Service Infrastructure and Programming Framework for Context-Aware Applications. In: Gellersen, H.-W., Want, R., Schmidt, A. (eds.) PERVASIVE 2005. LNCS, vol. 3468, pp. 98–115. Springer, Heidelberg (2005)
4. Gu, T., Pung, H.K., Zhang, D.Q.: A Service-Oriented Middleware for Building Context-Aware Services. Journal of Network and Computer Applications 28(1), 1–18 (2005)
5. Kieffer, S., Lawson, J.Y., Macq, B.: User-Centered Design and Fast Prototyping of an Ambient Assisted Living System for Elderly People. In: 6th International Conference on Information Technology: New Generations (ITNG 2009), pp. 1220–1225. IEEE Press, Los Alamitos (2009)
6. Nehmer, J., Becker, M., Karshmer, A., Lamm, R.: Living Assistance Systems: an Ambient Intelligence Approach. In: 28th international conference on Software engineering (ICSE 2006), pp. 43–50. ACM Press, New York (2006)
7. Ranganathan, A., Campbell, R.: A Middleware for Context-Aware Agents in Ubiquitous Computing Environments. In: Endler, M., Schmidt, D.C. (eds.) Middleware 2003. LNCS, vol. 2672, pp. 143–161. Springer, Heidelberg (2003)
8. Remagnino, P., Foresti, G.L.: Ambient Intelligence: A New Multidisciplinary Paradigm. IEEE Transactions on Systems, Man and Cybernetics, Part A: Systems and Humans 35(1), 1–6 (2004)

Group Behavior Recognition in Context-Aware Systems

Alberto Pozo, Jesús Gracía, Miguel A. Patricio, and José M. Molina

GIAA, Carlos III University, Spain
{alberto.pozo,jesus.garcia,miguelangel.patricio,
josemanuel.molina}@uc3m.es

Abstract. In most of the domains of the context-aware system the user make up a group and their behavior could be research with group behavior recognition techniques. Our approach try to take advantage of the context information to understand the user's behaviors like a group, and this information could be useful for other system to beat to the users. For this purpose it is exposed a new representation that concentrates all necessary information concerning relations two to two present in the group, and the semantics of the different groups formed by individuals and training (or structure) of each one of them.

Keywords: Group behavior recognition, context-aware systems, activity representation, computer vision.

1 Introduction

Pervasive computing is one of the most active fields of research. This discipline need the context information to beat user's behaviors, it need to understand and predict their context and the user's behavior.

For this reason the group behavior recognition techniques could collaborate in the domains where the users interact between them. Our approach tries to use the context information (especially the users position, but also others) to understand the users behavior.

Human activity analysis and behavior recognition has received an enormous attention in the last two decades of computer vision community. A significant amount of research has addressed to behavior recognition of one element in the scene.

Instead of modeling the activities of one single element, group behavior recognition deals with multiple objects and/or people, who are part of groups.

In behavior recognition there are two distinct philosophies for modeling a group; the group could be dealt as a single group (crowd) or as a composition of individuals with some shared objectives. On the pervasive computing domains the users are clearly differentiable, so the crowd perspective could not be appropriate. For this reason in this paper we focus the investigation in the second philosophy, where take place many distinguishable agents and how the context information could be useful for the task.

The present paper shows a new representation of the possible variables existed in the problem. This had been designed to put in order briefly the essential information of the system.

J. Cabestany, I. Rojas, and G. Joya (Eds.): IWANN 2011, Part II, LNCS 6692, pp. 645–652, 2011.

With the aim of achieve our project, it will rely on three levels of abstraction.

Firstly, a matrix will be established with the information of each binary relationship between any individual of the system. This matrix store one vector for each relationship with the features selected to the problem domain. It is important to emphasize that in many case the features selected will include the relative position vector.

In these terms and conditions, for each frame in the video, the outstanding information has being kept, inclusive the geometrical information.

Once being contained all the important information, the process continues in a second abstraction's level where the challenge is capturing the logical information implicated between the communication of individual and groups. For this reason it is necessary to make different combinations for representing every group of the system.

It is a relevant detail to remark that each individual can belong to a several groups at the same time, and the groups have the possibility to incorporate an undefined number of other groups or individual.

In the third level, a new representation is created to reduce the dimension of the problem. One of the important key in this type of domains is that the number of relations between the elements of the scene growth exponential in relation with the number of elements. For this reason, a new representation is created to save the essential information of each group without saving all the relations between each element. Instead of save all the possible edges in a graph, this approach only save important graph that can provide all the important information wasting less space.

The paper is organized as follows. Section 2 reviews related work. Section 3 describes the problem. Section 4 introduces our description. Conclusions are drawn in section 5.

2 Related Work

Context-aware system have became in very important field of research in the last years especially because the appearance of the handheld device. These devices need to know the context, even to predict their context.

In papers (1) or (2) we can see the need for recognize the device's context and understand it. There are a lot of research about the sensors and the way of store all the information. In (3) shows the important of understand the context and what requirements needs to have a context-aware system.

Often the context-aware systems are implemented on handheld device, wearable computers, etc. and their context depends on the user's behavior. So if we need to predict the device's context, we have to predict the users' behavior.

The users of a context-aware system rarely are isolated, so their behavior depends on the group's interactions, nearby users, etc.

Despite the fact that there is plenty of work on single object activities, (4) the field of group activities recognition is relatively unexplored.

Group behavior recognition is a complex task that can be observed under different points of view.

There are two big families of approaches, one logical and one geometrical.

The logical approaches (5) are focused in construct a context-free grammar to describe the group activity based in the individual activities of the group's member.

The main characteristic of this point of view is the important of the first level, the features extraction. They need a previous system that recognizes the activity of each element of the scene.

The geometrical approaches (6), (7) have a different point of view. The features extracted in this case are based on the coordinates of the elements of the scene.

This approaches use to have higher computational complexity and the number of the elements in the scene could become very important.

There are also approaches than combine both perspective, like (8) whose work recognize key elements of a sport (basketball) using the positions of the players. This approach needs to identify the key elements of the domain dealt, and these key elements could be different in many different situations.

One more general approach could be read in (9) where the trajectories of the players (in a Robocup mach) are coded to create set of patterns that identify each type of action.

3 Group Behavior Recognition in Context-Aware Systems

In pervasive computing the context is all the elements (and their relationships) that surround the system. These elements could provide useful information to the system, or could be necessary to predict their state in the nearby future to provide a good service to the user.

Our approach tries to use the context information (especially the users position, but also others) to understand the users behavior.

Group behavior recognition is composed by two steps: in the first one the features of the system should be extracted, and in the second one the features are used to recognize the behavior.

Handheld device, wearable computers, etc. usually have a lot of sensor that could provide information like position, orientation, loud, bright, etc. so in this paper we are going to focus in the second step, we use the extracted features by the devices to make the inference of the behavior.

3.1 General Description

In a general scene there is one area composed of many sub-areas and a number of groups that consist of some elements which could be users or object.

In a group, one element could be related with whatever other element of the group. Each element of the system and each relationship have a set of features (like positioning, color, shape, etc.) The features could suffer changes in time.

Each element of the system should belong to a group, and could belong to many groups at the same time.

It is important to emphasize that any element of the system must be in a group, so there are not isolated elements.

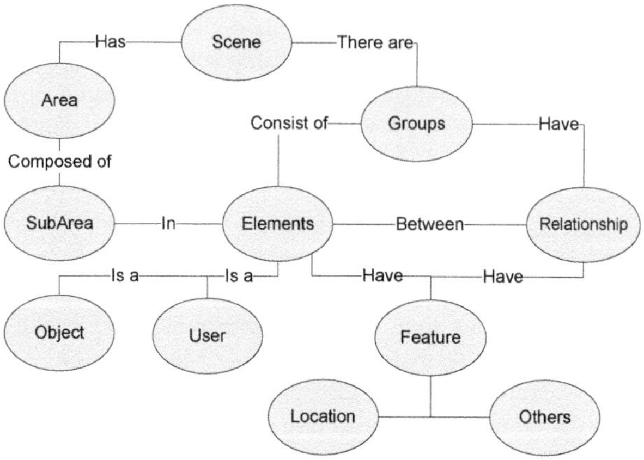

Fig. 1. General scene

3.2 Problem Description

Some of the general axioms of the problem describe above have been eliminated for more practical approach of the problem.

In our approach there is one sequence composed by a number of T instants, where are included a number of N elements (this number cannot change in time). The elements of the scene (all of them users) are distributed in a number of G groups, and each group is represented by a graph. (Group could be composed by two or more elements, and one element could be part of one or more groups).

Each node constitutes one user of the group and each edge constitutes one relationship, graphs are represented by his edges.

For each element and each edge we have a vector of features like positioning, that which is expressed by free vector in polar coordinate system.

Fig. 2 shows a scene with six elements conforming three groups. The definition of the groups is the semantic representation of the relations between the elements of the system.

The features selected to describe the elements of the scene will depend on the problem domain, and it will include its coordinates (in polar coordinate system or spherical coordinate system in case of 3-D positioning) and the coordinates of the free vectors that represents the edges of the graphs.

Fig. 2. Graphic representation of a system with six elements and three groups

For each element and each possible edge we save the features of the vector for each frame of the scene. One feature vector for each element and M free vectors for the edges, where $M = \frac{N*(N-1)}{2}$.

To describe the spatial feature relation between elements i and j in the frame t, there are two coordinates called $|r_{i,j,t}|$ and θ_t, with $|r_{i,j,t}| = \sqrt{(x_{j,t} - x_{i,t})^2 + (y_{j,t} - y_{i,t})^2}$ and $\theta_{i,j,t} = \tan^{-1}\frac{(y_{j,t} - y_{i,t})}{(x_{j,t} - x_{i,t})}$.

4 A Structured Representation for the Group Behavior Recognition Issue

Behavior recognition based on the positioning of each element of the group could be helped by the context information (provided by the device's sensors) obtaining better results in any situation.

However, the choice of the features (further away the positioning) it is dependent on the problem domain, so we need to select it in each case.

We propose a structured representation composed by three matrix called R, A and S. The first one save all raw data of the elements (and its relationships) in time (positioning and other dependent on domain features), while the second one represents the information about the semantic of the scene, composed by the number of groups founded and their makeup, and the third one represents the elements features and the important edges structure.

This structured representation contains the information about the features of each element of the scene, the features of the relations between the elements, and the group's structure information.

4.1 Features Vector

As we have wrote above, the structure representation is based on three matrixes. Two of this matrix (R and S) are composed by the features vector. This vector stores the features of one element or one relationship of the system in one definite instant. These features contain the geometrical information and other depending on the domain problem.

Each feature is storage in a natural number between one and eight. The first two features represent the geometrical information, where the first one (d) is the distance between the two elements of the relationship, or between the element and the pole (analogous to the origin of a Cartesian system). And the second one γ is the angle of this distance.

Element positions are calculated by $d_e = \left\lceil \frac{|r_e|}{|r_{eMax}|} \right.$. Where r_e is the distance between the element and the pole and r_{eMax} is the distance of the most remote element of the pole. Relative distance is calculated by the formula $d = \left\lceil \frac{|r_{ij}|}{|r_{max}|} \right.$. Where r_{ij} is the distance between the elements i and j, and r_{max} is the maximum distance between any elements of the graph. By definition d is a natural number between 1 and 8.

Direction between two elements of the graph (or to positioning one element) is defined by the formula $\gamma = \left\lceil \frac{|\theta_{ij}|}{\pi} \right.$. Where $_{ij}$ is the angle between the elements i and j (or the element and the pole). By definition γ is a natural number between 1 and

8. It is important to perceive that in spite of the graphs are not directed, to construct the reduced graph we have to distinguish between the same directions with different sense. So the possible directions are covered between - π and π radians.

All other features are calculated in the same way, using the max value of the feature, and the result is another natural number between one and eight.

4.2 Geometrical Information

All the features about the elements in the scene, and its relations are saved on matrix R, this information is used to construct the S matrix.

Matrix R is a three dimensional matrix with the information of each agent and each relationship presented at the scene.

A scene with N elements has $M = \frac{N*(N-1)}{2}$ possible edges that must be saved, and N elements features vectors.

Each vector of the matrix has P components (two or three for the geometrical information, and some more for the rest of the context information, depending on the problem domain).

The first ones N vectors represent the features of the N elements, and the next M vectors represent the features of the relationships between each element and other.

The R matrix has one row for each frame of the scene and N + M columns.

$$\begin{pmatrix} r_{11} & \cdots & r_{N1} & r_{(12)1} & \cdots & r_{(N-1N)1} \\ \vdots & \ddots & & & & \vdots \\ r_{1T} & \cdots & r_{NT} & r_{(12)T} & \cdots & r_{(N-1N)T} \end{pmatrix}$$

4.3 Semantics Information

The semantics information represents the associations between the elements of the scene to perform groups. One element could be part of many groups, and could be many groups.

This information makes it possible to create different associations between elements to grasp better the semantics context.

This semantics information is saved in a binary matrix with one row for each group, and one column for each element.

The matrix can only contain zeros or ones, which represent if this element forms part of the graph.

For example, in a scene with nine elements, and two groups, the matrix A could be like this one: $A = \begin{pmatrix} 1 & 1 & 1 & 1 & 1 & 0 & 0 & 0 \\ 0 & 0 & 0 & 0 & 1 & 1 & 1 & 1 \end{pmatrix}$. This matrix shows that there are two graphs, the first one composed by the elements: 1, 2, 3, 4, and 5; and the second one composed by the elements: 5, 6, 7, 8 and 9.

4.4 Structure Information

Matrices S define the structure of the graphs, there are one matrix for each graph.

Each S matrix has a number of T rows, and $M_g + N_g$ columns, where N_g is the number of elements of the group and M_g depends on N_g ($M_g = \frac{N_g*(N_g-1)}{2}$).

Each element of S is a features vector, describe in the section above. The selection of the important edges is made by the geometrical information of the features vector.

Each S matrix has the edges of the graph that defines its structure. If some edge have the same direction that another one and it is longer than the previous one, then this edge is not added to the matrix. One null value is added in this position.

Figure 4 shows the construction process: the shortest edge of the first element is added in (a). Then the second shortest is also added in (b). In (c) there is a shorter edge with the same direction (2) and the edge is not added. The process is repeated until all the elements are checked (d), (e) and (f).

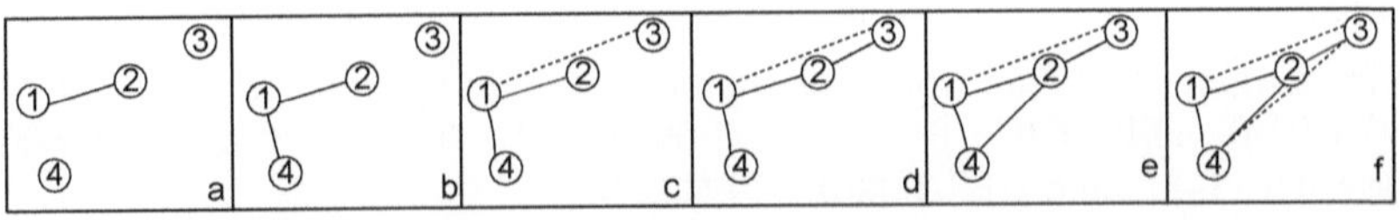

Fig. 3. Construction process

The matrix below shows the S of the graph in the Fig. 5b. First row represents the graph at the instant t = 0, and last row shows the graph at the instant t = T.

Fig. 5b shows an example of a graph with five elements where are presented the first two features (geometrical information). The first five columns represent the position of the elements, and the other represents the relationships. In the first frame the edges between the nodes 1-5, 3-4 and 3-5 are not defined because they have the same directions (and they are longer) than the edges 1-5, 3-2 and 5-2.

Then, in the frame T the graph's shape has changed, there are new relevant edges (like 3-5 and 3-4) and some relative distance have also changed.

$$\begin{pmatrix} 7^6 & 6^6 & 8^6 & 2^6 & 1^6 & 4^6 & 6^8 & 3^4 & -1^5 & 3^2 & 4^4 & 5^3 & -1^4 & -1^3 & 4^5 \\ \vdots & \vdots & \vdots & \vdots & \vdots & \vdots & \vdots & \vdots & \vdots & \vdots & \vdots & \vdots & \vdots & \vdots & \vdots \\ 7^6 & 6^6 & 8^6 & 6^6 & 5^6 & 4^6 & 5^8 & 4^4 & -1^5 & 4^5 & 6^8 & 5^2 & 6^4 & 7^3 & 5^5 \end{pmatrix}$$

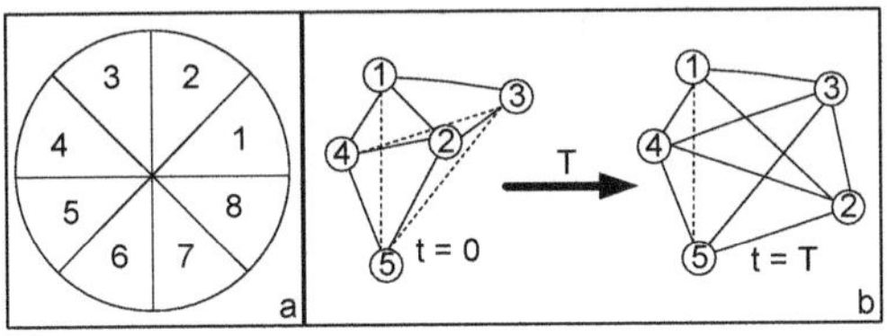

Fig. 4. a) Directions Code b) Example graph

5 Conclusions

In most of the domains of context-aware the users are not along and their behavior depends on the rest of the nearby users. Group behavior recognition could take advantage of the context information and it could be a good way to understand (and predict) the behaviors of the user on a pervasive computing system.

The structure of the information in context-aware systems is one of the most interesting fields of research, and it could be merging with the group behavior recognition systems.

Our approach had reduced the number of relations without loss information about the formation to realize the rezoning process. This approach is based in a novel structured representation of the important relations between the elements of the graphs.

The features selected in each case will depend on the problem domain, but in the most of the case the positioning system is available and provides a good source of knowledge.

Acknowledgements. This work was supported in part by Projects CICYT TIN2008-06742-C02-02/TSI, CICYT TEC2008-06732-C02-02/TEC, CAM CONTEXTS (S2009/ TIC-1485) and DPS2008-07029-C02-02.

References

1. Mäntyjärvi, Jani, Himberg, Johan and Huuskonen, Pertti: Collaborative Context Recognition for Handheld Device. Oulu : IEEE, (2003). Pervasive Computing and Communications, 2003.(PerCom 2003). Proceedings of the First IEEE International Conference on. pp. 161--168. 0769518931.
2. Mäntyjärvi, Jani and Seppänen: Adapting applications in handheld devices using fuzzy context information. Tapio. 4, Oulu : Elsevier, (2003), Interacting with Computers, Vol. 15, pp. 521--538. 0953-5438.
3. Baldauf, Matthias, Schahram, Dustdar and Rosenberg: A survey on context-aware systems. Florian. (2007), International Journal of Ad Hoc and Ubiquitous Computing, pp. 263--277.
4. Moeslund, T B, Kruger, V and Hilton: A survey of advances in vision-based human motion capture and analysis. A. 2-3, s.l. : Elsevier, (2006), Computer vision and image understanding, Vol. 104, pp. 90--126. 1077-3142.
5. Ryoo, M. S. and Aggarwal, J. K: Recognition of High-level Group Activities Based on Activities of Individual Members. s.l. : IEEE, (2008). Motion and video Computing, 2008. WMVC 2008. IEEE Workshop on. pp. 1--8.
6. Khan, Saad M. and Shah, Mubarak. Singapore: Detecting Group Activities using Rigidity of Formation. ACM, (2005). Proceedings of the 13th annual ACM international conference on Multimedia. pp. 403--406. 1595930442.
7. Ruonan, Li, Rama, Chellappa and Shaohua, Kevin Zhou: Learning Multi-modal Densities on Discriminative Temporal Interaction Manifold for Group Activity Recognition. New York : s.n., (2009), CVPR, pp. 1--8.
8. Perse, Matej, et al. 5, Ljubljana: A trajectory-based analysis of coordinated team activity in a basketball game. Elsevier, (2009), Computer Vision and Image Understanding, Vol. 113, pp. 612--621. 1077-3142.
9. Ramos, Fernando and Ayanegui, Huberto. Tracking behaviours of cooperative robots within multi-agent domains. Autonomous Agents. Tlaxcala : Vedran Kordic, (2010).

Context-Awareness at the Service of Sensor Fusion Systems: Inverting the Usual Scheme

Enrique Martí, Jesús García, and Jose Manuel Molina

Applied Artificial Intelligence Group, Universidad Carlos III de Madrid,
Av. de la Universidad Carlos III, 22, 28270, Colmenarejo, Madrid, Spain
emarti@inf.uc3m.es, jesus.garcia@inf.uc3m.es, molina@ia.uc3m.es,
http://www.giaa.inf.uc3m.es

Abstract. Many works on context-aware systems make use of location, navigation or tracking services offered by an underlying sensor fusion module, as part of the relevant contextual information. The obtained knowledge is typically consumed only by the high level layers of the system, in spite that context itself represents a valuable source of information from which every part of the implemented system could take benefit. This paper closes the loop, analyzing how can context knowledge be applied to improve the accuracy, robustness and adaptability of sensor fusion processes. The whole theoretical analysis will be related with the indoor/outdoor navigation system implemented for a wheeled robotic platform. Some preliminary results are presented, where the context information provided by a map is integrated in the sensor fusion system.

Keywords: sensor fusion, navigation, indoor, outdoor, context-aware, particle filter, software agent.

1 Introduction

Sensor Fusion is a process that consists in combining observations provided by several sensors about an entity of interest, so that the information finally obtained is better —in some sense— than what could be inferred by taking each of the sensors alone. This task is just a part of the more general concept of Data Fusion which, among other particularities, is not limited to sensing information. Location and tracking of dynamic objects [11][13][8][2][5] can be accounted as one of the most important applications of sensor fusion. Solving this problem requires a clear specification of, at least, what has to be estimated (variables of interest about the observed system), data provided by sensors, and how both system state and sensor readings are related. Fusion performance can be benefited of any additional information, such as a mathematical model of observed system dynamics. Fusion processes in real scenarios are, however, affected by a variety of external factors that cannot be accounted while modeling the problem, either because are subject to uncontrolled changes over time, or because they are even unknown to us.

J. Cabestany, I. Rojas, and G. Joya (Eds.): IWANN 2011, Part II, LNCS 6692, pp. 653–660, 2011.

To overcome that hurdle, fusion systems should be able to detect relevant changes in their environment, and adapt themselves for achieving the best performance: this is, context-aware sensor fusion. Most of the existing literature about context-aware applications featuring location and/or navigation follow this scheme in the opposite direction: fusion module performs location, and the obtained result is employed as a position-based context for higher-level applications [1][12].

The discipline of Data Fusion represents an example of fusion processes responsive to their environment. In the JDL model [9] for data fusion systems, the 4 level (Process Refinement) describes how to use the acquired information to feedback the lower levels by means of, for instance, sensor retasking or model modification. The goal of this paper is to analyze the direct application of contextual information to sensor fusion tasks. Many of the provided examples are based on the platform used in the experimental part, which was introduced in [10] and is briefly described in section 2. It consists in an autonomous robot that performs indoor and outdoor navigation using a variety of onboard and external sensors, enabling the scenario for representing a full scale fusion problem. Following that, section 3 conduct a theoretical analysis regarding the applicability of contextual information in sensor fusion processes. It begins covering the topic of modeling environment for context acquisition and processing. Immediately afterwards, a second part identifies the parts in a sensor fusion process where the obtained contextual information can be applied. Finally, some preliminary results using the described platform are presented in section 4. They show how sensor fusion can benefit from the use of contextual information either improving accuracy or reducing the computational burden of selected algorithms. Some remarks and conclusiones are given in the last section.

2 Sample Scenario

The scenario selected as reference for the analysis conducted in next section is the problem of combined indoor/outdoor navigation: estimation of position, orientation and dynamics of a robot, which is equipped with onboard sensors but also features communication capabilities with other entities that will serve as external sensors. From the architectural point of view, this navigation system is organized in layers in order to maximize its flexibility. Each tier plays a different role in the process of acquiring and transforming information into something useful for the final data sink: a Particle Filter which fuses all the information into the most likely estimation. One of the strongest reasons for selecting such a solution is to provide a reasonably complete scenario that does not limit the performed theoretical analysis.

Figure 1 contains a schematic view of the system. Information flows top-down in the diagram, with sensor data represented by small triangles. Upper levels are in charge of capturing information either by means of sensors physically attached to the platform or by exchanging data with external intelligent entities. The sensor abstraction layer is in charge of managing physical sensors as well

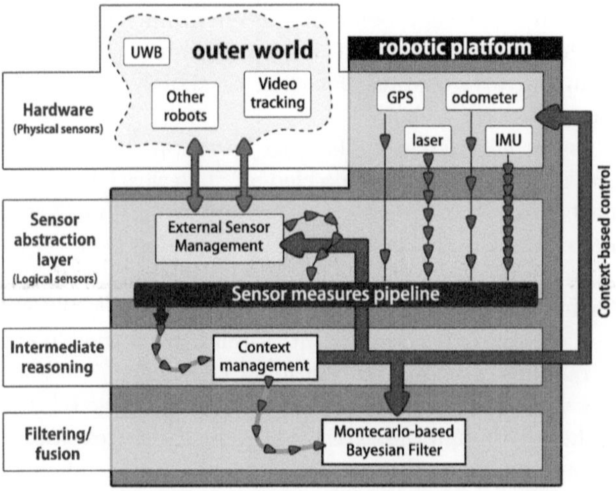

Fig. 1. Architecture of the proposed navigation system for an individual mobile platform

as providing a unified vision of external information sources. The applications of contextual information to sensor fusion described in section 3 involve acting over this layer.

The Intermediate reasoning layer receives and process raw sensor measures. The operations hosted by this level range from adapting sensing information to meet the various requirements of the filter, to more advanced inferences as deriving context information from the available readings. The diagram shows only the box corresponding to context reasoning for the sake of clarity. The last level contains simply a filtering-capable algorithm for integrating the incoming data.

3 Theoretical Analysis

This section is divided in two parts. The first one reviews different representations of environment for extracting contextual information, and the second details where and how this information can be used within a sensor fusion system.

3.1 Knowledge about Environment and Context Representation

Let us define the environment of an application as the multidimensional space where it operates, including other variables which an influence on the problem being solved. Those variables can be categorized in different ways, for instance, if we focus on what that knowledge is referred to, we can discern between information about the environment itself and information about the different entities populating it. Inside the first category we can account for several types of information attending to its nature.

If we are talking about either continuous or discrete valued variables that have a defined value for every point of the environment (i.e. fields), then the most straightforward representation is a map. Examples of this type of variables are ambient temperature, obstacle location or signal/noise level for a certain electromagnetic emission.

Some information which is not likely to be mapped, can be represented by statements instead (i.e. declarative knowledge). The weather is a perfect example of propositional context knowledge: sunny or rainy condition, current wind speed, etc. are factors to take into account in sensor fusion becasue they can affect the performance of some devices. Knowledge about external entities also lays in the field of statement-based information, but is a bit trickier because it tends to involve complex reasoning processes. Entity-related knowledge can be classified in two general families: feature and relational knowledge.

Among the many examples of feature knowledge we can cite identity, position, activity and utility. Although the extraction of such features can involve complex data structures and intrincated processing schemes —some of them still an open research field, as activity recognition—, they can be represented as value tuples or simple labels once have been determined. Relational knowledge describes the different interactions or links between entities. Ontologies for entity and graphs for groupal activity recognition can be enclosed in this category.

3.2 Applying Context to Sensor Fusion

The acquired contextual information can be injected in differents places of a sensor fusion system. The two principal insertion points are the set of sensors and the fusion algorithm. The first category, acting over the sensors, includes at least four uses of context information: sensor selection, modification of capture parameters, modification (correction) of raw sensing data, and finally sensing data augmentation (complete it with new information). The two first types are commonly known as sensor retasking in terminology of distributed multisensor data fusion. Nonetheless, they can be important features also in centralized, simple sensor fusion systems. For instance, selection plays an important role when dealing with redundant sets of sensors.

Sometimes direct observations can provide the required context, as in the case of fusing video and infrared sensors: in spite that both provide spatial information about non-occluded surfaces in the environment, poor lighting conditions discards a video camera as an effective sensor, while an infrared sensor can be affected by colors and reflection angles. Both effects can be detected using the sole video input. Another example is shown in figure 2. The motion of the rover in a rough floor cause vibrations which spoil the measures, as seen around second 9. On the other hand, some effects are not as easy to detect. A sustained magnetic interference can bias the readings of a magnetometer, but figuring out the existence of this problem requires further estimations processes and context data.

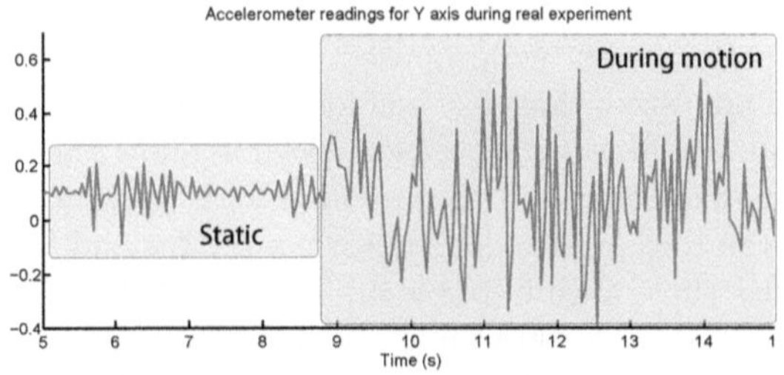

Fig. 2. Vibrations due to robot motion have a harsh effect on inertial unit measures

Regarding the modification of capturing parameters, we can cite changing the orientation of directional sensors such as PTZ video cameras. Existing algorithms for visual attention [7] and tracking can provide the information needed for the change of parameters.

The third option is to modify the sensed data to correct undesired effects. As an example, mobile entities can degrade the performance of map matching algorithms, because they affect laser readings. However, the context can be used to identify and remove the spurious beam hits. Finally, sensor data can be augmented by including additional data to be considered in the fusion process, as the confidence in a measure or a label indicating how to process it.

The second entry point for context information in sensor fusion systems is the proper fusion algorithm, by means of selection —change algorithm— and modification —manipulation of parameters—. Examples of algorithm selection can be employing a Particle Filter for indoor navigation, where walls and other obstacles make the problem highly nonlinear, but switching to a simpler and less costly approach such as least squares or a Kalman-like filter in open spaces. Multiple model systems and particle filters with adaptive population size constitute examples of algorithm modification.

4 Experiments on Navigation

This part of the paper gathers empirical results obtained with the platform described in section 2. A first subsection will thoroughly describe the configuration employed in the experiments, immediately followed by the obtained results along with an analysis on them.

4.1 Navigation System Setup

The experiments presented in this paper have been reproduced in laboratory from both simulated and real data. The real data has been obtained in controlled experiments where the robot was been equipped with a GPS sensor with meter-level precission and an inertial measurement unit (IMU). With simulation purposes, GPS measures are assumed to suffer a random gaussian-like noise with standard deviation of 1 meter.

The baseline navigation algorithm relies in a Particle Filter which performs loosely coupled fusion of the two proposed sensors. This approach will be compared with a similar system that also includes the information of a map. It must be noted that the available set of sensors do not allow the implementation of map matching techniques [6][3] that, provided with an almost perfect map, can result in an outstanding positioning performance. Instead, the map is used to discarding particles that move into a wall, as done in [4].

The system is tested in a very simple porch-like scenario. It presents obstacles to be mapped, while being an almost open space with avaialble GPS signal. The robot navigates in a relatively reduced space that, provided the low accuracy of GPS measures, makes the problem more difficult to solve.

4.2 Obtained Results

The conducted experiments compare navigation performance with and without a map of obstacles. Using standard sensing conditions, with update rates around 30 Hz for the IMU and 1 Hz for the GPS, the map-less navigation algorithm usually results in an average positioning error of 0.4 meters. This represents a 65% error decrease compared with the average 1.2 m error of bare GPS measures.

Figure 3 shows the filtered trajectory for one of the runs. The slash-marked path represents the true trajectory of the robot, while the track of dense small circles represent the Particle Filter estimation. GPS measures are marked as large circles, while the cloud of points with attached lines represent the position and orientation of particles at the end of the simulation. When using a map, the algorithm not only worsens its position estimation (0.6 meters, 50% improvement), but also spoils the continuity of the estimated trajectory with sudden jumps, as shown in figure 4. They are a direct effect of particle anihilation together with biased GPS measures. The real improvement, as in many sensor fusion systems, comes by the hand of degraded sensor performance. For the following experiments, degraded sensing conditions were assumed: IMU readings feature a noise characteristic of rough terrain, and GPS measures lower the update rate to 0.2

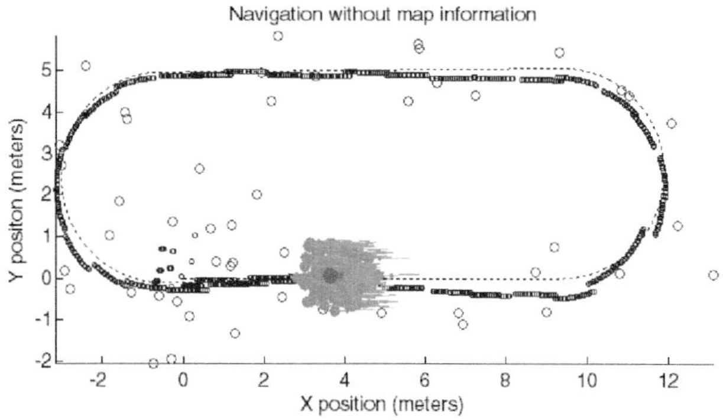

Fig. 3. Navigation without map, assuming best sensing conditions

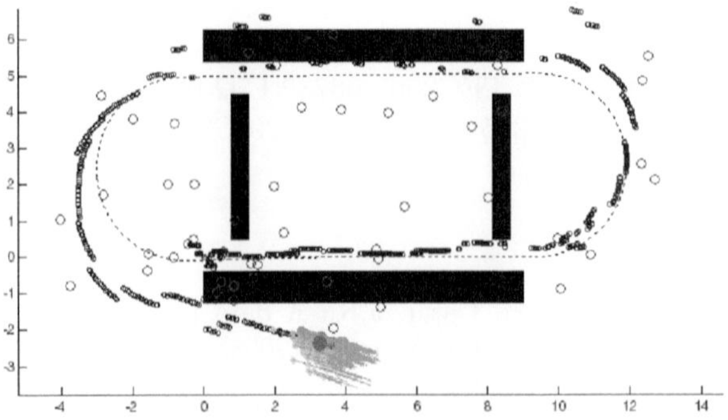

Fig. 4. Navigation with map, assuming best sensing conditions

Hz as in the case of occlusions. Under these conditions, taking into account map information leads to slightly better results. The baseline navigation algorithm have a mean position error of 1.2 meters, same as GPS measures, while including the map results in about 0.9 meters, although its stimate is still less smoother than the basic version.

In spite that the performed experiments are quite reduced, they show a fundamental fact: context information, when adecquately integrated into sensor fusion systems, can improve their robustness under conditions of degraded sensing performance. This is of vital importance for systems pretending to be autonomous and work unattended for long periods of time.

5 Conclusions

The contribution of this paper is two-fold. On the one hand, the theoretical analysis in section 3 tries to conciliate the worlds of Data Fusion and context-aware applications. It can be seen that all provided examples, which have been related to the sample scenario but can also be found in existing literature, solve problems using sensor fusion systems and context information. The problem is that authors are usually very focused in the context part and do not make use of those formalisms developed within Data Fusion field. Integrating both disciplines can lead not only to better results, but also to a faster progress thanks to not reinventing existing concepts. The other contribution is an scheme defining how artificial intelligence applications can feedback their sensor fusion modules in order to improve their results. Some preliminary experiments on indoor/outdoor navigation are also presented, where the simple use of a map provides the necessary context to improve location accuracy under degraded sensor performance. The obtained results are far from being spectacular in aboslute terms, but it is important to take into account that the goal was to test if even a very weak use of context information could serve to improve the performance of a sensor fusion system.

Acknowledgements

This work was supported in part by Projects ATLANTIDA, CICYT TIN2008-06742-C02-02/TSI, CICYT TEC2008-06732-C02-02/TEC, SINPROB, CAM MADRINET S-0505/TIC/0255 DPS2008-07029-C02-02.

References

1. Bernardos, A.M., Tarrio, P., Casar, J.R.: A data fusion framework for context-aware mobile services. IEEE, Los Alamitos (2008)
2. Castanedo, F., García, J., Patricio, M.A., Molina, J.M.: Data fusion to improve trajectory tracking in a Cooperative Surveillance Multi-Agent Architecture. Information Fusion 11(3), 243–255 (2010)
3. Dellaert, F., Fox, D., Burgard, W., Thrun, S.: Monte Carlo localization for mobile robots. In: Proceedings 1999 IEEE International Conference on Robotics and Automation (Cat. No.99CH36288C), pp. 1322–1328 (February 2001)
4. Evennou, F., Marx, F., Novakov, E.: Map-aided indoor mobile positioning system using particle filter. In:IEEE Wireless 2005 (2005)
5. Gómez-Romero, J., Patricio, M.A., García, J., Molina, J.M.: Ontology-based context representation and reasoning for object tracking and scene interpretation in video. Expert Systems with Applications 38(6), 7494–7510 (2010)
6. Gustafsson, F., Gunnarsson, F., Bergman, N., Forssell, U., Jansson, J., Karlsson, R., Nordlund, P.-J.: Particle filters for positioning, navigation, and tracking. IEEE Transactions on Signal Processing 50(2), 425–437 (2002)
7. Horaud, R., Knossow, D., Michaelis, M.: Camera cooperation for achieving visual attention. Machine Vision and Applications 16(6), 1–2 (2005)
8. Jing, L., Vadakkepat, P.: Interacting MCMC particle filter for tracking maneuvering target. Digital Signal Processing 20(2), 561–574 (2010)
9. Liggins, M.E., Llinas, J., Hall, D.L.: Handbook of Multisensor Data Fusion: Theory and Practice, 2nd edn. CRC Press, Boca Raton (2008)
10. Martí, E., García, J., Molina, J.: Opportunistic multisensor fusion for robust navigation in smart environments. In: Proceedings of CONTEXTS 2011 Workshop User-Centric Technologies and Applications, Springer, Heidelberg (2011)
11. Nemra, A., Aouf, N.: Robust INS/GPS Sensor Fusion for UAV Localization Using SDRE Nonlinear Filtering. IEEE Sensors Journal 10(4), 789–798 (2010)
12. Subercaze, J., Maret, P., Dang, N.M., Sasaki, K.: Context-aware applications using personal sensors. In: Proceedings of the ICST 2nd International Conference on Body Area Networks, p. 19 (June 2007)
13. Wendel, J., Trommer, G.: Tightly coupled GPS/INS integration for missile applications. Aerospace Science and Technology 8(7), 627–634 (2004)

Improving a Telemonitoring System Based on Heterogeneous Sensor Networks

Ricardo S. Alonso, Dante I. Tapia, Javier Bajo, and Sara Rodríguez

Department of Computer Science and Automation, University of Salamanca, Plaza de la Merced, s/n, 37008, Spain
{ralorin,dantetapia,jbajope,srg}@usal.es

Abstract. Information fusion helps telemonitoring systems to better unify data collected from different sensors. This paper presents last improvements made on a telemonitoring system aimed at enhancing remote healthcare for dependent people at their homes. The system is based on SYLPH, a novel platform based on a service-oriented architecture approach over a heterogeneous Wireless Sensor Networks infrastructure to create intelligent environments. Because of SYLPH, the system allows the interconnection of several sensor networks from different wireless technologies, such as ZigBee or Bluetooth. Furthermore, the SYLPH platform can be executed over multiple wireless devices independently of their microcontroller or the programming language they use.

Keywords: Information fusion, Distributed architectures, Context-awareness, Wireless Sensor Networks, Healthcare, Telemonitoring, Multi-agent systems.

1 Introduction

Telemonitoring systems allow patients' state and vital signs to be supervised by specialized personnel from remote medical centers. A telemonitoring system for healthcare needs to continuously keep track of context information about patients and their environment. The information may consist of many different parameters such as patients' location, their vital signs (*e.g.*, heart rhythm or blood pressure) or building temperature. Most of the context information can be collected by distributed sensors throughout the environment and even the patients themselves. In this sense, Wireless Sensor Networks (WSNs), such as ZigBee/IEEE 802.15.4 and Bluetooth, comprise a key technology to collect context information from users and their environment [1].

This paper presents the new changes introduced into a telemonitoring system aimed at enhancing healthcare for dependent people at their homes [2]. The system utilizes the SYLPH (*Services laYers over Light PHysical devices*) novel platform that integrates a SOA (Service-Oriented Architectures) approach with heterogeneous WSNs [3]. Unlike other SOA-WSNs architectures, SYLPH allows both services and services directories to be embedded into nodes with limited computational resources regardless of the radio technology they use. An essential aspect in this work is the use of Wireless Sensor Networks to provide the system with automatic and real-time information of the environment and allow them to react upon it. Therefore, these new

J. Cabestany, I. Rojas, and G. Joya (Eds.): IWANN 2011, Part II, LNCS 6692, pp. 661–668, 2011.

changes include the integration of n-Core [4], an innovative wireless sensor platform, to improve the context-awareness of the system.

The next section introduces the problem description and explains why there is a need for defining a new telemonitoring system. Then, it is described the basic components of the system, as well as the new changes introduced in the system to provide it with improved context-aware capabilities. In addition, it is explained some experiments made to evaluate the performance of the system applied to a real scenario, as well as the obtained results, comparing the previous version of the system and the new release presented in this paper. Finally, conclusions are depicted.

2 Problem Description

One of the key aspects for the construction of telemonitoring systems is obtaining information about the patients and their environment through sensor networks. This section presents the strengths and weaknesses of existing telemonitoring systems and discusses some of the problems of existent platforms aimed at integrating WSNs.

Biomedical sensors (*e.g.*, electrocardiogram, blood pressure, etc.) and automation sensors (*e.g.*, temperature, light, etc.) differ significantly in how they collect data. On the one hand, biomedical sensors obtain continuous information about vital signs that is important and should not be lost [5]. On the other hand, automation sensors obtain information at a lower frequency than biomedical sensors [1] because this information is generally less important than vital signs. In a telemonitoring scenario, it is necessary to interconnect WSNs from different technologies [6], so having a distributed platform for deploying applications over different networks facilitates the developers' work and the integration of the heterogeneous devices.

There are several telemonitoring healthcare developments based on WSNs [6] [7]. However, they do not take into account their integration with other architectures and are difficult to adapt to new scenarios [8]. This is because such approaches do not allow sensors and actuators to communicate directly with one another, and instead gather data in a centralized way. Excessive centralization of services negatively affects system functionalities, overcharging or limiting their capabilities [8]. A centralized model consists of a central node that gathers all the data forwarded by the nodes connected to it. One of the main problems in this model is that most of the intelligence of the system is centralized. Thus, it gathers the required data from the nodes and, based on such data, it decides what commands will be sent to the each node. That means that a node belonging to a certain WSN does not know about the existence of another node forming part of a different WSN in the same system.

Nonetheless, this model can be improved using a common platform where all the nodes in the system can know about the existence of any other node in the same system no matter the technology they use. This is achieved by adding a middleware logical layer over the existing application layers on the nodes. This way, a sensor node in a WSN can know about the existence of an actuator node in other WSN, so the sensor node can send a command to the actuator node directly at the application layer level. A service-oriented approach is adequate for being implemented in wireless sensor nodes as it allows distributing the functionalities of the system into small modules. Such small modules are ideal for being executed by devices with limited computational resources as wireless sensor nodes. The code executing in a certain node can invoke services offered by any other node in the system, regardless the latter node is in the same WSN or not. This way, the central node now only has to act as a gateway among the distinct WSNs connected to it. Thus, it has not to keep track of either the nodes in the system or the functionalities they offer.

There are different technologies for implementing WSNs, such as ZigBee or Bluetooth. The ZigBee standard allows operating in the ISM (Industrial, Scientific and Medical) band, which includes 2.4GHz almost all over the world [9]. The underlying IEEE 802.15.4 standard is designed to work with low-power and limited resources nodes [9]. ZigBee incorporates additional network, application and security layers over IEEE 802.15.4 and allows more than 65,000 nodes to be connected in a mesh topology network [9]. Another common standard to deploy WSNs is Bluetooth. Bluetooth allows multiple WPAN (Wireless Personal Area Network) or WBAN (Wireless Body Area Network) applications for interconnecting mobile devices or biomedical sensors. Bluetooth also operates in the 2.4GHz band and allows creating star topology networks of up to 8 devices, one acting as master and the rest as slaves, but it is possible to create more extensive networks through devices that belong simultaneously to several networks [1]. However, it is not easy to integrate devices from different technologies into a single network [8]. The lack of a common architecture may lead to additional costs due to the necessity of deploying non-transparent interconnection elements among different networks.

The SYLPH platform used for the telemonitoring system described in this paper tackles some of these issues by enabling an extensive integration of WSNs and providing a greater simplicity of deployment, optimizing the reutilization of the available resources in such networks. The SYLPH platform integrates a SOA approach for facilitating the distribution and management of resources (*i.e.*, services). SOA proposes a model based on a collection of services and a communication way between them. A service can be defined as a function that must be well-defined, self-contained, and non-dependent of the context or the state of other services [10]. Some developments try to reach integration between devices by implementing some kind of middleware, which can be implemented as reduced versions of virtual machines, middleware or multi-agent approaches [11]. However, these developments require devices whose microcontrollers have large memory and high computational power, thus increasing costs and physical size. These drawbacks are very important regarding WSNs, as it is desirable to deploy applications with reduced resources and low infrastructural impact, especially in healthcare telemonitoring scenarios.

There are developments that try to integrate WSNs and a SOA approach [8]. However, those developments do not consider the necessity of minimizing the overload of the services architecture on the devices. In contrast, our solution allows the services to be directly embedded in the WSN nodes and invoked from other nodes either in the same network or another network connected to the former. It also specifically focuses on using devices with small resources to save CPU time, memory size and energy consumption, which is very useful to design and construct smart environments. Furthermore, as previously mentioned, the system contemplates the possibility of connecting WSNs based on different technologies.

3 Telemonitoring System Description

This section describes the main features of the telemonitoring system designed and developed with the aim at improving healthcare of dependent people at their homes. This system utilizes WSNs for obtaining context information about users (*i.e.*, patients) and their environment in an automatic and ubiquitous way.

The system uses a network of ZigBee devices placed throughout the home of each patient to be monitored. The patient carries a remote control (a small ZigBee device embedded in a wristband) that includes an alarm button which can be pressed in case of emergency or the need for remote assistance. There is a set of ZigBee sensors that obtain information about the environment (*e.g.*, light, smoke, temperature, etc.) and react to changes (*e.g.*, light dimmers and fire alarms). In the previous version of the telemonitoring system [2], each ZigBee node included a C8051F121 microcontroller with 8KB of RAM and 128KB of Flash memory and a CC2420 transceiver, consuming only a few μA in sleep mode. In the new version of the telemonitoring system, these ZigBee devices have been substituted by new n-Core Sirius-A devices belonging to the novel n-Core platform [4]. Each n-Core Sirius-A 2.4GHz device includes an ATmega1281 microcontroller with 8KB RAM, 128KB Flash memory, an AT86RF231 transceiver and several communication ports (GPIO, ADC, I2C and USB/RS-232 UART) to connect to a wide range of sensors and actuators. There are also several Bluetooth biomedical sensors placed over the patient's body. Biomedical sensors allow data about the patient's vital signs to be acquired continuously. Each patient carries an Electrocardiogram (ECG) monitor, an air pressure sensor acting as respiration monitor, and a triaxial accelerometer for detecting falls. These Bluetooth devices use a BlueCore4-Ext chip with a RISC microcontroller with 48KB of RAM. All ZigBee and Bluetooth devices can offer and invoke services (*i.e.*, functionalities) within the network. There is also a computer connected to a remote healthcare center via Internet for forwarding possible alerts to caregivers and allowing them to communicate with patients. This computer acts as a ZigBee coordinator and is also the master of a Bluetooth network formed by the biomedical sensors as slaves. On the one hand, the computer works as a SYLPH Gateway so that it interconnects both WSNs. On the other hand, it runs a telemonitoring application based on the *Flexible User and ServIces Oriented multiageNt Architecture* (FUSION@) [12] to fuse information from the SYLPH sensor nodes and send commands to the actuator nodes.

Figure 1 shows an example of the system operation. In this case, a smoke sensor detects a higher smoke level than a previously specified threshold (1). Then, it invokes a service offered by the node which handles the fire alarm, making it to ring (2). At the same time, it also invokes a service offered by the computer that acts as both ZigBee master node and Internet gateway (3). Such gateway sends an alert through the Internet towards the remote healthcare telemonitoring center (4). At the remote center, the alert is received by a monitoring server (5), which subsequently queries a database in order to obtain the information relative to the patient (6) (*i.e.*, home address and clinical history). Then, the monitoring server shows the generated alert and the patient's information to the caregivers (7), which can establish a communication over VoIP (Voice over Internet Protocol) or by means of a webcam with the patient's home in order to check the incidence. The patient can also ask for assistance by pressing its manual alert button (using the personal remote control) or making a call through the VoIP terminal. In the example in Figure 1, the caregiver decides to request the monitoring server to start a voice and video communication with the patient's home (8). The monitoring server starts such a communication (9) through VoIP (10). As the gateway in the patient's home accepts it automatically (11), now the caregiver can see the patient and talk with him (12). Several webcams can be deployed through the patient's home to assure the chance of establishing the communication with the patient. If the patient is conscious, he can also talk with caregivers and explain the situation (13). If necessary, caregivers will call the fire department, send an emergency ambulance to the patient's home and give the patient instruction about how he should act.

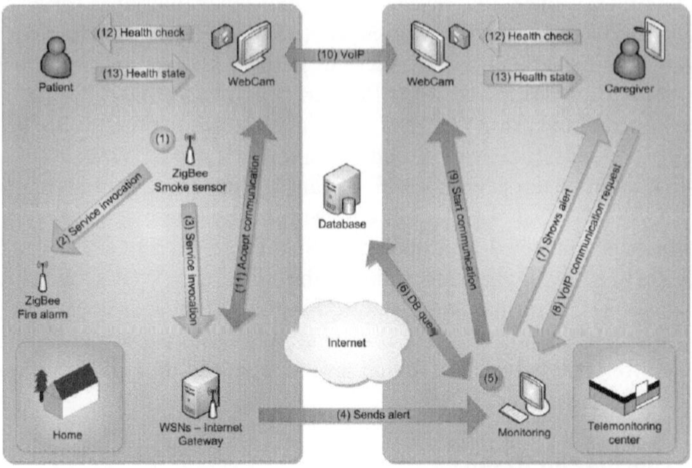

Fig. 1. Example operation of the telemonitoring system

As previously mentioned, the system implements a distributed architecture specially designed for integrating heterogeneous WSNs. This distributed architecture is called SYLPH (*Service laYers over Light PHysical devices*) [3]. It integrates a SOA approach over WSNs. The main objective of this proposal is to distribute resources over multiple WSNs by modeling the functionalities as independent services. SYLPH covers aspects relative to services such as registration, discovering and addressing. Some nodes in the system can integrate services directories for distributing registration and discovering services. SYLPH allows the interconnection of several networks from different wireless technologies, such as ZigBee or Bluetooth. In this case, the WSNs are interconnected through a set of intermediate gateways connected to several wireless interfaces simultaneously. Such gateways are called *SYLPH Gateways*. SYLPH implements an organization based on a stack of layers. Each layer in one node communicates with its peer in another node through an established protocol. In addition, each layer offers specific functionalities to the immediately upper layer in the stack. These functionalities are usually called *interlayer services*. The SYLPH layers are added over the existent application layer of each WSN stack, allowing the platform to be reutilized over different technologies.

The *SYLPH Message Layer* (SML) offers the upper layers the possibility of sending asynchronous messages between two nodes through the *SYLPH Services Protocol* (SSP), the internetworking protocol of the SYLPH platform. That is, it allows sending packets of data from one node to another node regardless of the WSN to which each one belongs. The *SYLPH Application Layer* (SAL) allows different nodes to directly communicate with each other using SSDL (*SYLPH Services Definition Language*) requests and responses that will be delivered in encapsulated SML messages following the SSP. The SSDL is the IDL (Interface Definition Language) used by SYLPH. SSDL has been specifically designed to work with limited computational resources nodes [3]. Furthermore, there are other interlayer services offered by the SAL for registering services or finding services offered by other nodes. In fact, these interlayer services call other interlayer services offered by the *SYLPH Services Directory Sub-layer* (SSDS). The SSDS creates dynamical services tables to locate and register services in the network. Any node that stores and maintains services tables is called *SYLPH Directory Node* (SDN).

As mentioned above, in SYLPH, a node in a specific type of WSN (*e.g.*, ZigBee) can directly communicate with a node in another type of WSN (*e.g.*, Bluetooth). Therefore, several heterogeneous WSNs can be interconnected through a SYLPH Gateway. A SYLPH Gateway is a device with several hardware network interfaces, each of which is connected to a distinct WSN. The SYLPH Gateway stores routing tables to forward SSP packets among the different WSNs with which it is interconnected. The information transported in the SSP header is enough to route the packets to the corresponding WSN. If several WSNs belong to the SYLPH network, there is no difference between invoking a service stored in a node in the same WSN or in a node from a different WSN.

3.1 Experiments and Results

Two experiments were realized to compare the performance of the new improved telemonitoring system with the previous prototype. As described before, the previous telemonitoring system used ZigBee nodes including each of them an 8-bit C51-based microcontroller with 8448B RAM, 128KB Flash memory and a ZigBee transceiver. On the other hand, the new system was formed by new n-Core Sirius-A 2.4GHz including an ATmega1281 microcontroller with 8KB RAM, 128KB Flash memory and an AT86RF231 transceiver [4]. Both systems were successively implemented in a healthcare institution in Salamanca (Spain). Both systems included a VoIP infrastructure connected between the remote center and the patients' homes. Patients were 20 elderly people with a relative risk of having a fall or home accident due to their limited mobility capabilities. In both systems each patient carried the three biomedical sensors previously described (id and fall detector, ECG and breath monitors), as well as a panic button to transmit alarms to the center in case of risk.

The patients selected had similar home characteristics (*i.e.*, 5 rooms, including bathroom and kitchen). In each home, 5 smoke sensors, 5 light sensors, 5 light dimmers and 1 fire alarm were installed. Both systems were subjected to observation during a period of four weeks in order to gather information and make comparisons. The data tracked were relative to the alerts registered by the system from the patients' homes. These alerts could come not only from the alarm button but also from any of the other sensors that constituted the telemonitoring systems. As a result, several risks sources, including the fall detector, the fire alarm or the heart pulse, were taken into account in the system data. The precise measured variables were: *average response time to incidents*; *average assisted incidents per day*; and *average number of false positives per day*. Table 1 illustrates how the new telemonitoring system reduced the average response time to incidents and reduced the false positives. Moreover, the new system allowed caregivers to detect some situations that the older system did not. This is because n-Core devices are more efficient and robust when implementing SYLPH platform due to their improved characteristics. The ZigBee stack implemented in n-Core devices are more robust than previous C-51 devices. This way, the deployed ZigBee network is more stable in the new telemonitoring system and frames transmissions have less faults and errors.

Table 1. Comparison between both telemonitoring systems

Factor	C51-based	n-Core
Average response time to incidents (minutes)	14.2	13.7
Average assisted incidents per day	3.1	3.4
Average number of false positives per day	1.6	1.2

4 Conclusions and Future Work

The system presented in this paper allows wireless devices from different technologies to work together in a distributed way in smart environments where information fusion is very important. Because such devices do not require large memory chips or fast microprocessors to exploit their functionalities, it is possible to create a more flexible system and reduce the implementation costs in terms of development and infrastructure support compared to other analyzed telemonitoring approaches [6] [7]. The distributed approach of this system makes it possible to add new components in execution time. In this respect, this model goes a step further in the design of information fusion scenarios (*e.g.*, e-healthcare). Furthermore, the integration of SYLPH and FUSION@ in the system facilitates the fusion of information coming from heterogeneous WSNs that can be thus managed by intelligent agents.

Future work includes the addition of new automation and biomedical sensors to the system to obtain additional context information. The suggestions and the necessities of patients and caregivers have been taken into account. In addition, some improvements are under development to enhance the overall system operation. An indoor Real-Time Location System based on the locating engine provided by the n-Core platform is intended to be implemented both in patient homes and medical centers. Patients will continue to carry the ZigBee wristbands as identification tags, and more ZigBee presence detectors will be present both in center and homes. Thus, if a patient suffers an accident at home, the system will warn caregivers about what room the patient is in, and activate a corresponding webcam. At the medical center, the system will keep track of the location of each patient, alerting the medical personnel if anyone leaves the center or accesses to a restricted area.

Acknowledgments. This work has been supported by the Spanish Ministry of Science and Innovation, Project T-Sensitive, TRA2009_0096.

References

1. Ilyas, M., Mahgoub, I.: Handbook of Sensor Networks: Compact Wireless and Wired Sensing Systems. CRC Press, Boca Raton (2004)
2. Corchado, J.M., Bajo, J., Tapia, D.I., Abraham, A.: Using Heterogeneous Wireless Sensor Networks in a Telemonitoring System for Healthcare. IEEE Transactions on Information Technology in Biomedicine 14, 234–240 (2010)
3. Tapia, D.I., Alonso, R.S., De Paz, J.F., Corchado, J.M.: Introducing a distributed architecture for heterogeneous wireless sensor networks. In: Omatu, S., Rocha, M.P., Bravo, J., Fernández, F., Corchado, E., Bustillo, A., Corchado, J.M. (eds.) IWANN 2009. LNCS, vol. 5518, pp. 116–123. Springer, Heidelberg (2009)
4. n-Core® Platform - Wireless Sensor Networks, `http://www.n-core.info`
5. Fazel-Rezai, R., Pauls, M., Slawinski, D.: A Low-Cost Biomedical Signal Transceiver based on a Bluetooth Wireless System. In: 2007 29th Annual International Conference of the IEEE Engineering in Medicine and Biology Society, Lyon, France, pp. 5711–5714 (2007)
6. Jurik, A.D., Weaver, A.C.: Remote Medical Monitoring. Computer 41, 96–99 (2008)
7. Varshney, U.: Improving Wireless Health Monitoring Using Incentive-Based Router Cooperation. Computer 41, 56–62 (2008)
8. Marin-Perianu, M., Meratnia, N., Havinga, P., de Souza, L., Muller, J., Spiess, P., Haller, S., Riedel, T., Decker, C., Stromberg, G.: Decentralized enterprise systems: a multiplatform wireless sensor network approach. IEEE Wireless Communications 14, 57–66 (2007)
9. Baronti, P., Pillai, P., Chook, V.W.C., Chessa, S., Gotta, A., Hu, Y.F.: Wireless sensor networks: A survey on the state of the art and the 802.15.4 and ZigBee standards. Comput. Commun. 30, 1655–1695 (2007)
10. Cerami, E.: Web Services Essentials: Distributed Applications with XML-RPC, SOAP, UDDI & WSDL. O'Reilly Media, Inc., Sebastopol (2002)
11. de Freitas, E.P., Wehrmeister, M.A., Pereira, C.E., Larsson, T.: Reflective middleware for heterogeneous sensor networks. In: Proceedings of the 7th Workshop on Reflective and Adaptive Middleware - ARM 2008, Leuven, Belgium, pp. 49–50 (2008)
12. Alonso, R.S., García, O., Zato, C., Gil, O., De la Prieta, F.: Intelligent Agents and Wireless Sensor Networks: A Healthcare Telemonitoring System. In: Demazeau, Y., Dignum, F., Corchado, J.M., Bajo, J., Corchuelo, R., Corchado, E., Fernández-Riverola, F., Julián, V.J., Pawlewski, P., Campbell, A. (eds.) Trends in PAAMS. Advances in Intelligent and Soft Computing, vol. 71, pp. 429–436. Springer, Heidelberg (2010)

Supporting System for Detecting Pathologies

Carolina Zato, Juan F. De Paz, Fernando de la Prieta, and Beatriz Martín

Department of Computer Science and Automation, University of Salamanca
Plaza de la Merced s/n, 37008, Salamanca, Spain
{carol_zato,fcofds,fer,eureka}@usal.es

Abstract. Arrays CGH make possible the realization of tests on patients for the detection of mutations in chromosomal regions. Detecting these mutations allows to carry out diagnoses and to complete studies of sequencing in relevant regions of the DNA. The analysis process of arrays CGH requires the use of mechanisms that facilitate the data processing by specialized personnel since traditionally, a segmentation process is needed and starting from the segmented data, a visual analysis of the information is carried out for the selection of relevant segments. In this study a CBR system is presented as a supporting system for the extraction of relevant information in arrays CGH that facilitates the process of analysis and its interpretation.

Keywords: CGH arrays, knowledge extraction, visualization, CBR system.

1 Introduction

Arrays CGH (Comparative Genomic Hybridization) [39] are a type of microarrays that allows analyzing the information of the gains, losses and amplifications [36] in regions of the chromosomes for the detection of mutations. These types of microarrays unlike expression arrays do not measure the expression level of the genes; this is the reason why its use and analysis differ from the provided by expression arrays. The data obtained by the arrays CGH allows detecting automatically the mutations that characterize certain pathologies [29] [25]. Moreover, this information is useful to cross it with genetic sequencing, facilitating the analysis of the genetic sequencings and the sequencing tasks [6].

Microarray-based CGH and other large-scale genomic technologies are now routinely used to generate a vast amount of genomic profiles. Exploratory analysis of this data is crucial in helping to understand the data and to help form biological hypotheses. This step requires visualization of the data in a meaningful way to visualize the results and to perform first level analyses [32]. At present, tools and software already exist to analyze the data of arrays CGH, such as CGH-Explorer [24], ArrayCyGHt [19], CGHPRO [7], WebArray [38] or ArrayCGHbase [27], VAMP [32]. The problem of these tools is that they follow a static processing flow, without the possibility of storing or selecting those techniques that suit the samples of each case best. Therefore, these tools do not permit to personalize the flow of actions for the extraction of knowledge or to store preferences that can be useful in future processes with similar needs. The tool that is presented incorporates automatic procedures that can carry out

J. Cabestany, I. Rojas, and G. Joya (Eds.): IWANN 2011, Part II, LNCS 6692, pp. 669–676, 2011.

the analysis and the visual representations, facilitating the extraction of information with the most suitable processing flow. This allows the revision of the information by personnel without a great statistical knowledge and guarantees the obtaining of a better analysis automatically.

The process of arrays CGH analysis is decomposed in a group of structured stages, although most of the analysis process is done manually from the initial segmentation of the data. The initial data is segmented [35] to reduce the number of gains or losses fragments to be analyze. The segmentation process facilitates the later analysis of the data and is important to be able to represent a visualization of the data. Normally, the interpretation of the data is carried out manually from the visualization of the segmented data, however, when great amounts of these data have to be analyzed, it is necessary to create a decision support process.

For this reason, in this work a CBR system is included to facilitate the analysis and the automatic interpretation of the data by means of the configuration of analysis flows and the incorporation of flows based on predefined plans. The execution flows include procedures for the accomplishment of segmentation, classification, extraction of automatic information and visualization. The classification process facilitates the diagnosis of patients based on previous data; the process of knowledge extraction selects the differentiating regions of sets of patients by means of statistical techniques. Finally, the visualization process facilitates the revision of the results.

This article is divided as follows: section 2 describes the arrays CGH, section 3 describes our system, and section 4 presents the results and conclusions.

2 CBR-CGH System

CGH analysis allows the characterization of mutations that cause several cancers. The relationship between the chromosomal alterations and the prognosis of illness is well established. Recently, conventional array-based expression profiling has demonstrated that chromosomal alterations are associated with distinctive expression patterns. The system proposed in this work focuses on the detection of carcinogenic patterns in the data from CGH arrays, and is constructed from a CBR system that provides a classification and knowledge extraction technique based on previous cases.

The CBR developed system receives data from the analysis of chips and is responsible of establishing the workflow for classifying individuals based on evidence and existing data. The purpose of CBR is to solve new problems by adapting solutions that have been used to solve similar problems in the past [21]. The primary concept when working with CBRs is the concept of case. A case can be defined as a past experience, and is composed of three elements: a problem description which describes the initial problem, a solution which provides the sequence of actions carried out in order to solve the problem, and the final state which describes the state achieved once the solution was applied. The way cases are managed is known as the CBR cycle, and consists of four sequential steps which are recalled every time a problem needs to be solved: retrieve, reuse, revise and retain. Each of the steps of the CBR life cycle requires a model or method in order to perform its mission.

The algorithm selected for the retrieval of cases should be able to search the case base and selects the kind of default problems according to the analyzed data. In our

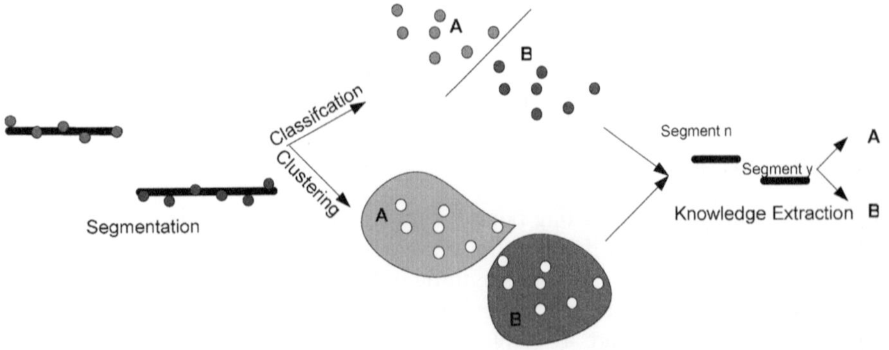

Fig. 1. Workflows in the classification, clustering, and knowledge extraction

case study, the system selects the workflows defined for each type of problem. The retrieved workflows are shown and the user selects one of them, then the activities are carried out. The revise phase consists of an expert revision for the proposed solution, and finally, the retain phase allows the system to learn from the experiences obtained in the three previous phases, consequently updating the cases memory.

The workflows set the sequence of actions in order to analyze the data. The kinds of default analysis are: clustering, classification and knowledge extraction. The figure 1 shows the available workflows and their activities since the initial state, for example a knowledge extraction process implies a segmentation and a clustering or classification activity.

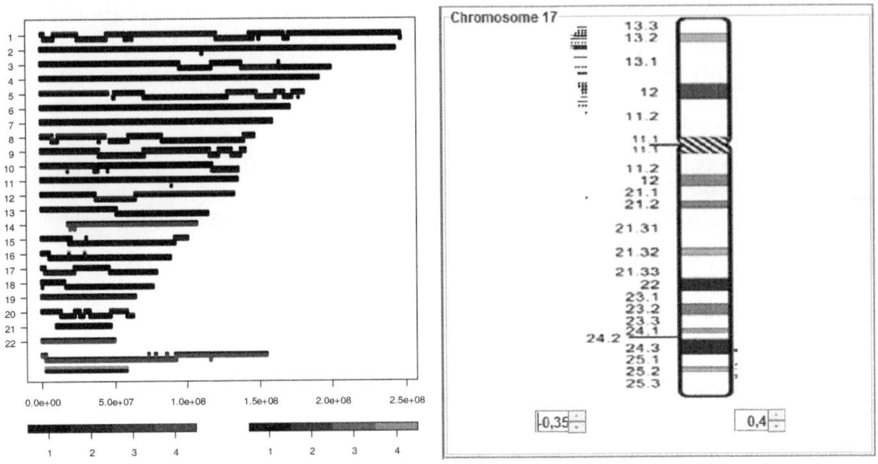

Fig. 2. Visualization of gains and losses using a) CGHcall and b) new method

In addition, a new visualization is provided to localized the mutations in an easier way, facilitating the identification of mutations that affects the gene codification among the large amount of genes. The figure 2a represents gains and losses using

CGHcall in R. The new visualization method is shown in the figure 2b, this visualization helps to locate the regions with mutations.

The system includes techniques for each of the activities (clustering, classification and knowledge extraction). Then, the applied algorithms in the steps are described.

2.1 Normalization and Segmentation

This stage constitutes the starting point for the treatment of the data and is necessary for the reduction of noise, the detection of losses and gains and the identification of breakpoints. The tool that is presented, through R Server, uses the package snapCGH [35], which allows both normalization and segmentation. Currently, many different segmentation algorithms are available, because of this, snapCGH incorporates software wrappers for several of these algorithms such as aCGH, DNACopy, GLAD and tilingArray. In [37][15] some comparisons between them can be found. The election of this package is due to the great acceptance, expansion and versatility, since it supplies many possibilities for the preprocessing.

2.2 Classification

The classification process is carried out according to a mixture of classifiers, although the system allows select a technique instead the mixture. A mixture of experts provide advances capacities by fusing the outputs of various processes (experts) and obtain the response more suitable for the final value [23] [28]. Mixtures of experts are also commonly used for classification and are usually called ensemble [41]. Some examples are the Bagging [5] or Ada-Boosting [11] algorithms. The classification algorithms can be divided in: decision trees, decision rules, probabilistic models, fuzzy models, based on functions, ensemble. The system selects these algorithms for each kind of method: decision rules RIPPER [8], One-R [16], M5 [17], decision trees J48 [31], CART [4] (Classification and Regression Trees), probabilistic models naive Bayes [10], fuzzy models K-NN (K-Nearest Neighbors) [1] and finally ensemble such as Bagging [5] and Ada-Boosting [11].

In order to calculate the final output of the system, RBF networks are used [13] [34]. The k cases retrieved in the previous phase are used by the classifiers and RBFs network as a training group that allows adapting its configuration to the new problem encountered before generating the initial estimation. The system presented in this article has a RBF network for each of the set of individual. Each of the RBF networks has as inputs the outputs estimated by the classifiers evaluated for the individual.

2.3 Clustering

Clustering techniques are typically broken down into the following categories [30] hierarchical, which include dendrograms [33], AGNES [18], DIANA [18], Clara [18]; neural networks such as SOM [20] (Self-Organized Maps), NG [26] (Neural Gas), GCS [12] (Growing Cell Structure; methods based on minimizing objective functions, such as k-means [14] and PAM [18] (Partition around medoids); or probabilistic-based models such as EM [2] (Expectation-maximization) and FANNY [18].

The provided methods are: in hierarchical clustering dendrograms [33], minimizing objective functions k-means [14] and PAM (Partitioning Around Medoids) [18] and in neural network SOCADNN (Self Organized Cluster Automatic Detection Neural Network) [3]. En el trabajo [3] se han realizado estudios sobre diferentes métodos de cluster y las ventajas que proporciona.

Hierarchical methods such as dendrograms do not require a number of clusters up front since they use a graphical representation to determine the number. Partition based methods as k-means and PAM, which optimize specific objective functions, have the disadvantage of requiring the number of clusters up front. Methods that are either hierarchical or minimize specific objective functions present certain deficiencies when it comes to recognizing groupings of individuals. ANN can adapt to the data surface, although they usually require additional time to do so. The SOM [20], have variants of learning methods that base their behaviour on methods similar to the NG [26]. They create a mesh that is adjusted automatically to a specific area. The ART networks can be considered as an alternative. The major disadvantage of these networks is the selection of the monitoring parameter [2] to determine the number of clusters. Another disadvantage is that the knowledge extraction is more complicated than in mesh-based networks, so learning is less evident.

2.4 Knowledge Extraction

Some techniques of the section 0 such as decision trees or rules, Bayesian networks or even rough sets could be applied in order to explain clusters or classifications although, the main objective in these problems is find maximum quantity of mutations that characterize a pathology. This information can be used in other studies as the sequencing of the concrete interesting regions with mutations. For this reason, statistical techniques are introduced in these activity for selecting the relevant segments. The introduced statistical techniques are broken down in non parametrics Kruskal-Wallis [42] and Mann-Whitney U-test [40] and parametrics ANOVA [9].

3 Results and Conclusions

In order to analyze the operation of the system, different data types of cancer, obtained from the data of the array CGH, were selected. In this case study we have 43 patients with GIST cancer, the data contain 4 kinds of pathologies: KIT Mutants, Wild-type, PDGFRA Mutants and BRAF Mutant, the pathology BRAF was removed because there was just one case with this illness. These data were previously classified, since the knowledge extraction is carried out from the previous classification. The data contain for each patients the kind of GIST and the segments with the gains and losses. The result of the relevant regions is shown in Table 1. Kruskal Wallis was applied for the extraction of this information, since the variables did not follow a normal distribution and therefore, a non-parametric test was required.

The figure 3 shows the highlighted region in the table 1. This region presents relevant differences among the detected GIST. In the box plots of the figure 3, PDGFRA doesn't have losses or it presents gains in the region where the others present losses or they don't have variations. We can validate the others regions in similar way.

Table 1. Total number of hits for the different classifiers

Chromosome	Start	End	Nclone	Wide
8	139136846	146250764	314	7113918
15	30686790	91341204	2425	60654414
23	91485305	91537583	3	52278
22	134661	49565815	1491	49431154
20	58058472	62363573	200	4305101
8	39535654	43647062	143	4111408
8	7789936	8132138	3	342202
8	11665221	39341523	879	27676302
3	137653537	163941171	784	26287634
15	56257	18741715	15	18685458
1	**9110683**	**24996793**	**548**	**15886110**
9	70803414	70803414	9	146631
20	47048133	58039998	342	10991865
15	20249885	30298095	302	10048210

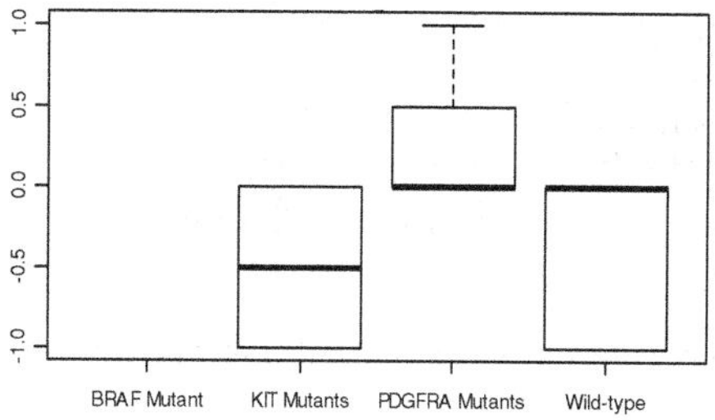

Fig. 3. Box plot for the region 9110683, 24996793

Although the system is still in a development phase, it is able to detect variations that allow characterizing different pathologies automatically. In addition, it permits the redefinition of execution flows, storing the sequence of actions that previously were considered satisfactory for its later use.

Acknowledgements. This work has been supported by the MICINN TIN 2009-13839-C03-03.

References

[1] Aha, D., Kibler, D., Albert, M.K.: Instance-based learning algorithms. Machine Learning 6, 37–66 (1991)

[2] Akhbardeh, A., Nikhil, Koskinenb, P.E., Yli-Harja, O.: Towards the experimental evaluation of novel supervised fuzzy adaptive resonance theory for pattern classification. Pattern Recognition Letters 29(8), 1082–1093 (2008)

[3] Bajo, J., De Paz, J.F., Rodríguez, S., González, A.: A new clustering algorithm applying a hierarchical method neural network. Logic Journal of IGPL (in Press)

[4] Breiman, L., Fried, J.H., Olshen, R.A., Stone, C.J.: Classification and regression trees. Wadsworth International Group, Belmont (1984)
[5] Breiman, L.: Bagging predictors. Machine Learning 24(2), 123–140 (1984)
[6] Brown, P.O., Botstein, D.: Exploring the new world of the genome with DNA microarrays. Nature Genetics 21, 33–37 (1999)
[7] Chen, W., Erdogan, F., Ropers, H., Lenzner, S., Ullmann, R.: CGHPRO- a comprehensive data analysis tool for array CGH. BMC Bioinformatics 6(85), 299–303 (2005)
[8] Cohen, W.W.: Fast effective rule induction. In: Proceedings of the 12th International Conference on Machine Learning, pp. 115–123. Morgan Kaufmann, San Francisco (1995)
[9] De Haan, J.R., Bauerschmidt, S., van Schaik, R.C., Piek, E., Buydens, L.M.C., Wehrens, R.: Robust ANOVA for microarray data. Chemometrics and Intelligent Laboratory Systems 98(1), 38–44 (2009)
[10] Duda, R.O., Hart, P.: Pattern classification and Scene Analysis. John Wisley & Sons, New York (1973)
[11] Freund, Y., Schapire, R.E.: Experiments with a new boosting algorithm. In: Thirteenth International Conference on Machine Learning, pp. 148–156 (1996)
[12] Fritzke, B.: A growing neural gas network learns topologies. Advances in Neural Information Processing Systems 7, 625–632 (1995)
[13] Fritzke, B.: Fast Learning with Incremental RBF Networks. Neural Processing Letters 1(1), 2–5 (1994)
[14] Hartigan, J.A., Wong, M.A.: A K-means clustering algorithm. Applied Statistics 28, 100–108 (1979)
[15] Hofmann, W.A., Weigmann, A., Tauscher, M., Skawran, B., Focken, T., Buurman, R., Wingen, L.U., Schlegelberger, B., Steinemann, D.: Analysis of Array-CGH Data Using the R and Bioconductor Software Suite. Comparative and Functional Genomics, Article ID 201325 (2009)
[16] Holmes, G., Hall, M., Prank, E.: Generating Rule Sets from Model Trees. In: Advanced Topics in Artificial Intelligence, vol. 1747/1999, pp. 1–12 (2007)
[17] Holte, R.C.: Very simple classification rules perform well on most commonly used datasets. Machine Learning 11, 63–91 (1993)
[18] Kaufman, L., Rousseeuw, P.J.: Finding Groups in Data: An Introduction to Cluster Analysis. Wiley, New York (1990)
[19] Kim, S.Y., Nam, S.W., Lee, S.H., Park, W.S., Yoo, N.J., Lee, J.Y., Chung, Y.J.: ArrayCyGHt, a web application for analysis and visualization of array-CGH data. Bioinformatics 21(10), 2554–2555 (2005)
[20] Kohonen, T.: Self-organized formation of topologically correct feature maps. Biological Cybernetics, 59–69 (1982)
[21] Kolodner, J.: Case-Based Reasoning. Morgan Kaufmann, San Francisco (1993)
[22] Brunelli, R.: Histogram Analysis for Image Retrieval. Pattern Recognition 34, 1625–1637 (2001)
[23] Lima, C.A.M., Coelho, A.L.V., Von Zuben, F.J.: Hybridizing mixtures of experts with support vector machines: Investigation into nonlinear dynamic systems identification. Information Sciences 177(10), 2049–2074 (2007)
[24] Lingjaerde, O.C., Baumbush, L.O., Liestol, K., Glad, I.K., Borresen-Dale, A.L.: CGH-explorer, a program for analysis of array-CGH data. Bioinformatics 21(6), 821–822 (2005)
[25] Mantripragada, K.K., Buckley, P.G., Diaz de Stahl, T., Dumanski, J.P.: Genomic microarrays in the spotlight. Trends Genetics 20(2), 87–94 (2004)

[26] Martinetz, T., Schulten, K.: A neural-gas network learns topologies. Artificial Neural Networks 1, 397–402 (1991)
[27] Menten, B., Pattyn, F., De Preter, K., Robbrecht, P., Michels, E., Buysse, K., Mortier, G., De Paepe, A., van Vooren, S., Vermeesh, J., et al.: ArrayCGHbase: an analysis platform for comparative genomic hybridization microarrays. BMC Bioinformatics 6(124), 179–187 (2006)
[28] Nguyena, M.H., Abbassa, H.A., Mckay, R.I.: A novel mixture of experts model based on cooperative coevolution. Neurocomputing 70, 155–163 (2006)
[29] Pinkel, D., Albertson, D.G.: Array comparative genomic hybridization and its applications in cancer. Nature Genetics 37, 11–17 (2005)
[30] Po, R.W., Guh, Y.Y., Yang, M.S.: A new clustering approach using data envelopment analysis. European Journal of Operational Research 199(1), 276–284 (2009)
[31] Quinlan, J.R.: C4.5: Programs For Machine Learning. Morgan Kaufmann Publishers Inc., San Francisco (1993)
[32] Rosa, P., Viara, E., Hupé, P., Pierron, G., Liva, S., Neuvial, P., Brito, I., Lair, S., Servant, N., Robine, N., Manié, E., Brennetot, C., Janoueix-Lerosey, I., Raynal, V., Gruel, N., Rouveirol, C., Stransky, N., Stern, M., Delattre, O., Aurias, A., Radvanyi, F., Barillot, E.: VAMP: Visualization and analysis of array-CGH, transcriptome and other molecular profiles Bioinformatics. Bioinformatics 22(17), 2066–2073 (2006)
[33] Saitou, N., Nie, M.: The neighbor-joining method: A new method for reconstructing phylogenetic trees. Mol. Biol. Evol. 4, 406–425 (1987)
[34] Savitha, R., Suresh, S., Sundararajan, N.: A fully complex-valued radial basis function network and its learning algorithm. Int. Journal of Neural Systems. 19(4), 253–267 (2009)
[35] Smith, M.L., Marioni, J.C., Hardcastle, T.J., Thorne, N.P.: snapCGH: Segmentation, Normalization and Processing of aCGH Data Users' Guide. Bioconductor (2006)
[36] Wang, P., Young, K., Pollack, J., Narasimham, B., Tibshirani, R.: A method for callong gains and losses in array CGH data. Biostat. 6(1), 45–58 (2005)
[37] Willenbrock, H., Fridlyand, J.: A comparison study: applying segmentation to array CGH data for downstream analyses. Bioinformatics 21(22), 4084–4091 (2005)
[38] Xia, X., McClelland, M., Wang, Y.: WebArray, an online platform for microarray data analysis. BMC Bionformatics 6(306), 1737–1745 (2005)
[39] Ylstra, B., Van den Ijssel, P., Carvalho, B., Meijer, G.: BAC to the future! or oligonucleotides: a perspective for microarray comparative genomic hybridization (array CGH). Nucleic Acids Research 34, 445–450 (2006)
[40] Yue, S., Wang, C.: The influence of serial correlation on the Mann-Whitney test for detecting a shift in median. Advances in Water Resources 25(3), 325–333 (2002)
[41] Zhanga, H., Lu, J.: Creating ensembles of classifiers via fuzzy clustering and deflection. Fuzzy Sets and Systems 161(13), 1790–1802 (2010)
[42] Kruskal, W., Wallis, W.: Use of ranks in one-criterion variance analysis. Journal of American Statistics Association (1952)

An Ontological Approach for Context-Aware Reminders in Assisted Living' Behavior Simulation

Shumei Zhang, Paul McCullagh, Chris Nugent, Huiru Zheng, and Norman Black

School of Computing and Mathematics, University of Ulster, BT37 0QB, UK
{zhang-s2,pj.mccullagh,cd.nugent,h.zheng,nd.black}@ulster.ac.uk

Abstract. A context-aware reminder framework, which aims to assist elderly people to live safely and independently within their own home, is described. It combines multiple contexts extracted from different modules such as activity monitoring, location detection, and predefined routine to monitor and analyze personal activities of daily living. Ontological modeling and reasoning techniques are used to integrate various heterogeneous contexts, and to infer whether a fall or abnormal activity has occurred; whether the user is in unhealthy postures; and whether the user is following their predefined schedule correctly. Therefore this framework can analyse behaviour to infer user compliance to a healthy lifestyle, and supply appropriate feedback and reminder delivery. The ontological approach for context awareness can provide both distributed context integration and advanced temporal reasoning capabilities.

Keywords: ontological modeling, temporal reasoning, context-awareness, reminder, behavior analysis.

1 Introduction

A healthy lifestyle can benefit people to maintain their health and reduce the risk of chronic diseases. Both the increasing numbers of elderly people within the population and the increased prevalence of related chronic diseases are placing economic burdens on health care systems on a global scale. Adherence to scheduled routines is important for maintaining a healthy lifestyle. Behavior analysis and reminder delivery can be utilized to encourage people to adhere to their predefined routines. Activity monitoring combined with activity-aware reminders may be able to improve lifestyle and possibly wellbeing. Especially, detections of abnormal activity (e.g. falls or other emergency alerts) can assist elderly people to live safely and independently at home, and potentially save healthcare costs.

The remainder of the paper is organized as follows: Related work is discussed in Section 2. Layered conceptual framework architecture is presented in Section 3. Section 4 is focused on the methodology for contexts modeling and reasoning. The experimental setting and initial experimental results are presented in Section 5. Discussion of the methodology and future work are presented in Section 6.

J. Cabestany, I. Rojas, and G. Joya (Eds.): IWANN 2011, Part II, LNCS 6692, pp. 677–684, 2011.

2 Related Work

Reminder systems normally deliver reminders according to a predefined routine based only on fixed times (Osmani *et al.* 2009). Scheduling can aid users in the management of their Activities of Daily Living (ADLs). For example, Levinson *et al.* (1997) presented a cognitive orthotic system that provided visible and audible clues about care plan execution using a handheld device. Early techniques that supplied reminders functioned in a similar manner to an alarm clock. The problems are that these reminder systems do not take into account what is the user's current status and whether this reminder will be useful or relevant to them at that particular point in time. In order to deliver appropriate reminders, systems are required to take into account: *where the user is, what the user is doing, and whether it is necessary to deliver a reminder*. Therefore, some studies have focused on context-aware reminders over the last decade.

Pollack et al. (2003) proposed Autominder, a location-aware cognitive orthotic system to assist people who suffer memory impairment. This system was embedded on a mobile robot, and the robot's on-board sensors could report which room the user was in, and Autominder could make decisions about whether and when reminders were given, according to the user's room location and their individual daily plan. Nevertheless, the approach of reminder delivery decided only by location, does not give information to estimate what the user is doing, what they have performed, and whether the reminder delivery provokes an undesirable interruption. For that reason, we believe that more contexts are required for inferring user's current status.

Challenges for context-awareness techniques are: how to integrate the different types of contexts, and how to infer the 'highest level' information based on the various contexts and relationships. Existing methods for context modeling vary in their expressive power, in their supportive ability for context reasoning, in their computational performance, and in the extensibility of context management. Strang *et al.* (2004) categorize context modeling approaches as: (1) Key-Value (2) Markup Scheme (3) Graphical (4) Object Oriented (5) Logic Based and (6) Ontology Based. The six approaches were evaluated for ubiquitous computing environments based on the requirements of distributed composition, partial validation, richness and quality of information, incompleteness and ambiguity, level of formality, and applicability to existing environments. Their experimental results illustrated that the ontology-based approach was the most expressive and best met the requirements.

Ontology based modelling and reasoning technologies have been adopted in pervasive computing and in particular for assistance with the completion of ADLs. For example, Kang *et al.* (2006) proposed a context-aware ubiquitous healthcare system, which used ontology based context management to derive abstractions from the data of the user's vital signals. Chen and Nugent (2009) used an ontology-based approach for activity recognition. They showed that ADL models were flexible and could be easily created, customized, deployed and scaled.

This study proposes a context-aware reminder framework, which integrates various contexts such as location, current activity, timeline and predefined schedule together to infer inconsistency between what the user is expected to do and what they are currently doing, and to make decisions about whether and when to issue reminders.

3 Framework Architecture

This reminding framework aims to monitor a user's ADLs and obtain the various contexts of when, where and what the user is doing, and hence to infer whether the user is in the correct place at the correct time and undertaking the pre-defined activity. It includes four different components and adopts a layered structure to connect each of these components as shown in Figure 1: data sensing (L1), context extraction (L2), context management (L3) and context-aware reminders (L4). The layered architecture facilitates extensibility and simplifies reusability.

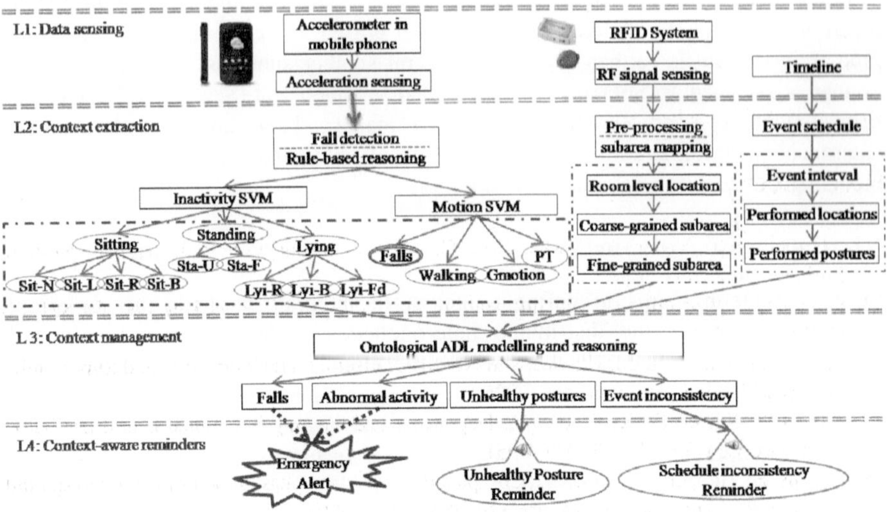

Fig. 1. Layered architecture of reminding framework, showing interaction of layers 1 to 4

In a preliminary study, two kinds of sensors were used for acquiring ADL specific data at the layer L1. A G-sensor embedded in the HTC smart phone was used to record acceleration for activity monitoring. A radio frequency identification (RFID) system was used for indoor location detection. Three modules in L2 were used to extract the 'high level' context: activity classification, location detection and event schedule design, and resulting high level context information stored. Further contexts integration and reasoning are performed in L3. Four types of 'higher level' contexts are inferred based on the ontological ADL modelling and reasoning: falls, abnormal activity (for example, lying in an inappropriate location), healthy or unhealthy postures and event consistency or inconsistency. Finally, feedback is delivered in the layer L4: emergency alert (fall down or abnormal activity), unhealthy posture reminder, and event inconsistent reminder.

4 Methodology

The challenges for delivering context-aware reminders are how to extract the precise high level contexts from the original sensing data, how to integrate the distributed heterogeneous contexts from different modules, and how to infer highest level contexts as feedback delivered to users. The activity classification and location detection algorithms have been described in (Zhang *et al.* 2009; 2010; 2011). This paper focuses on the context management and context-aware reminders based on three 'Context extraction' modules.

The activity monitoring module can extract the activity contexts for a total of 13 postures: sitting normal (Sit-N), sitting back (Sit-B), sitting leaning left and right (Sit-L and Sit-R); standing upright (Sta-U) and standing forward (Sta-F); lying right (Lyi-R), lying back (Lyi-B) and lying face down (Lyi-Fd), see Figure 1. The location detection module can extract the location contexts in three levels: room-based, and coarse-grained and fine-grained subareas within a room. The subarea size was divided according to their function. For example, the coarse-grained level is Bed, Desk, Sofa, Dining, and so on. Fine-grained level is further divided as several smaller subareas for some of coarse-grained areas, such as, Sofa area is fine-grained as which seat (Seat1, Seat2…). Context management is a process that is able to manage large amounts of distributed heterogeneous information in dynamic situations. Context management involves context modeling and context reasoning techniques.

4.1 Ontological ADL Modeling

Context modeling aims to define contexts, to integrate contexts, to explain the relationships among the distributed heterogeneous contexts, and to analyze the inconsistencies and incompleteness within the model. The ADL model was built in the Protégé-OWL environment. OWL ontologies consist of classes (concepts), properties, and instances (individuals). Classes can be interpreted as sets that contain instances. Properties are used to link instances. Instances represent objects in the domain in which we are interested. The ADL ontology is categorized into five main classes with 12 properties, as shown in Figure 2.

(1) Class *Timeline* (*temporal:Entity*) is a Protégé-OWL built-in class used to represent temporal durations and temporal constraints among the multiple classes. (2) Class *Person* describes the user's information using properties *hasName*, *hasAge* and *hasCondition*. (3) Class *SEvent* signifies the main tasks for a user's routines using seven properties *hasName*, *hasID*, *hasEvent*, *hasStartTime,* hasFinishTimeare, *performedIn* and *performedActivity*. (4) Class *Location* denotes where the user is at a given time using properties *hasName*, *hasStartTime*, *hasFinshTime*and *locateIn*. The location can be detected as which room, in addition to which part of a room. (5) Class *Activity* represents monitored user's activity postures with their timestamp using the properties *hasName*, *monitoredActivity*, *hasStartTime* and *hasFinishTime*.

Some of the 12 properties were specified to be the union of multiple classes. For example, classes of *Person, Activity, Location and SEvent* were specified as the union classes for the properties of *hasName*. This can be used to link the related instances from different classes. In addition, some of properties were defined with a corresponding inverse property. The inverse properties can link two different instances to each other. For example, the properties *performedActivity* and *monitoredActivity* are inverses that can be used to infer the relationships between the expected activity and monitored activity for a specified event as to whether or not it is consistent. In the similar way, the properties *performedIn* and *locatedIn* are inverse properties. It can link the expected location and detected location together for a specified event.

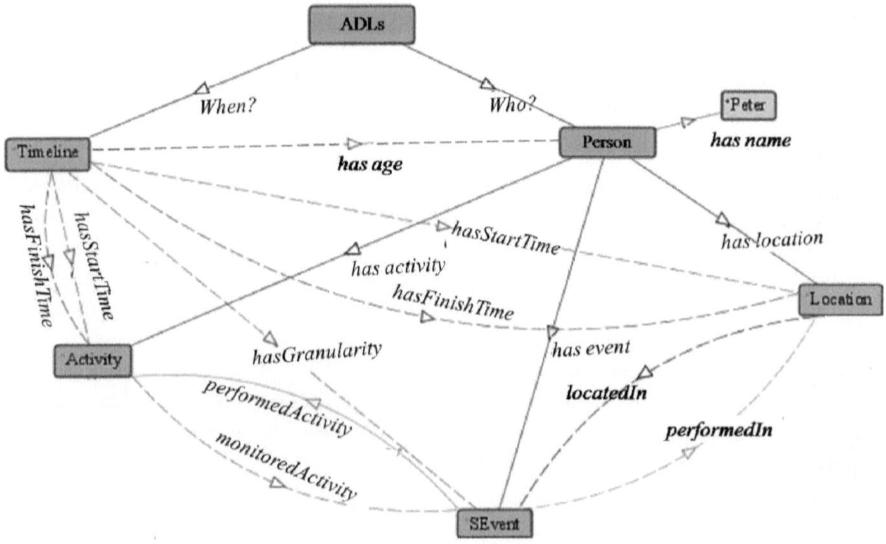

Fig. 2. ADLs ontological modeling and the relationships among the multiple classes

4.2 Ontological Context Reasoning

Context reasoning aims to infer 'highest level' information that the context-aware applications are interested in, based on the context model. The Semantic Web community has investigated the possibility of enhancing the expressive ontological languages through an extension with rules, and brought to the definition of logic languages. For example, semantic web rule language (SWRL) (Horrocks*et al.* 2004) and semantic query-enhanced web rule language (SQWRL) (O'Connor & Das 2009) have been widely used by studies. In this study, Ontological rule-based reasoning was performed by SWRL and SQWRL to infer inconsistency between monitored status and scheduled status based on the relationships among the personal activity posture, location and schedule. The relationships among the multiple classes represented using their properties were shown in Figure2. The *Timeline (temporal entity)* class links to the four classes (*Person*, *SEvent*, *Activity* and *Location*) through the corresponding time and duration properties. The *person* class links to the three classes (*Activity*, *SEvent* and *Location*). Additionally, the class *SEvent* links to *Location* and *Activity* using their corresponding inverse properties. If the property's value is consistent between each of the two pairs of inverse properties at the same time, it can be deduced that the person is following their schedule correctly. Otherwise, if one (location or posture) is not matched at a particular time, the system will infer an event 'inconsistent' result.

5 Experiments

The experiments used simulated scenarios to infer three types of reminders (as shown in Figure 1), based on corresponding rules in the Protégé environment. The rule-based reasoning involves activity relevant (whether a user has fallen or has unhealthy postures) and event relevant (consistency or inconsistency between the monitored status and expected status) queries. For conciseness, here we only use the activity relevant queries to illustrate the ontological reasoning.

Scenarios: *Assume that our user, Peter, had the coffee schedule from 10:00am to 11:00am. He went to the kettle area of the Kitchen and heated the water at 10:01. Then he sat down and drank the coffee in the dining area from 10:05 to 10:30. During this time his sitting posture was sitting normally for 15 minutes, sitting back for 5 minutes, and sitting leaning left for 5 minutes. Suddenly, he felt unwell, so he stood up with the intention to rest on the sofa. However, he fell down slowly on the floor when he tried to open the door at 10:30, until his neighbour Ian came to his house at 10:33, as Ian received an alert message.*

Activity relevant queries aim to infer whether the person has fallen, exhibited abnormal activity, or has adopted an unhealthy posture. The abnormal alert rule infers whether a person is lying, but not in the bed. For example, here, we pre-defined the abnormal activity and locations as:

abnormalA={*Lyi-R, Lyi-B, Lyi-Fd*} & *abnormalL*={*Kitchen, Bathroom, Stairs*}

The abnormal alert rule infers whether the *monitoredActivity* property has a value in *abnormalA* and the *locatedIn* property has a value in *abnormalL*. It is satisfied if both property values match the queries at the same temporal interval. Figure 3 illustrates how the *abnormalAlert* query was implemented using the SWRL rule with SQWRL queries, and the executing result is shown in Figure 4.

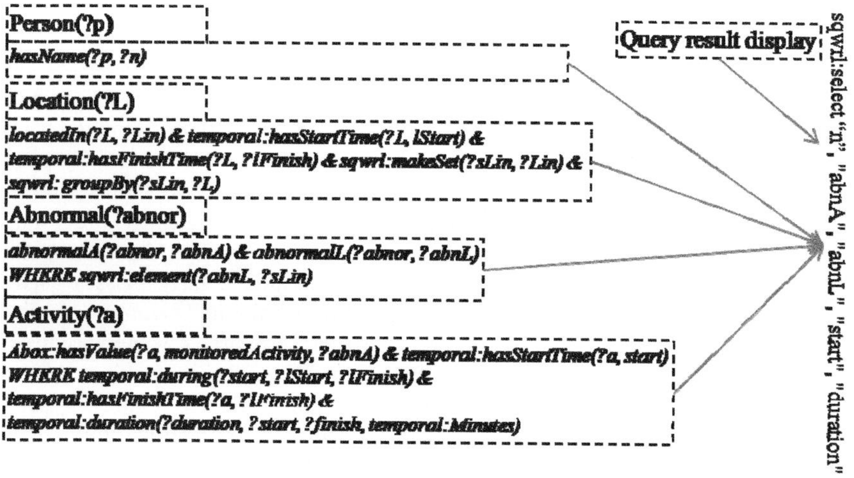

Fig. 3. SWRL rule visualization for *abnormalAlert* query

The *unhealthyPosture* rule infers whether a person is in an unhealthy posture. Here, five postures were defined as unhealthy:

$$unhealthyPosture \in \{Sit\text{-}B, Sit\text{-}R, Sit\text{-}L, Sta\text{-}F, Lyi\text{-}Fd\}$$

Therefore, if the *monitoredActivity* property has a value in *unhealthyPosture* for more than a predefined period of time, then an unhealthy posture reminder will remind the user to change their posture. For example, the experimental result in Figure 5 indicates that Peter was Sit-B and Sit-L for 5 minutes respectively in the Kitchen dining area.

In the initial implementation, the ontological queries were explained and demonstrated using simulated scenarios. Nevertheless, for a real behaviour analysis application, the user's ADLs can be analyzed for an appropriate period of time such as one week. Feedback can be provided to show when, where and for how long the user was correctly or incorrectly following his/her schedule.

SQWRLQueryTab | abnormalAlert

Person	Activity	Location	StartTime	finishTime	Duration(Min.)
Peter	Lyi-R	Kitchen.door	2010-05-12T10:30:00	2010-05-12T10:33:00	3

Fig. 4. Experimental result following execution to the *abnormalAlert* rule

SQWRLQueryTab | unhealthyPosture

Person	Location	Activity	startTime	finishTime	Dura
Peter	Kitchen.dining	Sit-B	2010-05-12T10:11:00	2010-05-12T10:16:00	5
Peter	Kitchen.dining	Sit-L	2010-05-12T10:25:00	2010-05-12T10:30:00	5

Fig. 5. Experimental result following execution to the *unhealthyPosture* rule

6 Discussion and Future Work

This paper proposed a context-aware reminder framework that utilized ontological ADL modeling and reasoning technologies. The ontological ADL model takes into consideration the wider contexts, which include five classes (Person, Timeline, Event, Location and Activity) with 12 properties. The reasoning is rule-based reasoning implemented using SWRL with SQWRL queries. Three reminders (fall / abnormal alert, unhealthy posture, and event inconsistent reminder) are extracted based on the relationships among the five classes defined by their properties in the ADL model.

The algorithms were evaluated using simulation to provide convincing example scenarios. The experimental results demonstrated that this remaindering framework has the ability to monitor and analyze personal behaviour, and provide feedback about whether the user has a healthy lifestyle (such as no unhealthy postures, following their schedule correctly), and whether falls or abnormal activities occurred during their daily living. The advantages of this ontology-based approach are that it is easier to integrate heterogeneous contexts from various context extraction modules (such as user identity, location, activity, timeline, schedule), to support the rule-based and temporal reasoning based on various relationships among the contexts. The limitation for the study is that the behaviour analysis is based on several distributed contexts, which were extracted using different technologies. Currently the contexts integration and feedback inference are performed offline, to investigate overall plausibility.

Future work will utilize embedded software and investigate system integration on a mobile platform (Smart phone with personal area network) to support delivery of relevant reminders in real-time.

References

1. Osmani, V., Zhang, D., Balasubramaniam, S.: Human activity recognition supporting context-appropriate reminders for elderly. In: 3rd IEEE International Conference on Pervasive Computing Technologies for Healthcare (2009)
2. Levinson, R.: The planning and execution assistant and trainer (PEAT). The Journal of Head Trauma Rehabilitation 12(2), 85 (1997)
3. Pollack, M.E., Brown, L., Colbry, D., McCarthy, et al.: Autominder: An intelligent cognitive orthotic system for people with memory impairment. Robotics and Autonomous Systems 44(3), 273–282 (2003)
4. Strang, T., Linnhoff-Popien, C.: A context modeling survey. In: Workshop on Advanced Context Modelling, Reasoning and Management (2004)
5. Chen, L., Nugent, C.: Ontology-based activity recognition in intelligent pervasive environments. International Journal of Web Information Systems 5(4), 410–430 (2009)
6. Kang, D.O., Lee, H.J., Ko, E.J., Kang, K., Lee, J.: A wearable context aware system for ubiquitous healthcare. In: IEEE Engineering in Medicine and Biology Society Conference, vol. 1, pp. 5192–5195 (2006)
7. Zhang, S., McCullagh, P., Nugent, C., Zheng, H.: A Theoretic Algorithm for Fall and Motionless Detection. In: 3rd IEEE International Conference on Pervasive Computing Technologies for Healthcare, pp. 1–6 (2009)
8. Zhang, S., McCullagh, P., Nugent, C., Zheng, H.: Activity Monitoring Using a Smart Phone's Accelerometer with Hierarchical Classification. In: 6th IEEE International Conference on Intelligent Environments, pp. 158–163 (2010)
9. Zhang, S., McCullagh, P., Nugent, C., Zheng, H., Baumgarten, M.: Optimal Model Selection for Posture Recognition in Home-based Healthcare. International Journal of Machine Learning and Cybernetics (Springer) (2011a)
10. Zhang, S., McCullagh, P., Nugent, C., Zheng, H.: Reliability of Location Detection in Intelligent Environments. In: Special Volume of Advances in Intelligent and Soft Computing, pp. 1–8. Springer, Heidelberg (2011b)
11. Horrocks, I., Patel-Schneider, P.F., et al.: SWRL: A semantic web rule language combining OWL and RuleML. W3C Member submission 21 (2004)
12. O'Connor, M., Das, A.: SQWRL: a query language for OWL. In: Fifth International Workshop on OWL: Experiences and Directions (OWLED) (2009)

Author Index

GPSR Compliance

The European Union's (EU) General Product Safety Regulation (GPSR) is a set of rules that requires consumer products to be safe and our obligations to ensure this.

If you have any concerns about our products, you can contact us on ProductSafety@springernature.com

In case Publisher is established outside the EU, the EU authorized representative is:

Springer Nature Customer Service Center GmbH
Europaplatz 3
69115 Heidelberg, Germany

Batch number: 10838023

Printed by Printforce, the Netherlands